Biology of Snail-Killing Sciomyzidae Flies

Written for academic researchers and graduate students in entomology, this is the first comprehensive analysis of sciomyzid flies.

Sciomyzid flies are important as prime candidates for the biological control of snails and slugs that help transmit diseases such as schistosomiasis or are important agricultural pests. They also can serve as a paradigm for the study of the evolution of feeding behavior in predatory insects. Starting with analyses of malacophagy in general and then in Diptera specifically, all important aspects of the Sciomyzidae are discussed, including behavior, ecology, life cycles, morphology, and identification. New behavioral and morphological classifications and hypotheses are proposed on the basis of unpublished information and a complete analysis of the extensive literature. Also included are keys to adults, larvae, and puparia and a checklist of world species, with information on geographical range and the location of type specimens. The accompanying DVD includes Clifford O. Berg's classic film on the biology of Sciomyzidae and biological control of snails.

LLOYD VERNON KNUTSON is an emeritus of the US Department of Agriculture's Systematic Entomology Laboratory at the Smithsonian Institution and is a Cooperating Entomologist of the Centro Nazionale per lo Studio e la Conservazione della Biodiversitá Forestale, Verona, Italy. He has studied the Sciomyzidae for more than 40 years and has published more than 100 papers on various aspects of their biology.

JEAN-CLAUDE VALA is Professor of Biology, Laboratoire de Biologie des Ligneux et des Grandes Cultures, at the University of Orléans, France. His more than 60 published works on the Sciomyzidae over the past 30 years cover field collections, life cycles, biology, classification, scanning microscopy of immature stages, and ecological adaptations relating morphology with larval habitats.

Biology of Snail-killing Sciomyzidae Flies

Lloyd Vernon Knutson
Salita degli Albito 29, Gaeta, Italy
Jean-Claude Vala
University of Orléans, Orléans, France

CAMBRIDGE UNIVERSITY PRESS
Cambridge, New York, Melbourne, Madrid, Cape Town,
Singapore, São Paulo, Delhi, Tokyo, Mexico City

Cambridge University Press
The Edinburgh Building, Cambridge CB2 8RU, UK

Published in the United States of America by
Cambridge University Press, New York

www.cambridge.org
Information on this title: www.cambridge.org/9780521867856

First published 2011

Printed in the United Kingdom at the University Press, Cambridge

A catalog record for this publication is available from the British Library

Library of Congress Cataloging-in-Publication Data

Biology of Snail-killing Sciomyzidae Flies / by Lloyd Vernon Knutson, Jean-Claude Vala.
p. cm.
ISBN 978-0-521-86785-6 (Hardback)
1. Sciomyzidae. 2. Snails–Biological control. 3. Snails as pests. I. Knutson, Lloyd,
II. Vala, Jean-Claude. III. Title.
QL537.S365B565 2010
595.77′4–dc22

2010016388

ISBN 978-0-521-86785-6 Hardback

This book is dedicated in remembrance of Clifford O. Berg, the pioneer of research on the natural history of Sciomyzidae.

Clifford O. Berg searching for sciomyzid larvae with a surface net of his design in Australia in 1961.

Clifford O. Berg, 1964. From Clarke (1987).

Contents

The plates will be found between pages 76 and 77

Foreword by Benjamin A. Foote

When I first arrived at Cornell University in the fall of 1950 to begin doctoral studies on the biology of snail-killing flies, little did I realize the pleasure awaiting me in the coming years. There was, of course, the continuing excitement of encountering new species and unraveling previously unknown life histories, but even more exciting was developing interactions with many dedicated dipterists, such as Lloyd Knutson and Jean-Claude Vala. The mentor of our group of young fly-chasers at Cornell was Clifford O. Berg, who had discovered, apparently almost by chance, the remarkable association between the larvae of Sciomyzidae and their gastropod prey. One of the early members of this group was Lloyd Knutson, who would go on to enjoy a prominent career with the US Department of Agriculture entomology staff at the United States National Museum in Washington, DC, and later in Europe. Lloyd's doctoral dissertation dealt with the life histories of many European species of the family, and he has continued to explore the mysteries of the Sciomyzidae to the present day. During his European ventures, Lloyd met in 1980 another active worker on the family, Jean-Claude Vala of France. These colleagues developed a highly productive collaboration that has continued to the present day and which has now resulted in the publication of this remarkable discourse on the family Sciomyzidae.

The Sciomyzidae is a moderate-sized family of 539 species scattered through all of the world's biogeographical regions. Its alpha taxonomy is in good shape as a result of the efforts of George Steyskal, Jean Verbeke, Rudolf Rozkošný, Robert Orth, Ted Fisher, and many other dedicated taxonomists. This publication builds on that earlier taxonomic work but focuses on the natural history of the Sciomyzidae.

The information presented in this volume brings together observations published in some 3000 research papers, including over 400 published in the last 20 years or so. As such, it allows workers interested in malacophagy, in general, and in the family Sciomyzidae, in particular, ready access to a large and diverse literature. Particular attention is given to larval feeding habits, life cycles, host and prey relationships, phenology and development, morphology of all life stages, biological control of agriculturally and medically important gastropods, systematic and evolutionary topics, and collecting and rearing methods. The authors point out areas where further research is needed and emphasize topics such as phylogenetic sequencing studies, where meaningful work is just now being initiated. Particularly valuable is the large number of figures and tables that summarize data collected from many sources, so workers can quickly acquire information in their area of interest. I am particularly impressed by the useful and workable keys to genera of adults, puparia, and larvae. Compiling such is a remarkable feat in itself.

It is notable that a family that was practically unknown biologically up to the early 1950s is now generally considered to be one of the best-known families of acalyptrate Diptera, with 240 of 541 (38%) recognized species having known larval feeding habits. Biological information is now available for 41 of the 61 genera. This is truly a significant accomplishment, particularly when it is recognized that the family has not generally been considered economically important. This book is a testament to those many dedicated individuals who laboriously worked out life cycles, described larval feeding habits, explored ecological relationships, and described and illustrated the structures of the eggs, larvae, and puparia.

All authors of scientific treatises hope that their contributions will stimulate interest in the subject and encourage workers to develop new insights and approaches. This volume certainly will fulfill those goals!

Benjamin A. Foote, Professor Emeritus,
Kent State University, Kent, Ohio, USA

Foreword by Rudolf Rozkošný

Study of Sciomyzidae became a great challenge and an opportunity for science beginning in the 1950s. New fields of basic research as well as practical applications began early after proposal of the hypothesis of the general malacophagy of the Sciomyzidae, formulated with remarkable insight by Clifford O. Berg. His hypothesis was based on his rearing of malacophagous sciomyzid larvae in Alaska. Berg well realized the opportunity for further theoretical studies of relationships between flies and molluscs and the potential value of malacophagous larvae as biocontrol agents of undesirable species of snails and slugs. He created very favorable conditions for a new scientific school at Cornell University in Ithaca, New York. Theses of his first postgraduate students were oriented especially to autoecological, phenological, and behavioral studies of individual species and genera of Sciomyzidae from different parts of the world. Berg and his students launched generalizations concerning the range of intimate relations among snail-killing flies and their prey and hosts and in the surprisingly different phenological manifestations within the family Sciomyzidae.

During the following 50 years an array of students at Cornell, Avignon, and other places published results of their studies, frequently in collaboration with colleagues from various parts of the world. Biological information is available for 203 of the known 539 species. This proportion is unusual among the acalyptrate flies, even when families of economic importance are considered.

The present monograph is a panoramic review of achievements in all branches of the science of Sciomyzidae. It includes chapters on the nature of malacophagy, an overview on natural enemies of Mollusca, and parts dealing with biological properties of, and relations among the species of the family (feeding habits and preferences, competition, phenology, reproduction, development and life cycles, ecology, morphology, physiology, behavior, population dynamics, genetics, systematics, biogeography, evolution, and biological control). History of research, notes on methodology, and a world checklist of Sciomyzidae complete this outstanding source of information. The authors present analytical evaluations of the huge amount of information available. They advance synthetic hypotheses in which they both participated recently. I mention especially, a new proposal of behavioral groups and trophic guilds, basic ideas about the evolution of the family, and an improved analysis of phenological types. They combine their deep and broad theoretical knowledge with personal experience in laboratory and field experiments performed not only in the temperate zone, but also in subtropical and tropical areas, with great success.

Sciomyzid larvae feed as aquatic predators of snails, pea mussels, and oligochaetes or as terrestrial parasitoids (and/or predators) of hygrophilous and terrestrial snails; a few attack slugs or consume snail eggs. The Sciomyzidae have colonized a great variety of ecosystems from various types of chiefly stagnant waters, moist habitats of semi-aquatic types, mesophytic woods, and even some xerothermic sites. The larval morphology shows differences in relation to those feeding habits. Larvae of Sciomyzini live chiefly as parasitoids in exposed aquatic, hygrophilous and terrestrial snails whereas larvae of Tetanocerini mainly include overt predators of aquatic molluscs and oligochaetes or terrestrial snails and slugs. Mature larvae of some *Tetanocera* spp. are able to immobilize their prey by injection of a neurotoxin produced in their salivary glands.

The methods used in the research have been as diverse as the investigated phenomena have been rich. Methods of rearing the larvae and laboratory and field experiments are elaborated in detail. Microstructures of eggs, larvae, and adults were studied by scanning electron microscopy. Crucial phases of the development of a few species were followed by video photography. Molecular analyses of DNA sequences as a tool for assessment of phylogenetic relationships are reviewed.

I had the good fortune to participate in the development of the modern study of Sciomyzidae, at least in the main contours, almost from the beginning. I was in correspondence with Clifford Berg from the 1960s and later with

Lloyd Knutson. Professor Berg offered me a one-year stay at Cornell University, but the occupation of what was then Czechoslovakia in 1968 interrupted such educational programs for many years. I met Lloyd Knutson during his first visit to Czechoslovakia in 1970. We discussed Sciomyzidae until the late hours. From that date until today, we have regularly exchanged information and have prepared a series of joint publications. Jean-Claude Vala and I initiated contact in the early 1980s when he started his studies on Mediterranean sciomyzids and began to prepare his Ph.D. thesis. In 1985, he invited me to participate in the defense of his dissertation, but I had to abandon my participation owing to duties at my university. Nevertheless, I have followed his scientific career during the subsequent years and met him when he visited Masaryk University in Brno in 2001. We discussed especially his new discoveries in feeding habits of some species attacking oligochaetes and new microstructures and sensilla on the body surface of larvae. We met at the Fifth International Congress of Dipterology in Brisbane, Australia, where Jean-Claude presented stunning new data on non-malacophagous larvae of Sciomyzidae.

So, my dear friends, it is really a pleasure to assist at the birth of your fascinating book. I am convinced that it will become a much-frequented source of updated information on all aspects of Sciomyzidae, not only for specialists and other dipterists, but also for professional as well as amateur researchers in different branches of biology.

Rudolf Rozkošný, Professor Emeritus,
Masaryk University, Brno, Czech Republic

Preface

We provide here the first review and analysis of the vast amount of information on this intensively studied group of flies, information gained over the past two and a half centuries that is widely scattered in publications in 16 languages, often in obscure journals, some critical studies being in as yet unpublished reports. We have crafted this work from the perspective of our own research on most aspects of Sciomyzidae during the past 4 + and 2 + decades, respectively, in North, Central, and South America, Western Europe, the Middle East, and Africa. We have sought to be comprehensive, not over-emphasizing archival detail but presenting enough appropriate detail when it is necessary to develop a conclusion, a hypothesis, a scenario, the germ of an idea, or a question. Often this has meant combining or relating significant bits of information from many studies, i.e., "bricks" that were not the main subject of a study but which resulted because the species under study provided the opportunity, and the observations then were included as related information. Mining the literature for such gems, unearthing buried bricks as it were, and relating them has been surprisingly rewarding prospecting.

We arrange the data, the "bricks" of Henri Poincaré (1902) and Marston Bates (1949), by subject, i.e., behavioral, ecological, and morphological, etc., not by species, life style, etc. While our approach has the advantage of enabling generalizations, it has the disadvantage of sometimes separating more or less related data gained on biological features of species *x*, at time *y*, at place *z*. A few species (*Ilione albiseta*, *Salticella fasciata*, *Sciomyza varia*, *Sepedon fuscipennis*, *S. sphegea*, *S. spinipes*, and *Tetanocera ferruginea*) have been particularly well studied. Thus it is useful to relate the discussions of these species under the various chapters, developing, so to speak, the architecture of these key species. But for many specific aspects, other species have been better studied. Relating data on those species to data on key species will be instructive. For these purposes, the text is extensively cross-referenced. The reader is encouraged to re-arrange the "bricks" to arrive at more sound generalizations, more pleasing edifices, than we present.

We allude to, ask questions about, and perhaps provide some mortar concerning mechanisms that might be responsible for the diversity of behavioral attributes and morphological features of Sciomyzidae. We are inclined to agree with Huston's (1994) conclusion in regard to the study of biological diversity in general: "The important question . . . is not which explanation for species diversity is correct, since virtually every explanation that has been proposed is important under some circumstances. The critical questions are which of the many potential explanations apply to a specific diversity pattern, whether any particular mechanism is the dominant explanation for a specific pattern . . . and whether there are any general rules about which mechanisms are likely to be important under particular environmental conditions, among specific groups of organisms, or at particular spatial and temporal scales." However, our essentially natural history approach to the study of Sciomyzidae, and some of our conclusions, e.g., the probable lack of competition within and among sciomyzids in nature, lead us to identify more with the autoecological than the demographic paradigm of ecological theory as distinguished by Hengeveld & Walter (1999).

We place special emphasis on ordering and arranging certain information (e.g., feeding behavior, feeding sites, phenology) to make the data easier to grasp, to provide food for thought, and to encourage further analysis and research. We do this at the risk of creating categories of the mind when they may not exist in nature, but with the pardoning admonition of one of our (LK's) instructors at Cornell in 1959, the late Howard E. Evans: "Look out the window – what do you see? Order, not chaos. Nature is ordered, not chaotic" (paraphrased from a 49-year-old recollection). We take the reader from an overview of malacophagy in general, through the major aspects of the biology of Sciomyzidae, building to sections on their potential use as an example of the evolution of behavior in parasitoid and predaceous insects, and finally to their use as economically sound and environmentally safe biocontrol agents. We attempt to relate the biology of Sciomyzidae to some major questions in behavior, ecology, and evolution, most for the first time.

Some say that a deficiency of modern science is a failure to respect, to make use of, "old" science. In fact, many critical theories, principles, etc. on which "new" science is based rely upon experiments and observations that were made only once and the results of which are questionable. This may be true of even pedestrian analyses such as ours on the Sciomyzidae. As we have plodded through a sizeable literature we have made extensive use of figures and tables published over the past century. Those were presented in various languages and formats. We have translated all of them into English and present them in a uniform format.

In the rich literature on the natural history of animals, there are relatively few treatises on flies, although these ubiquitous creatures make up perhaps as much as 10% of all described animal species. Most of the comprehensive books on flies concern those of pest/biocontrol importance (mosquitoes, black flies, house flies, tsetse flies, horse flies, tephritid flies), environmental significance (chironomid gnats), as genetic research subjects (drosophilids), or some such combination (coffin flies). One of the values of this book may be in providing some indications of subjects worthy of broader treatment in further comprehensive summaries of Diptera biology, morphology, etc. In reading recent manuals and handbooks on Diptera, we have seen little mention of many "unusual" features of some Sciomyzidae – feeding on aquatic oligochaetes, clams, snail eggs and slugs; hydrostatic sperm pump; loss of ptilinum and palpi; calcareous septum produced in snail shells by puparriating larvae, etc. – that may not be, in fact, so unusual among Diptera but which are better known in the Sciomyzidae. Study of such features in related families and other Diptera may be useful.

Much of this book is eighteenth-century biology from a twenty-first-century perspective. For that we make no excuse and hope that our effort may inform, instruct, and perhaps even incite further research on the biodiversity of flies and lesser organisms. We have included detail on some field study and laboratory experimental methods and methods of analysis to aid in their use in other studies.

You are invited to visit the Sciomyzidae website (www.sciomyzidae.info), which includes a nearly complete bibliography, directory of researchers, list of current research projects, needed research, lists of desiderata of material for research, and other resources useful to the dipterist.

L. V. Knutson,
Gaeta, Italy
J-C. Vala,
Orléans, France

Avant propos

It is indeed wonderful to consider, that there should be a sort of learned men who are wholly employed in gathering together the refuse of nature, if I may call it so, and hoarding up in their chests and cabinets such treasures as others industriously avoid the sight of. Addison (1770) cited in Stearn (1981).

There are no principles too deep, no speculations too lofty to find application in such creatures as flies. Williston (1896).

The flies, poor things, were a mine of observation. Levi (in Horvitz 2000).

On fait la science avec des faits comme une maison avec des pierres; mais une accumulation de faits n'est pas plus une science qu'un tas de pierres n'est une maison. Science is built of facts the way a house is built of stones; but an accumulation of facts is no more science than a pile of stones is a house. Poincaré (1902).

Facts form the raw material of science – the bricks from which our model of the universe must be built – and we are rightly taught to search for sound and solid facts, for strong and heavy bricks that will serve us well in building foundations, for clean and polished bricks that will fit neatly into ornamental towers. But while accumulating the bricks may be a contribution to science, we must take care that the pile does not become a hopelessly discouraging jumble. For science itself is not brick-making – it is, at the workaday technical level, bricklaying; and at the creative and artistic level, architecture, the designing of an edifice that will utilize all the bricks to the very best advantage.

The metaphor, of course, cannot be carried too far. The bricklayers and the architects of science are always acquiring strange, new, and beautiful bricks that make it necessary to tear their careful building down and start over. It is an unending, dreamlike game that seems to be limitless – the model of the universe will never be done, nor does any part of it seem to have a comfortable or dependable permanence. But still the bricks, as bricks, cannot be left in a jumbled pile, and we have the task of organizing them into some sort of a pattern, however transient. Bates (1949).

. . . flies are a topic like drains, not to be discussed in polite society, to be left to those strange people who cultivate a professional interest in them. It is a pity. Oldroyd (1964).

About the authors

LLOYD VERNON KNUTSON was awarded a Ph.D. in Entomology from Cornell University in 1963, his thesis research under Professor C. O. Berg being on the biology and immature stages of European Sciomyzidae. He was a National Science Foundation (NSF) supported Research Associate with Professor Berg during 1963–1968, much of the time spent in Europe conducting field and laboratory studies of Sciomyzidae. From 1969 to 1973 he worked with the US Department of Agriculture, Systematic Entomology Laboratory, at the Smithsonian Institution, in which position he was responsible for taxonomic research on Sciomyzidae and other Diptera. He was appointed Chairman of the USDA Insect Identification and Beneficial Insect Introduction Institute at Beltsville, Maryland, in 1973, where he served for 15 years, with one year on leave to the Ecology Program, Smithsonian Institution, and one year on leave as Director, Systematic Biology Program, NSF, Washington, DC. In 1988 he became Director of the USDA Biological Control of Weeds Laboratory in Rome, Italy, and in 1991 as Director of the USDA European Biological Control Laboratory when he was responsible for combining the USDA Rome, Italy, and Paris laboratories in Montpellier, France. He retired in 1997 in southern France and moved to Italy in 2002. He served as President of the Entomological Society of America (ESA) from 1989 to 1990. He is an Honorary Member of ESA as well as an Honorary Foreign Member of the Russian Entomological Society. He has conducted field work on Sciomyzidae in North, Central, and South America, in Europe from Crete to Finnish Lapland, and in Ghana, Iran, Israel, Nigeria, and Pakistan. He has published 92 papers on various aspects of Sciomyzidae, many of which have been in collaboration with colleagues throughout the world. Knutson has been a Cooperating Entomologist, Centro Nazionale per lo Studio a la Conservazione della Biodiversitá Forestale, Verona, Italy, since 2007.

JEAN-CLAUDE VALA studied relationships between aquatic snails and digenetic Trematoda larvae and between fish and monogenetic Trematoda under Professor Louis Euzet at Montpellier University, France, from 1971 to 1973. From 1973 to 1977 he taught parasitology in the Faculty of Sciences of Oran, Algeria. In 1977, he moved to the Faculty of Sciences of Avignon, France, where he taught courses in zoology (systematics, reproduction, physiology, ecology, and entomology). He was awarded a Ph.D. in Entomology from Montpellier University in 1985, his thesis being on the systematics, biology, and population dynamics of the Sciomyzidae of France. He has published 63 papers on Sciomyzidae, covering field collections, life cycles, biology, scanning microscopy of immature stages – particularly on sensilla – and ecological adaptations relating morphology with larval habitats. His graduate students on Sciomyzidae – Claire Haab, Mohamed Ghamizi, Sylvie Manguin, and Ghélus Gbedjissi – and colleagues Christine Brunel, Christine Caillet, Charles Gasc, and Jean-Marie Reidenbach conducted research with him on various aspects of larval feeding, behavior, phenology, and morphology of Sciomyzidae and elucidated several life cycles. He has conducted field work in France, Algeria, and Bénin. In addition to Sciomyzidae, he conducts research on the systematics of Tabanidae and the biology of Chironomidae, pests of rice in southern France. In 1989 his "Diptera Sciomyzidae Euroméditerranéens" was published in the *Faune de France*. From 1998 to 2000 he was the Director of the Department of Biology, Avignon University. Since then, he has been Professor of Biology, Laboratoire de Biologie des Ligneux et des

Grandes Cultures, Orléans University, France where he continues research on many aspects of sciomyzid biology. With his colleague Xavier Pineau, he also studies relationships between cultivated fields and their immediate environment and the colonization and establishment of insects in different types of fallow lands. The work is focused primarily on the biodiversity of Carabidae living on the soil as bio-indicators of degradations of biotopes. From 2006, with his colleagues in plant–insect relationships, he has participated in the project on an emergent pest, the woolly poplar aphid *Phloeomyzus passerini* (Signoret).

JEAN-CLAUDE particularly wishes to thank Professor Louis Euzet, University of Montpellier, and Professor Claude Combes, University of Perpignan and Académie des Sciences de Paris, who were his early mentors in scientific investigation and for their continued rigorous advice.

Acknowledgments

We acknowledge our longstanding colleagues Ben Foote and Rudolf Rozkošný for diverse and important advice, encouragement, and assistance over the past five decades. The classification of behavioral groups was developed in extensive communication with J. K. Barnes. For permission to use published figures and tables and to quote from their publications we thank the authors and publishers as noted in the text, legends and references. For the use of unpublished data we thank J. Abercrombie, S. L. Arnold, J. Bouniard, A. D. Bratt, C. Caillet, E. Colonnelli, J. B. Coupland, J. C. Deeming, K. Durga Prasad, L. G. Gbedjissi, M. Ghamizi, K. D. Ghorpadé, C. Haab, J. B. Keiper, R. J. Mc Donnell, T. Pape, J. B. Peacock, M. C. D. Speight, N. Vikhrev, and P. Withers. For the use of unpublished photos and figures we thank J. K. Barnes, J. Ebel, B. A. Foote, C. Gasc, C. and H. W. Hoffeins, S. A. Marshall, R. E. Orth, M. Thomas, and the late J. W. Stephenson.

We thank Marty Schlabach, A. R. Mann Library, Cornell University and staff of the Division of Rare and Manuscript Collections for recovering Professor C. O. Berg's classic film from the Cornell University Archives and preparing the DVD, and E. Brown for permission to reproduce it herein.

For checking names and classification of Mollusca, we thank C. Audibert, G. Manganelli, B. Métivier, and J-P. Pointier. For checking many names of insects in Index 2, we thank A. Freidberg, D. G. Furth, T. J. Henry, R. R. Kula, A. L. Norrbom, and N. E. Woodley. For obtaining literature references, we thank E. G. Chapman, H. V. Danks, C. Daugeron, E. Delfosse, J. Ebel, P. A. Espenshade, J. Guglielmi, the late M. Leclercq, M. Martinez, P. Mason, R. J. Mc Donnell, B. Merz, L. Munari, W. L. Murphy, D. Rodriguez, M. Schlabach, M. Sueyoshi, M. and C. Tauber, F. C. Thompson, H. Ulrich, and C. D. Williams. For assistance with translations we thank Agriculture Canada, M. Fargier, R. J. Gagné, L. W. Murphy, C. Sarré, R. Rozkošný, H. Ulrich, and J. Zuska.

Special thanks to the reviewers of parts of this manuscript: J. K. Barnes, D. A. Barraclough, E. G. Chapman, E. Colonnelli, R. H. L. Disney, B. A. Foote, A. Freidberg, M. Ghamizi, K. D. Ghorpadé, M. J. Gormally, K. C. Kim, S. A. Marshall, R. J. Mc Donnell, R. Meier, R. M. Miller, S. E. Neff, A. L. Norrbom, R. E. Orth, L. Papp, R. B. Root, R. Rozkošný, J. Sheahan, M. C. D. Speight, M. and C. Tauber, H. Ulrich, and C. D. Williams. We especially thank W. L. Murphy for an impeccable review of much of our final draft. Special thanks to the Cambridge University Press editorial team, D. Lewis, Commissioning Editor – Life Sciences, and his associates A. Evans, J. M. Jackson, A. Jones, L. Talbot, and L. Bennun, excellent copyeditor. We thank W. Cooper, former Editor for our initial contact with Cambridge University Press.

We greatly appreciate the gracious hospitality, assistance, and friendship of the literally hundreds of persons who have expedited our studies over the past, combined 80 decades.

Mara and Martine, thank you for your love, your care, and your patience over all these years of work.

Finally, we thank "our" flies, the Sciomyzidae.

"*Cela est bien dit répondit Candide, mais il faut cultiver notre jardian . . .*" [That is well said, answered Candide, but we must cultivate our garden . . .] Voltaire (1759).

1 • Introduction

All finite things reveal infinity.

Roethke (1964).

But the finite cannot be extended into the infinite.

Leonardo da Vinci.

Paris, 1846 and southwestern France, 1847. Then, a century later and continents apart, southwestern Alaska, 1950. In terms of space and time, these critical points are the most important in the early work on sciomyzid life cycles. The first description of the larva and puparium of a sciomyzid (that of *Tetanocera ferruginea*) was presented by Dufour (1847a), in a poetic style,

> *Vers la fin de l'automne de 1846, je découvris dans l'eau d'une mare, près de Saint-Sauveur [near Paris], au milieu des lemna et des callitriches, une larve dont . . . j'eus le bonheur, vivement senti, de la voir prospérer, se transformer en chrysalide . . . Là où l'oeil du vulgaire n'aurait certainement su voir qu'un fragment inerte de branche noircie par la pourriture, j'y voyais, moi, le berceau hermétique d'une nymphe tendre, emmaillotée, immobile, l'espoir de la prospérité de la Tétanocère . . . ce précieux conceptacle fœtal était appelé, par destination suprême, à braver la tempête pendant cinq mois de la plus mauvaise saison . . . à conserver sa vitalité malgré la glace qui pouvait l'ensevelir . . . Et, en définitive, l'éclosion de l'insecte ailé est venue, au printemps . . .*
>
> [Towards the end of fall of 1846, I discovered in the water of a pond close to Saint-Sauveur [near Paris], among *Lemna* and *Callitriche*, a larva for which I felt a deep happiness to see develop and change into a chrysalide [puparium] . . . Where a vulgar eye would have certainly seen only an inert fragment of twigs blackened by decay, I was seeing the hermetic cradle of a swaddled, still and tender nymph, the hopeful prosperity of the *Tétanocère* . . . this precious foetal container was called, by supreme destination, to brave the storm during five months of the worst season . . . to maintain its vitality despite the ice which could cover it . . . And, finally, the hatching of the winged insect at springtime . . .]

The first report of the larval food of a sciomyzid was presented by Perris (1850), who reared *Salticella fasciata* from larvae found in the terrestrial snail *Theba pisana* in southwestern France. He stated "*Cette larve . . . dévore . . . Helix pisana, probablement après qu'il est mort.*" ["This larva . . . devours . . . *Theba pisana*, probably after it had died"]. The first conclusive evidence that several sciomyzid larvae kill and consume gastropods was presented one century later by Berg (1953). Working in southwestern Alaska in 1950, he reared six species solely on snails (*Dictya expansa, Elgiva solicita, Sciomyza dryomyzina, Sepedon fuscipennis, Tetanocera ferruginea*, and *T. rotundicornis*) and "found a third-instar larva of *S. fuscipennis* eating a small *Lymnaea emarginata* in nature." The story of the original discovery by Berg in Alaska in 1950 was described in a personal account (Berg 1971a), in which he noted that he "stumbled onto a discovery." The noted wasp behaviorist, H. E. Evans (1985), in a vignette of the Cornell research, referred to Berg's discovery as "a prime example of serendipity – the gift for making fortunate discoveries accidentally" and noted the opportunities that the research offered for a study of evolution of behavior and for biocontrol.

In reviews of parasitoid and predatory insects, the Sciomyzidae generally have been treated inadequately. This has resulted, perhaps in part, because of the lack of a volume such as this that enables reviewers to encompass the extensive literature and because the behaviors of many species do not lend themselves easily to categorization. For example, Godfray (1994) does not mention Sciomyzidae in his *Parasitoids: Behavioral and Evolutionary Ecology* (disappointingly, despite his title, he refers only briefly to parasitoids other than Hymenoptera). Although until now no comprehensive review of sciomyzid biology has appeared, a few reviews of certain aspects have been published (e.g., Berg & Knutson 1978, Greathead 1981, Ferrar 1987, Barker *et al.* 2004) and some reviews/compilations (primarily taxonomic in nature) on a regional basis (e.g., Knutson 1970b, 1987, Vala 1989a,

Rozkošný 1984b, 1997a, 1998, 2002, Rivosecchi 1992). In developing this longer, holistic, world review and analysis we have had the advantages of space and time to consider all aspects more or less simultaneously. Thus we have had the advantages of relating to each other many aspects that have never before been treated in an integrated manner. We have placed special emphasis on modifying generalizations in light of recent data on Sciomyzidae and recent theoretical studies. Such an approach is useful for a group of insects such as the Sciomyzidae, about which studies of many aspects have been published, but usually quite separately (nearly 2000 publications since Linnaeus [1758] and more than 400 by more than 100 authors worldwide during the past 20 years). In addition to our primary objectives of review, synthesis, and analysis, including information from some unpublished documents, we provide work tools (e.g., keys to world genera, world checklist), information on methods of research, and suggestions for additional research that we hope will be stimulating and challenging to future researchers.

Because our study is worldwide in scope and as it is often instructive to compare features among species and genera from different zoogeographical regions or to relate features to regions, we occasionally indicate the regions after the species or genus names. Authors of valid species and genera of Sciomyzidae and Phaeomyiidae are given in the world checklist of species (Chapter 21); of mollusc species in Tables 1.2, 1.3, and 1.4. Authors of genera of molluscs are given in Table 1.1, and authors of other organisms are given in Index 3. In the figures and tables, first-, second-, and third-instar larvae and larval stadia are indicated by L1, L2, and L3. In the figures, scales are in millimeters unless otherwise indicated.

1.1 DEFINITIONS OF SAPROPHAGE, PARASITE, PARASITOID, AND PREDATOR, WITH EXAMPLES IN THE SCIOMYZIDAE, AND OF APO/PLESIOMORPHIC AND APO/PLESIOTYPIC

> A definition is the enclosing of a wilderness of ideas within a wall of words.
>
> Butler (1912).

Because the terms saprophage, parasite, parasitoid, and predator, and their adjectival forms, are critical to discussions throughout this presentation, it is important that we explain our usage at the outset. We prefer somewhat narrow definitions so as to provide as fine a level of discrimination among behaviors as possible and so as to provide the ability to describe different mixtures of behavior. We recognize that these are man-made divisions in a near continuum of behavior in nature. We often use the terms in conjunction as a kind of shorthand to indicate mixed and/or labile feeding behavior throughout the life cycle of some species, e.g., "saprophage–parasitoid–predator" for *Atrichomelina pubera*, or for a chronological sequence of behavior, e.g., "parasitoid-predator" for the many species that begin larval life feeding in a mollusc without killing it, but then become a true predator later.

Saprophage. A saprophage feeds in dead plant (saprophytic) or dead animal (saprozoic) matter. Cyclorrhaphous Diptera larvae feeding in dead material have been shown by Keilin (1912, 1915), Hartley (1963), Dowding (1967, 1968), and Roberts (1971) to have oral grooves leading into the mouth and ventral cibarial ridges in the pharynx, the latter serving as a filter mechanism to separate out micro-organisms and other particulate matter, which is the actual food. In the Sciomyzidae, only the plesiomorphic *Salticella fasciata* (Salticellinae) has been reared in the laboratory from hatching to pupariation and successful emergence solely on dead, decaying tissue of snails and other dead invertebrates (Knutson *et al.* 1970). It killed snails in laboratory rearings but in nature it appears to attack moribund snails (Coupland *et al.* 1994) (see Section 18.2). *Salticella fasciata* possesses weakly developed oral grooves and weakly developed ventral cibarial ridges (Knutson *et al.* 1970, Dowding 1971). The only record of a species of Sciomyzinae found in a dead non-gastropod in nature is that of *Atrichomelina pubera* found on the inner surface of a shell of a large unionid clam (J. B. Keiper, personal communication, 2006). Fourteen puparia of *A. pubera*, from seven of which adults had partially or entirely emerged, were attached to the inner surface of one valve of a *Lampsilis* sp. collected on the shoreline of a creek in Ohio. This probably was a dead individual washed up onto the shoreline, the flies then ovipositing onto it. *Atrichomelina pubera* and other Sciomyzidae appear to lack oral grooves and ventral cibarial ridges, but cross-sections of the pharyngeal region of most species have not been examined. All of the 46 reared Sciomyzini and many of the reared semi-terrestrial Tetanocerini complete development on the putrid, liquified tissues of the snails they have killed, beginning as early as the mid second stadium. Some Sciomyzini, e.g., *A. pubera*, have great flexibility in switching back and forth between saprophagous, parasitoid, and predatory feeding modes throughout larval development in the laboratory, depending upon the degree of crowding in the food snails and the relative sizes of the larva(e) and snails(s) attacked (Foote *et al.* 1960).

Table 1.1. Classification of genera of Mollusca mentioned in the text, tables, and figures

♦ Class BIVALVIA [LAMELLIBRANCHIA, PELECYPODA] (clams, mussels)
- ► Superorder Palaeoheterodonta
  - ■ Order Unionoida [Heterodonta]
    - Superfamily Unionoidea Rafinesque
      - Hyriidae Swainson
        - *Hyridella* Swainson
      - Unionidae Rafinesque
        - *Anodonta* Lamarck
        - *Fusconaia* Simpson
        - ***Lampsilis*** Rafinesque
        - *Quadrula* Rafinesque
- ► Superorder Heterodonta
  - ■ Order Veneroida
    - Superfamily Sphaerioidea Deshayes
      - Sphaeriidae Deshayes (fingernail clams, pea or orb mussels)
        - *Musculium* Link
        - *Pisidium* Pfeiffer
        - *Sphaerium* Scopoli
    - Superfamily Dreissenoidea Gray
      - Dreissenidae Gray
        - *Dreissena* van Beneden

♦ Class GASTROPODA
- Subclass Orthogastropoda [Prosobranchia, Streptoneura] (operculate snails)
  - ► Superorder Caenogastropoda
    - ■ Order Architaenioglossa [Mesogastropoda, Ctenobranchiata]
      - Superfamily Ampullarioidea Gray [Viviparoidea Gray]
        - Ampullariidae Gray [Pilidae Preston]
          - *Idiopoma* Pilsbry
          - *Marisa* Gray
          - *Pila* Röding
          - *Pomacea* Perry
        - Viviparidae Gray
          - *Campeloma* Rafinesque
          - *Filopaludina* Habe
          - *Viviparus* Montfort
      - Superfamily Cyclophoroidea Gray
        - Cochlostomatidae Kobelt
          - *Cochlostoma* Jan
    - ■ Order Neotaenioglossa
      - Superfamily Littorinoidea Gray
        - Littorinidae Gray
          - *Littoraria* Griffith & Pidgeon
          - *Littorina* Férussac
        - Pomatiasidae Newton
          - *Pomatias* Studer
      - Superfamily Cerithioidea Férussac
        - Cerithiidae Férussac
          - *Bittium* Leach *in* Gray
        - Pleuroceridae Fischer
          - *Goniobasis* Lea
          - *Oxytrema* Rafinesque
        - Thiaridae Troschel [Melaniidae Lamarck]
          - *Hubendickia* Brandt
          - *Melanoides* Olivier
          - *Melanopsis* Férussac
          - *Tarebia* H. & A. Adams
      - Superfamily Rissooidea Gray
        - Bithyniidae Troschel
          - *Bithynia* Leach
        - Hydrobiidae Troschel
          - *Fontelicella* Gregg & Taylor
          - *Hydrobia* Hartmann
          - *Hydrobioides* Nevill
          - *Lithoglyphus* Pfeiffer
          - ***Potamopyrgus*** Stimpson
          - *Tryonia* Stimpson
        - Pomatiopsidae Stimpson
          - *Lacunopsis* Deshayes
        - Rissoidae Gray
          - *Cingula* Fleming
        - Truncatellidae Gray
          - *Oncomelania* Gredler
  - ► Superorder Heterobranchia
    - ■ Order Ectobranchia
      - Superfamily Valvatoidea Gray
        - Valvatidae Gray
          - ***Valvata*** O. F. Müller
    - ■ Order [Subclass] Pulmonata [Euthyneura] (non-operculate snails, limpets, and slugs)
      - Suborder Basommatophora (aquatic snails and limpets)
        - Superfamily Acroloxoidea Thiele
          - Acroloxidae Thiele
            - *Acroloxus* Beck
        - Superfamily Chilinoidea H. & A. Adams
          - Chilinidae H. & A. Adams
            - *Chilina* Gray
        - Superfamily Lymnaeoidea Rafinesque
          - Lymnaeidae Rafinesque
            - *Acella* Haldeman [1]

Table 1.1. (*cont.*)

Austropeplea Cotton & Godfrey
Erinna H. & A. Adams
Fossaria Westerlund [1]
Galba Schrank
Lymnaea Lamarck
Myxas Sowerby
Omphiscola Rafinesque
Pseudisidora Thiele
Pseudosuccinea Baker [1]
Radix Montfort
Stagnicola Jeffreys [1]
Superfamily Planorbioidea Rafinesque
Physidae Fitzinger
Physinae Fitzinger
Physa Draparnaud
Physella Haldeman
Aplexinae Starobogatov
Aplexa Fleming
Planorbidae Rafinesque
Bulininae Fischer & Crosse [2]
Bulinus O. F. Müller
Ferrissia Walker
Indoplanorbis Annandale & Prashad
Planorbella Haldeman
Planorbinae Rafinesque
Afrogyrus Brown & Mandahl-Barth
Ancylus O. F. Müller [Ancylidae Rafinesque]
Anisus Studer
Bathyomphalus Charpentier
Biomphalaria Preston [*Australorbis* Pilsbry]
Drepanotrema Crosse & Fischer
Gyraulus Charpentier
Helisoma Swainson
Hippeutis Charpentier
Planorbarius Duméril
Planorbis O. F. Müller
Planorbula Haldeman
Promenetus Baker
Segmentina Fleming
Segmentorbis Mandahl-Barth
Taphius H. & A. Adams [*Australorbis* Pilsbry]
Trochorbis Benson
Tropicorbis Brown & Pilsbry
■ Order Stylommatophora (terrestrial snails and slugs)
Superfamily Achatinoidea Swainson
Achatinidae Swainson
Achatina Lamarck
Archachatina Albers
Limicolaria Schumacher
Superfamily Achatinelloidea Gulick
Achatinellidae Gulick
Achatinella Swainson
Superfamily Clausilioidea Gray
Clausiliidae Gray
Albinaria Vest
Balea Gray
Clausilia Draparnaud
Superfamily Cochlicopoidea Pilsbry
Cochlicopidae Pilsbry [Cionellidae Clessin]
Cochlicopa Férussac [*Cionella* Jeffreys]
Superfamily Enoidea Woodward
Cerastuidae Wenz
Euryptyxis Fischer
Enidae Woodward
Ena Turton
Zebrina Held
Superfamily Gastrodontoidea Tryon
Gastrodontidae Tryon [in part, Zonitidae Mörch]
Ventridens Binney & Bland
Zonitoides Lehmann
Oxychilidae Hesse [in part, Zonitidae Mörch]
Oxychilus Fitzinger
Aegopinella Lindholm [*Retinella* Fischer]
Pristilomatidae Cockerell [Vitreidae Thiele]
Vitrea Fitzinger
Superfamily Helicarionoidea Goodwin-Austen
Euconulidae Baker
Euconulus Reinhardt
Superfamily Helicoidea Rafinesque
Arionidae Gray
Arion Férussac
Bradybaenidae Pilsbry
Fruticicola Held [*Bradybaena* Beck]
Camaenidae Pilsbry
Obba Beck
Cochlicellidae Schileyko
Cochlicella Férussac
Helicidae Rafinesque
Arianta Turton
Cepaea Held
Cornu Born
Eobania Hesse [*Eubania* auct.]
Helicigona Férussac

Table 1.1. (*cont.*)

- - - *Helix* Linnaeus
    - *Otala* Schumacher
    - ***Theba*** Risso
  - Hygromiidae Tryon [Helicidae Rafinesque]
    - ***Candidula*** Kobelt
    - ***Cernuella*** Schlüter
    - *Euomphalia* Westerlund
    - ***Helicella*** Férussac
    - ***Hygromia*** Risso
    - *Monacha* Fitzinger
    - *Trochoidea* Brown
    - ***Trochulus*** Chemnitz [*Trichia* Hartmann]
    - *Xerolenta* Monterosato
    - *Xeropicta* Monterosato
    - *Xerotricha* Monterosato
  - Philomycidae Gray
    - *Pallifera* Morse
    - *Philomycus* Rafinesque
- Superfamily Limacoidea Lamarck
  - Agriolimacidae Wagner
    - ***Deroceras*** Rafinesque [*Agriolimax* Mörch]
  - Limacidae Lamarck
    - *Lehmannia* Heynemann
    - *Limax* Linnaeus
  - Vitrinidae Fitzinger
    - *Phenacolimax* Stabile
    - *Vitrina* Draparnaud
- Superfamily Parmacelloidea Fischer
  - Milacidae Ellis
    - *Milax* Gray
- Superfamily Polygyroidea Pilsbry
  - Polygyridae Pilsbry
    - *Mesodon* Rafinesque in Férussac
    - ***Stenotrema*** Rafinesque
    - ***Triodopsis*** Rafinesque
- Superfamily Punctoidea Morse
  - Charopidae Hutton
    - *Allodiscus* Pilsbry
    - *Charopa* Albers
    - *Flammulina* von Martens
    - *Phacussa* Hutton
    - *Ptychodon* Ancey
    - *Thalassohelix* Pilsbry
  - Endodontidae Pilsbry [in part, Discidae Thiele]
    - ***Anguispira*** Morse
  - Patulidae Tryon [in part, Discidae Thiele]
    - ***Discus*** Fitzinger
  - Punctidae Morse
    - *Laoma* Gray
    - *Punctum* Morse
- Superfamily Pupilloidea Turton
  - Chondrinidae Steenberg
    - *Abida* Turton
    - *Granaria* Held
  - Lauriidae Steenberg
    - *Lauria* Gray
  - Pupillidae Turton
    - ***Pupilla*** Fleming
  - Pyramidulidae Kennard & Woodward
    - *Pyramidula* Fitzinger
  - Valloniidae Morse
    - *Vallonia* Risso
  - Vertiginidae Fitzinger
    - ***Vertigo*** Müller
- Superfamily Rhytidoidea Pilsbry
  - Haplotrematidae Baker
    - *Haplotrema* Ancey
- Superfamily Streptaxoidea Gray
  - Streptaxidae Gray
    - *Gonaxis* Taylor
- Superfamily Succineoidea Beck
  - Succineidae Beck
    - ***Novisuccinea*** Pilsbry
    - ***Oxyloma*** Westerlund
    - *Quickella* Boettger [*Catinella* Pease]
    - ***Succinea*** Draparnaud
- Superfamily Testacelloidea Gray
  - Oleacinidae H. & A. Adams [Spiraxidae Baker]
    - *Euglandina* Crosse & Fischer
  - Subulinidae Fischer & Crosse
    - *Rumina* Risso
    - ***Subulina*** Beck

Notes: [ ], alternate names for molluscan taxa. ( ), common names. Bold face, genera known to be hosts/prey in nature. *Potamopyrgus* is not known to be attacked in nature, but laboratory tests are conclusive that this operculate snail is the prey of ***Neolimnia tranquilla***. [1] recognized as subgenera of *Lymnaea* by some specialists. [2] recognized as a subfamily of Planorbidae by some specialists.

Table 1.2. Snail species mentioned in the text, tables, and figures

SPECIES	REGION	HABITAT freshwater (A) marine, intertidal (M) terrestrial (T) semi-terrestrial (ST)
Abida secale (Draparnaud)	P	T
Achatina fulica Bowdich	AF	T
Achatinella stewarti var. *producta* Reeves	Hawaii	T
Acroloxus lacustris (L.)	P	A
Aegopinella nitidula (Draparnaud)	P	T
Aegopinella pura Alder	P	T
Afrogyrus oasiensis (Demian)	AF	A
Albinaria sp.	P	T
Ancylus fluviatilis O. F. Müller	P	A
Ancylus strictum Morelet	P	A
Anguispira alternata (Say)	N	T
Anguispira kochi (Pfeiffer)	N	T
Anisus leucostoma (Millet)	P	A
Anisus spirorbis (L.)	P	A
Anisus vortex (L.)	P	A
Aplexa hypnorum (L.)	N	A
Aplexa marmorata (Guilding) [*Stenophysa marmorata*]	NT	A
Arianta arbustorum (L.)	P	T
Austropeplea ollula (Gould) [*Lymnaea ollula*]	Hawaii	A
Austropeplea tomentosa (Pfeiffer)	A	A
Austropeplea viridis (Quoy & Gaimard)	O, A	A
Bathyomphalus contortus (L.)	P	A
Biomphalaria alexandrina (Ehrenberg) [*B. boisseyi* (Potiez & Michaud)]	AF	A
Biomphalaria glabrata Say	AF	A
Biomphalaria nigricans (Spix)	AF	A
Biomphalaria pfeifferi (Krauss)	AF	A
Biomphalaria schrammi (Crosse) [*B. paparyensis* (Baker)]	AF	A
Biomphalaria straminea (Dunker)	NT	A
Bithynia laevis Lea	O	A
Bithynia tentaculata (L.)	P	A
Bittium sp.	P, N	M
Bulinus africanus (Krauss)	AF	A
Bulinus forskali (Ehrenberg)	AF	A
Bulinus globosus (Morelet)	AF	A
Bulinus tropicus (Krauss)	AF	A
Bulinus truncatus (Audouin)	P, AF	A
Campeloma decisum (Say)	N	A
Candidula intersecta (Poiret) [*Helicella caperata* (Montagu)]	P	T

Table 1.2. (*cont.*)

SPECIES	REGION	HABITAT freshwater (A) marine, intertidal (M) terrestrial (T) semi-terrestrial (ST)
Candidula unifasciata (Poiret) [*Helicella unifasciata*]	P	T
Cepaea hortensis (O. F. Müller)	P	T
Cepaea nemoralis (L.) [*Trichia nemoralis*]	P	T
Cernuella virgata (da Costa)	P	T
Cingula sp.	P, N	M
Clausilia nigricans (Pulteney)	P	T
Clausilia bidentata (Ström)		
Clausilia dubia Draparnaud	P	T
Cochlicella acuta (O. F. Müller)	P	T
Cochlicella barbara (L.)	P	T
Cochlicopa lubrica (O. F. Müller)	P	T
Cornu aspersum O. F. Müller [*Helix aspersa*]	P	T
Discus cronkhitei (Newcomb)	N	T
Discus patulus (Deshayes)	N	T
Discus rotundatus (O. F. Müller)	P	T
Discus ruderatus (Studer)	P	T
Drepanotrema depressissimus (Moricand)	NT	A
Drepanotrema lucidum (Pfeiffer)	NT	A
Eobania vermiculata (O. F. Müller)	P	T
Erinna newcombi H. & A. Adams [*Lymnaea newcombi*]	Hawaii	A
Euconulus fulvus (O. F. Müller)	P	T
Euglandina rosea (Férussac)	AF	T
Euomphalia strigella (Draparnaud)	P	T
Euryptyxis latireflexa (Reeve)	P	T
Ferrissia sp.	N	A
Filopaludina martensi cambodiensis Mabille & Le Mesle	O	A
Filopaludina sumatrensis polygramma von Martens	O	A
Fontelicella californiensis Gregg & Taylor	N	A
Fruticicola fruticum (O. F. Müller)	P	T
Galba humilis (Say) [*Fossaria humilis, Lymnaea obrussa* (Say)]	N	A
Galba truncatula (O. F. Müller)	P	A
Gonaxis kibweziensis (Smith)	AF	T
Gonaxis quadrilateralis (Preston)	AF	T
Goniobasis sp.	N	A
Granaria frumentum (Draparnaud)	P	T
Gyraulus albus (O. F. Müller)	P	A
Gyraulus convexiusculus (Hutton)	O	A
Gyraulus crista (L.)	P	A
Gyraulus ehrenbergi (Beck)	AF	A

Table 1.2. (*cont.*)

SPECIES	REGION	HABITAT freshwater (A) marine, intertidal (M) terrestrial (T) semi-terrestrial (ST)
Gyraulus hiemantium (Westerlund)	O	A
Gyraulus laevis (Alder)	P	A
Gyraulus parvus (Say)	N	A
Helicella bolenensis (Locard)	P	T
Helicella itala (L.)	P	T
Helisoma anceps (Menke)	N	A
Helisoma duryi (Wetherby)	N	A
Helisoma tenue californiensis Baker	N	A
Helisoma trivolvis (Say)	N	A
Helix pomatia L.	P	T
Hippeutis cantori (Benson)	O	A
Hippeutis complanatus (L.)	P	A
Hippeutis umbilicalis (Benson)	O	A
Hubendickia siamensis Brandt	O	A
Hydrobia sp.	H	M
Hydrobioides nassa (Theobald)	O	A
Idiopoma ingallsiana (Lea)	O	A
Idiopoma pilosa (Reeve)	O	A
Indoplanorbis exustus (Deshayes)	O	A
Lacunopsis munensis Brandt	O	A
Lauria cylindracea (Da Costa)	P	T
Limicolaria sp.	AF	T
Lithoglyphopsis aperta Temcharoen	O	A
Lithoglyphus turbiniformis (Tryon)	O	A
Littoraria filosa (Sowerby)	A	M
Littoraria luteola (Sowerby)	A	M
Littorina littorea (L.)	N, P	M
Lymnaea maroccana Pallary	P	A
Lymnaea stagnalis (L.)	P	A
Marisa cornuarietis (L.)	NT	A
Melanoides tuberculata (O. F. Müller)	O	A
Melanopsis praemorsa (L.) [*M. algerica* (Pallary)]	P	A
Mesodon thyroideus (Say)	N	T
Monacha cartusiana (O. F. Müller)	P	T
Novisuccinea ovalis (Say) [*Succinea ovalis*]	N	ST
Obba sp.	O	T
Oncomelania formosana (Pilsbry & Hirase)	O	A
Oncomelania quadrasi (Moellendorff)	O	A
Oxychilus cellarius (O. F. Müller)	P	T
Oxychilus navarricus (Bourguignat)	P	T
Oxyloma decampi gouldi Pilsbry	N	ST
Oxyloma effusa (Pfeiffer)	N	ST
Oxyloma elegans (Risso) [*O. pfeifferi* Rossmässler]	P	ST

Table 1.2. (*cont.*)

SPECIES	REGION	HABITAT freshwater (A) marine, intertidal (M) terrestrial (T) semi-terrestrial (ST)
Oxyloma retusa (Lea)	N	ST
Oxyloma sarsii (Esmark)	P	ST
Oxyloma sillimani (Bland)	N	ST
Oxytrema carnifera (Lamarck)	N	A
Physa fontinalis (L.)	P	A
Physa venustula Gould	N	A
Physella acuta (Draparnaud)	P, N	A
Physella gyrina (Say)	N	A
Physella h. hendersoni (Clench)	N	A
Physella integra (Haldeman)	N	A
Physella virgata (Gould)	N	A
Pila ampullacea (L.)	O	A
Pila scutata Mousson	O	A
Planorbarius corneus (L.)	P	A
Planorbarius metidjensis (Forbes)	P	A
Planorbis planorbis (L.)	P	A
Planorbula armigera (Say) [*P. jenksii* (Carpenter)]	N	A
Pomacea bridgesi (Reeve)	O	A
Pomacea canaliculata (Lamarck)	O	A
Pomatias elegans (O. F. Müller)	P	T
Potamopyrgus antipodarum (Gray) [*P. jenkinsi* (Smith)]	New Zealand	A
Promenetus exacuous (Say)	N	A
Pseudisidora producta (Mighels) [*Lymnaea reticulata* Gassies]	Hawaii	A
Pseudosuccinea columella (Say) [*Lymnaea columella*]	N	A
Quickella sp.	P	ST
Radix auricularia (L.)	H	A
Radix auricularia rubiginosa Michelin	O	A
Radix balthica (L.) [*Lymnaea peregra* O. F. Müller in part, *Radix ovata* (Draparnaud)]	P	A
Radix gedrosiana (Say) [*Lymnaea gedrosiana* (Annandale & Prashad)]	P	A
Radix labiata (Rossmässler) ["*Lymnaea limosa*," *R. peregra* O. F. Müller]	P	A
Radix luteola (Lamarck)	O	A
Radix natalensis (Krauss)	AF	A
Rumina decollata (L.)	P	T
Segmentina hemisphaerula (Benson)	O	A
Segmentina nitida (O. F. Müller)	P	A
Segmentina trochoideus (Benson)	O	A

Table 1.2. (*cont.*)

SPECIES	REGION	HABITAT freshwater (A) marine, intertidal (M) terrestrial (T) semi-terrestrial (ST)
[*Trochorbis trochoideus*]		
Segmentorbis eussoensis (Preston)	AF	A
Stagnicola catascopium (Say) [*Lymnaea catascopium*]	N	A
Stagnicola elodes (Say)	N	A
Stagnicola emarginata (Say) [*Lymnaea emarginata*]	N	A
Stagnicola palustris (O. F. Müller)	H	A
Stagnicola palustris nuttaliana (Lea)	N	A
Stenotrema hirsutum (Say)	N	T
Subulina octona (Bruguière)	AF	T
Succinea californiensis Fischer & Crosse	N	ST
Succinea campestris Say	N	ST
Succinea luteola Gould	N	ST
Succinea putris (L.)	P	ST
Tarebia granifera (Lamarck)	O	A
Theba pisana (O. F. Müller)	P	T
Triodopsis tridentata (Say)	N	T
Trochoidea elegans (Draparnaud)	P	T
Trochulus hispidus (L.) [*Trichia hispida*]	P	T
Tropicorbis centimetralis (Lutz)	NT	A
Tryonia sp.	N	A
Valvata sincera Say	N	A
Vertigo genesii (Gredler)	P	T
Vertigo geyeri Lindholm	P	T
Vitrea sp.	P	T
Viviparus sp.	H, NT	A
Xerolenta obvia (Menke) [*Helicella candicans* (Pfeiffer)]	P	T
Xeropicta sp.	P	T
Xerotricha conspurcata (Draparnaud)	P	T
Zebrina sp.	P	T
Zonitoides arboreus (Say)	P	T
Zonitoides nitidus (O. F. Müller)	P	T

Species in boldface are known to be fed upon in nature; names in brackets are common synonyms.

Parasite. A parasite is an organism, animal or vegetal, that lives at the expense of another organism called the host. Three parasite groups can be distinguished. Ectoparasites live outside the host, which they grip with various mechanisms, e.g., rostrum, hooks, or claws. Some can be localized in cavities widely open to the outside, e.g., ears or nostrils. Mesoparasites invade the deep natural cavities of the host, e.g., genital parts, digestive tube, or respiratory system, all of which are in contact with the exterior. Endoparasites live inside cells, compact or liquefied tissues such as blood or lymph, and organs of the host. The parasite usually lives its entire or a relatively long portion of its life cycle in the host, causing some damage but consuming a small amount of the host. Parasites usually do not kill the host, are more or less host specific, and lack a free-living adult stage. We do not follow the

Table 1.3. Slug and semi-slug species mentioned in the text, tables, and figures

Species	Region
Arion fuscus (O. F. Müller)	P
Arion subfuscus Draparnaud	P
Deroceras laeve (O. F. Müller)	H
Deroceras reticulatum (O. F. Müller)	H
Deroceras sturanyi (Simroth)	P
Lehmannia poirieri (Mabille)	N
Limax maximus (L.)	P
Milax gagates (Draparnaud)	P
Pallifera sp.	N
Phenacolimax major (A. Férussac)	P
Philomycus carolinianus (Bosc)	N
Philomycus sp.	N

Table 1.4. Bivalve species mentioned in the text, tables, and figures

Species	Region
Anodonta cygnea (L.)	P
Anodonta sp.	H
Dreissena sp.	P
Fusconaia sp.	N
Hyridella menziesi (Gray)	SA
Lampsilis sp.	N
Musculium sp.	H
Pisidium casertanum (Poli)	H
Pisidium hodgkini (Suter)	SA
Quadrula sp.	N
Sphaerium novaezelandiae Deshayes	SA
Sphaerium occidentale Prime	N
Sphaerium partumeium (Say)	N
Sphaerium rhomboideum (Say)	N

definition of Price (1980), which includes phytophagous forms among insect "parasites." No sciomyzid is known to be a true endoparasite, although *Euthycera chaerophylli* is probably a mesoparasitoid of slugs.

Parasitoid. A parasitoid is an animal that like a parasite feeds (during the larval stage) in or on the host (but attacking one or more individuals) in an intimate manner for a relatively long period and is more or less host specific, but unlike a parasite and like a predator has a free-living reproductive stage and predictably kills the host and consumes most or all of its tissues. In some species of Sciomyzinae, several young larvae may feed together in a parasitoid manner in a living snail or slug for several days, especially under crowded conditions. Unlike the situation with parasitoid Hymenoptera, when the original host dies, they leave. Subsequently each attacks one or more additional snails or slugs in a predaceous manner. In the most intimate associations (e.g., *Sciomyza varia*, Barnes 1990) there is only one larva per snail and only one snail per larva. Most parasitoid Sciomyzidae are not 100% endoparasitic but keep their posterior spiracles exposed at the surface of the snail's or slug's tissues. Newly hatched larvae of some terrestrial species (e.g., *Tetanura pallidiventris*, Knutson 1970a) penetrate deeply into the snail, between the mantle and shell, apparently respiring cutaneously.

Berg (1964) provided a detailed discussion of definitions of parasite, parasitoid, and predator in regard to Sciomyzidae, arguing for the usage of the word parasitoid as originally proposed by Reuter (1913) for "*parasite-nartige Raubinsekten*" [parasite-like predators] and as accepted by Wheeler (1923), Bequaert (1925), Noble & Noble (1961), and many subsequent authors. Berg pointed out that according to some definitions of parasite and predator, genetically identical larvae from a common group of eggs have to be classified oppositely whenever one larva attacks a large snail where it finds enough food to satisfy its nutritional requirements, but another attacks a smaller snail that does not provide enough food. Yet another larva may enter the shell of a large snail and begin what seems destined to be a "parasitoid" life, only to be transformed suddenly into a predator because other larvae also chance to find the same snail. Knutson & Berg (1966) summarized their definitions, noting their use of the word parasitoid primarily as a comparative adjective to be used in a relative manner characterizing a range of feeding behavior between the parasitic and predaceous ends of the behavioral continuum. However, in subsequent years the use of parasitoid as a noun rather than an adjective became dominant in the sciomyzid literature and in the entomological literature in general. The use of the word parasitoid as a noun results in simpler, more direct sentence structure. It is used primarily as a noun in this book; other forms of the

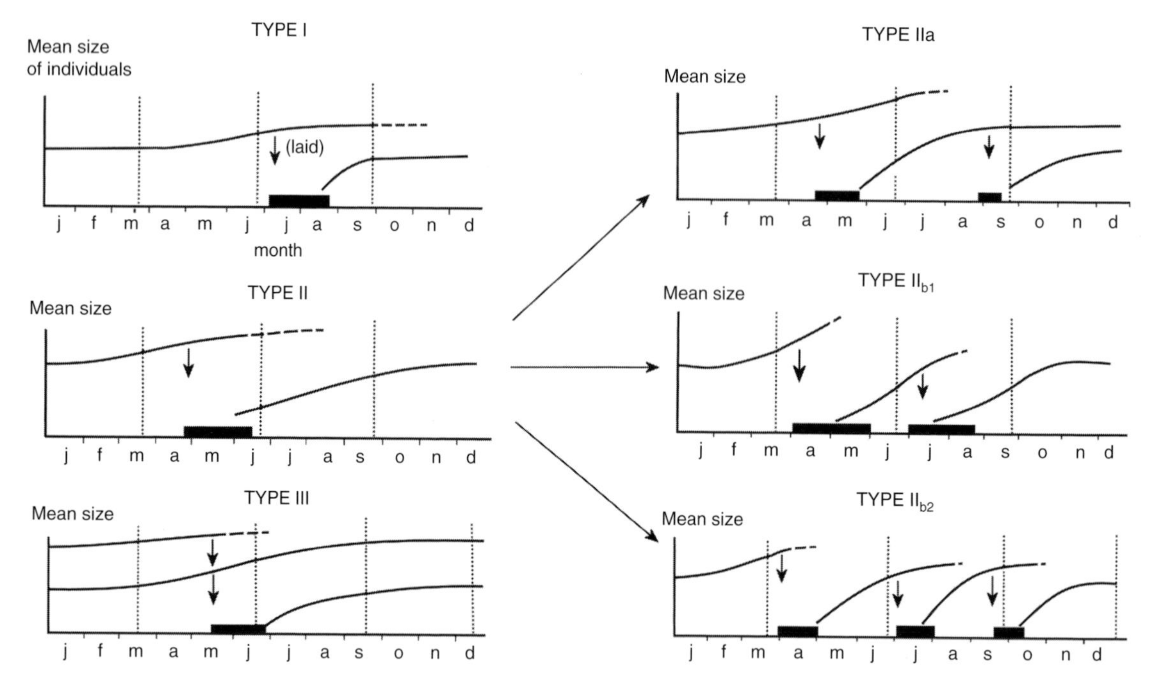

Fig. 1.1 Types of developmental cycles among freshwater snails. Short arrows, beginning of egg laying; black bars, duration of egg-laying period. From Russell-Hunter (1961).

word, such as "parasitoidize" and "parasitoidism," are clumsy and not used here. Askew (1971), Price (1975), Eggleton & Gaston (1990), Godfray (1994), and Yeates & Greathead (1997) also contrast the terms parasite and parasitoid. Our definition of parasitoid is most similar to that of Yeates & Greathead (1997) in their study of the evolutionary pattern of host use by bombyliid flies.

Among the Diptera, more subtle distinctions could be made between parasitoids (1) feeding within the host, that is, the endoparasitoid, which has almost all of its body buried in the flesh of the living host, but, as with nearly all parasitoid Sciomyzidae, keeps the posterior spiracles exposed, and (2) the mesoparasite that lives in internal cavities, i.e., cavities having contact with the exterior (the situation, perhaps, with *Euthycera chaerophylli*). Further specialization includes species (3) that live internally and produce a breathing tube (as in *Pelidnoptera nigripennis* of the sister group Phaeomyiidae), and (4) species that live, at least as first-instar larvae, between the mantle and shell (as appears to be the case with *Colobaea bifasciella, Sciomyza varia, Tetanura pallidiventris, Pherbellia schoenherri*, and possibly *Salticella fasciata*). Those more subtle distinctions appear to involve primarily adaptations to the respiration of the larvae. While they might be considered highly specialized adaptations, it is interesting that they are displayed by the more plesiomorphic Phaeomyiidae, Salticellinae, and Sciomyzini, in contrast to the apomorphic, strictly ambient-air breathing Tetanocerini, except early instars of some subsurface-feeding clam killers and *Hedria mixta*.

Predator. A predator quickly kills, in an overt manner, several prey in succession, feeding on fresh tissue only until satiated, then resting away from the prey. Usually it kills and feeds individually, but two to a few larvae may feed together. There is little prey specificity. Predators are represented in the Sciomyzidae by many freshwater Tetanocerini, and some semi-terrestrial and terrestrial Sciomyzini and Tetanocerini.

Apo/plesiomorphic and apo/plesiotypic. The terms plesiomorphic and apomorphic have been used widely in reference to the state (i.e., primitive or derived) of morphological characters. The terms also have been used in reference to behavioral and ecological attributes. However, with reference solely to behavioral and ecological attributes the suffix "morphic" is inaccurate usage of language. A few authors (e.g., Hennig 1965, J. F. McAlpine 1989) have used

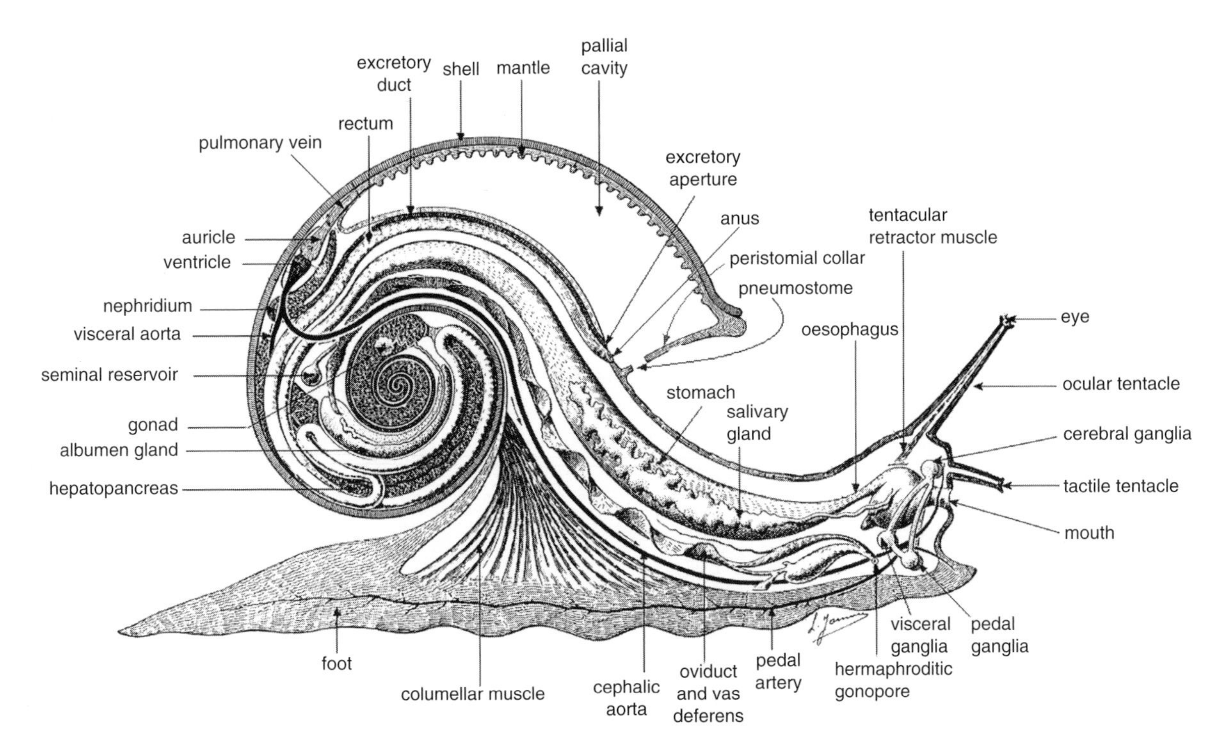

Fig. 1.2 Anatomy of a terrestrial snail, *Cornu aspersum*. From Jammes (1904).

plesio-oecal and apo-oecal when referring to behavioral and ecological attributes. However, these terms look and sound awkward, and, in fact, have not been used widely. Thus we use plesiotypic and apotypic when referring to behavioral/ecological attributes and to combinations of behavioral/ecological attributes and morphological characters. With regard to behavior and ecology, we use attributes (= features), not characters.

1.2 HOSTS/PREY OF SCIOMYZIDAE

Time flies like an arrow.
Fruit flies like a banana.

Lisa Grossman.

We use "host/prey" as a kind of shorthand in referring to the food of sciomyzid larvae in general discussions because of the range of behavior of the larvae and in recognition of the mixed and/or facultative parasitoid and predatory feeding behavior of the larvae of many species. We use "host" or "prey" when referring specifically to a parasitoid or predator, respectively. As the biology of the Sciomyzidae is considered, it will be useful to have some information on their hosts/prey at hand. A classification of the Mollusca adapted from Falkner *et al.* (2001) and showing the Orders, Superfamilies, and Families that include genera of known and likely hosts/prey is given in Table 1.1. The species of Mollusca mentioned in the text are given in Tables 1.2–1.4.

The family Lymnaeidae includes many freshwater and amphibious snails that are the sole or important hosts/prey of many Sciomyzidae. Species previously lumped together in the genus *Lymnaea* are now placed in the genera *Acella, Austropeplea, Fossaria, Galba, Lymnaea, Myxas, Omphiscola, Pseudosuccinea, Radix*, or *Stagnicola* by most malacologists. That classification is followed herein. Some specialists consider most of these genera as subgenera of *Lymnaea*. Currently, only the common Palearctic species *Lymnaea stagnalis* is maintained in the genus. In this text, for convenience we use "*Lymnaea*" when referring to these species in general or when the lymnaeid referred to was not identified to species and its generic placement was not specified according to the modern classification.

The types of developmental cycles among freshwater snails are shown in Fig. 1.1. The internal anatomy of a terrestrial snail is shown in Figs. 1.2 and 1.3, of a slug in Fig. 1.4, and of a non-operculate freshwater snail in Fig. 1.5.

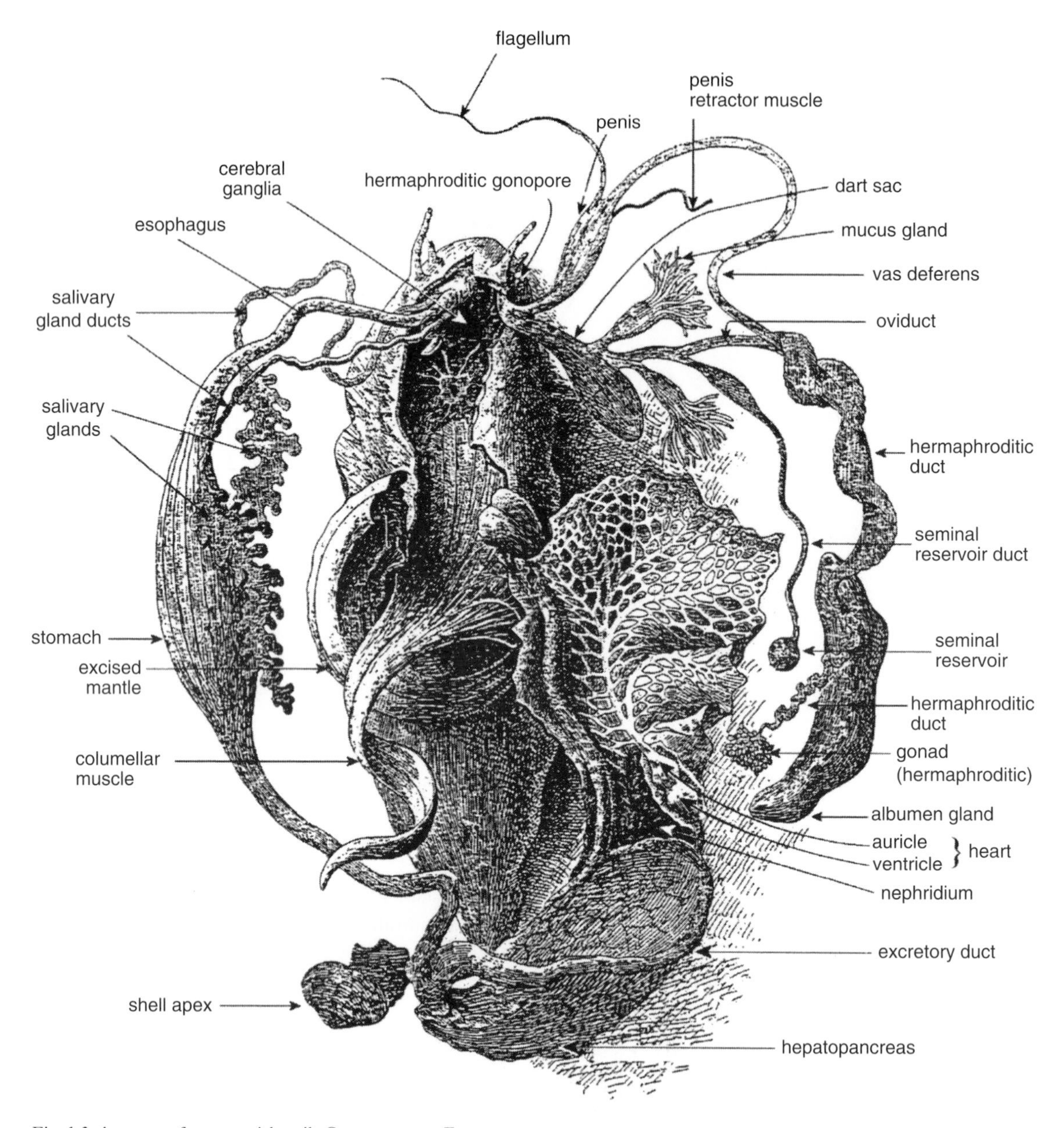

Fig. 1.3 Anatomy of a terrestrial snail, *Cornu aspersum*. From Jammes (1904).

These figures show somewhat different structures and have somewhat different terminology.

Understanding of the macro- and microhabitats, behavior, and population biology of the molluscan hosts/prey should contribute to an understanding of some of the adaptations of Sciomyzidae in habitat selection, food resource partitioning, phenology, and other aspects. However, few concurrent, detailed studies of Sciomyzidae and their hosts/prey have been conducted. Information on molluscs in the same situations where sciomyzids have been studied range from more or less general, non-quantitative observations reported in many publications on Sciomyzidae, through somewhat

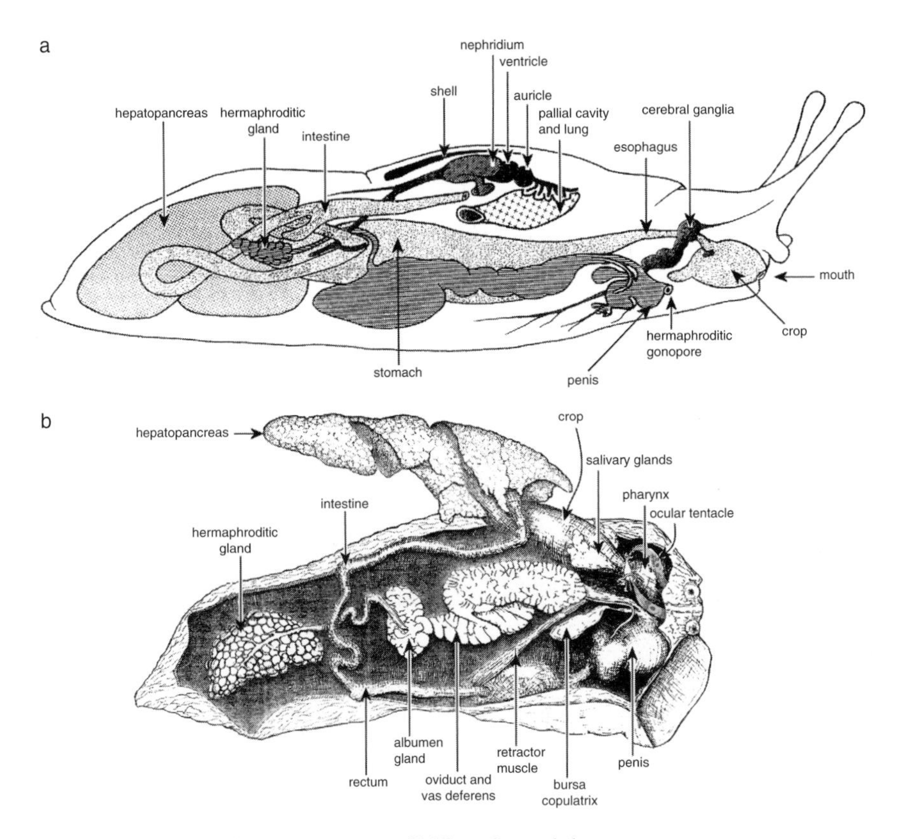

Fig. 1.4 (a) Anatomy of a slug, *Deroceras* sp. (b) View of expanded dissection of a slug, *Deroceras sturanyi*. Modified from Wiktor (2000).

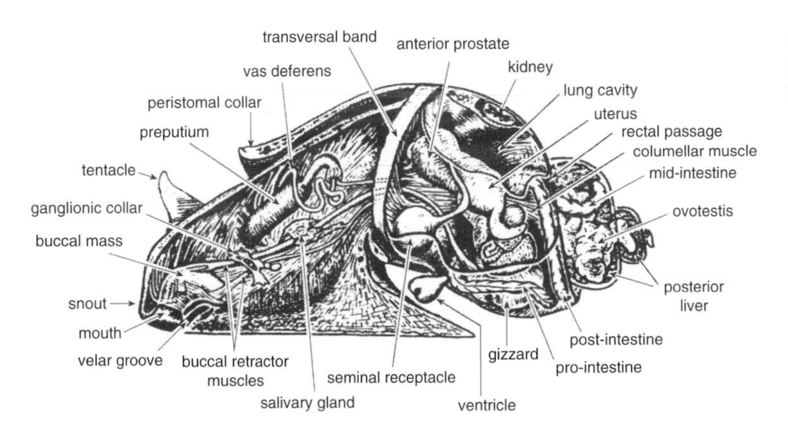

Fig. 1.5 Anatomy of a non-operculate freshwater snail, *Stagnicola catascopium* (Say). Modified from Walter (1969).

detailed quantitative field studies (e.g., Lynch 1965, Boray 1969, Hope Cawdery & Lindsay 1977), to a few quantitative experimental studies. Some aspects of quantitative studies, including the detailed studies of the freshwater snail predator *Sepedon fuscipennis* in the northeastern USA. (Eckblad & Berg 1972, Eckblad 1973a, Peacock 1973, Arnold 1978a, 1978b) and several aquatic and terrestrial species in southern France (Vala 1984b, Vala & Manguin 1987) also are reviewed in

Fig. 1.6 Dense population of the amphibious *Galba truncatula* on marginal mud beside drainage ditch in a low-lying marshy area in southern England. From Wright (1971).

Chapters 10 and 13. Information on "*Lymnaea*" spp., especially *Stagnicola palustris* in Europe and the similar *Stagnicola elodes* in North America, dominates these studies. Unfortunately, as Barnes (1990) noted, the taxonomic status of these species is unclear. He suggested that determinations be regarded as suspect, at least for North America. According to Burch (1982), several distinct species have been referred to as "*S. palustris*," and the true *S. palustris* may not occur in North America, where the taxonomy of the genus is uncertain; thus the name *Stagnicola elodes* is being used by some for a North American *palustris*-like snail. In the Palearctic, the species is polymorphic but is correctly identifed with the new name *Stagnicola palustris*.

It is of value to compare the biology of Sciomyzidae and that of their gastropod hosts/prey. Such comparisons of the well-known sciomyzid and gastropod faunas of western Europe are expedited by the database on 270 species of shelled Gastropoda (i.e., not including slugs) of western Europe presented by Falkner *et al.* (2001). The database is similar in structure to a well-developed database on European Syrphidae (Speight 2000). A similar database on western European Sciomyzidae is being developed (Knutson & Speight, unpublished data). The database on gastropods includes, in addition to 93 macrohabitat descriptors and 112 "microsite" (= microhabitat) descriptors, 31 morphological character categories, 21 range and endemism descriptors, and 62 traits descriptors. The traits descriptors concern sexes, reproduction, oviposition, number of eggs, number of live young, duration of egg development, main reproduction period, sexual maturity, longevity, food type, inundation tolerance, humidity preference, survival of dry period, light preference, hypogean activity, and intraspecific diversity. Falkner *et al.* (2001) described several procedures for use of their database: comparing sites, comparing stations within a site, predicting the fauna of a site, matching species distributions among samples and trait data, and delineating groups of

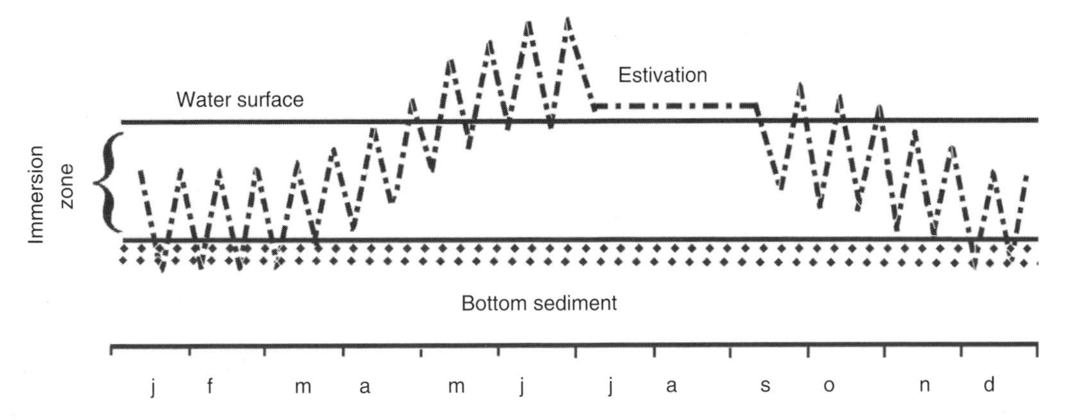

Fig. 1.7 Seasonal vertical displacement of *Galba truncatula* in western Europe. Modified from Rondelaud & Mage (1988).

species with similar characteristics. Examples of Correspondence Analysis (CA) of the latter two uses are presented. Comparisons of such CAs with CAs of behavioral features of Sciomyzidae may be useful in resolving a central question in sciomyzid biology, i.e., why are many groups of gastropods not utilized as hosts/prey by Sciomyzidae? Considerable detail on the macrohabitats and microhabitats of all freshwater snails of central New York, where extensive life-cycle studies of Sciomyzidae were conducted during the period 1953–1978, is presented by Harman & Berg (1971). Fisher & Orth (1983) summarized the behavior of molluscs in habitats in California, where these entomologists amassed enormous collections of Sciomyzidae.

1.2.1 Aquatic and semi-aquatic snails

The abundant and widespread snail species *Stagnicola palustris* and *S. elodes* in the Palearctic and Nearctic, respectively, have been most often recorded as natural hosts/prey of Sciomyzidae, have been used in many life-cycle studies in the laboratory, and have been used as the prey in critical laboratory and field experimental studies. Their habitat and biology are summarized below as examples of important aquatic hosts/prey (Eisenberg 1966, Harman & Berg 1971, Eckblad 1973a, 1973b, Russell-Hunter 1975, Jokinen 1983). These large species (maximum length about 29 mm) have a high degree of phenotypic plasticity and occur grazing on the surface film at the air–water interface in diverse microhabitats around the margins of almost all kinds of freshwater habitats. While most common in hard-water streams, small ponds, and lakes, they also thrive in shallow ephemeral aquatic habitats with rich organic substrates. In New York, *S. elodes* occurs in waters with a broad range of dissolved oxygen (0–14 ppm), carbon dioxide (0–44 ppm), and alkalinity (60–360 ppm as calcium carbonate) and a pH of 7.0–8.4 (Harman & Berg 1971). The various size classes have preferences for different water depths, with juveniles being eulittoral (occurring at the water surface) and adults most often in deeper water. The life cycle lasts about one year, with one reproductive period per year starting from early to mid April for 1–2 months (in New York; Eckblad 1973b). Under ideal conditions two or three generations may be possible during the spring and summer (in Michigan; Eisenberg 1966). During the reproductive period only two size classes are usually found; the adults and small juveniles of the year. During late spring the newly hatched snails are commonly found with the adults that have overwintered as small juveniles on the bottom of the water body. Newly hatched snails are about 1 mm long; the growth rate in nature was estimated to be 1/7 of that in the laboratory (Eisenberg 1966). Young individuals form

Table 1.5. Survival of stranded snails in Bool's Backwater, Ithaca, New York, collected on four dates during June 1970

Species	Percent alive (mean ± S.E.)
Stagnicola elodes	81.6 ± 4.78
Physella integra	11.5 ± 3.12
Gyraulus parvus	27.3 ± 6.91

Note: Modified from Eckblad (1973b).

an epiphragm, a more or less rigid sheet of dried mucus occluding the aperture of the shell, and estivate on drying mud, in mats of drying vegetation, and on vegetation above the waterline in dry periods, migrating when temperatures fall to basins that will be refilled with water as the season progresses.

Knowledge of horizontal and vertical distribution and movement of aquatic and semi-aquatic snails and their migration into and out of their microhabitats is especially important to considerations of them as hosts/prey of sciomyzids. A large number of both Sciomyzini and Tetanocerini with facultative or mixed parasitoid–predaceous–saprophagous behavior attack aquatic and semi-aquatic snails exposed on wet/damp surfaces (see discussion of these microhabitats in Chapter 10). The distinctions between exposed, stranded, and estivating snails in such microhabitats and the duration of time that such snails spend out of the water seem somewhat confused and/or imprecise in much of the sciomyzid literature. Greathead (1981) and Ferrar (1987) used "stranded snails" to describe the hosts/prey of one of the eight and nine behavioral groups of Sciomyzidae they recognized, respectively, albeit for rather different taxonomic assemblages of species and genera. The term stranded also has been used by other authors, e.g., Foote (1996a), who characterized, as a group, some *Tetanocera* spp. as predators of stranded snails in shoreline situations. Berg & Knutson (1978), in their classification of behavioral groups, noted that aquatic predators also are effective on "moist shores" but they did not use the term stranded. We feel that the term stranded has been used in too broad a sense. We define stranded snails as aquatic snails that are involuntarily and accidentally exposed in drying situations and with little opportunity to return to the water, as a result of slow, regular, annual seasonal changes in water levels or catastrophic, short-term climatic or other events that do not occur on a regular basis. This includes snails that die rather quickly in place if they do not return to the water and those that adapt by estivating, such as "*Lymnaea*" spp. that produce an epiphragm that prevents desiccation.

In contrast to being stranded, many freshwater snails are exposed on moist surfaces while foraging temporarily around their primary aquatic microhabitat for relatively brief periods throughout their season of activity. Semi-terrestrial snails also occur in such habitats. Some freshwater snails, e.g., *Galba truncatula*, are considered to be truly amphibious (Figs. 1.6, 1.7). Populations of these semi-terrestrial or amphibious snails seem to be heavily

Table 1.6. Two measures of horizontal dispersion for three snail populations in Bool's Backwater, Ithaca, New York during 1969–1970

Species	*Coefficient of dispersion	**k
Physella integra	9.43	6.289
Stagnicola elodes	12.98	2.725
Gyraulus parvus	67.33	0.180

Notes: *Average values of the variance/mean ratio for 20 sampling dates; **k values of the negative binomial distribution calculated by the regression method of Bliss & Owen (1958). Modified from Eckblad (1973b).

Table 1.7. Seasonal distribution of gastropods sampled (per m^2) in the Vallat-Blanc ravine near Avignon, southern France, January–November 1984

	Months										
Gastropod density per m^2	I	II	III	IV	V	VI	VII	VIII	IX	X	XI
Planorbids	4683	4077	1745	4878	3932	3012	2158	2493	3897	4514	4835
Lymnaeids	1082	1449	804	2013	1506	486	2102	1029	1506	2289	913
Ancylids	176	264	222	331	2116	12379	7942	3586	709	811	416
Total A	5765	5526	2549	6891	5438	3498	4260	3522	5403	6803	5748
Total B	5941	5790	2771	7222	7554	15877	12202	7108	6112	7614	6164

Note: Planorbid, *Bathyomphalus contortus*; Lymnaeids, *Radix balthica* and *Stagnicola palustris*; Ancylid, *Ancylus fluviatilis*; Total A, ancylid omitted; Total B, all four species. From Vala & Manguin (1987).

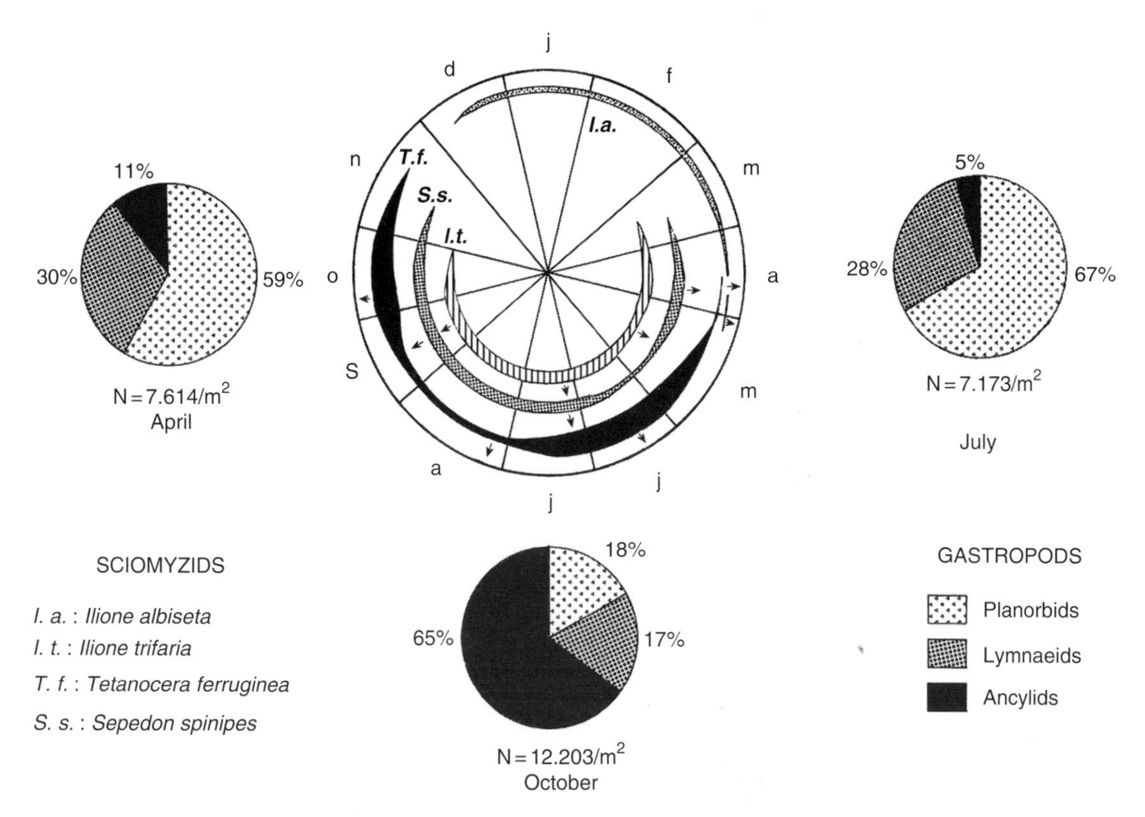

Fig. 1.8 Comparative chorologies of sciomyzids and gastropods in a typical aquatic habitat, the Vallat-Blanc ravine near Avignon, southern France. In center, periods of feeding by larvae of the four dominant sciomyzid species. Width of lines proportional to extent of feeding by larvae; arrows indicate emergence of adults. At periphery, percentages of gastropods during April, July and October. From Vala & Manguin (1987).

exploited by a large number of parasitoid/predaceous Sciomyzini and Tetanocerini (Behavioral Group 2). They also probably are exploited to some extent for brief periods by typical aquatic, predaceous Tetanocerini (Behavioral Group 11). See Chapter 3 for discussion of Behavioral Groups.

A few highly specialized parasitoid or parasitoid/predator Sciomyzini (Behavioral Group 3) attack estivating "*Lymnaea*" spp. or Planorbinae. At least the highly facultative predator–parasitoid–saprophage *Atrichomelina pubera* (Sciomyzini) (Behavioral Group 1) will feed on moribund snails as well as on healthy snails in such situations.

Eisenberg (1966) studied the regulation of density of enclosed and unenclosed populations of *Stagnicola elodes* in the broad marginal area of a shallow, 0.8–1.2 ha permanent pond in Michigan. Limited data were included on the five species (unidentified) of Sciomyzidae and other predators found during the study. No conclusions could be drawn on the effects, if any, of predation. The 28 enclosed areas

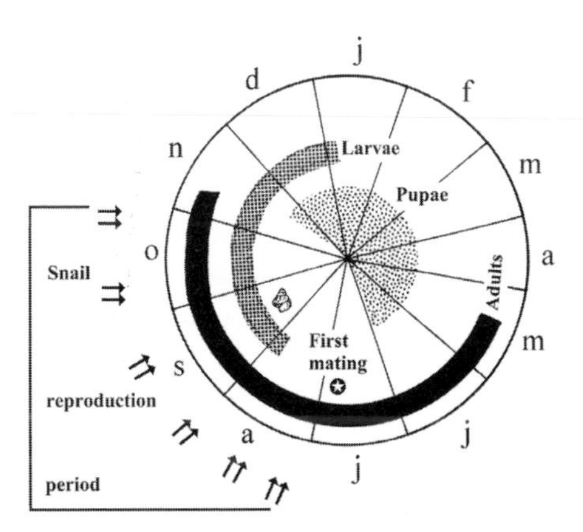

Fig. 1.9 Generalized phenological model of terrestrial Sciomyzidae and their prey in an oak grove near Avignon, southern France. From Vala (1984b).

(1 × 5 m) encompassed most of the type of habitat occupied by the snails and sciomyzid larvae, but migration of snails between exposed portions and the main body of the pond was not studied. The wealth of related data on the snail in this ideal habitat would be very useful in comparison with future concurrent studies of "*Lymnaea*" spp. and sciomyzids in similar habitats.

Eckblad (1973b) conducted population studies of three aquatic gastropods (*Stagnicola elodes, Physella integra*, and *Gyraulus parvus*) in an intermittent backwater habitat in New York, where he also conducted experimental predation studies. Many sciomyzid life cycles were studied by C. O. Berg and students at this locality. The experimental areas were similar in design to those in the study of Eisenberg (1966) (see above). The water level in this habitat annually drops beginning in May. As a result of a catastrophic event (removal of a down-stream barrier) there was little standing water from mid May to June 26, 1970, when the barrier was replaced. Although few adult *Stagnicola elodes* survived this period, survival of smaller individuals was much higher than among the other two species (Eckblad 1973b) (Table 1.5). Further concurrent studies are needed of sciomyzids and snails, focusing on the adaptations of the flies to such differential rates of survival of snail species.

Eckblad (1973b) provided information on the dispersion of the prey. In field experiments, he found that *Sepedon fuscipennis* larvae preferred *Stagnicola elodes* over *Physella integra* and *Gyraulus parvus*. He concluded that the apparent preference resulted from the fact that the *S. elodes* were more frequently at the water surface than were the other two species. In fact, *P. integra* can remain submerged throughout its life cycle. However, in laboratory experiments with *Tetanocera ferruginea*, Manguin *et al.* (1988b) found the opposite, with *Physella acuta* (80% at the surface) preferred to *Stagnicola palustris* (20% at the surface). In simple laboratory tests on vertical dispersion of *Gyraulus convexiusculus, Indoplanorbis exustus*, and *Radix auricularia* in Thailand, O. Beaver (1989) found that all three species preferred to be near the surface (66–100%). The results of both laboratory tests, however, may have been compromised by the oxygen content of the water in the experimental containers. Other studies of vertical dispersion include those of Cheatum (1934) and Pimentel & White (1959). Information on horizontal dispersion of the prey could be of interest in comparing susceptibility of prey species that aggregate with the susceptibility of those that do not. Eckblad (1973b) found that *G. parvus* was considerably more aggregated than either *S. elodes* or *P. integra* (Table 1.6), but other factors resulted in *S. elodes* or *P. integra* being more susceptible to predation by *S. fuscipennis*, especially because *S. fuscipennis* larvae were more commonly found at the surface film.

Lynch (1965) studied the seasonal abundance and variety of habitats of *Austropeplea tomentosa* quantitatively over a 4-year period in South Australia, including life cycle and population data on the aquatic predator *Dichetophora biroi*. Boray (1969) sampled populations of *A. tomentosa, D. biroi*, and *D. hendeli* over a 2-year period in New South Wales, Australia. Hope Cawdery & Lindsay (1977) presented quantitative data on populations of *Galba truncatula* and the semi-aquatic predator *Hydromya dorsalis* from a 3-year study in Ireland. Vala & Manguin (1987) obtained quantitative samples of *Radix balthica, Stagnicola palustris, Bathyomphalus contortus*, and *Ancylus fluviatilis* along 400 m of a 1-m wide ravine (that partially dried during the summer) near Avignon, southern France, during 1984 (Table 1.7). They correlated the marked changes in snail populations with periods of larval feeding of the four dominant species of aquatic, predaceous Sciomyzidae in the habitat (Fig. 1.8). The enormous number of molluscs that they found in this habitat is due to four reasons. First, the method was very comprehensive, and perhaps should be used in other studies. That is, a 30-cm diameter pipe was pushed down to the substrate, several times within each 1 m^2 area. The molluscs in these cores were sieved out, using sieves of various meshes, in the laboratory. Second, sampling was during the main reproductive season of the molluscs. Third, a maximum of small molluscs that would be important food for first-instar larvae were captured as a result of the sampling method and season of sampling. Fourth, the habitat was extremely eutrophic. The drainage ditches sampled were next to horticultural plastic tunnels that were sprayed with various fertilizers and some of these fertilizers flowed into the sampling sites.

1.2.2 Terrestrial snails and slugs

Vala (1984b) compared the period of reproduction of gastropods (*Pomatias elegans, Clausilia bidentata, Abida secale, Cornu aspersum, Cepaea nemoralis, Trochulus hispida, Lauria cylindracea, Phenacolimax major, Deroceras reticulatum*, and *Deroceras* sp.) in an oak grove near Avignon, southern France with the durations of the adult, larval, and pupal stages of six species of terrestrial Sciomyzidae that prey upon them (Fig. 1.9).

Other features of prey that determine their susceptibility include size as well as size in relation to biomass (Sections 7.2, 7.3), escape behavior and mucus production (Section 12.1), mobility, size of aperture, and shell morphology.

2 • Natural enemies of Mollusca

Natural enemies of Mollusca, especially freshwater and terrestrial Gastropoda, are found in a wide range of organisms, from viruses, bacteria, protozoans, and fungi, through helminths, other gastropods, and arthropods to fish, amphibians, reptiles, birds, and mammals. Because of the long-standing interest in the possibilities of biologically controlling populations of molluscs, much of this information is on organisms associated with freshwater snails of medical and veterinary importance and on terrestrial snails and slugs as agricultural pests. While the Sciomyzidae also offer possibilities for use in biocontrol, they are the only relatively large group of obligate insect natural enemies of Mollusca that have been studied as much from interests in basic biology and evolution of behavior as from practical interests.

2.1 GENERAL REVIEWS OF LITERATURE

> *Souvent, j'ai accompli de delicieux voyages, embarqué sur un mot dans les abîmes du passé, comme l'insecte qui flotte au gré d'un fleuve sur quelque brin d'herbe.* [Often, have I made the most delightful voyages, floating on a word down the abysses of the past, like an insect embarked on a blade of grass tossing on the ripples of a stream].
>
> Balzac (1832).

The first reviews of natural enemies of Mollusca were those of Cooke (1895, mainly enemies of terrestrial and marine species) and Gain (1896, enemies of British freshwater and terrestrial species). Other early reviews still of some interest include Schmitz (1917, Diptera), Keilin (1919, 1921, Diptera), Bequaert (1925, arthropods), Pelseneer (1928, diseases), Wild & Lawson (1937, primarily vertebrate predators in Britain), Fischer (1951, diseases), and Frömming (1956, central European terrestrial gastropods). Michelson (1957) reviewed 148 publications on predators, parasites, and other organisms, especially micro-organisms, associated with freshwater Mollusca, emphasizing schistosome-bearing snails. He concluded (as did Berg 1964) that the information published through 1956 gave "little rational basis for selecting a specific organism or group of organisms as biological control agents," and that vertebrates, except possibly fish, were the least likely possibilities. Michelson (1957) decried the lack of experimental and quantitative data on parasites and predators of freshwater snails, as did Greathead (1981) a quarter century later. In his book on the Giant African Snail (*Achatina fulica*) Mead (1961) provided an extensive review of the broad range of natural enemies of terrestrial snails, with emphasis on Coleoptera and Gastropoda, and on biological control of *A. fulica*. He was critical of the use of the predatory snails *Gonaxis kibweziensis* and *Euglandina rosea*, but somewhat optimistic regarding pathogenic micro-organisms. Berg (1964), in his "inquiry and challenge . . . on attitudes and policies of snail control," briefly reviewed some of the literature on organisms associated with freshwater snails of medical importance, noting the concentration on snail competitors and micro-organisms. In a generally overlooked paper, Petitjean (1966) reviewed natural enemies used in biological control, with especially detailed information on natural enemies of terrestrial snails, including a list of the 20 species of insect and snail natural enemies of *A. fulica* released in Hawaii to 1963. More than 46 species of micro-organisms and invertebrates associated with 25 species of slugs were included in a review of 42 publications (1921–1963) by Stephenson & Knutson (1966). Ten species were recorded as killing 14 species of slugs, with protozoans, brachylaemid flatworms, lung worms, lampyrid beetles, and Sciomyzidae as the most important natural enemies. Wright (1968) and Berg (1973a) included natural enemies of snails in their broad reviews of biological control of snail-borne diseases. Hairston *et al.* (1975) discussed the most promising natural enemies of schistosome-bearing snails and recommended that the World Health Organization "encourage research in the tropics . . . for sciomyzid flies or other obligatory snail predators." Ferguson (1977, a rare and obscure publication) considered predatory and competitive animals, ecological factors, and other approaches to control of schistosome-bearing freshwater snails, with many references, especially on other gastropods as competitors (*Helisoma* sp. and *Tarebia granifera*) and predators (*Marisa cornuarietis*). Godan (1979) focused on natural enemies of terrestrial gastropods as agricultural pests and on aquatic snails as intermediate hosts of flatworm diseases of man and domestic animals. She provided an especially complete review of protozoan and helminth parasites in terrestrial snails and slugs, other snails as predators and competitors of aquatic and terrestrial snails, and predaceous and parasitoid arthropods, especially Coleoptera, Phoridae, and Sciomyzidae. Brown (1980) succinctly reviewed vertebrate and invertebrate predators and other molluscs as competitors or predators of aquatic snails, especially hosts of *Schistosoma* species in Africa. Combes (1982, 1995) reviewed the sterilizing effects on their aquatic snail hosts by trematode parasites (especially Echinostomatidae) and antagonism between species of trematodes. The most complete, recent review is the 644-page book edited by Barker (2004), which focuses on natural enemies (bacteria to vertebrates) of terrestrial gastropods. Some of the 15 chapters also include information on aquatic gastropods.

2.2 MALACOPHAGOUS INSECTS OTHER THAN DIPTERA

Detailed studies have been made of a few obligate or predominantly malacophagous insects in addition to the Sciomyzidae and a few other species of Diptera, as noted below. No studies have been conducted of guilds of malacophagous insects in the original sense of Root (1967).

The most recent detailed reviews including other insects in addition to Sciomyzidae as natural enemies of aquatic and terrestrial gastropods are those of Baronio (1974, aquatic and terrestrial, 43 references 1919–1974), Ferguson (1977, aquatic, 16 references 1857–1971), Godan (1979, aquatic and terrestrial, 22 references 1931–1976), Mead (1961, terrestrial, especially biocontrol of *Achatina fulica*, 72 references 1824–1959), and 4 of 15 chapters in Barker (2004, especially the more recent publications on terrestrial species, nearly 1000 references 1841–2004). Rather than citing many of the original papers on insects, we refer to some of the detailed studies, papers missed in earlier reviews, and reports that are of interest in comparison to the biology of Sciomyzidae and/or methods of possible application in studies of Sciomyzidae. None of the above reviews covers all of the literature, although among them probably nearly all is included.

2.2.1 Hemiptera

At least four species of Belostomatidae (giant water bugs) prey upon freshwater and amphibious snails, by inserting their proboscis into the gastropods' foot or visceral mass. Voelker (1966, 1968) studied *Limnogeton fieberi*, a univoltine, apparently obligate but non-specific predator of freshwater snails in Egypt. He conducted quantitative experiments on feeding by the adults on nine species of operculate and non-operculate snails, and other aspects, and compared the behavior of this water bug with that of the opportunistic snail predator, *Diplonychus rusticus*, of the same family. *Belostoma flumineum* in North America is an important cause of mortality of freshwater snails (Kesler & Munns 1989). *Diplonychus annulatus* and *D. rusticus* have been studied as control agents of *Physella acuta*, *Pomacea bridgesi*, and *Radix luteola* in India (Raut & Saha 1989, Aditya & Raut 2001, 2002).

The Pyrrhocoridae (fire bugs) is primarily an Old World tropical, subtropical, and Australian family of about 300 species, the majority feeding primarily on plant seeds but some also scavenging opportunistically on dead invertebrates. Two genera, *Antilochus* and *Dindymus*, are predatory, with three of the 48 species of *Dindymus* (*D. albicornis*, *D. rubiginosus* var. *sanguineus*, and *D. versicolor*) being facultative predators of terrestrial snails (Jackson & Barrion 2004). The latter authors summarized the biology of *D. albicornis*, which inserts its proboscis through the shell of eggs and juveniles of the terrestrial *Obba* species in rainforest habitats in the Philippines. *Dindymus albicornis* fed for 30 minutes to 8 hours on a snail, depending on the size of the snail. The species is being investigated as a potential biocontrol agent of eggs of the rice pest *Pomacea canaliculata*. Species of *Naucoris* and *Corixa* (Corixidae) also have been reported to prey upon small planorbids in aquaria.

2.2.2 Coleoptera

The most recent, comprehensive review of malacophagous Coleoptera (treating Carabidae, Staphylinidae, Lampyridae, Drilidae, and Silphidae) is that of Symondson (2004), which covers primarily predators of terrestrial gastropods. The author noted that most reports to date are of a qualitative nature and concern species in Europe and North America. He stated,

> Most Coleoptera identified as gastropod predators are in fact polyphagous species, consuming variable proportions of non-gastropod prey in addition to slugs

> or snails. Even where gastropod consumption appears to be frequent, almost nothing is known about the relative importance of these invertebrates in relation to other prey, or about the factors that influence prey choice. Polyphagy applies even to certain species of carabids . . . that are morphologically adapted to feeding upon gastropod snails.

Symondson (2004) further noted that most work on the natural or classical biocontrol potential of tropical and Mediterranean species has focused on malacophagous species of lampyrids, drilids, and silphids, which are more or less specific to snails, rather than on the more generalist carabids, despite scant evidence of their having solved any major pest problem. While there has been great interest in polyphagous carabids and staphylinids in temperate areas, it is unlikely that any such species would be used in classical biocontrol because of concerns about the native, non-target biota. Symondson also concluded that the costs of culturing carabids would prevent their use in inundative biocontrol but that it might be possible in some instances to modify field designs and agricultural practices to augment natural populations of endemic predators.

2.2.2.1 CARABIDAE

The Carabidae (ground beetles), along with the Lampyridae, have received the most attention among snail- and slug-killing Coleoptera. Rather than detailing the behavior of carabids we will point out some of the major features and some types of studies and the methods used, many of which could be applied to studies of Sciomyzidae. Several major reviews of the biology of Carabidae have been published, the most recent and comprehensive one covering malacophagous carabids being that of Symondson (2004), who treated only those (the majority) attacking terrestrial gastropods, especially slugs. This author noted that,

> There is little doubt that carabids are the most common beetle predators of gastropods found in the majority of natural and man-made environments, in terms of both abundance and number of species. Even species in this [Symondson's] review that are thought to be gastropod specialists, or to have gastropods as a major component of their diets, are known to feed on many other prey items.

Symondson (2004) also wrote as follows. The diets of adult Carabidae have been extensively studied, but diets of larvae "have been seriously investigated for a very limited number of species. In general, larvae are more carnivorous and specialized than adults." This author continued, "Predation on snail[s] . . . appears to have led to the evolution of morphologically specialized carabid genera, characterized by the development in the direction of the two extreme and mutually exclusive forms, defined as cychrization and procerization." In the former, e.g., *Cychrus* and *Scaphinotus*, "the head and pronotum is narrow and elongated, with long, hooked, notched mandibles, adaptations that allow relatively large beetles to reach into, extract and consume the contents of small gastropod shells." The latter, e.g., *Carabus* (*Procerus*), "have evolved massive mandibles capable of crushing shells." Many other genera of Carabidae, including many that feed on slugs, lack such morphological adaptations.

Analyses of gut contents of Carabidae collected in the field have been highly successful in identifying the prey. In addition to visual inspection, a variety of biochemical and molecular methods, summarized by Symondson (2002a), could be applied to study of sciomyzid larvae. These include precipitin and enzyme-linked immunosorbent assays (ELISA), genus- and species-specific monoclonal antibody tests, isoelectric focusing, and use of PCR primers. Field and semi-field studies of Carabidae attacking slugs, especially, are much more detailed than those of Sciomyzidae attacking terrestrial snails and slugs. Sophisticated experimental techniques used in studying detection of slug prey by Carabidae also could be used in study of Sciomyzidae. The morphology of chemoreceptors of some Carabidae has been carefully studied, as have the defense mechanisms of slugs against predatory carabids and the ability of Carabidae to overcome such mechanisms.

2.2.2.2 DRILIDAE

This family of about 80 mainly tropical species seems to be restricted to preying upon terrestrial gastropods, primarily helicids. At least 19 original reports have been published (through 1969) on larvae of Drilidae (Baronio 1974, Symondson 2004). The latter author wrote,

> The larvae live for up to 4 years, following a highly specialized life history. The first-instar larva is an active predator with fully formed mandibles and legs, which subdues its prey by injection of neurotoxins and proteases. The larva then burrows inside the living gastropod and feeds on it slowly over a long period like a parasite. Eventually it turns into a maggot-like form with legs and mouthparts reduced to vestiges. It can

stay as a quiescent pre-pupa inside the shell of the gastropod prey for 4–5 years before pupating.

Cros (1926) provided a detailed study of *Drilus mauritanicus*, which attacks *Rumina decollata*. Plate (1951) presented details on the biology of Drilidae. Williams (1951) provided observations of drilid larvae attacking *Achatina fulica*. Snails up to 11.5 cm long were attacked by nearly mature larvae. A species from Kenya was studied in quarantine in Hawaii in 1948 for control of *A. fulica*, but was not released. Barker (1969) described the life cycle of *Selasia unicolor*, which is a predator of two species of *Limicolaria* (Achatinidae) in Nigeria. Schilthuizen *et al.* (1994) described the predatory behavior of first instars and adults of *Drilus* sp. against estivating *Albinaria* sp. (Clausiliidae) in Crete.

2.2.2.3 LAMPYRIDAE

Most of the approximatively 2000 species of Lampyridae are tropical. Although only a few species have been studied, most if not all of the voracious larvae, and many adults, are presumed to be polyphagous, feeding, in part, on terrestrial and (primarily) aquatic gastropods. Adult lampyrids feed on nectar, insects, and gastropods. The larvae prey upon invertebrates and use their hooked mandibles in a scissor-like motion to lacerate the tissue, beginning at the side of the foot of snails. The most detailed biological studies have been conducted on species in temperate regions, especially *Lampyris noctiluca*. Key studies of behavior include the following (obviously not pertinent to all species): Schwalb (1961) described the mucus-trail following behavior and poison production of *L. noctiluca*; Buschman (1984) found that larvae of *Pyractomera lucifera* captured snails both above and below the water surface and could remain submerged for many days; Copeland (1981) examined the inhibitory effects of midgut, hemolymph, head, thorax, and mandibles of two *Photuris* spp. on electrocardiogram recordings of the heart of *Limax maximus* and found that midgut and hemolymph were the most powerful inhibitors.

A few examples of Lampyridae as natural enemies of medically and agriculturally important gastropods have been noted in the literature; these were summarized by Mead (1961). Three species from Japan (*Luciola cruciata*, *L. lateralis*, and *Colophotia praeusta*) were introduced into Hawaii for control of aquatic snails. In Japan, they were observed to prey upon species of aquatic *Biomphalaria*, *Lymnaea*, *Melania*, *Oncomelania*, *Planorbis*, and *Tarebia*. *Luciola cruciata* was recorded eating 1.4–6.7 *Biomphalaria* (2.0–5.0 mm diameter) per week, considerably less biomass than most aquatic Sciomyzidae consume. The species did not become established in Hawaii. The terrestrial, nocturnal predator *Lamprophorus tenebrosus* from Ceylon was released in Hawaii in 1954 and in Guam in 1955 for control of *Achatina fulica*. Recoveries were made in Guam but not in Hawaii. During the course of development, male larvae consumed 20–40 achatinas, whereas females consumed 40–60. *Lampyris noctiluca* was imported into New Zealand from England for control of the terrestrial *Cornu aspersum*. *Phausis splendidula* and *L. noctiluca* also ate *Arion* slugs as well as terrestrial snails. The larvae attacked slugs more than twice their own length, several larvae attacking the same slug. Other Lampyridae recorded attacking gastropods include *Diaphanes* spp. feeding on *A. fulica* and other terrestrial snails in Ceylon. Jordan *et al.* (1980) noted observations of Lampyridae attacking *Oncomelania* spp. in Japan and *Bulinus* and *Lymnaea* spp. in southern Africa. Greathead (1981) noted that large numbers of larvae have been seen on floating mats of *Salvinia* (water-fern) in South America. Knutson & Ghorpadé (2004) suggested that the relatively lower diversity of Sciomyzidae in tropical areas as compared to northern temperate regions may result in part from competition with Lampyridae.

2.2.2.4 SILPHIDAE

The family Silphidae (carrion beetles), with 183 species worldwide, includes a small number of predators of terrestrial snails, primarily in north temperate regions. Adult *Ablattaria laevigata* and *Phosphuga atrata* in Europe bite the head and/or foot of a snail until it withdraws into the shell. Then there follows repeated biting and release of digestive proteases and an anal fluid by the predator (Baronio 1974). Species of *Ablattaria* have been found feeding on *Theba pisana* in nature in Israel and on the terrestrial snails *Monacha* sp., *Xeropicta* sp., *Candidula* sp., and *Zebrina* sp. in laboratory rearings (Symondson 2004).

2.2.2.5 STAPHYLINIDAE

Symondson (2004) summarized knowledge of malacophagy by Staphylinidae (rove beetles) as partly paraphrased below. Of the 32 340 species (slightly more than the number of described species of the biologically much better known Carabidae) from throughout temperate and tropical regions, only 2% of the species are known in the larval stage. Little is known about their life cycles, except that there is a remarkable range of behaviors: saprophagous, herbivorous, predatory, ectoparasitoid, mycophagous, pollen feeding, algal grazing, and ant and termite associating. Most of the information on the few slug and terrestrial snail feeders has come from studies in the northern hemisphere. Using antiserum and precipitin tests, Tod (1973) found the remains of *Deroceras reticulatum* in three species in the UK; however, such tests provide no way of distinguishing predation from scavenging. Mendis (1997) used a monoclonal antibody method to find remains of gastropod eggs in a number of small, field-inhabiting species. The Palearctic species *Ocypus olens* was accidentally introduced into southern California about 50 years ago. Initially it was regarded as a useful biocontrol agent of *Cornu aspersum* in gardens, nurseries, and orchards. An adult beetle (32.0 mm long) consumed 20 snails 10.0–20.0 mm in diameter in 3 weeks. Large larvae collected in a garden during April consumed 2–8 snails each during periods of 2–3 weeks in the laboratory, then pupated underground (Orth *et al.* 1975a, 1975b, Fisher *et al.* 1976). However, Wheater (1987) found that *O. olens* had difficulty attacking soft-bodied prey such as slugs and that *O. olens* fed more readily on dead or injured slugs than on living slugs. Wheater (1989) found no evidence of orientation to gastropod mucus and that direct contact was necessary to elicit a response. He considered that many *Ocypus* spp. probably are specialists on gastropods, but *O. olens* is a generalized predator. Sunderland (1996) even considered that *O. olens* might be beneficial to gastropod populations by removing parasitized and diseased individuals.

2.2.2.6 OTHER INSECTS

Other Coleoptera that have been reported as opportunistic predators of aquatic snails include larvae of *Hydrophilus* spp. (Hydrophilidae), adult *Dytiscus marginalis* (Dytiscidae), and some Cantharidae.

Mead (1961) referred to several reports of ants of the genus *Pheidologeton* robbing eggs of *A. fulica* and attacking newly hatched snails. Ferguson (1977) referred to *Solenopsis* sp. (fire ants) and *Paratrichina* sp. ants killing *Biomphalaria* sp. and *Marisa* sp. stranded above the waterline in Puerto Rico. Naiads of *Anax* sp. and *Aeschna* sp. (Odonata) have been reported to eat small planorbid snails in aquaria.

Of course, many of the above-mentioned natural enemies of gastropods likely are competitors and/or predators of larvae of Sciomyzidae but there is essentially no information in this regard. Results of simple laboratory experiments would be of interest.

3 • Malacophagy in Diptera

Several reviews of malacophagous Diptera have been published treating the Sciomyzidae and other families (Schmitz 1917, Keilin 1919, 1921, Bequaert 1925, Lopes 1940, Rozkošný 1968, R. Beaver 1972, Ferrar 1987, Coupland & Barker 2004).

The 69-page chapter on Sciomyzidae in Barker (2004) (by Barker – primarily a malacologist – and four sciomyzid specialists) is not limited to, but emphasizes, terrestrial gastropods. It includes the most detailed discussion to date of Sciomyzidae in relation to many aspects of modern biological control and ecological theory. Innovative sections include a revised cladistic analysis of genera, a detailed presentation of nine new "Eco-Groups" with ordination analyses of the distribution of genera in the Groups, an analysis of prey diversification with mapping of a cladogram of genera of Sciomyzidae onto a cladogram of phylogenetic relationships among families of potential molluscan prey, and discussions of many other aspects.

Diptera other than Sciomyzidae and Chironomidae associated with molluscs were extensively reviewed most recently by Coupland & Barker (2004). They included a detailed table that gives the gastropod hosts/prey (aquatic and terrestrial), geographical distribution, life strategy (parasitic, parasitoid, predator, plus suspected or facultative instances), and extensive literature references for the Calliphoridae (11 species), Rhinophoridae (1), Phoridae (16), Muscidae (6), Fanniidae (2), and Sarcophagidae (73). The only reference that we know of that they overlooked is that of the calliphorid *Pericallimyia greatheadi*, reared from the terrestrial snail *Euryptyxis latireflexa* in Oman (Deeming 1996).

Malacophagy means, literally, feeding on molluscs. In nature there is a near continuum of behavior by flies feeding on molluscs, from occasional saprophages to obligatory parasitoids and predators. This continuum is not concordant with cladistic lineages, except in the Sciomyzidae to some extent. All of the reported malacophagous Diptera are associated with Gastropoda (snails and slugs), except a few species in three genera of Sciomyzidae that are the only Insecta known to kill and feed on any freshwater mussels (Bivalvia), a few Chironomidae that are hematophagous endoparasites or commensals in some large freshwater mussels, and *Atrichomelina pubera* feeding on dead clams.

The 11 families of Diptera known to Keilin (1919) to feed upon living and/or dead molluscs were placed in four categories: (1) Parasitic (our categories 5, 6, 7, 10, 11, below) including Chironomidae and Calliphoridae; (2) Carnivorous (our categories 5, 6, 7, 8, 10, 11) including Anthomyiidae; (3) Epizoic (our category 9) including Phoridae; and (4) Saprophagous and Doubtful Parasitoids (our categories 2, 3, 4) including Sarcophagidae, Anthomyiidae, Phoridae, Ephydridae, Sciomyzidae, Dryomyzidae, Sepsidae, Sphaeroceridae, and Psychodidae. R. Beaver (1972) placed malacophagous Diptera in three categories: epizoics, parasitoids and predators, and saprophages. Coupland & Barker (2004) placed six families of Diptera other than Sciomyzidae (Calliphoridae, Rhinophoridae, Phoridae, Muscidae, Fanniidae, and Sarcophagidae) feeding on gastropods as "(i) saprophages, (ii) epizoic forms, (iii) parasitoids and predators," but in their list of 109 species they further characterized many as "suspected, facultative, possibly facultative, necrophagous, and pseudoparasitoids."

More precise delineation of the various types of malacophagy provides perspective in describing the behavior of Sciomyzidae, in comparing them with other Diptera, in evaluating their potential in biocontrol, and in developing an evolutionary scenario of Sciomyzidae. We recognize 11 categories of malacophagy found among 24 of the 110 ± families of Diptera of the world. The families that include some malacophagous larvae are indicated where they fit into these categories. For Sciomyzidae, the behavioral/host-prey/microhabitat categories in the classification of Behavioral Groups presented in Section 4.3 are indicated, as well as major differences in the feeding mechanisms of the larvae.

(1) Generally predaceous larvae that occasionally, opportunistically, and with no apparent preference kill freshwater, semi-terrestrial, and terrestrial gastropods. Some species of Tabanidae, especially *Tabanus* and *Hybomitra*, and perhaps some other predaceous families.

Table 3.1. Diptera reared from dead snails

1. Anthomyiidae
   Anthomyia confusana Michelson (P)
   Craspedochaeta pullula (Zetterstedt) (H)
   Homalomyia canicularis L. (P)
   Hylemyia vagans (Panzer) (P)
   Lasiomma octoguttatum (Zetterstedt) (H)
   Subhylemyia longula (Fallén) (H)
2. Calliphoridae
   Many genera and species. See Coupland & Barker (2004)
3. Carnidae
4. Chloropidae
   Cadrema pallida (Loew) (OC)
5. Curtonotidae
6. Dryomyzidae
   Dryomyza anilis Fallén (H)
   Dryomyza Fallén sp. (P)
7. Empididae
   Drapetis curvipes Meigen (P)
8. Ephydridae
   Allotrichoma Becker sp. (N)
   Discomyza incurva (Fallén) (P)[1]
   D. maculipennis (Wiedemann) (O)
   D. similis Lamb (AF)
   Hecamede albicans (Meigen) (H)
   Platygymnopa helicis Wirth (N)[1]
   Teichomyza fusca Macquart (P)
9. Fanniidae
   Fannia canicularis (L.) (H)
   F. pusio (Wiedemann) (N, NT)
   F. scalaris (Fabricius) (H)
10. Heleomyzidae
    Neolaria maritima (Villeneuve) (P)
11. Helcomyzidae
12. Lonchaeidae
13. Milichiidae
    Desmometopa inaurata Lamb (AF, A, NT)
14. Muscidae. See Coupland & Barker (2004)
    Alluadinella bivittata (Macquart) (AF)[1]
    Atherigona Rondani (*Acritochaeta*) sp.
    Hydrotaea armipes Fallén (H)
    H. occulata (Meigen) (H)[2]
    Muscina assimilis Fallén (P)[2]
    M. levida (Harris) (H)
    Ochromosca trifaria (Bigot) (AF)[1]
15. Otitidae
16. Piophilidae
    Parapiophila vulgaris (Fallén) (P)
17. Phoridae
    26 species in 9 genera, including some (e.g., *Spinophora maculata* (Meigen) P) reared only from dead snails. See Coupland & Barker (2004)
18. Platystomatidae
19. Psychodidae
    Philosepedon humeralis Meigen (P)[1]
20. Sarcophagidae
    Many genera and species in numerous genera (*Helicobosca* Bezzi, *Helicobia* Coquillett, *Sarcophaga* Meigen, etc.). See Coupland & Barker (2004)
21. Sciadoceridae
22. Sepsidae
    Nemopoda nitidula (Fallén) (H)
23. Sphaeroceridae (also see Chandler *et al.* 1978)
    Copromyza pedestris (Meigen) (P)[1]
    Leptocera clunipes Meigen (P)
    L. fontinalis (Fallén) (H)
    L. luteilabris Rondani (P)
    L. nana Rondani (P)
    L. palmata Richards (P)
    Sphaerocera pusilla (Fallén) (H)
24. Syrphidae

Note: [1]Species reared only from dead snails. The other species have been reared from various dead organic matter as well as from snails. Where no species are listed under a family, in most cases none was indicated in the original reference. Records from Chandler *et al.* (1978), Ferrar (1987), Coupland & Barker (2004), and other sources.
[2]Facultative predator of fly larvae in dead snails (Beaver, 1977).

(2) Generally saprophagous larvae that only occasionally and opportunistically feed on dead gastropods and are known to feed on other carrion and other organic matter. Species in many families; see Coupland & Barker (2004).
(3) Saprophagous larvae that also feed on other carrion and/or dung, etc., but show a preference for dead gastropods. See Table 3.1 and Coupland & Barker (2004).

Notes on categories 2 and 3: In his comprehensive review of feeding habits of Diptera larvae, Ferrar (1987) listed

23 families including species feeding on "carrion (including dead snails)." Table 3.1 lists these families, with reference to specific taxa, and additional species that we include from various other references. As indicated in the table, most species also have been reared from other dead organic matter, but a few have been reared only from dead snails. The levels of preference for dead gastropods by species in our Groups 2 and 3 are poorly known; no experimental studies on preference have been conducted. Elton (1966) noted that most saprophagous animals are relatively indiscriminate in their choice of food. However, food selection and feeding behavior of many malacophagous Diptera larvae likely are genetically controlled processes involving diverse sets of features, especially of physiology/nutrition, behavioral/sensory structural adaptations, microhabitat preferences, anatomy of feeding apparati, and associated morphological features. Further study of the Diptera feeding in dead snails, from an evolutionary perspective, would be of interest. We note that snail killing has evolved in only three of these families (Phoridae, Sarcophagidae, and Calliphoridae). In the nearly exclusively snail-killing family (Sciomyzidae) saprophagy is common during part of the larval life of many species, but only one species (*Salticella fasciata*) has been reared solely on a saprophagous diet.

In a quantitative study of Diptera reared from the terrestrial snail *Cepaea nemoralis* that were field collected or killed and then exposed throughout one year in Wales, R. Beaver (1972) found 16 species in the families Psychodidae, Empididae, Piophilidae, Heleomyzidae, Sphaeroceridae, Sarcophagidae, and Muscidae. Species representing our categories 2, 3, and 4 (see Chapter 6) were found. In an ecological analysis of these data, R. Beaver (1977) considered the Diptera breeding in dead snails as forming non-equilibrium "island" communities. He noted,

> The lack of specialization in breeding habits of the majority of Diptera found in the dead snails, and indeed of most Diptera found in carrion and dung, can perhaps also be related to the heterogeneity of the breeding sites in space and time. Since the habitat stability is effectively unity, and dispersal is necessary in each generation, a specialist in one type of carrion or dung would need very effective dispersal and a very predictable supply of new habitats in order to survive. It seems likely that this condition is rarely fulfilled in natural conditions.

He concluded, among other points, that the Diptera he studied are opportunistic colonists; biological interactions between the larvae in each snail are important in determining the community structure, diffuse competition being the most important, with size and dispersal capability also important; and that low habitat stability and dispersion of habitats in time and space are determining features, possibly accounting for lack of specialization. It would be of interest to analyze the origin of malacophagy in Sciomyzidae in regard to these conclusions about their extant, biologically equivalent ancestors.

Papp (2002) included "dead snails" along with droppings of forest animals, decaying fungi, *Vespa* nests, etc., in his study of the Diptera and Coleoptera guilds on "very small-sized feeding resources" in low montane forests in Hungary. He trapped adult flies on dead *Helix pomatia* and reared flies from traps baited with *H. pomatia* set out for 48-hour periods. Twenty traps of each type were set up at identical places on the same 3 days of July and August in 3 consecutive years (1995–1997). Of the 20 500 flies collected from ten types of feeding sources, 5013 individuals of 91 species were obtained from dead *H. pomatia*; in one series a maximum of 37 species was collected. Analysis of the data was not presented in this paper, as it was for four kinds of feeding sources (including Chao 2 and first-order jackknife analysis methods). In general, Papp (2002) concluded, "The quality, size, persistency and place of renewal of the sources, the potential size of each dipterous population, the flies' ability to find new sources and composition of the local fauna are all important factors in determining the actual frequencies of species found on extant sources." He noted,

> Although the primary texture of the forest community structure is formed by the more abundant forest species populations, those species in guilds on small-sized food sources put a colorful pattern on that texture. They are mostly rare and are probably insignificant for the main energy flow processes, but knowledge of their presence and life histories would seem to be indispensable for a complete understanding of ecosystem structures and diversity maintenance.

He also added, "I do not think that baiting or even manipulated baits would be proper tools for tests of general ecological relationships as was made by Kneidel (1984) and others." However, considering that the Sciomyzidae probably evolved from Diptera feeding saprophagously on dead gastropods, study, even by baiting, of those species of Diptera likely restricted to dead snails might be instructive in some ways in regard to the ecology and evolution of Sciomyzidae.

(4) Saprophagous larvae probably restricted to dead gastropods. See Table 3.1, R. Beaver (1972), and Coupland & Barker (2004).

(5) Larvae with a preference for killing/feeding in moribund terrestrial snails and probably feeding primarily on micro-organisms in liquefied tissues. Oral grooves and filter mechanism of ventral cibarial ridges present, ventral arch weakly toothed. Sciomyzidae, Behavioral Group 1 (*Salticella fasciata*). See Section 4.3 for behavioral groups of Sciomyzidae.

(6) Larvae that usually kill snails and/or slugs but which often can begin larval life feeding on dead gastropods; rarely, as *Atrichomelina pubera*, developing entirely on dead bivalve clams. These species spend most or all of their larval life feeding saprophagously on liquefied tissues. Sciomyzidae, Behavioral Group 1 (*A. pubera* and some *Limnia* species). Oral grooves absent, ventral cibarial ridges apparently absent, and ventral arch strongly toothed as in all Sciomyzidae except *Salticella fasciata*.

(7) Larvae restricted to killing snails and/or slugs; parasitoids and predators, many of which routinely feed in later larval life on decaying tissues of the host/prey. All Behavioral Groups of Sciomyzidae except 1, 5, 11, 13, and 14. Also, at least one species of Phoridae (*Megaselia fuscinervis*, Disney 1982), some Sarcophagidae, and some Calliphoridae (most detailed summaries by Ferrar 1975 and Coupland & Barker 2004). The most complete life cycles of malacophagous Sarcophagidae were presented by (1) Coupland & Baker (1994) for *Sarcophaga* (*Heteronychia*) *penicillata*, parasitoid on the terrestrial *Cochlicella acuta* in France, with comments on other Sarcophagidae for which there are fragmentary records and which require study to determine if they are obligate natural enemies of Gastropoda, and (2) by McKillup *et al.* (2000) and McKillup & McKillup (2000) for *Sarcophaga* (*Sarcorohdendorfia*) *megafilosia* and *S.* (*S.*) *meiofilosia*, parasitoids of the marine operculate snails *Littoraria filosa* and *L. luteola* in Australia. McKillup *et al.* (2000) referred to these two species as "what appears to be the first known insect parasitoids of a marine mollusk." They did not refer to *Hoplodictya setosa* (Neff & Berg 1962) and some *Dictya* spp. (Valley & Berg 1977) which, albeit predators, are the only other Diptera apparently restricted to killing coastal marine gastropods. The most complete life cycles of calliphorids parasitoid on snails with descriptions and figures of the immature stages are those of the Palearctic *Angioneura cyrtoneurina* by Čepelák & Rozkošný (1968) and *Melinda viridicyanea* by Keilin (1919).

(8) Larvae restricted to killing snails predaceously and routinely feeding only on fresh tissues. Sciomyzidae, many genera in Behavioral Groups 11, 13 except the Afrotropical species *Sepedonella nana* (Vala *et al.* 2000b) and *Sepedon* (*M.*) *knutsoni* (Vala *et al.* 2002) which are obligate predators of freshwater oligochaetes and *Sepedon* (*P.*) *ruficeps* (Gbedjissi *et al.* 2003) which is facultative for freshwater snails or freshwater oligochaetes.

(9) Ecto- and endoparasitic species and commensal species, often host specific, that usually do not kill the host. Phoridae: the apterous *Wandolleckia* spp., adults and larvae ectoparasitic inside the mantle cavity of *Achatina fulica* in Africa (Schmitz 1917, Bequaert 1925, Baer 1953, reviewed by Mead 1961). Chironomidae: the first publication on Chironomidae associated with molluscs appears to be the rather detailed paper by Barnard (1911), which has not been cited by most subsequent authors. He found larvae of a species tentatively identified as *Chironomus niveipennis* in *Radix balthica* in the fountains at Trafalgar Square, London, on many occasions. He experimentally infested snails several times. He noted, "The young larvae enter the pulmonary orifice, and burrow in the mantle, eventually reaching the liver . . . They inconvenience the host, but to what extent I do not fully know." Mathias & Boulle (1933) reported "*Lymnaea limosa*" parasitized by chironomid larvae. Rozkošný (1968) reviewed the early papers of Benthem-Jutting (1938) and Guibé (1942). He regarded "*Parachironomus varus varus*" as an ectoparasite, living in a cocoon on the shell of *Physa fontinalis* and feeding on the snails' integument. He regarded "*P. v. limanei*" as a true hematophagous endoparasite of *Lymnaea* sp., living on the surface of the snails' liver and feeding on blood. Armitage *et al.* (1995) reviewed most of the earlier literature and eight more recent papers on Chironomidae as parasites and commensals of Mollusca and other organisms. They overlooked the papers of Mathias & Boulle (1933), Pesigan *et al.* (1958), and Pringle (1960) (see below). Armitage *et al.* (1995) referred to "*Cryptochironomus* spp." as parasites of *Lymnaea*, *Radix*, and *Physa* snails and noted that the chironomid larvae "are in direct contact with the digestive glands and the gonads, and the infestation by all four larval instars and the relatively high content of hemoglobin and glycogen in the body of the chironomid larvae suggest that they are

truly parasitic." These authors noted the observation of Meier (1987) that infested *Radix* died soon after "*Parachironomus varus*" larvae left them. They also listed, as commensal species, *Eukiefferiella ancyla* on *Ancylus fluviatilis* in Sweden, and *Rheotanytarsus* spp. on *Oxytrema carinifera* in the USA and *Goniobasis* sp., an operculate snail in the USA. See also their review of chironomids associated with Bivalvia, below.

(10) Larvae restricted to the eggs of slugs during the entire life cycle. Phoridae: *Megaselia tertia, M. aequalis*, and *M. ciliata* (Robinson & Foote 1968, Disney 1977, 1979). Also larvae restricted to the eggs of snails and in later larval life to mature snails. Sciomyzidae: Behavioral Group 5, *Anticheta*. Chironomidae: curiously, the papers by Pesigan *et al.* (1958) recording the destruction of egg masses of *Oncomelania* sp. by *Polypedilum anale, P. kibatiense*, and *Chironomus acutus* in the Philippines and Pringle (1960) recording the invasion of egg masses of *Bulinus globosus* in Africa by larval chironomids have not been cited in subsequent reviews of Chironomidae or natural enemies of molluscs. The habits of these chironomids are not well known; it is unlikely that they are obligate snail-egg predators.

(11) Larvae restricted to killing Bivalvia. Sciomyzidae: Behavioral Group 14, *Renocera* (Foote 1976, Horsáková 2003), *Ilione lineata* (Foote & Knutson 1970), and *Eulimnia philpotti* (Barnes 1980a). The only other records of fly larvae associated with bivalve molluscs concern Chironomidae; most records were reviewed by Armitage *et al.* (1995). They noted that third- and fourth-instar larvae and pupae of *Baeoctonus bicolor* were found in the mantle cavity of two species of *Anodonta* in New Brunswick, Canada. These larvae construct tubes of particulate organic matter, normally attached to the anterodorsal surface of the gills. The authors stated, "The gills around each larval case were extensively damaged, suggesting that larvae actively fed on gill tissue . . . Despite the apparently parasitic nature of the larvae, it is unknown how seriously the host is affected by this parasitism and how widely the chironomid species occurs." Armitage *et al.* (1995) listed, as commensals, *Ablabesmyia* sp. on *Anodonta* spp., *Fusconaia* sp., and *Quadrula* sp. in the USA, and *Xenochironomus canterburyensis* on the lamellibranch bivalve *Hyridella menziesi* in New Zealand. Also, Beedham (1966) found a chironomid larva (*Glyptotendipes ? paripes*) between the shell and mantle of the large *Anodonta cygnea* in England, a likely case of opportunistic parasitism. Chandler *et al.* (1978) referred to larvae of the chironomid *Metriocnemus* sp. found in the mantle of the freshwater mussel *Dreissena* sp. as likely casual visitors.

As noted above for malacophagous insects in general, there have been no studies of guilds, in the sense of Root (1967), of malacophagous Diptera, except the papers by R. Beaver (1972, 1977) on Diptera breeding in dead snails. Although not using the term guild, the author emphasized aspects typical of guild studies: patterns of abundance of species and individuals, successional and seasonal changes in patterns of resource use, life-history strategies, composition of communities, etc. The reviews cited above and our classification of categories of malacophagy would seem to form a starting place for broader studies of malacophagy in Diptera.

4 • Life cycles

> Research is the process of going up alleys to see if they are blind.
>
> M. Bates.

The food and feeding behavior of sciomyzid larvae were discussed under eight Behavioral Groups by Berg & Knutson (1978) and under somewhat different Behavioral Groups, in each instance, by Greathead (1981, eight Groups), Ferrar (1987, nine Groups); Knutson & Vala (2002, 15 Groups), and Barker *et al.* (2004, nine Groups). Below we give a thumbnail sketch of the biology, followed by an expanded classification of Behavioral Groups, as in Knutson & Vala (2002).

4.1 OVERVIEW OF LIFE CYCLES

> We often fail to see what we are not looking for, even when it is clearly visible.
>
> Berg & Valley (1985a).

With information on the larval behavior of 203 of the 539 species (38% of the family) in 41 of the 61 genera, the Sciomyzidae is one of the best known families of Diptera. The flies pass through three typical larval instars, the brief fourth larval instar (or pre-pupal stage), pupal stage, pharate adult inside the puparium, adult, and egg stages (Figs 4.1 and 4.2). There are several distinct phenologies, with overwintering most commonly occurring in the pupal stage, but also, depending upon the species, in the adult, egg, or larval stage. Most species are multivoltine, but some are uni- or bivoltine. Larval habitats and food molluscs range from strictly freshwater to strictly terrestrial (Plate 3). The feeding behavior of larvae is exceedingly diverse, from saprophagous to predaceous to parasitoid, with many species having mixed or facultative behaviors. Freshwater, predatory behavior is best developed in the Tetanocerini, whereas terrestrial, parasitoid behavior is best developed in the Sciomyzini; the most intimately associated relationships in the latter are characterized by one larva per snail and one snail per larva. Species of Sciomyzini are muscoid, maggot-like, without adaptations for a freshwater existence, and feed on terrestrial, semi-terrestrial, or exposed freshwater snails. Their behavior ranges from saprophagous to parasitoid, with some predatory aspects in some species. Most Tetanocerini are adapted for a freshwater, predatory existence, each larva killing a dozen or more snails. A few feed on freshwater snails or fingernail clams beneath the surface for long periods. Tetanocerini in several genera are parasitoids/predators of terrestrial snails and/or slugs. *Sepedonella nana* and *Sepedon* (*M.*) *knutsoni* are obligate predators of freshwater oligochaetes (Plate 4e). *Sepedon* (*P.*) *ruficeps* is a facultative predator of freshwater snails and oligochaetes. Host specificity ranges from a few sciomyzids restricted to certain species or genera of terrestrial or semi-terrestrial gastropods or fingernail clams to the many freshwater and semi-terrestrial species attacking several genera of gastropods or a wide range of species and genera of gastropods in an ecological assemblage. Parasitoid larvae, slug killers, snail-egg feeders, clam killers, and *S. nana* and *S. knutsoni* are more host specific than are other larvae. Freshwater predators of snails feed on many genera in the laboratory. Relatively few have been found feeding in the field. The range of their natural prey may be much narrower in the field than in the laboratory. Puparia of some terrestrial and semi-terrestrial Sciomyzini are formed in the shell, but most sciomyzid puparia are formed in the litter, on the shoreline, or in the water; some of the latter are especially adapted for flotation. Most adults mate readily and frequently, the females laying many eggs in the microhabitat of the larvae. However, a few parasitoid Sciomyzini oviposit onto the shell of the host, e.g., *Pherbellia s. schoenherri* (Plate 4a). *Pelidnoptera nigripennis* (Phaeomyiidae), formerly included in the Sciomyzidae, is a parasitoid of diplopods (Figs 4.4, 4.5).

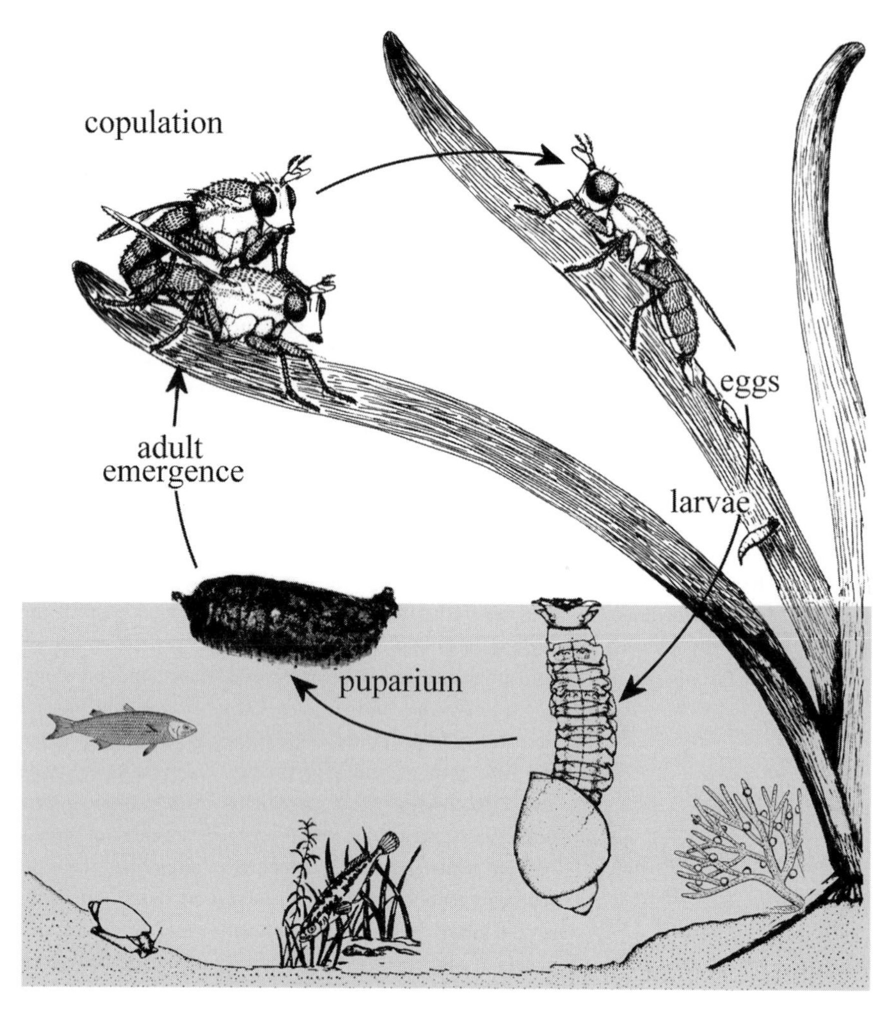

Fig. 4.1 Life cycle of a typical aquatic-predaceous sciomyzid, *Tetanocera ferruginea*. Modified from a drawing provided by B. A. Foote.

4.2 GENERALIZATION, SPECIALIZATION, AND LABILITY IN BEHAVIORAL ATTRIBUTES

> Yes, many things there are, which seem to be
>
> Perplexing, though quite falsely so, because
>
> They have good reasons which we cannot see.
>
> Dante.

Major behavioral attributes occurring throughout the life cycles are here characterized as generalized, specialized, or labile. Such characterization should aid an overview of behavior as background for our further detailed discussions. Such characterization also is useful in developing an evolutionary scenario. We are mindful that we are likely yet again taking the liberty of defining categories on paper where none exists in nature. In general, we agree with criticisms of rather rigid categorization, e.g., the statement by Tauber *et al.* (1984) that "it is not useful ... to view behavioral patterns as strictly learned (plastic) or strictly instinctive (stereotyped). The vast majority of response patterns have both genetic and environmental components," and "the productive approach is to elucidate the mechanisms underlying both components and to determine how they interact to adapt the organism to its environment." Such a mechanistic approach can be

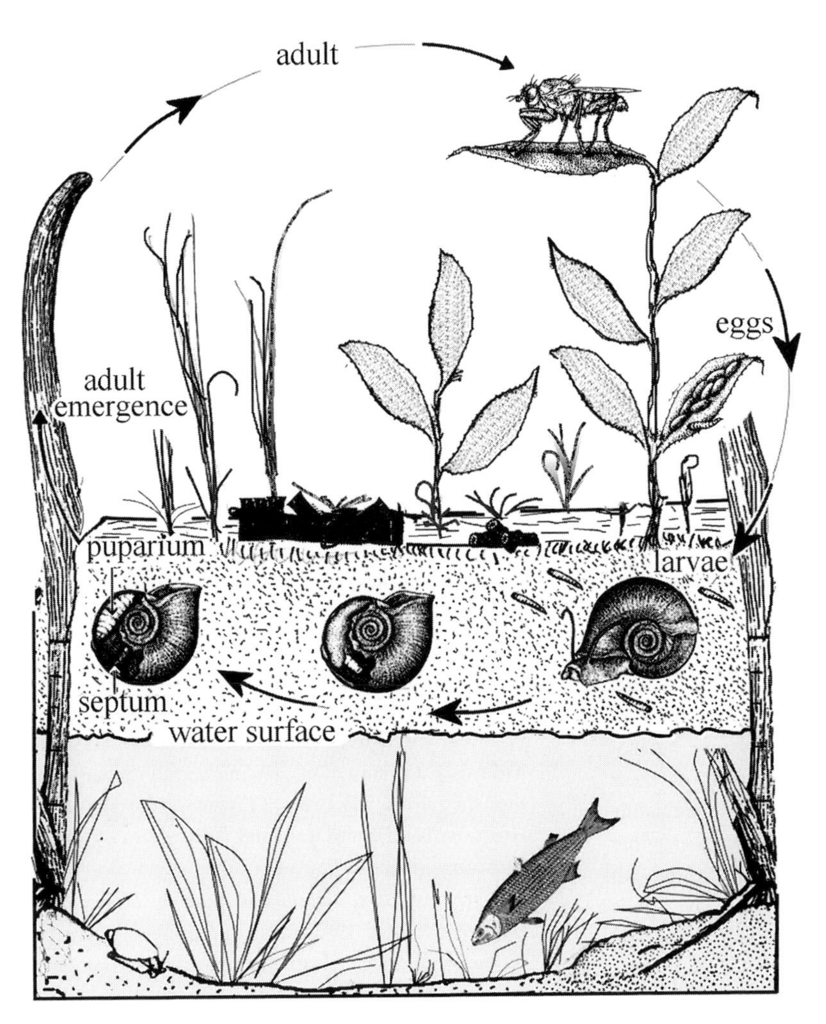

Fig. 4.2 Life cycle of a typical shoreline predator/saprophage, ***Pherbellia argyra***. Modified from Bratt *et al.* (1969).

applied on a case-by-case basis to specific behavioral attributes and of course is valuable. However, a broad analysis of behavior focusing on the products – not the processes – of adaptation, especially if conducted in a phylogenetic/cladistic setting, also is of value. Some authors, e.g., Futuyma & Moreno (1988) in their review of the evolution of ecological specialization, have used the terms generalized and specialized in a strictly ecological sense, as equivalents, respectively, of "eurytopic, polyphagous," etc. and "stenotopic, oligophagous," etc. For the purpose of developing an evolutionary scenario, this seems inadequate. Thus, we equate generalized with ancestral forms and specialized with derived forms. However, we do not use the terms as being based on the manifestations of *morphological* attributes in specific plesiomorphic or apomorphic genera as derived from a cladistic analysis.

We define as generalized those manifestations of an attribute that are (1) simple compared with other manifestations in the family, (2) likely not having resulted from extensive evolutionary pressures, and (3) similar to those of a presumed evolutionary form. Specialized is defined as manifestations that are more complex, probably have arisen as the result of substantial evolutionary pressures, and are not similar to those of a presumed ancestral form.

A few entire genera of Sciomyzinae, some species in genera scattered throughout the family, and – in a few species – different individuals from the same parents or even the same cohort, can be characterized as having labile behavior in regard to some, but not all, behavioral attributes. Lability (= plastic or facultative) is defined as different responses to different situations by the same genotype. Although the term "plastic" is more commonly used in the ecological literature, we prefer the term "labile" for several reasons. The noun form of facultative – facultativeness – is an awkward word. Plastic, of course, means something that can take any form, i.e., graded responses in a continuum of possible responses. That would be the correct usage, but plastic also is used to describe sets of non-graded, alternative responses. Lability in Sciomyzidae is seen especially in selection of microhabitat, selection of host/prey, and in larval feeding behavior. Notably, there is ability to begin feeding on dead or living tissue, to switch back and forth between feeding on dead or living tissue, to feed individually or communally, and there are differences in rate and amount of biomass consumed. Some lability also is seen in the presence or absence of diapause, voltinism, and in the overwintering stage of some species. Danks (2001) pointed out that "Plasticity is advantageous especially when the favorability of environments changes in the short term, so that a fixed response might prove inappropriate" and that "Plasticity is also necessary when habitats are heterogeneous on a small scale." At least the first of these conclusions applies to many Sciomyzidae, where labile feeding behavior is seen in the larvae of those species that live in damp situations, e.g., shoreline habitats that can become dry or can be inundated rather quickly. Labile feeding behavior is less common in the more aquatic and more terrestrial Sciomyzidae, in general. Although morphology varies considerably in the adults of many species of Sciomyzidae (e.g., color, size, chaetotaxy), polymorphism has not been noted, and no behavioral attributes are considered to be polytypic. Generalization, specialization, and lability also are discussed in Chapter 1 and Section 17.7. Below, G = generalized, S = specialized.

4.2.1 Adults

a. Pre-mating movements and posture normal, G; unusual, S.
b. Mating movements and posture normal, G; unusual, S.
c. No nuptial feeding, G; nuptial feeding, S.
d. In temperate climates do not overwinter as adults, G; overwinter as adults, S.
e. Overwintering adults in quiescence, S.
f. Overwintering adults in diapause, S.
g. Multivoltine, G; univoltine, S.

4.2.2 Eggs

a. Scattered singly or in small, disorganized groups on or slightly above the microhabitat substrate, G.
b. Laid side by side on vegetation, S.
c. Laid end to end on vegetation, S.
d. Laid on the shell of the host (a few Sciomyzini) or on intersegmental membranes of diplopods (*Pelidnoptera nigripennis*), S.
e. Laid on the soft tissues of the host, S.
f. Laid on eggs of the hosts/prey, S.
g. Diapause does not occur, G; occurs, S.

Notes:

(1) Coupland *et al.* (1994) speculated that *Salticella fasciata* oviposits on the host to avoid competition from other saprophagous flies. They concluded that oviposition on the substrate and larval ability to search for hosts/prey "seems to be a more recently derived feature of terrestrial sciomyzids." We consider oviposition on the host as a specialized manifestation. It occurs only in a few, specialized, parasitoid species of *Sciomyza*, *Colobaea*, *Tetanura*, and *Pherbellia* (all Sciomyzini); the specialized *Anticheta* (Tetanocerini); the labile *Atrichomelina* (Sciomyzini); and *Pelidnoptera nigripennis* (Phaeomyiidae).
(2) The eggs of *Ilione lineata* and possibly of *I. albiseta* (Knutson & Berg 1967) seem to require submergence in order to hatch. Hatching of eggs of *Euthycera cribrata* is often delayed several weeks until the chorion has been softened by rainfall or humidity – a specialized attribute (Reidenbach *et al.* 1989).
(3) Egg masses of a few species of *Sepedon* have been found in nature on vegetation ranging from 5 cm to 1.5 m above the substrate – a specialized attribute (Neff & Berg 1966).

4.2.3 Larvae

a. Essentially non-host specific, attack broad array of prey in an ecological assemblage, G; relatively host specific, S.
b. First instars able to begin feeding as a saprophage or on living hosts/prey, G; (a labile attribute of individual species).
c. First instars must attack living hosts/prey, then can feed saprophagously, S.
d. Almost entire larval development must be spent feeding on a series of freshly killed prey, S.

e. Able to switch nutrition during development between dead or living tissue, G (a labile attribute of individual species).
f. Entire larval period spent as an intimately associated parasitoid: one larva per host and one host per larva, S.
g. At least first instars must feed on embryonic snails, S.
h. Able to feed communally, G.
i. Usually feed individually, S.
j. Feed at water surface during most of larval life, S.
k. Feed beneath water surface, at least during early larval life, S.
l. Feed on damp substrates (shorelines, seepage areas, etc.), G.
m. Terrestrial, S.
n. Overwinter as developed larvae in diapause in egg membranes, S.
o. Overwinter as quiescent, mature larvae that pupariate in late winter to early spring, S.
p. Distinctive swimming behavior, S.
q. Not known to swallow air, G; swallow air, S.

Notes:

(1) Coupland *et al.* (1994) stated that "the more saprophagous than parasitoid nature ... of larval feeding in *S. fasciata* ... supports the view that it is phylogenetically close to the saprophagous dipteran ancestors of the Sciomyzidae ... but that its host specificity and oviposition on the host ... confuses its exact evolutionary placement." However, the cladistic analyses very reliably place the Salticellinae as the plesiomorphic sister-group of the Sciomyzidae.
(2) Larvae that feed on semi-terrestrial, "aerial" Succineidae or estivating Lymnaeidae are placed in category l because their hosts/prey probably fall to the damp surface of the habitat after being attacked.
(3) First-instar larvae of some slug-killing *Tetanocera* (Knutson *et al.* 1965, Trelka & Foote 1970), two *Euthycera* spp. (Knutson, unpublished data; Reidenbach *et al.* 1989), and some terrestrial snail-killing *Pherbellia* spp. (Bratt *et al.* 1969) do not search for the hosts but remain near the eggshells and ambush the passing hosts. Appleton *et al.* (1993) observed that first-instar larvae of *Sepedon neavei* suspend themselves from the water surface and attack snails passing below ("ambush" behavior, specialized attribute).
(4) Information on attributes h and i is based largely on laboratory rearings, thus artifices of the rearing conditions probably have resulted in more observations of communal feeding than actually occur in nature.

4.2.4 Pupae and puparia

a. Formed in shell (many Sciomyzini), in exoskeleton of diplopod (*Pelidnoptera nigripennis*, Phaeomyiidae), without calcareous septum or anterior end expanded, S.
b. Formed in shell, with calcareous septum, S.
c. Formed in shell, with anterior end expanded, S.
d. Formed in water, outside host/prey, S.
e. Formed on solid substrate, outside host/prey, G.
f. Overwinter in puparium, in quiescence, G.
g. Overwinter in puparium, in diapause, S.

Notes:

(1) (a, e). Site of pupariation labile in *Sciomyza aristalis* and *S. simplex*.
(2) (f, g). Overwintering in diapause labile in *Colobaea americana*, *C. bifasciella*, and *C. pectoralis*.

Another way of describing life cycles in Sciomyzidae is to typify some of the responses to the hosts/prey and environment in the various life-cycle stages. Danks (1999a, 1999b), primarily in regard to phenological and developmental aspects, considered responses as (1) fixed, (2) variable or programmed (modal and attenuated), or (3) flexible or opportunistic. These terms are compared in Table 4.1. It is reasonable to assume that in at least some species of Sciomyzidae various combinations of fixed, variable, and flexible responses exist. Fixed responses likely predominate in species living in more stable macrohabitats and microhabitats, e.g., truly aquatic species living in the water around emergent vegetation. Variable and flexible responses likely become more frequent in species living in less stable habitats, e.g., the seasonally but not closely predictably flooded temporary marshes of *Ilione albiseta*. The flexible response of Danks (1999a, 1999b) is more or less equivalent to the labile behavior as used above. While most of the various behavioral aspects and responses of Sciomyzidae can be termed specialized or fixed, it is particularly in feeding behavior and to some extent microhabitat selection and host/prey range that we see generalized behavioral aspects and flexible responses. From a phylogenetic viewpoint, the sum of generalized behavioral aspects are essentially equivalent to those of the ancestral and most primitive Sciomyzidae.

The concept of "trade off" perhaps can be applied to a behavioral aspect and to flexible responses. For example, the behavioral aspect of the ovipositing female consistently scattering eggs or placing them in a mass – an aspect not variable within a species – could be a trade-off between

Table 4.1. Types of life-cycle responses in insects

Type of response	Characteristics	Correlate	Example
1. Fixed	Set, unchangeable	Predictable conditions	Synchronized emergence
2. Programmed			
a. Modal	Grouped into subsets	Predictable seasonal change; also risk spreading	Diapause and non-diapause
b. Attenuated	Scattered, spread out	Risk spreading; widened resource use	Extended emergence
3. Flexible (opportunistic)	Opportunistic, usually cued by environment	Indicators for future conditions are reliable	Cued emergence time

Note: Modified from Danks (1999b).

(1) concentrating eggs and the resultant larvae in a snail-rich spot – assuming that the female can distinguish such spots – but exposing them to attack by hymenopterous parasitoids or (2) reducing the chances of attack by placing some eggs where snails are not as abundant – assuming that larval searching time is not as critical as adult searching time. The flexible response of larvae such as *Atrichomelina pubera* to feed gregariously or individually and the fixed response of larvae of intimately associated parasitoids such as *Sciomyza varia* to feed only individually could be trade-offs. That is, satisfying hunger immediately – thus saving the energy lost in further searching – against being out-competed and starved by more rapidly feeding, stronger larvae (plus possibly being subject to attack by those larvae). Many so-called trade-offs appear to have exceedingly obscure results.

4.3 CLASSIFICATION OF BEHAVIORAL GROUPS

> *Parmi les combinaisons que l'on choisira les plus fertiles sont souvent celles formées d'éléments tirés de domaines qui sont trés éloignés.* [Among chosen combinations the most fertile often are those formed of elements drawn from domains which are far apart].
>
> Poincaré (1908).

Knutson & Vala (2002) presented a more finely dissected classification of Behavioral Groups, 17 rather than the eight or nine previously recognized, to accommodate new behaviors that have been discovered recently. We believe it is important to deal with more or less discrete groups of species having very similar life styles in discussing behavior and especially in developing an evolutionary scenario. See Ferrar (1987) for a tabulation of most publications on life cycles through 1986 and Table 4.2 for documents produced since Berg & Knutson (1978).

Our current Behavioral Groups form an a posteriori classification of all biologically known Sciomyzidae of the world, each group or subgroup based on knowledge of the actual attributes of one to many species. The attributes are the kind of food eaten, the manner of killing and feeding, and the microhabitat. Thus, our Behavioral Groups are more similar to the "functional groups" rather than to the "guilds" of the ecological literature (Hawkins & MacMahon 1989). Unlike guilds in the sense of Root (1967), our Groups are not limited to sympatric species or restricted to members of a community.

In the classification below, where the feeding behavior is not entirely predaceous, parasitoid, or saprophagous but changes more or less regularly during the course of development, or is variable depending upon the presence of intraspecific competition, relative sizes of larva and host/prey, and/or microhabitat conditions, all behaviors are given, the predominant behavior first. In most of such cases, young larvae are more parasitoid and older larvae are more predatory and/or saprophagous. Genera and species of Tetanocerini are given in italics, Sciomyzini in bold, and Salticellinae are underlined. Species names are given where only one or a few species in a genus are known to exhibit the behavior. See also the categories of behavior based on feeding sites (Section 7.1) and posterior disc morphology (Section 14.1.3.3). In 1–15 below, A = Australian;

Table 4.2. Information on basic life cycles published since or not included in Berg & Knutson (1978), or in manuscripts in preparation, or in theses; not including studies of experimental nature

Species	Distribution	Behavior	Hosts/Prey	Reference
Colobaea (4 species)	P, N	Parasitoid Parasitoid/ predator	Freshwater snails	Knutson & Bratt, (unpublished data)
Coremacera marginata	P	Predator	Terrestrial snails	Knutson 1973
Dichetophora obliterata	P	Parasitoid/ predator	Terrestrial snails	Vala *et al.* 1987
Dictya floridensis	N	Predator	Freshwater snails	McLaughlin & Dame 1989
Dictya montana	N	Predator	Freshwater snails	Mc Donnell *et al.* 2007b
Dictya umbrarum	P	Predator	Freshwater snails	Willomitzer & Rozkošný 1977
Dictyodes dictyodes	NT	Predator	Freshwater snails	Abercrombie & Berg 1978
Eulimnia philpotti	SA	Predator	Fingernail clams	Barnes 1980a
Euthycera cribrata	P	Parasitoid/ predator	Terrestrial snails, slugs	Vala *et al.* 1983 Reidenbach *et al.* 1989
Euthycera stichospila	P	Parasitoid/ predator	Terrestrial snails, slugs	Vala & Caillet 1985
Hydromya dorsalis	P	Predator	Freshwater snails	Hope Cawdery & Lindsay (1977)
Ilione albiseta	P	Predator	Freshwater snails	Lindsay 1982, Lindsay *et al.* (2011)
Ilione lineata	P	Parasitoid/ predator	Fingernail clams	Knutson (unpublished data)
Limnia unguicornis	P	Parasitoid/ predator	Terrestrial snails, slugs	Vala & Knutson 1990
Neolimnia (8 species)	SA	Predator	3 on freshwater non-operculate snails 1 on freshwater operculate snails 4 on terrestrial snails	Barnes 1979b
Oidematops ferrugineus	N	Parasitoid	Terrestrial *Stenotrema hirsutum*	Foote 1977
Pherbellia griseicollis	H	Shoreline predator	Freshwater and terrestrial snails	Knutson 1988
Pherbellia inflexa	N	Parasitoid?	Terrestrial *Zonitoides* sp.	Foote 2007
Pherbellia limbata	P	Parasitoid	Terrestrial *Granaria frumentum*	Nerudová (Horsáková) & Vala (unpublished data)
Pherbina mediterranea	P	Shoreline predator	Freshwater snails	Vala & Gasc 1990a
Poecilographa decora	N	?	?	Barnes 1988
Protodictya (4 species)	NT	Predator	Freshwater snails	Abercrombie 1970
Renocera pallida	P	Parasitoid/ predator	Fingernail clams	Horsáková 2003 Knutson (unpublished data)
Renocera striata	H	Parasitoid/ predator	Fingernail clams	Knutson (unpublished data)

Table 4.2. (*cont.*)

Species	Distribution	Behavior	Hosts/Prey	Reference
Sciomyza dryomyzina	H	Parasitoid/ predator	Freshwater snails	Knutson 1988
Sciomyza testacea	P	Parasitoid	*Succinea* sp.	Knutson 1988
Sciomyza varia	N	Parasitoid	Estivating *Stagnicola elodes*	Barnes 1990
Sepedon knutsoni	AF	Predator	Freshwater oligochaete worms	Vala *et al.* 2002
Sepedon ruficeps	AF	Predator	Freshwater snails	Gbedjissi *et al.* 2003
Sepedon senex	O	Predator	Freshwater snails	O. Beaver 1989
Sepedon neavei, S. scapularis, S. testacea	AF	Predator	Freshwater snails	Barraclough 1983 Maharaj *et al.* 1992 Appleton *et al.* 1993
Sepedon trichrooscelis	AF	Parasitoid	Succineidae	Vala *et al.* 1995 Knutson 2008
Sepedon umbrosa	AF	Parasitoid/ predator	Terrestrial *Subulina octona*	L. G. Gbedjissi (pers. comm., 2003)
Sepedonea (9 species)	NT	Predator	Freshwater snails	Mello & Bredt 1978b Freidberg *et al.* 1991 Abercrombie 2000
Sepedonella nana	AF	Predator	Freshwater oligochaete worms	Vala *et al.* 2000 Vala & Gbedjissi (unpublished data)
Tetanocera (20 species)	H, N	Predator Parasitoid/ predator Predator Parasitoid/ predator	2 on shoreline snails 4 on Succineidae 10 on freshwater snails 4 on terrestrial snails, slugs	Foote 1996a Foote 1996b Foote 1999 Foote (2008)
Tetanoceroides (4 species)	NT	Predator	Freshwater snails	Abercrombie 1970
Tetanura pallidiventris	P	Parasitoid	Terrestrial snails	Knutson 1970a
Thecomyia limbata	NT	Predator	Freshwater snails	Abercrombie & Berg 1975
Trypetoptera punctulata	P	Parasitoid/ predator	*Lauria cylindracea* (viviparous terrestrial snail)	Vala 1986

AF = Afrotropical; H = Holarctic; N = Nearctic; NT = Neotropical; O = Oriental; P = Palearctic; SA = Subantarctic; WW = worldwide.

(1) Facultative, opportunistic, predators/parasitoids/ saprophages that can feed on dead, moribund, or living snails: *Salticella fasciata* (P) (terrestrial) (see also Section 18.2); or snails or clams. ***Atrichomelina pubera*** (N) (moist surfaces). See also Section 17.6.

(2) Predators/saprophages of non-operculate, primarily freshwater snails exposed on moist surfaces by receding or fluctuating water levels or while foraging or migrating, that is, most of the "stranded snail" situations of Greathead (1981) and Ferrar (1987) and the "shoreline" situation of Foote (1996a). ***Colobaea americana*** (N), ***C. pectoralis*** (P), ***C. punctata*** (P); ***Ditaeniella grisescens*** (P), ***D. parallela*** (N, NT);

at least 19 ***Pherbellia*** spp. (WW-SA); four ***Pteromicra*** spp. (N, P, AF); ***Sciomyza simplex*** (H); *Hydromya dorsalis* (P); *Perilimnia albifacies* (NT); *Pherbina coryleti*, *P. intermedia*, *P. mediterranea* (all P); *Protodictya nubilipennis*, *P. apicalis* (both NT); *Psacadina disjecta*, *P. verbekei*, *P. zernyi* (all P); *Shannonia costalis* (NT); *Tetanocera fuscinervis*, *T. silvatica* (both H).

(3) Parasitoids or parasitoids/predators more or less intimately associated with non-operculate freshwater snails estivating or otherwise exposed for a long period of time in temporary freshwater habitats (e.g., vernal ponds, marshes, swamps, playa lakes). ***Colobaea bifasciella*** (P); some ***Pherbellia*** spp. (WW-SA); ***Sciomyza varia*** (N).

(4) Parasitoids or parasitoids/predators more or less intimately associated with hygrophilous, semi-terrestrial Succineidae snails. ***Pherbellia s. schoenherri*** (P), ***P. s. maculata*** (N); ? ***Pteromicra anopla*** (N); ***Sciomyza aristalis*** (N), ***S. dryomyzina*** (H), ***S. testacea*** (P); *Hoplodictya spinicornis* (N); *Sepedon hispanica* (P, AF), *S. trichrooscelis* (AF); five *Tetanocera* spp. (P, N). ***Pherbellia s. schoenherri*** (P) also was found feeding on *Galba truncatula* in nature. See also Section 7.2 for detail on parasitoid species included in Groups 3 and 4.

(5) Obligate parasitoids/predators of exposed egg masses of freshwater Lymnaeidae or *Aplexa* (Physidae) or semi-terrestrial Succineidae snails during early larval life, followed by predation on juvenile to mature snails in damp or vernal situations. This niche is dominated by *Anticheta* spp. (N, P). *Hydromya dorsalis* (Group 2) and *Tetanocera ferruginea* (Group 11) also have been found occasionally in egg masses in nature. In laboratory studies larvae of 12 species in Groups 2 and 11 fed on eggs of freshwater snails. See Tables 5.1 and 5.2.

(6) Parasitoids intimately associated with terrestrial non-operculate snails. ***Oidematops ferrugineus*** (N); at least six ***Pherbellia*** spp. (WW-SA); ? ***Pteromicra steyskali*** (N); ***Tetanura pallidiventris*** (P).

(7) Predators/saprophages of non-operculate terrestrial snails. ***Pherbellia cinerella*** (P); *Coremacera marginata* (P); *Dichetophora obliterata* (P); *some Neolimnia* spp. (A); *Sepedon umbrosa* (AF); some *Tetanocera* spp. (H); *Trypetoptera canadensis* (N), *T. punctulata* (P). Some species, e.g., *S. umbrosa*, have some parasitoid aspects of behavior during early larval life.

(8) Predators/saprophages opportunistic on both terrestrial snails and slugs. *Euthycera arcuata* (N), *E. cribrata* (P), *E. stichospila* (P); *Limnia unguicornis* (P), ? *L. paludicola* (P); some *Tetanocera* spp. (H).

(9) Obligate ectoparasitoids/predators of slugs. *Tetanocera clara* (N), *T. elata* (P), *T. plebeja* (H) (third-instar *T. plebeja* also have been found in nature in semi-terrestrial *Oxyloma* snails and fed on terrestrial snails in laboratory rearings), *T. valida* (N). Ectoparasitoid slug feeders keep at least their posterior spiracles exposed; mesoparasitoids (below) live completely within the slugs.

(10) Obligate mesoparasitoids of slugs. *Euthycera chaerophylli* (P) (not reared beyond second stadium; up to about 30 days entirely within one living *Deroceras* sp. slug, without the posterior spiracles exposed to the exterior), and probably also *E. arcuata* (N).

(11) Predators of non-operculate snails at or just below the water surface, just above the surface on emergent vegetation, and occasionally those exposed on moist surfaces. Most *Dictya* spp. (N, P, NT); *Dichetophora biroi*, *D. hendeli* (both A); *Dictyodes dictyodes* (NT); four *Elgiva* spp. (N, P); five *Ilione* spp. (P); *Neolimnia repo*, *N. sigma*, *N. ura*. (all SA); *Protodictya chilensis*, *P. guttularis*, *P. lilloana* (all NT); most *Sepedon* spp. (WW-SA); *Sepedomerus caeruleus*, *S. macropus* (both NT); many *Tetanocera* spp. (N, P); four *Tetanoceroides* spp. (NT); *Thecomyia limbata* (NT); ten *Sepedonea* spp. (NT). Most larvae live at the water surface, with their posterior spiracles exposed, most of the time. Several aquatic predators habitually leave the water for moist surfaces when mature. Larvae of some species in this Group often have labile feeding behavior and might be placed as well in Group 2.

(12) Predators and predators/parasitoids of exposed and neustonic operculate aquatic snails.

a. *Hoplodictya setosa* (N) attacking *Littorina littorea* in strandline debris on Atlantic Ocean beaches.
b. *Dictya lobifera*, *D. oxybeles*, *D. pechumani* (all N) attacking saltmarsh operculates.
c. ***Pherbellia prefixa*** (N) attacking *Valvata* spp. exposed in freshwater marshes.
d. *Dictya fontinalis* (N); probably some *Ilione* spp. (P); *Neolimnia tranquilla* (SA) attacking neustonic freshwater operculate snails.

Note: The four species of Sciomyzidae in Groups 12 *a* and *b*, along with two species of Sarcophagidae, are the only Insecta known to be or very likely to be restricted to marine Gastropoda.

(13) Predators of non-operculate snails under the water surface, at least during the first part of larval life. Probably a few *Dictya* spp. (N, P, NT); *Hedria mixta* (N); *Ilione albiseta*; and *I. trifaria* (both P).

(14) Predators/parasitoids of fingernail clams. *Eulimnia philpotti* (SA); *Ilione lineata* (P); four *Renocera* spp. (H). All except *R. pallida* (P) feed beneath the water surface, at least during the first part of larval life.

(15) Predators of freshwater oligochaete worms. *Sepedonella nana*, *Sepedon knutsoni*, and *Sepedon ruficeps* (all AF) (the latter is opportunistic on oligochaetes and is primarily a predator of freshwater snails). [Note: larvae that feed listlessly on snails and other organisms provided in laboratory rearings (e.g., *Dictya ptyarion* [NA], Valley & Berg 1977) should be offered oligochaetes and other non-molluscans.]

The original Behavioral Groups of Berg & Knutson (1978) were not numbered to indicate any evolutionary sequence or relationships but were presented simply with the most common types first, followed by the unusual biologies (aquatic predators, then terrestrial parasitoids, scavengers, slug killers, egg or embryo eaters, subsurface foragers, killers of operculate snails, and clam killers). Subsequently, there have been four comprehensive proposals of Behavioral Groups (Greathead 1981, Ferrar 1987, Knutson & Vala 2002, Barker *et al.* 2004). The major differences among the earlier classifications and our classification concern scavengers (our current Group 1), aquatic predators (our Groups 11, 12, 13), and species feeding on "stranded" snails (primarily our Group 2). Our Groups are based on a combination of feeding behaviors, microhabitat, and hosts/prey, as were the previously described groups, without regard to the taxonomic or cladistic position of the flies. Our Groups are arranged, more or less, from least to most specialized and from those on damp substrates, to terrestrial species, to aquatic species.

Behavioral Groups have been discussed in relation to the species and genera of regional faunas in several works (Rozkošný 1984b, 1997a for northern Europe; Rozkošný 1998 for the Palearctic; Vala 1989a for Mediterranean Europe; and Rivosecchi 1992 for Italy). Notably, their Groups follow those of Berg & Knutson (1978), which they all cite, but none cites Greathead (1981), and only Rozkošný (1997a) cites Ferrar (1987). Chandler *et al.* (1978) listed the reared British species in nine groups.

Greathead (1981) presented a detailed summary table of the host preferences and habitats of the 25 genera published upon by then. He arranged the genera in eight Groups, differing primarily from that of Berg & Knutson (1978) by referring to their "scavengers" as his "Group 2, predators of stranded snails," including 11 genera, many more than did Berg & Knutson (1978). Moreover, he did not recognize a group of operculate snail killers but placed them in his Groups 2 and 1 ("Predators of aquatic snails"), separating the latter Group into three subgroups, i.e., those "a. hunting beneath the surface in still water, b. hunting in surface film in moving water, and c. predators in surface film of standing water." We consider his Group 1a (our Group 13) as too inclusive, i.e., it should not include all *Dictya* species. We do not recognize his Group 1b because we do not believe any of the reared Sciomyzidae are restricted to moving waters.

The appeal of recognizing a Group attacking "stranded" snails is indicated by Ferrar's (1987) apparently independent proposal of such a Group (e.g., he did not cite Greathead [1981] in his extensive bibliography). Ferrar's (1987) summary of Behavioral Groups essentially followed the original proposal of eight Groups by Berg & Knutson (1978), adding one group by separating parasitoids and predators of terrestrial snails from those attacking "stranded freshwater snails."

We do not recognize a Group restricted to stranded snails because we feel that "stranded" (= involuntarily, accidentally, and usually permanently separated from the primary, freshwater microhabitat) has been used in too broad a sense, as discussed in Section 1.2, and because it hides some behavioral, host/prey, and microhabitat adaptations and distinctions. Separating out these distinctions should be helpful in focusing further life-cycle studies. Many sciomyzid larvae, especially those in Groups 1, 2, 3, 4, 8, and 9, have been characterized as having "mixed" or "facultative" (= labile) feeding behavior. Although some species (e.g., *Atrichomelina pubera*, Foote *et al.* 1960) are truly labile in feeding behavior, at least some reports of such behavior probably result from artifices of laboratory rearings that mask the real behavior of the species. We separate out as Group 2 many species that are not truly freshwater or terrestrial but which feed on freshwater snails in damp habitats (referred to as "shoreline predators" by for example Foote 1996a, and including many species characterized as attacking "stranded" snails by Ferrar 1987). We further separate out *Succinea* feeders, obligate slug feeders, and opportunistic slug/snail feeders in the 17 Groups above.

The recent discovery that *Sepedon ruficeps* (Gbedjissi *et al.* 2003), an aquatic snail feeder according to previous studies (Knutson *et al.* 1967), can complete its larval development on freshwater oligochaetes as well as on snails (in laboratory rearings) calls into question the results of rearings of some Sciomyzidae exclusively on snails. Two other African species, *Sepedonella nana* (Vala *et al.* 2000b) and *Sepedon knutsoni* (Vala *et al.* 2002), are restricted to freshwater oligochaetes. Critically needed and of great importance are studies in which supposedly well-known species are reared under situations in which a broad range of hosts/prey are offered.

Although no sciomyzid larvae have been found feeding on freshwater operculate snails in nature, laboratory rearings indicate that such hosts/prey may be much more commonly utilized than earlier believed. Further research on predators of operculates is needed, especially in view of the possibility of biocontrol of operculate hosts of trematode diseases (see Section 5.1). In laboratory rearings, at least seven species of Sciomyzidae in five genera develop only on nine species of aquatic operculate snails (in nine genera of Cerithiidae, Hydrobiidae, Littorinidae, Rissoidae, Thiaridae, and Valvatidae) or strongly prefer them to a wide variety of non-operculate freshwater snails. The occurrence of the adults of most of these sciomyzid species primarily or only in the saltmarsh, freshwater vernal marsh, or spring-fed habitats of their hosts/prey further indicates the natural food preferences of their larvae.

Barker *et al.* (2004) identified 10 Ecomorphological Groups among the Phaeomyiidae (*Pelidnoptera*) and 35 biologically known genera of Sciomyzidae on the basis of ordination analysis of seven attributes of host/prey range and feeding behavior of larvae, four attributes of egg deposition site, six of habitat and microhabitat of larvae, 11 of larval morphology, and two of egg morphology. The Groups identified were Phaeomyiidae (Group 1) + Salticellinae (Group 2) + Sciomyzini (Groups 3–5) + *Anticheta* (Group 6) and Tetanocerini minus *Anticheta* (Groups 7–10) (Fig. 4.3). The authors noted that the dominant gradients correlated ($P < 0.01$) with Axis 1 of the ordination were oviposition site, chorion structure of the egg, ventral arch, mouthhooks, dorsal bridge, windows in cornua, cornual sinus, and pattern of integumentary spinules. The dominant gradients in Axis 2 were the degree of parasitoid oligophagy, extent of occupancy of dryland terrestrial habitat, accessory teeth, number of posterior spiracular disc openings, degree of development of interspiracular processes, and degree of development of posterior spiracular disc lobes. Further conclusions from Barker *et al.* (2004) relative to evolutionary significance of the groups are given in Section 17.1.

4.4 BEHAVIORAL EQUIVALENTS IN ZOOGEOGRAPHIC REGIONS

There are a large number, and some unusual cases, of behavioral equivalents among the Sciomyzidae in almost all of the zoogeographic regions. There have been long-term studies by many researchers over the past half-century in the Nearctic and Palearctic regions, where 66% of the species and 56% of the genera occur. There also has been an intense, 2.5-year period of research in New Zealand by J. K. Barnes and extensive field work in many parts of the Neotropics, especially by J. Abercrombie and C. O. Berg. The rich Afrotropical fauna is relatively poorly known biologically, as is that of the Oriental region. The Australian and Oceanic regions are the least studied. The distribution of behavioral equivalents by zoogeographic regions is shown in Table 4.3. We have omitted the Oceanic region since only three biologically unknown species have been recorded from that region.

The most extreme case of behavioral equivalency is *Colobaea bifasciella* (Knutson & Bratt, unpublished data) and *Sciomyza varia* (Barnes 1990), which are intimately associated parasitoids of estivating *Stagnicola palustris* and *S. elodes*, respectively, and are essentially the same in all behavioral features. Many other species are equivalent in food choice, microhabitat, and in some cases feeding behavior but show differences in various other features (the "trophic guilds" of Foote 1996a). The data in Table 4.3 are to a large extent a reflection of the greater number of species and genera and more extensive studies in the Nearctic and Palearctic than in other regions. However, the occurrences of striking equivalencies between the Holarctic and Afrotropical, and the Holarctic and Subantarctic, are surprising and suggest that other instances of behavioral equivalency may be found in other regions.

4.5 SUMMARY OF UNPUBLISHED AND RECENTLY PUBLISHED LIFE-CYCLE INFORMATION

The review by Berg & Knutson (1978) covered most publications on life cycles through almost all of 1977. The authors also had access to the theses and unpublished information of C. O. Berg's graduate students. Ferrar (1987) provided a summary and species-by-species list of much of the literature on life cycles and immature stages through 1985. Barker *et al.* (2004) gave an updated summary, with special reference to parasitoids and predators of terrestrial gastropods. The list in Table 4.2 refers to basic

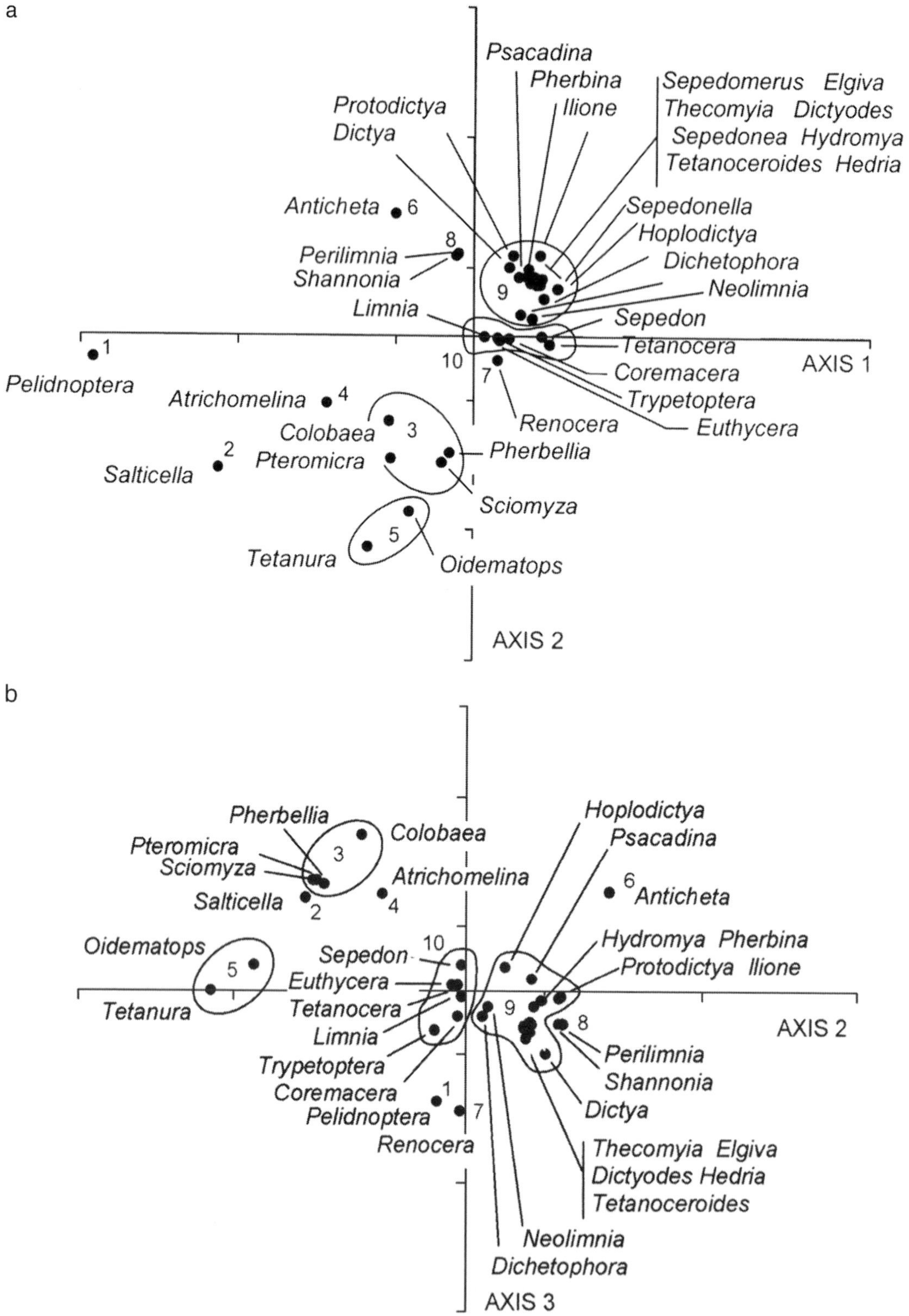

Fig. 4.3 Ordination patterns of 35 biologically known genera of Sciomyzidae (Salticellinae: 2, Sciomyzini: 3–5, Tetanocerini: 6–10), and Phaeomyiidae 1, based on analysis of 30 morphological, behavioral, and habitat features of immature stages (primarily larvae). (a) Axes 1 and 2; (b) axes 2 and 3. See Fig. 17.4. From Barker *et al.* (2004).

Table 4.3. Behavioral equivalents by zoogeographical region

Behavioral Categories	Zoogeographical regions						
	Palearctic	Nearctic	Neotropical	Afrotropical	Oriental	Australian	Subantarctic
1.	*Salticella fasciata*	*Atrichomelina pubera*	—	—	—	—	—
2.	16 spp. in 9 genera	10 spp. in 4 genera	4 spp. in 3 genera	—	—	—	—
3.	*Colobaea bifasciella*, several *Pherbellia* spp.	*Sciomyza varia*, several *Pherbellia* spp.	—	—	—	—	—
4.	5 + spp. in 4 genera	5 + spp. in 5 genera	—	*Sepedon hispanica* *S. trichrooscelis*	—	—	—
5.	*Anticheta* spp.	*Anticheta* spp.	—	—	—	—	—
6.	*Pherbellia* spp. *Pteromicra* spp. *Tetanura pallidiventris*	*Pherbellia* spp. *Pteromicra* spp. *Oidematops ferrugineus*	—	*Sepedon umbrosa*	—	—	—
7.	6 + spp. in 5 genera	3 + spp. in 2 genera	—	—	—	—	*Neolimnia* spp.
8.	5 + spp. in 3 genera	2 + spp. in 2 genera	—	—	—	—	—
9.	*Tetanocera* 3 spp.	*Tetanocera* 2 spp.	—	—	—	—	—
10.	*Euthycera chaerophylli*	—	—	—	—	—	—
11.	20 + spp. in 4 genera	20 spp. in 5 genera	20 + spp. in 7 genera	10 + spp. of *Sepedon*	5 spp. of *Sepedon*	3 spp. in 2 genera	3 spp. of *Neolimnia*
12a.	—	*Hoplodictya setosa*	—	—	—	—	—
12b.	—	*Pherbellia prefixa*	—	—	—	—	—
12c.	*Ilione* spp.?	4 *Dictya* spp.	—	—	—	—	*Neolimnia tranquilla*
13.	*Ilione* spp.?	Most *Dictya* spp. *Hedria mixta*	—	—	—	—	—
14.	*Renocera* spp. *Ilione lineata*	*Renocera* spp.	—	—	—	—	*Eulimnia philpotti*
15.	—	—	—	*Sepedonella nana* *Sepedon knutsoni* (*Sepedon ruficeps*)[1]	—	—	—

Note: [1]Feeds primarily on freshwater snails but will develop on freshwater oligochaetes when offered only that prey.

life-cycle papers published since late 1987 and a few earlier papers not cited by Berg & Knutson (1978). Listed in Table 4.3 is some unpublished information provided to us by various researchers on Sciomyzidae.

The life cycles of *Sepedonella nana* (Vala *et al.* 2000b) and *Sepedon knutsoni* (Vala *et al.* 2002) are of special interest as they are the first true sciomyzids found to have strictly non-malacophagous habits, i.e., feeding on freshwater oligochaete worms. The only other insects known to feed routinely on freshwater oligochaetes are at least two polyphagous Palearctic *Limnophora* spp. (Muscidae; Skidmore 1985) and some predaceous Chironomidae and Culicidae. Species of *Cryptochironomus* (Chironomidae) may be obligate predators of freshwater oligochaetes (Armitage 1968, Loden 1974, Armitage *et al.* 1995). *Cricotopus sylvestris* (Chironomidae), a detritivore and a pest of germinating rice seeds, is a facultative predator of oligochaetes (Loden 1974). In an analysis of the gut contents of *Thienemannimyia festiva* (Chironomidae), freshwater oligochaetes made up only 2.3% of all prey taken, while chironomids accounted for 81.5% and copepods for 7.3% (Tokeshi 1991). Campos & Lounibos (2000) found that in nature and laboratory rearings the predaceous larvae of *Toxorhynchites rutilus* (Culicidae) fed on freshwater oligochaetes as well as on 19 other taxa of aquatic prey in nine insect orders. In France during March, Keilin (1917) found, in wet leaf litter, numerous tiny oligochaetes with larvae of the common *Coenosia agromyzina* (Muscidae) and larvae of *Ula macroptera* (Tipulidae). He found that the *C. agromyzina* larvae preyed upon these oligochaetes and in their absence would attack each other, though usually without harmful effects. Skidmore (1985) found a puparium of *C. agromyzina* under moss on a rotten *Fraxinus excelsior* log; it produced a female about 2 weeks later at room temperatures. Ulrich & Schmelz (2001) reported upon repeated observations of adult Dolichopodidae of ten species in five genera preying upon Enchytraeidae (terrestrial oligochaetes) in woods in Germany. Three genera of Enchytraeidae were found at one of their study sites. They did not state that the Dolichopodidae preyed only on oligochaetes but noted, "While searching for Enchytraeidae, some of the flies met insects which would have been a suitable prey too, but apparently did not take notice of them. It appears that the worms were more attractive." They summarized the few other publications recording Dolichopodidae as predators of oligochaetes, including records of predation of Naididae on a muddy seashore by Dolichopodidae of various subfamilies. They described a piece of Baltic amber in which a dolichopodid and an enchytraeid were found together. Lopes (1979a, 1979b) noted that the sarcophagid "*Notochaeta aurata*" had been reared from an oligochaete. Species of *Bellardia*, *Calliphora*, *Onesia*, and *Pollenia* (Calliphoridae) and *Sarcophaga* (Sarcophagidae) are parasitoids of oligochaete earthworms (records summarized by Ferrar 1987).

In addition to the papers noted in Table 4.2, extensive new information has been published on the previously reared *Pherbellia s. schoenherri* (Vala & Ghamizi 1992), *Tetanocera ferruginea* (Vala & Haab 1984, Manguin *et al.* 1986, 1988a, 1988b; Manguin & Vala 1989, Foote 1999), *Salticella fasciata* (Coupland *et al.* 1994), *Coremacera marginata*, *Euthycera cribrata*, *Pherbellia cinerella*, *Trypetoptera punctulata*, and *Dichetophora obliterata* (Coupland & Baker 1995), *Sepedon sphegea* (Tirgari & Massoud 1981), and *Ilione albiseta* and *P. cinerella* (Gormally 1987a, 1987b, 1988a, 1988b). The biology of *Pelidnoptera nigripennis* (Phaeomyiidae, formerly placed in the Sciomyzidae), a parasitoid of millipedes, was published by Baker (1985) and Bailey (1989). Recent summaries of publications on the immature stages were provided for two of the largest genera, *Tetanocera* (Foote 1996a) and *Sepedon* and related genera (Knutson & Orth 2001).

4.6 LIFE CYCLE OF *PELIDNOPTERA NIGRIPENNIS* (PHAEOMYIIDAE)

As additional background, we note some of the major features of the genus *Pelidnoptera*, since until recently it has been considered one of the most primitive Sciomyzidae and since the life cycle and morphology of the immature stages of one species recently have been described. *Pelidnoptera* includes three western Palearctic species and a species from eastern Russia, Japan and Nepal (Sueyoshi *et al.* 2006, 2009). The genus was transferred to the new family Phaeomyiidae on the basis of five apomorphic characters of the adult by Griffiths (1972); this was supported by characters of the egg and larval stages of *P. nigripennis* described by Vala *et al.* (1990).

Pelidnoptera nigripennis occurs in forests and habitats of low bushes, but not in open grasslands, from central Fennoscandia to the Transcaucasus. Baker (1985) found eggs, larvae, and pupae of *P. nigripennis* (reported as *Eginia* sp., Eginiidae; included as a subfamily of Muscidae by some authors) on and in *Ommatoiulus moreleti* (Diplopoda: Iulidae) in a variety of terrestrial sites at 13 locations near Lisbon, Portugal. He obtained 40 puparia from several hundred *O. moreleti*, but none produced an adult. Bailey (1989) worked out the complete life cycle of

Fig. 4.4 Eggs of ***Pelidnoptera nigripennis*** on *Ommatoiulus moreleti*. Photo by J-C. Vala.

a

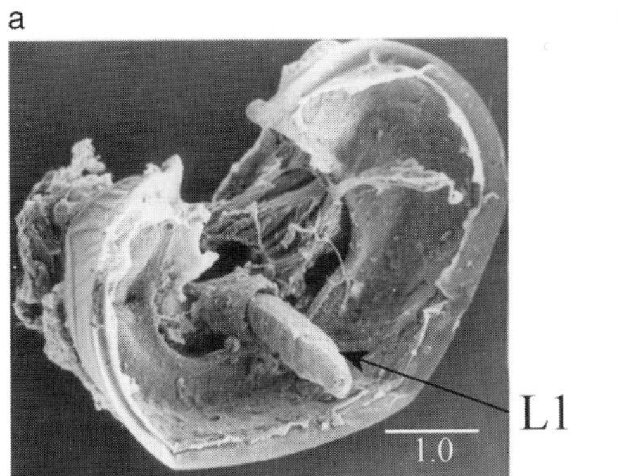

b

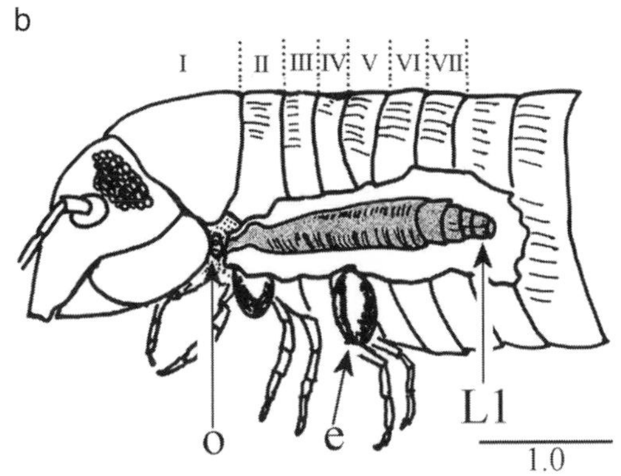

Fig. 4.5 ***Pelidnoptera nigripennis*** L1. (a) Scanning view of L1 *in situ* in *Ommatoiulus moreleti* after dissection (photo by J-C. Vala). (b) ***Pelidnoptera nigripennis*** L1 with most of its body within breathing tube which opens (o) through intersegmental membrane (stippled area); (e) egg. Head of millipede slightly extended. Segments I–VII numbered. Modified from Bailey (1989).

P. nigripennis from field and laboratory studies at the primary location (a forest of *Pinus pinaster* on coastal sand hills) used by Baker (1985), obtaining adults from 44 puparia. He considered the fly studied by Baker (1985) and recorded by Faes (1902) to be *P. nigripennis*.

Females lay 1–9 eggs (usually 1) on the anterior segments (usually 2–6) of stage 8 or larger of the host during the spring (Fig. 4.4). The first-instar larva penetrates through a unique soft area of the intersegmental membrane (Fig. 4.5). The endoparasitoid spends the summer as a first-instar larva within a respiratory tube, kills the host during the fall when in the third instar, and overwinters in the pupal stage inside the remains of the host. *Pelidnoptera nigripennis* is univoltine. The rates of parasitism of several thousand *O. moreleti* collected in nature ranged from 5% to 32%. In laboratory host-specificity studies, 15 species of millipedes from 5 families were exposed to *P. nigripennis* ovipositing females. The females oviposited onto ten species, but eggs adhered to only four, and only three species of *Ommatoiulus* were successfully parasitized. Only *O. moreleti* and *O. oliveirae* have been found parasitized in nature. The biology of *P. nigripennis* also was summarized in Vala *et al.* (1990). The species was released in 1988 in the Mount Lofty ranges, South Australia, for potential biological control of the accidentally introduced *O. moreleti*, but it did not become established.

Baker (1985) briefly summarized 11 publications that refer to various natural enemies of millipedes but overlooked all of the other papers on Diptera cited below, except Picard (1930). Among the Insecta, only about

15 species of Diptera and at least one species of Coleoptera seem to be parasitoids of millipedes. The published life cycles of all of the following species of Diptera are questionable and need to be verified. **Phoridae** – *Plastophora juli* reared from and adults attracted to and ovipositing onto "myriapods" in the USA (Knab 1913); *Megaselia elongata* reared from and adults attracted to *Iulus sabulosus* in France (Picard 1930); an unidentified species of *Megaselia* in France (Fage 1933); *Megaselia* (*Aphiochaeta*) *equitans* "riding on" a "large black millipede" in Yemen (Schmitz 1939); *Plastophora* sp. (Borgmeier 1963); and larvae of an undetermined species dissected from *O. moreleti* in Portugal (Baker 1985). **Muscidae** – two or three spp. of "Eginiidae" (Haase 1885, Ferrar 1987); *Neohelina* sp. reared from *Oncocladosoma* sp. in Australia (Bailey 1989). **Sarcophagidae** – *Spirobolomyia flavipalpis* and *S. singularis* reared from *Spirobolus* spp. in eastern USA (Aldrich 1916); *Sarcophaga* (*Pierretia*) *iulicida* bred from a larva found in *O. moreleti* in Portugal (Pape 1990a); *Blaesoxipha beameri* reared from "*Tylobolus* sp." (= probably *Hiltonius* sp. near *hebes* [= probably Spirobolidae] in California [Pape 1990b]). However, T. Pape (personal communication, 2004) questions whether the records of Sarcophagidae reared from millipedes represent cases of parasitoidism or scavenging, noting that he has "seen *Spirobolomyia* at rather high densities," causing him "to wonder if it can sustain itself entirely on millipedes." **Sphaeroceridae** – Adults of one species of *Limosina* attracted to and resting on "Iulides" in Africa (Roubaud 1916).

For some researchers (e.g., Marinoni & Mathis 2000) who consider malacophagy as one of the few autapotypic features of Sciomyzidae, the food choice alone might indicate the correct placement of *Pelidnoptera* in a separate family. Notably, Vala *et al.* (1990) confirmed the taxonomic separation of the Phaeomyiidae on the basis of morphological features of the immature stages: (1) egg chorion completely smooth; (2) postoral spinule band completely encircling segment I; and (3) number of sclerites in the cephalopharyngeal skeleton reduced, especially the ventral arch lacking (see Section 14.1). The description of the immature stages of *Eginia ocypterata* by Skidmore (1985) and much of the descriptions of Eginiidae by Ferrar (1987) very likely pertain to specimens of *P. nigripennis*.

R. Meier (personal communication, 1999) pointed out that "there is no obvious reason why feeding on Diplopoda should be plesiomorphic and feeding on molluscs apomorphic. It could be the other way around; after all there is no outgroup that could be used to polarize the character." He considered that the lack of the ventral arch in *Pelidnoptera* "is the main reason why it should be excluded from the Sciomyzidae." Interestingly, *Sepedonella nana* and *Sepedon knutsoni*, which are true but derived sciomyzids as based on adult characters, appear to be restricted to freshwater oligochaete worms but have a well-developed ventral arch (Vala *et al.* 2000b, 2002).

5 • Host/prey ranges and preferences

Si per multos annos viveres adhuc naturam unius festucae seu muscae seu minimae creaturea de mundo ad plenum cognoscere non valeres. [If one spends many years of one's life in the detailed study of a single grass, or fly, or any of the minimal creatures of the world, one would still not know all there is to know of them].

St. Bonaventure, as cited by Thompson (1917).

5.1 HOST/PREY RANGES

An overview of the food resources of Sciomyzidae in relation to those of cyclorrhaphous Diptera in general emphasizes the specialized nature of the family (Fig. 5.1). The larvae of all the 203 species of Sciomyzidae whose biologies have been studied to date, i.e., where there are enough biological data to place a species in a Behavioral Group, appear to be obligate killers and feeders on molluscs of one kind or another, i.e., "true malacophages," except for three species. The exceptions include one of the most plesiotypic sciomyzids, *Salticella fasciata*, which in nature seems to prefer moribund terrestrial snails (Coupland *et al.* 1994) but in the laboratory will kill snails and also feed on various dead invertebrates (Knutson *et al.* 1970), and two of the most apotypic species, *Sepedonella nana* and *Sepedon* (*Mesosepedon*) *knutsoni*, which appear to be restricted to freshwater oligochaete worms. Both of these species have been found in nature in empty tubes of the oligochaetes and also in tubes containing living worms (Vala *et al.* 2000b, 2002). The relatively apotypic *Sepedon* (*Parasepedon*) *ruficeps* is a predator of freshwater snails and feeds opportunistically on freshwater oligochaetes in laboratory rearings (Knutson *et al.* 1967b, Gbedjissi *et al.* 2003).

Extensive data on food ranges and preferences, especially from laboratory rearings, are included in many papers on life cycles, but there has been no comprehensive species-by-species summary except for the compilation by Godan (1979) for 30 species and the compilation by Greathead (1981) of 25 genera placed in eight Behavioral Groups. A tabulation of the general type of principal food preferences was presented for 34 genera by Ferrar (1987). A diagram (Knutson 1973) of the food of 26 genera of Sciomyzidae among 15 families of terrestrial gastropods and five families of freshwater molluscs shows the extensive range, both freshwater and terrestrial molluscs, of species within the two large genera *Tetanocera* (Tetanocerini) and *Pherbellia* (Sciomyzini). Knutson's diagram also points out the restricted range of the other two large genera, *Sepedon* and *Dictya* (Tetanocerini), to three families of non-operculate freshwater snails. The extensive natural range of food within *Pherbellia* and *Ditaeniella* is shown in a list (Bratt *et al.* 1969) of field and laboratory records of food snails and snails rejected in laboratory rearings for 25 of the 82 Nearctic and Palearctic species. Of the 63 species of snails in 18 families fed upon in nature and/or laboratory rearings, natural hosts are known for 15 *Pherbellia* and *Ditaeniella* species, 34 species of snails in 10 families being recorded. Foote (1977, 2007) listed 16 genera of terrestrial gastropods known to be preyed upon in nature by 25 North American species of sciomyzids in 10 genera, including both Sciomyzini and Tetanocerini; *Pherbellia albovaria* was recorded from the greatest number of genera (four). Seven of these 25 species of Sciomyzidae attack terrestrial gastropods in more than one genus, 12 being succineid snail or slug killers. Fisher & Orth (1983) presented a tabulation of 14 genera of molluscs known to be or probable natural hosts for the 49 species of sciomyzids in 13 genera in California.

Hosts and prey are included in the classification of Behavioral Groups in the previous section. An overview of the major categories of the hosts/prey of Sciomyzidae and Phaeomyiidae with the genera arranged as in the cladogram of Marinoni & Mathis (2000) is shown in Fig. 5.2. The general types of larval food are given in Fig. 5.3. Note that in the cladograms (Figs 5.2, 15.23, 15.24, 17.2–17.4), based on that of Marinoni & Mathis (2000) and the revision by Barker *et al.* (2004), (1) *Pelidnoptera* was used as the outgroup, (2) the Huttonininae were not included,

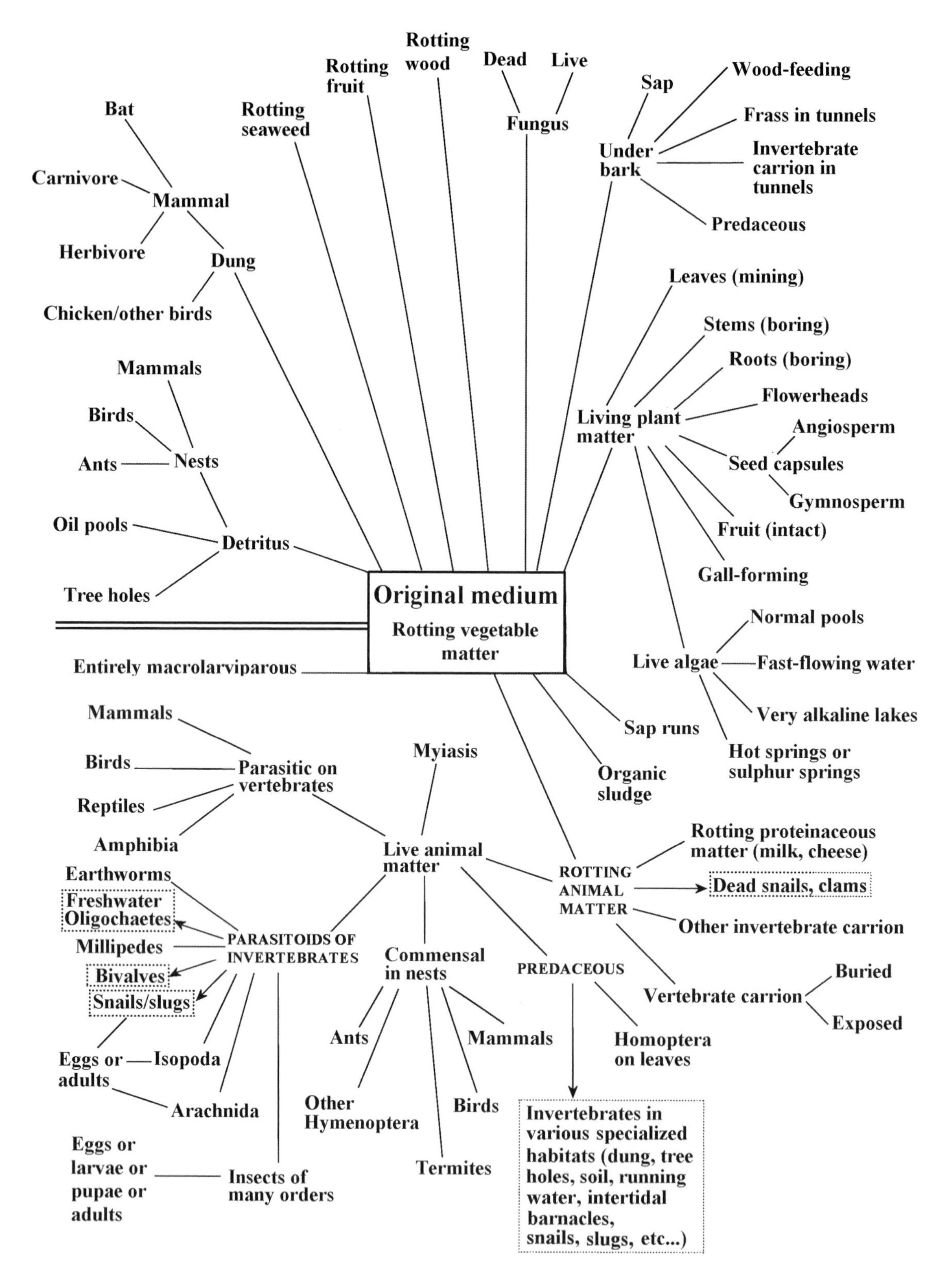

Fig. 5.1 Analysis of breeding media of Diptera Cyclorrhapha. Food resources of Sciomyzidae indicated by boxes. Modified from Ferrar (1987).

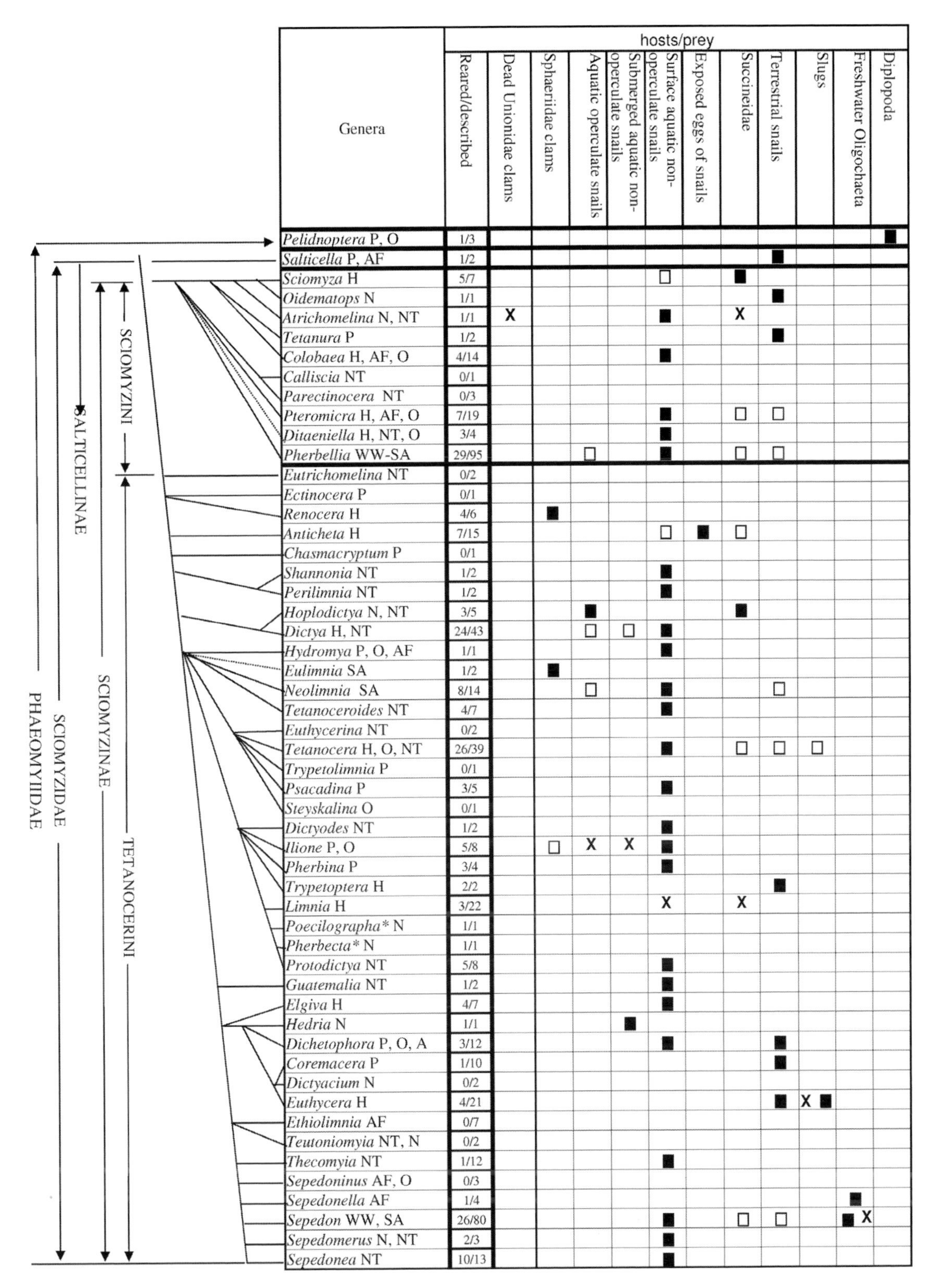

Genera	Reared/described	Dead Unionidae clams	Sphaeriidae clams	Aquatic operculate snails	Submerged aquatic non-operculate snails	Surface aquatic non-operculate snails	Exposed eggs of snails	Succineidae	Terrestrial snails	Slugs	Freshwater Oligochaeta	Diplopoda
Pelidnoptera P, O	1/3											■
Salticella P, AF	1/2								■			
Sciomyza H	5/7					□		■				
Oidematops N	1/1								■			
Atrichomelina N, NT	1/1	X				■		X				
Tetanura P	1/2								■			
Colobaea H, AF, O	4/14					■						
Calliscia NT	0/1											
Parectinocera NT	0/3											
Pteromicra H, AF, O	7/19					■		□	□			
Ditaeniella H, NT, O	3/4					■						
Pherbellia WW-SA	29/95			□		■		□	□			
Eutrichomelina NT	0/2											
Ectinocera P	0/1											
Renocera H	4/6		■									
Anticheta H	7/15					□	■	□				
Chasmacryptum P	0/1											
Shannonia NT	1/2					■						
Perilimnia NT	1/2					■						
Hoplodictya N, NT	3/5			■				■				
Dictya H, NT	24/43			□	□	■						
Hydromya P, O, AF	1/1					■						
Eulimnia SA	1/2		■									
Neolimnia SA	8/14			□		■			□			
Tetanoceroides NT	4/7					■						
Euthycerina NT	0/2											
Tetanocera H, O, NT	26/39					■		□	□	□		
Trypetolimnia P	0/1											
Psacadina P	3/5					■						
Steyskalina O	0/1											
Dictyodes NT	1/2					■						
Ilione P, O	5/8		□	X	X	■						
Pherbina P	3/4					■						
Trypetoptera H	2/2								■			
Limnia H	3/22					X		X				
*Poecilographa** N	1/1											
*Pherbecta** N	1/1											
Protodictya NT	5/8					■						
Guatemalia NT	1/2					■						
Elgiva H	4/7					■						
Hedria N	1/1				■							
Dichetophora P, O, A	3/12					■			■			
Coremacera P	1/10								■			
Dictyacium N	0/2											
Euthycera H	4/21								■	X ■		
Ethiolimnia AF	0/7											
Teutoniomyia NT, N	0/2											
Thecomyia NT	1/12					■						
Sepedoninus AF, O	0/3											
Sepedonella AF	1/4										■	
Sepedon WW, SA	26/80					■		□	□		■ X	
Sepedomerus N, NT	2/3					■						
Sepedonea NT	10/13					■						

Fig. 5.2 Major categories of hosts/prey of genera of Sciomyzidae and Phaeomyiidae arrayed above cladogram of Marinoni & Mathis (2000). Black square, usual hosts/prey; open square, other normal hosts/prey; X, facultative or suspected host/prey; *, host/prey unknown, reared from puparium. See Table 16.1 for abbreviations of zoogeographical regions given after genera.

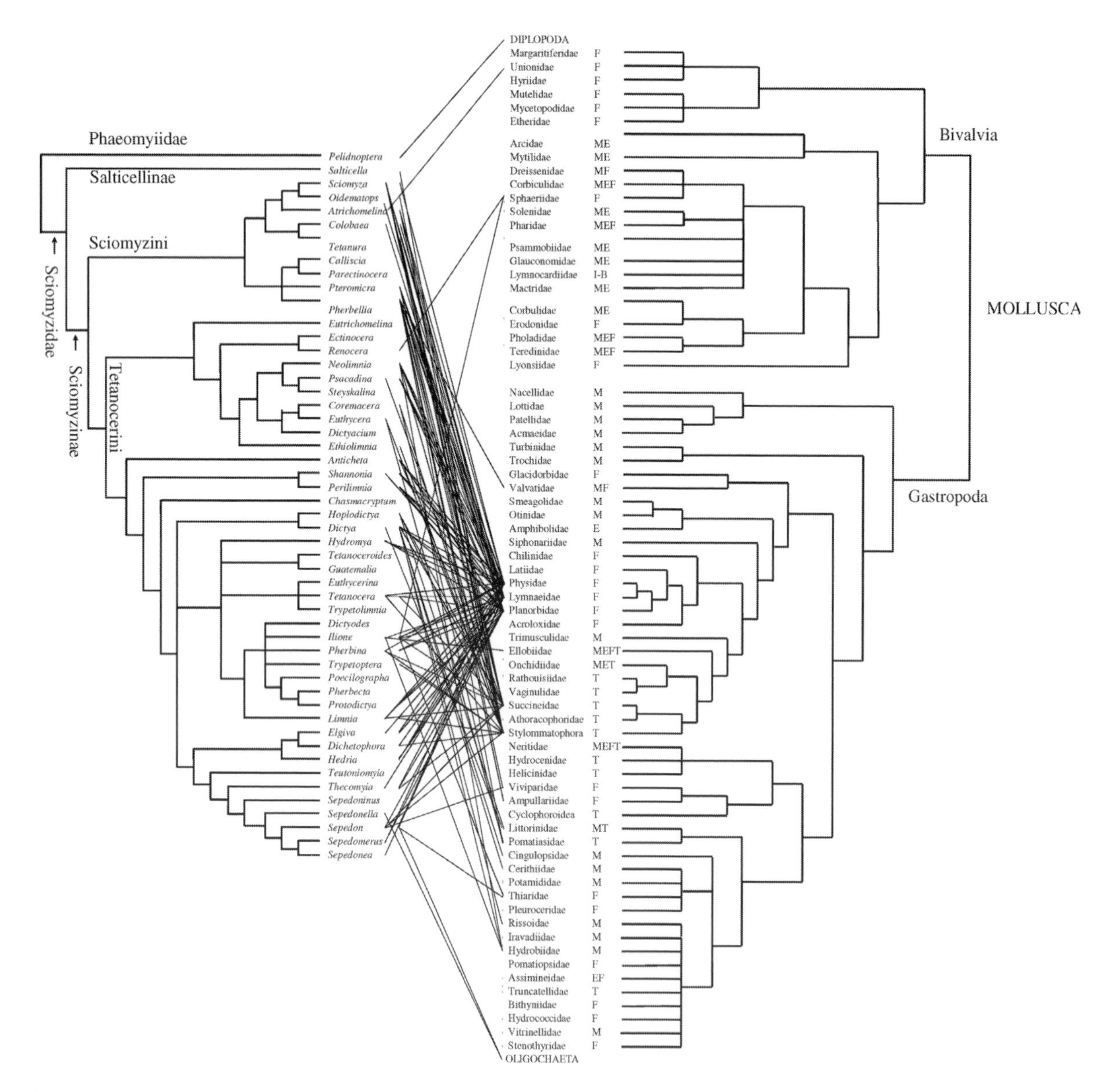

Fig. 5.3 Pattern of host/prey diversification in the biologically known genera of Sciomyzidae and Phaeomyiidae. Genera of Sciomyzidae as they appear in cladogram of Barker *et al.* (2004) are linked with a cladogram of known and potential molluscan hosts/prey (Diplopoda and Oligochaeta added here). The molluscan cladogram includes only those families with representatives in habitats that are or potentially could be used by Sciomyzidae. Stylommatophora is shown as a single terminal, but includes 71–92 families (most not attacked by Sciomyzidae). E, Estuarine waters; F, freshwater; I-B, inland brackish waters; M, marine littoral waters; T, terrestrial. Modified from Barker *et al.* (2004).

and (3) the following genera of Sciomyzidae were not included – *Apteromicra* (O), *Pseudomelina* (NT), *Neodictya* (P), *Neuzina* (NT), *Oligolimnia* (P), *Tetanoptera* (AF), and *Verbekaria* (AF) (all biologically unknown). The biologically known genera *Ditaeniella* (P, N, NT, AF) and *Eulimnia* (SA) were not included but have been interpolated by us, as shown by dotted lines. The numbers at branches of the cladogram are the numbers of characters found to separate the clades.

Barker *et al.* (2004) linked the phylogeny of molluscan clades that have representatives in freshwater, terrestrial, brackish water, and exposed marine littoral environments with a phylogeny of Phaeomyiidae and Sciomyzidae (Fig. 5.3) and noted the following points. The predominant pattern for the Sciomyzidae is association with the basommatophoran freshwater pulmonate families (Lymnaeidae, Physidae, and Planorbidae) which occur in shallow to moderately deep, permanent to ephemeral freshwaters throughout the world. Also widely utilized by the Sciomyzidae are the Succineidae, semi-terrestrial snails occurring throughout the world in diverse habitats, with many species found on moist soil or emergent vegetation in marshes, swamps, and at lake margins. Notably, these are the same freshwater systems occupied by the basommatophoran freshwater pulmonate families. The widespread availability of ecologically similar gastropod hosts/prey certainly has been a factor in the global distribution of the Sciomyzidae. In terrestrial habitats, the basal pulmonate families and a large number of more advanced Stylommatophora apparently are not attacked by Sciomyzidae. Prosobranchs are strongly underrepresented hosts/prey given their diversity in the habitats occupied by Sciomyzidae. The only terrestrial prosobranch gastropods attacked are European Pomatiidae. The only marine gastropods attacked are Littorinidae. Among the great diversity of the Bivalvia, only Sphaeriidae are attacked. Further conclusions from Barker *et al.* (2004) relative to the evolutionary significance of host/prey selection by Sciomyzidae are given in Section 17.1.

In general, the food range of malacophagous sciomyzids seems to be broad in most freshwater and terrestrial predator species and more restricted in terrestrial parasitoids, especially the few that oviposit onto the host. The food range also is more restricted in species that have unusual foods: operculate snails, snail eggs, slugs, and fingernail clams. Further information on species that oviposit onto the shell of the host snail is given in Chapter 6, and Sections 8.2, 9.2.2, and 18.2.

Although even freshwater predators probably have a certain degree of prey specificity, when the preferred prey is scarce it is likely that the predators can feed on unusual prey. A surprising instance in a laboratory rearing indicates this. P. Roberts (personal communication, 2005) provided a color photograph of a large third-instar larva together with a second-instar larva of *Pherbina coryleti*, a typical predator of freshwater snails, with their anterior ends deep into a hole at the umbo of a fingernail clam (*Sphaerium* sp.). The correspondent stated that the third-instar larva "appeared to make a hole in the shell," but this is highly unlikely – the shell probably was fractured during handling. Although only species of *Renocera*, *Eulimnia*, and *Ilione*, among all Insecta, are known to feed on fingernail clams, rarely have clams been provided in laboratory rearings of aquatic predators, as such prey seem to be well out of their range of food. The observation by P. Roberts is that of a laboratory-rearing artifact, but it raises questions about host/prey specificity.

The most intimately associated parasitoids, i.e., *Colobaea bifasciella*, *Oidematops ferruginea*, *Pherbellia schoenherri*, *Sciomyza dryomyzina*, *S. testacea*, *S. varia*, and *Tetanura pallidiventris*, appear to be restricted to one species or at most to one or two genera of closely related snails, even when exposed to other snails in laboratory rearings. However, while the Nearctic subspecies *Pherbellia schoenherri maculata* feeds only on the closely related succineids *Oxyloma* sp. and *Succinea* sp. (field and laboratory records; Bratt *et al.* 1969), the Palearctic subspecies *P. s. schoenherri* also will kill and feed on *Galba truncatula* in nature (Mc Donnell *et al.* 2004) (see Section 5.1). Newly hatched and young larvae, especially of terrestrial and semi-terrestrial species, seem to be more host/prey specific than are older larvae, at least in laboratory rearings. There are rather extensive field data for semi-terrestrial and terrestrial larvae, especially parasitoid species that remain in the snail for long periods, especially for those ovipositing onto or pupariating in shells. Most freshwater and terrestrial predators show broad prey ranges in laboratory rearings, probably much broader than their ranges in nature. There are few field records of the food snails of freshwater predators (Table 5.1), primarily because these larvae remain with the prey for a relatively short time, feeding on fresh tissue, and tend to leave the prey when disturbed (Mc Donnell *et al.* 2005a). Two aquatic predators of snails also have been found occasionally feeding in snail eggs in nature (Table 5.1) and several have fed on eggs in laboratory rearings (Table 5.2). See also Section 18.1.

A widespread impression seems to be that sciomyzid larvae are generally ineffective against operculate snails but information accumulating from laboratory rearings show that at least a few species likely are restricted in nature to operculates or strongly prefer them. The records are highlighted, with more detail, in Tables 5.3 and 5.4. It is likely that many species in Behavioral Group 2, those feeding on "stranded" snails of authors, the "shore-line predators" of Foote (1996a), especially the larvae that have strongly facultative behavior with ability to feed, in part, as saprophages, routinely feed on operculate snails that are out of the water and weakened, with the operculum not tightly closed. Further experiments are needed in this regard. The non-operculate terrestrial Palearctic snails of the

Table 5.1. Records of aquatic, predaceous Sciomyzidae larvae found feeding in snails and snail eggs in nature (not including ***Anticheta*** spp.)

Sciomyzidae	Snail (Snail eggs)	Location	No. – Instar	Date larva collected (L), died (D), pupariated (P), adult emerged (A)	Reference
Dichetophora biroi	*Austropeplea tomentosa*	Burbrook Stream, se. Adelaide, South Australia	1 – L?	?	Lynch (1965)
Dictya gaigei	*Lymnaea* sp.	Fish Point, Michigan	1 – L3	7.IX.71 (L) 13.IX.71 (P) 21–25.IX.1 (A)	Valley & Berg (1977)
Dictya umbrarum	*Lymnaea* sp.	nr. Umeå, Sweden	1 – L1	27.VII.67 (L) 11.VIII.67 (P)	Mc Donnell *et al.* (2005a)
Hydromya dorsalis	*Galba truncatula*	Glenamoy, Co. Mayo, Ireland	28 – L1 1 – L3	VI–VIII. 74, 75	Hope Cawdery & Lindsay (1977)[1]
	(*Lymnaea* sp.)	nr. Hillerød, Denmark	3 – L2	9.VI.64 (L) 21.VI.64 (P) 9.VII.64 (A)	Mc Donnell *et al.* (2005a)
	Stagnicola palustris	nr. Hillerød, Denmark	1 – L3	13.VI.64 (L) 19.VI.64 (P) 6.VII.64 (A)	Mc Donnell *et al.* (2005a)
Ilione trifaria	*Physella acuta*	nr. Montpellier, France	1 – L3	12.V.00 (L)	Mc Donnell *et al.* (2005a)
Sepedomerus macropus	*Austropeplea viridis*	Hawaii	"repeatedly"	?	Berg (1964)
Sepedon aenescens	*Gyraulus hiemantium*	Kagoshima, Japan	1 – L3	?	Nagatomi & Kushigemachi (1965)
Sepedon fuscipennis	*Stagnicola emarginata*	Anchorage, Alaska	1 – L3	23.VII.52 (L)	Berg (1953)
	Stagnicola elodes	Ithaca, New York	many L1, 2, 3	13.VI–30.VII.70	Eckblad (1973a)[2]
	Physella integra	"	"	"	"
	Gyraulus parvus	"	"	"	"
Sepedon sphegea	*Radix gedrosiana*	nr. Dezful, Iran	1 – L3	24.IV.73 (L) 26.IV.73 (P) 1.V.73 (A)	Knutson *et al.* (1973)
Sepedonea trichotypa	*Aplexa marmorata*	São Paulo, Brazil	1 – L3	26.V.67 (L)	Abercrombie (1970)
Tetanocera ferruginea	*Stagnicola elodes*	Shoshone, Idaho	1 – L?	22.V.59 (L)	Foote (1961b)
	Stagnicola elodes	Ithaca, New York	? – L3	?	Foote (1999)
	Helisoma sp.	"	? – L3	?	"

Table 5.1. (*cont.*)

	Stagnicola palustris	nr. Avignon, France	2 – L3	10.VI.86 (L) 16.VI.86 (P) 26.VI. (A)	Mc Donnell *et al.* (2005a)
	(*Lymnaea* sp.)	nr. Hillerød, Denmark	1 – L1	13.VI.64 (L) 11.VII.64 (P) 22.VII.64 (A)	”
	(*Lymnaea* sp.)	”	1 – L1	17.VI.64 (L) 14.VII.64 (III, D)	”
	Galba humilis	Cobleskill, New York	1 – L3	23.X.65 (L) 29.X. (D)	”
	Lymnaea stagnalis	Dangan, Co. Galway Ireland	1 – L3	14.XI.00 (L) 20.XI.00 (P)	”
	Lymnaea stagnalis	”	1 – L3	20.XI.00 (L) 1.XII.00 (P)	”
	Lymnaea stagnalis	Menlo, Co. Galway, Ireland	1 – 13	29.IX.00 (L) 11.X.00 (P) 31.X.00 (A)	”
	Bathyomphalus contortus	Dangan, Co. Galway, Ireland	1 – L3	2.X11.00 (L) 4.X11.00 (P)	”
	Physella acuta	nr. Avignon, France	2 – L3	10.VI.86 (L) 12, 14.VI. (P) 17, 18.VI.86 (A)	”
	Planorbis sp.	nr. Corfu, Corfu, Greece	1 – L3	20.IV.63 (L) 26.IV.63 (P) 11.V.63 (A)	”

Notes: A, adults emerged; D, died; L, larva collected; P, puparium formed; L1, L2, L3, first-, second- and third-instar larva.

[1]Snails captured, marked, released, recaptured.

[2]In screen cages in the field; see discussion.

[3]Modified from Mc Donnell *et al.* (2005a).

Table 5.2. Records of Sciomyzidae (other than *Anticheta* spp.) that fed on snail eggs in laboratory rearings

Sciomyzidae	Snail sp.	Larval instar that fed on eggs	Reference
Dictya brimleyi (N)	*Physella integra*	L1	Valley & Berg (1977)
Dictya floridensis (N)	*Physella h. hendersoni*	L1 – L3	McLaughlin & Dame (1989)
	Pseudosuccinea columella	L1 – L3 (pupariated)	"
Dictya hudsonica (N)	*Physa* sp.	L1	Valley & Berg (1977)
Dictyodes dictyodes (NT)	*Helisoma* sp.	L1	Abercrombie & Berg (1978)
Hydromya dorsalis (P, AF, O)	*Lymnaea* spp.	L1 – L3 (pupariated)	Knutson & Berg (1963a)
Psacadina verbekei (P)	*Lymnaea* sp.	L1 – L3	Knutson *et al.* (1975)
Psacadina vittigera (P)	*Lymnaea* spp.	L1 – L3	"
	Gyraulus sp.	L1	"
	Succinea sp.		
Sepedomerus macropus (NT, N)	*Austropeplea viridis*	L1 – L3	Chock *et al.* (1961)
Sepedon aenescens (O, P)	*Radix luteola*	L1 – L3 (pupariated & emerged)	K. Durga Prasad (personal communication, 1972)
Sepedon neavei (AF)	*Bulinus* sp.	L1	Barraclough (1983)
Sepedon plumbella (O, A, OC)	*Radix auricularia*	L1 – L3	Bhuangprakone & Areekul (1973)
	Indoplanorbis exustus		
Tetanoceroides fulvithorax (NT)	*Helisoma duryi*	L1	Abercrombie (1970)

family Clausiliidae have a structure, the clausilium, that functions somewhat like an operculum. Only larvae of *Pherbellia scutellaris* have been found feeding in nature in *Clausilia* (dead *C. nigricornis* and *C. dubia*). These snails have been killed and eaten by other *Pherbellia* spp. and *Tetanocera* spp. in laboratory rearings but were not accepted by species of several other genera of Sciomyzidae.

The extensive data on the broad range of snails killed and eaten during laboratory rearings of many species very probably give an incorrect impression of the actual host/prey range of some species. For example, in laboratory rearings *Pherbellia dorsata* killed and fed on more species of snails (28 species in 20 genera) than have been recorded for any other species of Sciomyzidae. However, all of the 41 larvae and five puparia of *P. dorsata*, studied rather extensively over a 2.5-month period at Snøgedam Pond, Hillerød, Denmark, were found only in *Planorbis planorbis*. Although this was the most abundant snail in the habitat, many other species, e.g., *Lymnaea* spp., were also abundant and widespread there (Bratt *et al.* 1969).

There also seem to be population differences in host specificity during laboratory rearings of some species. Such differences were seen, for example, in laboratory rearings of *Pherbellia similis*, many puparia of which were collected only in shells of *Planorbula armigera* at a locality near Kent, Ohio, and at two proximate localities near Geneva, New York. Larvae from the Kent population fed only on *P. armigera*; larvae from one of the Geneva localities fed only on *P. armigera*, *Helisoma trivolvis*, and *Gyraulus* sp.; and larvae from the other Geneva locality fed on all three of the above snails, plus *Physa* sp. and *Biomphalaria glabrata*, the latter a tropical species not occurring in the range of *P. similis* (Bratt *et al.* 1969).

5.2 EXPERIMENTAL STUDIES OF FOOD PREFERENCES

Extensive experimental studies have been conducted in the laboratory on host/prey ranges and preferences. Many of these studies include information on numbers and sizes of molluscs killed, biomass consumed, percentage survival on different molluscs, effects of prey and predator densities on rates of predation, effects of prey selection on rate of development, effects of water depth and temperature on predation, and competition. The latter subjects are reviewed in Chapters 7, 8, and 9.

5.2.1 Aquatic predators

Neff (1964) conducted the first laboratory experiments (also discussed, in more detail, by Berg 1964) on food preferences of aquatic predators, studying nine North American species

Table 5.3. Sciomyzidae larvae that killed and fed on operculate snails in laboratory rearings

Sciomyzidae	Habitat	Operculate snail prey	Preference for operculates	Pupariated on operculate diet/non-operculate diet	Reference
Dictya fontinalis (N)	Freshwater marshy meadows	*Fontelicella calforniensis* *Lithoglyphus turbiniformis* *Tryonia* sp.	2	Yes/yes	Fisher & Orth (1969b)
Dictya lobifera (N)	Coastal salt marshes	*Bittium* sp. *Cingula* sp. *Hydrobia* sp.	2	Yes/yes	Valley & Berg (1977)
Dictya oxybeles (N)	Coastal salt marshes	*Bittium* sp. *Cingula* sp. *Hydrobia* sp.	2	Yes/yes	"
Dictya pechumani (N)	Coastal salt marshes	*Bittium* sp. *Cingula* sp. *Hydrobia* sp.	2	Yes/yes	"
Hoplodictya setosa (N)	Coastal salt marshes	*Littorina littorea*	2	Yes/yes	Neff & Berg (1962)
Ilione trifaria (P)	Freshwater situations	*Hydrobia* sp. *Melanopsis praemorsa*	3	Yes/yes	Knutson & Berg (1967)
Neolimnia tranquilla (SA)	Freshwater situations	*Potamopyrgus antipodarum*	2	Yes/yes	Barnes (1979c)
Pherbellia prefixa (N)	Open,vernal freshwater marshes	*Valvata sincera*	1	Yes/no	Foote (1973)
Sepedon spangleri (O)	Freshwater situations	*Melanoides tuberculata* *Viviparus* sp. *Tarebia granifera* *Lithoglyphopsis aperta* *Lacunopsis munensis* *Hubendickia siamensis*	3	Yes/yes	Sucharit *et al.* (1976)
Tetanocera ferruginea (H)	Freshwater situations	*Bithynia tentaculata*	3	No/yes	Manguin *et al.* (1986)

Notes: Preference scale: 1, only operculates killed and eaten; 2, operculates strongly preferred; 3, operculates accepted but not preferred.

Table 5.4. Sciomyzidae larvae that fed on crushed operculate snails in laboratory rearings

Sciomyzidae	Habitat	Operculate snail	Reference
Sepedon aenescens (O, P)	Freshwater marshes and ponds	"Operculates"	O. Beaver *et al.* (1977)
		Melanoides tuberculata *Pila ampullacea* *Pila scutata*	Bhuangprakone & Areekul (1973)
Sepedon ferruginosa (O)	"	"Operculates"	O. Beaver *et al.* (1977)
Sepedon plumbella (O, A, OC)	"	"	"
Sepedon senex (O)	"	"	"
Sepedon spangleri (O)	"	"	"
Perilimnia albifacies (NT)	Various freshwater situations	*Campeloma decisum*	Kaczynski *et al.* (1969)
Shannonia meridionalis (NT)	"	"	"

in the genera *Dictya*, *Sepedon*, and *Tetanocera*, plus one Neotropical *Sepedomerus* species against 13 tropical species of snails in five non-operculate and one (*Oncomelania*) operculate genera that are intermediate hosts of *Schistosoma* spp. (Table 5.5). None of the snails is a natural prey, i.e., does not occur with the sciomyzids. Neff's experiments, in 300 ml jars each containing one larva and one snail on wet sand, consisted of 1100 "exposure days," each day equaling one snail and one larva per container per day. Because of the experimental design, comparisons of the effectiveness of larvae in a common genus could not be made. The study focused on numbers and sizes of snails killed and showed great differences in the vulnerability of the various genera of snails. All the sciomyzid larvae killed and ate the non-operculate, planorbid snails, whereas none killed the two operculate *Oncomelania* spp. In the tests against three species of *Bulinus*, *Tetanocera* larvae were notably more successful than *Sepedomerus*, *Sepedon*, or *Dictya* larvae.

When the freshwater predator *Sepedon plumbella* was used in choice tests in Thailand against six operculate species and three non-operculate species of snails, attacks were restricted to the non-operculates, except for eight individuals of the operculate *Filopaludina* spp. being killed (Bhuangprakone & Areekul 1973). Survival rates were 30%, 30%, and 25%, respectively, when larvae were reared on crushed *Filopaludina martensi cambodiensis*, *F. sumatrensis polygramma*, and *Melanoides tuberculata*, all operculate snails.

Eckblad's (1973a) enclosure experiments with *Sepedon fuscipennis* in New York remain the only experimental data showing prey species preference by an aquatic predaceous larva under field conditions. He found that small *Stagnicola elodes* were more susceptible than *Physella integra* or *Gyraulus parvus*. He related this to the fact that the *S. elodes* were more commonly found at the surface film than were the other two species. Hope Cawdery & Lindsay (1977) found 1.4%, 15.6%, and 29.4%, respectively, of about 100 *Galba truncatula* released in June–August 1974–76 in Wales infested by the shoreline predator *Hydromya dorsalis* when recaptured 96–144 hours later; apparently they did not sample other snails.

Sucharit *et al.* (1976) conducted experiments in Thailand with seven species of operculate snails and six species of non-operculates to test the range of prey of the aquatic predator *Sepedon spangleri*. Three experiments were conducted, consisting of 10 larvae and 20 snails of each species kept overnight in boxes with water to a depth of 5mm at 25 ± 3 °C with three replicates. Experiments 1 and 2 included two operculate species and three non-operculates. Experiment 3 included three operculate species. Different species were used in each of the three experiments. While 73.1% of the 360 non-operculate snails were killed, only 10.8% of the 280 operculate snails were killed (see Table 5.6 for species). The number of snails of the operculate *Hubendickia siamensis* killed (mean = 11.0) approached the number of non-operculates killed

Table 5.5. Summary of trials of ten species of sciomyzid larvae against snails of medical importance

	Sciomyzidae											
	Sepedon (*S.*) and *Sepedomerus* (*Sm.*)			*Tetanocera*			*Dictya*					
	S. armipes *S. fuscipennis* *S. tenuicornis* *Sm. caerulus*			*ferruginea* *vicina*			*brimleyi* *expansa* *pictipes* *stricta*			Totals		
Snails	Exp. days	Snails dead	Larvae dead	Exp. days	Snails dead	Larvae dead	Exp. days	Snails dead	Larvae dead	Exp. days	Snails dead	Larvae dead
Biomphalaria												
alexandrina	51	50	0	75	56	6	56	46	3	182	152	9
glabrata	154	147	8	92	91	13	52	51	6	298	289	27
nigricans	—	—	—	7	7	0	6	6	0	13	13	0
schrammi	—	—	—	17	17	2	3	3	0	20	20	2
pfeifferi	1	1	0	4	4	0	22	20	3	27	25	3
straminea	—	—	—	5	3	0	4	3	0	9	6	0
Planorbarius metidjensis	7	7	0	—	—	—	—	—	—	7	7	0
Tropicorbis centimetralis	—	—	—	4	4	2	4	3	1	8	5	3
Total	213	205	8	204	180	23	147	132	13	564	517	44
Bulinus												
africanus	11	0	4	88	72	11	58	17	18	157	89	33
globosus	—	—	—	72	53	10	—	—	—	72	53	10
truncatus	47	29	5	59	42	4	51	19	11	157	90	20
Total	58	29	9	219	167	25	109	36	29	386	232	63
Oncomelania												
quadrasi	17	0	4	—	—	—	44	0	6	61	0	10
formosana	39	0	10	40	0	2	52	1	4	131	1	16
Total	56	0	14	40	0	2	96	1	10	192	1	26

Note: Modified from Neff (1964).

(mean = 13.0). The operculate *Lacunopsis munensis* was more susceptible (mean = 5.7) than the non-operculate *Radix auricularia rubiginosa* (mean = 4.3).

When the freshwater predator *Sepedon senex* was used in laboratory choice tests in Thailand (O. Beaver 1989) including the non-operculates *Indoplanorbis exustus*, *Gyraulus convexiusculus*, and *R. auricularia rubiginosa*, larvae did not kill the last, which produced larger quantities of a more viscous mucus than the other two species, but they did kill and eat *R. auricularia rubiginosa* when offered only that species.

The most detailed laboratory experiments on prey preference and related aspects to date are those conducted in southern France on the polyphagous freshwater predators *Sepedon sphegea* (Haab 1984, Ghamizi 1985, unpublished Ph.D. theses) and *Tetanocera ferruginea* (Manguin *et al.* 1986, 1988a, 1988b, Manguin & Vala 1989).

Table 5.6. Ability of ***Sepedon spangleri*** L3 in killing 13 species of snails; divided into three groups of snails and three replications noted R I, R II, R III

Species of snails	N snails	Mean sizes of snails		N snails killed by 10 larvae			Average N snails killed
		Width (mm)	Length (mm)	R I	R II	R III	
Bithynia laevis	20	2.66	3.99	0	0	0	0.00
Melanoides tuberculata	20	2.65	7.35	0	7	3	3.33
Radix auricularia rubiginosa	20	2.66	5.20	2	4	7	4.33
Gyraulus convexiusculus	20	1.11	4.54	20	19	11	16.67
Segmentina hemisphaerula	20	1.59	3.83	20	20	13	17.67
All snails of experiment I	100			42	50	34	42.00
Viviparus sp.	20	4.68	5.62	1	2	1	1.33
Tarebia granifera	20	4.84	10.94	6	4	2	4.00
Segmentina trochoideus	20	1.13	2.59	12	17	10	13.00
Hippeutis umbilicalis	20	1.48	4.13	17	19	14	16.67
Indoplanorbis exustus	20	3.79	3.85	20	20	18	19.33
All snails of experiment II	100			56	62	45	54.33
Lithoglyphopsis aperta gamma race	20	1.28	1.90	1	0	0	0.33
L. aperta alpha race	20	2.49	3.86	1	3	0	1.33
L. aperta beta race	20	1.96	3.21	2	5	1	2.66
Lacunopsis munensis	20	4.07	4.65	10	5	2	5.66
Hubendickia siamensis	20	1.84	4.61	12	11	10	11.00
All snails of experiment III	100			26	24	13	21.00

Note: From Sucharit *et al.* (1976).

Manguin *et al.* (1986) conducted experiments in France with the aquatic predator *Tetanocera ferruginea* individually against three non-operculate snail species, one operculate snail, one limpet, and one fingernail clam species in 5.0mm water, at room temperatures, with ten replicates (Table 5.7). See Section 7.3 for detail on sizes and biomass of the operculate snails killed and eaten. Interestingly, the rate of success against non-normal prey (the limpet, operculate snail, and fingernail clam) in these single-prey systems in 5 × 17cm boxes was highest for first-instar larvae, but only two larvae feeding on limpets, *Ancylus fluviatilis*, of the 45 larvae provided non-normal prey eventually developed to the puparial stage. This is the only record of the larvae of any Sciomyzidae being reared successfully to pupariation on any species of limpet, although at a low rate of success. Another record of Sciomyzidae feeding to some extent on limpets is *Dictya ptyarion* feeding on *Ferrissia* sp. in laboratory rearings (Valley & Berg 1977). It is possible that some species of Sciomyzidae are restricted to or may feed extensively on limpets.

Manguin *et al.* (1988a, 1988b) and Manguin & Vala (1989) followed the suggestions for analyzing food preference made by Cock (1978) and others and as summarized by Hassell & Southwood (1978). The analysis consists of conducting functional response experiments using each prey type separately, then conducting further experiments in which various ratios of the different prey types are presented together. The authors utilized four autochthonous species of snails collected in the habitat of the adults (*Radix balthica*, *Stagnicola palustris*, *Physella acuta*, and *Bathyomphalus contortus*) and two tropical species that are important intermediate hosts of *Schistosoma* spp. (*Biomphalaria glabrata* and *Bulinus truncatus*). Each experiment, replicated ten times, included 20 snails of two species, associated in proportions

Table 5.7. Rate of success of predation by *Tetanocera ferruginea* against non-operculate snails (1–3), a limpet (4), an operculate snail (5), and a fingernail clam (6)

	Tested immature stadium			
Prey	L1	L2	L3	Pupa
1. *Bathyomphalus contortus*	100% (13)	92% (12)	92% (12)	77% (10)
2. *Stagnicola palustris*	100% (28)	54% (15)	46% (13)	36% (10)
3. *Physella acuta*	100% (31)	48% (15)	39% (12)	32% (10)
4. *Ancylus fluviatilis*	100% (15)	27% (4)	13% (2)	13% (2)
5. *Bithynia tentaculata*	100% (15)	0	0	0
6. *Pisidium casertanum*	100% (15)	0	0	0

Note: Numbers of each immature stadium tested shown in parentheses. From Manguin *et al.* (1986).

of 5/6, 8/12, 10/10, 12/8, and 15/5, respectively. Supplementary experiments were made with the four autochthonous snails in identical proportions (five of each). In all of the experiments, the sizes of the snails (length, or diameter for *B. contortus*) were proportional to the sizes of the larvae: 1.5 mm to $\geq$3.0 mm (first instar), 3.0–6.0 mm (second instar), and 6.0–10.0 mm (third instar).

All of the results, in terms of number killed or biomass consumed, showed that in the presence of two species of prey there were two trends of predation (Fig. 5.4). One trend is a species-independent type in the *S. palustris*/*R. balthica* and *B. glabrata*/*B. truncatus* systems (Fig. 5.4a, f; b, g). The rate of predation is proportional to the density of snails. In equal proportions, 10/10, an equivalent consumption of two prey species was observed ($P > 0.05$). The average curve of consumption (visually traced) thus corresponds to an X approximately centered at the point 10/10. The second trend is a species-dependent type. In this case, the larvae preferentially consume one prey species whatever its proportions are available. Then the abandonment of the second prey species is accentuated and its proportion decreases in the choice range. The curves of consumption thus are very asymmetric, their point of intersection being much more eccentric when predation is directed toward one of the prey ($P > 0.05$) (Fig. 5.4c–e, h–i). This accentuated preference in favor of *P. acuta* is significant in the association with *B. contortus* ($P < 0.001$), and the X curve of predation is not closed (Fig. 5.4e).

Expressed in biomass, consumption strongly increases during the course of development (Fig. 5.4f–j), with 9.4mg consumed by the first-instar larva, 56.0 by the second instar, and 268.5 by the third instar, with a geometric progression in the order of 5. In terms of biomass, consumption is similar for each prey species during the first stadium ($P > 0.05$). During the second stadium, consumption of *B. contortus* does not amount to more than 1.5% of the food consumed ($P > 0.01$), while biomass consumed of the other species is essentially equal. Only *S. palustris* is consumed to a lesser extent.

In the experiment grouping the four autochthonous species of snails provided in equal proportions (Fig. 5.8), the larvae adapt their attack according to the prey species and their own size, e.g., during the first stadium all of the snail species are killed and eaten avidly. This results primarily from factors inherent to each prey species. First are behavioral factors concerning the mobility of the prey species, enabling it to escape the predator. Surface-dwelling snails such as *Physella acuta* are more easily attacked than those such as *S. palustris* that utilize cryptic habitats, i.e., depths and shorelines inaccessible to the larvae. Second are morphological factors, i.e., a smaller shell aperture. This is the case for *B. contortus* vis-à-vis third-instar larvae. However, the presence of this species in the experiments contributes to limiting the mortality of first-instar larvae because of their small size and their weak ability to escape. Third are the weight factors of the prey related to the nutritive value of the species, of which the larger sizes have been better studied.

Manguin *et al.* (1988b) used the index *c* of Murdoch (1969) and *D* of Jacobs (1974) with some modifications to evaluate prey preference (30 replicates) (Figs 5.5–5.7, Table 5.8, Eq. 5.1). In their original formulation (Eq. 5.1a, b), these indices include the number of each prey present (*N1* and *N2*) and the number of each prey consumed (*P1* and *P2*). But, in the case of a strong

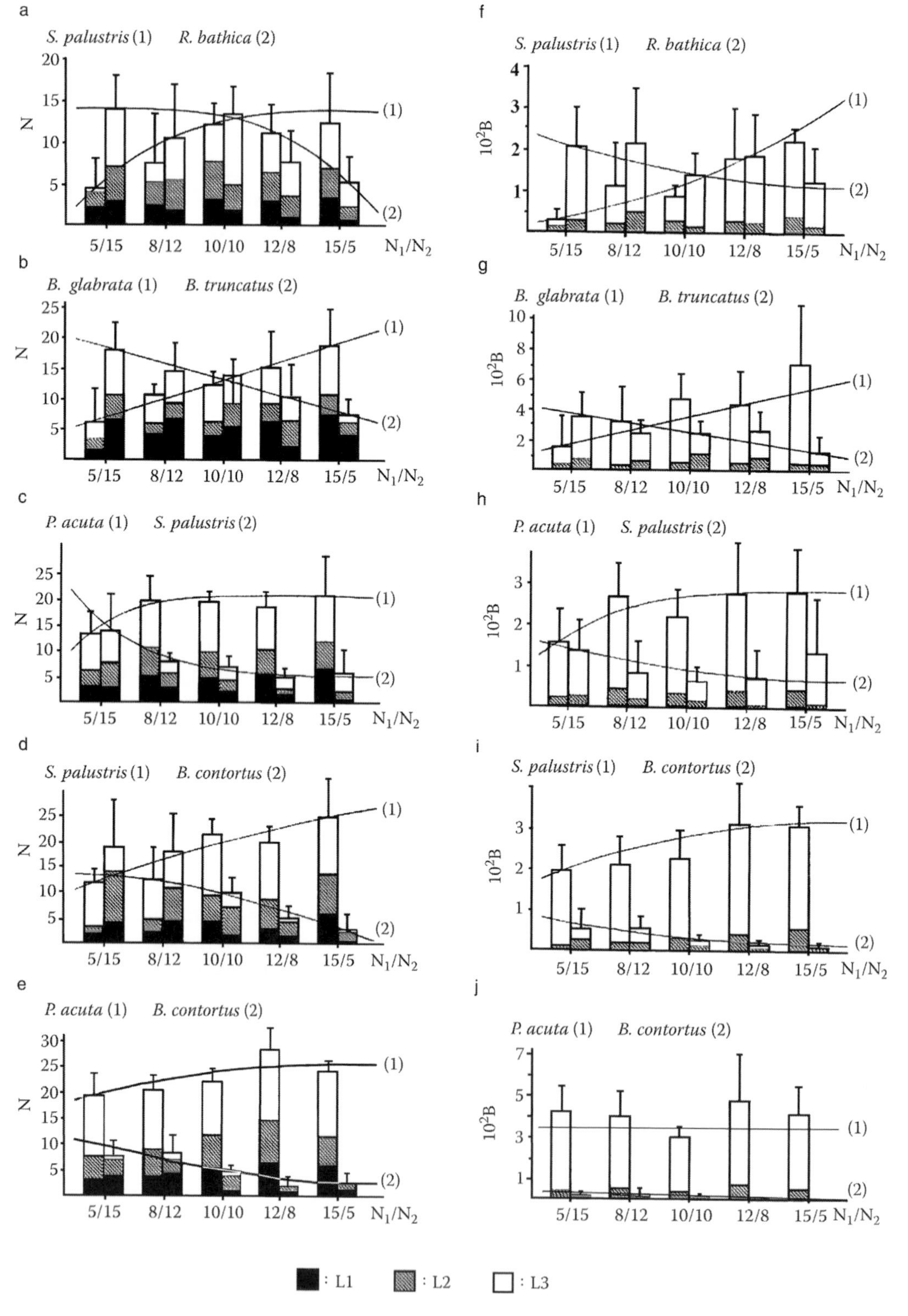

Fig. 5.4 Mean predation by *Tetanocera ferruginea* larvae (L1–L3 combined) according to relative densities of two snail species as prey. (a–e) mean predation in number (N) of consumed prey; (f–j) predation in biomass (B, in mg). Curve 1, visual curve of mean predation of first prey species; curve 2, visual curve of mean predation of second prey species. C.I., 5%. Snails: *Bathyomphalus contortus*, *Biomphalaria glabrata*, *Bulinus truncatus*, *Physella acuta*, *Radix balthica*, *Stagnicola palustris*. From Manguin *et al.* (1988a).

Table 5.8. Characteristics of correlation lines and preference indexes c' and M according to each larval stadium of *Tetanocera ferruginea* cited in Figs 5.6 and 5.7

Associated prey 1 + 2	L1	L2	L3
Physella acuta (1) *Stagnicola palustris* (2)	$1/Y = 0.35.\ 1/X + 0.007$ ($r = 0.34$)* $c' = 2.89$ $M = 0.44$	$1/Y = 0.39.\ 1/X + 0.006$ ($r = 0.61$)*** $c' = 2.59$ $M = 0.41$	$1/Y = 0.22.\ 1/X + 0.009$ ($r = 0.46$)** $c' = 4.47$ $M = 0.42$
Physella acuta (1) *Bathyomphalus contortus* (2)	$1/Y = 0.45.\ 1/X + 0.006$ ($r = 0.48$)* $c' = 2.21$ $M = 0.33$	$1/Y = 0.22.\ 1/X + 0.009$ ($r = 0.50$)* $c' = 4.48$ $M = 0.52$	$1/Y = 0.02.\ 1/X + 0.009$ ($r = 0.39$)* $c' = 50.30$ $M = 0.94$
Stagnicola palustris (1) *Bathyomphalus contortus* (2)	$1/Y = 0.60.\ 1/X + 0.006$ ($r = 0.36$)* $c' = 1.70$ $M = 0.20$	$1/YA = 5.56.\ 1/X - 0.097$ ($r = 0.54$)** $c' = 0.18$ $1/YB = 1.36.\ 1/X - 0.007$ ($r = 0.51$)** $c' = 0.73$ $M =- 0.10$	$1/Y = 0.20.\ 1/X + 0.009$ ($r = 0.51$)** $c' = 4.40$ $M = 0.64$
Stagnicola palustris (1) *Radix balthica* (2)	$1/Y = 0.40.\ 1/X + 0.007$ ($r = 0.10$) $c' = 2.40$ $M = 0.25$	$1/Y = 0.70.\ 1/X + 0.005$ ($r = 0.43$)** $c' = 1.48$ $M = 0.05$	$1/Y = 4.30.\ 1/X - 0.053$ ($r = 0.36$)* $c' = 0.23$ $M = - 0.28$
Biomphalaria glabrata (1) *Bulinus truncatus* (2)	$1/Y = 1.30.\ 1/X - 0.005$ ($r = 0.23$) $c' = 0.70$ $M = - 0.05$	$1/Y = 0.50.\ 1/X + 0.014$ ($r = 0.10$) $c' = 2.10$ $M = 0.17$	$1/Y = 0.90.\ 1/X - 0.001$ ($r = 0.74$)*** $c' = 1.10$ $M = 0.14$

Note: c' and M preference indexes; *, **, ***, d.f. 5%, 1%, and 1‰ respectively. $1/YA$, $1/YB$, part of curve under and above diagonal line which represents absence of preference (see Eq. 5.1, 5.2). From Manguin *et al.* (1988b).

preference, $P2 = 0$ and thus the two indices cannot be calculated. Murdoch (1969) proposed an equation (Eq. 5.1d) where X represents the percentage of a prey species in the habitat, and Y the percentage of the same species consumed. As the index c, deduced from the slope of regression, differs from that calculated in Murdoch's equation, Manguin *et al.* (1988b) established a new index called c' (Eq. 5.1e). So, if from 0 the values of c' approach 1, [1–3[, there is no consumption preference between tested snail species; if from 1 the variations of c' are approaching to 3,] 1–3[, there is only a weak preference. But, if the values are from 3 to infinity, [3 – infinity[, the preference is very strong. Graphically, when the curve deviates increasingly from the diagonal (Figs 5.5a–c; 5.6 and 5.7a, b), preference is in favor of one of the prey species. In absence of deviation or when the curve is more or less close to the diagonal line of the representative figure (Fig. 5.5d, e), there is no preference.

The experiments by Manguin *et al.* (1988a, 1988b) and Manguin & Vala (1989) on *Tetanocera ferruginea* concerned different combinations of two to six species of freshwater snails from the fly's habitat and two exotic species. In most associations (Fig. 5.5a–c), larvae of *T. ferruginea* preferred one species, whatever its relative density. In some associations preference changed during the course of development of the larvae. The prey characteristics determining choice were dispersal, mobility, size, and morphology of the snail species. The behavior was quantified in an index of preference. The authors concluded that larvae chose prey that provided the greatest return per unit of energy

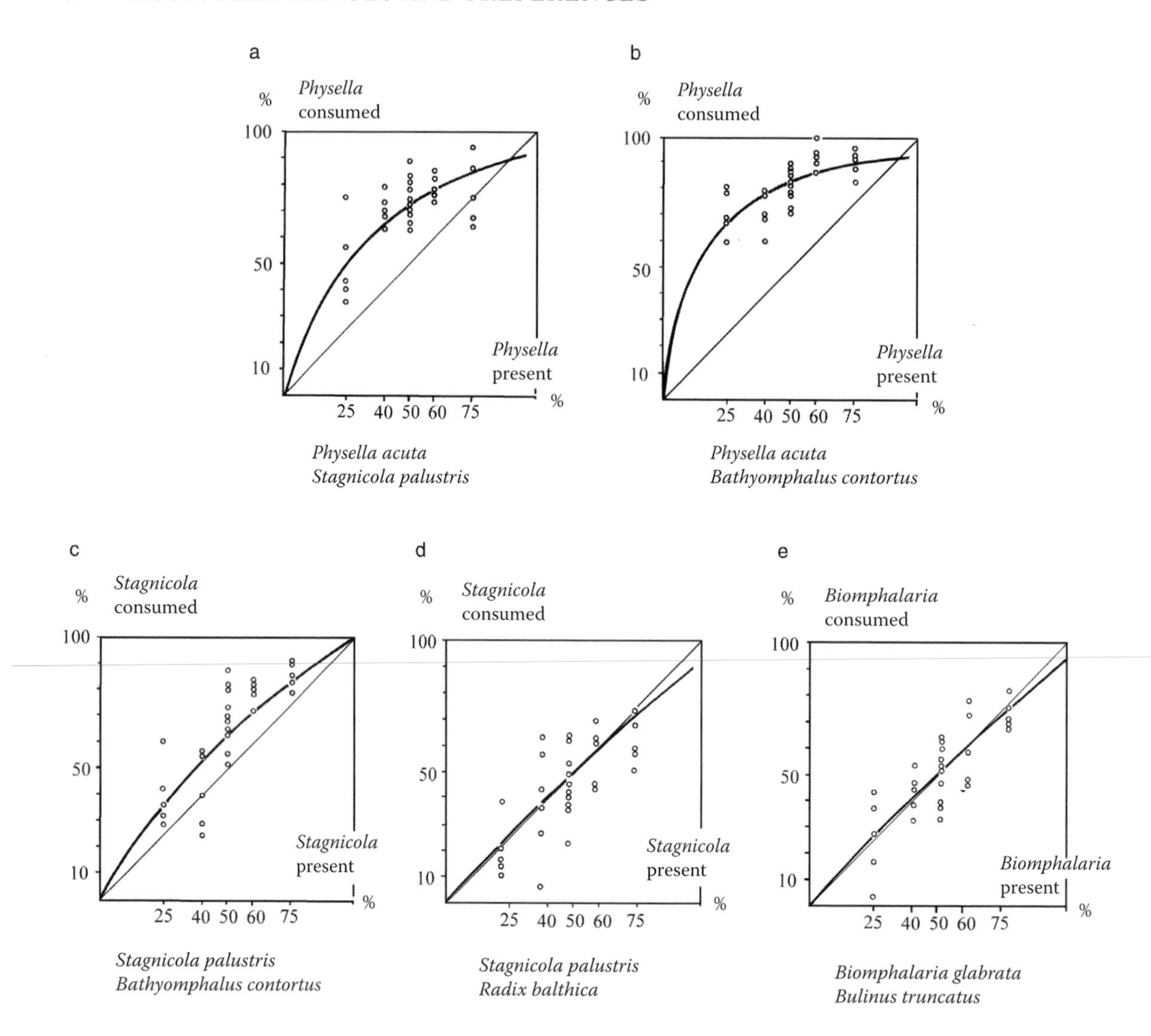

Fig. 5.5 Correlations of preference of ***Tetanocera ferruginea*** larvae (L1–L3 combined) according to relative proportions of pairs of snail prey offered. Correlations concern the interval comprised of 25–75% of presented prey (results extrapolated beyond these limits on figures). Diagonal line represents hypothetical null case of preference. When curve is localized in upper-left hand triangle, preference is more or less accentuated. See Table 5.8, Eq. 5.1, 5.2. From Manguin *et al.* (1988b).

expended. In the combination of two freshwater snails, with a total of 20 snails and in proportions of 25%, 40%, 50%, 60%, or 75% for one species, three behavioral types were revealed. When associated with *Physella acuta* and *Bathyomphalus contortus*, first-instar larvae obviously preferred *P. acuta*; the nutrition curve progressively withdrew from the diagonal line of non-preference. In association with *P. acuta* and *Stagnicola palustris*, larvae had a slight tendency to consume more *P. acuta*. This may be explained by the fact that the predator and prey live in the same microhabitat; larvae of *T. ferruginea* are surface predators and most often *P. acuta* crawl upside down on the water film, thus being in the path of the larvae (but see discussion of vertical dispersion of snails, Section 1.2). When the snails of the prey–predator association were *S. palustris* and *B. contortus*, no preference by first-instar larvae was observed. But second-instar larvae, which usually prefer *B. contortus*, modified their choice to *S. palustris* when the proportion of *S. palustris* surpassed 50% (Fig. 5.7a). This is considered an example of the "switching"

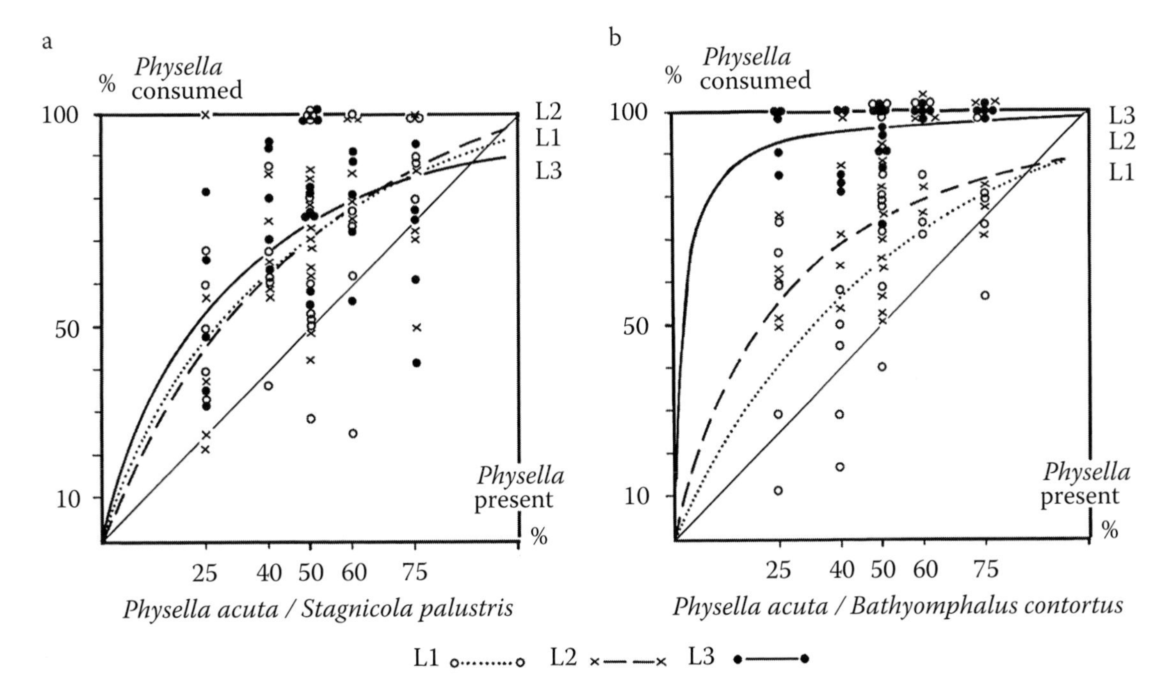

Fig. 5.6 Correlations according to each larval stadium, in both cases of strong preference of ***Tetanocera ferruginea*** larvae. Correlations concern the interval comprised of 25–75% of presented prey (results extrapolated beyond these limits on figures). Diagonal line represents hypothetical null case of preference. When curve is localized in upper-left hand triangle, preference is more or less accentuated. See Table 5.8, Eq. 5.1, 5.2. From Manguin *et al.* (1988b).

phenomenon *sensu* Murdoch (1969). During the third instar, the tendency towards consuming *S. palustris* preferentially remained and even increased. During the spring, populations of *B. contortus* are large. This species represents the first prey consumed by first-instar larvae hatching from eggs deposited by females issuing from overwintering puparia. Under natural conditions, freshwater predators probably adapt their prey-attack and their consumption according to the proportion and sizes of snail species available.

Contrary to Barraclough's (1983) laboratory finding that *Sepedon neavei* and *S. testacea* did not kill *Bulinus* snails in South Africa, Maharaj *et al.* (1992), also in South Africa, found that *S. scapularis* was an effective predator of both *Bulinus tropicus* and *B. africanus*, as well as of the immigrant *P. acuta*, with no clear preference shown by *S. scapularis* for any one of the three species. In single-predator species/multi-prey species experiments, *S. neavei* and *S. scapularis* proved to be polyphagous, with prey preferences changing among instars. *Sepedon neavei* attacked five species of non-operculate snails, and *S. scapularis* attacked nine species of snails in six genera (Appleton *et al.* 1993). When the researchers used two indigenous snail species, *B. tropicus* and *Biomphalaria pfeifferi*, and two immigrant species, *P. acuta* and "*Stenophysa* cf. *marmorata*," significantly more indigenous than immigrant snails were killed. When four indigenous snail species were used, the above two species along with *B. africanus* and *B. forskali*, all species of snails were killed by both sciomyzids, but significantly more *B. africanus* and *B. pfeifferi*, both medically important species, were killed than were the other two medically unimportant species.

5.2.2 Snail-egg/snail feeders

The only Insecta known to be restricted to eggs of Mollusca during at least the early part of larval development are five Nearctic and Palearctic species of *Anticheta* which attack snail eggs, and two species of Phoridae, *Megaselia aequalis* and *M. ciliata*, which attack slug eggs (Robinson & Foote 1968, Disney 1977). *Anticheta* spp. females lay their eggs only on snail egg masses. Larvae of some *Anticheta* spp. feed only on snail eggs throughout the three stadia, i.e., *A. analis*

(a)

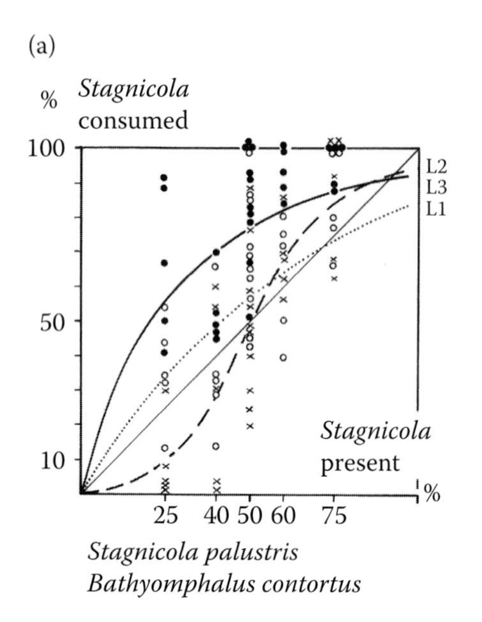

(b)

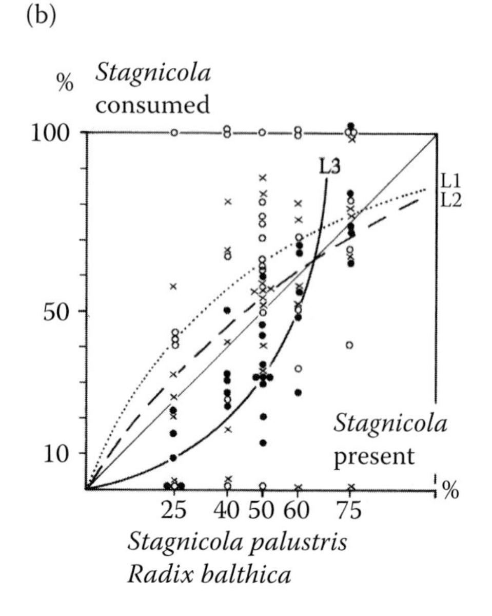

(c)

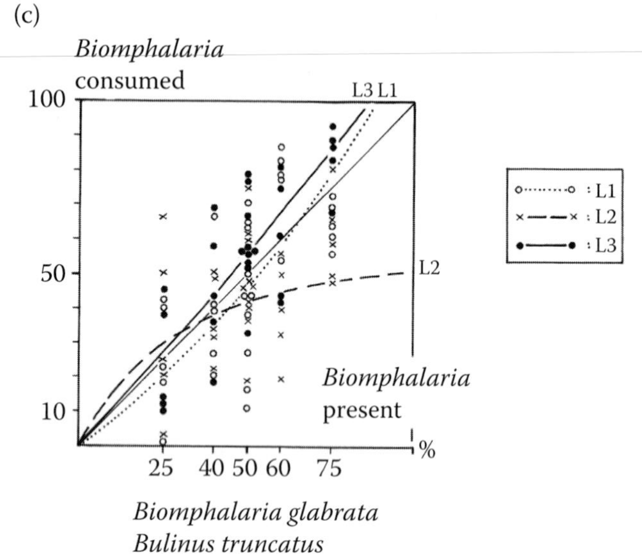

Fig. 5.7 Correlations concerning three weak and null cases of preference of *Tetanocera ferruginea* larvae of all three stadia. Correlations concern the interval comprised of 25–75% of presented prey (results extrapolated beyond these limits on figures). Diagonal line represents hypothetical null case of preference. When curve is localized in upper-left hand triangle, preference is more or less accentuated. See Table 5.8, Eq. 5.1. From Manguin *et al.* (1988b).

(Knutson 1966) on Lymnaeidae and *A. brevipennis* (Knutson 1966) and *A. borealis* (Robinson & Foote 1978) on Succineidae. Other species feed at first on snail eggs, then are capable of killing and feeding on juvenile to mature snails, i.e., *A. testacea* (Fisher & Orth 1964) on Succineidae and *A. melanosoma* (Knutson & Abercrombie 1977) on *Aplexa hypnorum* (Physidae). The only detailed host preference tests conducted to date are those of Fisher & Orth (1964) in California for *A. testacea*, which utilizes eggs and juvenile to mature snails of two succineid species, five freshwater non-operculates (*Helisoma tenue californiensis*, *Radix auricularia*, *Stagnicola palustris nuttalliana*, *Pseudosuccinea*

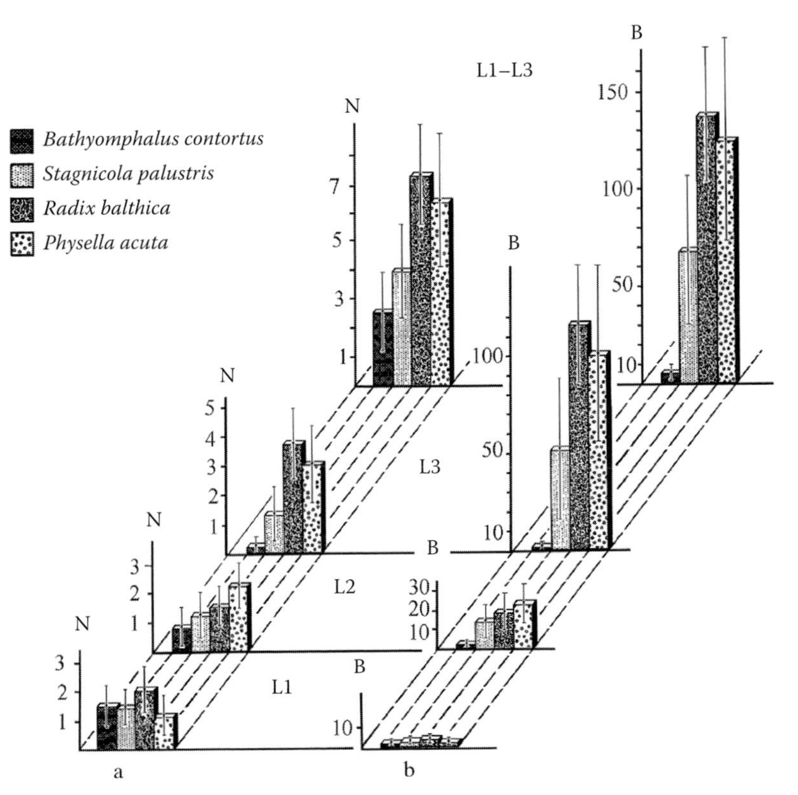

Fig. 5.8. Predation by ***Tetanocera ferruginea*** larvae (L1, L2, L3 and L1–L3 combined) in presence of four snail species as prey. (a) Predation in average number (N) of prey consumed by larvae; (b) average predation in biomass consumed (B, in mg). C.I. 5%. See Table 5.8, Eq. 5.1. From Manguin *et al.* (1988a).

columella, and *Physella virgata*), one terrestrial snail (*Cornu aspersum*), and one slug (*Lehmannia poirieri*). Although eggs of the succineids appear to be the natural oviposition site and food of young sciomyzid larvae, with older larvae feeding on juvenile to mature snails, sciomyzid eggs were laid on exposed eggs of all gastropod species, even those laid under water in nature, but to a lesser extent on *Helisoma*, *Cornu*, and *Lehmannia*. Larvae did not develop on gastropods of the latter three genera. See Section 18.1.3.4 for typical aquatic predators that have been found in gastropod egg masses in nature and have fed on gastropod eggs in laboratory rearings.

5.2.3 Terrestrial species

Field collections and host-specificity experiments with laboratory-reared *Salticella fasciata*, which oviposits only onto living, mature terrestrial snails, the larvae feeding primarily on dead tissue, were conducted by Knutson *et al.* (1970) and Coupland *et al.* (1994). Of 18 species of terrestrial snails in 15 genera collected in southern Wales, Knutson *et al.* (1970) found *S. fasciata* larvae only in *Cernuella virgata* and *Cepaea hortensis*. In the laboratory, however, females oviposited onto these two species as well as onto six other species in the genera *Arianta*, *Cochlicella*, *Fruticicola*, *Helicella*, *Monacha*, and *Theba*. In field collections in southern France (Coupland *et al.* 1994), only adults greater than 10mm diameter of the helicid species *Theba pisana* and *Cernuella virgata* were found bearing eggs. In their choice experiments, *S. fasciata* significantly preferred these snails as oviposition sites when kept in pairwise combinations with the co-occurring helicid species *Cochlicella barbara*, *C. acuta*, and *Cornu aspersum*. When maintained with similar sized *T. pisana* and *C. virgata*, ovipositing *S. fasciata* females showed no significant preference between species of snails. Of a collection of 100 individuals of these two species of snails made in southern France during October, 18 bore eggs of *S. fasciata* (Vala 1989a). Larvae of *S. fasciata* have been found in *T. pisana* collected in nature in southern France by Perris (1850, the first report of Sciomyzidae associated with molluscs); whether the snails were living or dead was not stated. They were also found by Mercier (1921) in living snails in southern France and in apparently living *Xerolenta obvia* by Povolný & Groschaft (1959) in Slovakia.

Larvae of five terrestrial predaceous species, *Coremacera marginata*, *Euthycera cribrata*, *Dichetophora obliterata*, *Pherbellia cinerella*, and *Trypetoptera punctulata*, were tested against nine species in eight genera of terrestrial snails, with 30 replicates each of one larva/one snail, in southern France (Coupland & Baker 1995, Coupland 1996a). All five sciomyzids each killed several snail species; *C. marginata* and *P. cinerella* killed all nine species, whereas *D. obliterata* and *T. punctulata* killed only five species each (Table 18.5).

5.2.4 Semi-terrestrial Succineidae snail feeders

Sixteen species of Sciomyzidae in the genera *Pherbellia*, *Pteromicra*, and *Sciomyza* (Sciomyzini) and *Hoplodictya*, *Limnia*, *Sepedon*, and *Tetanocera* (Tetanocerini) seem to be restricted, at least during early larval life, to juvenile to adult Succineidae of the genera *Oxyloma*, *Quickella*, and *Succinea*. Of worldwide distribution, succineids are moderately sized to large snails, characterized as semi-terrestrial, terrestrial-hygrophilous, aerial, or "amphibious," although none remains in water for prolonged periods. They are generally found on exposed vegetation or substrates around aquatic or damp habitats. Although many field collections of naturally infested succineids and limited host/prey trials rather conclusively prove the preference of the Sciomyzidae listed above to Succineidae, no rigorous host/prey preference experiments have been conducted to date. The succineid feeders range from parasitoids to predators; most species probably are specific only at the genus or family level. A few species, e.g., *Tetanocera arrogans* and *Hoplodictya spinicornis*, attacked and fed on a variety of aquatic and terrestrial snails during the second and/or third larval stadium. Knutson (2008) summarized the biology and relationships among succineid-feeding Sciomyzidae. A list of the species is included in Table 17.1.

6 • Host/prey finding

Il nostro geniale Federico Delpino [1869] *ebbe il torto di dichiarare che "uno dei più cospicui caratteri dei Ditteri è quello della loro stupidita*." [Our genial Federico Delpino [1869] was in the wrong when declaring that "one of the most prominent characters of the Diptera is their stupidity."].

Bezzi (1926).

The key factor of host/prey finding (see also Section 9.2.2, oviposition and egg production) in influencing the subsequent course of behavior during the life cycle is perhaps best described in the context of the classical theory of *r* and K strategies (Fig. 6.1). Species with *r*-selected traits exhibit high rates of increase and powers of dispersal that enable them to efficiently exploit patchy, ephemeral host populations. K-selected species predominate in perennial standing crop systems and exhibit lower rates of increase but high larval competitiveness (Force 1975). As shown in Fig. 6.1, most sciomyzid females, regardless of whether their offspring are terrestrial or aquatic, disperse their eggs in the microhabitat, with the newly hatched larvae searching for hosts/prey (*r*-strategy). Only five terrestrial parasitoids, perhaps one terrestrial parasitoid/predator, and at least five snail-egg feeders in damp habitats lay a limited number of eggs directly onto the host/prey (K-strategy). Although *r*-strategists are by far the most abundant and widespread in the Sciomyzidae, many are much rarer and more geographically limited than most K-strategists. Surprisingly, K-strategists have evolved primarily in the plesiomorphic subfamily Salticellinae and in the relatively plesiomorphic tribe Sciomyzini of the Tetanocerinae, with only five snail-egg-feeding *Anticheta* spp. and possibly *Dichetophora obliterata* in the Tetanocerini. K-strategists are known only in the Nearctic and Palearctic, while *r*-strategists occur throughout the world.

6.1 ADULTS

Information is limited concerning host/prey finding by adults, through microhabitat location or search for host/prey individuals. The ease of obtaining oviposition by field-collected females of most species kept under artificial conditions, such as small containers with substrates of damp cotton and often without molluscs, may indicate a low level of searching by females of most species. However, oviposition under such conditions may be due to stress, as is known for some other Diptera. The characteristic behavior of adults "patting" the surface in front of them with their front tarsi indicates searching behavior. The porrect and often elongate antennae characteristic of the family may be a morphological adaptation to enhance searching.

Gastropod feces and slime trails would seem to be likely materials to which ovipositing females could orient except in freshwater and damp habitats, where these materials likely are soon dissipated. Coupland (1996b) found that the terrestrial predator *Pherbellia cinerella* oviposited more frequently on substrates containing fresh feces of the terrestrial snail *Cernuella virgata* than on substrates containing snail mucus or water (control). *Cernuella virgata* and other terrestrial snails and slugs are known to return usually to the same resting site daily, and usually defecate before resting. Coupland proposed that larvae are more likely to find suitable prey upon emerging from the egg if female flies oviposit on the snails' feces.

Evidence has been presented that odors produced by freshwater snails attract adult sciomyzids. Large numbers of adults of the introduced species *Sepedomerus macropus* in Hawaii were captured with traps baited with freshly crushed *Lymnaea* snails. Fifty to 200 flies were captured during each of several half-hour periods, many more than could be collected by sweeping (Chock *et al.* 1961). Attempts to trap adult Sciomyzidae with crushed snails in the Ithaca, New York, area were not successful. DeWitt *et al.* (2000) noted that an act of predation on a snail causes it to release alarm pheromones that stimulate avoidance behaviors in nearby conspecific snail prey. Ovipositing sciomyzids perhaps also are sensitive to such pheromones produced by one or more snail species in a typical assemblage of prey. Feeding within a prey for a relatively long

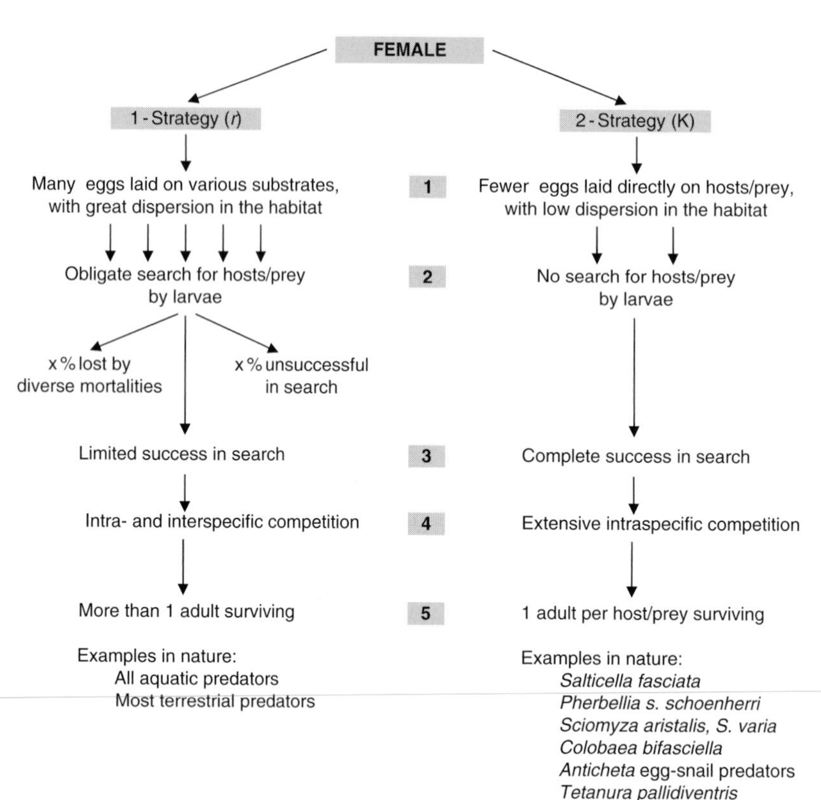

Fig. 6.1 Significance of methods of oviposition and host/prey finding in context of *r* and K selection.

time before killing it, especially by first- and second-instar larvae of many species of Sciomyzidae, may result in a prolonged period of release of alarm pheromones by the prey. This could result in escape, probably over a relatively short distance, by conspecific snail prey from cohort larvae and larvae of other sciomyzid species. However, it could serve as an olfactory beacon to a favorable oviposition site for the female flies. Observation of a natural population of snails when marked, laboratory-infested, conspecific snails are released among them would be of interest. Of course, at most times there are some dead snails, resulting from disease, predation, or age among a snail population and ovipositing flies may be attracted by the dead snails.

Only five sciomyzid species are known to oviposit routinely only on the shell of the host: *Pherbellia s. schoenherri* (but not the Nearctic subspecies *P. s. maculata*) (Verbeke 1960, Bratt *et al.* 1969, Moor 1980, Vala & Ghamizi 1992) and *Sciomyza aristalis* (Foote 1959a) on *Succinea* spp.; *Sciomyza varia* (Barnes 1990) and *Colobaea bifasciella* (Knutson & Bratt, unpublished data) on stranded or estivating *Stagnicola* and *Galba* spp.; and *Salticella fasciata* on terrestrial helicid snails (Knutson *et al.* 1970, Coupland *et al.* 1994). Eggs of *Atrichomelina pubera* were found on shells of freshwater snails in nature; in the laboratory they were laid on shells or on the moist substrate near the snails (Foote *et al.* 1960). Oviposition on the shell of the host might also be typical for the terrestrial predator *Dichetophora obliterata*, for which 50.7% of eggs laid in the laboratory were laid on *Helicella bolenensis* and *Theba pisana* snails included in the rearing containers (Vala *et al.* 1987). Five species of *Anticheta* oviposit only onto egg masses of *Aplexa*, *Galba*, *Stagnicola*, and *Succinea* (Fisher & Orth 1964, Knutson 1966, Knutson & Abercrombie 1977, Robinson & Foote 1978).

Tetanura pallidiventris is the only species in the family Sciomyzidae in which the ovipositor is highly modified (Figs. 9.26, 14.2a) and the only species known to lay eggs directly onto the flesh of snails, i.e., the retracted soft parts of resting *Cochlicopa lubrica* (Knutson 1970a). The extremely short incubation period of eggs of *T. pallidiventris* – within 12 hours of being laid, the shortest incubation period for any known species of Sciomyzidae – likely is correlated with the oviposition site, as the eggs certainly would be destroyed if the resting host became active. The thin, weakly sculptured chorion of the egg of *T. pallidiventris* also is likely a co-adaptation with the unusual oviposition site and incubation period. In nature, larvae and puparia of *T. pallidiventris* have been found in *C. lubrica*, *Aegopinella pura*, *Discus rotundatus*, and in *Clausilia* sp. in Poland (the latter an unpublished collection by A. Sulikowska, and determined by L. Knutson).

6.2 LARVAE

Searching by highly active, newly hatched larvae is the most usual method of host/prey location for the mainly predaceous freshwater or terrestrial species. From eggs laid on various substrates close to the mollusc microhabitats, sciomyzid larvae crawl to the water or substrate and aggressively attack their first prey. The fact that not all newly hatched larvae succeed in encountering a prey seems to be a major cause of mortality, partially compensated by the great number of eggs (200–600) laid by females of many such species. The manner of searching differs fundamentally between freshwater and terrestrial larvae. Most first-instar freshwater larvae must find a mollusc within the first few days following hatching in order to survive. Valley & Berg (1977) placed 57 newly hatched larvae of four Nearctic species of *Dictya*, aquatic predators, in containers supplied with water but no food and noted that they lived 3–8 days (average 4.8 days) without feeding. Once-fed, one-day-old larvae of *D. floridensis* that were subsequently starved lived 5–7 days (average 6.4 days). Different methods of locating prey by freshwater predaceous *Sepedon* spp. in South Africa were characterized as having "active" and "passive" searching (or "ambush" behavior) by Appleton *et al.* (1993). *Sepedon scapularis* larvae pursue snails, and while clinging to the shell probe for the edge, then penetrate the soft parts. *Sepedon neavei* suspends itself from the water surface by means of its hydrofuge interspiracular processes (float hairs) and attacks snails passing below. These general types of behavior have been described in many publications on the biology of aquatic predators.

Some terrestrial first-instar larvae, e.g., *Dichetophora obliterata*, can live for as long as 20 days after hatching before finding their molluscan food and without losing their attack potential (Vala *et al.* 1987). Newly hatched parasitoid larvae of two of the four slug-killing *Tetanocera* species (*T. elata* and *T. plebeja*) remain in place after hatching, with only their posterior end on the substrate and the body upright and motionless until a host slug passes by, whereupon they crawl onto it and then into protected recesses (e.g., beneath the mantle). Third-instar predatory larvae of these species initiate searching behavior upon contact with a slug (Knutson *et al.* 1965, Trelka & Foote 1970, Trelka & Berg 1977). Photographs of slug-killing larvae waiting for, searching for, and attacking hosts/prey were presented by Knutson *et al.* (1965) for *T. elata* (Fig. 9.28d), Trelka & Foote (1970) for *T. plebeja*, and Vala (1989a) for *Euthycera cribrata*. Bratt *et al.* (1969) also observed "ambush" behavior in first-instar larvae of some terrestrial species of *Pherbellia*. Slugs and terrestrial snails may be actively sought by *E. cribrata* but "usually the larvae remain poised for attack, with head erect. As soon as a slug encounters a larva, the larva rapidly attaches itself to the tegument using its mouth-hooks, climbs onto the prey and enters into the tissues . . . until only its posterior spiracles remain exposed" (Reidenbach *et al.* 1989). Similar behavior was noted for *Tetanocera melanostigma*, a parasitoid/predator of semi-terrestrial *Novisuccinea ovalis* snails (Foote 1996b); first-instar larvae placed in containers of moist sand,

> . . . moved slowly over the sand, stopping frequently and lifting the anterior parts of their bodies off the substrate and waving them to and fro. When they brushed against a *Succinea*, larvae became highly excited and attempted to crawl up onto the shell. Many then crawled onto the lateral surfaces of the expanded foot and attempted to insert themselves between the foot and collar forming the edge of the mantle.

This may represent a two-phase strategy (microhabitat and host/prey location), which is the most common method of host location in parasitoid Diptera (Feener & Brown 1997). Reidenbach *et al.* (1989) noted that development of the eggs of *E. cribrata*, a parasitoid/predator of slugs and terrestrial snails, usually required about 15 days in the laboratory, but hatching often was delayed several weeks until the eggs had been moistened. In nature, rainfall or high humidity, conditions that also favor the activity of their hosts/prey, likely induce hatching. The authors also noted that hatching in the laboratory is induced if the eggs are crushed by a passing mollusc, a feature that, in nature, would aid in synchronizing hatching with host/prey activity.

Geckler (1971) conducted laboratory experiments to determine the amount of time to first kill of *Helisoma anceps* snails by newly hatched larvae of the freshwater predator *Sepedon tenuicornis*, as a function of (1) snail density and (2) distance required to travel to first kill and snail density. Under the experimental conditions – snails

Table 6.1. Time to first kill by ***Sepedon tenuicornis*** L1 as functions of distance traveled and densities of 0.6 snails/cm^2 (a) and 1.2 snails/cm^2 (b)

Radial distance traveled (mm)	N	Mean time to kill (min)	± S.E.
(a)			
16	40	6.69	4.36
30	30	10.09	5.69
46	30	15.20	7.48
75	30	26.06	5.55
Regression line: Time to kill = 0.79–0.33 × (distance)			
(b)			
16	20	5.31	2.42
30	19	7.85	5.09
46	10	16.40	11.69

Note: Snails were *Helisoma anceps*. From Geckler (1971).

Table 6.2. Time to first kill by ***Sepedon tenuicornis*** L1 as functions of distance traveled and density of 0.3 snails/cm^2

Group	Radial distance traveled (mm)	N	Mean time to kill (min)	± S.E.
1	16	32	8.00	5.68
2	30	46	23.65	23.30
3	46	10	24.53	11.71
4	16*	11	8.75	4.83
5	30	11	10.45	4.41
6	46*	18	11.35	6.60

Note: *Larvae from the same rearing bottle on the same day. Snails were *Helisoma anceps*. From Geckler (1971).

and larvae placed on moist cotton in plastic boxes for density trials and placed in plastic rings of 16, 30, 46, and 75 mm diameter on moist cotton on a 20 cm diameter disk for distance plus density trials – Geckler found that the regression line relating time required to kill as a function of snail density was T (minutes) = 65.3–111.6 × (snail density). The regression line relating time to kill and snail densities of 0.6 and 1.2 snails/cm^2 as a function of distance from snails was T = 0.79–0.33 × (distance) (Tables 6.1 and 6.2). The author concluded that, "The time required for a larva to find and kill its first snail is highly variable and dependent, at least, on snail density and size and distance between larva and snail."

O. Beaver (1989) conducted detailed laboratory experiments on the searching ability of all three instars of larvae of the freshwater predator *Sepedon senex* attacking *Gyraulus convexiusculus* under more natural conditions in 92 mm diameter Petri dishes containing water 5 mm deep. The larvae and snails were at six prey densities of 1–12 snails per dish. She found that later-instar larvae had higher efficiencies of searching and detecting the prey than did young larvae. At similar distances from the prey, third-instar larvae moved faster toward their prey than did younger larvae. First- and second-instar larvae moved faster as they came closer to snail prey, indicating that the nearer they were to their prey the more accurately they could detect its position. However, Geckler (1971) noted, "Sometimes it appeared that larvae were being stimulated by the proximity of a snail but other times they would pass within 1/4 mm of a snail to kill one 1 cm away." O. Beaver (1989) also noted that larvae moved randomly at the beginning but more directly toward the snail as they located their prey. The increase in the rate of movement started from the release point, 92 mm away from the prey, indication that larvae can detect their prey at a distance of 92 mm or more. Third-instar larvae moved directly from the release point toward their prey at a constant rate of movement. O. Beaver concluded that not only the stage of development but also the distance between the predator and the prey influences searching time. The searching and handling time of larvae of all developmental stages varied with prey density. Increased prey density shortened searching time but slightly prolonged handling time because of interference by other snails. First-instar larvae spent more time in searching and handling their prey than did second- and third-instar larvae (Figs 6.2, 6.3).

In a quantitative experimental study, third-instar larvae of the slug parasitoid/predator *Tetanocera plebeja* appeared to follow fresh slime trails of *Deroceras laeve* and *D. reticulatum* only after recent contact with a slug, and they often followed the slime trails in the wrong direction (Trelka & Berg 1977). Coupland (1966b) found that both mucus and feces stimulated increased search behavior in first-instar larvae of the terrestrial predator *Pherbellia cinerella*. This study was the first proof of snail feces eliciting a response by any predator of snails.

Mc Donnell *et al.* (2007a), working with *Sepedon spinipes* and *Dictya montana* were the first to prove that larvae of aquatic sciomyzids could follow fresh snail mucus trails – of *Lymnaea stagnalis* and *Lymnaea* sp., respectively. The

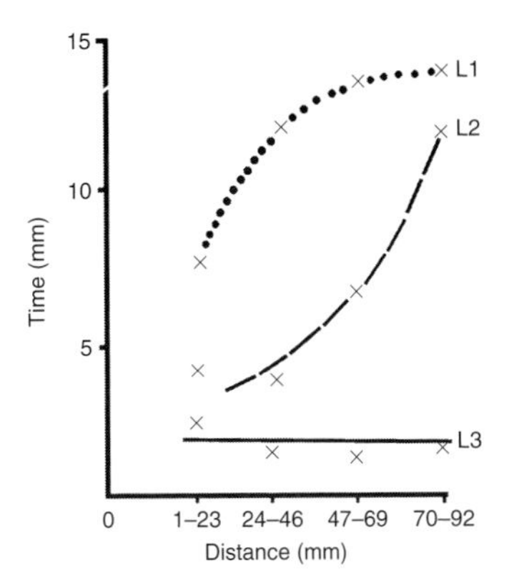

Fig. 6.2 Time (minutes) spent in moving 22 mm at four different distance intervals (x) between ***Sepedon senex*** larvae and *Gyraulus convexiusculus* prey. From O. Beaver (1989).

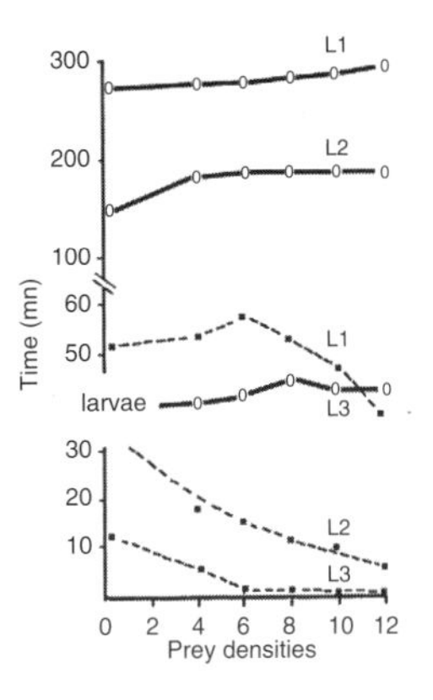

Fig. 6.3 Searching and handling time of ***Sepedon senex*** larvae at different prey densities. Square, searching; circle, handling. From O. Beaver (1989).

authors used Y-shaped mazes that consisted of a single stem and two arms, each of which were 3 cm long and 1 cm wide, separated by a 30° angle and lined with filter paper saturated with deionized water. Mucus was removed from the foot of three snails (approximately 8 cm long) with an artist's moist paintbrush and painted onto one of the arms (chosen at random). The other arm had no mucus trail and acted as a control. Neonates were acclimatized to the test environment by being placed on a saturated filter paper disk in a Petri dish for 15 minutes before testing. They were then placed on the Y-maze stem and encouraged forward by use of a stationary photo-optic light source. Larval response was measured as a strong response (SR: neonates followed the trail to its end), weak response (WR: larvae followed the mucus trail but deviated from it before reaching its end), or a non-response (NR: neonates did not follow the trail). Upon finding a mucus trail, larval searching behavior of both species switched from unstimulated to stimulated, the latter being characterized by an increase in larval velocity and the frequency of lateral head taps. In addition, significantly more neonates of *S. spinipes* and *D. montana* exhibited an SR than either a WR or an NR. Larvae of both of these species are aquatic predators, and their ability to follow mucus trails out of the water, in this case on filter paper, suggests that they have the potential to forage for food in semi-aquatic areas, e.g., shorelines. The authors also suggested that since gastropod mucus consists of both ubiquitous and species-specific mucus components (Skingsley *et al.* 2000) and since larvae of *S. spinipes* and *D. montana* are generalist aquatic predators, neonates likely utilize one or more of the ubiquitous constituents in detecting and following trails. Because the test larvae were naive, their trail-following behavior was likely to have been a genetic response. When trails were aged for 45 minutes, significantly more newly hatched larvae exhibited a WR than an NR or an SR. The authors speculated that an inability to follow aged trails effectively could, however, be advantageous in terms of locating prey efficiently. In any given snail habitat, the substrate is likely to consist of a mosaic of overlapping mucus trails of various ages; consequently the probability of locating prey would be reduced with increased ability to follow old trails. An acute sensitivity to fresh mucus, however, is more likely to result in successful snail location because the prey is likely to be in the immediate environment. P. Roberts (personal communication, 2005) also found that *Pherbina coryleti*, *Tetanocera arrogans*, and *T. ferruginea* followed fresh trails of the aquatic snail *Radix balthica* but showed weaker response to trails that had been aged for 15 minutes. Although slugs are not known to be normal prey of the semi-aquatic *P. coryleti*, it followed slime trails of *Arion fuscus*.

7 • Feeding behavior

When one tugs at a single thing in nature, he finds it attached to the rest of the world.

J. Muir.

7.1 FEEDING SITES

Characterization of Behavioral Groups by a combination of kind of host/prey, larval feeding behavior, and microhabitat as presented in the classification above (Section 4.3) has been the traditional approach in classifying behavioral diversity of sciomyzid larvae. Since there seems to be great diversity with regard to anatomical part, physiological state of the host/prey tissues, and nutritional condition of the food utilized by the larvae, at least from more or less gross observations, it is useful also to consider a classification based on precise feeding site. Only one detailed study has been conducted of the anatomical parts of snails fed upon by sciomyzid larvae (Barnes 1990). *Stagnicola elodes* snails were dissected after being fed upon by first-, second-, and third-instar larvae of *Sciomyza varia*, a specialized parasitoid intimately associated with its estivating host. First-instar larvae appeared to have abraded the posterior angle of the peristomial collar. Smaller second instars significantly damaged the surface of the peristomial collar and the mantle over the kidney and lung. Larger second instars consumed portions of the peristomial collar, mantle, and subtending kidney, resting with their anterior end over or in the albumen gland; at this stage the host was still alive. Both second and third instars appeared to prefer the albumen gland, a structure common to most non-operculate gastropods. Smaller third instars had inserted their cephalic segments into the albumen gland. Larger third instars destroyed portions of the intestine, uterus, seminal receptacle, and posterior prostate. The largest larvae ate all but the liver, ovotestis, gizzard, foot, vas deferens, preputium, and columellar muscle. Whereas the brains of snails and slugs are in positions relatively exposed to attack (Figs 1.2–1.5) and are attacked by some slug killers and many aquatic predators, they are avoided by the most parasitoid larvae, which must keep their hosts alive during much of their period of feeding.

Knutson & Berg (1967) presented photographs of the soft parts of a specimen of *Radix balthica* fed upon by *Ilione unipunctata*, illustrating the portions of the snail's body characteristically left uneaten (i.e., foot, mantle, and tentacles) by such freshwater predators. O. Beaver (1974b) noted that large snails attacked by aquatic and semi-aquatic predators may have only the digestive gland and reproductive organs eaten, even by large larvae, but small snails are usually eaten completely. Mass rearings of the freshwater predator *Dictya floridensis* to the puparial stage on freshly crushed and frozen crushed snails, in trays with water levels of 2.5 cm, were highly successful, with no significant differences in percentage pupariation between those feeding on live snails and those feeding on crushed snails (McLaughlin & Dame 1989). However, it is likely that in nature even freshwater predators feed on specific tissues, especially those tissues most readily reached while the posterior end of the predator is in contact with the ambient air.

Although general observations are scattered in life-cycle papers on the length of time individual larvae of the various behavioral groups spend feeding, limited quantitative data have been published. Lindsay (1982) Lindsay *et al.* (2011) conducted detailed experiments with third-instar larvae of the univoltine aquatic predator *Ilione albiseta* feeding in *Galba truncatula*, *Stagnicola palustris*, and *Radix balthica*. Larvae were starved 3–7 days, then placed in Petri dishes with snails. As each larva attacked, Lindsay noted the time it took for the larva to establish itself in the snail and the total feeding time. When the larva had finished feeding, it was weighed and the length of the snail was measured. The effect of snail size and larval size (weight), on total and post-establishment feeding time was then determined by use of regression analyses. His unique and important results and conclusions are paraphrased below.

Snail size was the paramount factor in determining the duration of feeding time. The equations for snail size against total feeding time and against post-establishment time are shown in Eq. 7.1a, b.

In the regression analyses of (a) larval weight against total feeding time and (b) larval weight against post-establishment time, analyses gave the equations as shown in Eq. 7.1c, d.

A multiple linear regression analysis conducted on snail size (x), larval weight (y), and total feeding time (z) for nine results chosen at random gave the equation $z = -28.26 + 43.78x - 0.11y$. This confirmed the linear regression results which showed that for third-instar larvae, snail size is the major criterion in determining the duration of feeding time. Fig. 7.1 shows the fitted regression line and data points for the regression of snail size against total feeding time. Lindsay (1982) Lindsay (*et al.* 2011) concluded that only a small proportion of time, during at least the third stadium, is occupied with actually feeding on snails. For example, a larva that eats five snails of an average length of 4 mm per week will spend less than 10 hours feeding on those snails in that week. He further concluded that this implies that, in the field, larvae can spend large amounts of time searching for prey and so could survive at fairly low snail densities. This is an important conclusion in regard to use of such Sciomyzidae as biocontrol agents.

Other observations indicate that feeding can be unusually voracious and rapid at high snail densities. Valley & Berg (1977) provided information on a third-instar larva of the aquatic predator *Dictya atlantica*. The prey species were not given but were of the families Lymnaeidae, Physidae, and/or Planorbidae and were 1.3–8.0 mm in greatest diameter or length. A surplus of snails apparently was available to the larva. The observed larva of this moderately sized species (7.2–13.0 mm long, range 5.3–14.2 mm for the genus *Dictya*) "ate 3 snails in 3 hours and 43 minutes, feeding for 56 minutes on the first victim, 82 minutes on the second, and 78 minutes on the third, with only 4- and 3-minute intervals between meals." The authors did not discuss whether this was a case of "wasteful feeding" (leaving tissue uneaten and/or defecating undigested tissue) noted in (2) below and in Section 7.3.

Feeding sites may be categorized as follows.

(1) Massive rupture of the hemocoel of freshwater snails by (usually) only one larva. Among many similar reports, Neff & Berg (1966) noted cessation of the heartbeat of freshwater snails in about 10 minutes after attack by many species of *Sepedon*. Foote (1999) noted progressive slowing of the heartbeat of a 10.2 mm diameter *Biomphalaria glabrata* during attack by a nearly mature third-instar larva of *Tetanocera ferruginea*, with heart action ceasing within 18 minutes. In snails having red hemolymph, this fluid is seen to pour out immediately upon attack by a predatory larva. Such snails never recover, even if the attacking larva is removed. Larvae occasionally are seen to leave the snail's flesh and engorge themselves on the hemolymph that fills the aperture (Neff 1964). Feeding strategies of second- and third-instar larvae of *Sepedon neavei*, a freshwater predator in South Africa, are shown in Fig. 7.2. These larvae feed voraciously for an hour or so, only on the fresh tissue that is within reach, often with much of the snail body mass remaining uneaten. They are in this sense "wasteful predators." They are also wasteful predators, especially during the third stadium, in that they apparently overfeed and defecate before digestion is completed. Eckblad (1976) noted that *Sepedon fuscipennis* in North America "were often observed to attack and kill snails encountered even when their guts contained snail tissue from a previous meal. The partially digested gut contents would commonly be voided as they began feeding on a newly killed snail." He quantified this by a coefficient of utilization of food (dry weight) for growth. O. Beaver (1974b) also commented on wasteful feeders and found that the mean ± 1 S.E. "food conversion ratio" (ratio of total wet weight of snails killed to total weight of larvae produced) of eight aquatic and semi-aquatic predators in Wales ranged from 6.1 to

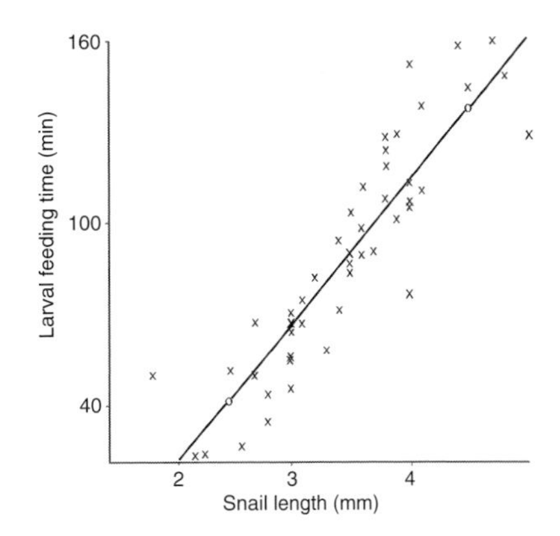

Fig. 7.1 Effect of snail size (length in mm) on total time (in minutes) spent feeding by ***Ilione albiseta*** L3. From Lindsay (1982), Lindsay *et al.* (2011).

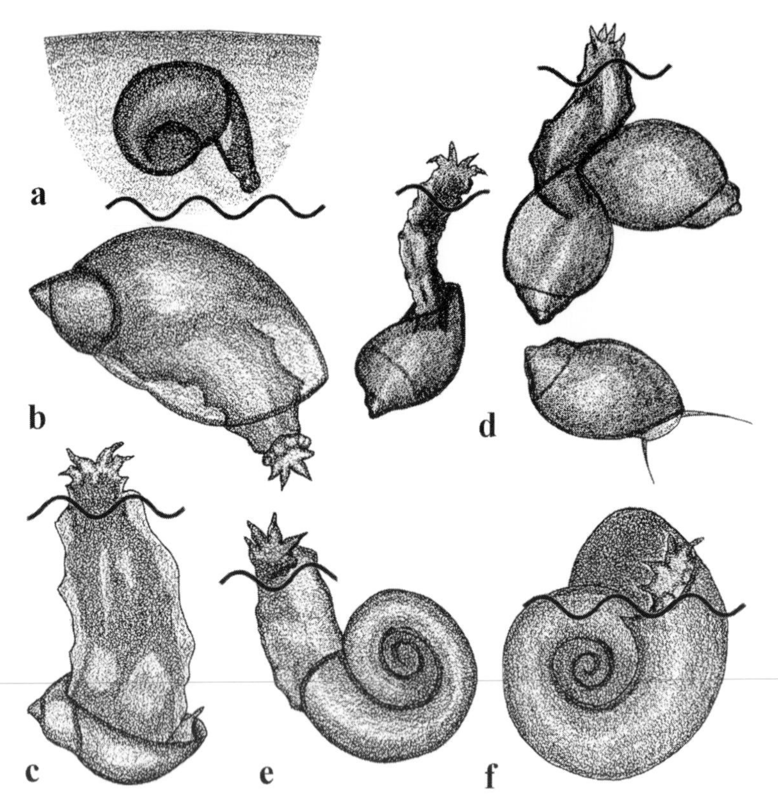

Fig. 7.2 Feeding strategies of ***Sepedon neavei*** L2 and L3. (a) L2 feeding on small *Biomphalaria pfeifferi* on Petri-dish wall, above water level. (b) L3 feeding in visceral hump of *Radix natalensis* out of water. (c) L2 probing foot of *Bulinus* sp. (above); foot withdrawn, rolling snail over, larva starts feeding while maintaining surface contact. (d) L3 feeding within overturned *Bulinus* sp., with posterior spiracles maintaining surface contact. (e) L3 moving into coiled shell of *B. pfeifferi*. (f) L3 entirely within coiled shell of *B. pfeifferi*; spiracular disc visible in aperture, where it maintains surface contact. Wavy lines indicate water level. Modified from Barraclough (1983).

31.7. The most wasteful feeders, such as *Ilione albiseta* and *Pherbina coryleti* – both univoltine predators that overwinter as third-instar larvae – had the highest ratios. We have noted this behavior in many freshwater predaceous sciomyzids. Further details on this aspect are given in Section 7.2. (**Behavioral Group 11**). Observations of snails dropping from the surface of the water after having been attacked and surviving, e.g., *Austropeplea tomentosa* attacked by *Dichetophora biroi* (Lynch 1965), probably are instances when the predaceous larva did not have time to rupture the hemocoel.

(2) Rupture of the hemocoel of freshwater (on a solid substrate), semi-terrestrial, or terrestrial snails by one or several larvae together. These larvae apparently feed indiscriminately on fresh soft parts within reach but then continue to feed on decaying, liquefied, putrid tissue until the shell is empty and subsequently attack another snail in the same manner if development is not complete. (**Behavioral Groups 2, 3, 7, 12**).

(3) Feeding on tissues of recently dead, dying, or moribund snails, the tissues being fresh at the beginning of feeding but putrid and liquefied during most of the feeding. In the case of *Salticella fasciata*, the only species in the family in which the larva is known to have ventral cibarial ridges, much or all of the food may consist of microorganisms filtered from the liquefied, rotting tissues (Dowding 1967, 1968, 1971). (**Behavioral Group 1**).

(4) Feeding, usually individually, in living snails on proteinaceous secretions (e.g., extrapallial fluid) and/or less vital soft parts for several days, then on various soft parts as described for *Sciomyza varia* (Barnes 1990), and finally on the liquefied tissues of the original snail attacked. The first-instar larvae of some of these species penetrate rather deeply into the snail, between the mantle and the shell, without their posterior spiracles exposed, thus apparently respiring by using dissolved oxygen in the liquid between the mantle and the shell. These larvae are usually host specific at the genus level of the host/prey. (**Behavioral Groups 3, 4, 6**) (Fig. 7.3).

(5) Similar feeding as in No. 4 but on slugs (and snails in the case of some species) and with a more predatory existence during stadia 2 and 3, i.e., a series of prey is attacked and fed upon for a short time. (**Behavioral Groups 8, 9**). The sites of attacks on slugs are

shown in Fig. 7.4 and Plates 3b, 4b–d. Comparisons of attacks on slugs by third-instar larvae of two species of *Tetanocera* showed that the site attacked influences the time required for the salivary gland toxin to immobilize the slugs, ranging from only five seconds to two minutes (Trelka & Berg 1977) (Table 7.1).

(6) Feeding in a mesoparasitoid manner, completely within the body of a slug, without the posterior spiracles exposed to the exterior. (**Behavioral Group 10**).

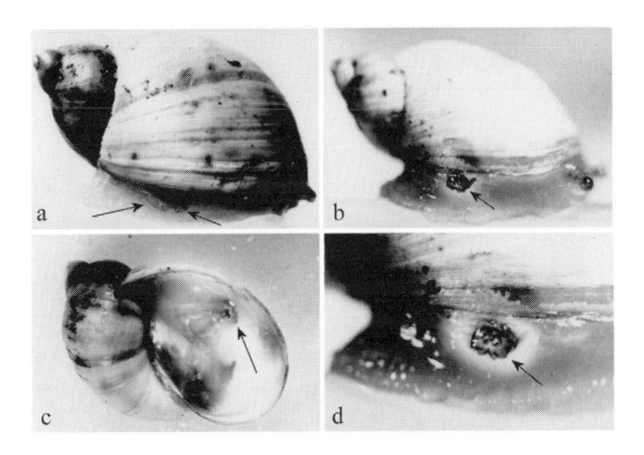

Fig. 7.3 Larvae (arrows) of ***Sepedon trichrooscelis*** feeding in *Succinea campestris*. (a) Two L2 near breathing pore of active *S. campestris*. (b) One late L2 or early L3 near breathing pore of active *S. campestris* (note extended eye stalks of still active snail). (c) One large L3 feeding in decaying tissue of dead *S. campestris*. (d) Posterior spiracular disc, same individual as shown in (b) at higher magnification. From Knutson (2008).

(7) Feeding on embryonic tissues within egg masses of *Aplexa*, *Physa*, *Stagnicola*, or Succineidae by young larvae, then, in the case of some species, predation on juvenile to mature snails by older larvae. (**Behavioral Group 5**).

(8) Feeding by young larvae on non-essential tissues of submerged fingernail clams (Sphaeriidae) or freshwater snails, then predatory feeding behavior by older larvae, with their posterior spiracles exposed to the air. (**Behavioral Groups 13, 14**).

(9) Preying upon freshwater oligochaete worms. (**Behavioral Group 15**).

The labile or mixed behavior and variable food of many aquatic and semi-aquatic species, based mainly on laboratory rearings, has been noted frequently. For example, *Sepedon neavei* was considered primarily an aquatic or semi-aquatic predator, but some first-instar larvae also fed on dead snails for 3–4 days, some fed briefly on embryonic *Bulinus*, and some did a bit of subsurface foraging and feeding (Barraclough 1983).

Additional methods might be used in studying the feeding behavior of Sciomyzidae. Allowing a larva to attack or feed very briefly on the host/prey, then removing the larva to see if the host/prey survives, would give indication of the production of a salivary gland toxin or destruction of a vital organ at the onset of attack/feeding. Some of the methods used in the study of the predatory behavior of other gastropod natural enemies (Carabidae, Drilidae, etc.), i.e., precipitin and enzyme-linked immunosorbent assays, genus- and species-specific monoclonal antibodies, iso-

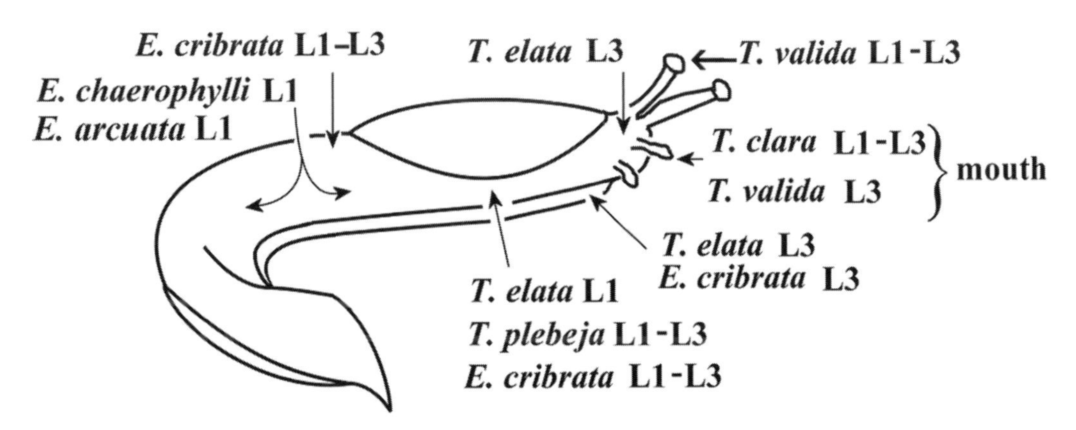

Fig. 7.4 Sites of attack by slug-killing *Tetanocera* spp. and *Euthycera* spp.

Table 7.1. Comparison of attacks by ***Tetanocera elata*** and ***T. plebeja*** and their effect on *Deroceras* slugs

	Usual site of attacks	Time required for immobilization with attacks at usual sites, mouthhooks inserted for entire period	Time elapsed until feeding was initiated	Time required for immobilization after 15-s attacks at posterolateral site
T. elata	Anteroventral part of foot	5–30 (15 ± 3.2)	15–60 (35 ± 4)	55 ± 9.4
T. plebeja	Posterolateral part of body	15–120 (54 ± 9.8)	60–240 (174 ± 30)	60 ± 10

Note: Data for each species based on samples of 20 L3. Time periods given all start at first insertion of mouthhooks. All values in seconds. From Trelka & Berg (1977).

electric focusing, and use of PCR primers likely could be used profitably in the study of sciomyzid larval feeding. See Section 2.2 and Symondson (2002b).

7.2 NUMBERS AND SIZES OF HOSTS/PREY ATTACKED (EXCEPT BY AQUATIC PREDATORS)

Except for several predators of freshwater snails, relatively little quantitative information is available on the numbers and sizes of hosts/prey killed and eaten during the course of development by individuals in the other major behavioral groups. The most detailed information available for groups other than aquatic predators of snails is presented below.

7.2.1 Snails on moist surfaces

There is a very broad range in the numbers and sizes, as well as species, of snails killed and eaten by larvae in the large group of behaviorally quite variable species of Sciomyzini and Tetanocerini that are neither strictly aquatic nor strictly terrestrial (Behavioral Group 2, see Section 4.3). Few rearings have been made of isolated larvae of this Behavioral Group, as there have been for larvae of most of the other Behavioral Groups. An isolated larva of *Pherbellia idahoensis* (adults 5.6–6.4 mm long) killed and ate three 12.0–15.0 mm diameter *Helisoma trivolvis* during the course of development. Several isolated larvae of *Pherbellia similis* (adults 3.6–4.4 mm long) ate 2–4 planorbid snails, 7.0–10.0 mm diameter. When one or two first-instar larvae of *Pherbellia argyra* (adults 4.3–5.4 mm long) attacked a single 10.0–15.0 mm diameter *Planorbis planorbis*, they developed to pupariation on the tissues of that one snail (Bratt *et al.* 1969). In laboratory rearings, the size and number of snails killed depends on the relative sizes of the snail host/prey and the larva, and whether larvae kill and feed in a solitary or gregarious manner. There are few data on larvae feeding gregariously in nature. Determination of the numbers and sizes of snails killed and eaten by many of these semi-terrestrial, semi-aquatic species with labile feeding behavior also is compromised by the fact that the larvae seem to be equally capable of feeding on dead, even liquefied tissues of a host/prey for extended periods or overtly attacking, killing, and feeding on the fresh tissues of a snail. The nutrition of many species is capable of alternating between fresh and decomposing tissue with apparent ease. Such larvae have remarkable resilience, spending long periods feeding or not feeding, are essentially non-host/non-prey specific, and the durations of the larval stages seem to be imprecisely fixed. Vala & Gasc (1990a) noted that while first- and young second-instar larvae of *Pherbina mediterranea* consumed few snails (but of a wide range in size), older second- and third-instar larvae behaved more like typical aquatic predators, killing and feeding on several snails, a total of about 16, 4.0–11.0 mm in length or diameter (Fig. 7.5). O. Beaver (1972b) noted that a *Limnia paludicola*, as "*L. unguicornis*," killed and ate 11 *Lymnaea stagnalis* and *Gyraulus albus*, 1.0–5.8 mm greatest diameter, during 100 days of larval life. *Limnia unguicornis* was recorded as an opportunistic predator/saprophage of slugs and terrestrial snails (Vala & Knutson 1990). Morphologically, some characters of the larvae of such species also are intermediate between those of typical terrestrial and typical aquatic species (see Figs 14.31, 14.32, Table 14.2).

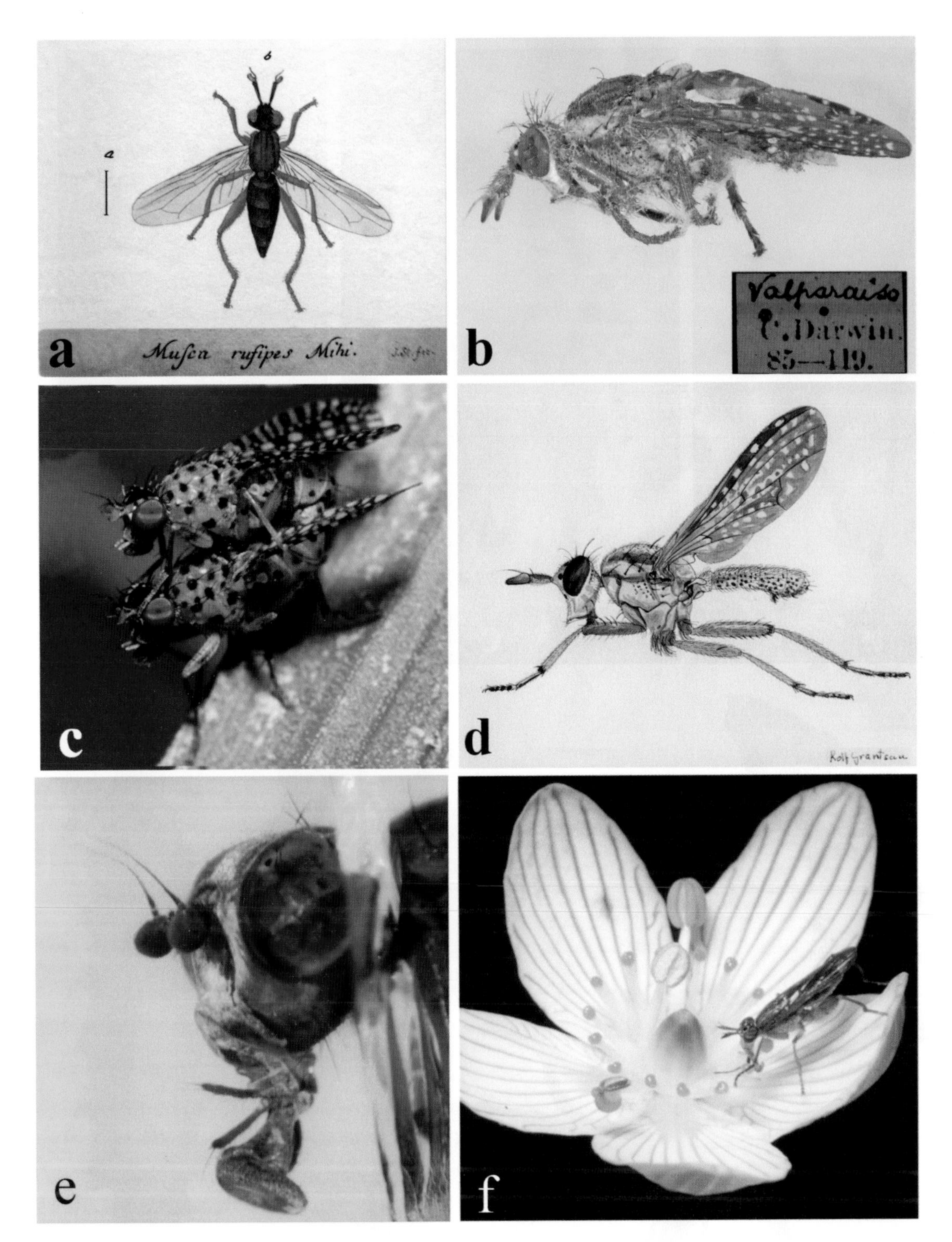

Plate 1. (a) "***Musca rufipes***" Scopoli (= ***Sepedon sphegea***), second known drawing of an adult Sciomyzidae (from Panzer 1798). (Note: the first is of the new species "*Musca Solicitus*" [= *Elgiva solicita*] by Harris, M. (1782?). In *An Exposition of English Insects*. . ., Decad IV, Tab. XXXIV, Fig 18. London: White & Robson.) (b) ***Protodictya chilensis*** female collected in Valparaiso, Chile, by C. Darwin (in BM [NH] collection). (c) Pair of ***Poecilographa decora*** mating (photo by S. A. Marshall). (d) ***Protodictya chilensis*** male (drawing by Rolf Grantsam). (e) ***Prosalticella succini***, fossil sciomyzid in Baltic amber, head, anterolateral (photo by C. & H. W. Hoffeins). (f) ***Sepedon*** sp. feeding on false nectaries of *Parnassia palustris* (Saxifragaceae) (photo by S. A. Marshall).

Plate 2. (a) ***Salticella fasciata***, L3, dorsal (photo by M. Thomas). (b, c) ***Pherbellia dorsata***, L3. (b) lateral; (c) dorsal. (d) ***Ilione lineata*** L1, dorsal (photos by J. Church). (e) ***Eulimnia philpotti***, L3, dorsal (photo by J. K. Barnes). (f, g) ***Tetanocera hyalipennis***, L3. (f) ventral; (g) dorsal (photos by L. Knutson).

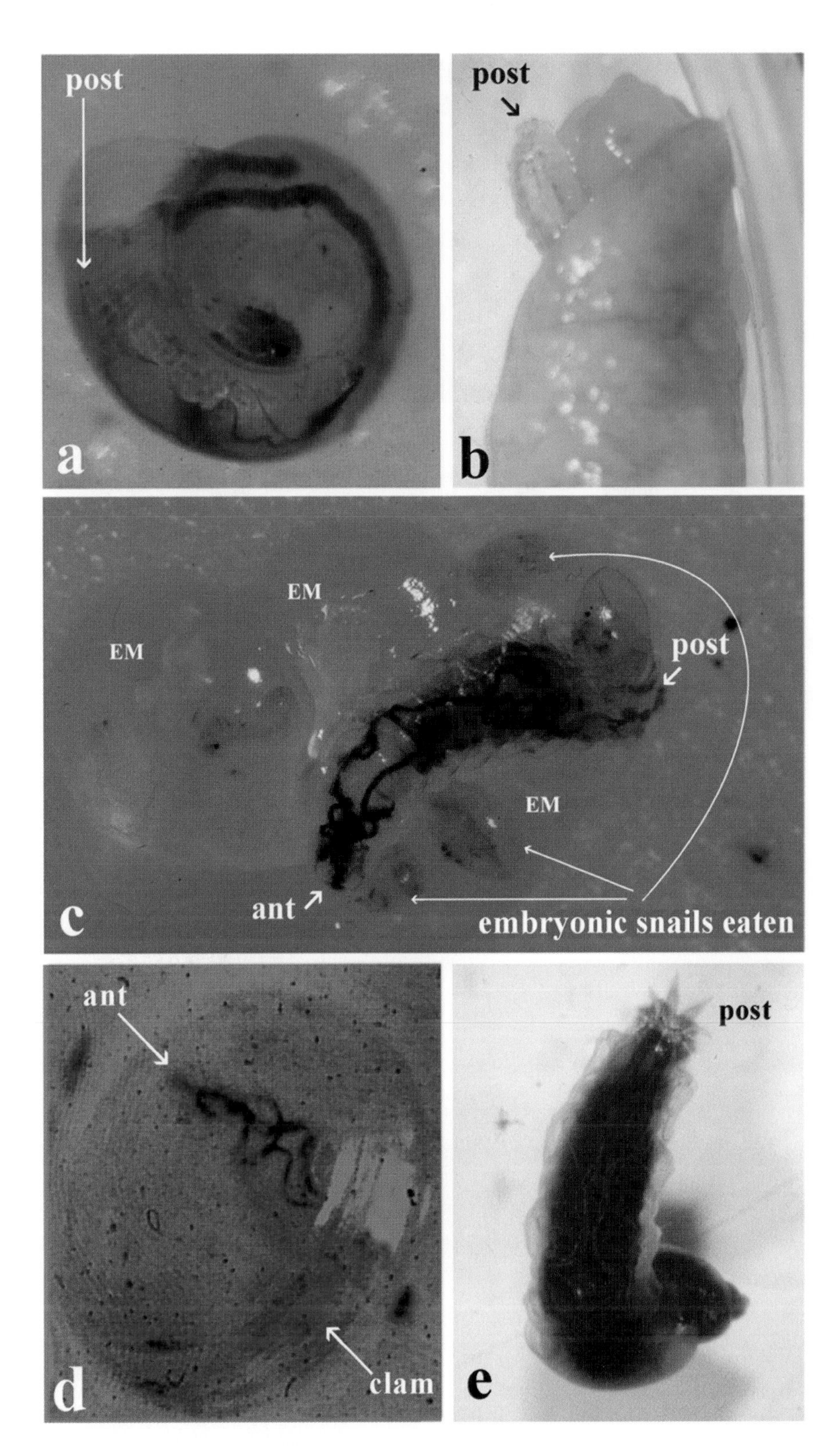

Plate 3. (a) ***Trypetoptera punctulata*** L3 attacking a terrestrial Helicidae. (b) ***Euthycera cribrata*** L3 in a *Deroceras* sp. slug; (c) ***Anticheta melanosoma*** L2 in egg mass of *Lymnaea* sp. (d) ***Ilione lineata*** L1 feeding in a fingernail clam; (e) ***Tetanocera ferruginea*** L3 eating *Physella acuta*. ant, anterior end of larva; EM, egg mass; post, posterior end of larva. (a, b, e: photos by J-C. Vala; c, d: photos by L. Knutson).

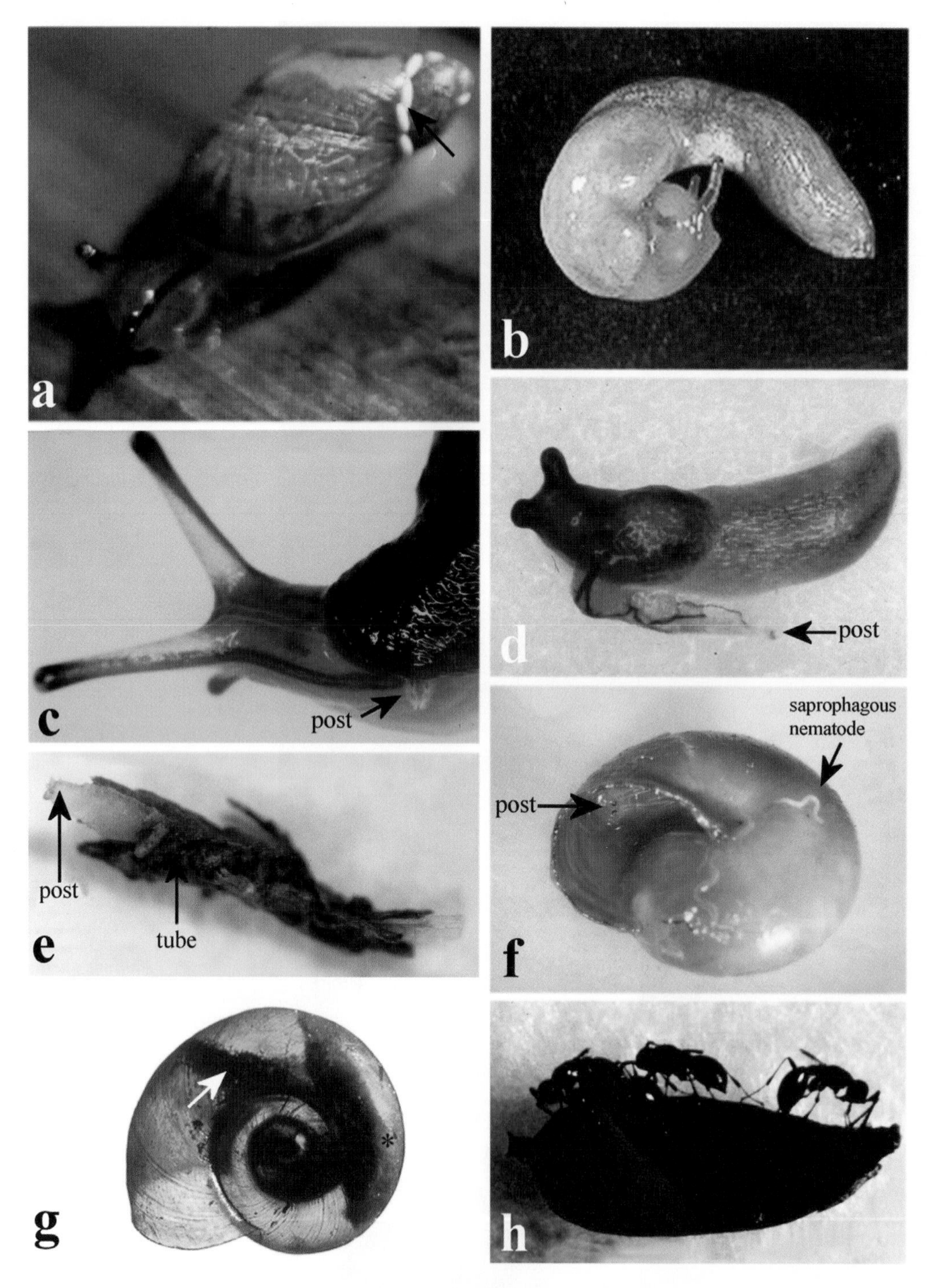

Plate 4. (a) eggs of ***Pherbellia s. schoenherri*** on shell of living *Oxyloma elegans* (photo by J-C. Vala). (b) *Deroceras* sp. attempting to eat attacking L1 of ***Euthycera chaerophylli***, spot of white mucus produced by slug where larva is penetrating into body (photo by L. Knutson). (c, d) ***E. cribrata*** L3 attacking *Deroceras reticulatum*. Arrows point to posterior end of larva (photos by J-C. Vala). (e) ***Sepedonella nana*** L3 preying on freshwater oligochaete *Aulophorus furcatus*. post, posterior end of ***S. nana*** larva; tube, case of oligochaete (from Vala *et al.* 2000b). (f) ***Trypetoptera punctulata*** L3 feeding on liquefied tissues of young *Candidula unifasciata*, containing saprophagous nematodes (photo by J-C. Vala). (g) puparium (star) of ***Pherbellia dorsata*** and calcareous septum (arrow) in shell of *Biomphalaria glabrata* (from Bratt *et al.* 1969). (h) *Phaenopria popei* (Hymenoptera: Diapriidae) ovipositing into puparium of ***Dictya*** sp. (from Knutson & Berg 1963b).

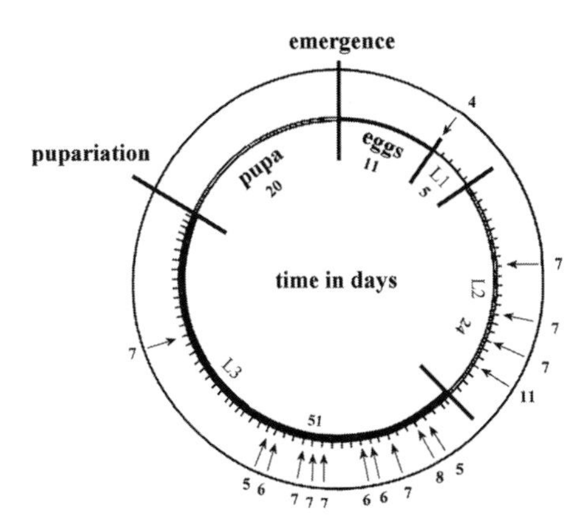

Fig. 7.5 Life cycle of an individual of ***Pherbina mediterranea*** showing duration of stages (in days, inner circle) and numbers of prey eaten. One day noted by one dash. Durations based on one example. Arrows mark days when a larva attacked a new prey, size of which is given at outer perimeter in mm. Modified from Vala & Gasc (1990a).

7.2.2 Terrestrial and estivating aquatic snails

Berg & Knutson (1978) included all terrestrial snail feeders, semi-terrestrial succineid snail feeders, and estivating aquatic snail feeders under one heading, "a heterogeneous group of terrestrial parasitoids," noting "their various degrees of specialization for parasitoid feeding." We separate all of these into our Behavioral Groups 1, 4, 6, 7, and 8. Berg & Knutson noted that for the relatively unspecialized Sciomyzini and Tetanocerini "each larva may kill three or four snails" and that the most highly specialized parasitoid Sciomyzini consume only one snail per larva.

We cite here the most precise of the extensive data available on examples of the broad range of predatory–parasitoid–saprophagous feeding among terrestrial species, beginning with the predators. Barnes (1979b) reared four species of *Neolimnia* (subgenus *Neolimnia*) primarily on ten species of the small terrestrial snails of the family Punctidae; most terrestrial snails in New Zealand are, in fact, small. He referred to them as overt predators, noting that a single larva may consume up to 60 or 70 snails before pupariating. Foote (2008) recorded a larva of the predator *Tetanocera phyllophora* killing 16 snails: two *Zonitoides arboreus* in the first stadium; two *Discus cronkhitei* in the second; and nine *D. cronkhitei*, one *Z. arboreus*, one *Stenotrema hirsutum*, and one *Anguispira kochi* in the third stadium.

In contrast to *Neolimnia* (*Neolimnia*) species and *T. phyllophora*, the predatory/saprophagous *Dichetophora obliterata* has an unusually low feeding rate, developing on just two individuals of the small species *Lauria cylindracea* and one *Helicella* sp. of 5.0 mm diameter (Vala *et al.* 1987). First- and second-instar larvae of the parasitoid/predator *Trypetoptera punctulata* develop on embryonic snails within the shell of the parent snail, the small (3.0–3.5 mm long) *L. cylindracea*. The third-instar larva kills and eats the parent snail, then attacks 12–15 other *L. cylindracea* or one larger *Helicella* sp., juvenile *Cornu aspersum*, or *Fruticicola* sp. (Vala 1986). Six (+?) larvae (2.0–5.0 mm long) found in nature in Ireland on at least two occasions in small (1.8–2.0 × 1.0–1.1 mm) *Vertigo geyeri* (as *V. genesii*) apparently were *T. punctulata* (Bhatia & Keilin 1937).

Bratt *et al.* (1969) noted that a second-instar larva of the parasitoid/predaceous *Pherbellia albovaria* found in a 4.0 mm diameter *Zonitoides arboreus* subsequently killed a 4.0 mm diameter *Discus patulus*, then a 5.0 mm diameter *Mesodon thyroideus*. Another second-instar larva of *P. albovaria* found in a 3.5 mm *D. patulus* subsequently killed a 6.0 mm diameter *D. patulus*. An isolated, laboratory-reared larva of this species killed and ate a total of five individuals of these three snail species during the 44 days of its development.

Adults of parasitoids and parasitoid/predators of terrestrial snails and estivating aquatic snails often differ greatly in size within a species, apparently reflecting great differences in the amount of food eaten by the larvae. For example, variation in the lengths of three structures on the left side of 54 field-collected adults of the terrestrial parasitoid *Pherbellia dubia* collected in Finland, Sweden, Denmark, England, France, and Austria is shown in Table 7.2. The shells of the hosts of those sciomyzids that pupariate within the shell also often differ greatly in size. Also, Barnes (1990) noted that adults of the highly specialized parasitoid *Sciomyza varia* emerged from field-collected shells of estivating *Stagnicola elodes* that were 7.0–19.8 mm long. He estimated that the shell-free dry weight of the snail tissue was 1.6–40.4 mg calculated from the weight-length regression equation for this snail from Eckblad (1971).

Whereas the sizes of adults resulting from parasitoid larvae that feed on only one snail appear to be determined by the size of the snail attacked, significant differences in sizes of adults of aquatic, predaceous Sciomyzidae also have been observed. Such differences in sizes do not seem to affect their reproductive potential. For example, Knutson & Berg (1964) noted that a very small male of the aquatic predator *Elgiva solicita* collected in Alberta, Canada,

Table 7.2. Size variation in field-collected adults of ***Pherbellia dubia***

	24 males		30 females	
Length of structure	Range (mm)	Mean (mm)	Range (mm)	Mean (mm)
Pleural suture	0.80–1.60	1.15	1.00–2.04	1.28
Femur	2.80–3.20	2.38 (*sic*)	2.00–3.80	2.64
Wing	3.20–4.80	3.66	3.36–6.64	4.38

Note: From Bratt *et al.* (1969).

copulated within one minute after being placed in a breeding jar with a female, twice his size, collected in Belgium. Denlinger & Žďárek (1994) summarized that free-living – and, we suggest, parasitoid – fly larvae probably show an early physiological commitment to metamorphosis. This has been shown experimentally by a precipitous decline in the juvenile hormone titre of some flies. While a fully fed third-instar larva of *Sarcophaga bullata* (Sarcophagidae) may weigh as much as 120 mg, as little as 3–10 hours of feeding by a newly molted third-instar larva can result in a mature larva as small as 20 mg being able to pupariate and produce an adult fly. The authors reasoned that such variability in size likely is an important adaptation to a limited and patchy food resource. Indeed, their molluscan food seems to be far more abundant than the Sciomyzidae species. However, the spatial distribution of many molluscs, especially snails in "terrestrial" situations, often is patchy.

Moor (1980) found that the small parasitoid *Pherbellia s. schoenherri* in Europe laid eggs on *Succinea putris* ranging in size from less than 10.0 mm to more than 17.0 mm long. The greatest percentage of successful larvae developed on the smallest snails, whereas the greatest number of larvae that became encapsulated by the host were found in the largest snails (Fig. 9.31). However, Bratt *et al.* (1969) found that if a larva of the North American subspecies *P. s. maculata* "began its development on a small snail [*Succinea* or *Oxyloma* spp.] (less than 10.0 mm long), it often ate all of the soft parts, left the empty shell of its original host, and killed another snail." They noted that first-instar larvae infested snails 5.0–15.0 mm in length and eventually killed even the larger snails. *Theba pisana* snails onto which the large parasitoid/saprophage *Salticella fasciata* oviposited in nature consisted only of large, second-year individuals, that is, breeding "adults," 11.0–19.0 mm in diameter (Coupland *et al.* 1994; Fig. 18.5).

7.2.3 Semi-terrestrial Succineidae snails

Information on the number and sizes of snails killed and eaten by succineid feeders, as information on host/prey preferences of this Behavioral Group, is quite limited, except for the well-studied parasitoid *Pherbellia s. schoenherri* (Table 9.3, Figs 9.30–9.34). *Sciomyza aristalis* and *S. testacea* also are intimately associated parasitoids of succineids, but they are less well studied than is *P. s. schoenherri*. Most other succineid feeders are predators with parasitoid features, especially during early larval life. Succineids found infested in nature (e.g., by *Tetanocera* spp., Foote 1996b) and attacked in the laboratory (e.g., by *Sepedon* spp., Knutson 2008) generally harbored one and sometimes up to 3 larvae. In laboratory rearings, up to 6 first-instar larvae of *Sciomyza dryomyzina* (Foote 1959a) and up to 10 *Hoplodictya spinicornis* (Neff & Berg 1962) infested one snail. When the initial host dies, these larvae leave the shell, each then attacking 2 or 3 more snails during the course of development. However, a large second-instar larva of *Tetanocera arrogans* found in a jar of field-collected terrestrial snails killed and ate one 7.0 mm *Succinea* sp., then molted, and during the third instar killed and fed on one 8.0 mm and one 9.0 mm *Succinea* sp., one 5.0 mm *Cepea nemoralis*, and one 6.0 mm *Trochulus hispidus* (Vala & Knutson, unpublished data). Knutson *et al.* (1967) noted that *Sepedon h. hispanica* larvae each killed and ate 2 or 3 large (about 9.0 mm long) *Succinea* spp. or 7–9 small (about 4.0 mm long) *Succinea* spp.

7.2.4 Slugs

The sizes of *Deroceras reticulatum* and *Milax gagates* slugs that were successfully infested and the numbers of different sizes killed by the three instars of the univoltine parasitoid–predator–saprophage *Euthycera cribrata*, which can develop on snails or slugs, were detailed by Reidenbach

Table 7.3. Effect of slug size on survival of prey and larval feeding period by ***Euthycera cribrata***

Larval instar	Size of slugs (mm) 1–5	6–10	11–15	16–20	21–25	26–30	31–35	36–40	41–45
S	2.6 ± 0.4	4.3 ± 0.4	5.0 ± 0.8	5.6 ± 0.6	6.7 ± 0.9	8.7 ± 0.9	10.2 ± 0.7		
L1								NI	NI
D	3.4 ± 0.4	7.4 ± 0.6	8.5 ± 0.3	9.4 ± 0.3	10.3 ± 0.5	11.2 ± 0.9	12.2 ± 0.6		
S	1.0 ± 0.0	2.0 ± 0.2	2.5 ± 0.2	4.0 ± 0.6	5.0 ± 0.5	5.4 ± 0.7	6.5 ± 0.9	7.0 ± 1.3	
L2									NI
D	1.6 ± 0.2	3.6 ± 0.4	4.2 ± 0.6	6.0 ± 0.7	8.2 ± 0.7	8.7 ± 0.9	10.4 ± 0.7	11.0 ± 0.8	
S		1.1 ± 0.1	1.3 ± 0.1	2.1 ± 0.2	2.8 ± 0.3	3.4 ± 0.4	3.6 ± 0.4	5.1 ± 0.5	5.2 ± 0.9
L3	NI								
D		1.7 ± 0.2	2.5 ± 0.2	3.8 ± 0.3	4.2 ± 0.4	5.1 ± 0.4	6.6 ± 0.8	8.0 ± 0.7	12.7 ± 0.9

Note: D, larval feeding period (days); NI, non-infested prey; S, prey survival (days). From Reidenbach *et al.* (1989).

Table 7.4. Effect of slug size on infestation rate and number of prey killed by ***Euthycera cribrata***

Larval instar		Size of slugs (mm) 1–5	6–10	11–15	16–20	21–25	26–30	31–35	36–40	41–45
	R	100%	100%	80%	75%	50%	40%	25%		
L1									*NI*	*NI*
	N	2–4	1–3	1–2	1–2	1	1	1		
	R	25%	65%	90%	85%	70%	46%	30%	21%	
L2										*NI*
	N	2–8	2–9	1–7	1–6	1–4	1–2	1	1	
	R		75%	88%	90%	75%	68%	50%	40%	34%
L3		*NI*								
	N		3–10	3–15	2–11	2–10	1–7	1–5	1–2	1

Note: R, rate of successful infestation; *N*, number of prey killed; *NI*, non-infested prey. From Reidenbach *et al.* (1989).

et al. (1989) (Tables 7.3, 7.4, Fig. 7.6, Eq. 7.2). Individual first-instar larvae were able to kill and develop successfully on slugs up to 35.0 mm long. Third-instar larvae killed slugs up to 60.0 mm long. In general, survival of larvae decreased as sizes of slugs increased. However, second- and third-instar larvae were less successful on slugs 1.0–10.0 mm long than on slugs 11.0–20.0 mm long. The most vulnerable slugs were 10.0–30.0 mm long. It was estimated that 15–25 slugs of various sizes were killed by each larva during the 2–3 months of development. Fewer slugs are killed during the much shorter period of larval development of multivoltine species of *Tetanocera*. *Tetanocera elata*, which is restricted to feeding on slugs, killed 5–9 slugs during 27–48 days (Knutson *et al.* 1965); *T. plebeja*, which also attacks terrestrial snails, killed up to 6 slugs 15.0–25.0 mm long during 12–20 days; *T. valida*, restricted to feeding on slugs, killed 3–7 during 13–20 days; and *T. clara*, restricted to slugs, killed 3–5 individuals, 8.0–35.0 mm long, in 20–33 days (Trelka & Foote 1970).

7.2.5 Fingernail clams (Sphaeriidae)

First-instar larvae of three species of *Renocera* in North America remained beneath the water surface within the

Table 7.5. Fingernail clams eaten by ***Renocera*** spp. larvae

	Species eaten	Maximum N eaten	Size range eaten (mm)
Renocera amanda			
L1	*Pisidium casertanum*	1	2.1–4.0
	Sphaerium occidentale		
	S. rhomboideum		
	S. partumeium		
L2	*P. casertanum*	6	2.0–6.2
	S. occidentale		
	S. partumeium		
L3	Same as L2	13	2.8–4.0
Renocera striata			
L1	Same as *R. amanda*	1	1.4–5.5
L2	Same as *R. amanda*	5	1.4–7.5
L3	Same as *R. amanda*	10	1.5–10.0
Renocera longipes			
L1	Same as *R. amanda*	4	1.2–1.9
L2	Same as *R. amanda*	11	1.2–4.2
L3	Same as *R. amanda*	11+	1.5–5.8

Note: From Foote (1976).

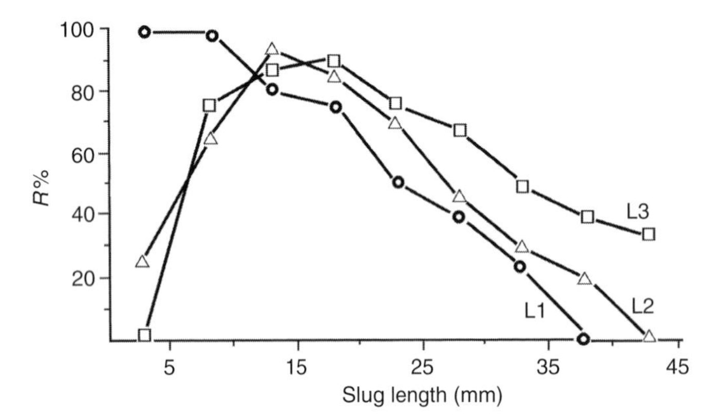

Fig. 7.6 Successful infestation rate (*R*) on various sizes of *Deroceras reticulatum* attacked by ***Euthycera cribrata***, L1, L2, and L3. See Table 7.4, Eq. 7.2. From Reidenbach *et al.* (1989).

mantle cavity of their clam host for 4–9 days before killing it (Foote 1976). They usually fed individually and killed only one clam during the first stadium. Second- and third-instar larvae preyed on a series of clams beneath the surface, with their posterior spiracles exposed at the surface film, killing the prey within 20–30 minutes and remaining with it for only about one hour. Individual larvae killed 14–35 clams during the three stadia, showing no preference for any size class, and attacked all four species of clams provided (Table 7.5). It was suggested that larvae lying beneath a closed clam release a muscle-relaxing toxin causing the valves to open, and that, in the case of *R. striata*, "Possibly the long, rather stiff bristles that project from the anal region of the larva irritated the inner surface of the mantle, thus keeping the valves slightly open." The strong spines on the pre-anal welt of second-instar larvae of *R. pallida*, the strongest spines on the surface of this larva, noted by Horsáková (2003), may have the same function. She found that larvae of *R. pallida* fed on exposed

Table 7.6. Random-age eggs of *Succinea californiensis*, *Oxyloma sillimani* (preferred host), and *Physella virgata* destroyed by larvae of ***Anticheta testacea*** compared with young and old eggs of *P. virgata* destroyed by larvae of ***A. testacea***

Larval instar	Factor	Host (eggs of random age)			Young and old eggs of *Physella virgata*	
		S. californiensis	*O. sillimani*	*P. virgata*	1- or 2-day-old eggs	7- or 8-day-old eggs
L1	Number of larvae	11	5	5	10	10
	Average eggs destroyed (N)	3.2	2.6	8.3	12.1	8.1
	Range of eggs destroyed (N)	2–5	1–4	5–13	7–16	6–10
	Larval development time (days)	2–3	2–3	2–6	2–3	2–3
L2	Number of larvae	5	5	6	10	10
	Average eggs destroyed (N)	5.6	6.0	34.7	37	15
	Range of eggs destroyed (N)	5–7	4–8	27–42	16–53	9–21
	Larval development time (days)	2–4	2–3	4–7	5–6	4
L3	Number of larvae	3	–	–	5	5 (survivors)
	Average eggs destroyed (N)	104	–	–	Incomplete data	283
	Range of eggs destroyed (N)	92–125	–	–	(90–154 + snails)	239–309
	Larval development time (days)	14–16	–	–	231	11–15
Total average eggs destroyed per larva (N)		112.8	–	–	–	306.1

Note: From Fisher & Orth (1964). *Hilgardia*, 36(1). © Regents, University of California.

Musculium sp., *Psidium* sp., and *Sphaerium* sp. One small clam was killed and eaten during the 5 days of the first stadium and, at most, 2 clams per day were killed and eaten during the 6 and 19–20 days of the second and third stadia, respectively. Thus, we estimate that each larva of *R. pallida* can kill about 50 clams during the course of its development. Larvae of *Ilione lineata* killed and ate 21–31 clams (1.5–4.0 mm greatest diameter) of three species of *Pisidium* (Foote & Knutson 1970, Knutson unpublished data). Larvae of *Eulimnia philpotti* in New Zealand preyed on at least *Sphaerium novaezelandiae* and *Pisidium hodgkini*, remaining submerged until late in the third stadium. First-instar larvae killed and fed on several clams (1.8–2.0 mm long), and it was conservatively estimated that each larva kills 50–100 clams during the course of development (Barnes 1980a).

7.2.6 Snail-egg/snail feeders

Fisher & Orth (1964) compared the number of eggs eaten by individually reared larvae of *Anticheta testacea*. They compared eggs of random age of *Physella virgata*, *Succinea californiensis*, and *Oxyloma sillimani* with 1–2-day-old and 7–8-day-old *P. virgata* eggs (Table 7.6). While first-instar larvae fed only on snail eggs, second- and third-instar larvae also killed and ate juvenile snails, but data on the numbers of snails killed were not given. Knutson (1966) noted that first-instar larvae of *A. analis* appeared to feed only on the gelatinous matrix of *Galba truncatula* egg masses, on which they were found in nature. When only one *A. analis* egg was laid on a *G. truncatula* egg mass, the larva developed to the third instar on the contents of that single mass and then fed in other egg masses. The contents of 6–9 masses were consumed by individual larvae during the three stadia; larvae did not attack juvenile snails. Larvae of *A. brevipennis*, found in egg masses of *Succinea* sp. in nature, fed on both *Succinea* sp. and *G. truncatula* egg masses but did not attack juvenile snails. Knutson & Abercrombie (1977) noted that a larva of *A. melanosoma* consumed 75 eggs in 11 egg masses of *Aplexa hypnorum* between hatching on August 6 and pupariating on August 20; some third-instar larvae also killed and ate juvenile and adult *A. hypnorum*. Robinson & Foote (1978) found that larvae of *A. borealis* consumed 4–7 eggs of *Oxyloma* during the first stadium, 3–11 during the second, and 105–110 during the third; only one larva (a second instar) killed a juvenile snail.

7.2.7 Oligochaete feeders

Two African sciomyzids appear to be obligate predators of freshwater oligochaete worms, *Aulophorus furcatus* of the family Naididae. The small (3.50–4.25 mm long) *Sepedonella nana* kills 10–21 worms (2.50–3.00 mm long, 0.25–0.30 mm diameter) during its 6–11 days of larval development (Vala *et al.* 2000a, 2000b). *Sepedon* (*Mesosepedon*) *knutsoni* (5.40–7.00 mm long) kills 50–120 worms during its 9–16 days of larval development (Vala *et al.* 2002; L. G. Gbedjissi personal communication, 2004). Both species were found in nature in empty and occupied oligochaete tubes. The aquatic snail predator *Sepedon* (*Parasepedon*) *ruficeps* in Africa also occasionally feeds on *A. furcatus* (Gbedjissi *et al.* 2003). Species of the major aquatic families of Oligochaeta occur throughout the world; thus oligochaete-feeding Sciomyzidae may be found in regions other than Africa.

7.3 EXPERIMENTAL STUDIES – AQUATIC PREDATORS

Almost all of the more rigorous experimental studies on larval feeding deal with aquatic predaceous Tetanocerini. While this section is not an historical treatment of experimental studies (see Section 19.5), there are a few historical aspects that help to clarify the subsequent discussions. The problems in analyzing and comparing research results on feeding behavior by Sciomyzidae probably are not unique to this family.

Previous reviews of the biology of Sciomyzidae have not included coverage of some of the earlier experimental studies (e.g., O. Beaver 1974b, 1974c) and some experimental studies are unpublished (e.g., Haab 1984, Ghamizi 1985). Also, some of the most complete studies (e.g., Manguin *et al.* 1986, 1988a, 1988b; Gormally 1988b, Manguin & Vala 1989, O. Beaver 1989, Mc Donnell *et al.* 2005b) were published after the review papers had been published. A further difficulty probably responsible for the lack of reviews of experimental studies of feeding behavior is the fact that several of these publications were in press during more or less the same period (1973–1974 and 1988–1989 papers) and were not cross-referenced.

Most of the experimental studies on aquatic predators such as *Ilione albiseta* and *Tetanocera ferruginea* included analyses of several aspects. Some aspects were discussed in relation to each other. The combinations of aspects varied among the publications. We attempt here to dissect out six of the main subjects concerning feeding behavior from most of the studies and to discuss them together, rather than separately, paper by paper. This approach has the disadvantage of losing some of the detail of relationships noted by the authors. In other papers, the factors analyzed are interactive or so closely related that they must be reviewed together, paper by paper. See also Section 9.3.4, in which numbers of snails killed and biomass consumed are discussed in relation to rates of development. The various kinds of experimental studies that have been conducted are distinguished in the subject index.

There also are difficulties in comparing many of the published results because of the different species and sizes of the flies and the snails used; differences in experimental designs and experimental conditions, especially temperature, photoperiod, numbers of larvae and prey used; and handling of the organisms. Other important variable factors include the size and type of containers used and, more importantly, the depth of water in the containers and the relative cleanliness of the containers (dead snail tissue and feces flushed out or not), which affect the state of health of the organisms.

In this discussion of aquatic predators, publications on experimental studies of the species listed below are analyzed. Experiments in some publications that are primarily qualitative life-cycle studies also are mentioned. General experimental conditions used in the studies are given below (a–m) to avoid repetition. Where conditions varied during a study, they are mentioned in the discussions. It is useful to compare the temperatures (19–28 °C) at which most benchtop life-cycle studies have been carried out to rates of development in the controlled temperature studies. The controlled studies are those of Bhuangprakone & Areekul (1973) – *Sepedon plumbella*; Barnes (1976) – *Sepedon fuscipennis*; Yoneda (1981) – *Sepedon aenescens*; Haab (1984) and Ghamizi (1985) – *Sepedon sphegea*; Vala & Haab (1984) and Manguin & Vala (1989) – *Tetanocera ferruginea*; Gormally (1988b) – *Ilione albiseta*; Maharaj (1991), Maharaj *et al.* (1992), and Appleton *et al.* (1993) – *Sepedon neavei* and *S. scapularis*; and Mc Donnell (2004), Mc Donnell *et al.* (2005b) – *Sepedon spinipes*.

The shallow water depth used in most experiments, usually 5.0 mm but varying from wet cotton to 5.0 cm except as noted below, clearly puts the predators at an advantage compared with what they would face under natural conditions. Most of the studies concern ten species of *Sepedon*, the second largest genus in the family, with 78 species. All species studied, except the univoltine *Ilione albiseta*, are multivoltine. Below, LD = light-dark photoperiod, RH = relative humidity. Abbreviations of zoogeographical regions are as in Table 16.1.

a. *Sepedon tenuicornis* (N), studied in North Carolina by Geckler (1971) in 8 × 6 × 2 cm plastic boxes with a layer of wet cotton, room temperatures of 19–25 °C, normal room lighting 8 am to 5 pm. Orth (1986) distinguished *S. gracilicornis* as a species distinct from *S. tenuicornis*, but there is hardly any doubt that Geckler's material was, in fact, *S. tenuicornis*, as the origin (North Carolina) is several hundred kilometers south of the known range of *S. gracilicornis*.
b. *Sepedon fuscipennis* (N), studied in New York by Eckblad (1973a). Laboratory experiments in various-sized containers with 5.0 mm or 5.0 cm of water as noted in the discussions, 21–24 °C room temperature, room lighting. Field experiments as noted in the discussions. Eckblad (1976): conditions not given. Also studied in New York by Barnes (1976) with special reference to the effects of temperature on survival (all stages) and rates of increase. Laboratory experiments at controlled temperatures of 15, 21, 26, 30, and 33 °C; LD 16:8; in small and large containers with 5.0 mm of water.
c. *Sepedon plumbella* (O), studied in Thailand by Bhuangprakone & Areekul (1973). Water depth uncertain (7.0 cm ?), in 19 × 33 cm glass jars at 25 ± 1 °C, lighting not given.
d. Four species of *Tetanocera* and one each of *Hydromya*, *Ilione*, *Pherbina*, and *Sepedon* (P and H), studied by O. Beaver (1974b, 1974c) in Wales. In Petri dishes lined with wet filter paper or with water at a depth of 5 mm, at 20–25 °C, lighting not given. Note that these studies included not only aquatic predators but some species in Behavioral Group 2 and that they included both univoltine and multivoltine species.
e. *Sepedon spangleri* (O), studied in Thailand by Sucharit *et al.* (1976). In 6.0 × 8.5 cm plastic boxes with water at a depth of 5.0 mm, room temperatures of 25 ± 3 °C, lighting not given.
f. *Sepedon aenescens* (O, P), studied in Japan by Yoneda (1981) at controlled temperatures of 15, 19, 22, 25, and 30 °C.
g. *Tetanocera ferruginea* (H), studied in France by Vala & Haab (1984) at LD 8:16 (13 and 20 °C), LD 16:8 (13 and 26 °C), LD 9:15 (20 °C), and LD 24:0 (20 °C), on wet filter paper. Also studied in France by Manguin *et al.* (1986) in 5 × 17 cm plastic boxes with water at a depth of 5.0 mm, other conditions not given. Also by Manguin *et al.* (1988a, 1988b) and Manguin & Vala (1989) at 20 ± 1 °C and LD 6:18.
h. *Sepedon sphegea* (P), studied in France by Haab (1984) at controlled temperatures of 12, 15, 20, 22, and 25 °C with LD regimes of 8:16, 12:12, and 16:8 in 5 × 15 cm dishes with water at various depths (5, 15, 30, and 50 mm) and by Ghamizi (1985) in 7 × 15 cm dishes with water to a depth of 30 mm, at room temperature and lighting. In these two exceptionally fine studies, unfortunately not published, protocols were developed for experimental approaches and data analyses and presentation in studies of predation, with emphasis on functional and numerical responses. These were based on the classical papers by many authors, e.g., Beddington, Hassell, Holling, Watt, etc. The rationales were especially well documented by Ghamizi (1985). Some protocols were further modified, extended, and applied to studies of *Tetanocera ferruginea* as noted above.
i. *Ilione albiseta* (P), individual and group cultures of field-collected and laboratory-reared larvae, and other stages studied by Lindsay (1982) Lindsay *et al.* (2011) in Ireland. Cultures kept outdoors and in an unheated room with no artificial lighting, in various-sized containers with or without a soil substrate and water to a depth of 5.0 mm. Also studied by Gormally (1988b) in Ireland. In 4.5 cm diameter (for first-instar larvae) to 8.5 mm diameter (for second- and third-instar larvae) Petri dishes with filter paper substrate and water at a depth of 5.0 mm, in water baths at various controlled temperatures of 14–26 °C and LD 16:8.
j. *Sepedon senex* (O), studied by O. Beaver (1989) in Thailand. Containers and water depths as noted in discussions; other conditions not indicated.
k. *Sepedon scapularis* and *S. neavei* (AF), studied by Maharaj (1991) in South Africa and published in part by Appleton *et al.* (1993). In 6.6 × 1.5 cm plastic Petri dishes with distilled water at a depth of 3 mm; in an insectary at 26 °C, 75% RH, and LD 12:12.
l. *Sepedon scapularis* (AF), studied in South Africa by Maharaj *et al.* (1992). In 8.2 cm diameter Petri dishes with distilled water at a depth of 3 mm (all other studies apparently used tap water; Gormally (1988b)

used 2-day-old conditioned tap water), at 28 °C and LD 12:12.

m. *Sepedon spinipes* (P), studied by Mc Donnell (2004), and Mc Donnell *et al.* (2005b) in Ireland at 14, 17, 20, 23, and 26 °C; LD 16:8; in Petri dishes, 3.5 cm diameter (for first instars) to 8.5 cm diameter (for second and third instars) with filter paper and water at a depth of 6 mm.

The related aspects of biomass killed, biomass consumed, and quantitative predation in single and multiple prey systems are discussed separately in Section 5.2. We discuss them separately partly because of the foci of the various studies but also because we want to highlight whether or not there are inherent differences among species relative to numbers/biomass killed/consumed and their configurations in different predator/prey systems. No conclusion is offered here, but the question seems well worth further experimentation, especially with regard to the selection of biocontrol agents.

A further intriguing question in analysis of data and conclusions on predation is the influence of "wasteful feeding," noted by several researchers, especially Eckblad (1973a, 1976), O. Beaver (1974b, 1974c), and Gormally (1988b) (also see Section 8.1). Can species be characterized as wasteful feeders or not, and what impact does wasteful feeding have on predation studies? Is there a difference between wasteful killing and wasteful feeding? We define the latter as resulting in defecation of undigested food, noted in some aquatic predaceous Tetanocerini, especially when they are disturbed. Or are these "labile behaviors"? Whether labile or not, what is the evolutionary significance? Notably, information on wasteful killing and feeding has been obtained from laboratory studies, where the environment is stable. They likely are much more significant in nature, where the environment is unstable (due to wind, rain, wave action, water flow, disturbance by other organisms), causing larvae to leave the prey prematurely.

7.3.1 Numbers and sizes of snails attacked, killed, and consumed

Geckler (1971), studying *Sepedon tenuicornis* and the planorbid *Helisoma anceps*, was the first to quantify the relationship between sizes of larvae and sizes of snails they killed. He found that, as observed by later authors, larvae can kill larger and larger snails as their own size increases (Fig. 7.7) and that 8.1 ± 2.5 (mean ± S.D.) snails of various sizes were killed

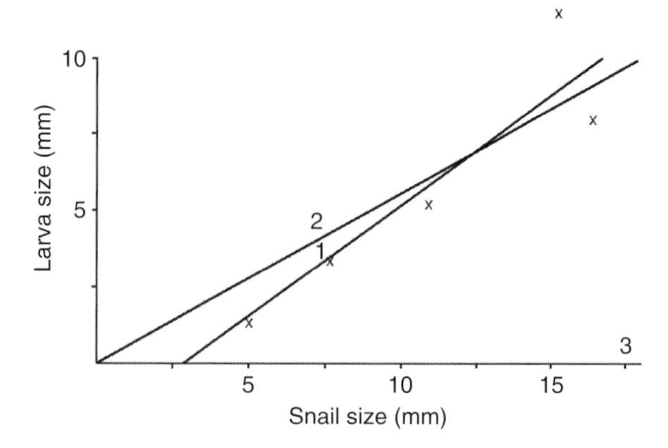

Fig. 7.7 Plot of calculated lines for 0.5 probability of kill of snails of different sizes by larvae of ***Sepedon tenuicornis*** of different sizes. Line 1, $y = 0.72\,x - 20$; line 2, $y = 0.56\,x$. From Geckler (1971).

and eaten to some (unspecified) extent during the course of development. The largest snails (>16.3 mm diameter) were killed by the largest larvae (>12.0 mm length) while the smallest snails (<3.0 mm) were killed only by the smallest, newly hatched larvae (<1.4 mm). The latter result probably was an artifact of the experimental conditions, since most subsequent researchers have observed that larvae of all instars will kill very small snails.

Sucharit *et al.* (1976) conducted laboratory experiments with *Sepedon spangleri* and 13 species of operculate and non-operculate freshwater snails. Twenty snails of each species were confined separately with ten larvae overnight in boxes. Of the non-operculate snails (six species of *Gyraulus*, *Hippeutis*, *Indoplanorbis*, *Lymnaea*, *Trochorbis*, and *Segmentina*) the average number of snails killed per box ranged from 13.0 to 19.3 overnight for the planorbid species (1.1–3.8 mm × 2.6–4.5 mm, mean width × length) and 4.3 for the *Lymnaea* species (2.7 × 5.2 mm).

Various aspects of predation by *Sepedon senex* against three non-operculate freshwater snails (*Gyraulus convexiusculus*, *Indoplanorbis exustus*, and *Radix auricularia rubiginosa*) (water 1.0 cm deep) were studied by O. Beaver (1989). Snail sizes and percentage killed per larval instar are shown in Table 7.7. She found that larvae killed snails selectively, depending upon the relative sizes of the larvae and the snails; that all instars killed snails with diameters as wide as about twice the length of the larvae; that the larvae more often selected small rather than large snails; and that mature larvae killed snails more efficiently than did immature larvae. In 10 replicates of experiments with 10 larvae and 100 *G. convexiusculus* (1.5–4.0 mm diameter) she recorded mean (±S.D.) consumption capacities (per sets of 10 larvae) of the first, second, and third instar as 6.0 ± 9.2, 4.8 ± 18.3, and 96 ± 6.2, respectively.

Sizes and numbers of the endemic, non-operculate, freshwater *Bulinus africanus* and *B. tropicus* and the immigrant, non-operculate, freshwater *Physella acuta* killed by *Sepedon scapularis* were studied in laboratory experiments by Maharaj *et al.* (1992) (single larvae and various numbers of snails). Small snails (<3.0 mm) were more frequently killed by all three larval instars than were larger snails. First-instar larvae killed few, if any, large snails (>7.0 mm), but second and third instars preyed effectively on all sizes. Third instars killed significantly more snails than did younger larvae. For larvae offered only one species of snail, the mean total number killed per larva during its entire development was 49 *B. africanus*, 45 *B. tropicus*, or 34 *P. acuta*. However, during prey specificity experiments in which all three snail species were offered as prey, an average of only 31 snails, irrespective of species, was killed by each larva during the course of development (Fig. 7.8; Tables 7.8, 7.9).

Table 7.7. Selective killing of some medically important snails by ***Sepedon senex*** larvae expressed as percentage killed

Larval instar	Larval length (mm)	Snail diameter (mm)	% killed
L1	1.39–1.53	(G) 1.39–1.65	9.87
		(G) 2.36–3.00	5.72
		(G) 3.80–3.82	0
L2	2.60–2.78	(G) 2.70–3.00	92.73
		(G) 3.60–3.90	89.47
		(I) 4.80–5.00	40.59
		(R) 4.17–5.00	0
L3	4.17–4.86	(I) 4.86–5.21	42.05
		(I) 5.70–8.00	38.04
		(R) 4.17–4.28	0
		(R) 6.05–7.00	0

Note: G, *Gyraulus convexiusculus*; I, *Indoplanorbis exustus*; R, *Radix auricularia rubiginosa*. From O. Beaver (1989).

From laboratory experiments on *Sepedon neavei* and *S. scapularis*, Appleton *et al.* (1993) (with reference to several of their published and unpublished studies) concluded that small snails (nine species of freshwater non-operculates) are the most vulnerable size class because they are preyed upon by all larval instars. First- and second-instar larvae attacked small (<3.0 mm diameter or length) and medium-sized (3.0–7.0 mm) snails more successfully than large snails (>7.0 mm), while third-instar larvae killed substantial numbers of all three size classes. *Sepedon scapularis* killed a mean of 28.9 ± 2.5 S.D. snails (3.2–4.4/day for 8–11 days), and *S. neavei* killed 35.0 ± 4.4 (1.3–1.6/day for 11–12 days).

Four of the most complete records of the number and sizes of freshwater snails killed and eaten during the course of development by individuals of freshwater predators are those by Foote (1999) for 16 individuals of *Tetanocera robusta* (total killed per individual: 12–26, size range 1.5–13.2 mm); by Knutson & Berg (1964) for 19 individuals of two species

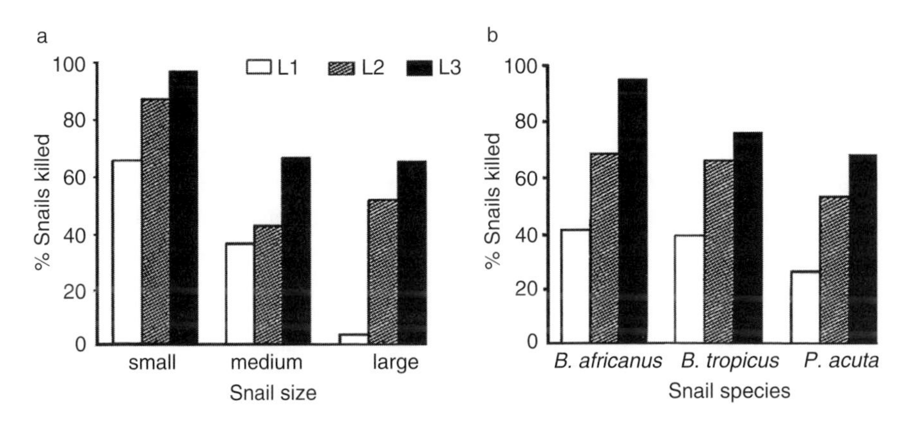

Fig. 7.8 Percentage of snails killed by three larval instars of *Sepedon scapularis* when exposed to (a) different size classes of snails and (b) different species of snails (*Bulinus africanus*, *B. tropicus*, and *Physella acuta*). From Maharaj *et al.* (1992).

Table 7.8. Mean total numbers of snails of different size classes and species killed by each instar of ***Sepedon scapularis*** larvae (N = 5 replicates in each case)

	Mean N of snails killed								
	Physella acuta			*Bulinus africanus*			*Bulinus tropicus*		
Size class	S	M	L	S	M	L	S	M	L
Instar									
L1	5.6	0.4	0.4	2.8	4.4	0.2	2.4	1.6	0
L2	8.4	0.6	1.8	9.2	3.2	3.0	6.8	4.8	4.2
L3	6.6	5.4	5.0	12.2	8.4	5.4	15.0	3.2	7.0
Total	20.6	6.4	7.2	24.2	16.0	8.6	24.2	9.6	11.2

Note: S, small (<3 mm); M, medium (3–7 mm); L, large (>7 mm). From Maharaj *et al.* (1992).

Table 7.9. Least significant differences (at P = 0.95) for number of snails killed by ***Sepedon scapularis*** larvae as result of three factors: (a) larval instar, (b) species of snail, (c) size of snail

Factor		N snails killed (mean ± S.E.)	Homogeneous groups
(a) Larval instar	L1	3.00 ± 0.41	a
	L2	3.56 ± 0.44	a
	L3	7.67 ± 0.67	b
(b) Species of snail	*Physella acuta*	3.80 ± 0.48	c
	Bulinus tropicus	5.00 ± 0.66	d
	Bulinus africanus	5.42 ± 0.64	d
(c) Size of snail	Small	7.58 ± 0.60	h
	Medium	4.67 ±0.36	g
	Large	1.98 ± 0.36	h

Note: Non-significant differences between groups represented by same letter in right-hand column. Modified from Maharaj *et al.* (1992).

of *Elgiva* (Table 7.10; total killed: 12–23, 1.0–13.0 mm), by Knutson & Berg (1967) for 10 individuals of *Ilione trifaria* (total killed: 28–40, 0.8–8.0 mm), and by Valley & Berg (1977) for 16 species of *Dictya* (Table 7.11; total killed: 12–53, 1.0–9.8 mm). See also Table 7.12. In their study of the natural history of 21 mostly very closely related species of Nearctic and Neotropical aquatic predatory *Dictya* spp., Valley & Berg (1977) reared 70+ individuals of 13 species in individual rearing containers and precisely recorded the numbers and sizes of the snails killed and eaten during each stadium, as well as the duration of the developmental periods. The data are highly comparable, the rearings having been conducted over a period of only a few years (primarily by Valley), by use of essentially the same rearing methods and under the same conditions (bench-top rearings at 20–28 °C) throughout. We have tabulated their results in Table 7.11. The snails killed and eaten were various combinations of *Galba humilis*, *S. elodes*, *Physella gyrina*, *P. integra*, *Physa venustula*, *P.* sp., *Biomphalaria glabrata*, *Drepanotrema depressissimus*, *D. lucidum*, *Gyraulus parvus*, *Helisoma trivolvis*, *H.* sp., *Promenetus exacuous*, and (by *D. bergi*) *Oxyloma* sp. All sciomyzids in Table 7.11 killed and ate species of Lymnaeidae, Physidae,

Table 7.10. Numbers and dimensions of snails killed by ***Elgiva solicita*** and ***E. connexa*** larvae from hatching to pupariation

Larva No.	N eaten by L1	N eaten by L2	N eaten by L3	N total eaten
Elgiva solicita				
Biomphalaria glabrata				
B1	3 (2.0–3.0)	3 (3.0–3.5)	6 (3.5–8.0)	12 (2.0–8.0)
B2	2 (1.5–2.0)	4 (3.0–4.5)	7 (4.0–10.0)	13 (1.5–10.0)
B3	2 (2.0)	6 (2.0–3.5)	6 (5.0–9.5)	14 (2.0–9.5)
B4	4 (1.0–1.5)	6 (2.0–2.5)	4 (2.0–8.0)	14 (1.0–8.0)
Helisoma trivolvis				
H1	3 (1.5)	7 (1.5–3.0)	5 (5.0–6.0)	15 (1.5–6.0)
H2	3 (1.5–2.0)	6 (2.0–2.5)	9 (2.0–10.0)	18 (1.5–10.0)
H3	3 (2.0–3.0)	4 (2.0–2.5)	7 (4.0–10.0)	14 (2.0–10.0)
H4	2 (2.5)	6 (2.0–2.5)	9 (2.5–7.0)	17 (2.0–7.0)
Physa sp.				
P1	4 (1.5–4.0)	5 (3.0–5.5)	8 (5.0–7.5)	17 (1.5–7.5)
P2	4 (1.5–5.0)	5 (2.5–4.0)	9 (4.0–8.0)	18 (1.5–8.0)
P3	4 (1.5–3.0)	6 (2.5–5.0)	7 (4.0–7.0)	17 (1.5–7.0)
P4	3 (1.5–2.5)	6 (2.0–3.5)	11 (3.5–7.5)	20 (1.5–7.5)
P5	2 (1.5)	4 (3.0–4.0)	13 (3.0–8.0)	19 (1.5–8.0)
Elgiva connexa				
Biomphalaria glabrata				
B1	6 (1.0–3.0)	6 (2.0–4.0)	5 (5.5–13.0)	17 (1.0–13.0)
B2	4 (1.0–3.5)	4 (2.5–3.0)	5 (4.0–13.0)	13 (1.0–13.0)
B3	4 (1.0–2.5)	5 (2.0–5.0)	4 (3.0–13.0)	13 (1.0–13.0)
B4	4 (2.0–4.0)	5 (1.0–5.5)	6 (4.0–11.5)	15 (1.0–11.5)
Physa sp.				
P1	4 (1.0–3.0)	4 (2.5–4.0)	11 (3.0–8.0)	19 (1.0–8.0)
P2	5 (1.5–3.0)	5 (2.5–4.5)	13 (2.0–8.0)	23 (1.5–8.0)

Note: Range in mm of snails eaten shown in parentheses. Greatest shell diameter for *Biomphalaria glabrata* and *Helisoma trivolvis*; length for *Physa* sp. From Knutson & Berg (1964).

and Planorbidae (except *D. texensis*, which destroyed and ate only Physidae and Planorbidae).

7.3.2 Biomass destroyed and consumed, and food conversion ratios

A major difficulty in reviewing the subject of number of snails killed (biomass destroyed) and biomass consumed and food conversion ratios has been the lack of distinction, by most authors, made between biomass destroyed and biomass consumed. Methods for determining biomass of snails by weight-length regression models are given by Eckblad (1971), O. Beaver (1974b), Manguin *et al.* (1986, 1988a), and Manguin & Vala (1989).

Geckler (1971) published the first experiments on biomass of snails destroyed. He found that with 34 larvae of *Sepedon tenuicornis* (a species similar in size and habits to *S. fuscipennis*) there was a great range in volume of *Helisoma anceps* killed; 241–931 mm^3 (mean = 544 ± 256 mm^3). Geckler did not distinguish between volume killed and volume consumed.

Eckblad (1973a) conducted a quantitative study of biomass consumed and energy transfer by larvae of *Sepedon fuscipennis* at four population densities of *Stagnicola elodes*. He estimated the weight of snail tissue killed by a larva during the course of development by a weight-length regression model and measured dry weight by percentage

Table 7.11. Ranges in numbers and sizes (in mm, greatest diameter or length) of snails killed and eaten by L1, L2, and L3 for ***Dictya*** spp. reared individually

	L1		L2		L3		
Dictya spp.	N snails	Size	N snails	Size	N snails	Size	L1–3 Total N snails eaten
adjuncta (8)	4–6	1.0–1.6	5–7	1.0–5.0	5–10	2.3–8.8	14–21
atlantica (?)	2–3	1.3–8.0	3–7	1.3–8.0	6–12	1.3–8.0	14–21
bergi (?)	3–6	3.0–5.7	3	3.3–6.0	4–6	6.0–8.8	ND
brimleyi (?)	5–8	<7.3	4–10	<7.3	5–10	<7.3	ND
expansa (11)	3–5	1.0–9.0	2–7	1.0–9.0	4–9	1.0–9.0	ND
floridensis (?)	ND	ND	ND	ND	ND	ND	20–22
gaigei (5)	4–6	1.3–2.5	2–5	2.9–4.5	8–12	3.2–9.1	17–21
hudsonica (?)	3–6	1.0–9.1	3–6	1.0–9.1	6–10	1.0–9.1	12–22
incisa (?)	2–7	1.1–2.1	5–9	1.6–3.3	6–16	2.0–7.0	18–26
lobifera (?)	ND	1.0–2.7	ND	2.1–8.1	ND	2.1–8.1	ND
matthewsi (6)	4–5	<13.0	3–5	ND	6–8	ND	13–17
oxybeles (?)	5–7	1.1–8.3	7–10	1.1–8.3	9–20	1.1–8.3	24–35
steyskali (9)	3–9	1.6–2.7	3–5	2.0–5.0	4–8	3.0–7.0	ND
stricta (5)	3–5	1.3–2.5	7–10	1.0–2.6	21–41	3.5–6.0	32–53
texensis (4)	2–4	1.9–3.0	ND	ND	6–8	5.0–9.8	12–16
umbroides (?)	3–6	2.0–3.0	2–6	ND	7–10	ND	ND
Minimum	2	1.0	2	1.0	4	1.0	12
Maximum	9	9.1	10	9.1	41	9.8	53

Note: Number of larvae of each species reared in parentheses after species name. ND, no data. Data from Valley & Berg (1977).

of ash content of tissue remaining in the snails killed. He determined calorific values of newly formed puparia and snail tissue with a Philipson Oxygen Microbomb Calorimeter. He found that prey density influenced the biomass consumed (Fig. 7.9) and that larvae utilized ingested snail tissue more efficiently when a lower snail biomass was ingested at lower snail densities (Fig. 7.10). Biomass consumed *(C)* equaled *Bk* (estimated total biomass of snails killed by a larva) minus *Br* (total snail tissue remaining after predation). The co-efficient of the utilization of food for growth *(U)* equaled $\leq$ *B* (increase in biomass of a larva) over *C*. The calculated mean ecological growth efficiency (net production in larval calories) over I (calories ingested) of 0.46 was somewhat higher than those published for most other invertebrate predators. Eckblad concluded that especially third-instar larvae overfeed, defecating before digestion is completed. Also, he concluded that the inverse relationships between utilization of snail tissue consumed and both biomass consumed and prey density seem to be important adaptations of this relatively specialized predator to prey density. Eckblad also provided an insightful discussion of numerical and functional response models.

In studies on intra- and interspecific competition among seven species of aquatic predaceous and semi-aquatic parasitoid/predaceous Tetanocerini, O. Beaver (1974b, 1974c) determined the total weight of *Lymnaea stagnalis* killed during the course of development of the larvae and their "food conversion ratios." She also compared predation rates, larval survival, rates of development, weight at which molting occurred, time intervals between molts, and sizes of adults produced, which are discussed elsewhere in this section and in other sections. The snail tissue, per se, was not weighed, but the shells of snails killed were measured and, to determine the relationship between weights and lengths, wet snails with shells of different lengths were weighed and a polynomial curve of the second degree was fitted to the data by

Table 7.12. Mean number of snails killed (± S.E.) during each larval stadium (L1, L2, and L3) of ***Ilione albiseta*** at five constant temperatures

Temperature (°C)	Stadium	N larvae	Mean N snails killed	Mean N snails killed/day	Average length of snails (mm)
	L1	18	5.5 ± 0.28[a]	0.42 ± 0.026[e, f, g]	2.9
14	L2	16	8.6 ± 0.66[a]	0.71 ± 0.054[k]	3.5
	L3	10	52.8 ± 5.60[a]	0.55 ± 0.027[l]	3.8
	L1	21	5.4 ± 0.24[b]	0.47 ± 0.035[h, i, j]	3.2
17	L2	21	7.1 ± 0.51[b]	0.76 ± 0.046[m, l]	3.9
	L3	8	45.6 ± 6.77[b]	0.50 ± 0.042[n, l]	4.4
	L1	26	6.0 ± 0.43[c]	0.68 ± 0.048[e, h]	3.3
20	L2	21	8.6 ± 0.63[c]	0.88 ± 0.086	3.5
	L3	7	40.4 ± 8.73[c]	0.43 ± 0.061[o]	4.0
	L1	28	5.7 ± 0.63[d]	0.72 ± 0.075[f, i]	3.3
23	L2	21	8.9 ± 0.88[d]	1.02 ± 0.065[k, m]	3.7
	L3	7	35.6 ± 7.75[d]	0.93 ± 0.068[l, n, o]	4.3
	L1	22	7.7 ± 0.88	0.83 ± 0.084[g, j, l]	3.0
26	L2	5	9.0 ± 0.71	0.92 ± 0.111	3.7
	L3	0	Larvae died	–	–

Note: Values with same superscript indicate significant differences between means at $P < 0.05$. Modified from Gormally (1988b).

computer using the method of least squares. Similar curves of larval weight against larval length were obtained for the sciomyzid species.

In the intraspecific studies O. Beaver found that the total weight of food killed depended on the duration of larval life and a high predation rate (Figs 7.11, 7.12). The mean ± 1 S.E. food conversion ratios were "reasonably" characteristic for each species and ranged from 6.1 ± 0.25 (*Sepedon spinipes*) to 31.9 ± 2.55 (*Pherbina coryleti*). Larvae that were "wasteful predators" and that killed more snails than needed had the highest food conversion ratios. She found that larvae did not feed less wastefully when the food supply was restricted, but Eckblad (1976) had found that larvae of *Sepedon fuscipennis* utilized food more efficiently at lower snail densities. In the interspecific studies, O. Beaver (1974c) found food conversion ratios of 2.7 (*Hydromya dorsalis*) to 27.1 (*Ilione albiseta*). The larvae tended to become more wasteful when near maturity (in agreement with Eckblad 1973a, 1976 for *S. fuscipennis*), and there was little interaction among species.

Vala & Haab (1984) studied the biomass of *Biomphalaria glabrata* – a tropical snail, not a natural prey of the sciomyzid species used in the experiments – and the sizes of snails consumed by each larval instar of *Tetanocera ferruginea* at several temperature and photoperiod regimes (LD 8:16 at 13 °C, 20 °C, and 26 °C; LD 9:15 at 20 °C; and LD 16:8 at 13 °C and 26 °C). The maximum amount of biomass consumed by all three instars was 1905 ± 838 mg at LD 16:8, 26 °C, and the minimum amount was 664 ± 160 mg at LD 8:16, 13 °C (Fig. 7.13). Younger larvae ate smaller snails and larger larvae ate larger snails. Third-instar larvae ate almost the entire size range offered (5.0–22.0 mm diameter) but not the smallest snails (2–4 mm diameter) (Fig. 7.14). See also Section 9.3.4, where their data are reviewed in relation to rates of development of the larvae under differences in these parameters.

Manguin *et al.* (1986) conducted experiments on both the number and biomass of three non-operculate freshwater snails (*Stagnicola palustris*, *Physella acuta*, and *Bathyomphalus contortus*) killed and eaten by *Tetanocera*

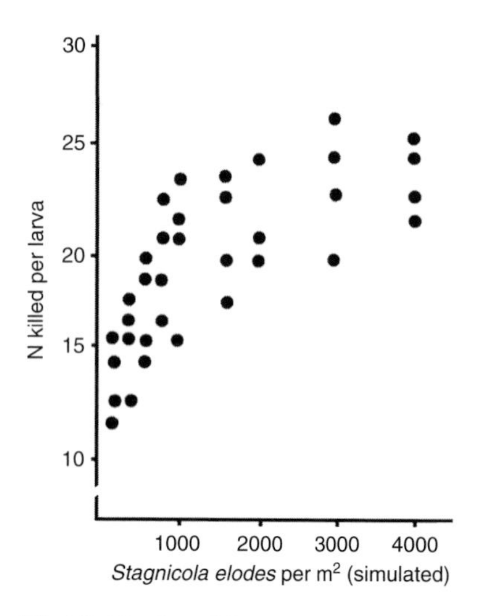

Fig. 7.9 Number of *Stagnicola elodes* (2–4.5 mm) killed during L1–L3 by ***Sepedon fuscipennis*** at nine prey densities increased at a decreasing rate when prey numbers were increased. From Eckblad (1973a).

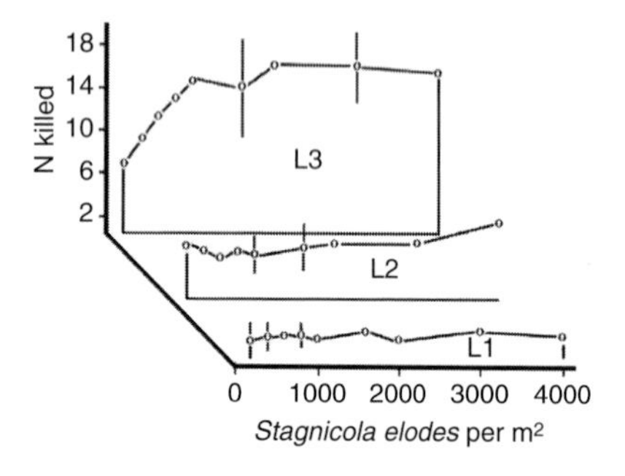

Fig. 7.10 Further analysis of data in Fig. 7.9 shows that functional response to prey density is largely a characteristic of L3. From Eckblad (1973a).

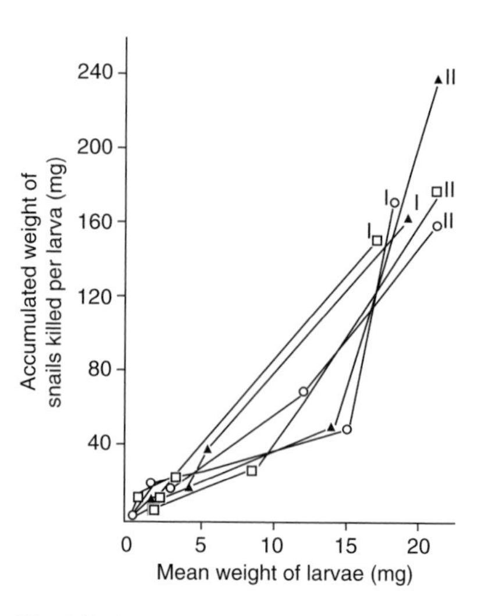

Fig. 7.11 Accumulated weight of *Lymnaea stagnalis* killed by ***Hydromya dorsalis*** L1–L3 in relation to larval weight. Square, one larva per dish; triangle, two larvae per dish; circle, four larvae per dish; I, II, replicates. From O. Beaver (1974b).

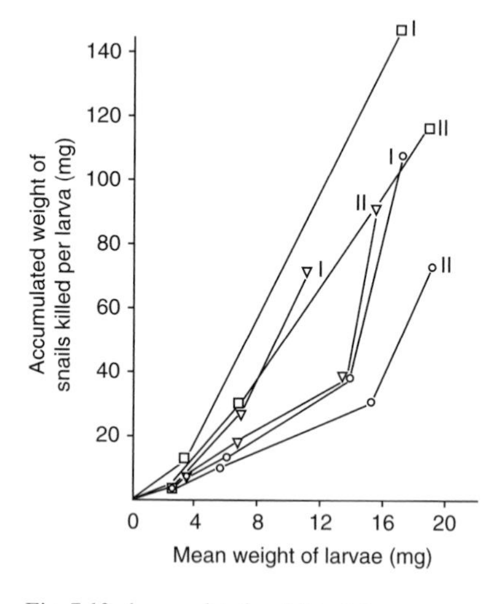

Fig. 7.12 Accumulated weight of *Lymnaea stagnalis* killed by ***Sepedon spinipes*** L1–L3 in relation to larval weight. Triangle, one larva per dish; circle, two larvae per dish; square, four larvae per dish; I, II, replicates. From O. Beaver (1974b).

ferruginea. They used the term "consommées" (consumed) but did not mention the amount of snail tissue killed and not eaten. [Note herein by Vala: in fact, all of the tissues were consumed.] Biomass of snails was determined by weight-length regression correlations. Experiments were conducted with snails provided as follows (length for *S. palustris* and *P. acuta*, diameter for *B. contortus*): 1.5–3.0 mm for first-instar larvae, 3.0–6.0 mm for second instars, and 6.0–10.0 mm for third instars, with 20 snails per predator available daily. They found that the larvae killed and ate a mean of 45 *B. contortus*, 28 *S. palustris*, or 30 *P. acuta*. Because the biomass of individual *B. contortus* is less than that of the other two species of similar dimensions, the larvae partly compensated by killing and eating more of them; differences in numbers and biomass killed and eaten were not significant when snails of similar biomass were used (Fig. 7.15). Manguin *et al.* (1986) also determined the rate of success of attack by larvae of all three instars against these snails and three other freshwater

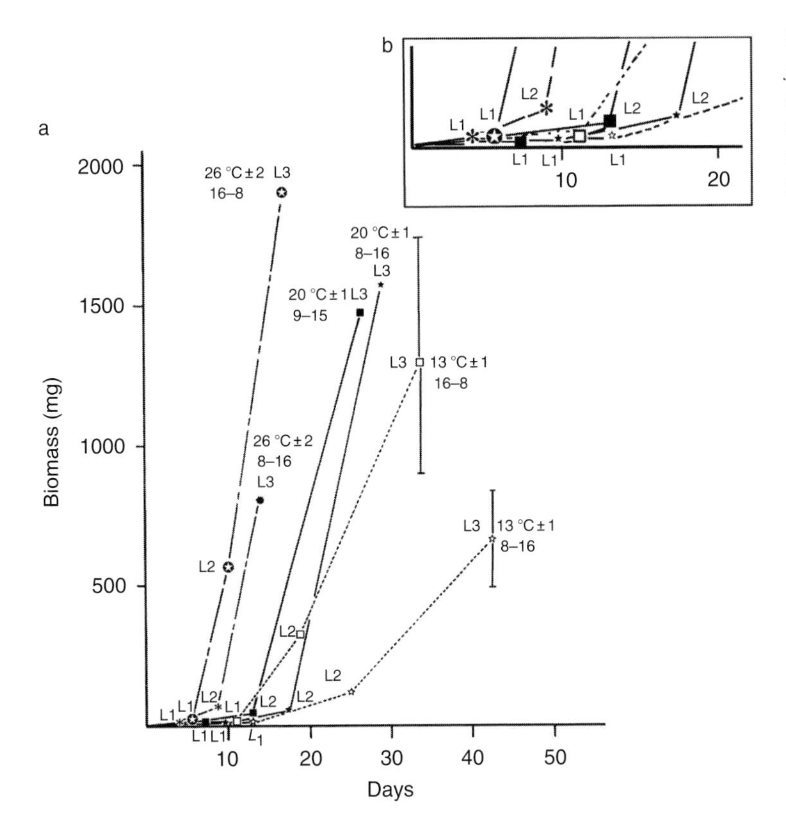

Fig. 7.13 (a) Biomass consumed by ***Tetanocera ferruginea*** L1–L3 under various controlled temperatures and photoperiods. Each symbol represents end of a stadium. (b) Enlargement of curves at beginning of feeding. Modified from Vala & Haab (1984).

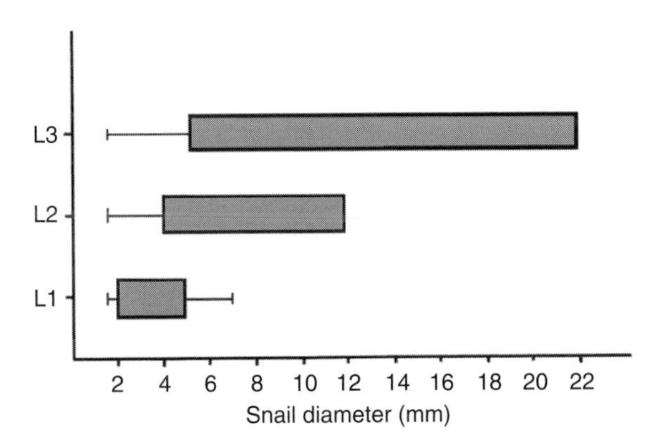

Fig. 7.14 *Biomphalaria glabrata* attacked by ***Tetanocera ferruginea*** L1–L3. Preferential sizes easily consumed by each stadium shown in rectangles. From Vala & Haab (1984).

molluscs (an operculate snail, a limpet, and a small bivalve) (see Table 5.7). They concluded that the velocity and aperture width of the snails affect success of the larvae.

Using the technique of Eckblad (1971), Gormally (1988b) compared the mean dry weight of *Radix balthica* attacked with those consumed per day by *Ilione albiseta*. See Section 7.2 on numbers of snails killed for rearing conditions. For first- and second-instar larvae biomass consumed was highest at 20 °C. For third instars and for the total larval duration period it was highest at 23 °C. The amount of tissue consumed increased significantly from the first to second instar and from the second to third instar at each temperature.

O. Beaver (1989) conducted laboratory experiments at room temperatures on *Sepedon senex*. One experiment to determine food consumption rate and growth rate (mg dry weight/larva/day) involved larvae from parents fed four different diets (honey and dry milk, honey and dry milk plus yeast or crushed snails, or both) and reared on both live and freshly crushed *Gyraulus convexiusculus*. Snail weights were obtained from shell diameter and dry weight relationship curves. Consumption rates ranged from 1.36 to 3.35 mg/day and were highest for larvae, from all parental diets, feeding on live snails. There was only about 1.0 mg per day difference between consumption rates of first- and third-instar larvae. Percentage conversion efficiency ranged from 0.5–0.8 for first-instar larvae,

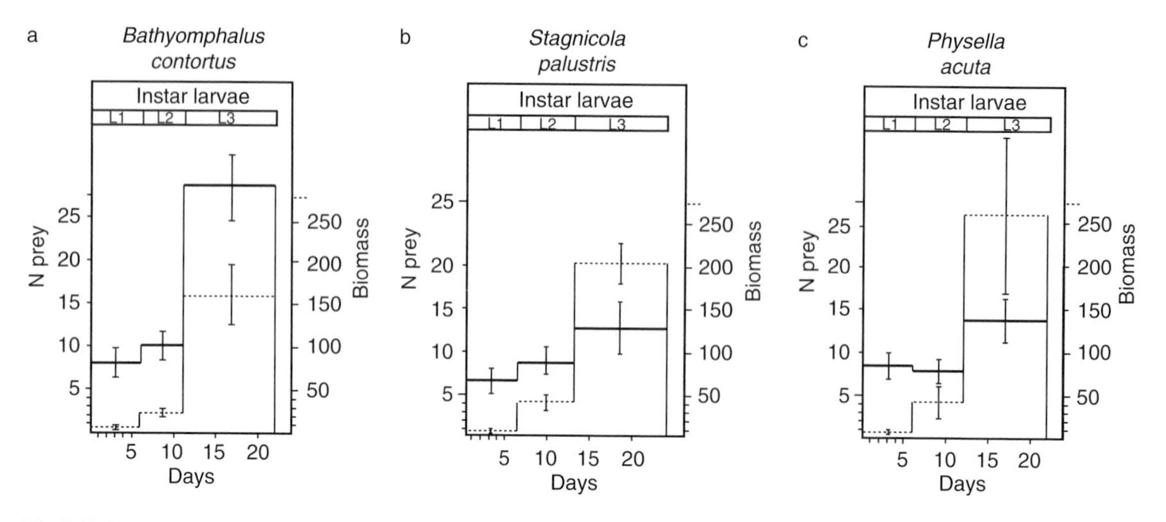

Fig. 7.15 Number (thick horizontal lines) and concordant biomass consumed (dotted horizontal lines) of snails killed by *Tetanocera ferruginea* L1–L3. (a) *Bathyomphalus contortus*; (b) *Stagnicola palustris*; (c) *Physella acuta*. Modified from Manguin *et al.* (1986).

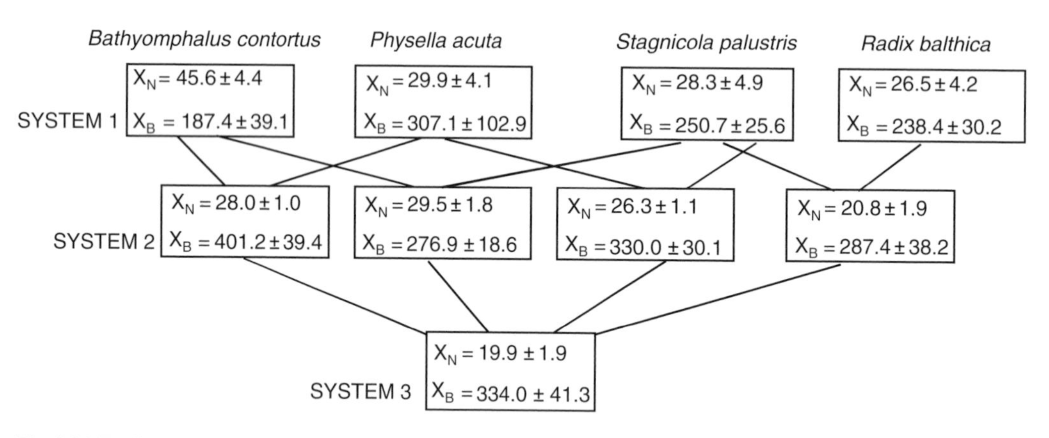

Fig. 7.16 Feeding of *Tetanocera ferruginea* larvae (combined L1–L3) with one (System 1), two (System 2), and four (System 3) prey species. N, Number of prey; B, corresponding biomass of prey. X_N and X_B, mean. From Manguin & Vala (1989).

7.0–15.0 for second instars, and 9.0–18.0 for third instars, the highest efficiencies being for larvae fed live snails. See also Section 9.3.4 and Fig. 7.15 for analyses of these data in regard to rates of development.

7.3.3 Prey and predator density and diversity, functional and numerical responses (see also Sections 9.3.4, 14.2.3.2.2, 18.1.3.1)

Manguin & Vala (1989) found that the foraging strategies of *Tetanocera ferruginea* changed as the diversity of prey species available increased. They used four species, *Radix balthica*, *Stagnicola palustris*, *Physella acuta*, and *Bathyomphalus contortus*, collected in the habitat of the fly in three experimental systems, each with a set of 20 snails. System 1 was monophagous; System 2 included two species (four combinations of the four snail species); and System 3 included all four species. An increased number of snail species available resulted in a decreased number killed, an increase in biomass consumed, and a shortened larval period (Figs 7.16, 7.17). They concluded that the larvae optimized their development by choosing a prey that provided the greatest return per unit of energy expended. They related their findings to the foraging theories of

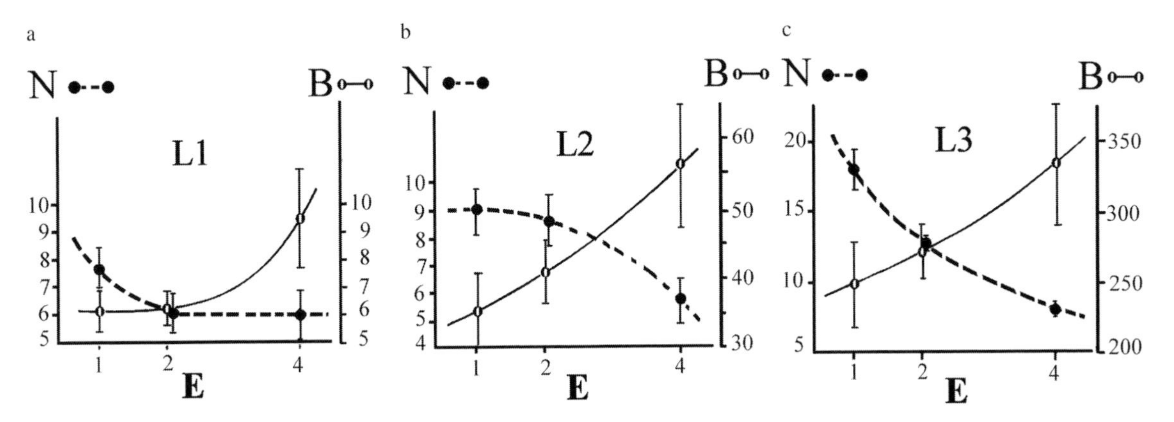

Fig. 7.17 Comparison between predation in number (N, left ordinate, dotted line, black points) and biomass (B, right ordinate, solid line, open circles) of *Tetanocera ferruginea* larvae in systems with one, two, and four prey species available (E). (a) L1; (b) L2; (c) L3. From Manguin & Vala (1989).

Emlen (1966), Hassell & Southwood (1978), and Luck (1984). They further concluded that the ability of the larvae to select the larger prey permitted them to kill sexually mature snails, thus increasing their effect on the snail population, a feature of significance in the possible use of the larvae as a biocontrol agent. In comparative experiments of 10 species of *Dictya*, *Sepedon*, and *Tetanocera* against 13 snail species that serve as intermediate hosts of *Schistosoma* spp., Neff (1964) found that the *Tetanocera* species, including *T. ferruginea*, were notably more successful against the snails than were the *Dictya* or *Sepedon* species.

Manguin *et al.* (1986, 1988a, 1988b) and Manguin & Vala (1989) did not comment on "wasteful feeding," the amount of snail tissue left behind after the predator is satiated. However, their conclusion that larvae select the larger prey fits with the observations of others that wasteful killing (and feeding?) is frequent, even in stable laboratory environments. That is, selection of larger prey at least provides the opportunity, in terms of food available, for wasteful killing.

In terms of the "functional" and "numerical" responses of Holling (1959), the snail mortality rates produced by aquatic predaceous sciomyzid larvae when their prey density increases is more or less typical of predators that are useful as biocontrol agents. That is, with increased prey density each larva kills more prey, i.e., the functional response, and the sciomyzid population increases, i.e., the numerical response.

Eckblad (1973a) found that the number of small (2.0–4.5 mm long) *Stagnicola elodes* eaten increased at a decreasing rate as prey density increased in laboratory experiments on *Sepedon fuscipennis* (water depth 5.0 mm, room temperature 21–24 °C). Larvae killed a mean of 14 snails at a prey density of $200/m^2$ and 24 snails at $4000/m^2$ (Figs 7.9, 7.10). While third-instar larvae displayed a strong functional response, first- and second-instar larvae showed no significant functional response. Eckblad compared three predictive equations for the total number of snails killed per larva in shallow water and found that a simple power function may give a better predictive equation than the predator models of either Holling (1959) or Watt (1959). Mc Donnell (2004) found a strong functional response for all instars of *Sepedon spinipes* at five temperatures.

In laboratory experiments with 10 larvae per 400 *S. elodes* per 1 m^2 and 1000 m^2, Eckblad (1973a) found no evidence that predator density affected predation rate. In field experiments, larval densities were manipulated in 18 enclosed 0.6 × 4.6 m quadrats perpendicular to the shoreline. This was an arrangement similar to Eisenberg's (1966) study of the regulation of density in *S. elodes* populations; see Section 1.2. The population density of smaller (<4.5 mm) *S. elodes* was reduced when predator density was increased, but populations of *Gyraulus parvus*, *Physella integra*, and larger *S. elodes* were not significantly reduced; see Chapter 13.

In experiments with *Sepedon sphegea* and the tropical planorbid *Biomphalaria glabrata*, Haab (1984) noted that predation by larvae is dependent on density, which fits the type II functional response described by Holling (1959).

For all larval stages of *Sepedon senex* reared in Thailand on *Gyraulus convexiusculus*, O. Beaver (1989) found that the number of snails eaten per larva per day increased with snail density. Maximum daily consumption of first-, second-, and third-instar larvae were 3, 6, and 10 snails of

2.0–3.0 mm diameter and 1, 3, and 4 snails of 4.0–5.0 mm diameter, respectively. Mature larvae had greater ability to kill snails at all prey densities studied. For all instars the average rate of predation increased with prey density but decreased with predator density. The ability of larvae to kill snails at all prey and predator densities increased with the age of the larvae. She found that the functional response curves of first- and second-instar larvae were of Type I and that of the third instar were of Type III, as suggested by Holling (1959) at some predator/prey densities and prey sizes, but for all stages the response was between Types I and II at other densities and sizes.

In laboratory experiments with *Sepedon neavei* and *S. scapularis*, Maharaj (1991) found that predation on *Bulinus tropicus* by *S. neavei* was affected by prey density, but that predation on *B. tropicus* by *S. scapularis* was not. In tests using *B. tropicus* at estimated densities of 312, 833, and 1667 snails per m^2, *S. neavei* killed increasing numbers of snails (21.2 ± 3.8, 23.4 ± 3.4, and 26.3 ± 2.8, respectively). *Sepedon scapularis* killed fewer snails than did *S. neavei* and with no density-dependent variation, the mean being 18.4 ± 2.6.

Mc Donnell (2004) investigated the impact of snail availability, either one snail per day or excess snails per day, on predation capacity, predation rate, larval stage duration, and larval survival for *Sepedon spinipes* at five constant temperatures (14, 17, 20, 23, and 26 °C) and LD 16:8. He found that percentage larval survival was comparable between the two feeding regimes (>75%) and that median larval stage duration was shorter by only one day when excess snail food was available at all temperatures. This is in contrast to Geckler (1971), who found that duration of the larval stage of *Sepedon tenuicornis* was extended more than 2 weeks when only one snail was provided daily as food. In addition, Mc Donnell found that predation capacity and rate were significantly greater when excess prey was provided. In fact, at 23 and 26 °C, twice the number of snails were killed when they were in excess than when they were not. The author related his results to foraging theory and postulated that larvae of *S. spinipes* may be able to feed saprophagously when the snail density is low.

7.3.4 Rates of predation

Information on rates of predation is combined with other factors in several publications noted above. Some additional observations and conclusions follow. Eckblad (1973a), in laboratory and field experiments with *Sepedon fuscipennis* and *Stagnicola elodes*, found that predation rates were more influenced by prey density and water depth than by differences in levels of prey aggregation and predator density. In intra- and interspecific competition studies of eight primarily aquatic predaceous species, O. Beaver (1974b) found that the weight of food killed per day by a single larva increases from the first to the third instar. Predation rates were not correlated with adult size or with more predatory or more parasitoid habits among the species studied. These included strictly predatory species to those with mixed parasitoid to predatory habits. Predation rates of the various species differed little in the first-instar larvae but varied considerably in later instars (Figs. 7.18, 7.19).

7.3.5 Survival of predators

Predation and survival also are treated with other factors in the above sections, and a few additional points are noted here. From laboratory experiments on snail vulnerability and larval success, Geckler (1971) concluded that newly hatched larvae of *Sepedon tenuicornis* are "in a precarious position for survival" and that differences of little more than 1.0 mm in diameter of *Helisoma anceps* can significantly alter their chances of success. In laboratory experiments on *Sepedon senex* with *Gyraulus convexiusculus* (10 larvae with 100 snails in 45 cm diameter bowls, with dead snails and dead larvae replaced daily), O. Beaver (1989) maintained water depth at 10 cm, much deeper than in experiments by other researchers, but found that rates of survival were similar to those found by other authors

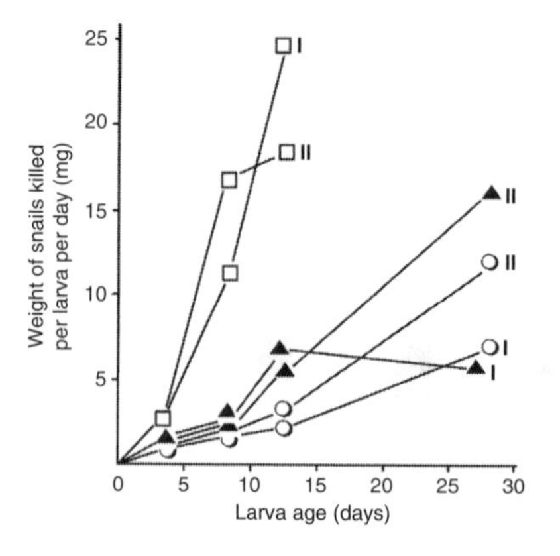

Fig. 7.18 Predation rate of ***Hydromya dorsalis*** larvae in relation to larval age in experiments with *Lymnaea stagnalis*. Square, one larva per dish; triangle, two larvae per dish; circle, four larvae per dish; I, II, replicates. From O. Beaver (1974b).

among the instars: 6.0% ± 9.2% for first instars, 48.0% ± 18.3% for second instars, and 96.0% ± 6.2% for third instars. The percentage survival of first-instar larvae in these experiments was in the same range as the percentage of snails killed by first-instar larvae in experiments on sizes of snails killed, conducted in 1.0 cm of water. These results conflict with those of Eckblad (1973a) (see Section 7.3.3).

Survival of *Sepedon scapularis* reared on two endemic, non-operculate freshwater snails, *Bulinus africanus* and *B. tropicus*, and one immigrant species, *Physella acuta* (3 mm water, 28 °C, LD 12:12) were found to be significantly influenced by instar, size of snail attacked, and species of prey provided (Maharaj *et al.* 1992) (Fig. 7.20).

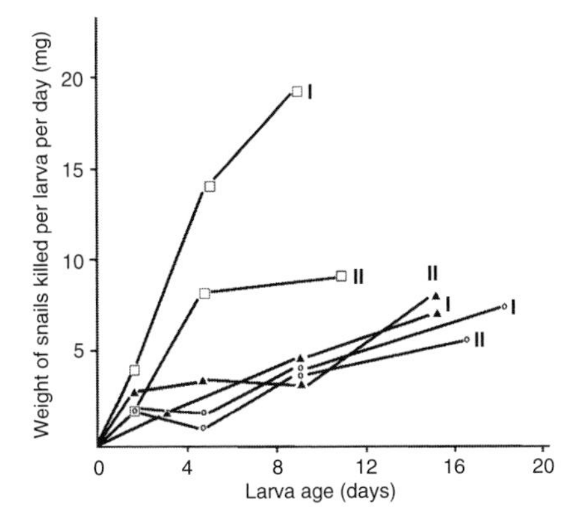

Fig. 7.19 Predation rate of ***Sepedon spinipes*** larvae in relation to larval age in experiments with *Lymnaea stagnalis*. Square, one larva per dish; triangle, two larvae per dish; circle, four larvae per dish; I, II, replicates. From O. Beaver (1974b).

Survival was affected by larvae becoming trapped in snail mucus, snail feces entangling larval interspiracular processes (both considered to result, perhaps, from the experimental conditions), and prey avoidance behavior. However, survival of first-instar larvae likely is influenced by various factors affecting different species differently. For example, Gormally (1988b) found that survival of unfed first-instar larvae of the univoltine *Ilione albiseta* (most of which hatch during the fall) decreased as temperature increased, from a mean of 28.4 days at 14 °C to 11 days at 26 °C, and that first-instar larvae could survive as long as one month without food.

Mc Donnell (2004) found different results for unfed larvae of the multivoltine *Sepedon spinipes*: a median of eight days at 14 °C and of three days at 26 °C. In addition, he found that larval survival could be prolonged significantly at all temperatures by feeding neonates on one snail (*Lymnaea stagnalis*, 2.0 mm in length). In fact, at 20, 23, and 26 °C, larval longevity doubled when each larva was provided with one snail meal. Similar results also were observed outdoors, during July in Ireland, where the life-span of larvae increased from 7 to 13 days when each neonate was allowed to feed on one *Galba truncatula* (2.0 mm). The author discussed the importance of these results in relation to biocontrol attempts; see Section 14.2.3.2.2.

7.3.6 Effects of water depth and temperature

Eckblad (1973a) tested predation rates and larval survival of *Sepedon fuscipennis* in 3 × 3 × 2 factorial laboratory trials at depths of 5.0 mm and 5.0 cm of water and prey densities

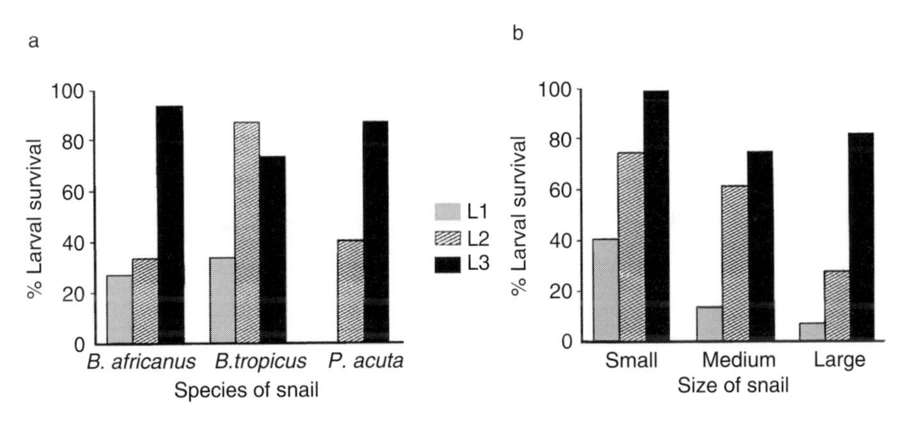

Fig. 7.20 Survival of ***Sepedon scapularis*** L1–L3 when feeding on (a) different species of snails (*Bulinus africanus*, *B. tropicus*, and *Physella acuta*). (b) different size classes. From Maharaj *et al.* (1992). Note: no L1 survived after attempting to attack large *P. acuta*.

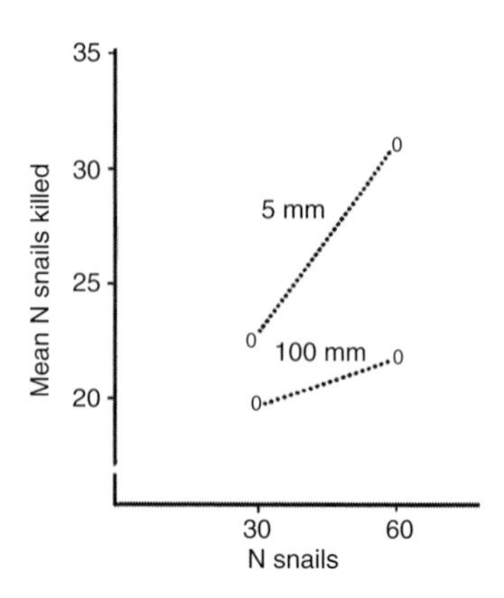

Fig. 7.21 Interaction between numbers of *Stagnicola elodes* (2–4.5 mm long) and water depth (5.0 and 100 mm) on number of snails killed by ***Sepedon fuscipennis*** L1–L3. From Eckblad (1973a).

of 20, 40, and 60 *Stagnicola elodes* in containers of 32, 128, and 513 cm^3 containing one first-instar larva each. The number of snails killed increased with decreased water depth and higher prey densities. The mean survival was 52.2% at 5.0 mm and 40.0% at 5.0 cm, the difference being caused primarily by differential survival of first-instar larvae. Concurrent with the laboratory experiments, two consecutive experiments were carried out in field enclosures of 1256 cm^3 at water depths of 5.0 mm and 10.0 cm and at densities of 30 and 60 snails in a 2 × 2 factorial, and in enclosures of 78, 312, and 1256 cm^3 with water at a depth of 5.0 mm and the same snail densities in a 3 × 2 factorial. Fifteen first-instar larvae were placed in each chamber. The first field experiment confirmed the laboratory trials (Fig. 7.21), with mean survival 36.7% at 5.0 mm and 18.3% at 10.0 cm. The variability in survival associated with water depth masked the influence of prey density in the 2 × 2 factorial field test and in both laboratory tests.

Yoneda (1981) determined the number of snails eaten per day by *Sepedon aenescens* at various controlled temperatures. He found that at 30 °C, 2.7 *Hippeutis cantori* and 3.3 *Physella acuta* were eaten per day, while only 1.4–1.8 were eaten per day at 15, 19, 22, and 25 °C.

The effect of temperature on number of snails killed, biomass consumed, and survival of *Ilione albiseta* was studied at 14, 17, 20, 23, and 26 °C and LD 16:8 by Gormally (1988b). *Radix balthica*, 2.5–3.5 mm in length, were provided to first-instar larvae and *R. balthica*, 3.5–4.5 mm, were provided to second- and third-instar larvae. The mean total number of snails killed per larva was high: 66.9, 58.1, 55.0, and 50.2 at 14, 17, 20, and 23 °C, respectively (no significant difference). There was a significant difference in the mean number of snails killed per day during the third stadium between 23 °C and 14, 17, and 20 °C. Third-instar larvae maintained at 26 °C died (Table 7.12). With regard to wasteful feeding, Gormally found that the difference between the amount of snail tissue attacked and the amount consumed per day was less at 23 °C than at the lower temperatures during the third instar and for total larval duration.

8 • Competition

> I have a strongly held skepticism about any strongly held beliefs, especially my own.
>
> Lapparent (cited by Coates, 2003).

8.1 INTERSPECIFIC COMPETITION

The extent to which interspecific competition occurs among animals has been investigated for many years (e.g., Andrewartha & Birch 1954). The central role of interspecific competition in organizing questions in ecological theory and research approaches has been reviewed by Hengeveld & Walter (1999) and Walter & Hengeveld (2000). They were led to distinguish two incompatible ecological paradigms. The first is the dominant demographic paradigm, largely based on competition. The second is the poorly developed autoecological paradigm, largely based on the proposition that "the primary actions of organisms involve their responses to the environment that surrounds them" (Walter & Hengeveld 2000). These authors (above papers and others cited therein) concluded "the ecological and evolutionary significance of interspecific competition is of questionable significance."

The fact that most Sciomyzidae are multivoltine and that none is known to have a larval feeding duration of more than several months (most only a few weeks) – i.e., there being no need for complex, more-than-one-year life cycles – may be one indication that their food supply is reliable. This in turn indicates that sciomyzids compete little for food, the only resource likely to be in short supply for Sciomyzidae, and that they compete only rarely.

However, if there is competition for food among species of Sciomyzidae and between Sciomyzidae and other malacophagous organisms, then the process should be considered in the context of temporal terms (phenology) and spatial terms – large scale (geographical distribution) and small scale (habitat). Obviously, many opportunistic predatory insects share food resources with Sciomyzidae, but there are relatively few obligate predatory or parasitoid malacophagous insects that might compete with Sciomyzidae. Why is this? Knutson & Ghorpadé (2004) speculated that Lampyridae in the tropics might be especially important competitors. Only in the case of *Atrichomelina pubera* have other Diptera (Sarcophagidae, Phoridae, Ephydridae, and Piophilidae) been reared from snails collected in nature that also contained Sciomyzidae. *Atrichomelina pubera* is a highly facultative species of the tribe Sciomyzini that can feed, on a day-to-day basis, in a parasitoid, predatory, or saprophagous manner on a broad ecological assemblage of exposed aquatic and terrestrial snails (Foote *et al.* 1960) and also occasionally on large, dead clams (J. B. Keiper, personal communication, 2006). R. Beaver (1977) found co-existence in the same snail to be frequent among most of the ten most common species of the Diptera (representing six families, but not including Sciomyzidae) breeding in dead snails. The fact that there are no records of different species of sciomyzid larvae feeding together in the same host/prey individual in nature also indicates a lack of competition. In laboratory experiments, Appleton *et al.* (1993) observed larvae of *Sepedon neavei* and *S. scapularis* feeding simultaneously on the same snails without any visible interference. No experimental studies have been conducted in nature on interspecific competition among Sciomyzidae, and only two laboratory studies have been conducted (O. Beaver 1974c, Ghamizi 1985; Ghamizi & Vala, unpublished data).

Adaptive differences in phenology perhaps evolved in response to competition, but may serve to lessen or avoid competition. Some of these aspects were analyzed in detail by Berg *et al.* (1982) and are reviewed in Section 9.1, with the addition of Groups 5b and 6. The large-scale (continental) geographical ranges of many sciomyzid species in the same trophic/microhabitat guild overlap broadly or are essentially the same. In terms of small-scale range, that is, at the same general site (from a few square meters to several hundred square meters in extent), adults of many different species of various genera usually are found. Often several species can be collected together in a few sweeps of a collecting net.

The larvae are distributed primarily according to microhabitat preference and the more or less concordant distribution of their food resource, as described in Section 4.1. Larvae of members of the same trophic/microhabitat guild often are found in close proximity, and many species more or less overlap in terms of feeding and other behavior. These facts of natural history might lead one to assume that there is a certain level of interspecific competition. Particularly in the publications by B. A. Foote, there is emphasis on resource partitioning as a result of competition, e.g., among certain North American *Tetanocera* species (Foote 1996a, 1996b, 1999). However, for the ten North American species of aquatic predaceous *Tetanocera* that he studied intensively in the field and laboratory, Foote (1999) concluded that there does not seem to be much partitioning of the food resource but there are distinct differences in geographical distribution, habitat distribution, and especially in voltinism and the period of the year when larvae are feeding. He referred to the idea of "the ghost of competition past" in regard to these ten species. He suggested that during past times of increased environmental stress, e.g., drought, when aquatic habitats were dramatically reduced and when snail populations were significantly lower, partitioning of Sciomyzidae could have resulted from intense competition for food.

In most sites, snail populations are enormous compared with sciomyzid populations. Such observations, however, may sometimes be an artifact of selection of study sites. Sites often have been chosen largely because Sciomyzidae, and thus their hosts/prey, are abundant. However, it is possible that there are sites where food is limited and competition occurs.

Ghamizi & Vala (unpublished data) found in laboratory rearings of *Tetanocera ferruginea*, *T. arrogans*, *Sepedon sphegea*, and *S. spinipes* that the main result of interspecific competition is the much smaller size of emergent adults (see Section 8.2.2). In all of their observations, *T. ferruginea* was always the most voracious species, and its larvae grew more rapidly than did larvae of the other species. However, *T. arrogans* is not a behavioral equivalent of the other species but is a parasitoid/predator of Succineidae.

In nature, interspecific competition seems to be related to three main factors: the behavior of the species of molluscs that are potential prey; slow or fast developmental rate of larvae; and the univoltine or multivoltine characteristic of the life cycle. Some species, e.g., *Ilione albiseta*, apparently avoid competition with other sciomyzid species because their larval cycle occurs largely during the winter. When spring begins in temperate areas, competition should be low in permanent aquatic habitats because most species of snails are in a reproductive stage then. During this season, many sciomyzid larvae are found in great numbers attacking different sizes of prey, particularly first-instar larvae, which more easily attack the abundant small snails (Vala & Manguin 1987). Following the summer and the drying of some temporary aquatic habitats, competition may be more intense because the snail and fly populations have become more or less concentrated.

O. Beaver (1974c) conducted a laboratory study of interspecific competition among eight species of sciomyzids in five genera in Wales, using primarily *Lymnaea stagnalis* as the prey and pairs of species (A and B) with "similar habits and food preferences" at ratios of 1A:4B, 2A:2B, and 4A:1B larvae. All were placed under the same rearing conditions, similar to those in her study of intraspecific competition among the same species, but in 5mm of water, with food sufficient for two larvae in containers of four larvae, and sufficient for three larvae in containers of five larvae. She referred to all sciomyzid species as aquatic predators. In fact, they occur in quite different microhabitats – three are shoreline predators (*Tetanocera silvatica*, *T. fuscinervis* [as "*T. unicolor*"], and *Hydromya dorsalis*) and five are aquatic predators (*Tetanocera ferruginea*, "*T. punctifrons*" [= *T. montana* ?, voucher specimens not seen by us], *Ilione albiseta*, *Pherbina coryleti*, and *Sepedon spinipes*). The author did note, "*T. silvatica* and *T. unicolor* are known to have more amphibious and parasitoid tendencies than *H. dorsalis*, while *S. spinipes* is a more aquatic and predatory species." It also should be noted that whereas all eight species feed on freshwater non-operculate snails in laboratory rearings, information on their natural prey is limited. The natural prey are known only for *H. dorsalis* (*Galba truncatula* and *Stagnicola palustris*) and *T. ferruginea* (*Stagnicola elodes*, *S. palustris*, *Galba humilis*, *L. stagnalis*, *Physella acuta*, *Helisoma* sp., *Planorbis* sp., and *Bathyomphalus contortus*). Furthermore, three species (*I. albiseta*, *T. montana* ?, and *P. coryleti*) are univoltine, overwintering as larvae, whereas the other species are multivoltine, overwintering in the puparium (except *S. spinipes* as adults). This also affects Beaver's conclusions because larvae of some of the species pairs are not present at the same time of year in nature and likely are not direct competitors. However, the early-season species may make inroads into the prey population, thereby decreasing the food supply for the later species. Further studies on interspecific competition should encompass these points. Range and mean total larval period, maximum weight of larvae forming puparia, weight of snails killed and food conversion ratios, and minimum and maximum growth rates of larvae forming puparia were analyzed, and the results were

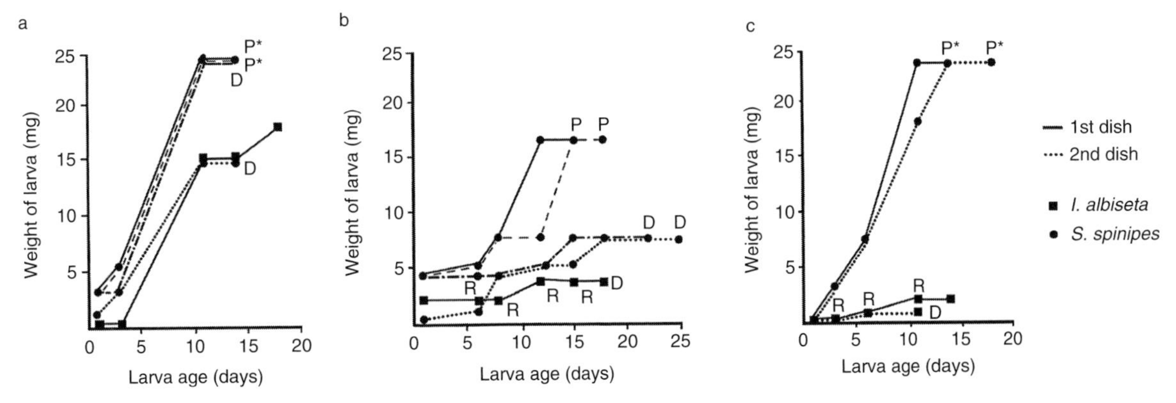

Fig. 8.1 Growth curves of ***Ilione albiseta*** and ***Sepedon spinipes*** L1–L3. (a), (b) One ***I. albiseta*** and four ***S. spinipes*** per dish. (c) Two ***I. albiseta*** and two ***S. spinipes*** per dish. D, larva died; R, larva died and was replaced by another of weight indicated; P, larva pupariated but no adult emerged; P*, larva pupariated and later produced an adult. From O. Beaver (1974c).

compared with her previous experiments on intraspecific competition among the same species (O. Beaver 1974b). She concluded that, in general, growth is related to the total number of other sciomyzid larvae present rather than their species. Some interactions involving growth rate and amount of food eaten were evident, especially in the experiments including *S. spinipes*, showing that species to have a competitive advantage over *I. albiseta* (Fig. 8.1) and *P. coryleti* and showing *H. dorsalis* to have some competitive advantage over *T. fuscinervis* and *T. silvatica*.

8.2 INTRASPECIFIC COMPETITION

Pour bien savoir les choses, il faut savoir le détail.
[To be known well, things must be known in detail].
La Rochefoucauld (1665).

8.2.1 Semi-terrestrial and terrestrial parasitoid Sciomyzini and Tetanocerini

Many of the species in this group spend much of later larval life feeding in the liquefied, putrid tissues of the host snail. The absence of ventral cibarial ridges and oral grooves in these larvae indicate that they are not filter-feeding on the micro-organisms developing in this "nutrient" but instead are ingesting the decaying tissues. Although infestation of a snail by several larvae is common in laboratory rearings of many Sciomyzini and Tetanocerini, solitary feeding seems to be the rule in nature.

In Sciomyzini, multiple infestations (of one species) have been found in nature only in the case of *Atrichomelina pubera*, *Pherbellia s. maculata*, and *P. seticoxa*. At least two puparia of *A. pubera* were found in many field-collected shells of *Helisoma anceps* (Foote *et al.* 1960). Fourteen puparia of *A. pubera* (from which six adults had emerged) were found attached, together, to one valve of a large, dead *Lampsilis* clam in nature; obviously the larvae had fed together (J. B. Keiper, personal communication, 2006). Two larvae of *P. s. maculata* were found in an *Oxyloma* sp. snail in nature. Of the many shells of *Helisoma trivolvis* collected in nature, many harbored two *P. seticoxa* puparia and a few contained three, four, or five puparia (Bratt *et al.* 1969). Infestations of 20–40 first-instar larvae of *P. seticoxa* feeding in one large *Helisoma* (10mm diameter) were common in laboratory rearings (Bratt *et al.* 1969). In laboratory rearings up to eight *A. pubera* larvae fed in one snail (Foote *et al.* 1960).

In the Tetanocerini, four of 35 *Succinea* sp. collected in nature were each infested by one larva of *Tetanocera melanostigma*, one *Succinea* sp. had three *T. melanostigma* larvae, and two of 100 *Oxyloma effusa* each contained one *T. melanostigma* larva. Of about 2000 *Oxyloma* sp. collected in nature, only 12 were infested with *T. rotundicornis*, and of these succineids only a few contained two larvae (Foote 1996b). Egg masses of *Aplexa hypnorum* collected in nature bore one to four eggs each of *Anticheta melanosoma* (Knutson & Abercrombie 1977). Of 26 egg masses of *Galba truncatula* collected in nature, 15 bore one egg of *Anticheta analis*, nine bore two eggs, and two bore four eggs

(Knutson 1966). Some behavioral and/or physiological mechanisms seem to have evolved among Sciomyzidae to severely limit multiple feeding. Such mechanisms are known among other families of Diptera. Bryant & Hall (1975) have shown that larval conditioning of the food medium occurs in *Musca domestica*, limiting the number of larvae feeding together.

Intraspecific competition among feeding larvae is documented fairly well in results of laboratory rearings for a few terrestrial Sciomyzini. These species, especially those that oviposit onto estivating aquatic snails or semi-terrestrial or terrestrial snails, have the most intimate relationships with the host known in the family. They are: *Sciomyza varia*, host: *Stagnicola elodes*; *S. aristalis*, host: *Novisuccinea ovalis*; *Pherbellia s. schoenherri*, hosts: *Oxyloma elegans* and *Succinea putris*; *Colobaea bifasciella*, hosts: *Stagnicola palustris*, *Galba truncatula*; and *Tetanura pallidiventris*, hosts: *Clausilia* sp., *Cochlicopa lubrica*, *Aegopinella pura*, and *Discus rotundatus*. However, the precise mechanisms of exclusion of supernumerary larvae are imperfectly known. In these species, one or more eggs are laid only on the shell of the host, or on the soft tissue by a highly modified ovipositor in the case of *T. pallidiventris*. The host does not die until about one week after the larvae or larva enter(s) it. Only one larva develops in the host even if several eggs are laid on or in it and even if the host is small and other potential hosts are available. The puparium is formed in the shell and usually is specially adapted. Exceptions include *P. s. schoenherri* and *P. s. maculata*, which more often form the puparium outside, and *P. s. maculata*, which often does not oviposit onto the host.

As early as 1958, the essentials of the life cycle of *S. varia* (as *Pteromicra inermis*) were presented at a scientific meeting and then published by Berg *et al.* (1959). Foote (1959a) published the complete life cycle of *S. aristalis*. Berg (1961) proposed the idea of one larva per snail and one snail per larva in outlining the life cycles of *S. varia*, *S. aristalis*, and *P. s. schoenherri*. Much more information subsequently was published for *P. s. schoenherri* (Verbeke 1960, Moor 1980, Vala & Ghamizi 1992) and *S. varia* (Barnes 1990).

Berg *et al.* (1959) noted that in laboratory rearings they found 1–25 eggs of *S. varia* on each shell without once finding more than one living larva in each. They speculated that the larvae "conserve the available food for their own use by killing all other larvae that try to enter the shells they occupy." With regard to *S. varia*, *S. aristalis*, and *P. s. schoenherri*, Berg (1961) stated that "the first larva to enter may have some way of eliminating competitors." Barnes (1990) found that when up to 14 eggs were laid on one shell, other first-instar larvae often were found dead near the established larva and "It appears that the first larva to enter the shell effectively excludes all others, no matter how many eggs are present on the shell." For *C. bifasciella* (Knutson & Bratt unpublished data), a behavioral equivalent of *S. varia*, snails bearing eggs were isolated in 1 × 5 cm glass vials to enable close observation and to prevent loss of the larvae on the substrate. When more than one egg was on a shell, all larvae entered the snail, but all except one larva left within one day, the unsuccessful larvae being easily found in the tube. Some *Cochlicopa lubrica* infested by first-instar larvae of *Tetanura pallidiventris* in the laboratory remained alive for several months. When dissected after death of the snail, a dead first- or second-instar larva was found in each. Single larvae found in living *C. lubrica* and dead *Aegopinella pura* in nature eventually pupariated and some produced adults, and a puparium found in a *Discus rotundatus* shell in nature produced an adult (Knutson 1970a).

Knutson (1970a) noted encapsulation of larvae of *Tetanura pallidiventris* by *C. lubrica*, and Moor (1980) carefully studied encapsulation of larvae of *Pherbellia s. schoenherri* by *Succinea putris*. In these species, the survival of only one larva per snail may result from the ability of one of a cohort of attacking larvae to survive encapsulation by developing faster rather than due to direct attack by other larvae.

Larvae of the Nearctic subspecies *P. s. maculata*, which does not routinely oviposit onto shells or pupariate in shells, have been found in both *Succinea* spp. and *Oxyloma* sp. in nature. Multiple infestations by first- and second-instar larvae were common in crowded rearing containers, up to six larvae in a 15 mm *Succinea* sp., but third-instar larvae seldom shared a snail, even when crowded. Larvae often attacked a second snail, and there was no evidence of competitive exclusion of larvae (Bratt *et al.* 1969).

Many other Sciomyzidae (Salticellinae, Sciomyzini, and Tetanocerini) attacking semi-terrestrial and terrestrial snails and slugs have more or less parasitoid habits, especially during early larval life. However, none has the complete repertoire of parasitoid features noted for the five species above. Only in two other species is it more or less proven that there is only one larva per snail. The female *Salticella fasciata* (Salticellinae) lays eggs (usually one to six per snail) only in the umbilici of moribund terrestrial *Theba pisana* and *Cernuella virgata* snails in nature, and only one larva develops on the decaying tissues of the host (Knutson *et al.* 1970, Coupland *et al.* 1994). First-instar larvae of *Trypetoptera punctulata* (Tetanocerini), whose eggs

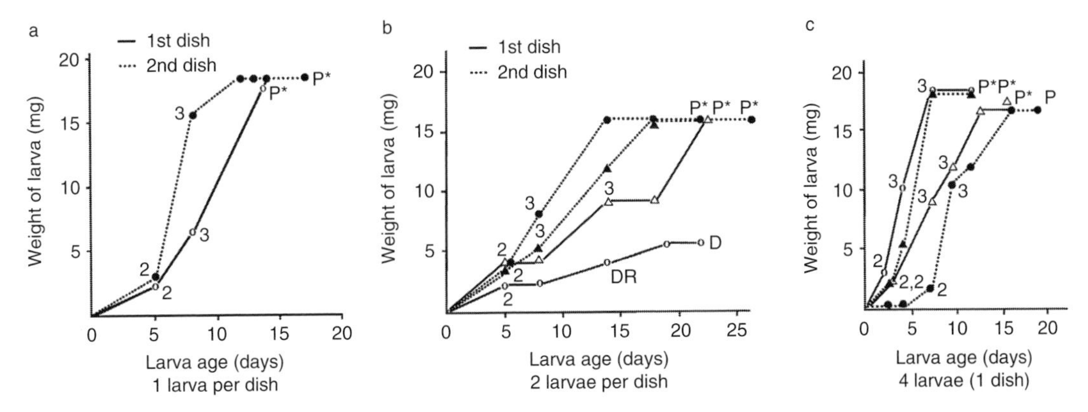

Fig. 8.2 Growth curves of ***Sepedon spinipes*** L1–L3. (a) One larva per dish. (b) Two larvae per dish. (c) Four larvae in one dish. D, larva died; R, larva died and was replaced by another of weight indicated; P, larva pupariated but no adult emerged; P*, larva pupariated and later produced adult. 2, molting from L1 to L2. 3, molting from L2 to L3. Note: in Fig. c the lines indicate four larvae in one dish, not separate dishes. From O. Beaver (1974b).

are not laid on the shell, feed on small juveniles inside the pallial cavity of the viviparous terrestrial parent *Lauria cylindracea*. The fastest developing larva occupies the first whorl of the shell, appearing to obstruct the posterior spiracles of other larvae feeding within. Second- and third-instar larvae attack a series of up to 15 snails, apparently excluding other larvae in the same way (Vala 1986).

Larvae of some *Pherbellia* spp. that, at least in laboratory rearings, feed together in a single snail sometimes have dark, paired, comma-shaped marks on the integument. These marks probably are scars resulting from the larvae slashing at each other with their paired mouthhooks; this may be a mechanism to force dispersal (e.g., *P. dorsata*, Plate 2b, c). J. B. Coupland (personal communication, 2000) noted that he has seen similar scars on larvae of saprophagous species of *Sarcophaga*.

8.2.2 Aquatic and semi-aquatic predators

In many species, in laboratory rearings, multiple larvae simultaneously attack an individual snail, but when the tissues have been consumed, each larva then attacks one or more subsequent snails if it is still in the process of developing. O. Beaver (1974b) conducted intraspecific competition experiments in Wales for eight aquatic and semi-aquatic species in five genera with densities of one, two, or four larvae and *Lymnaea stagnalis*, similar to her experiments on interspecific competition (O. Beaver 1974c). When more than one larva was placed in a dish, one or two of them always grew faster than the others. The

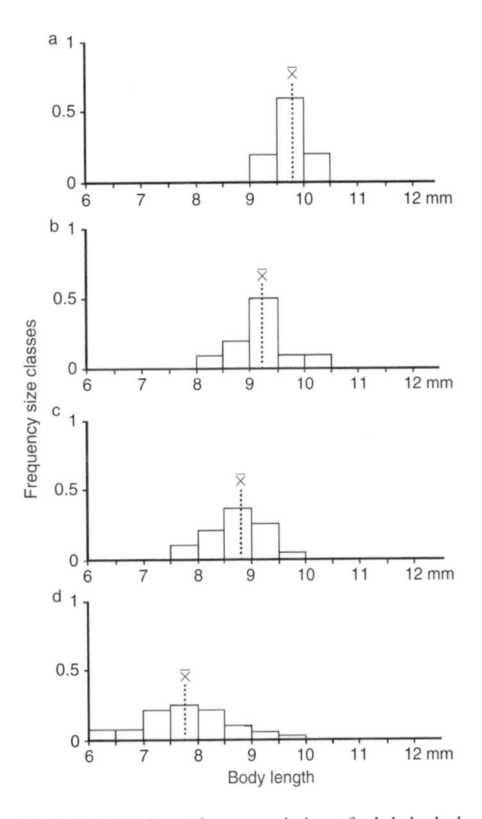

Fig. 8.3 ***Sepedon sphegea***, variation of adult body length as function of larval intraspecific competition. Data when 1 (a), 5 (b), 10 (c), 20 (d) L3 are in presence of 25 *Radix balthica* as prey. x, mean of adult length in each case. Modified from Ghamizi (1985).

slower-growing larvae were more likely to die than the faster-growing larvae, and they tended to die at a lower weight and were less likely to pupariate. As the intensity of competition increased, the mean time spent in each instar by a species usually increased (Fig. 8.2). Her results also are of value to discussions of predation (Chapter 7) and development (Chapter 9).

Ghamizi (1985, unpublished thesis) conducted a series of detailed laboratory experiments on intra-specific competition among larvae of the aquatic predator *Sepedon sphegea* in France. Among his results, he showed that the size of adults varied according to the degree of intraspecific competition. In nature, the length of *S. sphegea* (base of antenna to end of abdomen) is about 10–12mm, with the females being longest. In a series of experiments, Ghamizi (1985) associated 25 *Radix balthica* and 1, 5, 10, and 20 larvae, respectively (equals 40, 200, 400, and 800 predators/m^2). When competition was low (Fig. 8.3a) adults were about the same size as those collected in nature, i.e., 9.8 ± 0.6mm long, and the population was distributed only in a few size classes. Under intense competition the sizes of adults decreased gradually, with a deviation to the lower values. There also were more size classes, with a reduced frequency of adults in each class (Fig. 8.3b–d). These results showed that significant intraspecific competition has direct effects on the adult sizes. O. Beaver (1974b, 1974c) also noted a relationship between the larval weight determined just before pupariation and the size of emergent adults.

Competitive displacement in feeding also may occur among individuals of some species in two of the three genera that include the only Sciomyzidae that are restricted to subsurface fingernail clams (Sphaeriidae). For *Renocera striata* (as *R. brevis*) in North America, Foote (1976) noted that "Probably the normal situation in nature is for 1 larva to remain throughout the 1st stadium in 1 clam and for each clam to be infested by only 1 larva." Barnes (1980a) described the life cycle of the New Zealand fingernail-clam killer, *Eulimnia philpotti*, larvae of which are subsurface predators throughout almost all of their life cycle and which apparently feed individually. He noted, "nearly all larvae that left their original host without completing the first stadium died without being able to invade a second clam."

9 • Phenology, reproduction, and development

> Almost every insect which you see has undergone a transformation as singular and surprising, though varied in many of its circumstances. That active little fly, now an unbidden guest at your table, whose delicate palate selects your choicest viands, one while extending his proboscis to the margin of a drop of wine, and then gaily flying to a more solid repast from a pear or a peach; now gamboling with his comrades in the air, now gracefully currying his furled wings with his taper feet, was but the other day a disgusting grub, without wings, without legs, without eyes, wallowing, well pleased, in the midst of a mass of excrement.
>
> Kirby & Spence (1846).

The approach used to report life-cycle phenomena by various sciomyzid researchers throughout the world has been quite similar in style and terminology, thus enhancing the comparative value of the reports. The rather consistent and similar approach used by C. O. Berg and his 14 graduate students and by J-C. Vala and his six graduate students and their co-authors, which accounts for most of the publications on life cycles and also that used in this book, is similar to that recommended by Danks (2000) (Figs. 9.1, 9.2). We note the following differences from Danks (2000).

(1) We do not recognize a free "prepupal stage" after larval feeding ends, as shown in Danks's generalized scheme of the visible stages of development in a typical endopterygote insect (Fig. 9.1). See also Fig. 9.45. We use the term prepupa as used by Denlinger & Ždárek (1994), i.e., the stage between pupariation and larval-pupal apolysis. While most sciomyzid larvae form the puparium (pupariate) when feeding ceases, some continue as mature larvae outside the snail shell for several days to several weeks before pupariating. This may include a "wandering phase," noted for some other brachycerous larvae, or a quiescent larval or larval/pupal phase, of overwintering in the case of long durations.

(2) For Sciomyzidae, we have no evidence to indicate that the period between mating and oviposition is the main time of dispersal (as shown in Fig. 9.2). Dispersal of Sciomyzidae could well occur between hardening of the exoskeleton and mating, during the oviposition period, during seasonal or aseasonal times when the habitat and snail populations are stressed, and/or also could result largely by dispersion of floating puparia by seasonal flooding for many aquatic and semi-aquatic species.

(3) We do not describe one complete generation as being from 50% of oviposition from one generation to the next, but rather as from emergence from the puparium of one generation to emergence of the next.

Most Sciomyzidae appear to have simple life cycles in terms of development, responding to only a few influencing elements such as temperature, photoperiod, food, moisture, rainfall, etc., whereas a few appear to have complex life cycles subject to a complicated series of controls. However, for even the best-studied species – *Ilione albiseta*, *Sepedon fuscipennis*, and *Tetanocera ferruginea*, all aquatic predators – the nature and organization of the integrated systems that probably are involved have not been studied extensively enough in a coordinated manner under controlled conditions for us to apply this statement globally. Much of the information on sciomyzid life cycles could be criticized as resulting from piecemeal investigations. As Danks (1991) concluded, "a majority of entomologists continue to analyze the individual elements of life cycles, such as diapause responses to photoperiod, in isolation," and "For life cycles subject to complicated series of controls, such analysis provides only static glimpses of ongoing processes of development and tends to be of limited help for the interpretation of life cycles in nature." Indeed, the life cycles of many insects can be exceedingly complex, for example, no less than 18 alternate life-cycle pathways are known for the egg stage alone of the Australian grasshopper *Chortoicetes terminifera* (Wardaugh 1986).

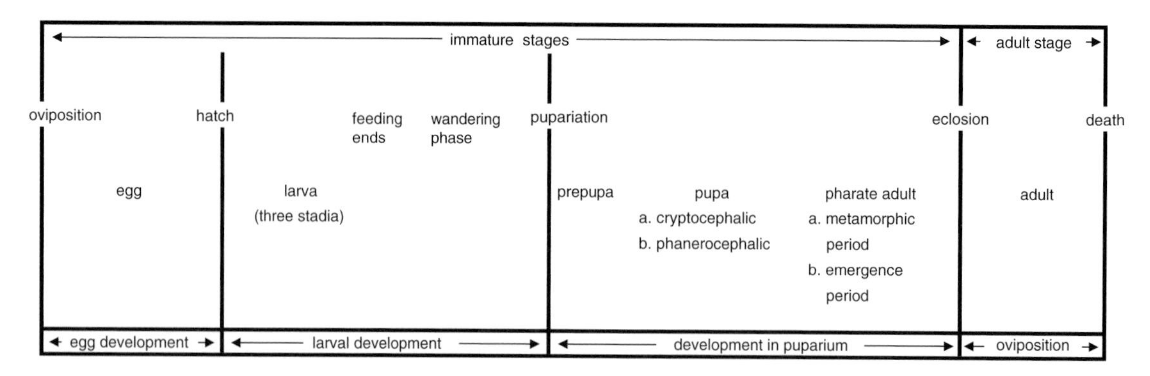

Fig. 9.1 Diagram of visible stages of development in a typical endopterygote insect, emphasizing immature stages. Modified for Schizophora from Danks (2000). See also Fig. 9.45.

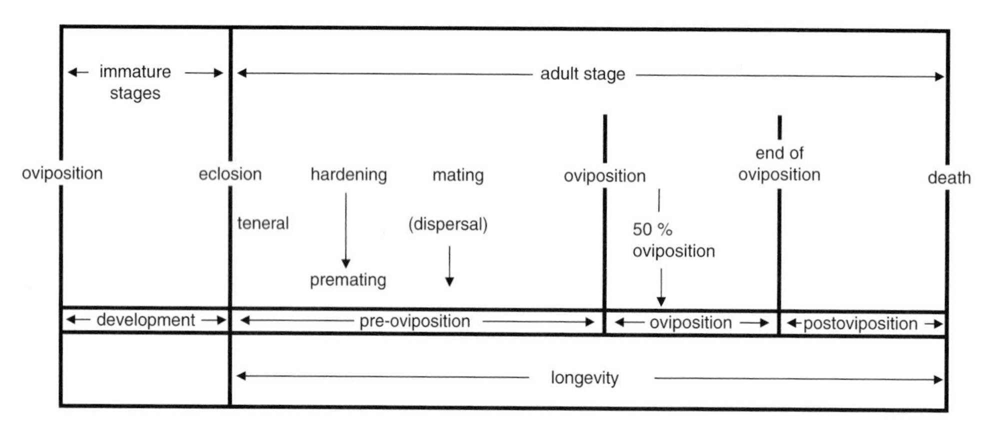

Fig. 9.2 Diagram of visible stages of development in a typical oviparous insect, emphasizing adult stage. Modified from Danks (2000).

The most obvious differences in development among Sciomyzidae are in (1) voltinism – most are multivoltine, many are univoltine, and a few appear to be bivoltine, (2) overwintering stage – most in the puparium, some as adults, a few as larvae, and very few in the egg membranes, and (3) the stage in which diapause, if present, occurs. Diapause is most common within the puparium, less so in the adult, and uncommon in the egg or larval stage, with pupae of some multivoltine species apparently having the ability to diapause or not to diapause. Further studies are needed of the token stimuli responsible for the induction, duration, and termination of diapause; diapause in only some individuals of a population; variation in rates of development; and the effects of photoperiod, temperature, moisture, and larval instar. Such studies would be enhanced if carried out in the framework for analyzing complex life-cycle pathways as proposed by Danks (1991). This involves flowcharts that identify the options available to a given species and that show how the individual elements are controlled. Especially prime subjects among the Sciomyzidae would be those species that show mixed phenological behaviors as noted for *Tetanocera vicina* by Berg *et al.* (1982). Other prime subjects would be species living in habitats that experience profound seasonal changes, such as temporary aquatic habitats (e.g., *I. albiseta*) and other habitats that experience different conditions from year to year. Since the timing of some of these seasonal and annual changes are largely unpredictable, complex systems of development that allow great flexibility of response might be expected in the species occurring in such habitats. We note that it has often turned out that as life cycles are more intensively studied, they are found to be more complex.

9.1 PHENOLOGY

> If we again consider each species in different climates we shall find obvious varieties both as regards size and form; all are influenced more or less strongly by climate. These changes only take place slowly and imperceptibly; the great workman of Nature is Time: he walks always with even strides, uniform and regular, he does nothing by leaps; but by degrees, by gradations, by succession, he does everything; and these changes, at first imperceptible, little by little become evident, and express themselves at length in results about which we cannot be mistaken.
>
> Buffon (1749) (translation quoted from Dendy, 1914).

Parallel to the diverse adaptations for habitat and host/prey selection and for feeding behavior in Sciomyzidae are the diverse phenological adaptations for the effective use of time in a seasonal sense (Danks 1993). This timeliness enhances larval feeding and thus development of as many generations per year as appropriate to the constraints of weather, habitat conditions, and host/prey availability.

Phenology has been a focal point in publications on life cycles of Sciomyzidae. Since this aspect is critical in working out further life cycles, in understanding the evolution of species, and in biocontrol efforts, it is discussed here in some detail. See Section 9.1.7 for background information on dormancy, in general. The data available as of 1982 on many species living in "cool temperate latitudes" of North America and Western Europe (essentially above 40° N. Lat. in North America and above 30° N. Lat. in Western Europe) and a few in New Zealand were analyzed by Berg *et al.* (1982), with the addition of much new information. They established five groups of voltinism and overwintering, noted below. The subject was one of Prof. Berg's special interests, as seen in his personal but scientific discussion in the 1982 paper, a *Festschrift* for the late C. W. Sabrosky. Many of the data analyzed were on the many species easily available in the Ithaca, New York area where Berg and his students had carried out field and laboratory studies throughout the year during the previous 30 years, along with data from less extensive field work in Alaska, Montana, Ohio, and Pennsylvania, as well as from field work in Western Europe and New Zealand.

9.1.1 Duration of life cycles

There are no clear indications that any sciomyzid has a life cycle extending more than one year. However, there are a few species in which members of the same cohort (i.e., larvae developing from eggs laid on the same day) may include individuals, during late summer to fall, that overwinter, apparently in quiescence, in the puparium or that overwinter as adults. In regard to the overwintered adults, these possibly have a life cycle that is up to a few months longer than one year. This would depend on the survival of the overwintered adults. This likely is the case in the late fall generation of the multivoltine *Pherbellia cinerella* at low elevations in mid to southern latitudes of its range in Europe. It is known to overwinter, apparently in quiescence, in the puparium in northern Europe (Bratt *et al.* 1969) and as an adult in southern Europe (Vala 1984b). The capture, marking, and release of a female *Sepedon fuscipennis* at Ithaca, New York, in August and her recapture in a reproductively active condition the following May (Arnold 1978b) also may indicate, again depending upon survival of adults, a life cycle slightly longer than one year in that and other northern hemisphere *Sepedon* species. Furthermore, some species in Phenological Group 6 may have a life cycle slightly more than one year.

9.1.2 Overwintering

Berg *et al.* (1982) summarized that multivoltine species overwinter primarily in the puparium, but some overwinter as adults, and univoltine species overwinter as embryonated eggs, partly grown larvae, or in the puparium. They concluded that multivoltinism with overwintering in the puparium is the most common and widespread behavior. They recognized that some species, especially in their Group 5, are labile and have the ability to develop seasonally in alternate ways.

It is not known if Sciomyzidae survive freezing temperatures by being freezing tolerant, i.e., ability to tolerate formation of extracellular ice in the body, or freezing intolerant, i.e., ability to prevent freezing by depressing the supercooling point of their body fluids through synthesis of antifreeze or cryoprotectant compounds such as glycerol and removal of extra-cellular nucleating agents. It is likely that both general mechanisms function within the family. Essentially nothing is known about some elements involved in overwintering in cold areas, for example, water relationships and cold injury. However, survival seems to be rather high in all four overwintering stages (primarily in the puparium but also by many species as adults and a few as embryonated eggs or free larvae).

There is little evidence that overwintering larvae in temperate climates feed during winter. Feeding during the winter probably is slight since food in the gut would act as ice nucleation sites when the temperature drops to freezing. However, Knutson *et al.* (1975) reported that second- and third-instar larvae of the aquatic predator *Pherbina coryleti* killed and ate snails at 5 °C during laboratory rearings. Other evidence indicates that *P. coryleti* is a univoltine species that overwinters as third-instar larvae. Foote (1999) reported that a first-instar larva of *Tetanocera latifibula* consumed two small *Physella* while being held in a refrigerator, without mentioning the temperature.

Puparia of many aquatic and semi-aquatic predators in cold-winter areas are often found at the water surface during periods of ice melt in the winter; those of most species probably survive being frozen in the ice cover. Foote (1999) collected six puparia of *Tetanocera ferruginea* on February 10 at Ithaca, New York, when ice covered the marsh habitat. As the ice sagged under the considerable weight of the collector,

> . . . water collected on top of it, and six puparia. . . were found floating in this ice. Undoubtedly, the water that was forced up onto the ice as the ice was depressed must have escaped through the holes that had formed around the stems of emergent vegetation, as melting occurred first around these dark objects. Puparia typically float into contact with any object which thus breaks the surface film, and they must have been concentrated in such situations before the marsh froze.

Adults emerged from these puparia in the heated laboratory. Valley & Berg (1977) collected a puparium of *Dictya atlantica* "while still frozen in a thin crust of ice on 22 March" in central New York. It produced an adult on March 29 in the heated laboratory. As another example, we collected a puparium of *Tetanocera arrogans* on March 8, 2006 at 1295 m in a small roadside marsh in southern Italy by pushing the ice cover below the surface; a male emerged 12 days later after being held at room temperature.

While most sciomyzid larvae that pupariate out of the water probably seek sheltered overwintering sites where changes and extremes of temperature are buffered, some early emerging species may seek sites that are among the first to be warmed during the early spring. A long, non-feeding pause during the third larval stadium of some sciomyzids, e.g., *Coremacera marginata* and some other terrestrial predators in Phenological Group 5b (Vala 1984b), may be an adaptation for wandering to select exposed sites for pupariation and overwintering, i.e., sites that warm up quickly during the spring rather than deeply protected sites. Selection of such overwintering sites is known to be the case in many arctic insects (Danks *et al.* 1994). As noted above, the extended third stadium also may be important in providing sufficient time for the photoperiod-sensitive structures in the brain of the larva to gather information on the seasonal change in photoperiod and thus to trigger diapause mechanisms.

Many semi-aquatic and terrestrial Sciomyzini, including both univoltine and multivoltine species, routinely form their puparia in the shell of the host/prey snail. This behavior, especially in those species of *Pherbellia* and *Colobaea* that also produce a calcareous septum occluding the opening of the shell (all northern hemisphere species), may be an adaptation for overwintering as well as for protection, with the shell serving as a sort of cocoon. The routine ejection of decayed tissue remaining in the shell by such pupariating larvae may to be remove sources of ice nucleation as much as to remove sources of growth of harmful fungi and bacteria or material that might attract natural enemies of the pupa.

In regions with cold winters, many of the species that overwinter as adults, for example, most *Sepedon* and *Elgiva* species, *Pherbellia schoenherri*, and *P. cinerella* in the southern parts of its range, can be collected on warm days during the winter. Their apparent "basking" behavior likely enhances the rate of egg development.

9.1.3 Classification of Berg *et al.* (1982)

The five groups proposed by Berg *et al.* (1982) with the representative species included by them are summarized below. These groups as originally diagrammed by Berg *et al.* (1982), with the additional Group 5b proposed by Vala (1984b), are shown in Fig. 9.3. The new Group 6, proposed here, is shown in Fig. 9.5. Vala (1989a) presented another format for portraying phenology (Fig. 9.4) that allows more detail to be included.

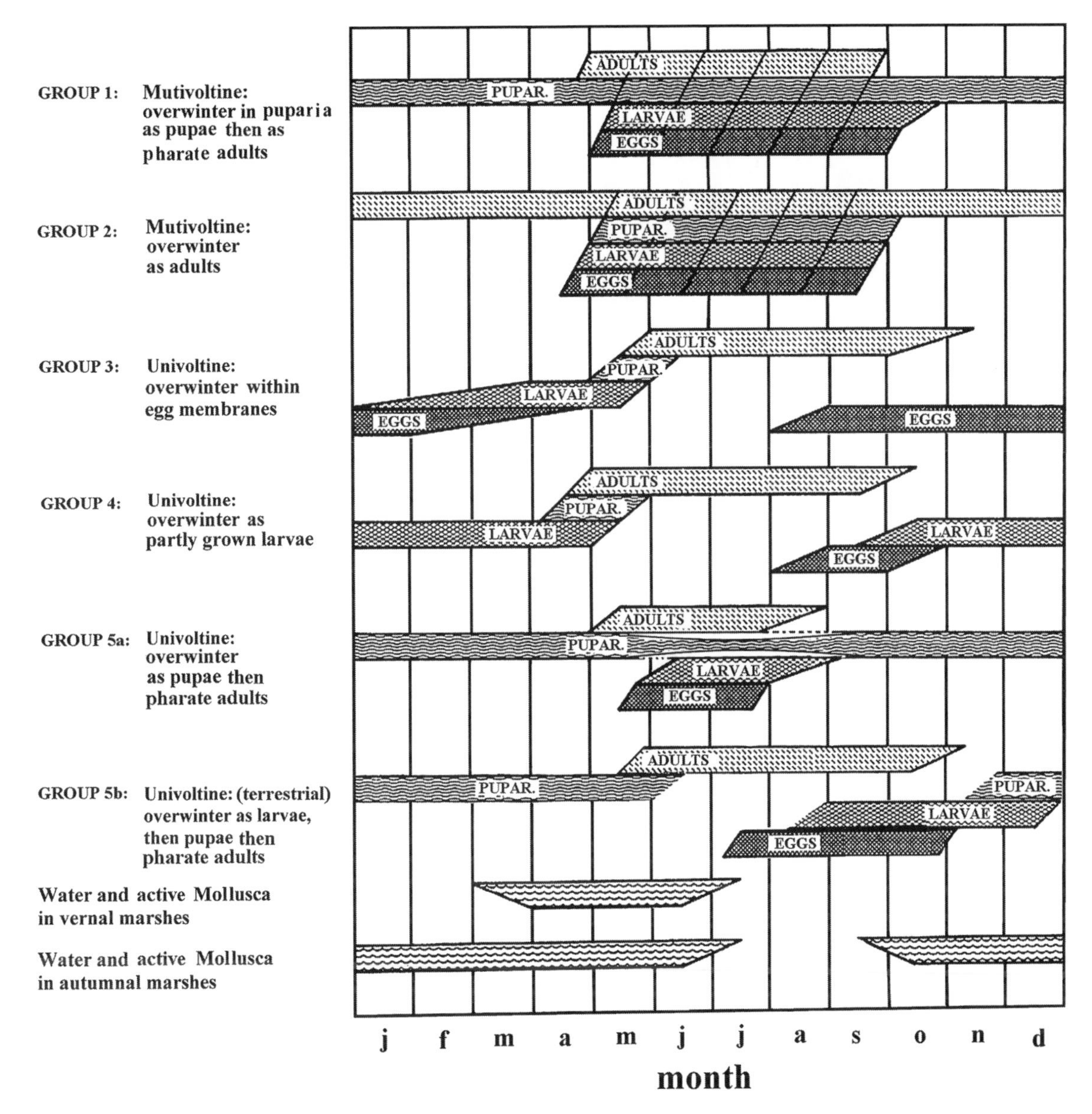

Fig. 9.3 Seasonal occurrence in North Temperate zone of stages in life cycles of phenological groups 1, 2, 3, 4, 5a, 5b of Sciomyzidae, and of water, snails, and fingernail clams in temporary ponds and marshes. Modified from Berg *et al.* (1982).

Group 1: *Multivoltine species overwintering in the puparium as diapausing or quiescent pupae or pharate adults.* The puparial stage is found throughout the year. The overwintering stage ranges from very young, unpigmented pupae to pharate adults in the puparium. Pupae or pharate adults of some species are in diapause; those of others are simply quiescent. Adults emerge during early spring and fly between April and October, producing three to five successive generations. Larval stages last from May to November. The first generation is often concomitant with reproduction of gastropods in the habitat. Included are many aquatic and terrestrial species of both tribes of Sciomyzinae.

Group 2: *Multivoltine species overwintering as diapausing adults.* Reproductive diapause, at least in some species, is corroborated in the female with reduced ovaries and accessory

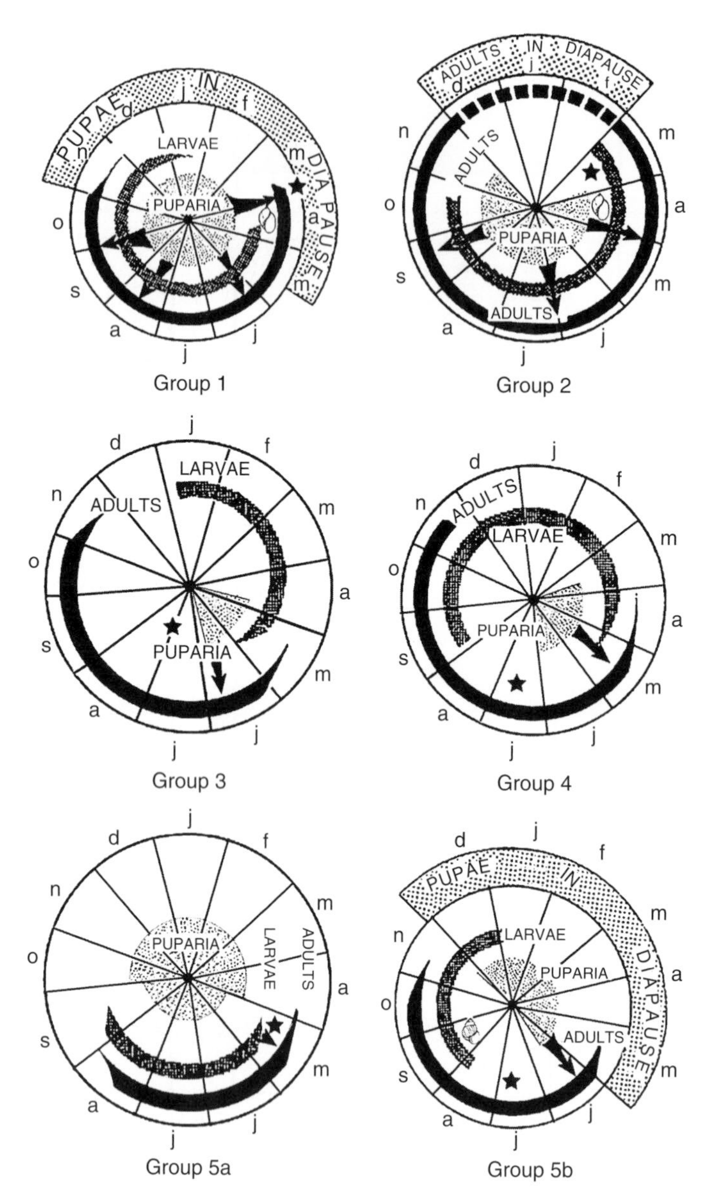

Fig. 9.4 Phenological groups of Sciomyzidae with addition of Group 5b. Star, approximate time when first eggs are laid; arrow, main emergence of adults. Modified from Vala (1989a).

glands and hypertrophied fat bodies, and in the male with slightly developed testes. The generations succeed one another during spring and summer as in Group 1. Included are many aquatic predators of the genera *Sepedon* and *Elgiva*, the terrestrial parasitoid *Pherbellia schoenherri*, the terrestrial predator *P. cinerella* in southern parts of its range, and possibly *Psacadina*. Three sciomyzids have been collected on snow: *Pherbellia schoenherri maculata* (November 20 at Plummer, Minnesota; Bratt *et al.* 1969); *P. s. schoenherri* (December 31 at Baerum, Akershus, southern Norway; Hågvar & Greve 2003), and *P. cinerella* (December 28, 1979 near Avignon, southern France; Vala 1984b). But, see reference in Section 9.3.2 to the possibility that *Sepedon spinipes* overwinters as a quiescent larva, pupa, or adult.

Group 3: *Univoltine species overwintering within egg membranes.* The first larval stadium, within the egg

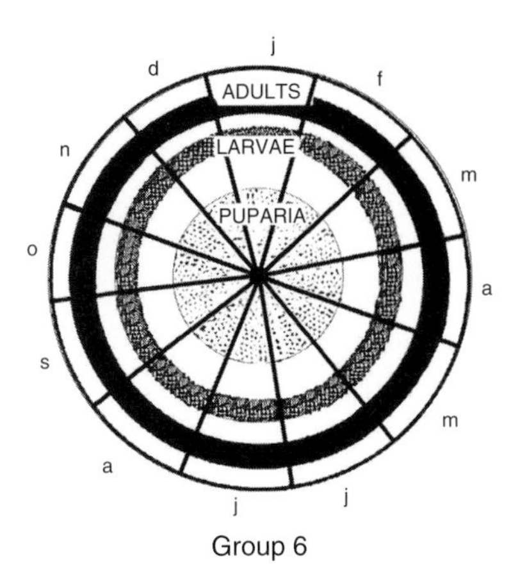

Fig. 9.5 Phenological Group 6, proposed mainly for tropical Sciomyzidae.

membrane, undergoes diapause, extending this stage to several months. Adults experience a reproductive diapause during spring and early summer. Included are *Tetanocera latifibula*, *T. montana*, *T. loewi*, *T. soror*, possibly *Ilione albiseta*, and *Hedria mixta*.

Group 4: *Univoltine species overwintering partly in the larval stage.* Adults experience an estival diapause, after which eggs are laid. They hatch promptly, and larvae begin to develop before winter. This Group was based primarily on *Tetanocera vicina*, along with *T. plumosa* and *T. obtusifibula*. Some other species in this Group show minor variations or have Group 3 or 4 features, depending upon local conditions. *Pherbina coryleti* adults mate in spring and early summer, but oviposition is delayed for several months. *Eulimnia philpotti* in New Zealand mate and oviposit during spring and early summer, and the incubation period is short. Berg *et al.* (1982) concluded, on the basis of laboratory studies and limited field data, that *Ilione albiseta* and *I. lineata* follow pattern 3 or 4, depending upon the availability of food and water, and that they follow pattern 4 in warmer latitudes. However, subsequent rearings through the complete life cycles of *I. lineata* (Knutson unpublished data) and *I. albiseta*, with extensive field data on the latter (Lindsay 1982, Lindsay *et al.* 2011), have essentially confirmed these species as typical members of Group 4.

Group 5: *Univoltine species overwintering as pupae.* Puparia are formed from June until fall, depending upon the species; most have a pupal diapause lasting until the following spring. Adults are active and oviposit from May through August. Included are many species of *Anticheta*, *Renocera*, and *Pherbellia* living in seasonally aquatic sites and feeding on snail eggs, fingernail clams, and aquatic snails, respectively, and six species of *Pherbellia*, *Oidematops*, and *Tetanura* attacking terrestrial snails. Berg *et al.* (1982) referred to this group as "a heterogeneous assemblage of species that apparently have become univoltine in response to quite different evolutionary pressures." We believe that the terrestrial species in this group are better placed in Group 5b (see below).

9.1.4 Additions and modifications

Several species in cool-temperate latitudes that were subsequently reared fit into the above groups, as noted below.

Group 1 – Four Nearctic *Tetanocera* preying upon succineid snails show no evidence of diapause in the pupal stage (Foote 1996b). Barnes' (1990) studies of *Sciomyza varia*, one of the most specialized parasitoids, confirmed Berg's (1964) supposition that the species overwinters as diapausing pupae and is multivoltine.

Group 2 – Laboratory rearings (Knutson 1988) indicate that *Pherbellia griseicollis*, a non-host-specific, semi-terrestrial predator/saprophage, may belong to this group but probably is univoltine.

Group 5 – Puparia of the poorly known *Poecilographa decora* collected during June produced adults a week or two later. Adults were collected in early July. Mating and oviposition quickly followed, but the incubation period was prolonged; first-instar larvae did not feed on the wide variety of molluscs offered to them (Barnes 1988).

More recently, relatively long-term ecological studies have been conducted throughout the year in southern France around Avignon by Vala *et al.* and around nearby Montpellier by Coupland (both at about the same latitude as Ithaca, New York). This is a Mediterranean area, usually with long, hot, dry summers but wet, cool winters. The semi-aquatic predator *Pherbina mediterranea* fits well in Group 4 (Fig. 9.6a), not in Group 3 as stated by Vala & Gasc (1990a). The flight period of the multivoltine terrestrial predator *Pherbellia cinerella*, widely distributed from Swedish Lapland through the Near and Middle East to Afghanistan, apparently is dependent on local climatic conditions, with overwintering in the quiescent pupal stage in northern parts of the range (Group 1) (Bratt *et al.* 1969) and in the southern parts of the range as adults that become active when the temperature increases above 10 °C for several days (Group 2) (Vala 1984b, 1989a). Vala (1989a) collected

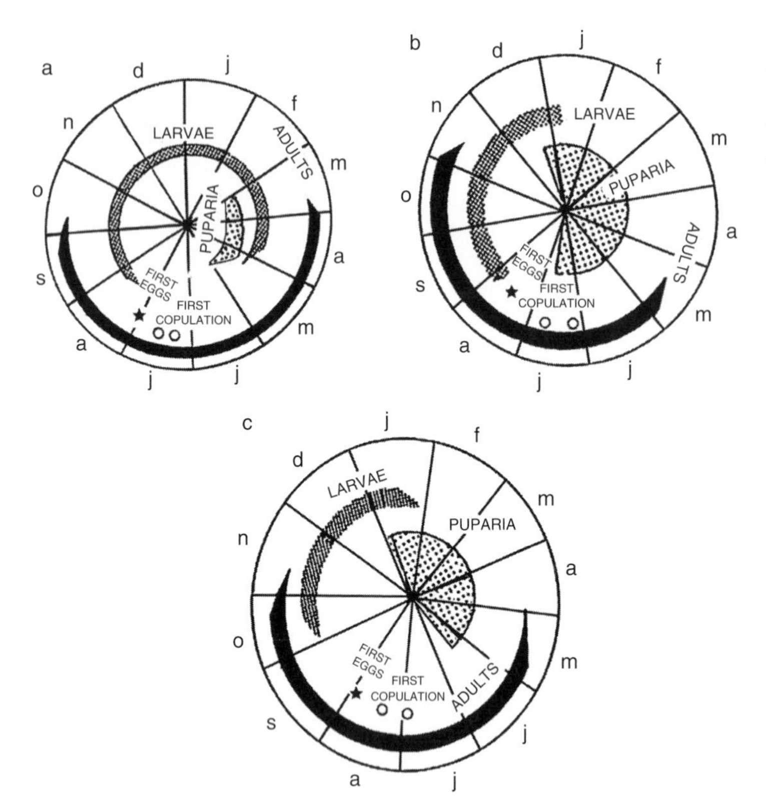

Fig. 9.6 (a) Phenology of ***Pherbina mediterranea*** (from Vala & Gasc 1990a). (b) Phenology of ***Euthycera stichospila*** (from Vala & Caillet 1985). (c) Phenology of ***Trypetoptera punctulata***. From Vala (1986).

adults during all months of the year (1979–1984) near Avignon, but Coupland & Baker (1995) did not find them during July and August (1991–1992) about 100 km southwest of Avignon. Two males were collected on December 28, 1979 near Avignon when the ground was covered with 40 cm of snow (Vala 1984b). The successful and widespread but poorly known *Limnia unguicornis*, a predator/saprophage on semi-terrestrial snails and slugs, apparently is univoltine and overwinters as a diapausing pupa (Vala & Knutson 1990).

Several univoltine parasitoids/predators/saprophages of terrestrial snails studied in southern France do not fit well into Group 5 of Berg *et al.* (1982). These are *Coremacera marginata* (Knutson 1973), *Dichetophora obliterata* (Vala *et al.* 1987), *Euthycera cribrata* (Vala *et al.* 1983), *E. stichospila* (Vala & Caillet 1985, as *E. leclercqi*; Fig. 9.6b), and *Trypetoptera punctulata* (Vala 1986; Fig. 9.6c). These species are univoltine with exceptionally long pre-oviposition periods, larval period from late summer or early fall to mid-winter, and overwintering is completed as diapausing pupae. Vala's (1984b) 5-year sampling of adults of these species in an oak forest near Avignon and Coupland & Baker's (1995) 2-year sampling of adults of all except *E. stichospila* 100 km southwest of Avignon also provide phenological information. Vala (1984b) proposed Group 5a for Nearctic and northern Palearctic species originally included by Berg *et al.* (1982) in Group 5. He proposed Group 5b for southern Palearctic species having phenologies like the five species described above. We also include the six terrestrial Nearctic and Palearctic species placed in Group 5 by Berg *et al.* (1982) in Group 5b, and possibly *Salticella fasciata*, which they placed in Group 1.

Vala & Manguin (1987) studied the seasonal dynamics as well as species richness and diversity of 14 species of aquatic, semi-terrestrial, and terrestrial species in and around a temporary aquatic habitat near Avignon over a 3-year period (Fig. 9.7). Figures of seasonal variation in flight period and abundance of eight multivoltine species (Fig. 9.8) and four univoltine species (Fig. 9.9) in several kinds of habitats near Avignon were presented by Vala (1989a).

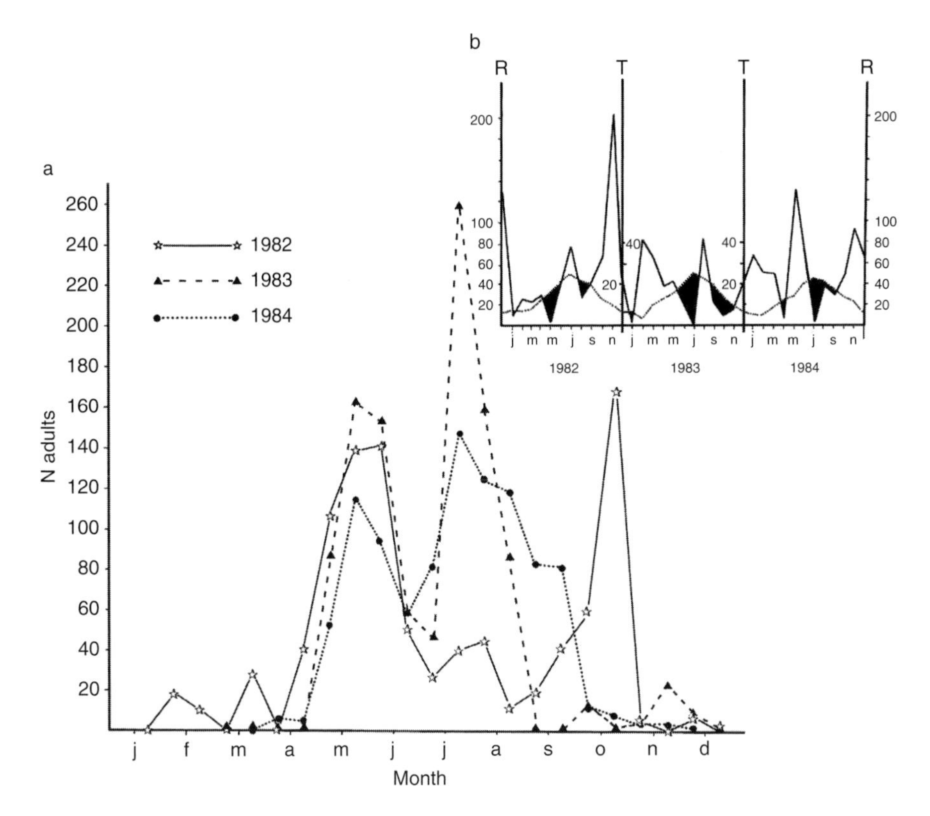

Fig. 9.7 (a) Seasonal variation of 14 species of Sciomyzidae in and around temporary aquatic habitat near Avignon, southern France during 1982–1984; (b) Ombrothermic curves during this period in this area; R, rainfall (solid line) in mm; T, temperature (dotted line) in °C. From Vala & Manguin (1987).

In southern California (32°–36° N. Lat., Mediterranean climate), Fisher & Orth (1983) made monthly collections between 1962 and 1966 at four locations (elevations of 76, 259, 1372, and 2730 m). Eight of the 17 species occurring there (of 49 species known from California) probably breed continuously throughout the year in southern California and other warm parts of their ranges. Adults of these species were collected during nearly every month in southern California: *Atrichomelina pubera* (February 19 – December 7), *Pherbellia nana nana* (March 10 – December 22), *P. trabeculata* (February 13 – November 13), *Ditaeniella parallela* (January 14 – December 27), *Dictya montana* (January 1 – December 31), *Hoplodictya acuticornis* (January 9 – December 27), *Sepedon bifida* (January 14 – December 27), and *S. pacifica* (January 16 – November 13). These species are all widespread across North America or at least from the Mississippi River Valley westward, except the two *Sepedon* species, which have more restricted western ranges, and *P. trabeculata*, which ranges from southwestern USA to Guatemala. Of the species that range into cooler areas, all are known to pass the winter there in the puparium, except for the two *Sepedon* species, which probably overwinter as adults.

Rozkošný (1997a) summarized the phenology of northern European Sciomyzidae and placed the species in the five groups of Berg *et al.* (1982). Capture records of 87 species of Palearctic Sciomyzidae from museum specimens were used by Soós (1958) to attempt to determine flight periods and numbers of generations per year. He recognized three groups: (1) eurychrone species with several generations and an unbroken flight period; (2) stenochrone species with short, distinct flight periods and one generation per year; and (3) non-eurychrone species but with several generations per year. Some authors do not agree with his conclusions in regard to the numbers of generations per year, for certain species, but Rozkošný (1962) quantitatively collected 28 of these species (261 specimens) in four localities in Moravia by sweep net, and his data supported Soós's proposal. See Chapter 13.

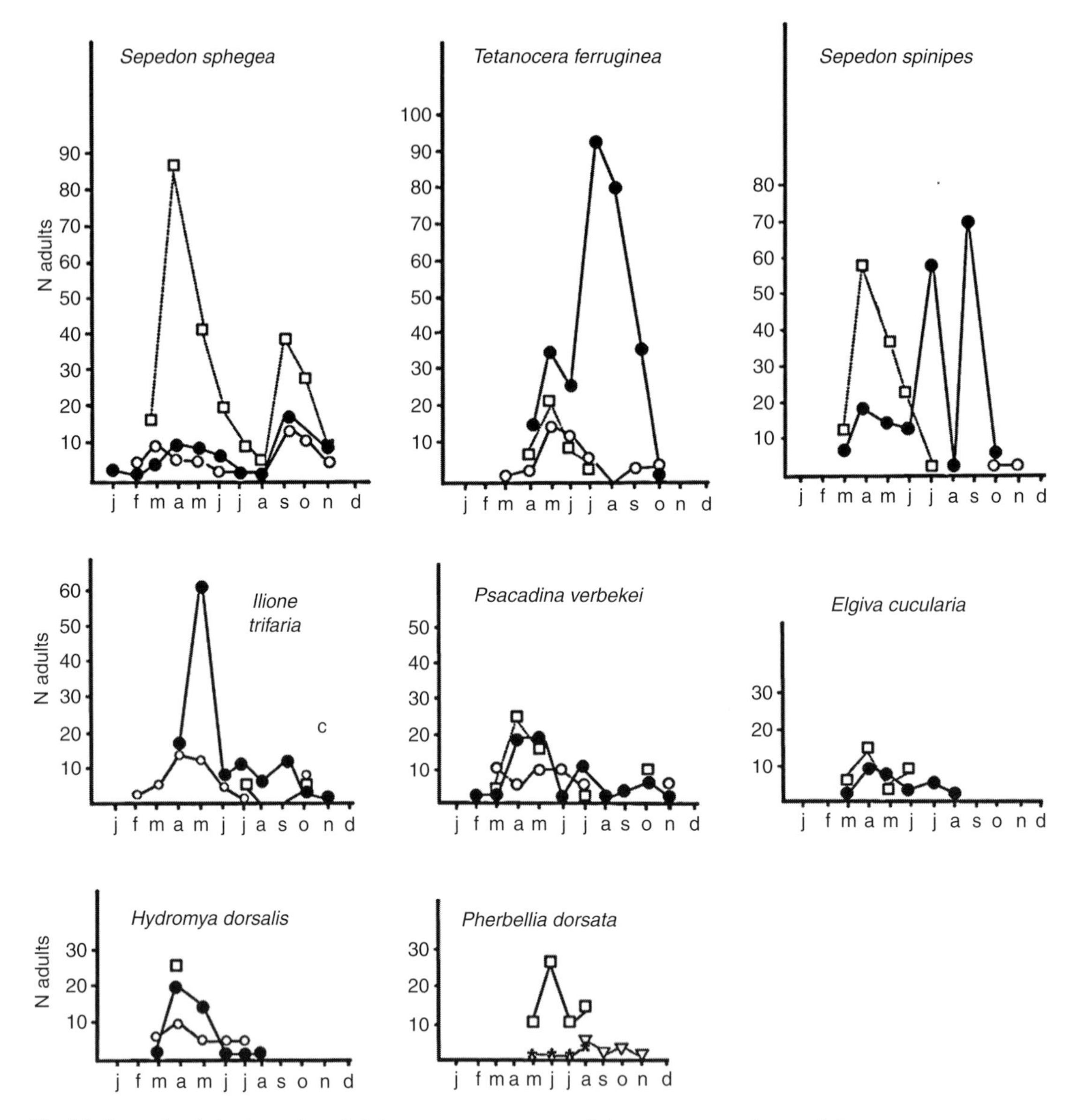

Fig. 9.8 Seasonal variation in number of adults of some multivoltine Sciomyzidae in various habitats near Avignon, southern France. Square, marshes; black circle, temporary canal; open circle, permanent canal; triangle, cultivated pasture; star, natural meadow. From Vala (1989a).

9.1.5 Phenologies in south temperate latitudes

Neolimnia was studied throughout the year, over 2½ years, in New Zealand (Barnes 1979b). Three species of the subgenus *Pseudolimnia* preying upon aquatic non-operculate snails and one attacking an aquatic operculate snail were found to be multivoltine, without developmental or reproductive diapause. Adults, all three larval instars, and puparia were collected during all seasons; thus these species fit into our new Group 6 (see below), which otherwise includes mostly tropical species. Four species of the subgenus *Neolimnia* are predatory on terrestrial snails; adults appear suddenly, during November, but overwintering and voltinism are unknown.

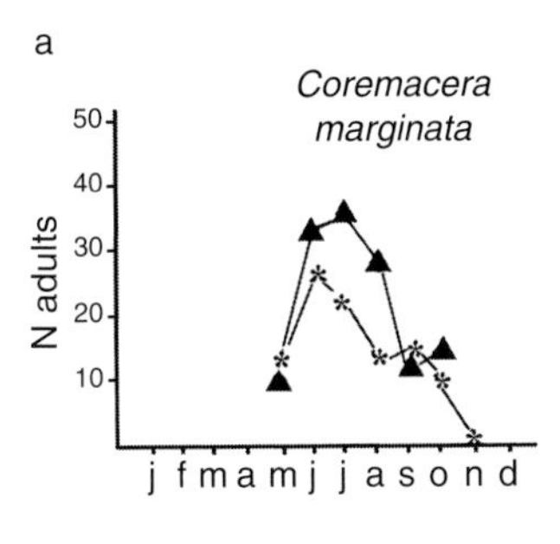

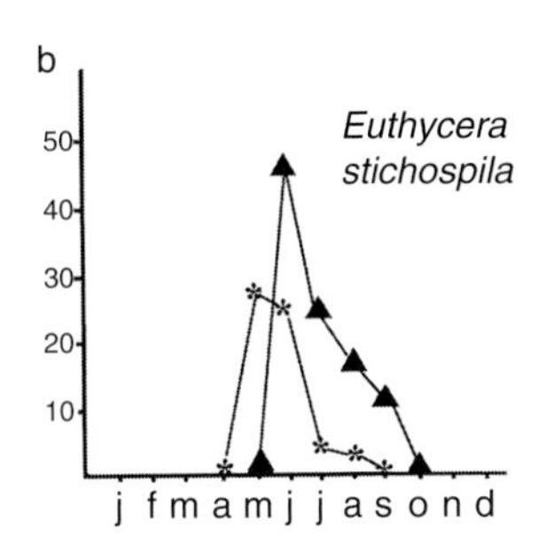

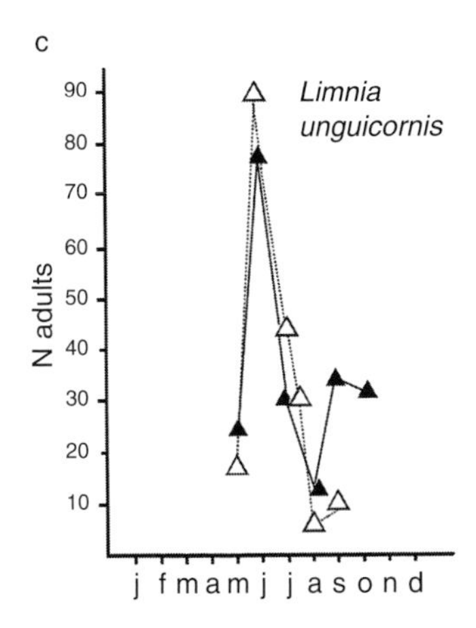

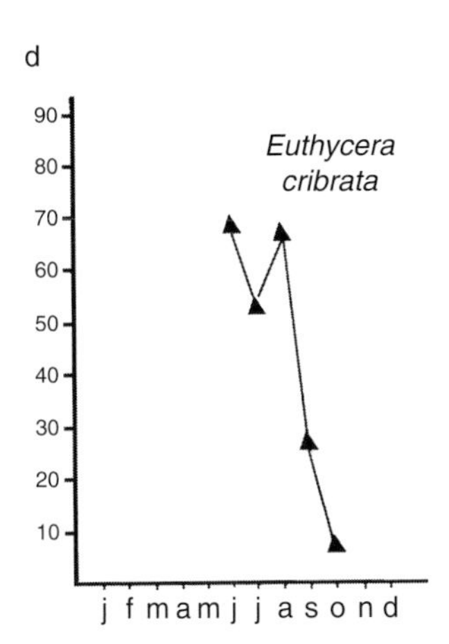

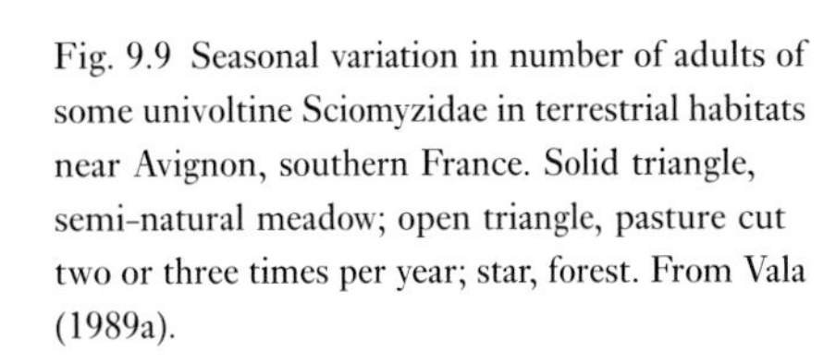
Fig. 9.9 Seasonal variation in number of adults of some univoltine Sciomyzidae in terrestrial habitats near Avignon, southern France. Solid triangle, semi-natural meadow; open triangle, pasture cut two or three times per year; star, forest. From Vala (1989a).

Two cool-adapted Neotropical species, *Perilimnia albifacies* (western and southern South America) and *Shannonia meridionalis* (central and southern Chile), both Tetanocerini predators/parasitoids/saprophages of aquatic snails, seem to have similar phenologies, breeding continuously, lacking diapause in any stage, and adults being active throughout the year (Kaczynski *et al.* 1969). These species also fit into Group 6.

In New South Wales and South Australia (Mediterranean climates), field studies over several years of the aquatic predators *Dichetophora biroi* and *D. hendeli* (Boray 1964, 1969; Fig. 9.10 and Lynch 1965; Fig. 9.11), indicated that their phenologies are considerably labile. Larvae were found primarily during spring, shortly after *Austropeplea tomentosa* snail populations peaked and especially where water levels were decreasing. Adults were found throughout the year. The entire life cycle took only 21 days.

Barraclough (1983) collected adults of five species of *Sepedon* (the freshwater predators *S. neavi* and *S. testacea*, the biologically unknown *S. pleuritica*, and two unidentified species) at three sites south of Pietermaritzburg (Natal), near the southeastern coast of South Africa. All five sciomyzid species were collected only during the warmer months, especially 1–2 months after the annual rains (April 11 – June 19 and January 15 – February 18) at an impounded section of the Mpushini River near Pietermaritzburg, with diverse vegetation and where the freshwater snails *Physella acuta*, *Bulinus* sp., *Radix natalensis*, and *Biomphalaria* sp. were abundant. Many flies were collected during April (+50 per 100 sweeps in a 6 × 6 m area). The numbers rapidly decreased after mid May, and sciomyzids were absent by mid June. The seasonal occurrence of adults was quite different 20 km closer to the coast in a less arid and warmer region where three species, including *S. neavei*, were collected in large numbers between May 4 and June 19. At a shaded, muddy pool (an impounded area below a dam wall) only *R. natalensis* was present in large numbers. *Bulinus* sp. and *Biomphalaria* sp. were also present.

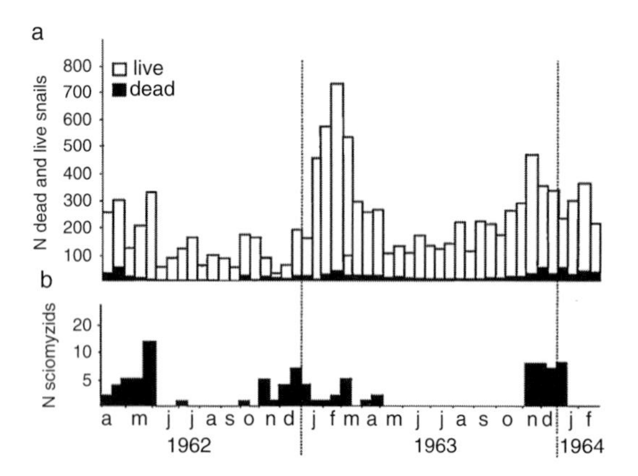

Fig. 9.10 Seasonal variation in populations of (a) *Austropeplea tomentosa* and (b) larvae of ***Dichetophora hendeli*** and ***D. biroi*** at Hampton, central tableland, New South Wales, Australia. From Boray (1969).

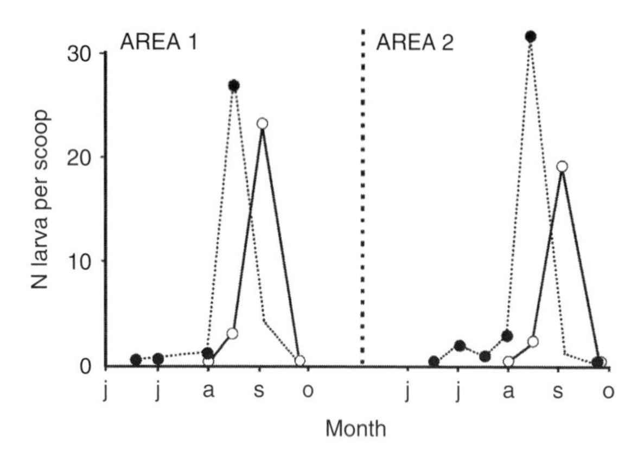

Fig. 9.11 Seasonal variation in populations of *Austropeplea tomentosa* and larvae of ***Dichetophora biroi*** in two experimental areas at Murray Lakes, South Australia. Black circle, number of snails per scoop; open circle, number of larvae per scoop. From Lynch (1965).

9.1.6 Phenologies in tropical zones and proposal of a new phenological group

Since most information on tropical Sciomyzidae pertains to aquatic and semi-aquatic predators of freshwater, non-operculate snails, here we refer almost entirely to these Behavioral Groups, with some remarks on terrestrial and semi-terrestrial forms included in the summary (Section 9.1.6.2) of information from the Afrotropical region.

Stereotyped phenology characterizes aquatic and semi-aquatic predators in tropical zones. They appear to be multivoltine, with a variable number of generations per year (perhaps 4–12) which are not discrete but are successive and overlapping. In laboratory rearings, these species show no indication of diapause, develop promptly, have a short pre-oviposition period, a long oviposition period, and short egg, larval, and pupal periods. For these similarly behaving species from warm areas of the Neotropical, Afrotropical, and Oriental regions we designate the new Group 6. As noted in Section 9.1.3, a few temperate zone New Zealand and Neotropical species also seem to fit into this Group.

As in temperate zones, life cycles of sciomyzids in tropical zones (essentially the area between the southern and northern hemisphere annual isotherms of 18 °C during the coldest months) must be adapted to the seasonal changes in their snail host/prey life cycles and populations, in this case to the changes primarily resulting from the wet/dry seasonal phenomena. Except for the Afrotropical *Sepedon ruficeps* (Gbedjissi *et al.* 2003), little study has been conducted of such adaptations among tropical Sciomyzidae. The extensive literature on the ecology of tropical snails, especially disease-transmitting species, provides a baseline of information relative to the study of seasonal adaptations of sciomyzids. A better understanding of these adaptations will be an important key to the utilization of Sciomyzidae as biocontrol agents in tropical zones.

Information is available from laboratory rearings of about 35 species, limited field collections of larvae and puparia, and collection records of adults in warm areas of the Neotropical, Oriental, and Afrotropical regions. Few year-round field studies have been conducted, and no summary has been published of phenologies of sciomyzid species in tropical zones. It is also difficult to generalize about number of generations per year, life-cycle periods, and stages adapted to survive inclement conditions for species in tropical latitudes. Such areas lack the very strong, area-wide seasonal changes in temperature and photoperiod characteristic of mid to extreme northern latitudes. As noted, seasonal adaptations probably are primarily in response to wet and dry periods, but these occur at different times of the year in different tropical areas, to variable extents, and they vary considerably from year to year in most areas (Walter 1985). The effects of aseasonal and localized habitat deterioration resulting from unpredictable dry conditions also needs to be considered (Tauber *et al.* 1984).

During dry seasons, adults of species whose larvae feed on aquatic and semi-aquatic snails possibly go into diapause or quiescence or migrate to residual wet places with the generations continuing essentially uninterrupted. When one considers that the adults of many species of tropical aquatic predators have been collected throughout the year, the latter scenario seems likely to be the more common. Larvae developing in aquatic to moist situations obviously are threatened by the onset of the dry season, and lacking ability to move significant distances, other than to follow retreating shorelines, may survive by pupariating and remaining in diapause or quiescence until favorable moist conditions resume. However, Bratt *et al.* (1969) noted that larvae of *Ditaeniella parallela* (southern Canada to Costa Rica), a parasitoid-predator of aquatic snails in moist situations, "were the most active and resistant to dessication [sic] of any [of 28 species of *Ditaeniella* and *Pherbellia*] reared. If not provided with enough snails, they often left the rearing dishes and wandered over the [dry] laboratory tables . . . Many larvae were found alive more than 1 meter from the rearing dishes." It seems unlikely that the egg stage is involved in dry-season survival in tropical zones.

Not only is there a lack of information about Sciomyzidae in regard to dry-season survival, but as Tauber *et al.* (1998) emphasized, "Of the major physical factors that influence insect seasonal ecology, moisture is least understood and least appreciated." They hypothesized that "moisture influences insect life cycles via one or more of three mechanisms – as a token stimulus for diapause [we add, also likely as a token stimulus for the termination of diapause], modulator of developmental or reproductive rates, or behavioral cue for vital seasonal events," similar to the effects of the much better-known phenomena of temperature and photoperiod regimes in northern latitudes. While the three experimental paradigms described by Tauber *et al.* (1998) pertain primarily to insects living in soil or other non-living solid media and which pass the dry season in a state of *larval* diapause, some of these paradigms might be adapted to the more free-living Sciomyzidae, with all life stages more or less exposed to ambient conditions. These authors also noted that in addition to elucidation of life cycles of individual species and gaining information that may be useful in selection of biocontrol agents, comparative studies on the mechanisms of response to moisture, if conducted within a phylogenetic setting, could help elucidate evolutionary pathways of adaptation to long-term climatic changes. This could be especially instructive in regard to the explosive speciation of the primarily aquatic predaceous genus *Sepedon* in the Afrotropical region and the unusual development of many endemic genera of aquatic predators in the Neotropical region.

As noted above, knowledge of the hosts/prey also will be important in improving our understanding of the phenology of tropical Sciomyzidae. Particular attention should be paid to comparison of permanent and temporary aquatic sites in the same area, as done, for example, in the studies of *Sepedon ruficeps* by Gbedjissi *et al.* (2003) in Bénin. Systematically collecting substrata from clearly defined snail microhabitats, then searching the material for puparia or holding the material for emergence of adult flies, as done by Przhiboro (2001) in the St. Petersburg, Russia, area would be important. Emergence traps also should be useful. Collecting snails estivating in cracks in the mud of drying aquatic habitats and from other protected sites, then examining them for larvae and puparia, not only may help elucidate life cycles but may help to identify species of Sciomyzidae that could be especially important as biocontrol agents, since such snails obviously escape attempts at control by molluscicides. Comparison of dates of collections of adults in areas where there has been extensive, year-round collections with climate diagrams, such as those of Walter (1985) also will be of interest.

A summary of seasonal data from the three major tropical regions is given below. This information is based largely on collections of adults whose larvae are mainly predators of aquatic snails. These data usually are associated with meteorological data. The two most likely scenarios that would explain adult collection data are as follows. One might expect that if the species has been estivating in puparia or as adults, then the adults would be most common for a short period after the beginning of the wet season, as snail populations build up, the moisture serving as a behavioral cue for emergence from puparia and for adult activity. With the short pre-oviposition and incubation periods recorded for tropical species in laboratory rearings, there would seem to be no advantage for emergence and adult activity at the very beginning of the wet season, when snail reproduction is just beginning. Alternatively, one might expect adults to be more common later during the wet season as a result of the build-up of their populations during the time when prey are most available. While such data on adult collections will be useful, especially if obtained by routine, regular sampling throughout the year over a several-year period, along with more precise meteorological data on specific habitats and data on snail populations, routine collections for immature stages throughout the year will be more conclusive.

9.1.6.1 SOUTHERNMOST NEARCTIC AND NEOTROPICAL SPECIES

Life-cycle data are available for 17 species of sciomyzids occurring in Florida and tropical Central and South America. Most are aquatic predators. These include *Dictyodes* (Abercrombie & Berg 1978), *Dictya* (Valley & Berg 1977), *Protodictya* (Abercrombie 1970), *Sepedomerus* (Neff & Berg 1966), *Sepedonea* (Neff & Berg 1966; Abercrombie 1970, 2000; Bredt & Mello 1978; Knutson & Valley 1978), *Thecomyia* (Abercrombie & Berg 1975), three semi-aquatic predators/saprophages: *Protodictya hondurana* (Neff & Berg 1961), *Ditaeniella parallela* and *Pherbellia trabeculata* (Bratt *et al.* 1969), and one parasitoid/predator on succineids: *Hoplodictya spinicornis* (Neff & Berg 1962). There is also biological information on ten species of three genera essentially restricted to the south temperate zone of South America: *Tetanoceroides* (Abercrombie 1970) and *Perilimnia* and *Shannonia* (Kaczynski *et al.* 1969).

A series of collections, primarily by sweep net, were made of the common (160 specimens) *Thecomyia longicornis* and the rare (12 specimens) *T. papaveroi* in a 0.7 ha^2 area of protected forest near Belém, Brazil on 49 dates throughout 1977 and 1978 (Knutson & Carvalho 1989). It was concluded that *T. longicornis* breeds continuously throughout the year. The seasonal distribution was compared with monthly temperatures and rainfall over the 10-year period (Fig. 9.12). Most adults were collected during two peaks just before (December) and at the height (March) of the rainy season. *Thecomyia longicornis* ranges across northern South America from Belém to central Peru, and north to northeastern Nicaragua. *Thecomyia papaveroi* is known only from Belém and another locality about 800 km from the mouth of the Amazon River. The only reared species of *Thecomyia*, *T. limbata*, bred continuously in the laboratory, without diapause. The entire life cycle required only 33–59 days; adults were collected (southern Brazil, Paraguay to Venezuela and Panama) during all months (most commonly, October to January); larvae were collected in May and June; and puparia were collected in May (Abercrombie & Berg 1978).

Mello & Bredt (1978b) made monthly collections of adults of five species of known or likely aquatic predators (*Sepedonea barbosai*, *S. canabravana*, *S. telson*, *S. guianica* as *S.* "*vau*" (*nomen nudum*), and *Sepedomerus bipuncticeps*) between 9 and 11 am during 1975 and 1976 by sweeping shoreline vegetation at Lagoa das Pedras, Formosa Co., 162 km northeast of Brasilia, Brazil. Graphs of their data on the three most common species are shown in Fig. 9.13. As in most data from Africa, population peaks occurred as the period of heaviest rainfall declined, but in some years adults of some species were found during every month.

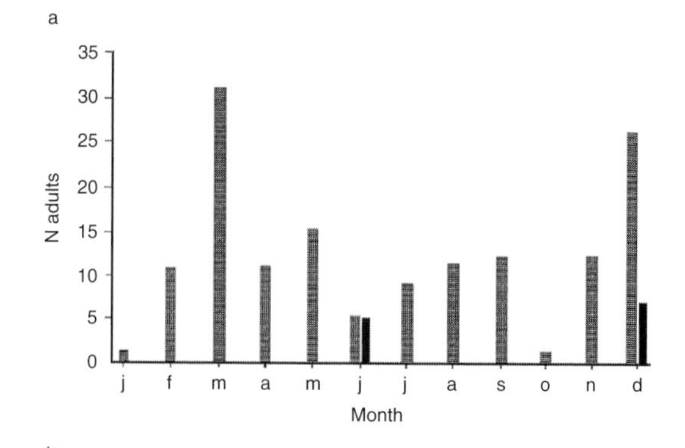

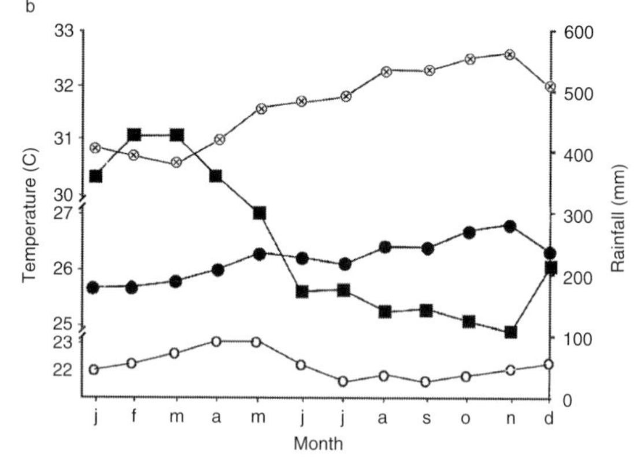

Fig. 9.12 (a) Seasonal variation of adults of ***Thecomyia longicornis*** (gray bars) and ***T. papaveroi*** (black bars) during 1961–1971, 1977–1979 at Belém, Pará, Brazil. (b) Black circle, average; open circle, minimum; circle with x inside, maximum monthly temperature; black square, average rainfall 1969–1979. From Knutson & Carvalho (1989).

9.1.6.2 AFROTROPICAL SPECIES

Seasonality of populations of aquatic Sciomyzidae in western Africa seem to be correlated also with the type of habitat, i.e., permanent or temporary aquatic as well as with rainfall and temperature. Gbedjissi *et al.* (2003) sampled aquatic snails and the common and abundant aquatic predator *Sepedon ruficeps* in permanent and temporary sites near the coast in Bénin, western Africa, from

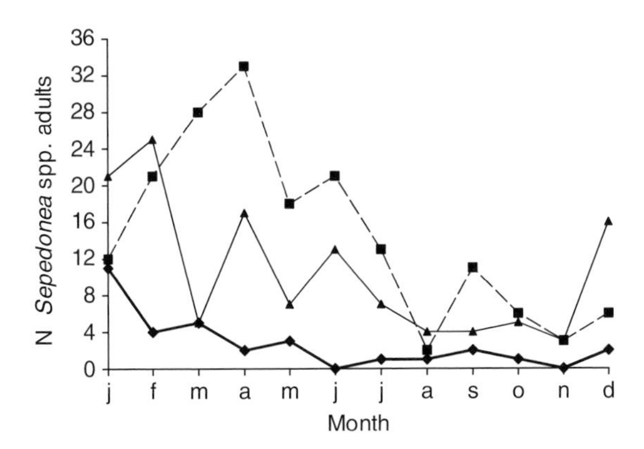

Fig. 9.13 Seasonal variation in population of adults of *Sepedonea barbosai* (triangle), *S. canabravana* (square), and *S. guianica* (diamond) collected near Brasilia, Brazil during 1975 and 1976. From Mello & Bredt (1978b).

March 1996 to January 1997 (Fig. 9.14). In a permanent site (Cotonou), *S. ruficeps* was present throughout the year, with populations peaking in September (Fig. 9.14b). In a temporary site (Cocotomey), adults were collected July through December, with populations peaking in October (Fig. 9.14c). Rainfall was highest during June, and at the permanent site surface temperature of the water was lowest during August through September (Fig. 9.14a). As would be expected, populations of *Biomphalaria pfeifferi* and *Radix natalensis* – two of the likely natural prey – at the permanent site were lowest before the rainy period and highest following the rainy period (Fig. 9.14d). Quite different peaks of abundance were seen for *Sepedonella nana*, a predator of aquatic oligochaetes (Fig. 9.15). For some tropical regions there probably is a period of limited egg and larval development during periods of drought and limited adult activity during periods of rainfall.

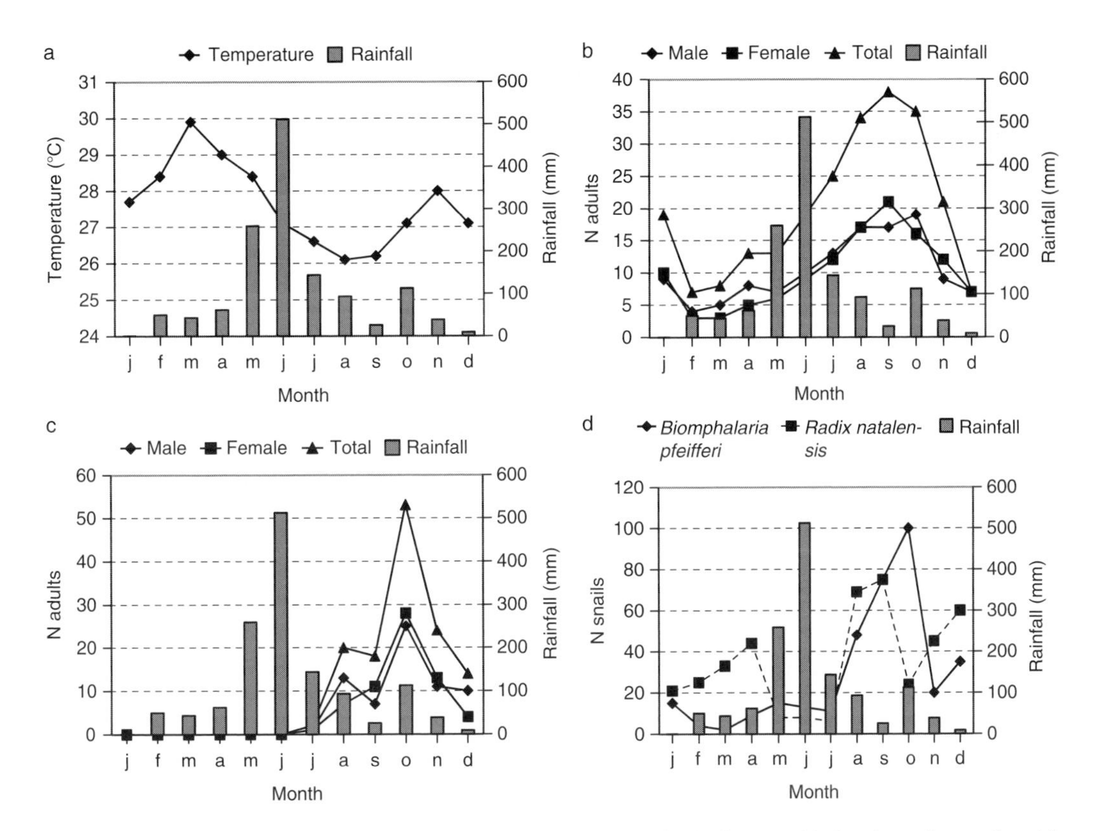

Fig. 9.14 Seasonal variation in populations of *Sepedon ruficeps* adults and snails in two habitats in Bénin during 1997. (a) Rainfall and temperature of water surface in permanent aquatic habitat in Cotonou. (b) *Sepedon ruficeps* collected in permanent aquatic habitat in Cotonou. (c) *Sepedon ruficeps* collected in temporary aquatic habitat in Cocotomey. (d) Population changes of *Biomphalaria pfeifferi* and *Radix natalensis* in permanent aquatic habitat in Cotonou. Modified from Gbedjissi *et al.* (2003).

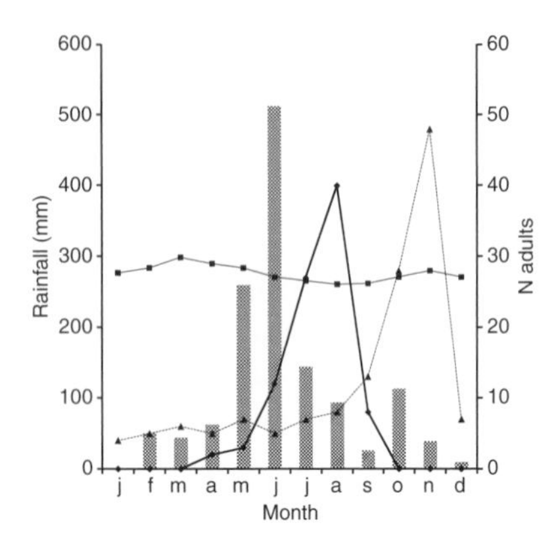

Fig. 9.15 Seasonal variation in populations of adults of ***Sepedonella nana*** collected at permanent (Akasso, triangle) and temporary (Agnavo, diamond) aquatic habitats in coastal Bénin during 1998 with rainfall in mm (histogram) and temperature in °C (square) in Cotonou (1997). Modified from Vala *et al.* (2000b).

Other inconclusive, conflicting information is available from central Africa. Verbeke (1963) collected 787 specimens of 11 species of *Sepedon* and related genera December 1950 to September 1952 in Garamba Park, northeastern Democratic Republic of Congo. He compared the collection data with average monthly rainfall, and concluded that his "Sepedoninae" (taxonomic group including the 11 investigated species), (1) fly throughout the year, (2) maximum abundance is during the *dry* season (November through March), and (3) the curve of abundance of Sciomyzidae is inverse to that of precipitation (Fig. 9.16). However, half of the specimens referred to by Verbeke (1963) were *Sepedon trichrooscelis*, subsequently shown by Vala *et al.* (1995) and Knutson (2008) to be atypical for *Sepedon* in being a parasitoid/predator of semi-terrestrial Succineidae and the closely related *S. lippensi*, also probably a semi-terrestrial species. Also, only 14 of the 399 specimens of *S. trichrooscelis* and *S. lippensi* were collected during the five months of April through August (the rainy season), indicating that at least these species do not breed

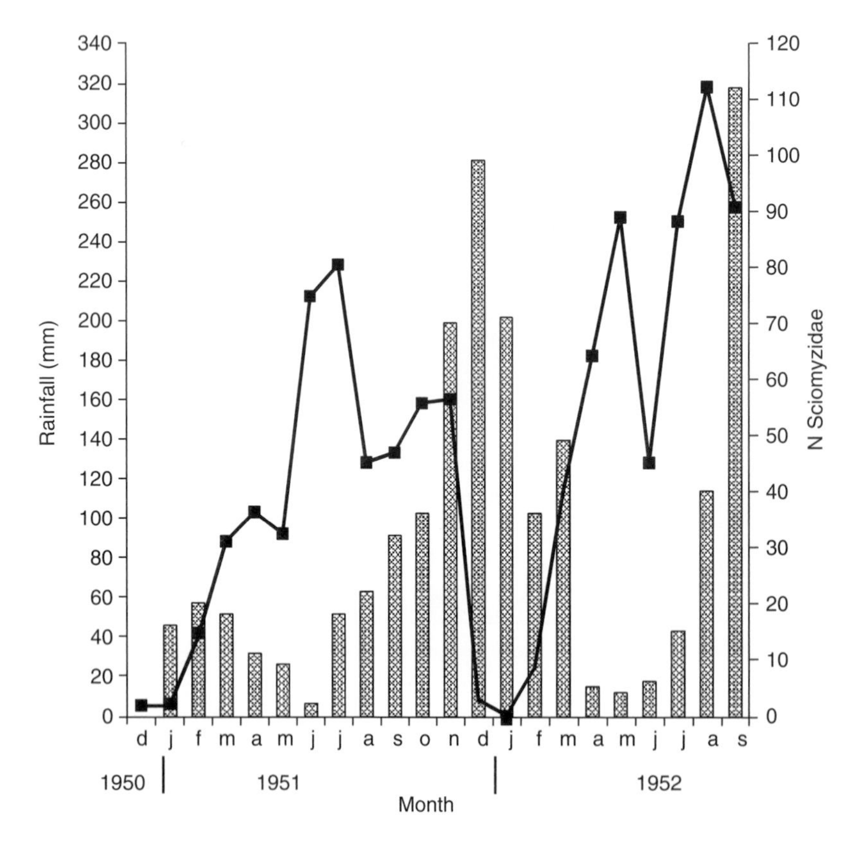

Fig. 9.16 Seasonal variation of adults of 11 species of ***Sepedon*** and related genera captured in Garamba National Park (northeastern Democratic Republic of Congo). Histogram, adults captured, December 1950 to September 1952; line, rainfall. Modified from Verbeke (1963).

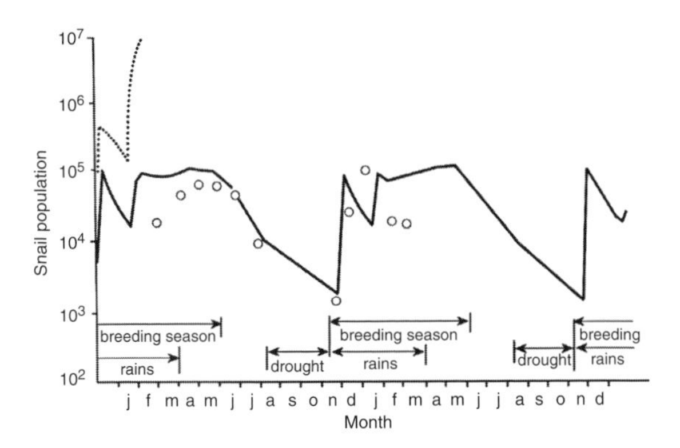

Fig. 9.17 Seasonal variation in populations of *Bulinus globosus* predicted by mathematical model compared with observations on natural population in Zimbabwe (after Jobin & Michelson 1967). Open circle, natural population; solid line, predicted population assuming adverse effect of crowding to begin at density of 1 snail per 100 l; dotted line, model prediction in absence of crowding effect. From Brown (1980).

continously. However, Vala *et al.* (1995) and Knutson (2008) concluded that *S. trichrooscelis* is multivoltine, based on laboratory rearings in which the entire life cycle required only 30 days and 25–40 days, respectively, and on limited field collections of immatures.

The rather extensive but partly conflicting information on number of adult sciomyzids collected in tropical areas plotted against rainfall may be, in some cases, due to artifacts of collecting, along with natural dispersal of the flies. That is, during rainy periods the flies probably are more dispersed, and more difficult to collect, leading to lower numbers captured. During dry seasons, the aquatic species probably are concentrated in wet spots and terrestrial species in shady spots, thus easier to collect in large numbers. Also, fewer man-hours are spent collecting during rainy seasons as opposed to drier seasons. As noted, the type of habitat, permanent or temporary aquatic, probably is important.

Rainfall, the breeding season, and population density of the aquatic snail, *Bulinus globosus*, which is likely one of the major prey of *Sepedon* species in Central Africa, are shown in Fig. 9.17. This snail is present in much of Africa south of the Sahara (Brown & Kristensen 1993), mostly south of 12° S. Lat. in West African areas.

9.1.6.3 ORIENTAL SPECIES

In tropical areas of the Oriental region, the biologies of only five species, all aquatic predators in the genus *Sepedon*, have been studied; there has been no analysis of data on flight periods. In laboratory rearings started with flies collected in Thailand and Indonesia (O. Beaver *et al.* 1977), these species had short pre-oviposition periods, long oviposition periods, immature stages developed promptly, the entire life cycle required only 14–26 days, and there was no indication of diapause. Adults were collected in all months, and at least some were most abundant shortly after the beginning of the rainy season. Of the five, *S. aenescens*, which is closely related to the Palearctic *S. sphegea*, ranges into Japan and there adults overwinter in diapause, with ovaries poorly developed, ovarioles small and narrow, but fat bodies not enlarged and flies active and feeding on warm days (Nagatomi & Kushigemachi 1965, ChannaBasavanna & Yano 1969).

9.1.7 Dormancy (diapause and quiescence)

Diapause and quiescence in relation to overall phenology are treated also in previous parts of this section. As in other aspects of behavior, the Sciomyzidae are diverse in displaying diapause in the adult, egg, larval, and pupal stages, according to the species, and in displaying quiescence.

Much of the information on diapause and quiescence in Sciomyzidae is incomplete, based primarily on limited field collections and temperature (but usually not photoperiod) controlled laboratory studies. The three species that have been the most extensively studied and those in Phenological Group 5b are discussed below. As background for the reader, we relate some of the information on Sciomyzidae to one of the major reviews of dormancy (Tauber *et al.* 1984).

Dormancy was defined by these authors as a seasonally recurring period during which development and reproduction are suppressed and the organism is in a state of quiescence (torpor) which does not involve preparatory hormonal or physiological changes and is terminated as soon as conditions become favorable *or* in a state of diapause which is initiated by token stimuli, involves preparatory hormonal and physiological changes, and growth and reproduction are suppressed even if favorable conditions prevail. While they noted that not all insects undergo dormancy, total lack of dormancy is rare and "the generalization that tropical insects do not undergo dormancy or diapause is clearly questioned."

As in many reports on diapause in insects, discussions of diapause in Sciomyzidae often refer to "obligatory" diapause (diapause expressed in every individual of a generation, regardless of environmental conditions) and "facultative" diapause (diapause averted under appropriate environmental conditions). Tauber *et al.* (1984) considered these terms of limited usefulness, obligatory diapause having "served largely as a catch-all for univoltine species for which the sensitive stage(s) and the diapause-controlling environmental stimuli have not been determined" in contrast to facultative diapause "in which the controlling environmental factors are known and alterable."

Tauber *et al.* (1984) also stated that "An important limitation on the usefulness of the photoperiodic response curve [in regard to studies on dormancy] is that it is based on stationary light-dark cycles that do not occur in nature; it therefore does not take into account the considerable influence of changing day length." We would extend this thought by noting that the overall success of an overwintering, multivoltine species in making the greatest use of the time available (essentially, maximization of the number of generations per year and population increase, thus increased opportunity for genetic recombination) is largely dependent on early emergence from the overwintering stage. Emergence of Sciomyzidae clearly is dependent to some significant extent on the overwintering site. Since, in the northern hemisphere, south-facing slopes receive significantly more insolation than north-facing slopes, exposure to the trajectory of the sun as well as the duration of and changes in day length also probably are important in the build-up of Sciomyzidae populations. Since the immature stages obviously do not have the ability to change north/south-facing slope positions, we predict that rather simple sampling studies during the period of emergence, first mating, and first egg-laying will show larger populations of adults on south-facing rather than north-facing slopes. Since adults can easily move from north/south-facing slopes over a short time, emergence traps situated in both sites will be the important sampling devices.

"Chilling" (usually exposure to 5 °C for various periods) has most often been used in attempts to terminate diapause in laboratory rearings of Sciomyzidae (and many other insects). However, as Tauber *et al.* (1984) noted, "It is evident in most cases that temperature acts to regulate the *rate* of diapause development . . . not as a specific signal to terminate diapause." They concluded that while "aestival diapause generally requires a specific stimulus for termination . . . hibernal diapause . . . which often ends "spontaneously" sometime during winter, may not require a terminating stimulus" and that "in general, diapause has ended . . . when token stimuli no longer prevent growth and development and when thermal and other responses have returned to the non-diapause state."

The significance of quiescence in the survival of Sciomyzidae under inimical conditions is indicated by the cold, heat, and drought hardiness of many species. The recorded spring-to-fall flight period of many univoltine species and limited capture-mark-release-recapture data for a few multivoltine species show that adults of many species clearly survive frequent aseasonal environmental changes. Also, we have often transported many species, of all stages, in small containers with little semblance to natural conditions over long distances and for a week or more, between laboratory locations, with essentially no care, and have been surprised to re-initiate rearings with the obviously stressed individuals.

Tauber *et al.* (1984) noted an "intimate association between seasonal migration and dormancy," that "free-living insects often move from their site of reproduction to another site to undergo dormancy," and that "Insects entering diapause often undergo color and/or morphological changes [polyphenism]" that provide "protection against seasonal exigencies." There are no records of seasonal polyphenism in Sciomyzidae. However, Zuska & Berg (1974) convincingly related variation in light and dark color forms of two species of *Tetanoceroides* in temperate South America to their north–south geographical distributions. The authors considered the variation as cases of behavioral thermoregulation. The only indications of migration associated with dormancy among Sciomyzidae are a few observations of a few adults of some species, e.g., *Sepedon sphegea*, collected in the fall in non-typical habitats rather far from their known, typical habitats.

The majority of life-cycle studies of sciomyzids have been conducted with samples from populations from 43–47° N. Lat. in North America (Ithaca, New York; Kent, Ohio; and Moscow, Idaho) but from 44–60° N. Lat. in Europe (Avignon, southern France; Harpenden, southern England; Hillerød, Denmark; and near Helsinki, Finland). Also, population samples from latitudes far south or far north of these have been used for many life-cycle studies in laboratories situated at 45–50° N. Lat. in Europe and North America. Day lengths between the extremes of 43–60° N. Lat. differ throughout the year by as much as 3 hours (Fig. 9.18). Thus, especially for relatively broadly north–south distributed species, one could expect differences between European and North American populations of Holarctic species and differences between Palearctic and Nearctic species that, on the whole, appear to belong to the same phenological group.

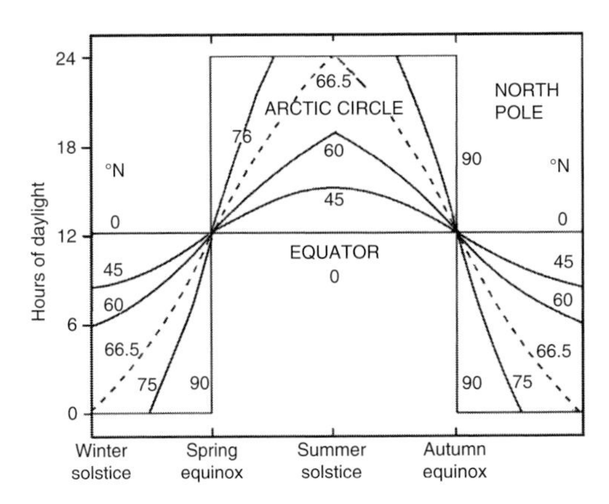

Fig. 9.18 Annual variation in day length (sun above horizon) at selected latitudes. Day lengths in high arctic indicated by zone above *c*. 75° N. Lat. From Danks (1993).

Since most information on diapause and quiescence in Sciomyzidae concerns the overwintering stage in cool, temperate areas we briefly summarize this by stage. Most Sciomyzidae in cool, temperate climates – both terrestrial and aquatic, parasitoid and predaceous, univoltine and multivoltine, Sciomyzini and Tetanocerini – overwinter in the puparium. Most species overwinter in a state of quiescence but many in diapause. Results from rearings of *Colobaea americana*, each larva of which feeds on one *Gyraulus parvus* and pupariates and overwinters in the shell, demonstrate the controlling influence of temperature on the development of many species in Group 1. Whereas five consecutive generations (duration of each, 19–25 days) were produced beginning in January in the laboratory, several puparia collected in nature on November 6, 1963 were held at 5 °C for about 2 years (until December 20, 1965) and produced adults 9–19 days after being removed to room temperature. This is the longest period of time that any stage of any species of Sciomyzidae has been kept alive (Knutson & Bratt unpublished data). Part of the population of pupae in the fall go into diapause or quiescence and part produce another generation of adults in some primarily univoltine *Anticheta*, *Colobaea*, and *Pherbellia* spp. and in some late summer to early fall generations of some multivoltine *Tetanocera* spp. (and in cultures of the latter reared under short day conditions). This is perhaps an adaptation to ensure survival in the event of an early winter and also to take advantage of a prolonged season favorable for development of the larvae. Palearctic and Nearctic species of *Sepedon* (multivoltine), *Elgiva* (multivoltine), and possibly *Psacadina* (probably univoltine) and *Pherbellia schoenherri* (multivoltine) overwinter as diapausing or quiescent adults. In Mediterranean latitudes *Pherbellia cinerella* (multivoltine) overwinters as quiescent adults but in northern latitudes in the puparium; other species may have similar latitudinal differences. Overwintering as pharate first-instar larvae in the egg membranes is exhibited by some univoltine Tetanocerini: *Hedria mixta*, four species of Palearctic and Nearctic *Tetanocera*, and possibly by part of the populations of *Ilione lineata* and *I. albiseta*. These species also display adult estival diapause, except for *H. mixta* and some Nearctic populations of *T. montana*. Some univoltine Tetanocerini overwinter as partly grown to mature larvae in diapause: three Nearctic *Tetanocera*, *Pherbina coryleti*, and *Ilione albiseta*.

Much of the information on diapause and quiescence published to 1980 was reviewed by Berg *et al.* (1982) in establishing their five Phenological Groups, with much new information. Here, we focus on recent and experimental studies and offer a few interpretations of the data. Detailed experimental studies under temperature and photoperiod controlled laboratory conditions, in outdoor screen cages, and field data throughout the year are available for only three Tetanocerini. These are (references given below): *Sepedon fuscipennis*, multivoltine with facultative adult diapause; *Tetanocera ferruginea*, multivoltine with facultative pupal diapause; and *Ilione albiseta*, univoltine with overwintering usually as larvae, primarily third instars, and with estival adult diapause.

Also, limited laboratory experiments under controlled temperature and photoperiod were reported for three Nearctic *Tetanocera* spp. which are multivoltine with pupal diapause (Trelka & Foote 1970), *Tetanocera vicina* which is univoltine with overwintering as diapausing and quiescent larvae and estival adult diapause, and *Dictya* spp. which are multivoltine with overwintering as quiescent pupae in various stages of development (Berg *et al.* 1982). Survival and subsequent reproductive capacity of adults of the multivoltine *Dictya* spp., *Elgiva solicita*, *Sepedon armipes*, and *Pherbellia* s. *maculata* were studied to some extent, along with detailed studies of *S. fuscipennis*, in outdoor screen cages maintained for periods of 21, 23, and 64 days during the winter in Ithaca, New York (Berg *et al.* 1982). Brief experiments involving only exposure to various temperatures, but often with examination of the state of development of pupae by dissecting puparia, are reported in some life-cycle papers. The effect of photoperiod, the primary inducer of diapause in northern latitudes, has been relatively neglected, except for *I. albiseta*, *S. fuscipennis*, and *T. ferruginea*. The effect of moisture and rainfall in tropical areas is unknown.

Sepedon fuscipennis. Diapause has been studied most extensively, from field collection data, outdoor experiments, and controlled laboratory experiments in the multivoltine, aquatic predaceous *S. fuscipennis*, which undergoes facultative imaginal diapause. The information from the several studies fortunately is quite comparable since all were conducted at Ithaca, New York (42°26′ N. Lat.). This species and several other Nearctic and Palearctic *Sepedon* were first found to overwinter as adults by Neff & Berg (1966); all puparia collected during the winter failed to produce adults. Arnold (1978b) found that a female captured, marked, and released in August was reproductive when recaptured the following May. Adults survived in outdoor screen cages, without mating or laying eggs from October to April. Berg *et al.* (1982) summarized experiments on survival and oviposition. Adults collected January 30 survived a 21-day exposure in outdoor screen cages with the temperature dropping to a low of –19 °C. Pairs that survived a 64-day exposure during February to early April when the temperature dropped as low as –22 °C subsequently produced viable eggs. Winter survival of males seems necessary in order to carry viable sperm through the winter; overwintered females isolated from males did not lay viable eggs (Berg *et al.* 1982).

Females collected during June in 1974, 1975, and 1976 ceased oviposition in late August to mid September each year when the photoperiod was decreasing at the maximum rate, despite different weather patterns (Arnold 1978b; Fig. 9.19). Of 16 females collected in late August and held at 21 °C and LD 15:9 only two oviposited by January but of 35 collected a couple of weeks later and held at 25 °C and LD 17:7, 29 oviposited by the beginning of November (Barnes 1976). A rapid population decline during August–October was found by quantitative sweep-net collections and mark-release-recapture methods (Eckblad & Berg 1972).

Barnes (1976) reared a large number of adults at LD 16:8 and 8:16 and at 18, 21, 26, and 30 °C. He transferred some reared under long day to short day, and determined the condition of fat bodies, testes, and ovaries of adults subjected to the various regimes. He concluded that the imaginal diapause is facultative, influenced by both temperature and photoperiod in the sensitive and responsive freshly emerged females, and is initiated by late August. He found that diapause could be induced experimentally

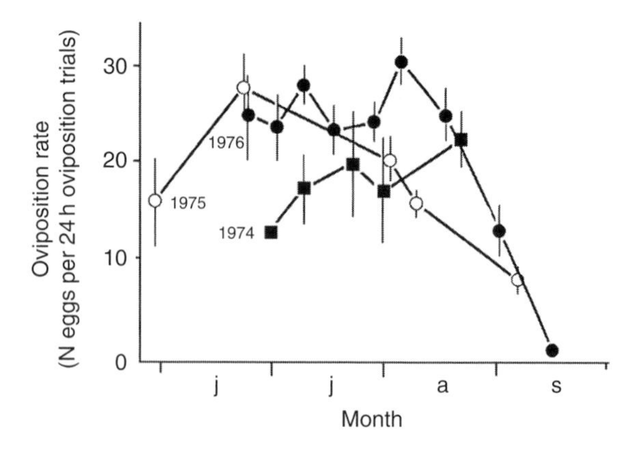

Fig. 9.19 Oviposition rate (mean number of eggs laid per female per day) by ***Sepedon fuscipennis*** (captured at Ithaca, New York) that oviposited during 24 h oviposition trials in 1974 (square), 1975 (open circle), and 1976 (black circle). Vertical bars are ± 1 S.E. Only one female oviposited on each of first sampling date in 1974 and last date in 1976. Modified from Arnold (1978b).

by low temperatures, even during long photoperiods, as was also found by Vala & Haab (1984) for induction of diapause in pupae of *Tetanocera ferruginea*. Barnes (1976) noted that the conditions under which the immature stages developed may moderate diapause in the adult, and that it is not known if diapause is induced in mature adults. Diapause was easily terminated by transferring laboratory-reared females held for about 1 month at relatively low temperature (21 °C) and short day to higher temperature (30 °C) and long day conditions, but the effects of temperature and photoperiod on females remaining at lower temperatures for longer periods is not known (but for temperature, see experiments on outdoor exposure in screen cages by Berg *et al.* 1982). Examination of fat bodies, testes, and ovaries in non-diapausing and diapausing adults showed well-developed fat bodies and smaller testes, ovaries, and accessory glands in the latter (Fig. 9.20). However, ChannaBasavanna & Yano (1969) found that in *Sepedon aenescens* (which undergoes reproductive diapause during the winter in Fukuoka, Japan), whereas the ovaries of diapausing flies were "ill-developed, with ovarioles very small and hardly recognizable," the fat bodies were not well developed. They concluded that the low fat content explains their observations that the diapausing adults are active and feed when temperatures are slightly high during the winter, which is unusual for insects in imaginal diapause. Barnes (1976) also included extensive information on the effect of temperature on development, survival, and oviposition.

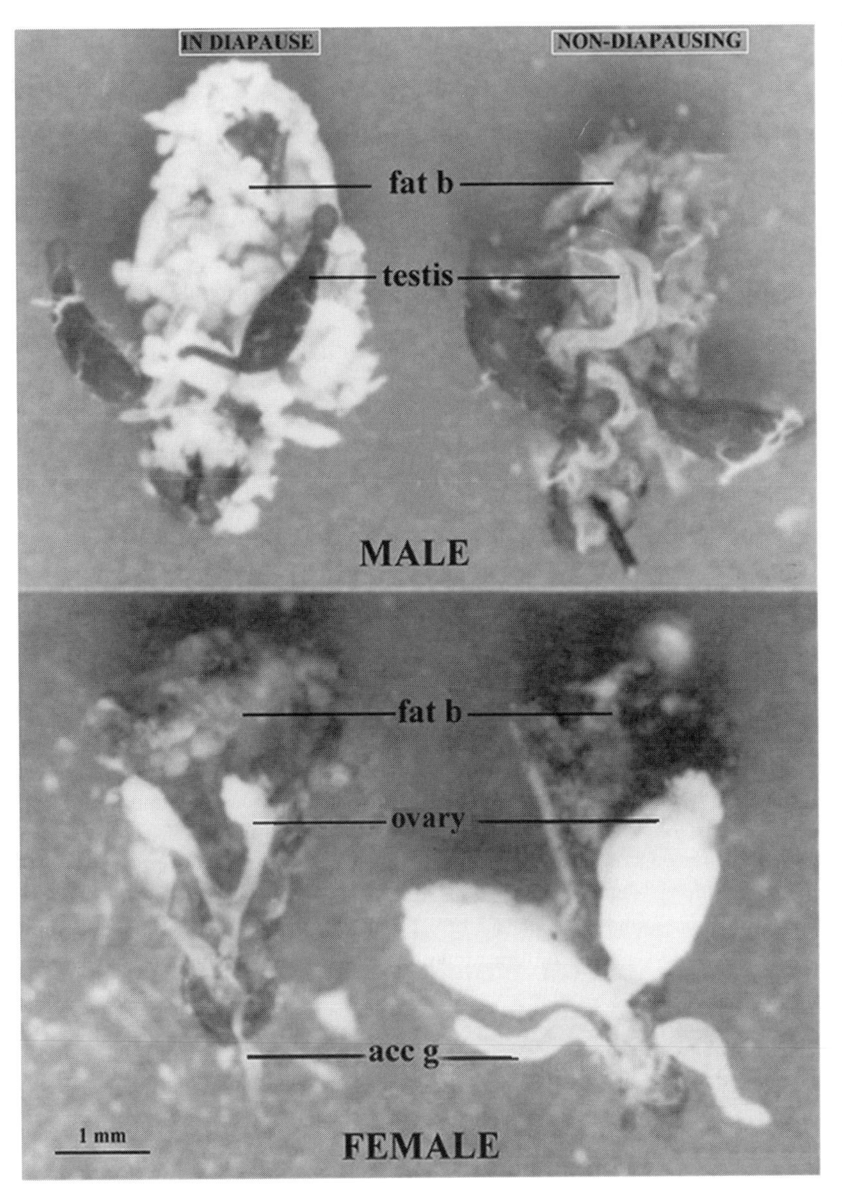

Fig. 9.20 Dorsal view of dissections of abdomens of male and female *Sepedon fuscipennis*. Fat hypertrophy (fat b) and reduced size of testes, ovaries, and accessory glands (acc g) apparent in diapausing flies (left) when compared with non-diapausing flies (right). Note: much of fat bodies cut away to make reproductive systems visible in photographs. Modified from Barnes (1976).

Arnold (1978b) validated some of the laboratory studies of Barnes (1976) by examination of field-collected females. The reproductive status and output of each female was assessed by (1) examination of abdomens after marking but before feeding during the 1–3 days females were held in field cages before release, (2) collection of all eggs each female laid in captivity, and (3) counting those eggs laid in the 24-hour oviposition trials beginning after marking and feeding. The percent of females that oviposited or had eggs visible through the abdominal integument (for 20–77 females collected on 6–11 dates, each, during May–September of 3 consecutive years) rose to about 80–95% shortly after the days of longest photoperiod and dropped to near 0% by mid September; the oviposition rate (during 24-hour oviposition trials) was similar (Fig. 9.21).

Tetanocera ferruginea. Fewer field data but from a wider set of localities and laboratory data from rearings under a more diverse set of temperature and photoperiod regimes are available for this multivoltine aquatic predator which

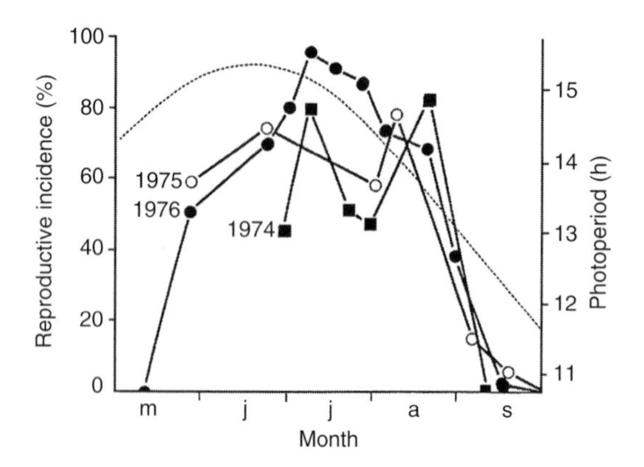

Fig. 9.21 Reproductive incidence (fraction of females either ovipositing or having eggs visible through abdominal integument) in captured *Sepedon fuscipennis* females in 1974 (square), 1975 (open circle), and 1976 (black circle), and photoperiod at latitude of study site (42°30′ N. Lat.), Ithaca, New York. Except for two first dates in 1974 (n = 11 and 10, respectively) and first two dates in 1976 (n = 2 for each), percentages based on 22–77 females. From Arnold (1978b).

overwinters as pupae in diapause. Berg (1953) noted that pupae formed during early September by larvae collected during mid August at Anchorage, Alaska (61°10' N. Lat.) remained in diapause for 4–5 weeks despite exposure to laboratory temperatures and adults emerged after the puparia were held at 3 °C for 7 weeks. At lower latitudes (Ithaca, New York and Kent, Ohio [41°10' N. Lat.]) larvae have been found as late as November 29; puparia collected from early December to mid April contained unpigmented, undeveloped pupae; and puparia collected February 10 to early April produced adults 9–11 days after being brought into the laboratory (Berg *et al.* 1982, Foote 1999). They did not study the effects of photoperiod, but suggested that the uniformity in stage of development of overwintering pupae indicates that diapause is fixed genetically at this stage. A few records from Europe (Vala & Knutson unpublished data) also indicate multivoltinism with overwintering in the puparium. Adults emerge and mate early during the spring: females collected March 27 at Valencia, Spain and Corfu, Greece and April 6 at Montpellier, France and kept without males laid viable eggs. Third-instar larvae have been collected between April 20 and October in central and southern Europe. Puparia that produced adults have been collected as early as March 25 (Moravia) and June 20 (at the Arctic Circle in Sweden).

Vala & Haab (1984) reared this species in the laboratory at Avignon, France (about the same latitude as Ithaca, New York, but with a Mediterranean climate) at the following photoperiod and temperature regimes: LD 8:16 at 13, 20, and 26 °C; LD 9:15 at 20 °C; LD 16:8 at 13 and 20 °C; and LD 24:0 at 20 °C; they did not include specific data from field collections. They found that diapause is induced by photoperiod. Diapause is depressed if the temperature exceeds 20 °C and photoperiod is more than LD 9:15, but proceeds despite high temperatures if the photoperiod is LD 8:16 or less (Fig. 9.22). In comparing their results with the data of Berg (1953) from Alaska, they concluded that diapause is obligatory or facultative, depending on climatic conditions.

Ilione albiseta and *I. lineata*. Understanding of diapause and/or quiescence in these univoltine, freshwater predators which spend the early larval life beneath the surface remains unresolved. Knutson & Berg (1967) considered that they could overwinter as pharate larvae in the egg membranes or as partly grown larvae depending on the availability of water and their prey during the winter. *Ilione albiseta* is discussed in detail in Section 9.3.3.3 on development of species representative of the Phenological Groups.

Phenological Group 5b. As noted in Section 9.1.2, Vala (1984b) proposed a new Phenological Group, 5b, based on his rearings and field studies in southern France of several univoltine, terrestrial, predators/saprophages: *Dichetophora obliterata*, *Euthycera cribrata*, *E. stichospila*, *Trypetoptera punctulata*, and *Coremacera marginata* (see also Knutson 1973 for *C. marginata*). These are characterized by overwintering of larvae from September until midwinter, completion of overwintering in pupal diapause, and an estival reproductive diapause of adults. Unlike the original Group 5 of Berg *et al.* (1982) (univoltine species overwintering in puparia) puparia are not found throughout the year. However, that aspect of the Berg *et al.* classification seems to result simply because of the long and variable duration of overwintering puparia, some of which overlap with the summer generation of puparia. Induction of diapause in such species raises some interesting questions. It is known that diapause can be induced in larvae living in microhabitats where light levels are lower than in the general environment (e.g., larvae of the chironomid *Metriocnemus knabi* and the mosquito *Wyeomyia smithii* living in water collected within the pitcher plant; Saunders 2000). Consider that the brain of the larva is the site of photoperiod sensitivity. How then can photoperiod be responsible for inducing pupal diapause in terrestrial and semi-terrestrial sciomyzid larvae that spend larval life

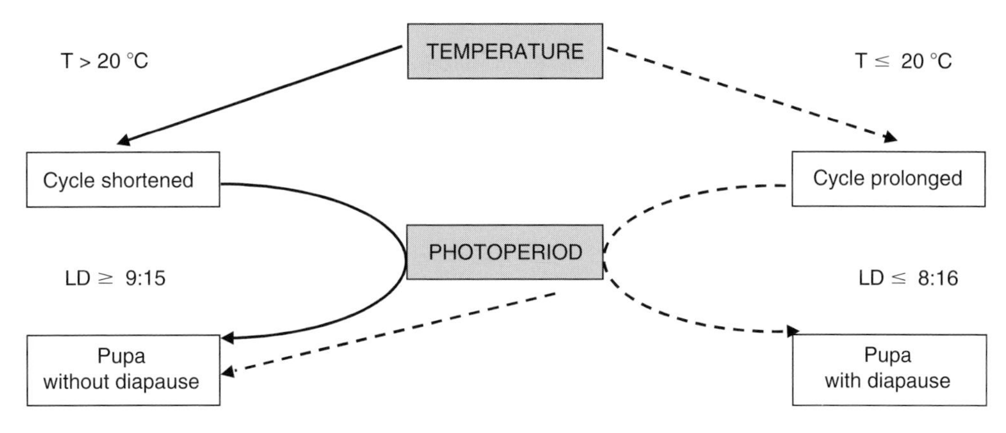

Fig. 9.22 Influences of various combinations of temperature and photoperiod on life cycle of *Tetanocera ferruginea*. Modified from Vala & Haab (1984).

essentially in total darkness with at least their anterior ends buried in the decaying tissues inside a snail shell? Perhaps, as has been noted in at least some of these species, it is due to a prolonged pre-puparial wandering phase in which the larvae do not feed and thus are exposed to some level of light. Such an adaptation to allow for exposure for photoperiod induction would of course be a maladaptation in regard to exposure to predators and parasites, although no parasitic Hymenoptera have been reared from Sciomyzidae with this kind of phenology. Even freshwater and shoreline, predaceous larvae which spend much of their life exposed at the surface live in situations with low light levels. We note that while most of the body integument is heavily pigmented in most of these larvae, the anterior three segments (surrounding the brain) are not pigmented (e.g., *Tetanocera hyalipennis*, Plate 2f, g).

9.1.8 Evolutionary implications of voltinism and overwintering stage

Berg *et al.* (1982) speculated that multivoltine species overwintering in the puparium is probably the most primitive pattern of overwintering strategy. Diapause is controlled by complex mechanisms. Tauber *et al.* (1984) summarized that three major types of genetic mechanisms help to determine diapause characteristics. Since diapause is an evolved adaptation, it cannot be considered the most "primitive" form of behavior. It therefore follows that the most primitive pattern of overwintering is as quiescent, not diapausing, pupae. Berg *et al.* (1982) further speculated that evolution of phenological diversity has resulted in partitioning of resources in time and space; species whose larvae are active from fall to spring being able to exploit food resources with little competition from other Sciomyzidae. They considered that seasonal restriction of larval feeding by some univoltine species enables them to move into temporary aquatic habitats less favorable for multivoltine species.

The supposition by Berg *et al.* (1982) that multivoltinism with overwintering in the puparium is the primitive life-cycle pattern in Sciomyzidae seems reasonable from several aspects. It is not only the most common pattern in Sciomyzidae, typical of the sister group Phaeomyiidae and most of the plesiomorphic genera of Sciomyzidae, but also the most common in the many families of Muscomorpha–Brachycera (puparium-forming Diptera) in general. Assuming that Sciomyzidae originated in a temperate area, i.e., climates where the majority of species and genera occur today, overwintering in the protected puparium is more likely than in the other, less protected stages. We note the remote origin of puparium formation at the base of the Muscomorpha–Brachycera lineage, the profound genetic basis that puparium formation must have, and the absence of a single reversal to the morphology of the more primitive pupal case of the Orthorrhapha. Thus we conclude that the selective advantages of the protective puparium likely became genetically associated with behavioral mechanisms to survive winter and other conditions (e.g., dry weather) inimical to development and survival. Furthermore, multivoltinism has the obvious selective advantage of greater reproductive potential than univoltinism.

Other ways of surviving inimical conditions have evolved only sporadically in a few genera of Sciomyzidae, but broadly in *Tetanocera*. These are: Group 2 – as adults: only in one species of the Sciomyzini genus *Pherbellia*, possibly in one species of the Tetanocerini genus *Psacadina*, *Elgiva solicita*, and the decidedly apomorphic Tetanocerini genus *Sepedon*; Group 3 – as developed larvae in egg membranes: in the Tetanocerini genera *Ilione*, *Hedria*, and a few *Tetanocera*; and Group 4 – as young larvae: in the Tetanocerini genera *Ilione*, *Pherbina*, *Eulimnia*, and a few species of *Tetanocera*. The evolution of phenologies other than multivoltinism with overwintering in the puparium likely have their origins, as suggested by Berg *et al.* (1982) as specialized adaptations allowing temporal and spatial partitioning. While it is difficult to show that this has resulted from competition, it obviously results in less competition.

In three of the four most successful genera (*Pherbellia*, 95 species; *Tetanocera*, 39 species; and *Dictya*, 42 species), Group 1 phenology is the most common scenario. The other highly successful genus (*Sepedon*, 78 species) displays only Group 2 and 6 phenologies. The largest genus essentially restricted to temperate areas, *Tetanocera*, has the greatest diversity of phenologies (Groups 1, 3, 4, 5) but does not display the phenologies (Group 2, 6) of the genus, i.e., *Sepedon*, that is best developed in tropical areas. Although the apotypic genus *Sepedon* is also well represented in the Holarctic region (19 species in the Nearctic, 8 in the Palearctic) most of the reared species there belong to Group 2.

9.2 REPRODUCTION

> *Achever le cycle de l'année sous la forme adulte, se voir entouré de ses fils aux fêtes du renouveau, doubler et tripler sa famille, voilà certes un privilège bien exceptionnel dans le monde des insectes.* [To complete the annual life cycle in the adult form, to be surrounded by its own offspring, to double or triple its family, is a very exceptional privilege in the insect world].
>
> Fabre (1921).

The major patterns of phenology and the closely associated variations in many aspects of reproduction and development that are often part of these patterns are central to considerations of population biology and applications in biological control. Thus some information presented in Chapters 13 and 18 is closely related to the present section. Also, co-adaptations in phenology, reproduction, and development often are closely linked with microhabitat preferences and life styles of the larvae. These features have been described routinely in many publications on life cycles.

9.2.1 Courtship and mating behavior

Chemical, visual, tactile, and olfactory mating cues have been noted for a few species of Sciomyzidae, but for most species there is no overt courtship behavior and most mating is of the "assault type" (Foote 1977), with rather stereotyped body positions, as noted in many papers on life cycles. Despite the fact that many Sciomyzidae have elaborate wing patterns, courtship display involving wing movements, except in *Sepedon* spp. (which lack wing pattern) noted below, has not been observed. Female Sciomyzidae have not been observed rejecting males, except for a few superficial observations. Male mating preference for female survivorship, as analyzed for species of the related Sciomyzoidea families Dryomyzidae and Coelopidae (e.g., Dunn *et al.* 2001), would be of interest.

Most adults mate within 1–15 days after emergence from puparia, usually within about one day, although many *Sepedon* females resist the advances of males for several days after emergence (Neff & Berg 1966). Most adults mate frequently and often daily, e.g., up to 37 consecutive days for *Salticella fasciata* (Knutson *et al.* 1970). Mating tends to decrease as females age. The duration of each mating is from a few seconds to 2–3 hours.

Mating trophallaxis has been documented in seven families of Acalyptratae (Freidberg in Sivinski *et al.*, 1999), including seven species of *Sepedon* in the Nearctic, Afrotropical, and Oriental regions (Green 1977; Yano 1978; Barraclough 1983; Berg & Valley 1985a, 1985b). Some males release a nearly clear liquid anally that the females feed on during copulation (Fig. 9.23a–d). Other males are considered to secrete an opaque nuptial food orally. Some appear to regurgitate food to the females by direct labellar contact and some attract and nourish potential mates with food (e.g., dead snails) that they discover and defend. A *Sciomyza dryomyzina* male was seen to apply his labellum to the back of the female's head during mating (Knutson 1988). In *Neolimnia* spp. the male's proboscis was extended but did not touch the female's head (Barnes 1979b); in neither case did there appear to be a drop of liquid on the labellum.

Nuptial feeding by the female of *Sepedon aenescens* was illustrated with color photographs by Sung-Yang Lee

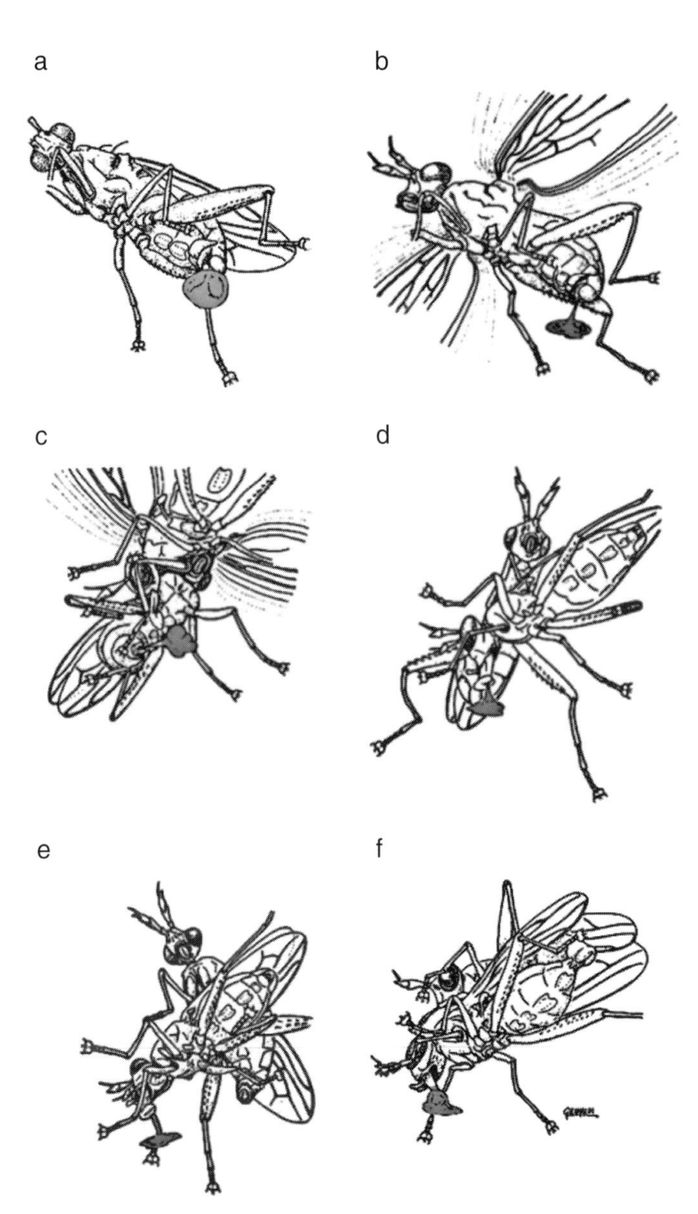

Fig. 9.23 Nuptial feeding by ***Sepedon floridensis***. Drawings made from motion-picture film of pair of flies standing on side of glass breeding jar. (a) Male initiating precopulatory behavior by depositing droplet of fluid. (b) Male vibrating wings. (c) Female approaching droplet beneath male and vibrating wings. (d) Female feeding, male mounting her. (e) Male changing position. (f) Pair copulating, female feeding. Modified from Berg & Valley (1985a).

(in Green 1977), who believed the substance was produced orally. Berg & Valley (1985a) presented drawings, from photographs, of *Sepedon floridensis* males secreting the food anally and the female (while *in copula*) feeding on it. They described their observations of *S. aenescens* behaving in the same manner (Fig. 9.23e, f). Barraclough (1983) presented a drawing of a pair of *Sepedon neavei* with their probosces in contact while *in copula*, suggesting that the male may be passing regurgitated food to the female (Fig. 9.24). Berg & Valley (1985a, 1985b) reviewed the observations reported by Green (1977), Barraclough (1983), and the unpublished observations of O. Beaver on *Sepedon senex* and *S. plumbella* and those of K. Durga Prasad (personal communication, 1972, 1973) on *S. aenescens*, *S. ferruginosa*, and *S. plumbella*; these authors considered the food to be produced orally by the male. Berg & Valley (1985a, 1985b) overlooked the report of Yano (1978) who recorded that a male of *Sepedon senex* in Thailand produced 2–3 mm diameter masses of a "milky white foam like substance" at its labellum and then the

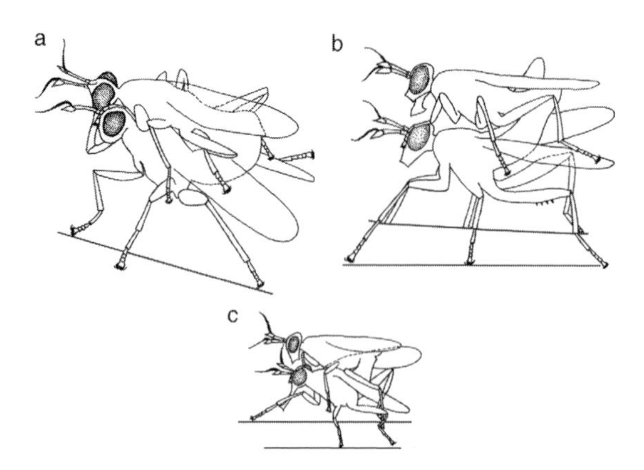

Fig. 9.24 *Sepedon* pairs copulating. (a) *Sepedon neavei*. (b) *Sepedon testacea*. (c) *Sepedon pleuritica*. Modified from Barraclough (1983).

same fly ate them during a 30-minute period. Yano (1978) also reported that a *S. aenescens* (as *S. sauteri*, sex undetermined) was seen feeding on a "sticky milky white substance on [a] leaf" in Japan.

Valley & Berg (1977) noted that their laboratory observations indicate that females of *Dictya* can distinguish between males of their own and other species,

> When mounted by a male of a different species, a female either invaginated the tip of her abdomen or extended it downward. Both actions prevented the males from establishing genitalic contact. Males that mounted females of a different species usually left them again in less than a minute. The consistency of this behavior strongly indicates that the female selects the proper mate.

These observations may be particularly useful in attempting to rear species of *Dictya*, most of which can be identified only by characters of the male genitalia, and may be useful in rearing closely similar species in other genera. The fact that a male of *Dictya pictipes* mated successfully with a grossly malformed female (body flexed through almost 90°) led O'Neill & Berg (1975) to speculate that, at least in that genus, tactile cues were not as important as olfactory cues.

Valley & Berg (1977) noted that a female *Dictya stricta* collected with adults of *D. brimleyi* was placed with a male of the latter species. On one occasion the male mounted the female but she directed her abdomen downward and the male left. When the male *D. brimleyi* was replaced with a male *D. stricta*, the pair mated within four hours. These authors also mentioned that a female *D. floridensis*, collected with a male *D. expansa*, were placed in the same rearing jar. During the 91 days they were together they were never observed to mate but "the frayed wings of both adults may have indicated that mating had been attempted and struggles had ensued."

Pre-mating wing movements have been observed in some species. The pre-mating behavior of 16 species of *Sepedon* studied by Neff & Berg (1966) is different than that of other genera,

> Prior to mating, the male moves back and forth in front and to either side of the female in a path that describes an arc of about 180°. As he moves in this path, the forelegs are raised and lowered nervously and a bobbing motion of the abdomen occurs. His wings are extended laterally from their resting position over the abdomen, vibrated rapidly for several seconds, and then returned to the resting position. The frequency of these wing vibrations increases as long as the female remains still and presumably receptive; however, if she backs away and flicks her wings several times in quick scissors-like movements, the male ceases his display. If the female remains still and gives no wing motion, the male circles behind her slowly and mounts her from behind.

In the slug killers *Tetanocera plebeja* (wing plain) and *T. clara* and *T. valida* (wing patterned) the female spreads its wings in front of the male before mating (Trelka & Foote 1970).

With minor differences in many species and a few major differences noted below, the copulatory posture of most sciomyzid species consists of the male, with wings closed, astride the female and facing in the same direction, his front tarsi along the mesal margins of her compound eyes or, in some *Sepedon* spp., grasping her antennae; in some *Colobaea* spp. grasping the anterodorsal angles of her thorax. His mid tibiae rest on the costal margin of her wings, and his hind tarsi grasp her abdomen. The copulatory posture of *Poecilographa decora* is rather typical of the family (Plate 1c). In *Salticella fasciata* (Salticellinae), one fore leg of the male is held outstretched and the mid tarsi rest on the substrate (Knutson *et al.* 1970; Fig. 9.25). The female *Tetanura pallidiventris* (which several authors have placed in a separate subfamily) flies over the anterior end of the male, lands behind him, and the male quickly turns toward her. The male then climbs onto the female's back, facing toward her posterior end. He quickly turns to face in the same direction as the female and the genitalia are coupled. Then his front tarsi rest on the humeral angles of her thorax, his middle tarsi grasp the ventral surfaces of her uplifted and partially outstretched wings, and the apices of his hind tibiae grasp the sides of her postabdomen (Knutson 1970a). These changes in orientation probably are associated with the fact that the terminal segments of the female abdomen are (uniquely within the family) twisted about 60° from the longitudinal axis, apparently an adaptation for laying eggs on the retracted tissues within the shell of a living snail. Rivosecchi (1992) presented habitus figure of *T. pallidiventris* male and female postabdomens *in copula* and a detailed figure showing the conjoined internal genitalic structures of a pair *in copula* (Fig. 9.26). Kaczynski *et al.* (1969) noted that in some mating pairs of *Perilimnia albifacies* and *Shannonia meridionalis*, females resting head downward on sticks in the rearing jars were mounted by males in that position. They then walked down the stick until they appeared to be standing on their head on the moss substrate, remaining in this position until copulation ended an hour later. The authors noted that this behavior has been observed in other (un-named) genera of Sciomyzidae. Photographs of pairs *in copula* have been presented for *Tetanocera plebeja* (Trelka & Foote 1970), *S. fasciata* (Knutson *et al.* 1970), *Hedria mixta* (Foote 1971), *Tetanocera* sp. (Barker *et al.* 2004), and *Anticheta borealis*, *T. plebeja*, and *T. plumosa* (Marshall 2006). Drawings of pairs *in copula* have been presented for *Sepedon neavei*, *S. pleuritica*, and *S. testacea* (Barraclough 1983) and *S. floridensis* (Berg & Valley 1985a). Females are often active, i.e., feeding, grooming, or flying short distances while coupled with the males. Females of some species shake their body if they are not receptive to males.

Fig. 9.25 Pair of *Salticella fasciata* mating. From Knutson *et al.* (1970).

Females of many species retain viable sperm in their seminal receptacles for months after mating. For example, a female of the multivoltine aquatic predator *Dictya floridensis* collected on March 27 and kept without a male laid 1162 viable eggs over the following 91 days (Valley & Berg 1977), and two females of the univoltine, terrestrial parasitoid-predator *Dichetophora obliterata* collected on May 6 did not oviposit until 64 and 72 days later, laying 155 and 78 viable eggs, respectively (Vala *et al.* 1987).

9.2.2 Oviposition and egg production (see also Chapter 6 on host/prey finding)

Here we comment mainly on site and pattern of egg deposition, number of eggs produced, hatching stimuli, and oviposition rates. Information, some from studies under controlled conditions, on the two latter aspects and pre-oviposition period, oviposition period, incubation period, and survival are included with information on development of other stages in the next section. As noted in Chapter 6,

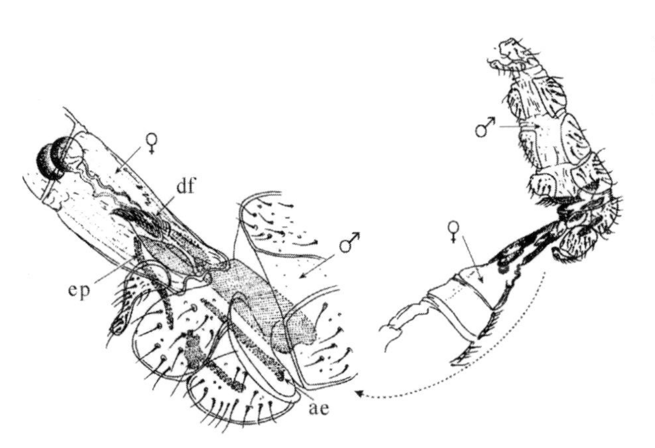

Fig. 9.26 *Tetanura pallidiventris*. Male and female terminalia *in copula*. ae, aedeagal apodeme; df, distiphallus; ep, epiphallus. From Rivosecchi (1992).

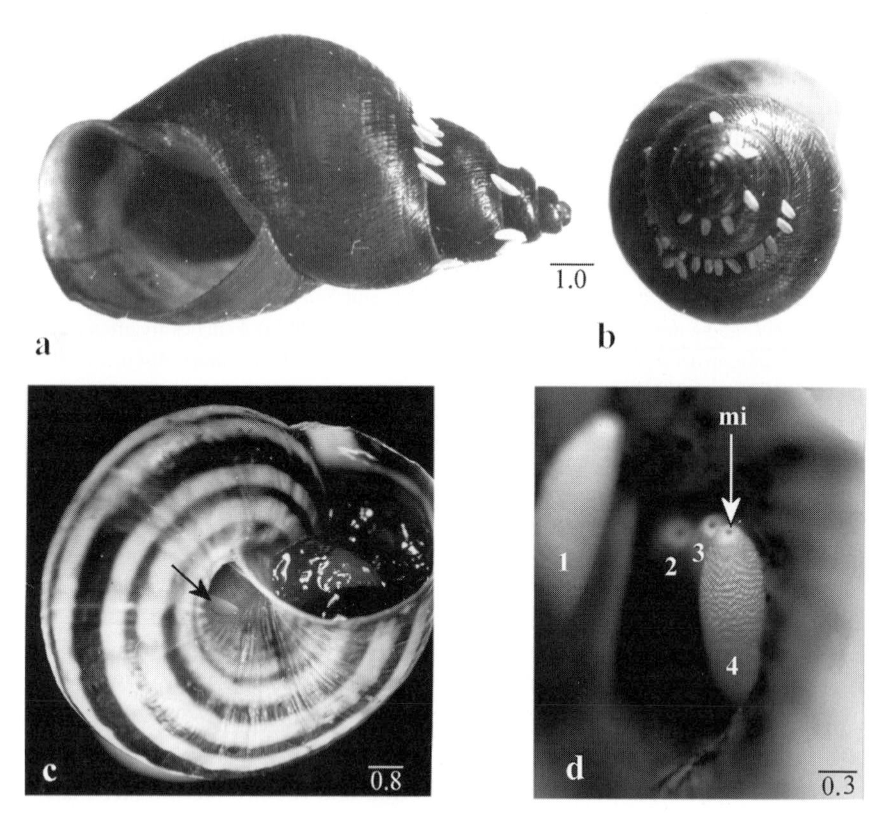

Fig. 9.27 Eggs of Sciomyzidae laid on snails.
(a), (b) ***Colobaea bifasciella*** eggs on estivating *Stagnicola palustris* in lateral and apical views. In nature, usually only one or two eggs are laid on a snail (from Knutson & Bratt unpublished data). (c) ***Salticella fasciata*** egg (arrow) in umbilicus of *Cernuella virgata* (from Knutson *et al.* 1970). (d) Eggs of same in umbilicus of *Helicella bolenensis* collected in nature (photo by J-C. Vala). mi, micropyle; 1–4, eggs. Scale bars in mm.

only five species – one *Pherbellia*, two *Sciomyza*, one *Colobaea* (Sciomyzini), and one *Salticella* (Salticellinae) (Fig. 9.27) – are known to lay their eggs only on the shell of the host snail. *Tetanura pallidiventris* (Sciomyzini) oviposits only onto the flesh of the host. Five species of *Anticheta* (Tetanocerini) oviposit only onto egg masses of the host. In laboratory rearings, the terrestrial predator *Dichetophora obliterata* (Tetanocerini) laid 51% of its eggs on the terrestrial snails *Helicella bolenensis* and *Monacha cartusiana*, but eggs have not been found on snails in nature (Vala *et al.* 1987). All other species, including terrestrial, semi-terrestrial, and aquatic species, at least in laboratory rearings, lay most of their eggs on damp to dry substrates in the rearing containers. Of these, eggs of only the following species have been found in nature: *Sepedon aenescens*, 1–37 eggs in masses on leaves of rice and other plants (Nagatomi & Kushigemachi 1965); *S. spinipes* (on a dry leaf), *S. fuscipennis* (in masses of 12–25) and *S. armipes* and *S. pusilla*, on *Typha* sp. leaf margins (Neff & Berg 1966); *S. sphegea*, in groups of 5–7 on *Lemna trisulca* (Gercke 1876, questionable record); *S. fuscipennis* (816 eggs in 93 masses), *Elgiva solicita* (as *E. sundewalli* Kloet and Hinks) (86 eggs in 41 masses), and *Tetanocera* sp. (98 eggs in 19 masses) on *Bidens* sp., *Alisma* spp., grasses, and other vegetation (Juliano 1981); and *Dictya expansa* (Valley & Berg 1977, site not given).

To determine the natural oviposition sites more precisely, it would be useful to place gravid females in cages over microhabitats in nature, with subsequent search for eggs. Also, holding gravid females in tall cages with vegetation in the laboratory would be of interest. In laboratory rearings, eggs of most species are scattered, singly or in small, unorganized groups, indiscriminately on wet to damp moss or cotton, the sides and gauze covers of breeding containers, included vegetation (Fig. 20.2a), and the shells of crushed snails included as food for the adults.

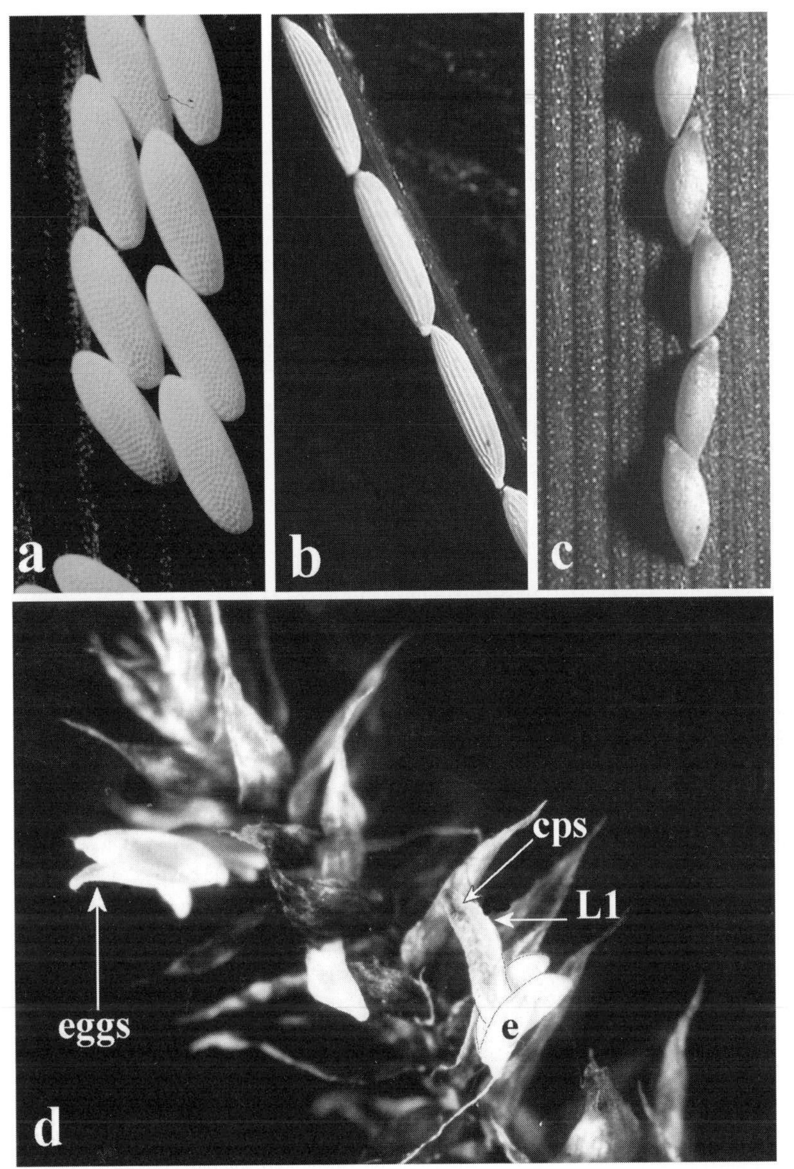

Fig. 9.28 Eggs of Sciomyzidae on vegetation. (a) Eggs of ***Ditaeniella parallela*** (from Bratt *et al.* 1969). (b) Eggs of ***Hoplodictya setosa*** (photo by E. C. Bay). (c) Eggs of ***Renocera*** sp. (photo by B. A. Foote). (d) Eggs and L1 of ***Tetanocera elata*** on moss (photo by L. Knutson); cps, cephalopharyngeal skeleton; (e) egg membrane from which larva hatched.

Eggs of most species are firmly attached and must be moistened to be removed. Eggs of *Colobaea bifasciella* on *Lymnaea* shells remained attached after prolonged storage in 70% ethanol. Eggs of most *Sepedon* are laid side by side in orderly masses on upright grasses, sedges, or other longitudinally ribbed vegetation, each mass composed of a single row of horizontally placed eggs (up to 37 in nature, larger egg masses in the laboratory) (Fig. 11.1). Eggs of the *S. pusilla* group are laid singly or in groups of two or three on leaves and sporophytes projecting from the moss covering the bottom of breeding jars (Neff & Berg 1966). Those of *Dictyodes dictyodes* are usually placed side by side in groups of 10–12 (Abercrombie & Berg 1978). Eggs of *Dictya* usually are laid side by side or nearly so to produce horizontal, vertical, or diagonal rows (Valley & Berg 1977). Those of *Protodictya hondurana* are usually arranged in vertical rows, lying end to end (Neff & Berg 1961). In the genus *Neolimnia*, females of aquatic predators of the subgenus *Pseudolimnia* usually lay their eggs side by side but those of terrestrial predators of the subgenus *Neolimnia* are deposited individually (Barnes 1979b).

Neff & Berg (1966) noted that the hatching sequence of *Sepedon* eggs was not in the same sequence in which they were laid. Abercrombie & Berg (1978) noted this also for *D. dictyodes* and that in a single cluster of ten eggs the incubation period ranged from 3 to 5 days. While deposition of eggs singly, scattered in the microhabitat of the larvae, would seem to increase survival, deposition of eggs in masses may indicate that adults of these species are particularly adept at locating microhabitats for the larvae. This could be a trade-off for the disadvantage of greater exposure to parasitoid wasps and other natural enemies.

Concentration of eggs in masses might depend on the ability of the female to select microsites where incubation will proceed the most rapidly (for multivoltine species). Another possibility is selection of microsites where a long incubation period will correspond with the seasonal development of the microhabitat suitable for the hatching larvae, for example, *I. albiseta* in temporary aquatic situations. Cues for site selection in the latter case are obscure but could be plant species that are abundant in such areas. We can envision elegant experiments in this regard.

Lindsay (1982) Lindsay *et al.* (2011) provided quantitative data on egg masses and oviposition sites from laboratory rearings for *Ilione albiseta*, a univoltine, aquatic predator most common in situations that are dry during the summer (Table 9.1). Females oviposit during the summer, the eggs have a long incubation period and hatch in late summer to fall when the habitats fill with water. Gormally (1988a) found for this species that a greater percentage of eggs were laid singly (93.6–94.8%) at 23 and 20 °C than at 14 and 17 °C (80.8–81.1%). He demonstrated a strongly statistically significant association between temperature and eggs being laid singly or in groups, significant at the 0.1% level. Barnes (1976) found that egg rows produced by a single *Sepedon fuscipennis* vary from 1 to 30 eggs, and row size is influenced by temperature (Table 9.2). Field-collected females consistently produced larger egg masses (40+) than laboratory-reared females.

Eggs of the slug killers *Tetanocera plebeja*, *T. elata*, and *Euthycera chaerophylli* are scattered singly or in groups of two to seven on vegetation near the substrate. The newly hatched larvae remain in place or crawl up to 45 mm upward, raise the anterior third to half of the body at 30–60° angles, intermittently wave the anterior end about for several minutes, and attack a passing slug (Knutson *et al.* 1965, Trelka & Foote 1970). Eggs and larvae of *T. elata* are shown in Fig. 9.28d.

The number of eggs laid on snail hosts by females of intimately associated parasitoids is more limited than the number of eggs laid by predaceous species. In some cases, egg production for one female varied between only 90–100 for *Pherbellia s. schoenherri* (Vala & Ghamizi 1992). Even less egg production was noted for *Sciomyza aristalis* according to Foote (1959a) who recorded 32 eggs laid by one laboratory-reared female in 2 months and 32 and 39 laid by two wild-caught females over a 16-day period. Complete fecundity records for *Sciomyza varia* were 81, 67, and 106 eggs laid on snails during, respectively, 18, 37, and 38 days of life (Barnes 1990). However, a laboratory-reared *Salticella fasciata*, which oviposits onto helicid snails, but which might be more of a saprophage than a parasitoid, produced 389 eggs over 151 days, with the mean life-time fecundity of ten females being 205 ± 14 eggs (Coupland *et al.* 1994). The complete daily oviposition record for one laboratory-reared *S. fasciata*, kept with a male and provided with five fresh *Helicella itala* each day, is shown in Fig. 9.29 (265 eggs in 65 days). Three wild-caught *Anticheta testacea* each laid 5–10 eggs on egg masses daily for 50 consecutive days, after which counts were discontinued (Fisher & Orth 1964). Low numbers of eggs produced by some species that oviposit onto the host probably result from the oviposition target being the host. Thus mortality due to searching for food by newly hatched larvae is reduced.

A great deal of information is available on all aspects of the life cycle of *Pherbellia s. schoenherri*, a multivoltine parasitoid of *Oxyloma* spp. and *Succinea* spp. which overwinters as adults (Verbeke 1960, Rozkošný 1967, O. Beaver 1972a, 1973, Moor 1980, Vala & Ghamizi 1992). The latter two papers provided detailed information on oviposition. The biology of the North American subspecies, *P. s. maculata*, which unlike *P. s. schoenherri* does not routinely oviposit on or pupariate in the shell of the host, was detailed by Bratt *et al.* (1969) and Berg *et al.* (1982). Moor (1980) studied the seasonal fluctuation of parasitism of *Succinea putris* by *P. s. schoenherri* in nature near Basel, Switzerland (Fig. 9.30), and the number of eggs found on snails of five size classes (Fig. 9.31). She also studied survival of larvae and especially encapsulation of larvae by the host (see Section 12.1). Vala & Ghamizi (1992) made quantitative collections of adults of *P. s. schoenherri* and *Oxyloma elegans* throughout the year (May 1984 to June 1985) near Avignon, southern France (Fig. 9.32). They also conducted laboratory experiments on oviposition at different densities of the host, placing one female or pair of adults with 1, 2, 4, 5, 10, 15, 25, or 35 *O. elegans* (Fig. 9.33). They extensively

Table 9.1. Number of eggs per mass and oviposition sites of ***Ilione albiseta***

Mass	1	2	3	4	5	Total
N eggs	477	152	24	16	5	674
%	70.77	22.55	3.56	2.37	0.74	

a. Number of eggs laid per mass

	Vegetation			Surfaces		
	Grass	Others	Total	Upper	Lower	Total
N eggs	408	183	591	204	257	461
N cultures	12	12	12	12	12	12
	$\chi^2 = 20.8$; d.f. = 11; $P < 0.05$			$\chi^2 = 9.9$; d.f. = 11; $P > 0.1$		

b. Oviposition sites used

	Live vegetation	Dead vegetation	Total
N eggs	81	460	541
%	14.97	85.03	

c. Oviposition site preferences

Note: From Lindsay (1982), Lindsay *et al.* (2011).

Table 9.2. Egg row size of ***Sepedon fuscipennis*** at various temperatures

		Calculated average egg row sizes for individual females		
Temperature (°C)	N	Mean	C.V.	Maximum
15	17	5.5	32.4	9.2
21	59	6.6	32.9	14.0
26	54	5.1	34.2	11.3
30	52	5.0	37.3	10.6
33	24	3.7	46.9	8.5

Note: C.V., coefficient of variation. From Barnes (1976).

analyzed the functional response based on their laboratory data according to the equations of Holling (1959), Rogers (1972), and Arditi (1983) (Tables 9.3, 9.4, Fig. 9.34).

In the above studies it was assumed that *Pherbellia s. schoenherri* is host specific to *Succinea* and *Oxyloma* (Succineidae) snails. Bratt *et al.* (1969) found that in laboratory rearings *P. s. maculata* attacked only *Succinea* sp. and *Oxyloma* sp. and did not attempt to feed on the aquatic snails *Helisoma* sp., *Lymnaea* sp., and *Physa* sp. However, O. Beaver (1972a) found that females of *P. s. schoenherri* with a choice of *Succinea putris* and *Galba truncatula* laid a few eggs also on the latter and that, in the absence of *S. putris*, larvae attacked living *G. truncatula*. She concluded that "*G. truncatula* may form an alternative for the larvae in nature if *Succinea* spp. are rare or absent." Mc Donnell *et al.* (2004) collected aquatic snails and sciomyzid larvae by dragging a net through aquatic vegetation, then put the snails (only) into a pail of water. In this way, they found a third-instar larva of *P. s. schoenherri* feeding in a *G. truncatula* on July 28. The larva did not feed further in the laboratory, pupariated on July 31, and a male emerged August 12. No eggs were found on the *G. truncatula* shell. Furthermore, J-C. Vala and M. Ghamizi (unpublished data) found that in laboratory rearings *P. s. schoenherri* did not lay

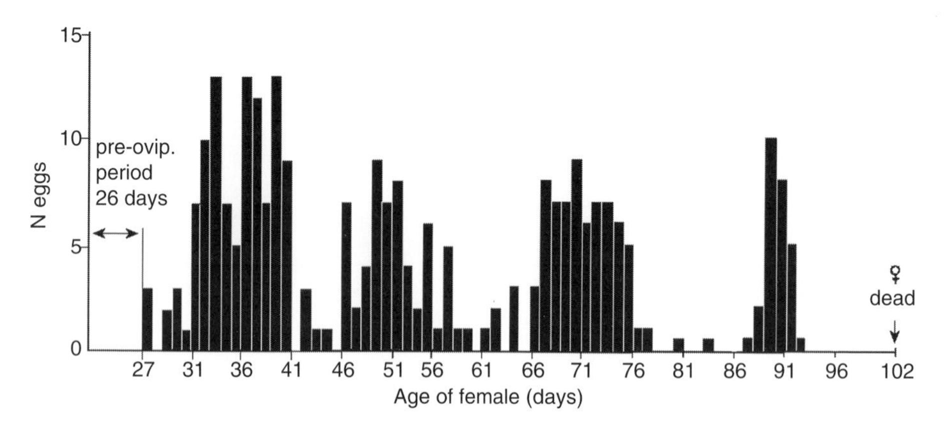

Fig. 9.29 Complete daily oviposition record for one laboratory-reared female of ***Salticella fasciata***. Modified from Knutson *et al.* (1970).

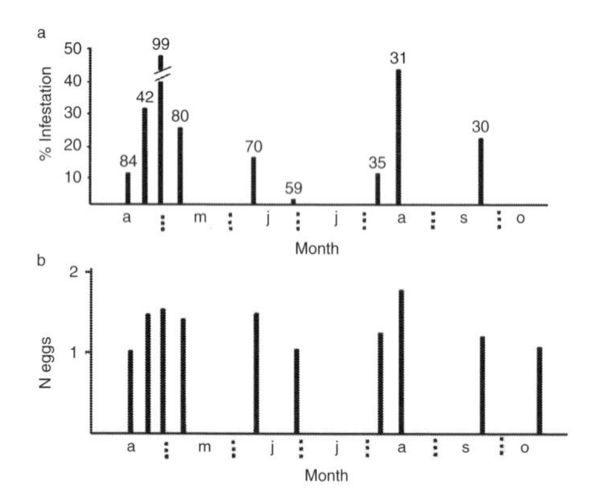

Fig. 9.30 Seasonal variation in infestation of *Succinea putris* by ***Pherbellia s. schoenherri***. (a) Percentage of examined snails infested (numbers above columns are total number of examined snails); (b), number of eggs per snail. Modified from Moor (1980).

eggs in the absence of snails but immediately oviposited when *Oxyloma elegans* were included and they laid eggs also on and developed on the terrestrial snail *Helicella* sp. after several days when only that species was included in the breeding jars.

Information on pre-oviposition, oviposition, and incubation periods and rates; number of eggs laid; and percentage hatching is provided in the many papers on life cycles. Most of these data are from rearings at ambient laboratory conditions. O. Beaver (1973) summarized data from her studies (1972a, 1972b) conducted at laboratory temperatures, of the life cycles of 17 species in Wales. Fecundity data from rearings under a variety of controlled conditions are presented for *Ilione albiseta*, *Sepedon fuscipennis*, *S. sphegea*, and *Tetanocera ferruginea* in Section 9.3, with summaries of development for the genera *Sepedon* and *Tetanocera*. Below we comment on a few other aspects.

O. Beaver (1973) discovered that oviposition curves for the longest-lived females reared in the laboratory seemed to follow an S-shaped curve, and she concluded that many of the oviposition curves of wild-caught females could be interpreted as segments of this curve. Some curves might be truncated at the left-hand end because the flies had been collected part-way through their lives and already had laid some eggs in the field; other curves might be truncated at the right-hand end because the flies died prematurely. Figure 9.35 shows oviposition curves for four wild-caught females of the univoltine, aquatic predator *Pherbina coryleti* that overwinters in the puparium, and for one laboratory-reared female (indicated by L); the dates are dates of collection of the wild-caught four and date of emergence for the laboratory-reared female. Gormally (1988a) found similar curves for 40 females of *I. albiseta*, a univoltine aquatic predator that overwinters as larvae, reared at five constant temperatures, >70% RH, and LD 16:8 (Fig. 9.36). Vala (1986) found similar curves for three wild-caught females of the univoltine terrestrial predator *Trypetoptera punctulata* that overwinters as mature larvae and later in the puparium (Fig. 9.37). See also Mc Donnell *et al.* (2005b) for *Sepedon spinipes*.

Most female sciomyzids produce several eggs daily during their lives (see above and Section 9.3). However, some univoltine species that overwinter as larvae and

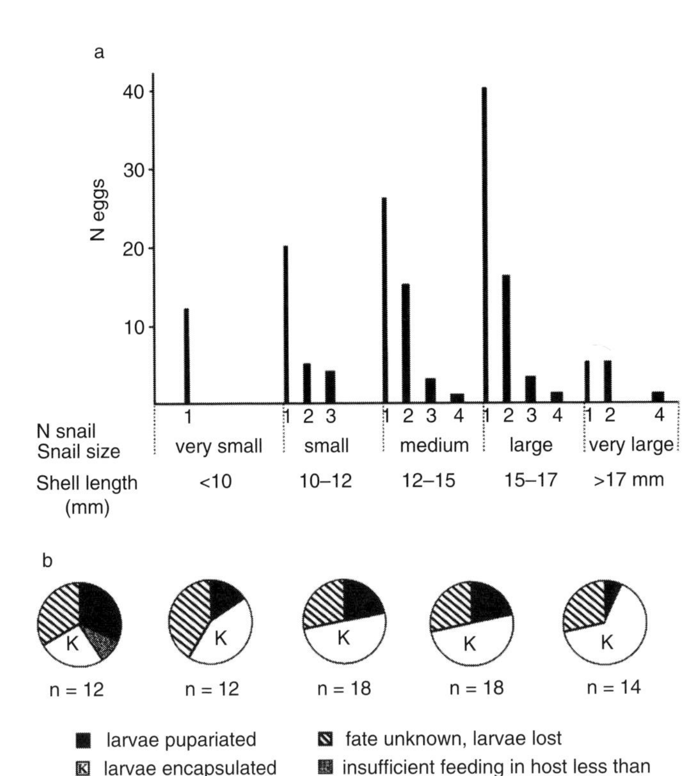

Fig. 9.31 (a) Distribution of numbers of eggs of ***Pherbellia s. schoenherri*** per individual of *Succinea putris* in various size classes of snails. (b) Fate of larvae hatching from eggs. From Moor (1980).

emerge from puparia during the spring, e.g., *I. albiseta*, oviposit intermittently (Fig. 9.38).

Barnes (1976) found that *Sepedon fuscipennis*, a multivoltine, aquatic predator that overwinters as adults, oviposits mostly during the daily photophase. Arnold (1978b) found on a seasonal basis that the peak of reproductive incidence for this species occurred during June–August (40–95%) when the number of pre-reproductive females had fallen to a low value relative to the number of older females in the population. He found that oviposition ceased at the same time each year, when photoperiod was shortening at the maximum rate (Fig. 9.21).

Eggs of *Ilione lineata* must be completely submerged for one to several weeks to hatch (Knutson & Berg 1967). Submergence in an oxygen-deficient medium may be necessary, in part, for eggs of *I. albiseta* to hatch (Gormally 1985b). The need for eggs to be moistened or submerged to hatch does not necessarily indicate a subsurface life style, but may simply be an adaptation that ensures hatching at a time of the year when the environment is wet or inundated and aquatic prey are available. Barnes (1979b) found that eggs of the four species of the terrestrial snail predators *Neolimnia*, subgenus *Neolimnia* in New Zealand hatched readily when kept under high humidity or were moistened but if they were not given sufficient moisture within 3 or 4 weeks of being laid they became desiccated and the unhatched larvae died. He considered that this behavior ensures that the larva will emerge during moist periods when its snail prey is most active. Neonate larvae can be dissected from the egg membranes and will then develop successfully for some but not all of the several species of univoltine, aquatic-predaceous *Tetanocera* that overwinter as first-instar larvae in the egg membranes and also from the eggs of other univoltine *Tetanocera* and *I. albiseta* that overwinter as larvae hatching from eggs laid in late summer to fall (Knutson & Berg 1967, Foote 1999). The hatching stimuli for these species are poorly known and the factors of temperature, photoperiod, oxygen depletion, and others that are probably involved require further experimental studies. The observation that eggs of aquatic species have aeropyles at both poles but that most terrestrial species have aeropyles only on the posterior pole suggests whether a species is aquatic or terrestrial. Submergence, and dissection of neonate larvae from egg membranes, may be useful aids in rearing species that

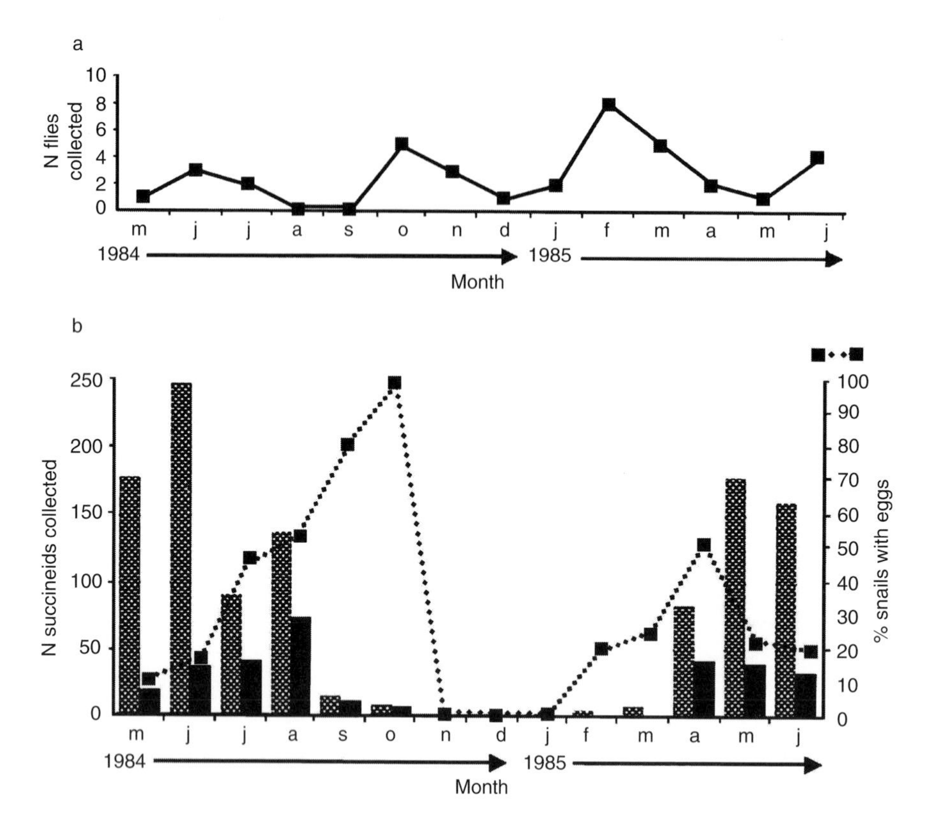

Fig. 9.32 Seasonal variation of ***Pherbellia s. schoenherri*** adults (a) and *Oxyloma elegans* (b) in habitat near Avignon, southern France. Solid bars, snails bearing eggs of the sciomyzid; hatched bars, non-infested snails; dotted line, total snails with eggs during the year. Modified from Vala & Ghamizi (1992).

have thus far resisted attempts to rear them. See also Mc Donnell *et al.* (2005b) for *Sepedon spinipes.*

The process of hatching has been described only for the aquatic predators, *Dictya* spp. (Valley & Berg 1977). When larvae are fully formed, they move to the posterior end of the egg, lunge forward repeatedly, and pierce the membranes near the micropyle with their mouthhooks, the entire process requiring less than 2 minutes.

9.3 DURATIONS OF LIFE-CYCLE ASPECTS AND STAGES AND RATES OF DEVELOPMENT OF IMMATURE STAGES

> The leading idea which is present in all our researches, and which accompanies every fresh observation, the sound which to the ear of the student of Nature seems continually echoed in every part of her works. Is Time! Time! Time!
>
> K. H. Adler.

9.3.1 General considerations

The duration and rates of development of the life stages are key subjects for understanding the overall life cycles, are critical to population studies, selecting species for further in-depth studies, and in selecting biocontrol agents. Of course, development needs to be considered in the context of food consumption, the phenology of the species, and environmental conditions.

Most data on the durations and rates of development are based on rearings conducted at diverse room temperatures, under a wide range of relative humidities, varying natural and artificial light conditions, in a wide variety of containers with different substrates and moisture treatments, and diverse provisioning of hosts/prey. Data on durations and rates of development in most publications have been given as the range from the slowest to fastest. These data also are influenced by the fact that in laboratory rearings there usually was an over-abundance of food

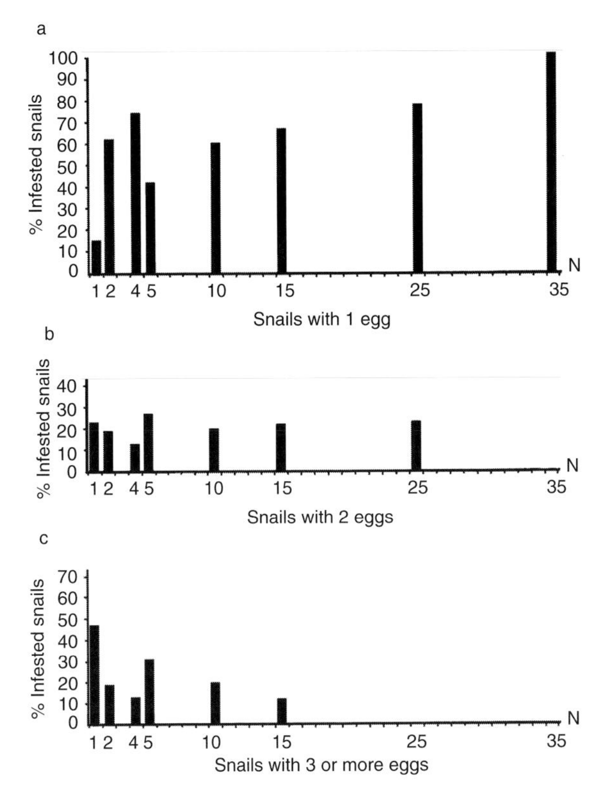

Fig. 9.33 Percentage of *Oxyloma elegans* with eggs laid by one ***Pherbellia s. schoenherri*** in laboratory rearings according to number of snails provided. (a) Snails bearing 1 egg; (b) Snails bearing 2 eggs; (c) Snails bearing 3 or more eggs. From Vala & Ghamizi (1992).

readily available (but in some cases likely not the preferred hosts/prey), little exposure to inimical conditions, no exposure to natural enemies, but probably considerable exposure to "laboratory" pathogens, and little expenditure of energy by the larvae in searching for food. Most rearings have been initiated with one to several females and carried out with up to a few hundred eggs and larvae, and fewer pupae. Only in some cases have rearings of the same species been repeated over several years. Many laboratory observations have been complemented by occasional collections of life stages in nature, but generally not in a regular manner throughout the year. Data on species reared in the same laboratory, by the same researcher using the same methods, are likely to be somewhat more comparable. For example, a large number of species from several regions were reared by C. O. Berg and his students between 1954 and 1978 in Ithaca, New York and in many locations in Europe, South America, and New Zealand; by B. A. Foote and his students in Kent, Ohio from 1961 to present; and by J-C. Vala and his students in Avignon and Orléans, France and in Bénin, 1980 to present.

9.3.2 Summary of development in *Pherbellia* and *Sepedon*

Despite the deficiencies noted above, at least gross comparisons, e.g., whether a species is univoltine or multivoltine, the overwintering or estivating stage, relative rates and duration of development, and presence or absence of diapause or quiescence can be made from rearing data, most obtained under non-controlled conditions. We summarize these data below for two of the largest and best-known genera, *Pherbellia* and *Sepedon*. Another speciose genus, *Tetanocera*, includes a large number of species reared by many researchers under diverse conditions, representing all Phenological Groups except Groups 2 and 6, and thus cannot be succinctly summarized. A detailed review of development in *Tetanocera* has been prepared by Vala & Knutson (unpublished data). The summaries below, based on rearings under a variety of ambient conditions in the laboratory, show broad ranges. Species representing the extremes, with their habitat, if relevant, are given in parentheses.

Pherbellia. The largest genus in the family, with 95 species. Behavioral Groups 2, 3, 4, 6, and 11b. Phenological Groups 1 and 2 with at least six terrestrial species in Group 5a and a few tropical species in Group 6. Bratt *et al.* (1969) presented the basic life cycles of 25 Nearctic, Palearctic, and Holarctic species, including both the typical semi-terrestrial species and six terrestrial species. O. Beaver (1972a) added biological information on *P. cinerella*, *P. s. schoenherri*, and *P. ventralis*. Moor (1980) and Vala & Ghamizi (1992) provided further extensive detail on the intimately associated parasitoid of *Oxyloma* spp. and *Succinea* spp., *P. s. schoenherri*. Foote (1973) reared *P. prefixa*, the only known predator/parasitoid *Pherbellia* species attacking operculate snails, in Montana. Knutson (1988) reared *P. griseicollis*, a typical parasitoid/predator of aquatic snails, in northern Sweden. Nerudova-Horsáková & Vala (unpublished data) reared the terrestrial parasitoid *P. limbata* in the Czech Republic, and Foote (2007) reared the terrestrial parasitoid *P. inflexa* in Montana. Detailed information was provided on overwintering of *P. similis* in New York by Berg *et al.* (1982); on seasonal aspects of *P. cinerella*, a terrestrial predator in Phenological Group 2 by Coupland & Baker

Table 9.3. Comparisons of results in terms of functional response in laboratory experiments on number of eggs laid on succineids in associations of one ***Pherbellia s. schoenherri*** and N *Oxyloma elegans*

Observed values		Calculated values		
N succineids	± C.I. with 95%	Holling	Rogers	Arditi
1	0.76 ± 0.22	0.72	0.75	0.76
2	1.31 ± 1.24	1.13	1.19	1.27
4	1.64 ± 0.44	1.58	1.65	1.89
5	2.36 ± 0.56	1.72	1.78	2.09
10	2.50 ± 1.19	2.08	2.11	2.63
15	3.00 ± 1.02	2.27	2.25	2.86
25	2.75 ± 1.32	2.38	2.37	3.08
35	2.66 ± 2.45	2.43	3.18	–
χ^2		0.66	0.60	0.22

Note: Observed values and their calculated values obtained from equations proposed by Holling (1959), Rogers (1972), and Arditi (1983). C.I., confidence interval. From Vala & Ghamizi (1992).

Table 9.4. Experimental data for determination of Holling's disc equation and Rogers' random parasite equation

N	1	2	4	5	10	15	25	35
n	17	32	14	11	6	6	8	3
Na	0.76	1.31	1.64	2.36	2.50	3.00	2.75	2.66
σ(Na)	0.43	0.69	0.74	0.80	1.04	0.89	1.48	0.58

Note: n, number of experiments; N, number of succineids present in each experimental box; Na, average number of succineids infested (eggs laid onto shells) per day; σ(Na), deviation. From Vala & Ghamizi (1992).

(1995) and Vala & Manguin (1987); and on effects of temperature on development of *P. cinerella* by Gormally (1985b, 1987a).

Overwintering stage: most species in the puparium; *P. griseicollis* and *P. s. schoenherri* as adults, *P. cinerella* in either stage; *P. scutellaris* as third-instar larvae.

Generations per year: one or two in some terrestrial species; up to ten in many semi-terrestrial species in regions where the breeding season is interrupted by cold or other unfavorable conditions, up to 19 where breeding is continuous.

Maximum adult longevity: 21 (*P. albovaria*, terrestrial) to 277 days (*P. griseicollis*, semi-terrestrial).

Pre-mating period: 1–15 days, usually 1–2.

Pre-oviposition period: 2–50 days (*P. knutsoni*, terrestrial). The many species attacking aquatic snails usually have periods of only a few days, except the boreal, semi-aquatic predator *P. griseicollis* apparently must be exposed to long periods of cold (overwintering as adults) before mating. For the terrestrial predator *P. cinerella*, 9–36 days were given by Bratt *et al.* (1969) and 4–21 days were given by Coupland & Baker (1995).

Incubation period: 1–17 days (species attacking exposed aquatic snails and semi-terrestrial snails have the shortest periods, usually 1–5. Six terrestrial species have periods of 4–9, 5–6, 5–13, 8–10, 8–12, and 8–12).

First larval stadium: 1–12 days (*P. dubia*, terrestrial).

Second larval stadium: 1–10 days (*P. dubia*, *P. scutellaris*, terrestrial).

Third larval stadium: 2–165 days (*P. albocostata*, terrestrial). There are limited differences in the first and second stadia but considerable variation in the third stadium, with the terrestrial species in Phenological Group 5a having long third stadia (17–21).

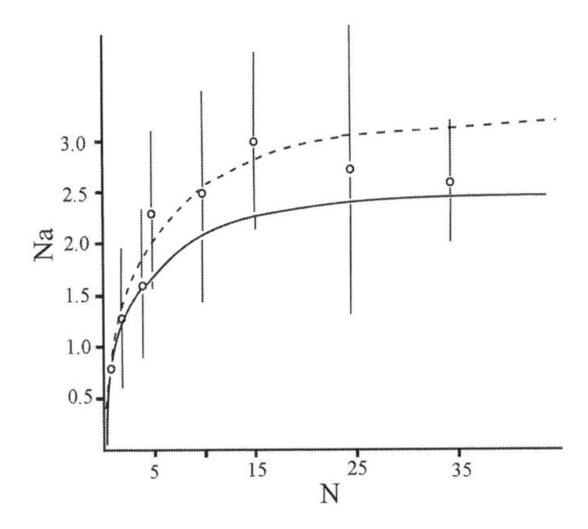

Fig. 9.34 Curves of functional response of type II obtained in association between ***Pherbellia s. schoenherri*** and *Oxyloma elegans* according to equations proposed by Rogers (1972) (solid line) and Arditi (1983) (dashed line). Na, number of snails found daily bearing eggs of the fly; N, number of snails present in each situation. From Vala & Ghamizi (1992).

Total larval life: 4–173 days (*P. albocostata*). Most commonly, only 7–15 days for species attacking exposed aquatic and semi-terrestrial snails.

Puparial period: 6–21 days for species breeding continuously. Six terrestrial species overwinter as diapausing pupae and one (*P. scutellaris*) as a third-instar larva in diapause.

Oviposition to adult emergence: 15–62 days for multivoltine species, 63–285 for univoltine species.

Sepedon. This is the second largest genus of Sciomyzidae (78 spp.), occurring in all except the Subantarctic region, and especially numerous in the Afrotropical region. Most species are aquatic predators (Behavioral Group 11) in Phenological Group 2, but two Afrotropical species are parasitoid/predators of succineids (Behavioral Group 4), one Afrotropical species is a parasitoid/predator of terrestrial snails (Behavioral Group 7), two Afrotropical species is an obligate predators of aquatic oligochaetes (Behavioral Group 15) and one Afrotropical species is a facultative predator of aquatic snails and oligochaetes. Neff & Berg (1966) studied the life cycles of 11 Nearctic species and two Palearctic species, and O. Beaver (1972b) described the life cycle of the Palearctic *Sepedon spinipes*. Mc Donnell (2004) and Mc Donnell *et al.* (2005b) carried out extensive studies of all stages of this species under controlled laboratory conditions. A summary of the data published on North Temperate zone species is presented below, followed by a summary of data on the 12 reared species from the southern hemisphere.

Overwintering stage: primarily as an adult. Two first-instar larvae of *S. spinipes* found on November 9 in Wales by O. Beaver (1972b) developed in the laboratory to the third instar by the end of November and fed very little from December to March; one pupariated April 24 and the other died May 23. Also on November 9 she found a puparium that produced an adult the following year. She concluded that this species are overwinter as an adult, a larva, or in the puparium.

Generations per year: 2 (*S. borealis*, *S. neili*, *S. pusilla*) to several.

Maximum adult longevity: usually 3–5 months for summer generations. *Sepedon* spp. in the Temperate zone overwinter as adults except *S. spinipes* may overwinter in the puparium. The winter generation of adults is especially long-lived in laboratory studies, e.g., *S. pusilla* – 224 days, *S. sphegea* – 226, *S. fuscipennis* – 285, and *S. neili* – 351. For *S. bifida* reared at a constant temperature of 20 °C and 50% RH, females lived 80–140 days and males, 52–131. Arnold (1978b) reported that a female captured, marked, and released in August 1974 (Ithaca, New York) was reproductive when recaptured in May 1975.

Pre-mating period: usually several days for females but only 6–12 hours for males.

Pre-oviposition period: for summer generations, 4–24 days, average 9.6; for *S. bifida* reared at a constant temperature of 20 °C and 50% RH, 19–28, average 21. Much longer for last-generation-of-the-year adults that go on to overwinter, e.g., 70–76 for *S. sphegea*, 147 for *S. anchista*, and 240–279 for *S. neili*, one of the three species with only two generations per year.

Incubation period: range within fastest to slowest developing species: 3–4 to 3.5–9 days.

First larval stadium: fastest to slowest range: 2–3.5 to 3–6 days.

Second larval stadium: fastest to slowest range: 2–4 to 4–6 days.

Third larval stadium: fastest to slowest range: 4–6 to 8–10 days.

Total larval life: fastest to slowest range: 8–13.5 to 15–26 (estimated) days.

Puparial period: fastest to slowest range: 5–7 to 10–14 days.

Oviposition to adult emergence: fastest to slowest range: 24–38 to 43.5–71 days (for summer generations).

In the southern hemisphere seven Afrotropical species of *Sepedon* have been reared (Knutson *et al.* 1967a,

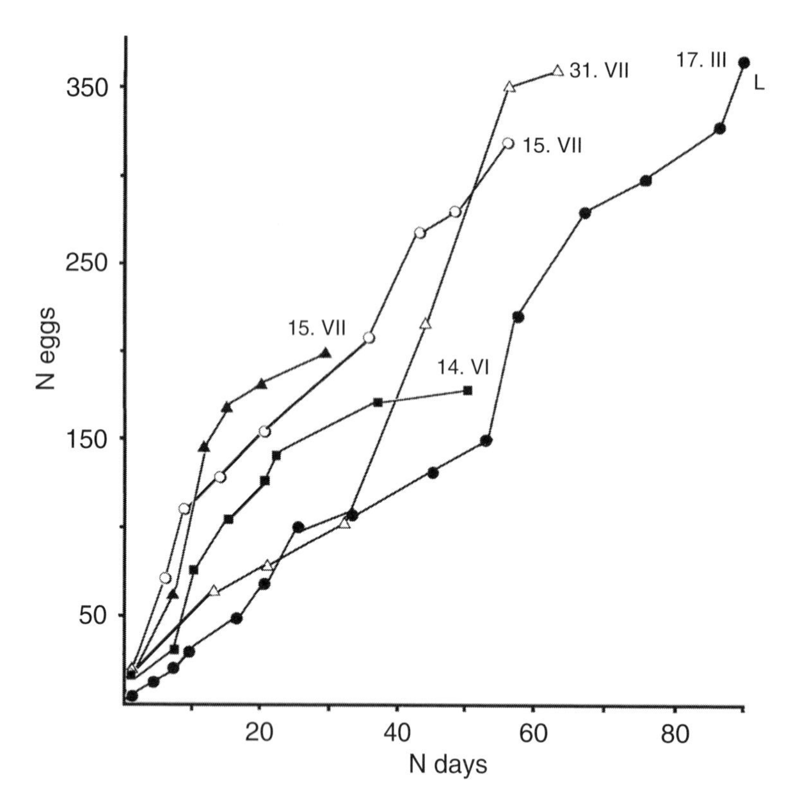

Fig. 9.35 Cumulative numbers of eggs laid by four field-collected (dates are dates of collection) and one laboratory-reared (L, date of emergence) ***Pherbina coryleti***. On abscissa, 0 is the day first eggs were laid. Modified from O. Beaver (1973).

Barraclough 1983, Maharaj *et al.* 1992, Appleton *et al.* 1993, Vala *et al.* 1995, 2000b, 2002, Gbedjissi *et al.* 2003, Knutson 2008) and five Oriental species have been reared (O. Beaver *et al.* 1977 and papers cited therein; O. Beaver 1989). Most of the data are from laboratory rearings and collections of adults, with few data on field collections of immature stages.

We summarize below the data on ranges of most of the developmental periods of Afrotropical aquatic predators from Barraclough (1983) and Appleton *et al.* (1993), and of Oriental aquatic predators from O. Beaver *et al.* (1977). Included are our calculations of the number of generations per year theoretically possible assuming continuous breeding throughout the year, and our estimation of the pre-mating period. We have not included the data of Bhuangprakone & Areekul (1973) for *S. plumbella* because their rearings were with crushed snails only. The extensive data of O. Beaver (1989) on rearings of *S. senex* from parent flies fed four different diets are shown in Table 9.20.

Estivating stage: unknown, probably as adults and in puparia.

Generations per year: 8–14.

Maximum adult longevity: 267 days (*S. spangleri*); usually 30–60 days.

Pre-mating period: estimated at 2–3 days.

Pre-oviposition period: 2–23 days (*S. aenescens*).

Incubation period: 1–7 days (*S. testacea*).

First larval stadium: 2–11 days (*S. spangleri*).

Second larval stadium: 2–9 days (*S. spangleri*, *S. testacea*).

Third larval stadium: 2–13 days (*S. spangleri*), 14 (*S. testacea*).

Total larval life: 8 days (*S. scapularis*) – 30 days (*S. testacea*, calculated).

Puparial period: 3 days (*S. senex*) – 22 days (*S. ferruginosa*).

Oviposition to adult emergence: 25 days (*S. scapularis*) – 44 days (*S. testacea*).

9.3.3 Development of species representative of phenological groups

There have been studies of development under controlled conditions, with different temperatures and photoperiods,

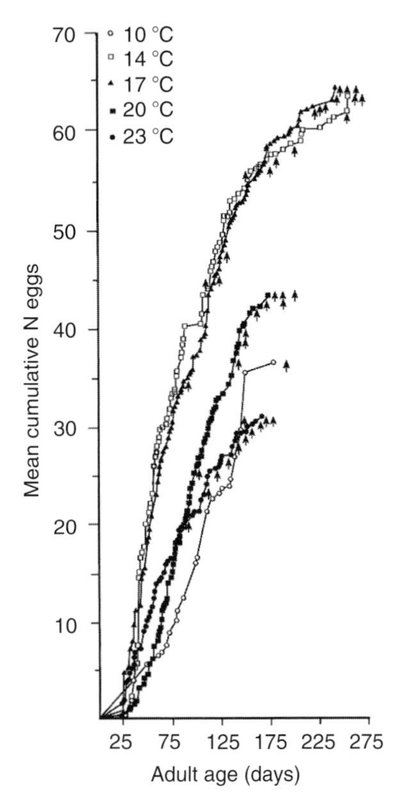

Fig. 9.36 Mean cumulative number of eggs laid each day by laboratory-reared *Ilione albiseta* at five constant temperatures. Points, days on which eggs were laid; arrows, days on which females died. From Gormally (1988a).

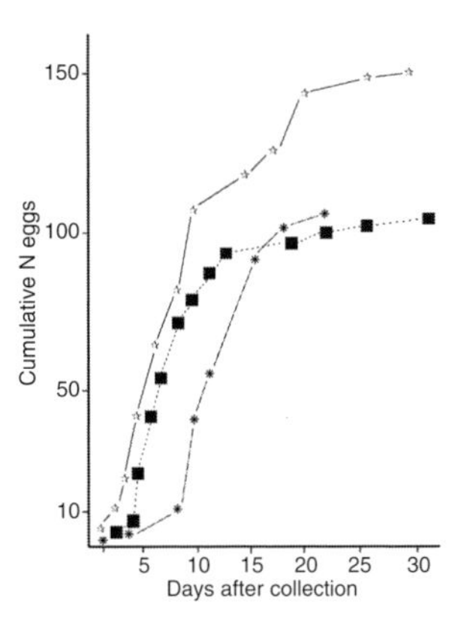

Fig. 9.37 Cumulative numbers of eggs laid by three field-collected ***Trypetoptera punctulata***. From Vala (1986).

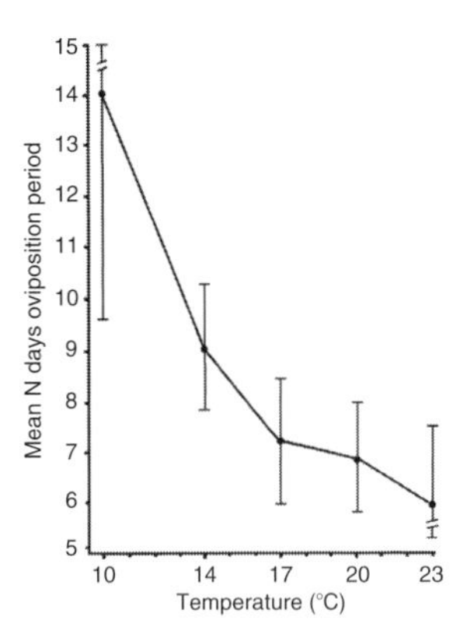

Fig. 9.38 Mean number of days of oviposition period for laboratory-reared *Ilione albiseta* held at five constant temperatures. Vertical lines, 95% confidence limits. From Gormally (1988a).

of only seven species of Sciomyzidae, representing four of the seven Phenological Groups. These are, Group 1: multivoltine species overwintering in puparia (*Tetanocera ferruginea* and *Pherbellia cinerella* [also overwinters as adults]); Group 2: multivoltine species overwintering as adults (*Sepedon fuscipennis*, *S. aenescens*, and *S. spinipes*); Group 4: univoltine species overwintering as larvae (*Ilione albiseta*); and Group 6: tropical multivoltine species breeding continuously (*S. plumbella*). A few other species have been reared under one or more sets of controlled conditions, with the mean as well as the range of development recorded, but without comparative results given (e.g., rearings of *Neolimnia* and *Eulimnia* in New Zealand at 15 and 20 °C, LD 16:8 by Barnes 1979b, 1979c; 1980a, respectively). Below we summarize rates and duration of development for representatives of Groups 1, 2, 4, and 6 that have been studied under controlled conditions.

9.3.3.1 GROUP 1: MULTIVOLTINE SPECIES THAT OVERWINTER IN THE PUPARIUM. *TETANOCERA FERRUGINEA*, AQUATIC PREDATOR AND *PHERBELLIA CINERELLA*, TERRESTRIAL PREDATOR

Although there are only two studies of development of *Tetanocera ferruginea* under controlled conditions (Vala & Haab 1984, Manguin & Vala 1989), the natural history has been studied extensively (Foote 1999, Vala & Knutson [unpublished data], and papers cited therein). There is some information from overwintering experiments (Berg *et al.* 1982), and prey choice and consumption have been detailed much more extensively for this species than for any other sciomyzid (Manguin *et al.* 1986, 1988a, 1988b, Manguin & Vala 1989). Vala & Haab (1984) reared

Table 9.5. Duration of each stadium of *Tetanocera ferruginea* at various combinations of temperature and photoperiod

	Photoperiod (LD)	8:16			9:15	16:8		24:0
	Temperature (°C)	13 ± 1	20 ± 1	26 ± 2	20 ± 1	13 ± 1	26 ± 2	20 ± 1
Duration of stadia (days)	Egg	14.7 ± 1.6 N = 20	5.17 ± 0.1 N = 41	3.0 N = 16	4.2 ± 1.7 N = 36	13.0 ± 1.7 N = 23	4.0 ± 0.5 N = 14	4.9 ± 1.2 N = 13
	L1	13.2 ± 0.6 N =13	9.8 ± 1.6 N = 14	4.0 N = 15	6.5 ± 1.1 N = 20	11.0 ± 1.2 N = 30	5.6 ± 0.8 N = 13	6.0 ± 0.5 N = 31
	L2	12.2 ± 0.6 N = 12	7.7 ± 1.4 N = 7	3.9 ± 0.3 N = 15	6.7 ± 1.4 N = 15	7.9 ± 1.4 N = 30	4.6 ± 0.7 N = 13	6.1 ± 0.6 N = 33
	L3	17.3 ± 1.4 N = 11	12.6 ± 1.6 N = 5	6.1 ± 0.3 N = 16	11.5 ± 1.6 N = 6	14.9 ± 1.0 N = 30	7.4 ± 2.4 N = 11	7.4 ± 0.7 N = 34
	L1–L3	42.5 ± 1.8 N = 10	29.0 ± 1.4 N = 5	13.9 ± 0.6 N = 14	26.5 ± 2.4 N = 6	33.8 ± 1.9 N = 30	17.0 ± 3.2 N = 11	19.4 ± 1.2 N = 29
	Pupa	82.5 ± 10.7 N = 14	92.2 ± 5.5 N = 31	14.9 ± 2.0 N = 13	22.5 ± 3.4 N = 6	32.0 ± 1.8 N = 30	11.6 ± 2.0 N = 7	20.0 ± 2.2 N = 33

Note: From Vala & Haab (1984).

T. ferruginea on *Biomphalaria glabrata* at the following temperatures and light regimes, and at a constant humidity of ± 100%: LD 8:16 at 13 and 20 °C, LD 16:8 at 13 °C and 26 °C, LD 24:0 at 20 °C, and LD 9:15 at 20 °C. In addition to showing the development periods for eggs, larvae, and pupae (Table 9.5), they showed developmental rates of larvae in terms of duration and amount of biomass consumed (Fig. 7.13). Manguin & Vala (1989) conducted rearings at LD 16:8 and 20 °C using three experimental systems, each with a set of 20 snails: 20 each of (1) only one species (*Bathyomphalus contortus*, *Stagnicola palustris*, *Physella acuta*); (2) two species in equal numbers of ten each (*B. contortus* with *S. palustris*, *B. contortus* with *P. acuta*, *S. palustris* with *P. acuta*; and *S. palustris* with *Radix balthica*); and (3) four species (five individuals of each species). They found that the length of time required to complete development was significantly longer when one species of snail was available (22 days) compared with four species available (19 days) (Table 9.6, Fig. 9.39).

Duration of the egg, larval, and pupal stages for the well-studied *Pherbellia cinerella*, which belongs to this phenological group except that adults can overwinter in the southern part of the range, at LD 16:8 and 14, 17, 20, 23, and 26 °C were provided by Gormally (1985b, 1987a). He found that the mean duration of the egg stage decreased from 14.73 days at 14 °C to 4.42 days at 26 °C and percent hatching was greatest at 14 °C. As temperatures increased, mean larval durations decreased, but the percentage of larvae pupariating also decreased (100% at 14 °C, 33.3% at 26 °C); there was a trend for shorter mean pupal durations at higher temperatures. Mc Donnell (2004) provided additional information. Gormally (1985b) also provided some information on the duration of the egg stage of the poorly known terrestrial species, *Limnia unguicornis*, which may belong to this Phenological Group.

9.3.3.2 GROUP 2: MULTIVOLTINE SPECIES THAT OVERWINTER AS ADULTS. *SEPEDON FUSCIPENNIS*, *S. AENESCENS*, AND *S. SPINIPES*, AQUATIC PREDATORS

Neff & Berg (1966) presented the basic life cycles for *Sepedon fuscipennis* and *S. spinipes*. *Sepedon fuscipennis* is one of the best-studied species of Sciomyzidae from the point of view of development. Eckblad & Berg (1972) provided information on adult longevity and larval survival for summer populations. Barnes (1976) reared this species at constant temperatures of 10, 15, 18, 21, 26, 30, and 33 °C

Table 9.6. Comparison of duration of larval stadia of *Tetanocera ferruginea* and number and biomass of snails eaten in System 1 (one prey species), System 2 (two prey species), and System 3 (four prey species)

	System 1	Comparison of systems 1 and 2	System 2	Comparison of systems 2 and 3	System 3	Comparison of systems 1 and 3
	N = 40		N = 40		N = 40	
Duration of larval stadia (days)						
L1	6.2 ± 0.3	NS	6.2 ± 0.3	NS	5.6 ± 0.7	NS
L2	6.0 ± 0.5	NS	5.5 ± 0.4	NS	6.0 ± 0.5	NS
L3	10.6 ± 0.8	**	7.4 ± 0.4	*	8.1 ± 0.6	**
L1–L3	22.4 ± 1.7	**	19.1 ± 0.8	NS	19.7 ± 1.1	**

Note: NS, differences not significant (P > 0.005); * differences significant (P < 0.05); ** differences highly significant (P < 0.01). From Manguin & Vala (1989).

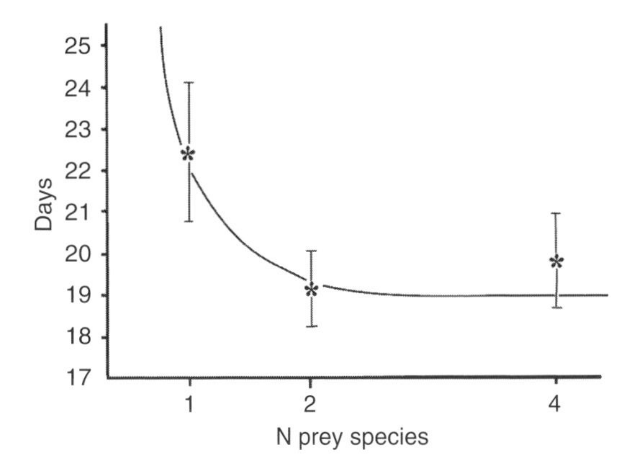

Fig. 9.39 Duration of larval stadia of *Tetanocera ferruginea* according to number (1, 2, or 4) of prey species available. Modified from Manguin & Vala (1989).

at 70% relative humidity and LD 16:8 to determine the effect of temperature on fecundity, development, survival, and diapause. Females usually oviposit from a few days after emergence until death (Table 9.7). Rarely is a period of more than 2 or 3 days passed in which a given female lays no eggs. Total fecundity is quite variable among females at any particular temperature. In general, long-lived females lay more eggs than short-lived females. The maximum number of eggs laid by a female was 2129. Rates of egg-laying for individual females during short periods may reach 50 or more eggs/day (Table 9.8, Fig. 9.40). Sustained rates of 25 or 30 eggs per day are not uncommon in the 10–30 °C range. Developmental rates were greatest at 30 °C (Fig. 9.41; Table 9.9). Total survival of the immature stages was greatest at 26 °C (37%) (Table 9.10). Adult longevity increased as temperatures decreased down to 15 °C. Facultative diapause in the overwintering adults is characterized by cessation of ovarian development and spermatogenesis, and hypertrophy of the fat bodies. Both temperature and photoperiod influence diapause induction in the adult stage. Berg *et al.* (1982) conducted extensive overwintering experiments, providing information on adult survival and reproductive capacity of overwintered adults.

Yoneda (1981) provided information on development at five constant temperatures from 15 to 30 °C, for *Sepedon aenescens* (Table 9.11). This species is closely related to *S. sphegea*, whose biology was studied by Neff & Berg (1966), and which is fairly well established as a member of Phenological Group 2. Yoneda found no significant difference between rates of development of larvae that ate *Physella acuta* and larvae that ate *Hippeutis cantori*. The calculated thresholds of temperatures for development were 13.5 °C in the egg stage, 8.0 °C in the larval stage, 11.1 °C in the pupal stage, and 10.3 °C for completing the whole immature period. The thermal constants in each developmental stage were about 36, 160, 78, and 268 day-degrees, respectively.

Mc Donnell (2004) and Mc Donnell *et al.* (2005b) studied the development of all stages of *Sepedon spinipes* under controlled conditions at 14, 17, 20, 23, 26 °C; LD 16:8. An interesting aspect of these studies is that two feeding regimes – limited (only one snail per day) and excess – with *Radix balthica* as prey were used. Both the mean and

Table 9.7. Pre-oviposition periods of ***Sepedon fuscipennis*** at constant temperatures

Temperature (°C)	15	18	21	26	30	33
N	19	13	54	57	49	25
Mean[a] (days)	30.6	14.2	9.8	7.4	5.4	9.9
C.V.	14.3	6.7	20.2	19.9	21.5	15.3

Note: [a]Geometric means. C.V., coefficient of variation. From Barnes (1976).

Table 9.8. Oviposition data for ***Sepedon fuscipennis*** at constant temperatures

		Fecundity (eggs/female)			Oviposition rate (eggs/day)		
Temperature (°C)	N	Mean[a]	C.V.	Max.	Mean	C.V.	Max.
15	17	125.5	16.3	374	3.0	51.7	9.4
21	59	321.2	19.2	1801	6.3	33.4	17.3
26	54	304.3	19.5	2129	9.6	36.3	33.0
30	52	236.8	23.2	1180	12.5	45.7	41.7
33	24	57.3	30.0	374	5.8	75.2	20.8

Note: Females that did not oviposit excluded from this analysis.
[a]Geometric means; C.V., coefficient of variation. From Barnes (1976).

Table 9.9. Parameter estimates for sigmoid functions relating ***Sepedon fuscipennis*** egg, larval, and pupal developmental rates to temperature

Life stage	C	k_1	k_2
Egg	0.01972	3.94247	−0.17514
L1	0.01480	3.58117	−0.18249
L2	0.00872	3.81923	−0.18783
L3	0.00560	3.69824	−0.18181
Pupa	0.00959	3.96639	−0.19286

Notes: L1, L2, and L3 refer to developmental rates for larvae from eclosion to end of first, second, and third stadia, respectively. C, asymptote of the curve; k_1, k_2, empirical constants. See also Fig. 9.41. From Barnes (1976).

median duration of the pre-oviposition period tended to decrease with increasing temperatures. The duration of the oviposition period was more variable; median duration was longest at 14 °C and shortest at 23 °C. Median total fecundity was greatest at 20 °C (670 eggs) and least at 23 °C (213 eggs). The mean and maximum longevity of laboratory-reared adults decreased as temperatures increased; the median was 218 days at 14 °C and 75 days at 26 °C. Egg hatching was completed at 13, 9, 8, 6, and 5 days, respectively, at the given temperatures. In general, both mean and median duration of the incubation period decreased as temperatures increased. Percentage hatch was greatest at 23 °C (78%) and least at outdoor temperatures (June and July) (53%). For larvae the median duration was significantly longer at 14 °C and 17 °C than at 23 °C and 26 °C regardless of the feeding regime. Median

Table 9.10. Survival of immature stages of ***Sepedon fuscipennis*** at constant temperatures

	Eggs	Larvae	Pupae	
Temperature (°C)	Mean ± S.E.	Mean ± S.E.	Mean ± S.E.	Total
10	.829 ± .016	ND[a]	ND	
15	.939 ± .015	.094 ± .025	.890 ± .075	.079
18	.875 ± .028	.267 ± .045	.932 ± .028	.218
21	.909 ± .019	.331 ± .058	.956 ± .021	.288
26	.955 ± .017	.420 ± .088	.922 ± .021	.370
30	.924 ± .017	.346 ± .029	.921 ± .023	.294
33	.542 ± .052	.466 ± .049	.676 ± .032	.171

Note: [a]Only two larvae of 350 pupariated; these puparia were misshapen and very small, and no adults emerged. ND, no data. From Barnes (1976).

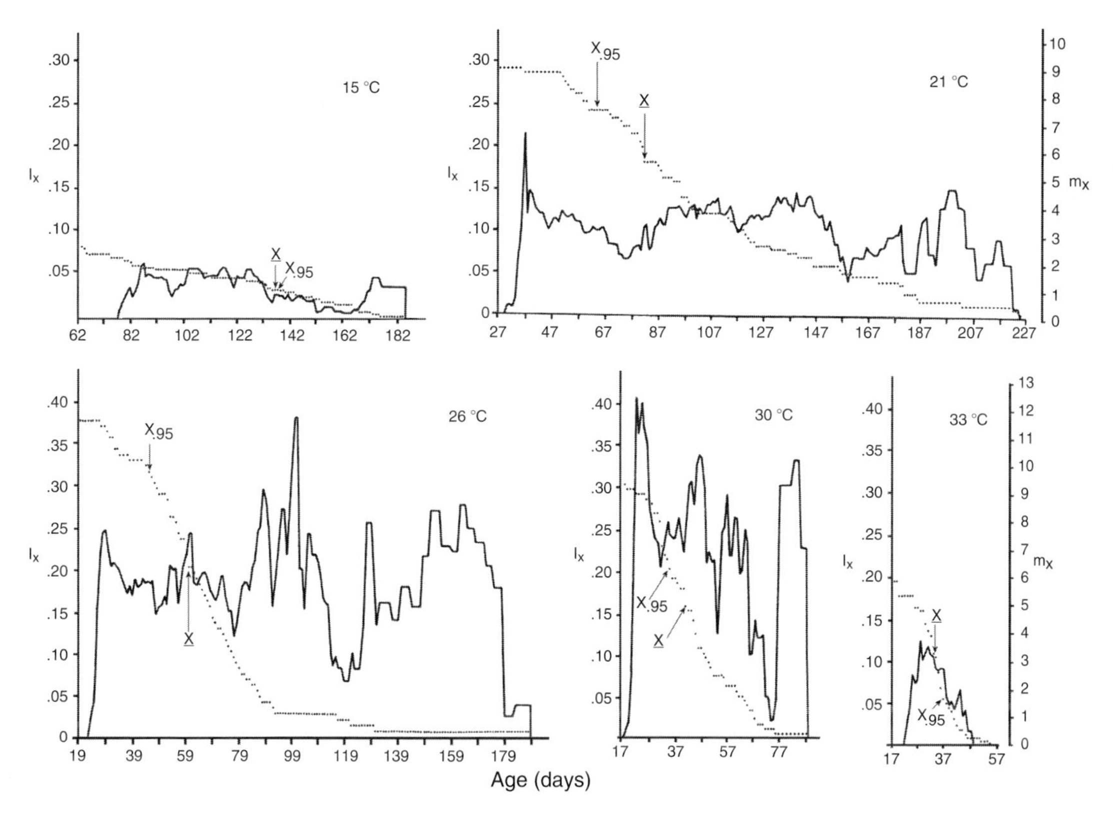

Fig. 9.40 Survivorship (l_x, dotted line) and fecundity (m_x, solid line) curves for ***Sepedon fuscipennis*** adults at various constant temperatures. $\underline{X}$ and X, age of female. From Barnes (1976).

Table 9.11. Duration of development of ***Sepedon aenescens*** at various temperatures

	Duration of development (days)			
Temperature (°C)	Eggs	Larvae	Pupae	Total
15	11.8 ± 0.8	25.1 ± 1.1	19.1 ± 1.0	56.1 ± 1.3
19	5.0 ± 0.0	14.8 ± 1.0	9.0 ± 0.1	28.7 ± 0.9
22*	5.0 ± 0.0	10.8 ± 0.5	7.6 ± 0.7	23.3 ± 0.6
25	4.0 ± 0.0	9.6 ± 0.5	5.8 ± 0.4	19.4 ± 0.5
25*	4.0 ± 0.0	9.0 ± 0.6	6.0 ± 0.6	19.0 ± 0.8
30	2.0 ± 0.0	7.5 ± 0.5	4.1 ± 0.5	13.5 ± 0.5
30*	2.0 ± 0.0	7.3 ± 0.5	3.9 ± 0.2	13.2 ± 0.4

Note: Values are mean ± S.D. Snail prey was *Hippeutis cantori* for rows marked with an asterisk and for others were *Physella acuta*. Modified from Yoneda (1981).

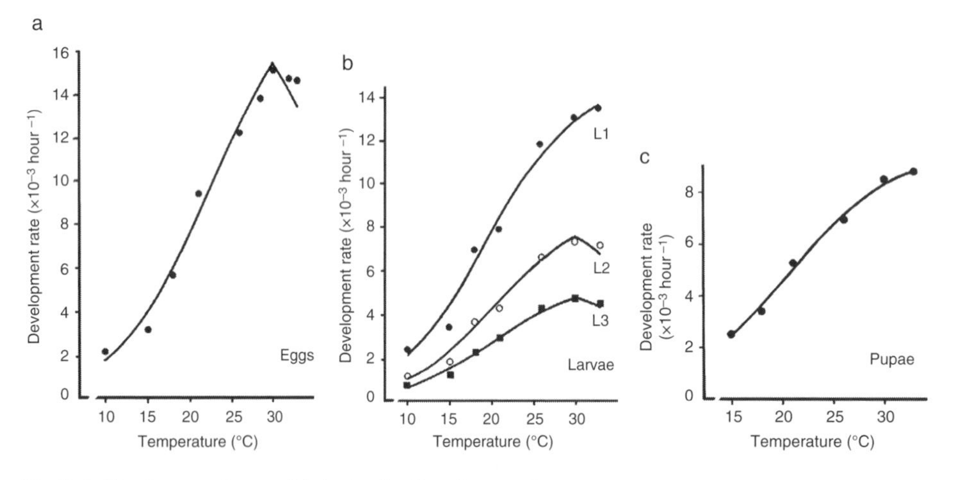

Fig. 9.41 Developmental rates of ***Sepedon fuscipennis*** eggs (a), larvae (b), and pupae (c). Rates calculated by taking reciprocal of developmental time. Solid lines represent fitted sigmoid functions with parameter estimates as given in Table 9.9. From Barnes (1976).

duration of the larval period was significantly shorter with excess snails than with limited snails at each temperature except 14 °C. Emergence of adults from puparia for both limited and excess feeding regimes was greatest at 14 °C and least at 17 °C. As temperature increased (14–26 °C) the mean and median pupal stage duration decreased for both feeding regimes. When larvae were supplied with both limited and excess food, a significant difference was observed in median pupal stage duration between all temperatures except between 14/17 °C, 17/20 °C, 20/23 °C, and 23/26 °C. The median duration of the pupal stage when larvae were fed on excess snails was significantly shorter at 17, 20, and 26 °C than when provided limited food.

9.3.3.3 GROUP 4: UNIVOLTINE SPECIES THAT OVERWINTER AS PARTLY GROWN LARVAE. *ILIONE ALBISETA*, FRESHWATER PREDATOR

This aquatic predator, occurring primarily in vernal and fall sites has been the subject of extensive studies of

Table 9.12. Data on collections of 204 larvae of ***Ilione albiseta*** from traps in a *Galba truncatula* habitat (water-logged drainage ditches) at Glenamoy, Co. Mayo, Ireland by month, instar, and months of pupariation and emergence

	Oct.	Nov.	Dec.	Jan.	Feb.	Mar.	Total
1980–81	39	53	29	15	3	4	143
1981–82	20	15	10	12	4	0	61

a. Larval field collections by month (1980–81 and 1981–82)

	L1	L2	L3	Total
1980–81	0	11	132	143
1981–82	2	16	43	61

b. Larval field collections by instar (1980–81 and 1981–82)

	March	April	May	June	July	Total
N Pupariated	4	53	17	5	0	79
N Emerged	0	0	22	33	0	55

c. Month of pupariation and emergence

Note: From Lindsay (1982), Lindsay *et al.* (2011).

development under controlled conditions (Gormally 1985b, 1987a, 1988a, 1988b) and there are also extensive data on phenology, especially overwintering of larvae (Lindsay 1982, Lindsay *et al.* 2011). Because it is a key reference species and is of interest from biocontrol aspects, its development is reviewed here in some detail. Knutson & Berg (1967) first worked out the basic life cycle, initiated with material from Greece and Italy. O. Beaver (1972b) conducted a limited rearing in Wales and included the species in her laboratory studies of intraspecific and interspecific competitions of eight species in five genera (1974b, 1974c). Lindsay (1982) Lindsay *et al.* (2011) conducted much more extensive rearings and experiments over a 3-year period in Ireland, providing quantitative data on most aspects of the life cycle. On the basis of some data in Knutson & Berg (1967), Berg *et al.* (1982) speculated that *I. albiseta* also can overwinter as developed larvae within egg membranes (Phenological Group 3) but this has not been supported by the later studies.

Overwintering only in the larval stage is strongly supported by the data of Lindsay (1982) Lindsay *et al.* (2011), who was very successful in collecting larvae of all instars in the field in Ireland. During the fall–winter–spring collecting period of 1980–1981 no first instars were found and the last second instar was found on December 2. During the 1981–1982 collecting period the last first instar was found on October 13 and the last second instar on December 4. Most larvae (175 third instars of a total of 204 larvae) were found from the beginning of October to the end of December (Table 9.12). No larvae were found after March in either 1980–1981 or 1981–1982 and no larvae were found before October 1981 despite collections being carried out from September of that year. It is surprising that so few larvae were found in February and March since the evidence from the laboratory cultures indicates that the larvae generally do not begin to pupariate until April. One possible reason is the evidence from the laboratory cultures that mature third-instar larvae spend most of their time out of water and thus are more difficult to collect.

The first puparia obtained in the laboratory from field-collected larvae were formed at the end of March and the last puparium from which an adult successfully emerged was formed on May 21, 1981. None of the puparia formed after that date produced an adult. The mean duration of the pupal stage was 48.5 ± 7.7 days, but the duration varied with pupariation date as the later-formed puparia had much shorter pupal stages. Figure 9.42a shows the

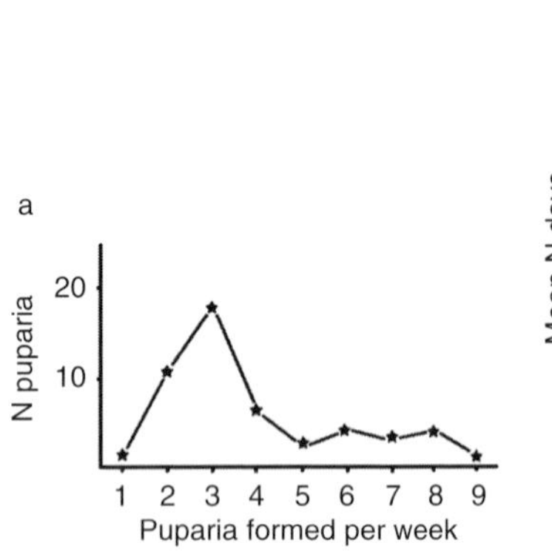

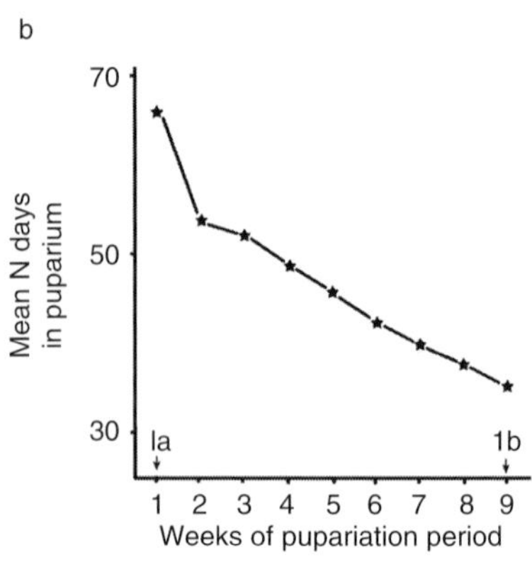

Fig. 9.42 (a) Numbers of puparia of ***Ilione albiseta*** (from 143 larvae field collected October–March, 1980–1981 in Co. Mayo, Ireland) formed (in laboratory) per week during March, April, and May. Including only puparia (55) that produced adults of total formed (79). (b) Effect of pupariation date on mean duration of puparial stage of 55 individuals of ***I. albiseta*** (same individuals as in a). Arrows indicate last week of March (1a) and last week of May (1b). Modified from Lindsay (1982), Lindsay *et al.* (2011).

numbers of puparia formed per week in 1981 and Fig. 9.42b shows the mean duration of the pupal stage for puparia formed in each week of the pupariation period.

Oviposition, the duration of the egg stage, and factors stimulating hatching are of particular interest in this species. Knutson & Berg (1967) noted that adults that emerged from puparia in spring and early summer were sexually inactive but those collected in late summer mated and the females oviposited mid August to mid November, with hatching delayed 33–76 days. They found that immersion in water resulted in hatching of some eggs but immersion was not necessary for all eggs. O. Beaver (1973) confirmed some of these observations, found great variability in oviposition rate, percentage of eggs hatching, and incubation period. She suggested that the differences in durations of the egg stage may be adaptations to survival in the fly's variable and unpredictable, primarily temporary, aquatic habitats.

All the above rearings were conducted at ambient laboratory temperatures or, in part, outdoors. Gormally (1988a) compared the considerable differences among the fecundity and oviposition data in the earlier papers, noting the need to define optimal laboratory rearing conditions for continuous rearings, especially for biocontrol purposes. *Ilione albiseta*, with *T. ferruginea* and *S. spinipes*, was one of the three "fundamental" species in a study of the sciomyzid community (14 species) of a temporary aquatic site in southern France (Vala & Manguin 1987).

The following series of studies of the effects of temperatures on *I. albiseta* (14, 17, 20, 23, and 26 °C at LD 16:8 and > 70% RH) were carried out by Gormally in Ireland: (1) fecundity and longevity of adults, studied also at 10 °C and ambient temperatures of 8.2–24 °C (Gormally 1988a); (2) duration of egg stage, studied also at 2–3 °C and also including *Pherbellia cinerella* and *Limnia unguicornis* (Gormally 1985b); and (3) duration of larval and pupal stages, including also *P. cinerella* (Gormally 1987a). Gormally (1988b) also conducted studies under these conditions of the effects of temperature on predation and survival (see Chapter 7) and oxygen consumption by larvae (see Chapter 18).

Gormally (1988a) found that the length of the pre-oviposition period is not affected by the temperature at which adults are maintained, and suggested that it could be dependent on the temperatures at which their larvae had been reared. Most eggs were laid at 14–17 °C and the mean oviposition period declined progressively above and below 14 °C. There was no significant difference between oviposition rates or pre-oviposition periods at each temperature, but the mean number of days between egg laying for each female was significantly greater at 10 °C than at 17, 20, and 23 °C and greater at 14 °C than at 23 °C (Table 9.13). Adults lived significantly longer at 14 and 17 °C than at 20 °C (Table 9.14).

In Gormally's experiments, eggs were placed in individual compartments of aged tapwater, covered with plastic film, and submerged in water baths at controlled temperatures. The effects of constant temperatures on durations of the egg stage are shown in Table 9.15. The minimum duration was 6 days at 7 °C, and the maximum, 205 days at 2–3 °C. At 2–3 °C and 14 °C, the percentage of eggs hatching was very gradual and extended over a long period (14–205 days). He found a high percentage of hatching of eggs within 24 hours after being held for 5, 10, 15, and 20 days on wet filter paper and then immersed for 24 hours in a 0.1% solution of ascorbic acid (used to chemically lower the level of oxygen) at 23 °C (Table 9.16). Gormally suggested that when the temporary

Table 9.13. Oviposition data for *Ilione albiseta* at constant and ambient temperatures (± S.E.)

Temperature (°C)	N adult pairs	Mean N eggs laid	Infertile eggs (%)	Mean pre-oviposition period (days)	Mean oviposition period (days)	Mean oviposition rate
10	3	32 ± 10.5	16.7	68.0 ± 7.2	84.3 ± 13.5	0.40 ± 0.14
14	10	57.7 ± 6.5	12.3	39.6 ± 4.5	131 ± 15.8	0.50 ± 0.08
17	10	57.8 ± 5.0	12.5	42.1 ± 6.1	128 ± 13.1	0.51 ± 0.64
20	10	40.2 ± 9.6	17.9	60.1 ± 5.8	77.9 ± 11.6	0.52 ± 0.10
23	9	27.7 ± 7.3	9.2	57.9 ± 11.7	51.3 ± 12.2	0.99 ± 0.32
Ambient (8.2–24)	9	47.7 ± 3.8	0.7	41.7 ± 8.2	117.9 ± 10.8	0.44 ± 0.06

Note: From Gormally (1988a).

Table 9.14. Longevity (days) of *Ilione albiseta* adults at constant and ambient temperatures (± S.E.)

	N		Mean longevity		
Temperature (°C)	male	female	male	female	Mean total adult longevity
10	2	3	166.5 ± 27.5	160.0 ± 19.0	162.6 ± 12.3
14	10	10	184.6 ± 14.6	194.0 ± 16.3	189.3 ± 10.7
17	10	10	156.0 ± 12.6	202.0 ± 20.3	179.3 ± 12.7
20	10	10	159.7 ± 9.6	161.0 ± 9.7	160.4 ± 6.7
23	9	9	137.3 ± 8.4	140.3 ± 10.1	138.8 ± 6.4
Ambient (8.2–24)	9	9	128.3 ± 15.9	148.4 ± 10.9	138.4 ± 9.7

Note: From Gormally (1988a).

Table 9.15 Duration of egg stage of *Ilione albiseta* at constant temperatures

Temperature (°C)	N eggs used	Mean duration (days)	S.E. of mean	% Hatch
2–3	232	102.84	4.05	47.8
14	146	88.53	5.00	53.4
17	178	67.23	4.07	46.6
20	82	63.40	3.87	51.2
23	155	53.16	1.05	61.7
26	59	34.05	2.65	62.7

Note: From Gormally (1985b).

aquatic habitats began to fill up with water in the fall (and remained full during the winter), reducing conditions would ensue, causing a gradual decrease in oxygen content, which would stimulate hatching. However, he noted that as most eggs eventually hatch without the stimulus of an oxygen-free medium, as shown by his constant temperature experiments, hatching may be triggered by some physiological process such as food resource depletion. He concluded that, "Clearly all conditions including changing temperatures and oxygen depletion occur in the field and it is possible that they all have a partial effect on the duration of the egg stage in the field." Thus the effects of submergence on hatching are not resolved for *I. albiseta*.

Pritchard *et al.* (1996) used the data of Gormally (1985b) on the development of eggs of *I. albiseta*, *L. unguicornis*, and *P. cinerella* in their study of the development of aquatic insect eggs in relation to temperature and strategies for dealing with different thermal environments. They found that *I. albiseta*, unlike the nine species of Culicidae and two species of Tipulidae included in their study, had a significantly positive slope of average thermal reaction norm, which is the relationship between the number of day-degrees required to complete egg development and temperature. This indicates adaptation to cold temperatures. *Limnia unguicornis* also showed a positive slope and *P. cinerella* showed a negative slope (warm adapted) but the data for those two species were not significant.

Table 9.16. Comparison of effects of 24 h immersion in water and in 0.1% ascorbic acid solution on hatching of conditioned ***Ilione albiseta*** eggs at 23 °C

	Water		Ascorbic acid (0.1 %)	
Age of eggs (days)	N	% Hatch	N	% Hatch
5	0	0	0	0
10	0	0	12	65
15	0	0	14	70
20	0	0	10	50

Note: From Gormally (1985b).

Knutson & Berg (1967) found even greater evidence of the need for submergence in rearings of *I. lineata* where almost 100% hatching, under the surface, was obtained within 2–10 days when embryonated eggs where submerged. The first and second stadia of this species is spent entirely submerged, feeding on fingernail clams.

In Gormally's experiments (1987a), mean duration of first- and second-instars of *I. albiseta* was shortest at 23 °C but at 26 °C the duration increased again. There was 100% survival of first-instar larvae at all temperatures. Mean third-instar larval duration decreased from 104 days at 17 °C to 27.5 days at 26 °C. There was considerable variation in the duration of all larval stages, particularly during the third stadium, at all temperatures except 23 °C where total duration was significantly shorter. Total percentage larval survival was greatest at 17 °C and least at 26 °C. The ranges of duration of the stadia were: first, 5–20; second, 5–19; and third, 17–195 (Table 9.17). Mean pupal duration decreased as temperature increased and this occurred also under outdoor conditions (Table 9.18). The duration of the pupal stage was the least variable part of the life cycle.

9.3.3.4 GROUP 6: TROPICAL, MULTIVOLTINE SPECIES THAT BREED CONTINUOUSLY. *SEPEDON PLUMBELLA*, AQUATIC PREDATOR

Sepedon plumbella is the most common species of Oriental *Sepedon*, ranging from northern India and southern China to the southeastern coast of Australia and New Caledonia. It has a distinctly more southern range than the other common *Sepedon* in the Orient, *S. aenescens* (Phenological Group 1) which extends from southern India and the Philippines to northeastern China and northern Japan. Adults of *S. plumbella* have been collected commonly during all months but there are no records of immature stages collected in nature.

Table 9.17. Effects of temperature on duration of three larval instars of ***Ilione albiseta***

	N			Mean duration of instars (days) ± S.E.			Survival %		
Temperature (°C)	L1	L2	L3	L1	L2	L3	L1	L2	L3
14	21	20	14	13.1 ± 0.4	12.5 ± 0.3	99.7 ± 13.1	100	100	71.4
17	17	17	11	11.4 ± 0.8	9.3 ± 0.4	104.0 ± 20.1	100	100	72.7
20	28	24	13	9.0 ± 0.5	10.4 ± 0.6	92.3 ± 10.7	100	95.8	53.9
23	28	25	18	8.4 ± 0.3	8.7 ± 0.6	35.5 ± 5.7	100	88	44.4
26	23	22	6	9.8 ± 0.5	10.2 ± 1.5	27.5 ± 4.5	100	27.3	33.3

Note: Modified from Gormally (1987a).

Table 9.18. Effects of temperature on duration of *Ilione albiseta* pupal stage and adult emergence

	N			Mean duration of pupal stage (days) ± S.E.			Emergence (%)	
Temperature (°C)	a	b	c	a	b	c	a	b
14	10	–	9	28.3 ± 0.9	–	28.3 ± 0.9	90	–
17	13	–	10	22.0 ± 0.5	–	22.0 ± 0.5	76.9	–
20	8	5	10	15.0 ± 0.8	17.4 ± 0.5	16.0 ± 0.6	57.1	100
23	8	7	12	11.8 ± 0.2	11.9 ± 0.1	11.8 ± 0.1	62.5	100
26	2	8	8	Died	11.0 ± 0.2	11.0 ± 0.2	0	100

Note: Reared at (a) constant temperatures, (b) ambient temperatures and placed at constant temperatures on pupariation, (c) reared at (a) and (b) combined, resulting in adult emergence. From Gormally (1987a).

Bhuangprakone & Areekul (1973) reared *S. plumbella* in Thailand on wet filter paper in Petri dishes in a constant-temperature room (25 ± 1 °C), without mention of the lighting regime. Adults were exposed to ambient temperatures (24–32 °C). We note that 25 °C is the highest or nearly the highest temperature at which the other species were reared under constant temperatures, and in Bhuangprakone & Areekul's rearings only freshly crushed snails were provided to the larvae. Six of the nine species of crushed snails were operculates, and in preference tests of third-instar larvae none of these was killed by the larvae but three non-operculates (*Gyraulus convexiusculus*, *Radix auricularia*, and *Indoplanorbis exustus*) were readily attacked and eaten. There were only slight differences in duration of stages in larvae developing successfully on the three non-operculate species and three of the operculate species, but there was longer life of the adults and a higher rate of oviposition during rearings on the operculate, *Melanoides tuberculata*. Comparative data with three other tropical, Oriental species of *Sepedon* and *S. aenescens* reared at 25 ± 3 °C and ambient temperatures by O. Beaver *et al.* (1977) (Table 9.19) show only a few differences in rates and durations. No diapause or quiescence has been observed in any of these species, except for indications of diapause in overwintering adults of *S. aenescens* in Japan (ChannaBasavanna & Yano 1969). More detail on *S. senex* reared on various combinations of live and crushed snails and from parents fed various diets are given in the next section.

9.3.4 Effects of diet, biomass consumption, and other factors on rates and durations of development

There have been indications in many publications on life cycles that protein in the form of crushed snails added to a diet consisting of a paste of honey, dried milk, and brewer's yeast increases adult fecundity and longevity. This was quantified as early as 1961 by Chock *et al.* for *Sepedomerus macropus* reared in Hawaii for mass releases against *Austropeplea ollula*. O. Beaver (1989) carried out detailed experiments in Thailand at room temperatures on diets of adults and larvae of *Sepedon senex*. Two types of experiments were conducted. In the first, adults were reared on four different diets: (A) honey – dried milk at a ratio of 3:5; (B) honey – dried milk – yeast at 3:5:2; (C) honey – dried milk – crushed snail at 3:5:1; and (D) honey – dried milk – yeast – crushed snail at 3:5:2:1. In the second type, larvae from adults fed with each diet were used to investigate the effect of larval food on growth and development. In one set larvae were fed with live snails, in another they were fed with freshly crushed snails. The incubation period of eggs, growth rate, developmental periods of the larvae, and the pre-oviposition period of progenies from flies fed with high-protein food were shorter than those of larvae from parents fed with low-protein food (Table 9.20, Fig. 9.43). Larvae from flies fed with live snails had shorter developmental periods and higher growth rates than those fed with crushed snails. Larvae from flies fed with higher-protein food had higher growth rates and higher conversion efficiencies than those fed with low-protein food (Tables 9.21, 9.22). The larvae consumed more tissue of live snails than of crushed snails. Conversion efficiency and growth rates of larvae fed with live snails were higher than of those fed with crushed snails.

The publication on *S. plumbella* by Bhuangprakone & Areekul (1973) referred to in the previous Section also includes comparative data on rates and durations of development and survival rates of larvae provided different snails. Some data are also found in the publication of Yoneda (1981)

Table 9.19. Range (R) and mean (M) ± S.D. of periods of incubation, larval instars, puparial, and pre-oviposition of Oriental species of *Sepedon*

	Incubation period		L1		L2	
Species	R	M ± S.D.	R	M ± S.D.	R	M ± S.D.
S. ferruginosa	1–6	3.16±0.73	2–8	3.74±1.45	2–8	3.78±1.38
S. plumbella	1–5	2.86±0.66	2–7	3.38±0.79	2–6	3.48±0.70
S. senex	1–7	3.09±0.79	4–5	4.38±0.48	3–5	3.83±0.69
S. spangleri	1–10	3.38±1.20	2–11	4.89±1.46	2–9	3.84±1.35
S. spangleri[1]	3–4	3.55±0.51	4–5	4.35±0.49	2–4	2.95±0.76
S. aenescens	1–6	2.65±0.93	2–5	3.03±0.78	2–5	3.48±0.78
S. aenescens[2]	–	4.00±0.00	–	3.30±0.46	–	2.80±0.29
	L3		**Puparial period[3]**		**Pre-oviposition period**	
	R	M ± S.D.	R	M ± S.D.	R	M ± S.D.
S. ferruginosa	2–10	4.90±1.58	5–22	7.21±1.61	2–22	13.84±6.10
S. plumbella	3–9	4.66±1.14	5–14	7.27±1.94	5–20	8.54±3.97
S. senex	3–4	3.40±0.49	3–6	5.00±1.00	3–18	10.75±5.34
S. spangleri	3–12	6.08±2.38	5–17	8.06±3.38	2–10	4.70±1.78
S. spangleri[1]	3–5	4.45±0.61	8	8.00±0.00	3–5	4.60±0.86
S. aenescens	2–5	3.46±0.72	3–8	5.31±1.04	7–23	14.50±5.68
S. aenescens[2]	–	3.70±0.48	–	6.00±0.64	–	–

Notes: [1]Data of Chandavimol *et al.* (1975), rearings at 25 ± 3 °C on *Biomphalaria* in Thailand.
[2]Data of Yoneda (1981), rearings at 25 °C on *Hippeutis cantori* in Japan.
[3]Period from puparium formation to emergence of adult. All results in days. From O. Beaver *et al.* (1977).

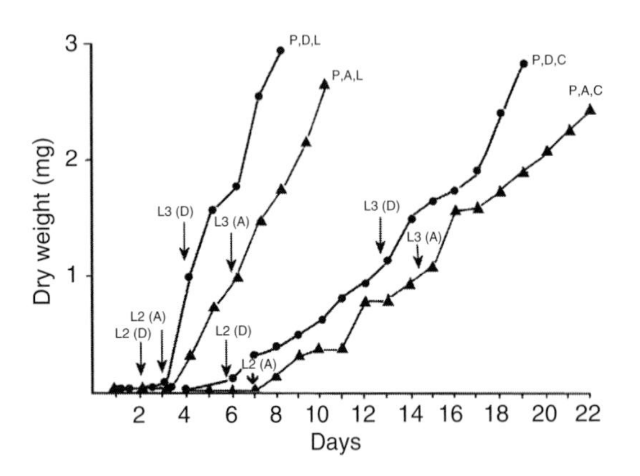

Fig. 9.43 Growth rates and developmental periods of *Sepedon senex* larvae (from parents fed with diets A (honey + dried milk) and D (honey + dried milk + yeast + crushed snail) fed with live and crushed *Gyraulus convexiusculus*. Arrows indicate dates of molting into L2 and L3. A, D, Parents' diet; L, larvae fed with live snails; C, larvae fed with crushed snails; P, pupariation date. Modified from O. Beaver (1989).

for *Sepedon aenescens*. Barraclough (1983) noted some differences in the duration of larval life (28–35.5 days) for the freshwater predator *Sepedon neavei* reared on four genera of non-operculates in South Africa.

The range and mean duration of the three larval stadia, weight at which ecdysis occurs, time periods between ecdyses, and other aspects of development of eight species of aquatic predators in five genera were studied as part of experiments on inter- and intraspecific competition in Wales by O. Beaver (1974b, 1974c) (see Chapters 4 and 8). The duration and rate of development of larvae (primarily third instar) of the aquatic predator *Sepedon sphegea* according to species of prey, density of prey, biomass consumed, and density of the predator was studied in detailed experiments on the predator–prey dynamics by Haab (1984) and Ghamizi (1985) (unpublished theses).

Ghamizi (1985) presented the results of laboratory experiments on predation of *Physella acuta*, *Radix balthica*, and *Stagnicola palustris* by *Sepedon sphegea* larvae. For base-line data, he determined the maximum consumption

Table 9.20. Developmental periods (days) of ***Sepedon senex*** from parents fed four different diets, A–D, when larvae fed with live and freshly crushed *Gyraulus convexiusculus*

		L1	L2	L3	Pupal period	Pre-oviposition period	Incubation period	
Food of parent flies	Food of larvae	M ± S.D.	M ± S.D.	M ± S.D.	M ± S.D.	M ± S.D.	M ± S.D.	Sex ratio (M:F)
A	l.sn.	3.30 ± 0.48	2.80 ± 0.42	3.80 ± 0.42	8.80 ± 1.31	8.60 ± 0.89	3.00 ± 0.68	
	c.sn.	7.25 ± 0.71	7.38 ± 0.92	6.75 ± 0.71	11.88 ± 1.89	12.33 ± 2.52	3.05 ± 0.64	5:3
B	l.sn.	3.11 ± 0.33	2.44 ± 0.53	3.78 ± 0.44	7.66 ± 0.87	8.60 ± 0.84	2.68 ± 0.48	
	c.sn.	6.71 ± 0.76	7.28 ± 0.98	6.71 ± 0.75	10.29 ± 0.95	11.33 ± 1.53	2.80 ± 0.41	3:4
C	l.sn.	3.00 ± 0.00	2.10 ± 0.20	3.60 ± 0.52	6.40 ± 0.70	8.60 ± 0.55	2.63 ± 0.49	
	c.sn.	6.33 ± 0.87	7.11 ± 0.33	6.33 ± 0.87	9.28 ± 1.54	9.75 ± 1.50	2.78 ± 0.52	4:5
D	l.sn.	2.00 ± 0.00	2.00 ± 0.00	3.50 ± 0.53	5.40 ± 0.70	8.80 ± 1.10	2.56 ± 0.51	
	c.sn.	5.86 ± 0.69	7.00 ± 0.00	6.28 ± 0.95	9.28 ± 1.49	9.65 ± 0.58	2.72 ± 0.46	4:3
D[a]	l.sn.	2.30 ± 1.70	2.78 ± 0.78	3.62 ± 1.25	5.24 ± 0.51	8.87 ± 0.01	2.50 ± 0.61	
	c.sn.	5.24 ± 0.13	6.25 ± 0.73	6.78 ± 0.86	7.32 ± 1.01	9.02 ± 0.51	3.07 ± 0.41	NR
D[b]	c.sn.	4.89 ± 1.46	3.84 ± 1.35	3.40 ± 0.49	5.00 ± 1.00	0.75 ± 5.34	3.09 ± 0.79	NR

Note: A, honey + dried milk; B, honey + dried milk + yeast; C, honey + dried milk + crushed snail; D, honey + dried milk + yeast + crushed snail; [a] Reared in 5 cm diameter dish with 1 cm depth of water. [b] Data from O. Beaver *et al.* 1977; l.sn., live snail; c.sn., crushed snail; M:F, male:female; NR, not recorded. From O. Beaver (1989).

Table 9.21. Food consumption and growth rate (mg dry weight/larva/day) of ***Sepedon senex*** larvae from parents fed four different diets, A–D as in Table 9.20

		L1		L2		L3	
Food of parent flies	Food of larvae	Food consumption	Growth rate	Food consumption	Growth rate	Food consumption	Growth rate
A	l.sn.	2.25	0.012	3.04	0.35	3.21	0.47
	c.sn.	1.34	0.0055	1.72	0.131	0.131	0.182
B	l.sn.	2.25	0.013	3.15	0.40	3.32	0.57
	c.sn.	1.36	0.006	1.83	0.132	2.21	0.236
C	l.sn.	2.25	0.016	3.18	0.46	3.32	0.57
	c.sn.	1.36	0.0064	1.87	0.134	2.25	0.296
D	l.sn.	2.34	0.020	3.19	0.48	3.35	0.59
	c.sn.	1.37	0.0069	1.91	0.136	2.44	0.309

Note: l.sn., live snail; c.sn., crushed snail. From O. Beaver (1989).

for each larval instar, rearing larvae separately with unlimited food. The rearings were conducted at a temperature of 20.0 ± 1.0 °C, a photoperiod of 12:12, and in water at a depth of 3 cm. In days, using the first-mentioned prey, the median development data were 4.8 ± 0.2 (L1), 3.7 ± 0.2 (L2), 5.4 ± 0.3 (L3), 13.8 ± 0.3 (L1–L3) and 9.9 ± 0.37 (pupae). These results corroborate the values cited by Neff & Berg (1966) and Haab (1984) for *S. sphegea*. By varying the prey density, the author showed the importance of the density and behavior of the prey. Many experiments (each repeated five or ten times) were conducted with prey densities equivalent to 300 m^2 (5), 900 m^2 (15), 1500 m^2 (25), 2100 m^2 (35), and 3000 m^2 (50) (number of prey is indicated in parentheses).

With *Physella acuta*, Ghamizi (1985) found that the effect on total duration of the larval cycle (L1–L3) was highly significant, as low as 18.6 ± 4.5 to 12.4 ± 2.1 days ($P < 0.001$, I. C. 95%) when prey density increased (Table 9.23). These reductions of number of days concerned mainly the first- and third-instar larvae, specifically, from 9.00 ± 0.10 to 6.20 ± 2.30 and 6.00 ± 2.80 to 3.20 ± 0.60, respectively. On the contrary, during the second stadium only slight variations were seen, from 3.60 ± 1.20 to 3.00 ± 1.00 ($P > 0.05$). These data showed that the duration of the pupal stage was essentially independent of the prey density fluctuations. When prey density gradually increased, the duration of larval life tended to the median development rate when larvae were reared with unlimited food. These results also show one of the numerical aspects of development of *Sepedon sphegea*, i.e., its predation behavior being, in part, a function of prey density. Therefore, Ghamizi (1985) carried out some studies on variation in predation by using only third-instar larvae with different snail species. He found that the variation in number of prey eaten during the third stadium also decreased when *Stagnicola palustris* or *Radix balthica* were used as prey.

Table 9.22. Conversion efficiency (growth rate: food consumption rate) of ***Sepedon senex*** larvae from parents fed four different diets, A–D as in Table 9.20

		Conversion efficiency (%)		
Food of parent flies	Food of larvae	L1	L2	L3
A	l.sn.	0.5	12.0	15.0
	c.sn.	0.4	8.0	9.0
B	l.sn.	0.6	13.0	16.0
	c.sn.	0.4	7.0	11.0
C	l.sn.	0.7	14.0	17.0
	c.sn.	0.5	7.0	13.0
D	l.sn.	0.8	15.0	18.0
	c.sn.	0.5	7.0	13.0

Note: l.sn., live snail; c.sn., crushed snail. From O. Beaver (1989).

Overall, as a function of increased numbers of prey, Ghamizi (1985) showed that there was a gradual increase in daily numbers of snails consumed during each larval stadium of *S. sphegea* (Fig. 9.44a, c, d). This is one characteristic of the functional response of larvae; their predation differs as a result of differences in the density of available prey. Concerning only the first- and second-instar larvae, there were two results. In terms of number of snails consumed the total is nearly the same between these two larval stadia – the food consumption curves are parallel (Fig. 9.44a), but there is a very sharp difference in their daily consumption which is marked by an inversion in the position of the curves of daily consumption (Fig. 9.44c). Since the second-instar larva has a shorter duration (Fig. 9.44b), its consumption is greater. The first-instar larva lives longer and its curve of duration of development is in a superior position with regard to that of the second-instar larva (Fig. 9.44b). But, for the level of daily consumption, the curve of the second-instar larva occupies the higher position (Fig. 9.44c), attesting to an intake of food that is much more elevated since its duration of life is two or three times less. Concerning the third-instar

Table 9.23. Duration of each immature stadium of ***Sepedon sphegea*** as function of density of *Physella acuta* as prey

Physella acuta		Duration of each immature stadium (in days)				
Density m^2	N present	L1	L2	L3	L1–L3	Puparia
300	5	9.00 ± 1.40	3.60 ± 1.20	6.00 ± 2.80	18.60 ± 4.50	6.00 ± 2.60
900	15	7.40 ± 1.60	3.00 ± 1.00	4.40 ± 1.60	14.80 ± 1.50	6.00 ± 1.70
1500	25	6.60 ± 2.70	3.60 ± 1.80	4.20 ± 1.12	14.40 ± 3.10	6.40 ± 1.80
2100	35	6.20 ± 2.10	3.20 ± 0.60	3.80 ± 0.12	13.20 ± 3.30	6.60 ± 1.60
3000	50	6.20 ± 2.30	3.00 ± 1.00	3.20 ± 0.60	12.40 ± 2.10	6.20 ± 1.40

Note: Data from Ghamizi (1985).

Table 9.24. Total and daily consumption by ***Sepedon sphegea*** L3 as function of species of prey

Prey characteristic		Prey species					
Density m2	N present	*Physella acuta* Consumption		*Radix balthica* Consumption		*Stagnicola palustris* Consumption	
		Total for one L3	Daily for one L3	Total for one L3	Daily for one L3	Total for one L3	Daily for one L3
300	5	5.60 ± 0.70	1.40 ± 0.22	5.60 ± 0.70	0.90 ± 0.31	3.60 ± 0.80	0.58 ± 0.20
600	10	7.20 ± 2.30	1.58 ± 0.20	5.60 ± 0.70	1.00 ± 0.29	5.00 ± 1.70	0.86 ± 0.12
900	15	**8.60 ± 2.5**	1.88 ± 0.11	6.20 ± 1.20	1.28 ± 0.11	5.20 ± 1.80	0.92 ± 0.11
1200	20	8.40 ± 1.60	1.98 ± 0.21	6.80 ± 1.80	1.38 ± 0.12	4.80 ± 1.20	1.00 ± 0.22
1500	25	7.80 ± 2.30	1.94 ± 0.32	**7.20 ± 1.80**	1.58 ± 0.21	**5.60 ± 0.70**	1.42 ± 0.20
2100	35	8.00 ± 1.70	2.12 ± 0.26	7.20 ± 1.20	1.70 ± 0.22	5.40 ± 1.30	1.44 ± 0.32
3000	50	8.40 ± 2.31	2.30 ± 0.22	7.40 ± 1.31	1.88 ± 0.12	5.60 ± 0.70	1.72 ± 0.33

Note: Bold indicates levels of prey where maximum consumption begins. Data from Ghamizi (1985).

larva, the total number of prey consumed is more than the sum of the two preceding stadia (Fig. 9.44d). Its consumption rose from 1.40 *Physella acuta* per day at a density of 300 per m^2 to 2.30 per day when the density attained 3000 prey per m^2 (Table 9.24).

Predation by third-instar larvae of *S. sphegea* varied according to the species of mollusc utilized. Ghamizi (1985) noted that the total was 5–11 with *P. acuta*, 5–9 with *R. balthica*, and 3–7 with *S. palustris* (Table 9.24). For each of the species of prey, the variation in total consumption was significantly different as a function of their densities ($F = 79.2$; $P < 0.001$) and the species of prey ($F = 49.2$; $P < 0.001$). The curves of predation (Fig. 9.44d) present a similar profile between the three species of prey, with an increase in the number of snails consumed, jointly with that of the density of prey present. However, the levels of the curves show variation due to the behavior of each of the prey species. *Physella acuta*, a species that lives near the surface of the water and is more easily attacked, was more extensively preyed upon than *S. palustris* which is rather benthic and less accessible to the larvae. *Radix balthica*, which moves easily between the surface and the bottom of the aquatic habitat, occupies an intermediate position in the predation by the larvae. The rearings were conducted in containers with tap water to a depth of 3 cm. Nevertheless, the general aspect of the curves corresponds to the Type II functional response, which tends toward an asymptote, as described by Holling (1959). At elevated densities, the numbers of molluscs consumed were equivalent to the results obtained in the experiments under conditions of excess food.

Temperatures at which larvae develop can determine size and weight of larvae, pupae, and adults. Yoneda (1981) found that the weight of pupae of *S. aenescens* was less for larvae reared at 15 and 30 °C compared with those reared at 19, 22, and 25 °C, while length and width did not show any special tendencies. Length of the wing of adults was shortest for those reared at 30 °C (6.1 ± 0.29 mm) and longest at 19 °C (6.9 ± 0.31). Gormally (1988b) measured the length of the pharyngeal sclerite of second-instar larvae of *Ilione albiseta* reared at 14, 17, 20, 23, and 26 °C and found that the length decreased from 536.44 m at 14 °C to 502.50 m at 26 °C. Lengths at 14 and 17 °C differed significantly from those at 23 °C.

Mc Donnell *et al.* (2005a) found that the mean and median weight of newly formed puparia of *Sepedon spinipes* tended to be greater at lower temperatures (14/17 °C) than at higher temperatures (23/26 °C) for both limited and excess feeding regimes for the antecedent larvae. Mc Donnell (2004) found that larval size of *S. spinipes* may also be strongly influenced by temperature and prey density. He presented the mean and median length of the cephalopharyngeal skeleton of second-instar larvae reared at 14, 17, 20, 23, and 26 °C when fed limited and excess snails. For both feeding regimes, the mean cephalopharyngeal skeleton length decreased as temperature increased. When provided with limited food, the cephalopharyngeal skeleton was significantly longer at 14 °C than at the higher temperatures. But under excess food conditions there was no significant difference in cephalopharyngeal length between the experimental temperatures.

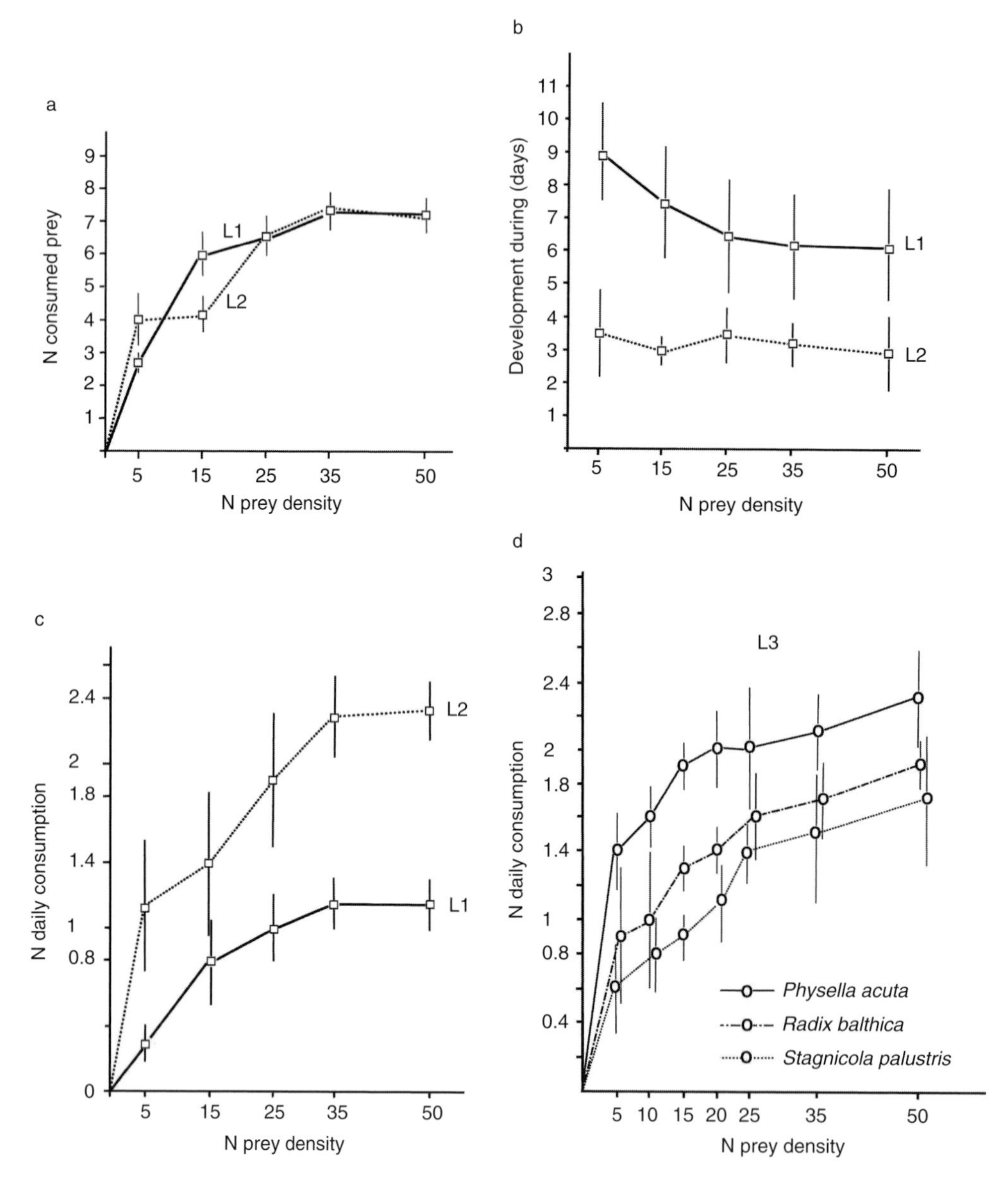

Fig. 9.44 (a–d) Development and number of prey consumed by larvae of ***Sepedon sphegea*** following prey density and species of snail prey. (a) L1 and L2, number consumed with *Physella acuta* as prey; (b) L1 and L2, duration of development (days) as function of prey density; (c) L1 and L2, daily consumption with *Physella acuta* as prey; (d) L3, daily consumption with three snail species. From Ghamizi (1985).

9.3.5 Post-feeding larval phase to adult eclosion

For recent reviews of the post-feeding larval phase in higher Diptera, see Darvas & Fónagy (2000) and Denlinger & Ždárek (1994). The latter authors provided a very useful timetable of the major developmental events, from the beginning of the third stadium to eclosion (Fig. 9.45). They noted that most of the experimental work on the stages and phases comprising this period has been on relatively few species, especially of the Calliphoridae, Drosophilidae, Glossinidae, Muscidae, Sarcophagidae, and Tephritidae. This limitation is true of the biology of higher flies in general.

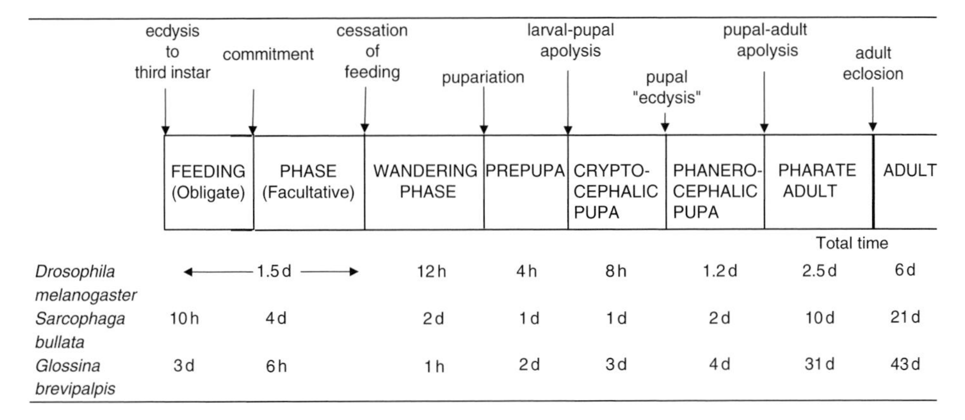

Fig. 9.45 Timetable of major developmental events in cyclorrhaphous Diptera from beginning of L3 to adult eclosion. Arrows indicate landmarks for beginning and end of each developmental stage. Times (d, days; h, hours) below each stage are approximate durations of each interval at 25 °C. From Denlinger & Žďárek (1994).

9.3.5.1 WANDERING PHASE (POST-FEEDING LARVAL PHASE)

While most Sciomyzini seem to pupariate very shortly after feeding stops, in many publications on life cyles of Tetanocerini it has been noted that mature third-instar larvae stop feeding for a day or more before pupariating. Obviously, at least a short, non-feeding period is needed by all larvae to complete digestion of the last meal and to empty the gut before pupariation. Then, the occurrence of a post-feeding "wandering phase" during which the larva searches for a pupariation site is unclear with regard to most Sciomyzidae. Of course, the 31 species in eight genera of Sciomyzini that routinely pupariate in the shell of the host snail do not wander.

Darvas & Fónagy (2000) concluded that "There is no wandering phase when the place of pupariation or pupation is where the larva develops" and "In some cases . . . of cyclorrhaphous species (e.g., Calliphoridae, Sarcophagidae, Muscidae, etc.), especially when a species developed in very wet or aquatic environment (usually necrophagous, coprophagous or saprophytic species) a wandering (= exodus [from the food]) phase starts before pupariation." This generalization is not very applicable to Sciomyzidae. Some aquatic predaceous Tetanocerini larvae pupariate in place; others seem to wander. Some semi-terrestrial to terrestrial species that can feed in a predaceous or saprophagous manner as mature larvae pupariate in place outside the shell whereas others seem to wander. For example, some multivoltine aquatic predators (e.g., *Elgiva* spp.) pupariate 12–72 hours after feeding stops, their puparia are strongly adapted for flotation, and the puparia are found in the water. Other univoltine aquatic predators (e.g., *Ilione albiseta*) that overwinter as mature larvae leave the water and have been found in moist cryptic locations, but their puparia also are adapted for flotation, which is particularly important when the habitat becomes flooded during the following spring. Neff & Berg (1966) did not give any information regarding the duration of a non-feeding pause before pupariation for the 15 species of aquatic predaceous *Sepedomerus*, *Sepedon*, and *Sepedonea* (Tetanocerini) that they reared. However, they found puparia of two species, *Sepedon tenuicornis* and *S. spinipes* glued to emergent vegetation as high as 1 m above the water surface. Obviously, the larvae had to wander to reach those pupariation sites.

In regard to the aquatic Tetanocerini that overwinter as free larvae (e.g., *I. albiseta*) and terrestrial Tetanocerini that overwinter as larvae then later in the puparium but outside the shell (e.g., *Coremacera marginata*) the distinction between wandering and simply overwintering as larvae (or as larvae then in the puparium) is especially unclear. The phases seem to meld together and probably are highly influenced by the micro-climatological situation. Of such species, *I. albiseta* has been studied the most thoroughly. Results were inconclusive (Knutson & Berg 1967) in two preliminary laboratory rearings of a few individuals at room temperature and natural lighting. The authors reported that two larvae (parents collected in Belgium) reared during October did not feed during the 7 and 8 days before pupariation. Beaver (1972b) reported that several larvae (parents collected in Wales) reared from August to December did not feed 10–15 days before pupariation but others killed and ate a large *L. stagnalis* a day before pupariation.

That *I. albiseta* is univoltine and overwinters as third-instar larvae is clearly shown by extensive information on collection of immature stages: one larva (stadium not given) in Greece during April (Knutson & Berg 1967); several larvae (number and stadia not given) in Wales during November (Beaver 1972b); 21 larvae, mostly third instars, collected in Ireland during March and April (Gormally 1987b); 204 larvae, mostly third instars, collected in Ireland between October and March (Lindsay 1982, Lindsay *et al.* 2011); 414 newly formed puparia collected April 6–11 in central Italy (Knutson & Berg 1967); and 12 puparia found during March and April in Ireland (Gormally 1987b). Other extensive seasonal information, e.g., collections of adults, time of mating, oviposition, egg-hatching, etc. in these publications support the conclusions.

In his extensive study of the biology of *I. albiseta* in Ireland, Lindsay (1982) described the larval wandering or overwintering pre-pupariation pause in detail, based on field-collected and laboratory-reared larvae. Among his collected larvae, based only on larvae which eventually became adults, the pause was highly variable, ranging from 73–155 days with a mean of 109.7 ± 22.4 days. During some of the longer pauses the larvae attacked one to three snails at widely separated intervals. This was in contrast to an attack rate of two or three snails per week before the pause began. Thus the author felt that the pause could be most accurately measured from the time that regular killing of snails ended. Observations from his reared larvae confirmed the extreme variability and duration of the pause for *I. albiseta*. Some larvae which hatched from eggs during July entered the pause by the beginning of December while others which had hatched during December and January were still killing snails as of April 12. This suggests that the date of egg hatch has a major effect on the pause, as it determines the environmental conditions to which the larvae will be exposed. It was also noted that during the pause larvae generally left the water or were found floating at the water surface. They tended to be much less active and reacted more sluggishly to disturbance than earlier instar larvae. Gormally (1987a) also noted that third-instar larvae were commonly found out of the water in his cultures. At least the first few days of this pause likely is a wandering period since mature larvae were found in moist, cryptic places and in refuge traps, that is, shallow dishes level with the surface of depressions where water collects in habitats that tend to dry during the fall. But most of the pause probably is simply an overwintering phase, as indicated by the observation that during some of the longer pauses larvae killed snails at widely separated intervals. The effect of the pause on larval metabolic rate is an area of future research.

Gormally (1988b) noted that at temperatures of 14, 17, and 20 °C the pre-pupariation pause for third-instar *I. albiseta* larvae ranged from 7 to 35 days, but not more than 7 days at 23 °C. Mc Donnell *et al.* (2005b) found that for *S. spinipes* in general the mean and median duration of the non-feeding pause (before ecdysis for first- and second-instar larvae and before pupariation) decreased as temperature increased from 14 to 26 °C, for larvae provided limited (one snail per day) and excess food. For third-instar larvae, the pause ranged from 3.18 ± 1.87 SD days at 14 °C to 0.75 ± 0.61 days at 26 °C.

Bratt *et al.* (1969) did not comment on a post-feeding, pre-pupariation pause among the 25 species of *Pherbellia* and three species of *Ditaeniella* they reared. These Sciomyzini included semi-terrestrial predators/saprophages and terrestrial parasitoids. However, they described an enlightening experiment they conducted with third-instar larvae of *D. parallela*, a species that can feed as a predator or saprophagously during the second and third stadium. Larvae were reared in a glass-sided observation chamber having a 20 mm layer of moss at the bottom. Seventeen larvae pupariated at depths of 1–16 mm, average 8 mm. The puparia (3.7–5.6 mm long) were formed at angles of 45°–95° from the horizontal, with the anterior end up and the anterior spiracles just below the surface for individuals that pupariated at the average or shallower depths. They also reported that larvae of this species were the most resistant to desiccation of any of the 28 species they reared, noting that many larvae were found wandering over the laboratory tables, more than 1 m from the rearing dishes when insufficient food was provided; larvae of other species that escaped from rearing containers were found dead a few cm away. Their observations seem to indicate a wandering phase in some individuals of this species.

Neff & Berg (1966) noted that larvae of *Sepedon fuscipennis* "become distinctly yellow because of the increased size and color of fat bodies beneath the transparent integument from 24 to 36 hours before formation of the puparium." Such development of fat bodies probably is typical in mature larvae and probably occurs shortly before, during, and shortly after evacuation of the gut after the last meal. The observation of these authors indicates, thus, a non-feeding, pre-pupariation pause of about 24 to 36 hours, in general, for the multivoltine, adult-overwintering species of *Sepedon* and other genera. The observation also could be important in further studies of wandering by larvae in that it could provide an indication of the cessation of feeding, especially for species of Sciomyzini and terrestrial Tetanocerini that have an unpigmented integument and aquatic Tetanocerini (e.g., *Elgiva* spp. and other genera) that have a lightly pigmented, translucent integument.

A major difficulty in resolving the question of wandering by sciomyzid larvae as based on the available literature stems from the fact that most rearings have not been conducted in arenas large enough to allow larvae to wander, but in small containers that preclude wandering. However, the rather common observation that larvae of many species, including aquatic predators, pupariate in the uppermost parts of such containers also indicates wandering.

For further study of a possible wandering phase among sciomyzid larvae it is useful to note some of the conclusions from experimental studies of other Diptera that feed saprophagously, some of which exhibit well-defined larval wandering. The comments by Denlinger & Ždárek (1994) that if a larva enters diapause (e.g., *Lucilia sericata* and *Calliphora vicina* [both Calliphoridae]), "it may remain in a wandering-like stage for up to 10 months" seems to reflect a lack of distinction between the phenomena of wandering and of overwintering as a larva. The authors noted that wandering is under weak, genetically determined hormonal controls responding to environmental stimuli. It ranges from less than one hour in some Glossinidae to over one week in some Sarcophagidae, and the duration may vary widely among closely related species. Larval wandering has been shown to be a "gated" event (occurring during a specific period of time, i.e., as a circadian rhythm) at dusk and during the night for species of Drosophilidae, Calliphoridae, and Sarcophagidae (Richard *et al.* 1986). The latter authors noted that the selective advantages of a gate for larval wandering are not clear, and suggested that it might be to avoid undue desiccation and diurnal predators, but, on the other hand would result in exposure to nocturnal predators.

9.3.5.2 PUPARIATION AND PUPATION

We suggest that if sciomyzid larvae do, in fact, wander and are also similarly gated, a selective advantage might be to find a pupariation site of appropriate high temperatures and/or degree of wetness. Such favorable pupariation sites would allow the most rapid development within the puparium and subsequent early eclosion of adults. In turn, this would result in the greatest number of generations that could be produced, thus the greatest amount of genetic recombination, especially for multivoltine species with overlapping generations. Thus the end result in the most ideal scenario would be the fullest and most efficient exploitation of resources. For uniquely adapted flies such as Sciomyzidae, larval wandering and pupariation site, seemingly insignificant features, become of critical significance.

Another selective advantage of wandering could be to allow the brain of the larva (the photosensitive receptor site) to receive the appropriate low light stimuli needed to initiate estival or hibernal diapause or quiescence. This also is important in the sequence of events in the fore-going scenario. For aquatic and semi-aquatic Tetanocerini, the fact that the first few anterior segments of these otherwise generally darkly pigmented larvae, e.g., *Tetanocera hyalipennis* (Plate 2f, g), are not pigmented would argue for the latter reason. Whatever the selective advantage(s) might be, it is unlikely that the generally highly adaptive sciomyzid larvae would simply pupariate wherever they find themselves upon the termination of feeding, unless that site, e.g., within the shell of the host/prey, resulted in other advantages, such as protection from predators, flotation during flooding, etc.

Pupariation in the species of Sarcophagidae, etc., that have been studied is a non-gated process that can occur at any time of day or night, because, in these species, it occurs beneath the surface (Denlinger & Ždárek 1994). However, pupariation of Sciomyzidae occurs in exposed situations, beneath the surface, or within snail shells, depending on the species. Thus the literature on timing of pupariation may not pertain to Sciomyzidae. However, there are no data on the diel timing of pupariation of Sciomyzidae. As for other aspects of the larval post-feeding to adult eclosion part of the life cycle, video-camera surveillance of larvae in appropriately sized and designed arenas, with a range of aquatic to terrestrial situations, deep to shallow substrates of various densities, various vegetation, and exposure to light of changing day-length and angles of incidence is needed.

The durations of the four "phases" occurring within the puparium – prepupal, cryptocephalic and phanerocephalic pupal, and pharate adult (Figs 9.1 and 9.45) – have been more or less treated as simply the pupal or puparial stage in most publications on Sciomyzidae. This amalgamation of phases has been due primarily to the practical difficulty of observing the developing animal within the darkly pigmented puparium of most species and the lack of dissection of puparia after they have been formed until the emergence of the adult.

The cryptocephalic (invaginated head) pupal phase follows the prepupal/pupal molt. At this stage, only the prothoracic pair of spiracles are supposedly functional and the abdomen is clearly still larval. In the following phanerocephalic (everted head) pupal phase the head and thoracic appendages are fully everted. During the gradual course of the metamorphic phase of the pharate adult the organism is essentially motionless. The eclosion phase begins with active movement of the fully developed adult (Darvas & Fónagy 2000).

Bratt *et al.* (1969) found that two puparia of *Pherbellia dubia* (univoltine, overwintering in the puparium) formed on 4 July contained well-developed but unpigmented pupae 2 and 3 days later. The translucent puparia allowed Knutson & Berg (1964) to determine the prepupal stage to be about 24 hours for *Elgiva solicita* and 9–48 hours for *E. connexa* (both multivoltine, overwintering as adults).

Berg *et al.* (1982), focusing primarily on overwintering and phenology, noted that the developmental stages of some *Tetanocera* and *Dictya* species (both in their Group 1, multivoltine species overwintering in the puparium) that survive the winter range from very young pupae through older pupae to pharate adults. Their phenological conditions ranged from mere quiescence to true diapause. Berg *et al.* (1982) presented details from indoor and outdoor experimental studies on these species. Further studies of development within the puparium of Sciomyzidae are needed, especially in regard to respiration and the effects of environmental conditions.

9.3.5.3 ECLOSION

Eclosion of the adult from the puparium is a gated event, occurring close to dawn in carefully studied species of Calliphoridae, Drosophilidae, Muscidae, and Sarcophagidae, but in mid afternoon in Glossinidae (Denlinger & Žďárek 1994). The authors considered eclosion near dawn to be an advantage because wing expansion is more effectively accomplished at high humidity and/or because synchronization of a population of diurnally active flies is an advantage for their subsequent activities of mate location, mating, initial feeding, etc. It is not known if eclosion in Sciomyzidae is a gated event. Such information would be useful in allowing more precise timing during surveillance of puparia about ready to produce adults.

The process of eclosion in Sciomyzidae has been studied in detail only for the highly unusual *Sepedon spinipes* (R. J. Mc Donnell & L. Knutson, unpublished data). Notably, this species and the 112 other members of *Sepedon* and five related genera are the only members of Schizophora (45 000 ± species), except the highly modified, parasitic Nycteribiidae, that lack a ptilinum and ptilinal fissure. The balloon-like, armored, enervated, and musculated ptilinum, which protrudes through a fissure on the top of the head of eclosing Schizophora, and then is withdrawn into the head, is fundamental to eclosion behavior, except in *Sepedon* and related genera. In the ptilinum-less *S. spinipes*, and assumedly in the related species, eclosion (recorded by time-lapse videophotography) is effected by protrusion of the mouthparts and antennae and movements of the lower face. Eclosion in these genera perhaps is aided by the rows of large hooks on the labellum, also known only in *Sepedon* and related genera. However, labellar microteeth are present in the unrelated genus *Dictya*, which has a ptilinum and ptilinal fissure. The total duration of eclosion in *S. spinipes* ranged from 121–201 seconds, followed by the normal walking, wing expansion, grooming, etc. of newly emerged flies.

10 • Macrohabitats and microhabitats, guild structures and associations, threatened species, and bio-indicators

> Flee the advice of those speculators whose reasons have not been confirmed by experience.
>
> Leonardo da Vinci.

10.1 MACROHABITATS AND MICROHABITATS

> *En Camargue ... Au mois d'août ... De place en place, les étangs fumaient au soleil comme d'immense cuves, gardant tout au fond un reste de vie qui s'agitait, un grouillement de salamandres, d'araignées, de mouches d'eau cherchant des coins humides.* [In the Camargue ... during August ... here and there, marshes were steaming in the sun like enormous tanks, keeping remnants of life right at their bottom, like, for example, the hustle and bustle of salamanders, spiders, and aquatic flies looking for damp spots].
>
> Daudet (1866).

We use the terms "macrohabitat" for the general type of aquatic, terrestrial, or intermediate situation, e.g., pond, marsh, swamp, backwater of stream, different types of woods, meadow, etc., where adults occur and "microhabitat" for the more precise locations within a macrohabitat, e.g., surface of pond, emergent vegetation in a marsh, shoreline situations, sunlit patches in shady broad-leaved woods, wet spots in *Equisetum* meadows, etc., where immature stages develop.

Our definition of macrohabitat is in the broad, traditional sense, equivalent to "biotope," i.e., the characteristics of areas inhabited by species, not in the strongly different autoecological sense of, e.g., Walter & Hengeveld (2000), the living conditions specific to individuals of a species. Walter & Hengeveld (2000) considered that,

> the environmental context of organisms is species specific and "habitat" must be defined in relation to use of the environment by the subject organisms. Habitat therefore relates to specific neurophysiological and behavioral interactions between individual and environment (e.g., Hüber 1985) and is thus species specific.

As information on the environmental requirements of species of Sciomyzidae is refined, a shift to the autoecological usage of habitat perhaps may be desirable.

Description of some species of Sciomyzidae as "eurytopic" may in fact be inaccurate, such species being found in a variety of macrohabitats (traditional sense) *not* because they are broadly adaptive to the environment, but because the variety of situations all meet their macrohabitat (autoecological sense) requirements; the flies thus not being influenced by features of the situation that do not impinge upon them either negatively or positively.

Oldroyd (1964, 1970) pointed out that larvae of Diptera are seldom fully terrestrial or fully aquatic, but most live in moist situations. He suggested that rather than "aquatic" and "terrestrial" we should speak of "watery" and "earthy" microhabitats. This distinction applies well to the microhabitats of most Sciomyzidae. While a few sciomyzid larvae live beneath the water surface most of their lives, many live at the water surface, i.e., they are "neustic" or "periphytic" around emergent vegetation, many live on moist (wet to damp) surfaces, and even the larvae feeding on snails in xeric habitats are in a "watery" microhabitat, i.e., the fresh to liquified tissues of the host/prey. Most so-called terrestrial larvae seem to spend little time outside the moist confines of the host/prey shell. Exceptions are for the relatively short duration of the predatory phase (late second and third stadia) of slug killers and the protracted third stadium of a few univoltine snail predators (e.g., *Coremacera marginata*) that appear to have a prepuparial pause or "wandering" phase, then overwinter, first as mature larvae and then in the puparium. Many of the microhabitats of sciomyzid larvae are relatively distinct, discrete, very similar from one geographical area to another, and can be described in a few meaningful words. However, for the many species that live between truly aquatic and truly terrestrial microhabitats a wide variety of somewhat nebulous terms has been used, e.g., semi-terrestrial, semi-aquatic, humid, hygrophilous, seepage areas and, especially, "shoreline" areas. The hosts/prey in these situations are stranded, foraging, estivating, migrating, wandering, or otherwise exposed on moist surfaces. We consider these nebulous terms more or less equivalent, but certainly there are some differences among them in vegetation, mollusc fauna, diel and seasonal moisture levels, exposure to sunlight, etc. The actual ranges of species within the ecological gradients of such habitats and in ecotones are not well known, although many larvae and ovipositing females appear to exploit a rather broad range.

The terminology that has been used in describing sciomyzid habitats, in general, lacks precision and uniform usage. For example, while "turlough" is used to describe ephemeral water bodies considered unique to Ireland (Ryder *et al.* 2005, Williams *et al.* 2009), the description of them seems similar to the widespread "sloughs" in the midwestern USA (Whiles & Goldowitz 2001). Keiper *et al.* (2002) noted that the various types of wetlands are difficult to define precisely, and in at least the USA there is no generally accepted classification scheme. Glossaries prepared for databases used for inventory and monitoring biodiversity should provide useful guidelines. Macro- and microhabitat definitions for a database on Sciomyzidae, similar to the well-developed databases on western Palearctic Syrphidae (Speight *et al.* 2000), are being developed (Speight & Knutson, unpublished data).

10.1.1 Publications on macro- and microhabitats

Most information on Sciomyzidae in macrohabitats and microhabitats is from northern and southern temperate latitudes, where the majority of species have been reared. The generalizations below are derived from our field studies in Europe, North America, Africa, and South America and from publications on life cycles and faunistic studies. Many of these papers include descriptions of macrohabitats, often with lists of Sciomyzidae, molluscs, and vegetation present. More detail on habitats or summaries of habitat distribution are given in other papers, as noted in the list of major papers below, most of which concern collections of adults, and in the following discussion. It is instructive to compare studies of the macro- and microhabitats of the Sciomyzidae with those of their hosts/prey. For shelled Gastropoda of western Europe, Falkner *et al.* (2001) is an extremely valuable tabulation of macrohabitats (93 descriptors) and microhabitats (112 descriptors) of 270 species.

10.1.1.1 PALEARCTIC

Western Russia. Przhiboro (2001) described the physical characteristics, vegetation, and substratum of the "zone of waterline", i.e., "from 10 cm above the water level to 5 cm below this level" of five sites (ranging from 30–40 cm wide to 50 m wide) on the shores of two small lakes near St. Petersburg. He obtained adults of 11 species in seven genera from substrate samples (obviously containing puparia) held in the laboratory for emergence. Further collections of this nature, from an even wider variety of and even more narrowly defined microhabitats, from aquatic to terrestrial, would be useful, as they aid in pinpointing the microhabitat of the puparia, as well as providing information on the overwintering stage and seasonal aspects and also could give some indications of the hosts/prey.

Northern Europe. Summary of 81 species in 23 genera (Rozkošný 1984b).

Sweden. Summary of 53 species in 17 genera (Knutson 1970b).

Germany. One aquatic and one terrestrial site in a bog in Schleswig-Holstein where 15 species were collected by sweep net and 20 in Malaise traps over a 2-year period (Kassebeer 2000). Kassebeer (2001a) collected 1421 specimens of 68 species in 21 genera during 102 excursions over a 4-year period in 28 diverse locations in Berlin and Brandenburg.

Italy. Descriptions and photographs of ten types of macrohabitats of adults, with notes on vegetation, associations of species of Sciomyzidae, altitudinal distribution, and diel periodicity (Rivosecchi 1992). Distribution of adults of eight species in macrohabitats in two locations in central Italy (Rivosecchi & Santagata 1978). Distribution of adults of 29 species in five kinds of macrohabitats in 72 locations in central Italy, with a diagram of altitudinal distribution (Rivosecchi & Santagata 1979).

Northern France. Five microhabitats along a transect in an aquatic to terrestrial situation where 18 species were collected in yellow pan traps and sweep nets over a 3-year period (Vala & Brunel 1987; Table 10.1, Fig. 10.1).

Southern France. Terrestrial habitats (Coupland & Baker 1995; Table 10.2). Temporary aquatic macrohabitat where seasonal aspects of 14 species of Sciomyzidae and their hosts/prey were studied quantitatively over a 3-year period

Table 10.1. Collections of 18 species of Sciomyzidae in yellow pan traps across an aquatic to terrestrial habitat transect near Amiens, northern France

	Microhabitats						
Species	A	B	C	D	E	Total	Collection periods
1. *Anticheta analis*	1	1				2	13/V to 20/VI
2. *Coremacera marginata*				1		1	11 to 18/VII
3. *Elgiva cucularia*	19	10	1			30	2/V to 8/VIII
4. *Elgiva solicita*	14	6				20	20/VI to 22/VIII
5. *Hydromya dorsalis*		1				1	28/III to 25/V
6. *Ilione albiseta*	1					1	18 to 25/VII
7. *Ilione lineata*	1					1	1 to 8/VIII
8. *Limnia unguicornis*	2	5				7	20/VI to 22/VIII
9. *Pherbellia dorsata*	2					2	18/VII to 1/VIII
10. *Pherbellia s. schoenherri*	1					1	25/VII to 1/VIII
11. *Pherbina coryleti*	3	1	2	1		7	4/IV to 12/IX
12. *Psacadina verbekei*	14	5	7		1	27	4/IV to 29/VIII
13. *Sciomyza testacea*	1					1	27/VI to 7/VII
14. *Sepedon sphegea*	4	2			1	7	4/VII to 26/IX
15. *Sepedon spinipes*	1	2	2	1	1	7	11/IV to 20/X
16. *Tetanocera arrogans*	26	2				28	13/VI to 19/X
17. *Tetanocera elata*		10				10	30/V to 8/VIII
18. *Tetanocera ferruginea*	118	24				142	28/III to 26/X
	Total	208	69	12	3	3	295
	Percentage	70	24	4	1	1	100

Note: A–E, microhabitats: see Fig. 10.1. From Vala & Brunel (1987).

(Vala & Manguin 1987; Fig. 9.7, Table 10.3). Also, an oak grove where the phenologies of six species of Sciomyzidae and their hosts/prey were studied quantitatively over a 6-year period (Vala 1984b).

England. Aquatic macrohabitats in Suffolk (Disney 1964). The snail fauna and vegetation of Ivinghoe Hills, near Whipsnade Zoo, Bedfordshire, where the eurytopic *Pherbellia cinerella* was collected abundantly, was described in detail by Stratton (1963).

Southern Wales. While collecting *Salticella fasciata* on the sand dunes at Tenby for their studies of the life cycle, Knutson *et al.* (1970) sampled the snails and vegetation in the microhabitats. The distribution of 3637 specimens of 18 species of snails from 35 quadrats collected along three transects in a 20 × 300 m area during July and September and the frequency of 12 species of angiosperms were analyzed by Stratton (1970). The area now is a commercial park for the amusement of the general public, not for dipterists. Drake (1988) described a coastal grazing marsh where 20 species, including 10 considered notable or rare, were collected with 510 other species of Diptera in 65 samples swept from marginal vegetation during June–July; *Elgiva solicita* was collected in 46 of the samples and *Sepedon spinipes* in 42.

Ireland. Summary of the collection records of the 40 species then known (now 57 species) with notes on macrohabitats of many (Chandler 1972). Nine aquatic and semi-aquatic sites in Co. Sligo where ten species, especially *Ilione albiseta*, *Limnia unguicornis*, and *Pherbellia cinerella*, were studied (Gormally 1987b). Detailed descriptions, emphasizing hydrology and plant species and heights, of a temporary aquatic habitat (turlough) in Co. Mayo where Sciomyzidae and other Diptera were sampled, the data being treated to extensive statistical analyses (Ryder *et al.* 2005, Williams *et al.* 2009).

Table 10.2. Total number of Sciomyzidae caught in each of five terrestrial habitats near Montpellier, southern France, February 1991–January 1993

Habitat	*Pherbellia cinerella*	*Euthycera cribrata*	*Coremacera marginata*	*Trypetoptera punctulata*	*Dichetophora obliterata*
Pasture	3051 (83)	1 (0.5)	2 (0.5)	3 (1)	0
Low scrubland	312 (8)	39 (24.5)	327 (68)	8 (2)	13 (9)
Forest (riverine)	20 (0.5)	120 (75)	150 (31)	351 (97)	125 (91)
Dunes	325 (8.5)	0	1 (0.5)	0	0
Total	3688	160	480	362	138

Note: Percentage of total catch shown in parentheses. From Coupland & Baker (1995).

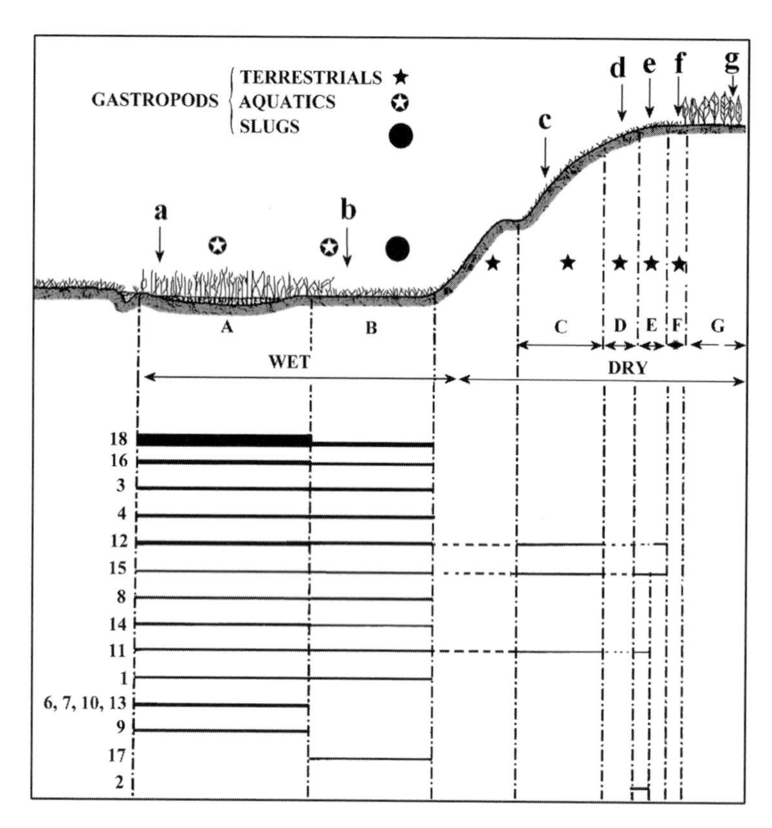

Fig. 10.1 Collections of 18 species of Sciomyzidae in yellow pan traps across an aquatic (A) to terrestrial (G) transect in a habitat near Amiens, northern France. (a–g) locations of traps. See Table 10.1 for species 1–18. Thickness of horizontal lines indicates relative numbers of flies captured. From Vala & Brunel (1987).

10.1.1.2 NEARCTIC

New York. Aquatic site where population dynamics of *Sepedon fuscipennis* were studied (Eckblad 1973b).

Connecticut. Lotic and lentic macrohabitats where dispersal of *S. fuscipennis* was studied (Peacock 1973, Table 13.1).

California. Twelve sites in northern, central, and southern California where about 24 000 adults of 49 species in 13 genera were collected between 1962 and 1976 (Fisher & Orth 1983).

Idaho. Nine aquatic sites where life cycles were intensively studied, with a summary of physiographic, biotic, and climatic regions in the State (Foote 1961b).

Table 10.3. Sciomyzidae collected in temporary aquatic habitat at Rochefort-du-Gard, southern France, during 1982–1984

Species (i)	F_i	n_i	Annual q_i 1982	1983	1984	Overwintering stadium
1. *Tetanocera ferruginea*	84	41	549	323	280	in puparium
2. *Ilione albiseta*	55	27	138	155	428	L3
3. *Ilione trifaria*	55	27	23	176	33	?
4. *Sepedon spinipes*	55	27	70	167	106	Adult
5. *Sepedon sphegea*	43	21	26	11	70	Adult
6. *Psacadina verbekei*	43	21	30	69	17	Adult
7. *Hydromya dorsalis*	33	16	24	8	10	Larva?
8. *Elgiva cucularia*	22	11	3	18	7	Adult
9. *Tetanocera arrogans*	20	10	14	21	0	in puparium
10. *Pherbellia cinerella**	14	7	59	1	0	Adult and in puparium
11. *Coremacera marginata**	12	6	7	2	4	Larva/in puparium
12. *Pherbina coryleti*	10	5	4	5	4	Larva
13. *Euthycera cribrata**	2	1	3	0	0	Larva/in puparium
14. *Pherbellia s. schoenherri**	2	1	0	0	1	Adult
$\underline{Q}$			950	1056	990	

Note: Each species (i) is characterized by F_i, frequency of presence; n_i, number of sweeps with species *i*; q_i, number of collected adults with species *i*; $\underline{Q}$, total number of collected adults of all species. **Asterisk** indicates terrestrial species frequently collected, but not breeding in this habitat. See Eq. 10.1, Fig. 10.6. Modified from Vala & Manguin (1987). See Bratt *et al.* (1969) for information on overwintering of *P. cinerella* in the puparium.

Nebraska. Five temporary and permanent aquatic sites – sloughs in a large wet meadow complex – emphasizing hydrology, where emergence production of Sciomyzidae, eight other families of Diptera, and three families of Trichoptera was studied (Whiles & Goldowitz 2001; Table 10.4).

Alaska. General summary of the four biotic provinces (Foote *et al.* 1999).

The occurrences of 25 Nearctic, Palearctic, and Holarctic species of *Pherbellia* and three *Ditaeniella* species in 20 types of aquatic and terrestrial macrohabitats were tabulated as typical or occasional and five sites in Denmark, Wales, and the USA were described in detail (Bratt *et al.* 1969). The macrohabitat in Zealand, Denmark where many species were reared and studied intensively is shown in Fig. 10.2. See also the papers reviewed in the following section on Guilds and Associations (Section 10.2).

10.1.1.3 OTHER REGIONS

There is limited information on habitats in other regions scattered in papers on life cycles. The distribution of 787 specimens of 11 species of *Sepedon*, *Sepedonella*, and *Sepedoninus* collected during 1949–1952 in 16 types of habitats in Garamba Park, northern Democratic Republic of Congo, were tabulated by Verbeke (1963). The greatest number and diversity of sciomyzids were collected along rivers and in marshes. Of the species collected, one of the most numerous, *Sepedon trichrooscelis*, is a parasitoid/predator of Succineidae. Habitats of the aquatic predators *Sepedon neavei* and *S. testacea* in South Africa were described in detail by Barraclough (1983). Habitats in Bénin, where *Sepedon trichrooscelis*, the oligochaete predator *S. knutsoni*, and the freshwater snail predator *S. ruficeps* were studied, were described by Vala *et al.* (1995), Vala *et al.* (1994), and Gbedjissi *et al.* (2003), respectively. Some information on habitats in South Africa for several genera was provided by Miller (1995).

10.1.2 Kinds of macro- and microhabitats of Sciomyzidae

10.1.2.1 ADULTS

The only extensive comparative information on the ranges in elevation at which adult Sciomyzidae have been

Table 10.4. Taxa contributing greatest amounts to annual (a) insect emergence abundance and (b) production at five sites in wet meadows in Nebraska, USA

Study site					
M13 [94] (ephemeral)		WR1 [158] (ephemeral)		M12 [296] (intermittent)	
(a) Abundance					
Family (%)	N	Family (%)	N	Family (%)	N
Chironomidae (33%)	152.1	Tipulidae (28%)	103.2	Culicidae (95%)	23 021.1
Sciaridae (27%)	125.0	Chironomidae (19%)	70.6	Chironomidae (1%)	331.4
Sciomyzidae (8%)	38.0	Sciaridae (17%)	65.2	Ceratopogonidae (1%)	211.9
Dolichopodidae (8%)	38.0	Muscidae (16%)	59.8	Muscidae (1%)	146.7
Total	456.4	Total	374.9	Total	24 124.1
(b) Emergence production					
Family (%)	mg	Family (%)	mg	Family (%)	mg
Sciomyzidae (45%)	60.6	Muscidae (35%)	64.7	Culicidae (90%)	4590.4
Muscidae (20%)	27.6	**Sciomyzidae** (25%)	45.9	Muscidae (3%)	171.9
Sciaridae (12%)	16.0	Tipulidae (12%)	22.3	**Sciomyzidae** (2%)	80.5
Tipulidae (10%)	13.9	Chironomidae (7%)	12.6	Baetidae (1%)	55.4
Total	134.7	Total	183.4	Total	5099.4

Study site			
M11[331] (intermittent)		WR2 [365] (permanent)	
(a) Abundance			
Family (%)	N	Family (%)	N
Chironomidae (32%)	635.7	Chironomidae (96%)	1553.9
Culicidae (19%)	364.0	Ceratopogonidae (1%)	21.7
Limnephilidae (11%)	219.0	Sciaridae (1%)	16.3
Ceratopogonidae (7%)	141.3	Tipulidae (1%)	10.9
Total	1957.7	Total	1619.1
(b) Emergence production			
Family (%)	mg	Family (%)	mg
Limnephilidae (57%)	560.0	Chironomidae (94%)	239.8
Culicidae (14%)	137.9	Ephydridae (3%)	7.0
Sciomyzidae (11%)	112.6	Tipulidae (2%)	4.7
Leptoceridae (5%)	50.3	Sciaridae (1%)	2.1
Total	982.3	Total	255.7

Note: Numbers in parentheses show percentage of contribution of each family to total at that site. Values in parentheses after study site code numbers are annual hydroperiods in days. Abundance, no. m^2/year; production, mg dry matter m^2/year. Slightly modified [headings] from Whiles & Goldowitz (2001).

a

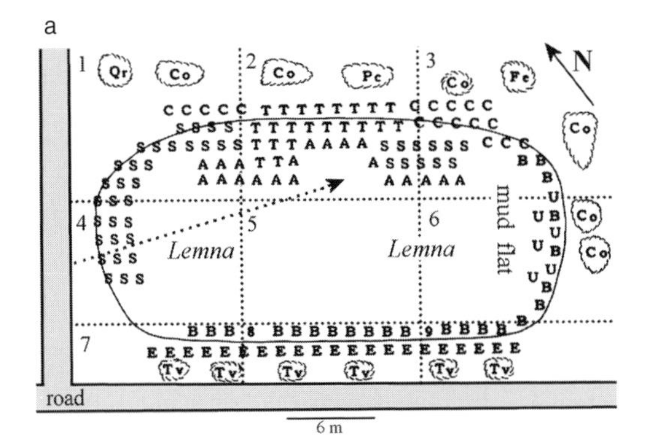

b

c

Fig. 10.2 Habitat in Hillerød, Zealand, Denmark. a, Distribution of vegetation. b, Photo (taken 1960) along transect indicated by arrow in a. Plants: A, *Alisma* sp.; B, *Bidens tripartita*; C, *Carex vesicaria*; Co, *Crategus rhipidophylla*; E, *Equisetum arvense*; Fe, *Fraxinus excelsior*; Pc, *Prunus cerasifera*; Qr, *Quercus robur*; S, *Sparganium simplex*; T, *Typha latifolia*; Tv, *Tilia vulgaris*; U, *Oenanthe aquatica*. Sciomyzidae collected: ***Pherbellia argyra*, *P. dorsata*, *P. dubia*, *P. obtusa*, *Pteromicra nigrimana*, *Colobaea pectoralis*, *Sciomyza simplex*, *Tetanocera ferruginea*, *T. hyalipennis*, *Pherbina coryleti*, *Sepedon sphegea*, *S. spinipes*, *Elgiva solicita*, *E. cucularia*,** and ***Limnia unguicornis***. Snails present: *Planorbis planorbis* (quadrats 4–6), *Anisus vortex* (1), *Galba truncatula* (all quadrats), *Succinea* sp. (quadrats 1–3), *Zonitoides arboreus* (all quadrats), *Cepea nemoralis* and *Trochulus hispidus* (quadrats 1–3) (from Bratt *et al.* 1969). c, Same habitat 30 years later; x, benchmark location (photos by L. Knutson).

captured was provided by Fisher & Orth (1983) for the 40 species in 13 genera occurring in California and by Rivosecchi & Santagata (1979) for 29 species in central Italy (Toscana to Abruzzo and Lazio). Based on 1027 specimens collected at 200–1400 m at 72 locations, the latter authors found *Pherbellia annulipes*, *P. pilosa*, *Euthycera stictica*, and *E. zelleri* only at 200 m, with *Ditaenella grisescens* the only species restricted to above 1400 m. However, collections elsewhere in Europe show somewhat different and broader ranges of elevations for most species. The data of Fisher & Orth (1983) are based primarily on their extensive collections throughout California during 1962–1976, plus specimens in museum collections and a few published records. A total of 24 000 specimens was examined. Surprisingly, only five species (*Dictya incisa*, *Pherbellia oregona*, *P. vitalis*, *Pteromicra pectorosa*, and *Tetanocera loewi*) were found only at the lowest elevations (below 305 m), with *D. montana* found at the very lowest (–58 m) but also up to 2621 m. All of the rest of the species were found at moderately high elevations (1067 + m). The species found at the highest elevations were *Limnia severa* and *T. plumosa* (2743 m), *Pherbellia schoenherri maculata* (2789 m), and *Sepedon pacifica* (3048 m). The last species was found at the greatest range of elevations (3 to 3048 m). Fourteen other species were found at ranges of elevation from less than 10 m up to between 1387 and 2743 m.

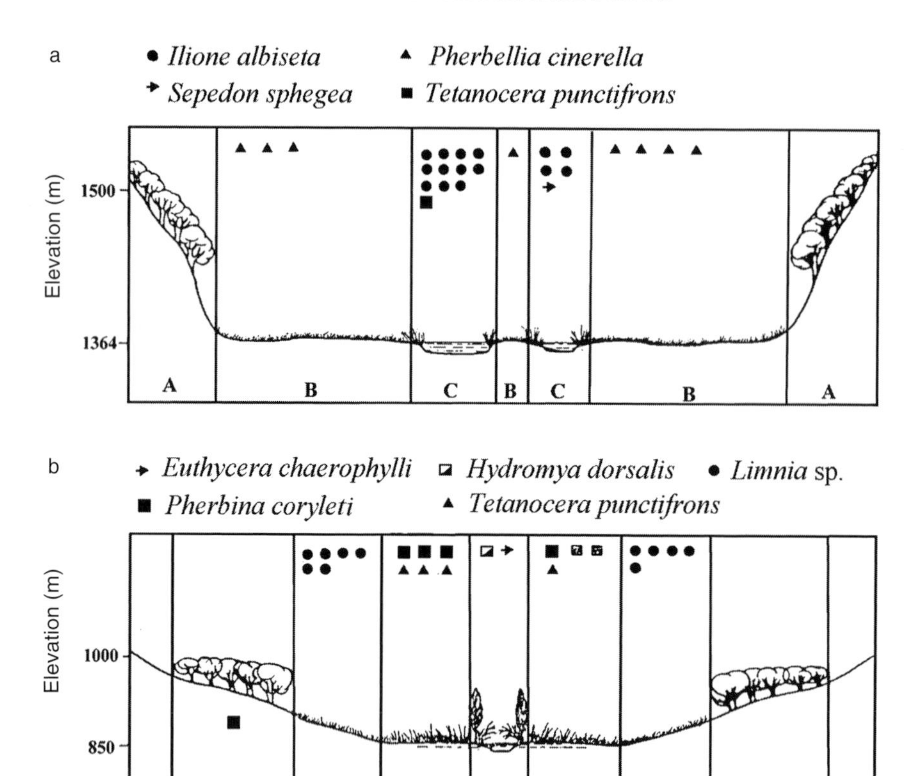

Fig. 10.3 Microhabitats of Sciomyzidae along transects in two habitats in eastern central Italy. (a) Voltigno tableland around Lake Sfondo (Gran Sasso d'Italia, Abruzzo), August 20, 1977. A, Beechwoods; B, prairie; C, lakeshore. (b) Valley of R. Varri (Abruzzo), May, 1977; A, pasture; B, beechwoods; C, humid pasture; D, *Juncus* marsh; E, aspen grove and shrubs covering river like a tunnel. Each symbol corresponds to five specimens. From Rivosecchi & Santagata (1978).

Sciomyzidae occur in a great variety of macrohabitats, from truly terrestrial to truly aquatic, and the slow-flying adults are usually, but not always, closely associated with the microhabitat of the larvae. Adults of some aquatic species have been found in mesic terrestrial situations as much as 100 m from aquatic habitats in the late summer to fall, e.g., *Sepedon sphegea* near Avignon, France. This is apparently the result of dispersion due to moving between patchy habitats, seasonal effects, and/or a build-up of the population. Specimen label data may indicate a rather broad macrohabitat range for many adult Sciomyzidae, but can be misleading as to the generally more restricted microhabitat range of the larvae. For example, one male (genitalia examined by Knutson) and one female of *Tetanocera phyllophora* are labeled "U. S. S. R., Siberia, Teletskoya Lake, pans in lake drift line, 16–20. VI. 1991, S. A. Marshall" (Guelph Univ. Collection). However, larvae of this species were found in the terrestrial snail *Euconulus fulvus* in nature and developed on various terrestrial snails, but not aquatic snails, in laboratory rearings (Vala & Knutson, unpublished data). Diagrams of macrohabitats showing the microhabitat distributions of some species are shown in Figs 10.1, 10.3, 10.4. Szadziewski (1983) collected only 13 species of Sciomyzidae in his study of saline flies of Poland in eight inland, coastal, and marine macrohabitats (90 000 adults and larvae collected, representing 516 species in 55 families). Only *Pherbellia cinerella*, *Pherbina coryleti*, and *Limnia unguicornis* were numerous and all 13 species were considered as haloxene, i.e., occurring more often and more numerously in non-saline macrohabitats.

However, *Pherbellia mikiana* adults have been collected by sweeping closely over *Salicornia* sp. growing just a meter from the edge of the sea and along the margin of commercial salt-production ponds in Malta, and it is known from the littoral on Elba Island (Munari 1986). Also, *Calliscia*

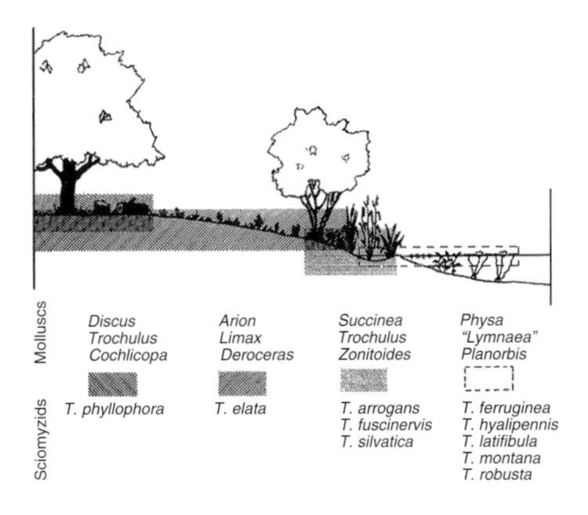

Fig. 10.4 Microhabitats of *Tetanocera* spp. and gastropod hosts/prey in Sweden. Modified from Knutson (1970b).

callisceles adults were found (photographed, the superb photos identified by L. Knutson) by S. A. Marshall (personal communication, 2007) far out in the intertidal zone of coastal Chile with flies of the family Canaceidae, which are almost exclusively intertidal in habitat. The only report of a sciomyzid collected in a cave (at the entrance) is one adult of the slug-killer *Euthycera chaerophylli* in Moravia (Bezzi 1907).

10.1.2.2 AQUATIC LARVAE

Most aquatic sciomyzid larvae live in various still-water habitats – ponds, marshes, backwaters of streams, protected bays of lakes, slowly flowing irrigation ditches, etc. – where they feed primarily on non-operculate snails at the water surface, foraging on vegetation, or exposed on wet shorelines. A few species live beneath the surface of the water during early larval life, feeding on fingernail clams (three *Renocera* spp., *Ilione lineata*, and *Eulimnia philpotti*), or non-operculate snails (*Hedria mixta*, probably some *Ilione* species, and some *Dictya* species that can spend long periods, at least during early larval life, beneath the surface). Quite a few species live in seasonal marshes, pools, and seepage areas (e.g., *Anticheta* spp., *Ilione albiseta*, *Renocera amanda*, several *Pherbellia* spp., and some *Tetanocera* spp.).

Whereas the "micro-altitudinal" distribution of Sciomyzidae across wet–dry habitat transects has been emphasized, their horizontal distribution along apparently uniform, level habitats has not. The occurrence of Sciomyzidae within such habitats that appear to be more or less uniform often is rather patchy. This probably is due primarily to the patchy distribution of the hosts/prey. For example, Wright (1971) commented on the patchy distribution of snails in drainage ditches traversing a low-lying marshy area in southern England (Fig. 10.5),

Fig. 10.5 An apparently uniform habitat, a drainage ditch crossing marshy lowland in southern England – but in which freshwater gastropods have a very patchy and seasonally variable distribution. From Wright (1971).

> The water in these ditches is slow-moving to stagnant and frequently choked with vegetation, providing an ideal habitat for most of the British species of freshwater gastropod molluscs. The distribution of the snails in the ditches is very patchy and subject to marked seasonal variation. The specific components of the mollusc fauna in a ditch such as the one shown may be quite different over distances of thirty to forty yards.

Quantitative information on Sciomyzidae, over seasons but also throughout the day and from day-to-day, in relation to the less vagile snail species complexes and populations would be of interest.

While describing in detail the macrohabitats and microhabitats of larvae of *Perilimnia albiseta* and *Shannonia meridionalis*, Kaczynski *et al.* (1969) came to conclusions about the quality of adjacent open water that may apply to the large number of semi-terrestrial ("shoreline") Sciomyzini and Tetanocerini. These two cool-adapted, western and southern Neotropical species feed as predators, parasitoids, or saprophages, depending on the food available and intensity of intraspecific competition, on freshwater snails. They occur in a very broad range of macrohabitats from sea level to 3800 m with adjacent waters being flowing or stagnant, shaded or unshaded, and ranging,

> from very soft, poorly buffered, and poor in dissolved nutriments in mountainous seepage areas, to hard, highly nitrogenous, rich in dissolved organic matter, and appreciably salty on the coastal plains and the deserts of Coquimbo and Atacama Provinces in Chile.

They noted that the larvae are broadly tolerant to their immediate microhabitats, but more importantly are not even exposed to the full range of the water qualities of the macrohabitats, and thrive wherever they find a permanently moist area supporting populations of the broad range of acceptable snails. Eckblad (1973b) presented a chemical analysis (16 characteristics) of an intermittent backwater in Ithaca, New York where populations of *Sepedon fuscipennis* and its snail prey were extensively studied, but these were not discussed in relation to the fly or the snails. The water quality of a temporary freshwater habitat in southern France where the sciomyzid and snail communities were studied was briefly characterized by Vala & Manguin (1987). Water chemistry for five riparian sloughs in Nebraska where Sciomyzidae were included in a study of insect emergence production was detailed by Whiles & Goldowitz (2001).

No species is known to be restricted to a madicole microhabitat although these situations are exploited by some eurytopic forms. The impression that *Hydromya dorsalis* is limited to madicole microhabitats (Knutson & Berg 1963a) was proved incorrect by later findings of the larvae in diverse aquatic and semi-aquatic microhabitats. However, some Sciomyzidae prefer unusual habitats. *Hoplodictya setosa* preys on the operculate *Littorina littorea* in open salt marshes and in strandline debris on Atlantic Ocean beaches (Neff & Berg 1962). *Dictya*, with only one species in the Palearctic but 34 in the Nearctic and eight in the Neotropics, appears to have radiated into diverse habitats in North America. Adults of at least three species are found only in open, coastal salt marshes in North America where their larvae apparently feed on small operculate snails (Valley & Berg 1977). At least two are restricted to or are common in spring macrohabitats. *Dictya fontinalis* apparently feeds on operculate snails in a spring habitat in southern California (Fisher & Orth 1969b). *Dictya texensis* larvae were found in a thermal (34–35 °C) spring/stream habitat in Virginia, where they fed on *Physa* sp. (Robinson & Turner 1975). But *D. texensis* has been found in many diverse habitats, including other thermal water situations, across the USA (Valley & Berg 1977). *Tetanocera robusta* larvae and puparia have been found in hot springs (24 °C) in Iceland (Nielsen *et al.* 1954) but also in many other kinds of aquatic habitats. Adults of the very common aquatic predator *Sepedon ruficeps* were collected on February 8 in a large *Typha–Phragmites* marsh fed by rapidly flowing channels issuing from the sulphurous Bati Hot Springs near Nazareth, Ethiopia (Knutson *et al.* 1967a) but occurs in many other kinds of habitats.

Important relationships between emergent vegetation, the surface tension of the water, and the air need to be considered in characterizing the microhabitats of neustic and perineustic aquatic larvae and puparia of Sciomyzidae. It is likely that adaptations to physical conditions of this interface, at least in part, are responsible for some external morphological features of larvae and puparia. These include microstructure of the surface of the integument, size of posterior spiracular disc lobes and body tubercles and length of posterior interspiracular processes. Also, length and posterior or dorsal direction of the postanal portion of segment XII has recently been pointed out by Chapman *et al.* (2006). The tendency for aquatic larvae and puparia to be found most commonly next to emergent vegetation has been mentioned in several papers on life cycles. Arnold (1978a) pointed out that larvae of *Sepedon* and *Dictya* are found in the surface film of still, shallow, vegetation-choked waters where larvae of *Anopheles* mosquitoes live. He referred to publications (Renn 1943, Hess & Hall 1943, Rozeboom & Hess 1944) where this phenomenon was carefully studied in relation to the micro-distribution of *Anopheles* larvae. Other references on the effect of surface tension on mosquito larvae include Renn (1941) and Russell & Rao (1941). Rozeboom & Hess (1944) found that the ratio of the length of the "intersection line" (where air, water, and emergent vegetation meet) to open water in 1 m^2 quadrats was highly correlated with the density of *Anopheles* larvae in stands of aquatic vegetation. Singh & Micks (1957) conducted experiments on the effects of surface tension on pupal development and adult emergence of five species in three genera of mosquitoes. They found that none of the species could emerge when the surface tension was 41 dynes/cm or less or 78 dynes/cm or more, and that the differences in the behavior of the three genera towards the surface tension seemed to be in accord with their natural habitat.

Arnold (1978a) suggested that in studies of populations of larvae and puparia the distribution of vegetation should be utilized to achieve maximum sampling precision. Keeping in mind that there is the general observation that adult Sciomyzidae usually do not venture far from the microhabitats of their larvae, it is of interest to note that Peacock (1973) concluded in regard to adults of the aquatic predator *Sepedon fuscipennis*,

> the density of vegetation, rather than the size of the emergent stand, the plant diversity, or the presence of preferred plants, was probably more important in providing suitable lotic habitats for sciomyzids and their hosts.

The presence of seven species of *Sepedon* in three stream and three impoundment habitats near Pietermaritzburg, South Africa, could not be correlated with vegetation type or height (density was not mentioned) in 2–5-month surveys by Appleton *et al.* (1993). The interior of dense stands of vegetations, e.g., of *Phragmites*, are not especially productive collecting sites for adult sciomyzids, probably because of the difficulty of sweeping therein. However, such situations may in fact be prime sites for further quantitative/experimental sampling of populations of larvae by cutting the vegetation in plots several centimeters above the water surface to enable ease of sampling. The water surface-tension phenomenon also is intuitively understood by collectors. That is, pushing emergent vegetation below the surface, thus breaking the tension, and looking for floating larvae and puparia is an effective collecting technique.

Obviously, other factors are involved in adaptation to aquatic microhabitats, including (1) the presence of snails grazing on the biofilm on emergent vegetation, (2) the relative safety of sheltered compared with open situations, (3) the conservation of energy in crawling/swimming around emergent vegetation as opposed to moving about in open water, and (4) the opportunity to grasp emergent vegetation with the well-developed anal proleg and ventral abdominal tubercles for support while feeding. The morphology of the posterior spiracular disc and interspiracular processes have been compared among some genera in developing ecomorphological classifications of larvae, as detailed in Section 14.1.3. The size and the morphology of the anal proleg and ventral abdominal tubercles and their cuticular armature and the relative length and posterior or dorsal direction length of the postanal portion of segment XII also deserve study.

A trend towards increasingly intimate associations with the hosts/prey appears to be correlated with the ecological shift from aquatic to terrestrial habitats (Foote 1977). There is also a strong correlation between voltinism and habitat among species feeding on aquatic snails. A seasonal restriction of need for food by larvae enables some univoltine species, e.g., *Sciomyza varia*, several species of *Tetanocera* in the Nearctic and Palearctic, *Hedria mixta*, *Pherbina coryleti*, and some *Ilione* spp., to colonize ephemeral aquatic habitats.

Let us attempt to visualize the ideal habitats and microhabitats of aquatic and semi-aquatic Sciomyzidae, for the purposes of identifying study sites, sites for collection of biocontrol agents, and sites likely comparable to those where Sciomyzidae originated and evolved. Current patterns of the distribution of diverse and successful genera suggest areas: (1) of basic soil substrates supporting large and diverse populations of non-operculate, aquatic and semi-aquatic snails; (2) in temperate climates with rather regular rainfall and temperature patterns; (3) where there are eutrophic, large ponds and sheltered bays of lakes without significant wave action or water flow; (4) in a poorly drained region where such major habitats are more or less contiguous and relatively stable over geological periods of time. Narrowing the focus, relatively open and well-insolated areas with patches of shade, protected from strong winds, are indicated. The ideal microhabitats seem to be among moderately dense and diverse herbaceous emergent vegetation on very gently sloping south-facing banks, and in a few centimeters of water to the more or less permanently wet, muddy shoreline, a few centimeters from the water's edge.

10.1.2.3 TERRESTRIAL LARVAE

Terrestrial species in North America and Europe are found in mesic woods, meadows, and around marshy habitats. In Europe, but apparently much less so in North America, many species, e.g., *Salticella fasciata*, some *Euthycera* spp., some *Coremacera* spp., and *Dichetophora* spp., are found in dry, open, rocky and grassy, exposed situations. It would be interesting to compare the distribution of Sciomyzidae in the five "mediterranean" shrublands worldwide, noting the species endemic to these areas and those present but with broader distributions. Some species, e.g., *Trypetoptera punctulata*, live primarily in exposed ecotone situations, between forests and scrub or grassy areas.

Aquatic and semi-aquatic Sciomyzidae seem to be able to breed in moderately disturbed situations, but not where the substrate is strongly disturbed by cattle around the edges of water bodies, drainage/irrigation ditches disturbed by humans, etc. However, a few terrestrial species apparently can develop in somewhat less natural and more highly disturbed situations, but not in highly disturbed cropland when agricultural activities are underway. *Salticella fasciata*, the larvae of which develop during fall and over the winter, clearly can utilize such areas when human activity is at a low level. The persistent population studied by Coupland *et al.* (1994) in southern France occupied "typical Mediterranean waste ground" (a disused vineyard). Adults and larvae were collected September 7–14 on disturbed seaside dune slacks in southern Wales; adults were collected on golf links in September in southern England and from "wheat field stubble" in the Paris, France area on June 14 (Knutson *et al.* 1970). *Pherbellia cinerella* apparently has successfully adapted to grazed pasture in southern France (Coupland & Baker 1995). Speight (2001, 2004) as noted below, found 23 species in Malaise and emergence traps in disused to disturbed sectors of a farm in Co. Cork, Ireland. *Pherbellia cinerella* was trapped only in the production sector. The slug killer *Tetanocera elata* was the most abundant species, being found in emergence traps in production, infrastructure, and disused sectors, including pasture, silage grassland, and cereal strips. Knutson *et al.* (1965) noted that the habitat distribution of this very common and widespread species is very broad, like that of its preferred host/prey *Deroceras reticulatum*. They noted an infestation rate of 14% in a sample of *D. reticulatum* taken from a disused horticultural plot in southern England. Vala & Knutson (1990) captured and released adults of the common and widespread slug/snail parasitoid/predator *Limnia unguicornis* twice each month between May and October, in a semi-natural meadow and an adjacent meadow where the vegetation was harvested two or three times per year, in southern France. The numbers, of a total of about 350 flies, were about the same in both locations, except in the harvested location they declined substantially below that of the semi-natural location beginning mid July (Fig. 13.8).

10.1.3 Further research needs

Long-standing questions about the microhabitats of the many so-called, ill-defined "shoreline" species could be resolved by laboratory experiments. Also, some of the key life-cycle events could be identified from such experiments. These should be long-term, replicated in large, artificial microhabitats under controlled conditions of temperature, overhead light passage as well as duration, moisture, rainfall, and simulation of natural physical conditions such as shoreline inclination, substrates, north/south exposure, water level, water quality, etc. Larvae and snails should be released together, in various cohorts of species and numbers of individuals, in the aquatic, shoreline, and terrestrial sections of such artificial microhabitats. Then sampling of the dispersed snails to pinpoint micro-sites where predation occurs, time-lapse videophotography to gain information on the larval searching arena, and recovery of puparia to pinpoint pupariation sites are possibilities. Draining water from such artificial microhabitats to simulate seasonal changes may prove to be especially informative. Follow-up studies in nature on the species utilized would further clarify some of the central issues in understanding the adaptive radiation of Sciomyzidae.

Many features of microhabitat distribution have not been investigated. For example, in northern latitudes, are larvae of species that live on snails exposed on moist banks of ponds, and puparia of those species, more abundant on the south-facing banks where temperatures are higher than on north-facing banks? Sunshine warms shallow water and the surface of the ground by as much as 20–30 °C above the ambient temperature (Danks *et al.* 1994). This warming effect likely enhances the diel as well as seasonal activity of larvae, the "basking" of overwintering adults and thus their sexual activity and development of eggs, and probably results in earlier emergence from puparia during the spring.

10.2 GUILDS AND ASSOCIATIONS

There have been few analyses of aspects of these subjects, either of Sciomyzidae as members of broader guilds and associations including other mollusc feeders, or within the family. In fact, there have been no studies of guilds including Sciomyzidae in the original sense of Root (1967), that is, of a group of sympatric species involved in competitive interactions in exploiting the same sorts of environmental resources in a similar way, regardless of their taxonomic relationship. This is despite the fact that there is a wealth of baseline information, especially for many species and genera of aquatic predators in North America and Western Europe. Although the niches occupied by Sciomyzidae are diverse in terms of macro- and microhabitat, seasonality, taxonomic complement of hosts/prey, larval feeding site, and nutritional state of food resource, relatively few other Insecta occupy these niches. See Chapter 3 for a review of malacophagy among insects. Surprisingly, the niches where most competition likely occurs (predation of non-operculate snails in open water and predation/saprophagy of exposed aquatic and semi-aquatic snails) are those niches where, in terms of diversity of species and genera and in numbers of individuals, the Sciomyzidae have been most successful, worldwide. Species of several families of aquatic insects (e.g., Belostomatidae, Hydrophilidae, and Notonectidae) are opportunistic predators of snails and other organisms in open water. Species of several other families, especially Diptera, feed in dead or dying snails, and a few species of Calliphoridae, Phoridae, and Sarcophagidae are obligate predators or parasitoids of gastropods. Some Coleoptera (species of Carabidae, Drilidae, Lampyridae, Silphidae, and Staphylinidae) are more or less restricted to preying on terrestrial snails or slugs. No insects other than three genera of Sciomyzidae are known to be restricted to fingernail clams (Sphaeriidae), no insects except *Sepedonella nana* and *Sepedon knutsoni* are known to be restricted to freshwater oligochaetes, and few other insects are restricted to snail egg masses.

We suggest that in almost all habitats, the disparate sizes of populations of Sciomyzidae (comparatively few individuals) and the populations of their prey/host molluscs (comparatively many individuals), along with a low degree of host specificity by larvae of most sciomyzids has resulted in a lack of formation of significant guild structures. This is despite relatively strong microhabitat preference by most Sciomyzidae. In a series of publications on the biology of 20 of the 39 species of Nearctic *Tetanocera*, the fourth largest genus in the family and biologically diverse (with representatives in six of the 15 behavioral categories), Foote (1996a, 1996b, 1999) commented on presumed guilds and he associated habitat and food resource partitioning.

Sciomyzidae have been included in some synoecological studies. In a study of the community (47 families) of Diptera in a lowland oak–ash forest in southern Moravia, Czech Republic, 22 species of Sciomyzidae were fairly numerous. They were included in the dominant (species and individuals) group of predators, but overall did not appear among the +5% dominant or 2–5% subdominant species (Vaňhara 1981). In a study of the community of Diptera (50 families) in a forest-steppe, a beech–oak forest, and the forest edge near Brno, Czech Republic, only four species of Sciomyzidae were found, and in very low numbers (Rozkošný & Vaňhara 1993). Coupland & Baker (1995) made a quantitative study of five terrestrial species in five pasture sites, five riverine forest sites, five lowland scrub sites, and five dune sites near Montpellier, southern France, using standardized sweep-net collections made monthly from February, 1991 to January, 1993. Their data on a total of 3688 individuals collected are shown in Table 10.2. A spatio-temporal survey of two phytophagous groups of insects (46 spp. of Agromyzidae, Diptera and 77 spp. of Tenthredoidea, Hymenoptera) and two predaceous Dipteran groups (77 spp. of Dolichopodidae and 12 spp. of Sciomyzidae) was conducted between March 28 and June 27 in nine marshy locations in the adjacent Acon and Somme valleys in northern France by yellow-pan trapping (Brunel 1986). Sciomyzidae were not included in the

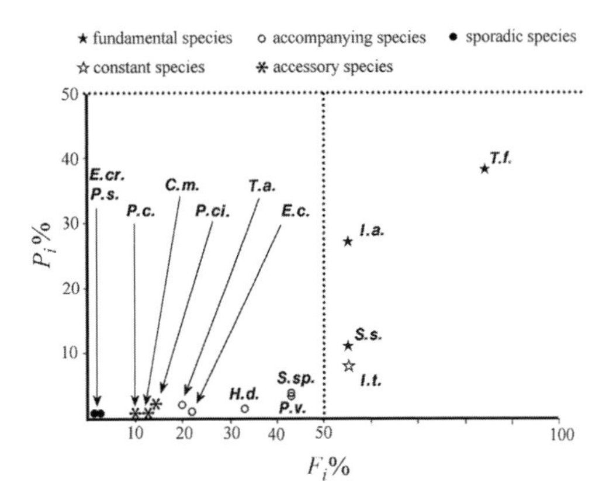

Fig. 10.6 Ecological position of Sciomyzidae (see Table 10.3 for species) in temporary aquatic habitat in Rochefort-du-Gard, southern France, according to frequency of presence (F_i) and relative abundance (P_i). See Eq. 10.1. From Vala & Manguin (1987).

factorial analyses of correspondence of chorology and plant phenology, but were included in the indices of site similarities (index of Ochiai 1957). Sciomyzidae were collected at all seven sites, with S indices ranging from 0.93 in sites I_B and I_C to 0.24 in sites II_B and II_A, i.e., 40% of the sciomyzids captured in I_B and I_C.

Studies of assemblages of Sciomyzidae in an oak grove (six species) (Vala 1984b) and in a temporary aquatic habitat (14 species) (Vala & Manguin 1987) near Avignon, southern France, based on quantitative, routine sweeping, focused on species richness, diversity, phenology, and synchrony between the sciomyzid species and their mollusc hosts/prey. Data on the 2996 specimens of the 14 species collected and then released in a temporary aquatic habitat near Avignon during 1982–84 by Vala & Manguin (1987) (Table 10.3, Eq. 10.1) were especially carefully analyzed, and the methods could be applied well to other studies. The frequencies of presence were calculated and grouped according to the classification of Bigot & Bodot (1972–1973) (Fig. 10.6). The specific richness was analyzed, the Shannon & Weaver (1948) index was used to analyze diversity, and the Daget (1979) index was used to display population equilibration (Fig. 10.7). The species composition and seasonal changes of 32 species collected daily in Malaise traps during April–October in three wooded and one pond habitat in Slovakia were analyzed for dominance, site similarity, alpha-diversity, and evenness (Roller 1995). Although only one site (a mixed forest, Cephalantero–Fagenion community) produced many flies (672 of the

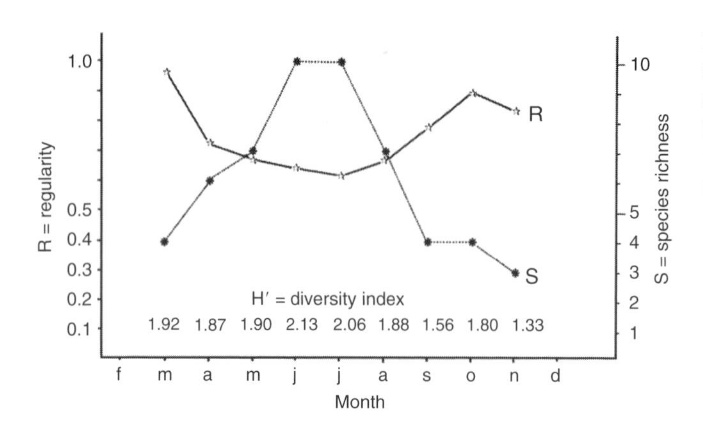

Fig. 10.7 Variation in regularity (R), species richness (S), and Shannon & Weaver diversity index (H′) for species of Sciomyzidae collected in temporary aquatic habitat in Rochefort-du-Gard, southern France during 1984. See Table 10.3. Modified from Vala & Manguin (1987).

775 collected at all four localities), the methods used are of interest. Two species, the generally uncommon *Pherbellia pallidiventris* (448 specimens) and the common slug killer, *Tetanocera elata* (101 specimens) were dominant there. *Pherbellia pallidiventris* (36 specimens) also was the dominant species of 20 species in 10 genera collected in a Malaise trapping study of Sciomyzidae from April to October 1993 in a terrestrial situation in Schleswig-Holstein (Kassebeer 1999b).

King & Brazner (1999) included Sciomyzidae in their study of numerical abundance, biomass, and taxonomic composition of insects in four coastal wetland areas along the eutrophic to oligotrophic, shallow, northwestern shore of the 193 × 30 km Green Bay of Lake Michigan, Wisconsin. They trapped insects flying at just above (to 24 cm) the surface of three vegetation zones with floating sticky traps during the spring and summer of 1995. They estimated abundance and biomass, using length–mass regression equations for biomass, and described their various statistical analyses. They presented their data by family (26) in 8 orders of insects, except by subfamily and tribe for Chironomidae. For Diptera, the 14 families were identified to "morphospecies." The Sciomyzidae, found in all four areas and all three vegetation zones (most common in the "dense emergent" and least common in the "open water-submergent" zones), ranked eleventh in abundance and fourth in biomass of the 14 families. Sciomyzidae (14 species) were among the most species-rich families, along with the Chloropidae (28 species) and Ephydridae (15 species), found in weekly net samples made between May 6 and September 29, 1989 in stands of *Carex lacustris* and *C. stricta* in a small marsh in northeastern Ohio (Foote 2004).

Whiles & Goldowitz (2001) carried out a very detailed field-experimental analysis of insect emergence abundance and productivity in two ephemeral, two intermittently wet, and one permanent "sloughs" (wet meadows) in the Platte River wetlands of midwestern USA. These kinds of situations are prime habitats for many species of aquatic and semi-aquatic species of Sciomyzidae. Similar habitats are present in much of the northern hemisphere and perhaps are represented in southern hemisphere temperate areas. The authors provided extensive information on ten physical characteristics of each of the five sites. They summarized the abundance, emergence production, species richness, Shannon index diversity, and number of unique taxa for each site. Production was presented in terms of mean dry mass per family. The authors summarized that the "Sciomyzidae (primarily *Pherbellia* and *Elgiva*), Muscidae, and Tipulidae . . . . which are relatively heavy-bodied dipterans, were important contributors to emergence production at most sites, particulary those with shorter annual hydroperiods" (Table 10.4). M. R. Whiles (personal communication, 2004) informed us that the species of Sciomyzidae that he was able to identify were *Atrichomelina pubera*, *Pherbellia* spp., and *Elgiva* sp. The last, based on our knowledge of the distribution of Sciomyzidae, certainly was *E. solicita*. Unfortunately, voucher specimens were not preserved. The need for preservation of voucher specimens from ecological, molecular, etc. studies was highlighted by Knutson (1984), but remains until today a serious omission in such studies.

Eight species of Sciomyzidae were included with 39 other families of Diptera (determined to family) sampled in a grazed ephemeral wetland in Ireland (Ryder *et al.* 2005). Vegetation height appeared to play a more important role in determining the sciomyzid species composition than plant species composition. Mc Donnell & Gormally (2000) had come to the same conclusion studying sciomyzids in the same habitat, and they also suggested that disturbance as a result of grazing by cattle appeared to negatively impact sciomyzid species richness.

The most modern synoecological studies including Sciomyzidae that give formulae for and/or references to publications on indices and statistical tests used are those of Vaňhara (1981, 1986; dominance, alpha-diversity, equitability), Vala & Manguin (1987; alpha-diversity, equitability, specific richness), Rozkošný & Vaňhara (1992, 1993; alpha-diversity, equitability, Jaccard index [coefficient of similarity]), Roller (1995; alpha-diversity, dominance, site-similarity, evenness), While & Goldowitz (2001; diversity, productivity, relationships), Ryder *et al.* (2005; canonical correspondence analysis [CCA], Sorensen similarity index), and Williams *et al.* (2009; CCA).

10.3 THREATENED SPECIES AND BIO-INDICATORS

Marcus:
Alas, my lord, I have but kill'd a fly.
Titus:
But? How if that fly had a father
and mother?
How would he hang his slender
gilded wings
And buzz lamenting doings in the air!
Poor harmless fly,
That, with his pretty buzzing melody,
Came here to make us merry!
And thou hast kill'd him.

Shakespeare (1564).

Many species of Sciomyzidae are known from only a few specimens each. Many other species are considered rare, but in some cases this is due to artifacts of time of collecting, habitats searched, and/or collecting methods. For example, we have only 70 date-locality records of the seemingly rare *Tetanura pallidiventris*, a boreo-alpine, terrestrial parasitoid extending from Ireland to Japan. This species is univoltine, has a short flight period, with the adults from a population of overwintering pupae disappearing shortly after emergence. All collection records we have seen range only from June 5 (Estonia) to August 5 and 6 (French Pyrenees). However, on two of the seven dates and sites where we collected it, *T. pallidiventris* was one of the most common flies present. Hundreds of adults were seen on the upper surfaces of almost all vegetation in a mixed woodlot in southern Finland on July 4, but none could be found on July 11, or during later visits. Chandler (1972) recorded seven collections of this species from Ireland between June 25 and July 11 and noted that "it was very numerous and large numbers of adults were seen at rest or running about" on leaves of ground-level plants and logs on June 30 in a beech woods in Killarney, Co. Kerry, Ireland.

In recent years, publications on collections in local habitats more frequently have included notes on the status of species as threatened or not, becoming localized, decreasing, etc., or expected to be collected but not collected because of habitat destruction (e.g., Speight 1994). Six species of Sciomyzidae occurring in lowland wet grasslands in Wales are noted in the Red Data Book as 2 or 3 on a 1–3 scale (Drake 1988). Almost half of the 31 scarce and rare species of Sciomyzidae (of a total of 66 British species) were recorded minimally on one or more of 24 such sites in England as analyzed by this author. The scarce and threatened species of Great Britain were listed by Falk (1991). The Sciomyzidae (50 species) were included along with "Lower Brachycera," Scathophagidae, Syrphidae, and Tipulidae in the 15 groups of invertebrates (6989 species) used in the preliminary attempts to develop an invertebrate assemblage classification (ISIS) for conservation interests in England (Webb & Lott 2006). Their provisional classification of assemblage types for freshwater wetlands consisted of five "broad assemblage types" (macrohabitats) and 20 "specific assemblage types" (microhabitats). The authors discussed habitat classification in general and identified major impediments and needs in terminology, site evaluation, etc. to use of invertebrates in conservation.

Vaňhara (1981, 1986) and Vaňhara & Rozkošný (1997) noted a distinct reduction in the numbers of species and specimens of Sciomyzidae in the Horní les forest floodplain in Moravia over a period of two decades (1971: 14 species and 87 specimens; 1981: five and 12; 1991: three and four). This apparently was due to water management practices. Rozkošný (1999b) recorded 47 species in 17 genera from detailed, long-term surveys of the 83 km^2 UNESCO Pálava Biosphere Reserve in southern Moravia. He noted that most Sciomyzidae are "very narrowly restricted to wetlands and thus are jeopardized by a general loss of suitable habitats." He suggested that in this reserve the snail-egg feeding *Anticheta* species are endangered or vulnerable, as is *Pherbellia limbata*, a xerothermic species living on steppe-like slopes on limestone; *Ilione lineata* and *Tetanocera montana* were considered rare.

Speight (2001, 2004) inventoried the Sciomyzidae (17 species) and Syrphidae (73 species) of a case-study farm in

Co. Cork, Ireland during 1994–2000 and considered the faunas in relation to the total potential role of the farm and its component habitats in supporting the regional faunas of the two families. There are 57 species of Sciomyzidae known from Ireland, 28 of which are recorded from Co. Cork. For the Syrphidae, similar percentages of the Co. Cork fauna (60%) and Irish fauna (40%) were found on the farm. The main groups of nine habitats were the production area (*c.* 30 ha), infrastructure habitats (*c.* 5 ha), disused sector (*c.* 5 ha), and transitory set-aside areas. Both the infrastructure and disused sectors included various aquatic habitats as well as terrestrial habitats. Most of the specimens were collected by 27 Malaise traps set out in all sectors, April–September 2000 and twenty 1 m^2 emergence traps set out only in production and set-aside sectors, April–August 2000. Of the 17 species of Sciomyzidae, 182 specimens were collected in the Malaise traps, with five species in the production sector, four in the infrastructure sector, all 17 in the disused sector, and none in the set-aside. The author concluded that the solitary specimen of *Tetanocera elata* collected in an emergence trap in a pasture field of more than 1 ha on the farm is potentially indicative of a large population of the species developing in one of the pasture fields. He speculated that a "naive estimate" of specimens produced per ha would be 2500 for a species collected just once in one trap. This is consistent with the general knowledge that this very successful species is one of the most common, abundant, and widespread sciomyzids in the Palearctic. The most abundant species collected were the slug-killer, *T. elata* (61 specimens), and the fingernail clam-killer *Ilione lineata* (39). The other species, both aquatic and terrestrial, ranged from 1 to 17 specimens, with six species represented by unique specimens. M. C. D. Speight (personal communication, 2003) noted that further examination of samples "confirmed the development of *T. elata* in the fields, with the highest numbers emerging from pasture but some from silage grassland and even a few from 'cereal strips' – stands of cereals installed to provide a seed supply for dwindling farmland bird species." *Tetanocera ferruginea* and *T. hyalipennis* also emerged from the field surface traps, and an additional species, *Pteromicra angustipennis*, was collected by Malaise traps.

The author discussed the species of flies found, those predicted to occur, their habitat associations, and conservation value, especially in detail for the Syrphidae. That family is better represented and there is more information on its distribution and conservation value in Ireland than for the Sciomyzidae. All of the sciomyzid species found would be predicted to occur on the farm on the basis of the habitats present, but also other species should be present. Only *Elgiva solicita* and *Tetanocera punctifrons* were considered potentially of some "conservation value." In general, the sciomyzid situation paralleled what was found for the Syrphidae. If the disused sector land was brought into production it would be predicted that the sciomyzid fauna would be reduced to eight species. There would be a further reduction to five species if the infrastructural land were converted to production. All would disappear if the entire farm was converted to crop production. The unimproved, seasonally flooded, oligotrophic *Molinia* grassland/acid fen habitats of the disused sector were considered the most important for maintenance of the sciomyzid fauna, and one of the areas most threatened by development. The results of the overall study in general were discussed in relation to conservation values and management of farmland.

The above study was followed up by analyses of data from 1 and 4 m^2 emergence traps set up in all of the farm habitats except six considered irrelevant or not suitable to the trap designs, plus one newly developed habitat, an artificially produced overland flow wetland (Speight 2004). The emergence trap data largely corroborated the predictions of habitat occupancy recorded from the farm by the earlier Malaise trapping. A total of over 500 specimens were collected by both methods. There were 968 trapping units (1 unit = 1 m^2 of ground surface trapped for 1 month) for the emergence traps. Six species of Sciomyzidae, all recorded from emergence traps with two of them also from Malaise traps, were added to the 17 recorded from the previous Malaise trap study. This amounted to 44% of the then known Irish sciomyzid fauna, including all except five species known from Co. Cork but not from the farm. The five species were predicted to occur on the farm.

In total, 18 of the 23 species were collected by emergence traps; thus they could be assumed to be breeding successfully on the farm. Extensive notes on habitats occupied and on numbers of flies collected by both methods were included. *Pherbellia cinerella* was recorded only from emergence traps on production land. All other species were recorded from disused land, with seven of these also recorded from one or the other of the other two main habitat categories. *Tetanocera elata* was the only species appearing in emergence traps in all three (production, infrastructure, and disused land). It was the only species recorded repeatedly by both Malaise and emergence traps from production land. It occurred at the greatest number of sites, and Malaise and emergence trap collections were concordant at all except one site. While ten specimens of *Pteromicra angustipennis* were collected in Malaise traps, no fewer than 28 were collected in emergence traps.

Disused land was shown to play the dominant role in supporting the farm's sciomyzid fauna, with patches of it including wetland habitats being particularly important. It was predicted that 55% of the sciomyzids breeding on the farm would disappear if the disused land were converted to crop production or grassland. The methodologies used in these two studies likely could be applied elsewhere.

Williams *et al.* (2009) sampled adults of seven species of Sciomyzidae in an ephemeral wetland in Ireland and described the physical and vegetational aspects of the habitat in detail. Their results provided strong evidence for high microhabitat specificity in Sciomyzidae and indicated a major influence, primarily, of vegetational structure (height) and, secondarily, of hydrological regime on the sciomyzid assemblage. They noted "extreme specificity along the environmental gradient," and commented on this aspect as a factor to be considered in use of Sciomyzidae as biocontrol agents and their value as bio-indicators. They concluded that "sciomyzid communities are best thought of as tracking hydrological regimes at a fine spatial scale rather than the heterogeneity in vegetation zones per se." They suggested that "sensitivity to both hydrological regime and the influence of vegetation structure make this family particularly suitable to studies of sustainable grazing management of wetland habitats."

Haslett (2001) included Sciomyzidae, Syrphidae, and Muscidae in a study of spatial heterogeneity of habitats, especially patch content and border complexity as related to biodiversity and conservation of Diptera in Austria. With regard to Sciomyzidae, he focused on patch content and land use along a 70 km stretch of the Salzach River where 22 species were collected in Malaise traps at 26 sites, grading from old, dry woodlands to woodland/meadow mosaics to early successional wetland and river edge. Unfortunately, the species names and numbers of specimens were not given (apparently included in an unpublished environmental assessment). The species were grouped according to primary type of snail host/prey, (1) terrestrial, (2) aquatic or terrestrial, and (3) aquatic according to such data in Vala (1989a). The number of species per site was one to ten. The data were subjected to TWINSPAN cluster analysis as an "ensemble view of the mosaics" but the analysis did not address patch borders, the latter subject being analyzed in relation to the data on Syrphidae and Muscidae. The author found the results "extremely helpful to the Environmental Assessment Exercise concerned," and that the relative abundance of the species of Sciomyzidae could be used to characterize habitat mosaics and are sensitive to human land use. The author concluded that if it is true "that patch borders generally have an influence on the "absorbing power" of habitat mosaics, then borders need to be incorporated as a new axis in the definition of a habitat." Among the Sciomyzidae, we suggest adults of the terrestrial *Trypetoptera puncutala* and *Tetanura pallidiventris*, semi-terrestrial (shoreline) *Pherbellia* spp., and the aquatic *Sepedon sphegea* as optimal subjects in the study of border effects.

Fisher & Orth (1983) collected Sciomyzidae extensively, primarily with a suction machine, throughout California between 1962 and 1976, capturing 24 000 adults of 49 species in 13 genera. R. E. Orth & R. J. Mc Donnell (personal communication, 2002) noted that during collections in and around Riverside, southern California in 2001, no Sciomyzidae were found at the many localities that were very productive two decades earlier, apparently due primarily to habitat modification resulting from invasion of a Palearctic tall cane, *Arundo donax*. Also, population decline of molluscs and Sciomyzidae may be due to acid rain in areas of low calcium in the substrates. Similar declines have been noted in Ohio by B. A. Foote (personal communication, 2006).

> If all mankind were to disappear tomorrow, the world would regenerate back to the rich state of equilibrium that existed 10 000 years ago. If insects were to vanish, the terrestrial environment would collapse into chaos.
>
> E. O. Wilson.

11 • Natural enemies

Comme tous les êtres vivant, les insectes son soumis aux attaques de toute sorte d'ennemis qui empêchent leur multiplication à l'infini et conduit à un équilibre entre les différentes espèces animales et végétales.
[Like all living beings, insects are subjected to attacks from all sorts of enemies that prevent them from multiplying endlessly, which helps to lead to a balance between all the various animal and vegetal species].

Gruner & Riom (1977).

The following information leads us to the conclusion that natural enemies would not be significant limiting factors to the use of Sciomyzidae as biocontrol agents except, possibly, by parasitoid Hymenoptera not host-specific. A total of 218 records of 96 species and morphospecies in 47 genera of natural enemies reared from at least 64 species in 23 genera of Sciomyzidae have been tabulated in detail by R. J. Mc Donnell & L. Knutson (unpublished data).

11.1 PATHOGENS

The adults of many Sciomyzidae are restricted to damp, shady situations ideal for the development of pathogens. Notably, adults are fastidious, spending much time in cleaning movements of the legs across the body. Diseased adults or immatures rarely have been collected in nature, although the larvae of many species, especially terrestrial snail feeders in the second and third stadia, spend several days to a month or more inside the shells of their host, feeding on decaying tissue. Mechanisms for avoiding disease and preventing development of competitors in such a favorable nutrient seem likely. A mechanism by which intra- and interspecific competitions are avoided by *Musca domestica* and which also may provide protection against disease organisms has been studied by Bryant & Hall (1975). They showed that within a few hours after a batch of eggs hatch, larval conditioning of the medium occurs and ovipositing females avoid the conditioned medium.

In rearing *Salticella fasciata* on terrestrial snails, Knutson *et al.* (1970) noted "the thick, whitish film of bacteria that characteristically covers the exposed tissues of dead and decaying non-infested snails within several hours does not develop in snails that have larvae feeding in them; the presence or absence of a bacterial film readily distinguishes infested from non-infested snails." However, saprophagous nematodes often are seen in terrestrial snails in which mature, terrestrial, primarily parasitoid-predatory sciomyzid larvae are feeding in a saprophagous manner, apparently unimpeded by the nematodes as they complete development (Plate 4f).

When small rearing containers of larvae of aquatic species are not changed daily or when dead snails are not removed and the containers not flushed with clean water, an apparent viral disease may appear, resulting in bright red Malpighian tubules, a flaccid body with bulging weak spots in the cuticle, and death.

The only records of fungal pathogens of adults are *Hirsutella citriformis* (Hypocreales) attacking *Sepedon aenescens* (as "*Sepedon sphegea*") as well as a delphacid and a psyllid (Rombach & Roberts 1989), and a pair of *Dictya pictipes* collected on July 12 in northeastern USA that died one day later due to *Entomophthora* sp. (Valley & Berg 1977).

Entomopathogenic nematodes have emerged from a few field-collected adults (Fig. 11.1b). Parasitism by unidentified Nematoda was noted in two females of *Tetanocera obtusifibula* collected on July 26 in Idaho; four worms emerged from one female on June 29 and two from the other on June 30; both females died one day after emergence of the worms (Foote 1999). Valley & Berg (1977) provided some detail on the behavior of a male *Dictya steyskali* collected on June 9, and a female reared from a field-collected puparium on May 11 (both collected in New York) that harbored nematomorph worms. Verbeke (1948) recorded a species of Gordidae reared from a female *Ilione albiseta* in Belgium. "Sciomyzidae" were recorded as hosts of free-living marine nematodes of the genus *Pseudocella* (Leptosomatidae) on the coast of Kamchatka and from the Kara Sea by Platanova (1985, 1988) (determinations questionable).

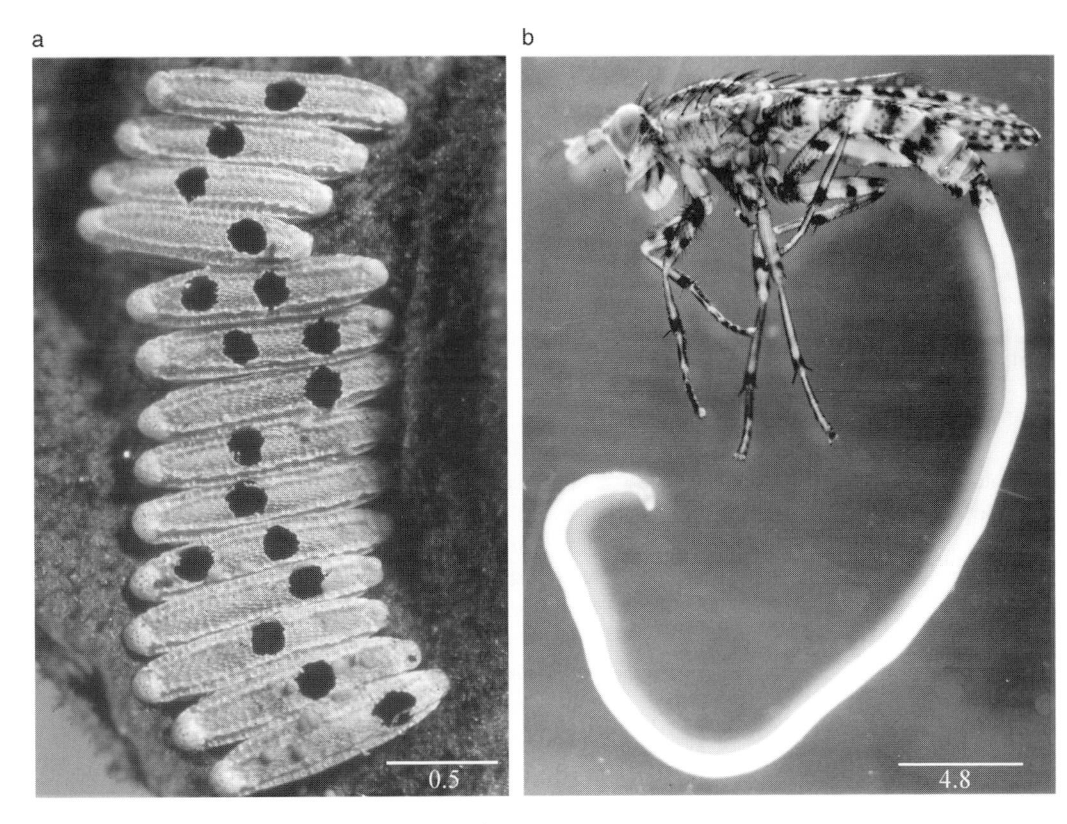

Fig. 11.1 (a) Eggs of ***Sepedon americana*** parasitized by *Trichogramma* sp. (from Neff & Berg 1966). (b) Emergence of adult nematode from anal aperture of ***Dictya*** sp. female (photo by E. C. Bay).

As many as 30 minute, round, brownish "sacs" occasionally have been found in the abdomen of some adult sciomyzids during the course of macerating the abdomen in KOH for study of the genitalia. These are resistant to strong caustic solution. They may be microsporidia or eggs of parasitoid Hymenoptera but no parasitoid Hymenoptera have been reared from adult Sciomyzidae.

11.2 PARASITOIDS

Eggs of some species, especially the *Sepedon* species that lay eggs in groups, are attacked in nature by parasitoid Hymenoptera of the family Trichogrammatidae (Fig. 11.1a). At least five identified and four unidentified species of *Trichogramma* and *Trichogrammatoidea* have been reared from six species of *Elgiva*, *Sepedon*, and *Tetanocera*. Juliano (1981) studied extensively, at four sites near Ithaca, New York, *Trichogramma julianoi* (which parasitizes *Sepedon fuscipennis* and *Elgiva solicita*) and *Trichogramma* sp. near *californicum* (which parasitizes *S. fuscipennis*, *E. solicita*, *Tetanocera* spp., and pyralid moths). *Trichogramma julianoi* was most active in early and late summer; *Trichogramma* sp. near *californicum* was most active in mid and late summer. Another undescribed species, similar to *T. semblidisi*, that occurs in the same habitats is not known to parasitize *S. fuscipennis* in nature but did in laboratory rearings; its primary hosts are Stratiomyidae. Mortality of *S. fuscipennis* eggs due to *Trichogramma* spp. varied significantly over the summer and reached 43.6% in late July and early August (Fig. 11.2). Of the eggs of various freshwater insects (Sciomyzidae, Ephydridae, Syrphidae, Stratiomyidae, Tabanidae, Pyralidae, Sialidae, Corydalidae, Dytiscidae, Chrysomelidae, and Coccinellidae), Sciomyzidae (up to 31.4% parasitization) and Stratiomyidae (at least 30% parasitization) were the hosts most heavily parasitized by *Trichogramma* spp. In laboratory trials, *Trichogramma* sp. near *californicum* parasitized all of these. Juliano (1982) offered eggs of *S. fuscipennis* < 1, 1–2, 2–3, and 3–4 days old to

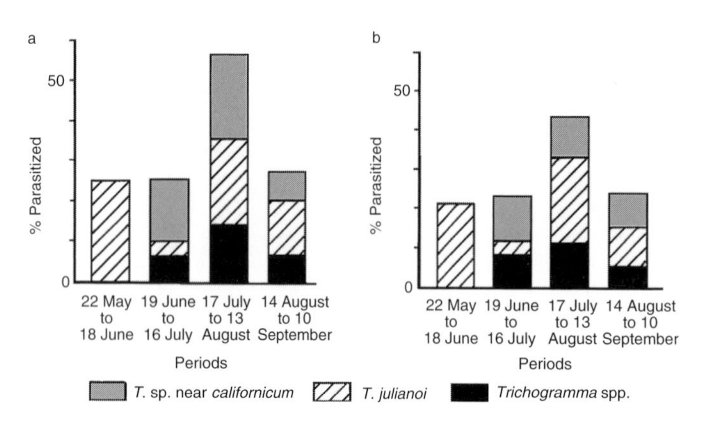

Fig. 11.2 Percent parasitization of eggs of ***Sepedon fuscipennis*** by *Trichogramma* spp. during four periods. Data from nine sites. (a) Egg masses; (b) individual eggs. From Juliano (1981).

inexperienced female *Trichogramma* sp. near *californicum* in a laboratory experiment. Quoting from his abstract,

> Exposure period (2 h) and number of host eggs offered at 1 time (10) were held constant. Percentage of hosts parasitized, total number of parasitoid progeny found in a group of hosts, percentage of parasitized hosts yielding adult parasitoids, and survivorship of parasitoid progeny to adulthood all decreased significantly with increased host age. Younger hosts yielded more than 1 adult parasitoid more often than did older hosts. Over 50% of the adult parasitoids emerging from hosts < 1, or 1–2 days old were female. Hosts 2–3 days old produced only males.

Some *Sepedon* species have been considered as important alternate hosts of *Trichogramma* parasites of rice-stem borers in Southeast Asia (Nagatomi & Kushigemachi 1965, Yano 1968, 1975, 1984), maintaining populations when egg masses of the borers are not present. Of 134 egg masses of *S. aenescens* containing 2123 eggs collected on June 3 in Kyushu, Japan, 117 masses (87.3%) and 1395 eggs (65.8%) were parasitized by *Trichogramma japonicum* (Nagatomi & Kushigemachi 1965).

At least 15 identified species and 40 unidentified species of parasitoid Hymenoptera of the families Braconidae (*Aphaereta*, *Aspilota*, and *Phaenocarpa*), Pteromalidae (*Eupteromalus*, *Spalangia*, *Trichomalopsis*, and *Urolepis*) and Ichneumonidae (*Atractodes*, *Eriplanus*, *Exolytus*, *Hemiteles*, *Mastrus* [?], *Mesoleptus*, *Orthizema*, *Phygadeuon*, and *Theroscopus*) that oviposit into larvae and emerge from puparia of flies have been reported from 33 species in 13 genera, including both terrestrial and aquatic Sciomyzidae, primarily in the Nearctic and Palearctic. Many of the Ichneumonidae that have been reared from field-collected puparia, especially in North America, have not been identified or published on as they belong to genera in need of taxonomic revision. There is a large amount of reared material in the sciomyzid collections at Cornell University and at the Smithsonian Institution. William K. O'Neill began doctoral research at Cornell on the biology and taxonomy of ichneumonid parasitoids of sciomyzids, but died in 1974 before completing the study.

In a field study of larval survivorship by Eckblad & Berg (1972), ichneumonid parasites accounted for 3.9% mortality of *Sepedon fuscipennis* puparia; total mortality during two seasons was 13%. In southern France, Vala & Manguin (1987) showed that mortality of *Sepedon sphegea* was significant in the spring generation of the flies with 86% of puparia collected in a temporary aquatic habitat on May 10 parasitized by *Mesoleptus ripicola*. Two collections of 59 puparia of *Atrichomelina pubera* produced 12 flies, 28 wasps (24 of them Ichneumonidae) and 19 failed to open (Foote *et al.* 1960). Close correlations in seasonal activity of seven parasitoid wasp species (Braconidae: *Phaenocarpa antichaetae*; Ichneumonidae: *Mastrus* [?] sp., *Mesoleptus* spp., *Phygadeuon* spp.), reared from puparia of the snail-egg feeding *Anticheta melanosoma*, and seasonal activity of that sciomyzid, and the food snail (*Aplexa hypnorum*) were described by Knutson & Abercrombie (1977). The seasonal activities of all organisms were associated with changes in the water level of the ephemeral, vernal pond habitat. From the overwintered puparia there was sequential emergence; first the flies, then the braconids, then the ichneumonids (Fig. 11.3). From a total of 161 overwintering puparia of *Tetanocera ferruginea* collected between February 28 and April 5 on five occasions at two sites near Ithaca, New York, 108 flies and ichneumonids of six species (four *Theroscopus* spp., one

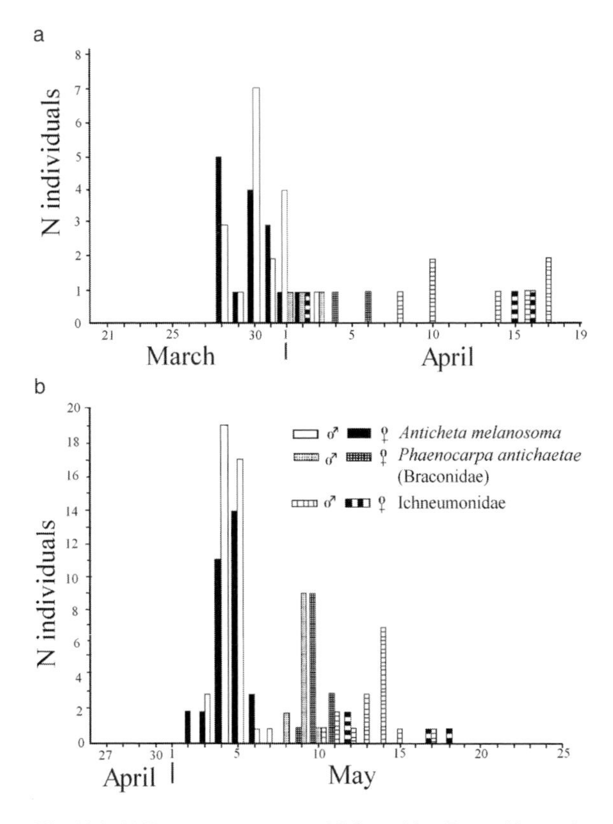

Fig. 11.3 (a) Emergence patterns of Sciomyzidae, Braconidae, and Ichneumonidae from 56 overwintering puparia of ***Anticheta melanosoma*** collected March 21, 1966. (b) Emergence patterns from 134 overwintering puparia of ***A. melanosoma*** collected April 27, 1966. Both collections made at Geneva, New York. Male that emerged May 7 was ***A. borealis***; all other flies and puparia were ***A. melanosoma***. From Knutson & Abercrombie (1977).

Mesoleptus sp., and one *Phygadeon* sp.) emerged (21% parasitization), with parasitization of the individual collections ranging from 16–33% (Foote 1999). In each sample, the wasps emerged 2–6 days after the last emergence of the flies.

Further data on parasitoid natural enemies of Sciomyzidae could be obtained by exposing laboratory-reared immature stages in suitable containers in the field, then retrieving them and holding them in the laboratory for emergence of parasitoids. This technique was successfully used by O'Neill (1973) who obtained the new species *Trichopria atrichomelinae* (Diapriidae) from puparia of laboratory-reared *Atrichomelina pubera* exposed for 7 days in *A. pubera* habitats.

Parasitoid Hymenoptera with broad host ranges are probably the most important biotic limiting factor for Sciomyzidae. Several of the Ichneumonidae and Braconidae species reared from sciomyzid puparia have been reared also from puparia of other families of Diptera. Fisher & Orth (1964) found that *Phygadeuon* sp. reared from puparia of the aquatic predator *Sepedon praemiosa* and exposed to young pupae of the egg-feeder *Anticheta testacea* readily oviposited in and completed development on the latter. A few parasitoid Diapriidae, *Trichopria* spp. (Knutson & Berg 1963b, O'Neill 1973) (Plate 4h) and *Spilomicrus* spp. (Abercrombie 1970, Barnes 1979b), seem to be specific to the pupal stage of some aquatic and terrestrial Sciomyzidae. O'Neill (1973) reported extensive field collections and laboratory studies of two species of *Trichopria*.

Seasonality of development of larvae may be related, in part, to parasite pressure, with those species whose larvae develop during the late fall to early spring escaping attack by parasitic Hymenoptera. Berg *et al.* (1982) noted,

> Although the Sciomyzidae are so heavily attacked by parasitoid Hymenoptera that collections of puparia formed in late spring, summer and fall often yield more Hymenoptera than flies, we have never reared a parasitoid wasp from any puparium of species in these groups [Groups 3 and 4 – univoltine species that overwinter within egg membranes or as partly grown larvae].

11.3 PREDATORS

Adult sciomyzids are relatively solitary, secretive, inconspicuous, do not congregate, and are not often collected on flowers (thus their name: scio-myz = shade-fly). This suite of features, except their slow flying, would seem to protect them from many opportunistic predators. Adults seem to be unusually sensitive to their substrate and surroundings; their near-continual probing the substrate and area in front of them with their fore tarsi is a field identification character for the family.

During mass rearings of the introduced aquatic predator *Sepedomerus macropus* in containers outdoors in Hawaii, predators "exacted a heavy toll" (Chock *et al.* 1961). Odonate naiads accidentally introduced into the containers attacked larvae, *Pheidole megacephala* ants robbed the containers of larvae and puparia, and the hemipteran *Mesovelia mulsanti* preyed heavily on first-instar larvae. In a study of larval survivorship of *Sepedon fuscipennis* in the field over a 2-year period in Ithaca, New York, 1000 larvae were kept in floating cylinders in a typical habitat of the species; only one instance of predation was observed, when an immature *Notonecta* sp. attacked and fed on a third-instar larva (Eckblad & Berg 1972). Nymphs of various Odonata and larvae of *Tropisternus* sp. (Hydrophilidae) collected in the same study area were successful predators of *S. fuscipennis* under laboratory conditions. *Tetanocera* species, generally abundant in a large nature reserve near Rome, Italy were absent in areas where *Gambusia* spp. fish were disseminated for control of mosquito larvae (Rivosecchi 1992). O. Beaver (1989) noted that in laboratory trials in which predaceous Odonata nymphs and four species of fish were kept overnight with larvae of *Sepedon senex* all of the larvae were consumed. A mature larva of the tabanid *Hybomitra schineri* killed and ate 37 second- and third-instar larvae of *Tetanocera ferruginea* in the laboratory during 21 days before pupating (Knutson 1965). Ten species of wasps, flies, and Orthoptera have been recorded as predators of adult Sciomyzidae (Mc Donnell & Knutson, unpublished data). Survey of collections of predaceous odonates, wasps, Asilidae, and Empididae for specimens with sciomyzids as prey would be of interest. Vala (1984b) attributed decreases in the populations of six species of terrestrial sciomyzids in an oak grove in Rochefort-du-Gard, southern France after the peak of emergence during April to be partially due to "spiders and Asilidae." While Sciomyzidae obviously are preyed upon by birds and fish there are only four actual records (Mc Donnell & Knutson, unpublished data).

12 • Defense mechanisms

When the moon shall have faded out from the sky, and the sun shall shine at noonday a dull cherry-red, and the seas shall be frozen over, and the ice-cap shall have crept downward to the equator from either pole, and no keels shall cut the waters, nor wheels turn in mills, when all cities shall have long been dead and crumbled into dust, and all life shall be on the very last verge of extinction on this globe; then, on a bit of lichen, growing on the bald rocks beside the eternal snows of Panama, shall be seated a tiny insect, preening its antennae in the glow of the worn-out sun, representing the sole survival of animal life on this our earth, – a melancholy "bug."

Holland (1903).

12.1 MOLLUSC DEFENSE MECHANISMS AGAINST SCIOMYZIDAE

Morphological and behavioral adaptations have evolved in gastropods in response to the four primary types of predation they experience – shell entry (e.g., sciomyzid larvae, decapods), shell crushing (e.g., fish, birds), shell boring (e.g., some Mollusca), and whole-shell swallowing (e.g., fish, birds). Shell shape does not seem to be a factor in the success of attacking sciomyzid larvae as it is for some larger entry-type predators such as decapods that more successfully attack rotund snails than snails with a narrower aperture (DeWitt *et al.* 2000). While, in general, smaller sciomyzid larvae prey on smaller snails and larger larvae prey on larger snails, their attenuate, small diameter, telescoping anterior segments allow even large larvae to attack small snails.

Most mollusc hosts/prey of Sciomyzidae seem relatively vulnerable, unlike the highly mobile and agile or heavily armored adult insect hosts (e.g., Hymenoptera, Diptera, and Coleoptera) of several groups of dipterous parasitoids (e.g., Phoridae, Pipunculidae, Pyrgotidae, and Tachinidae). Some snails produce alarm pheromones when attacked (see references in DeWitt *et al.* 2000), like the social or colonial insect hosts of insect parasitoids that are well known to have pheromones or behaviors to which the attacker can adapt and counter (Feener & Brown 1997).

Damage to the shells of snails fed upon by sciomyzid larvae has been reported only by Appleton *et al.* (1993). They figured 6–7 mm shells of the aquatic *Biomphalaria pfeifferi* and *Bulinus tropicus* massively "broken during attacks by a 3rd instar *Sepedon neavei* larva," but did not state that they actually saw the larva breaking the shells. We believe that it is likely that some other predator, a contaminant in their rearing, was responsible. P. Roberts (personal communication, 2005) stated that third-instar larvae of *Pherbina coryleti* made a rather large hole in the umbo of two individuals of *Sphaerium* sp.; but despite numerous attempts the behavior was not replicated. The shells probably were at least fractured accidentally.

The operculum of salt marsh and freshwater prosobranch snails in six families (see Chapter 4) does not always hinder the attack of seven species of the aquatic predators *Dictya* and *Neolimnia*, or the parasitoid/predator *Pherbellia inflexa*. The larvae strongly preferred or could develop only on these snails, which seem to be the natural prey, in laboratory rearings. They could not kill such snails when the operculum was closed, but readily killed them when the soft parts were exposed. The opercula of most of these snails are coriaceous and pliable and can be bent aside easily. All of the snails tested were small (± 6 mm greatest dimension), except *Littorina littorea* (13–15 mm long). Larvae of 11 species of aquatic predators in the genera *Dictya*, *Sepedon*, and *Tetanocera* could not kill prosobranch snails in the genera *Oncomelania*, *Bithynia*, *Filopaludina*, and *Pila* from the Orient with hard, calcareous opercula (Berg 1964, Neff 1964, Bhuangprakone & Areekul 1973); *S. aenescens* also could not develop or developed poorly on crushed snails of the latter four genera. In trials of ten species of Nearctic aquatic predators of the genera *Sepedon*, *Tetanocera*, and *Dictya* against *Oncomelania formosana* and *O. quadrasi*, Neff (1964) noted that,

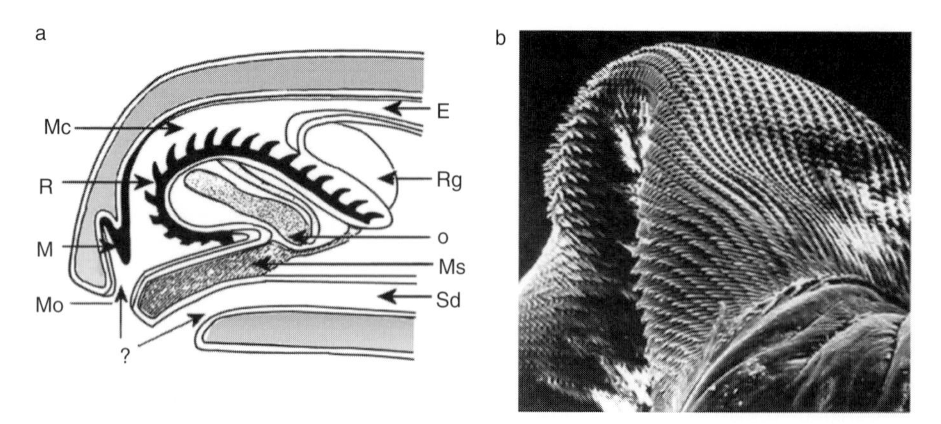

Fig. 12.1 (a) Longitudinal section through the head of a slug. M, maxilla; Mc, mouth cavity; Mo, mouth opening; Ms, muscle of radula; o, odontophore cartilage; E, esophagus; R, radula; Rg, radular gland; Sd, salivary gland duct opening; ?, possible points of penetration by sciomyzid larvae. (b), Radula of *Deroceras reticulatum*. Modified from Godan (1979).

> Operculum closure has enough force to injure the anterior end of the larva. After such an encounter a larva usually failed to feed or carry out further attacks. Subsequent examination disclosed injury to the larva's feeding mechanism (i.e. the cephalopharyngeal apparatus and its musculature).

However, Barnes (1979b) found that the aquatic predator *Neolimnia tranquilla* in New Zealand killed and consumed the operculate snail *Potamopyrgus antipodarum* in preference to non-operculate snails. The operculum of *P. antipodarum* is not coriaceous and is inflexible. Although the larvae could not attack snails with closed opercula, attacks usually were successful when the soft tissues were exposed, "No larva was ever found to be injured by being wedged between the inflexible operculum and the shell of the snail."

The membranous epiphragm of some estivating snails might seem to offer protection, but it usually has a weak spot or open breathing pore. Such estivating *Stagnicola* and *Galba* spp. are the obligate hosts of the behavioral equivalents *Colobaea bifasciella* (Knutson & Bratt unpublished data) and *Sciomyza varia* (Barnes 1990). Also, the radula in the mouth of gastropods would seem to provide protection. Surprisingly, the larvae of at least two species of Sciomyzidae habitually attack their hosts by penetrating into what appears to be the mouth opening (*Pallifera* and *Philomycus* slugs by *Tetanocera clara* [Trelka & Foote 1970] and terrestrial *Subulina octona* snails by *Sepedon umbrosa* [L. G. Gbedjissi, personal communication, 2004]). However, these larvae may in fact penetrate into the opening of the salivary gland duct. Penetration in that place, rather than into the mouth opening, would probably result in pushing the radular muscle mass upwards, thus closing the mouth opening and preventing the radula (Fig. 12.1) from contact with the invading larva. Also, production of toxin by two other slug-killing *Tetanocera* species, which attack the surface of the slug's body, is well-known (Trelka & Berg 1977) and larvae that penetrate the "mouth" may also produce immobilizing toxin.

Aquatic snails under attack by sciomyzid larvae often retract, release air, and drop from the surface, similar to their avoidance responses to leech predation described by Wilken & Appleton (1991). Lynch (1965) noted that *Austropeplea tomentosa* which dropped to the bottom after being attacked by *Dichetophora biroi* normally survived. Maharaj *et al.* (1992) described avoidance behavior of snails attacked by *Sepedon scapularis*. Snail predator-avoidance behavior mediated by sensing predation on nearby conspecific snails is a well-known phenomenon (DeWitt *et al.* 1999) but has not been examined in snail/sciomyzid associations. Mollusc mucus (see summary by Skingsley *et al.* 2000) is species-specific and consists of a variety of lactins, charged mucopolysaccharides, glycoproteins, proteins, uronic acid, sialic acid, hexosamine, and other molecules in an aqueous carrier medium. Release of mucus to the surface from various glands is in the form of membrane-encased mucus granules that are ruptured by various cues, including mechanical shearing. The mucus provides information about the mollusc's sexual condition and direction of movement to conspecifics, and the latter

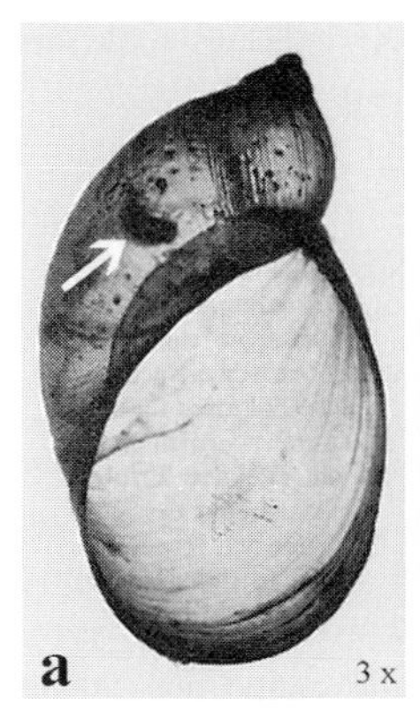

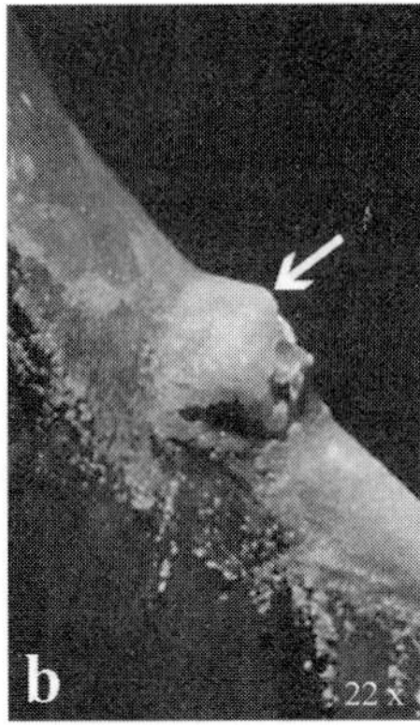

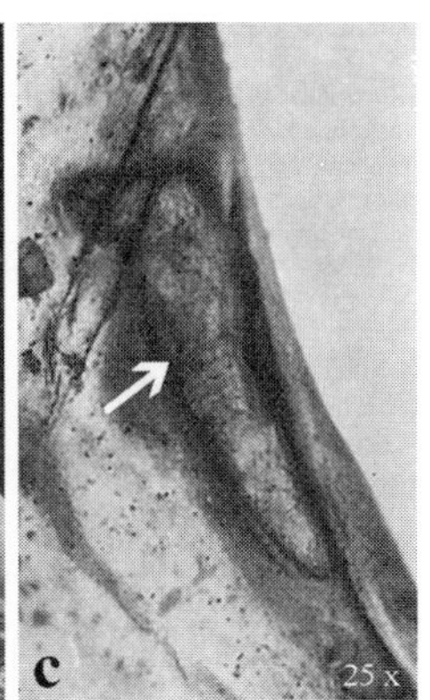

Fig. 12.2 (a) Shell of *Succinea putris* with encapsulated larva of ***Pherbellia s. schoenherri*** in position 4. (b) Encapsulated larva in position 2. (c) Encapsulated larva in position 3. Compare positions with Fig. 12.3. From Moor (1980).

is also sensed by some predators, notably Lampyridae (Schwalb 1961). A special defense mucus is produced by some snails, particularly terrestrial snails.

When attacked, terrestrial and aquatic snails retreat into the shell and often release sticky mucus but except for operculate species the soft parts are still exposed. This mucus might be, in part, egg jelly as DeWitt *et al.* (2000) recorded being released by *Physa* sp. when they were penetrated by the decapod *Orconectes obscurus*. Some terrestrial and semi-terrestrial sciomyzid larvae have developed mechanisms to enable them to counter the sticky mucus coat of their hosts/prey. When *Deroceras* slugs are attacked by first- and second-instar *Tetanocera elata* they produce a whitish mucus abundantly, but this does not deter the larva from crawling relatively rapidly on top of the relatively quickly moving slug (Knutson *et al.* 1965). However, first-instar larvae of *Tetanocera plebeja* often are entrapped in the mucus of their *Deroceras reticulatum* hosts/prey (Trelka & Foote 1970). Aquatic larvae are also often impeded by mucus. Mortality of larvae, especially first-instar larvae, of the aquatic predator *Sepedon senex*, caused by retraction and production of frothy mucus by aquatic snails, especially *Radix auricularia rubiginosa*, was detailed by O. Beaver (1989). Maharaj *et al.* (1992) observed larvae of all stages of the aquatic predator *Sepedon scapularis* to die in the mucus secretions or as a result of their interspiracular processes becoming entangled with snail feces. *Pseudosuccinea columella*, an exotic, invasive aquatic snail in southern Africa, apparently uses its foot as a shield, preventing the larva from moving beneath the shell and wraps its foot around the larva, enveloping it in mucus and immobilizing it.

Except for the encapsulation response of terrestrial and semi-terrestrial snails attacked by *Tetanura pallidiventris* and *Pherbellia s. schoenherri* noted below, the snails attacked by the other three most intimately associated parasitoid Sciomyzidae, *Sciomyza varia*, *S. aristalis*, and *Colobaea bifasciella*, do not show any defensive behavior and do not produce viscous mucus when attacked. The larvae of these species penetrate between the soft parts and the shell, in some cases so far into the snail that the posterior spiracles are not exposed in the aperture, and remain in place feeding on the living and often active snail for several days. The terrestrial and semi-terrestrial hosts of some other Sciomyzidae with somewhat similar parasitoid behavior only during early larval life (e.g., *Sepedon hispanica*, Knutson *et al.* 1967a) also remain alive and active for a few days after being attacked.

Direct attack by hosts/prey has been observed in the case of *Deroceras* slugs that curl upon themselves to eat a *Euthycera chaerophylli* (Plate 4b) and *E. cribrata* larvae attempting to penetrate into the slug. Carnivorous snails have been observed to ingest attacking larvae: *Pherbellia dorsata* larvae by the non-natural hosts *Gonaxis* spp. and *Achatinella* sp. (Davis *et al.* 1961) and *P. dubia* by the natural hosts *Aegopinella* sp. and *Oxychilus* sp. (Bratt *et al.* 1969). Since sciomyzid larvae, except probably *E. chaerophylli*, are not mesoparasitoids but feed primarily on muscle, gland, and other tissue incapable of an encapsulation response, encapsulation seems to be rare. Encapsulation of *P. s. schoenherri* by *Succinea putris* has been studied in detail (Moor 1980; Figs 12.2, 12.3). She noted high levels of encapsulation in all size classes of snails parasitized, with the highest level (about 65%) in the largest snails and the lowest level (about 25%) in the smallest snails (Fig. 9.31). Also, the finding of first- and second-instar larvae of *Tetanura pallidiventris* shallowly embedded in a hard, calcareous matrix on the inner wall of all shells of a host snail, *Cochlicopa lubrica*, containing larvae or puparia

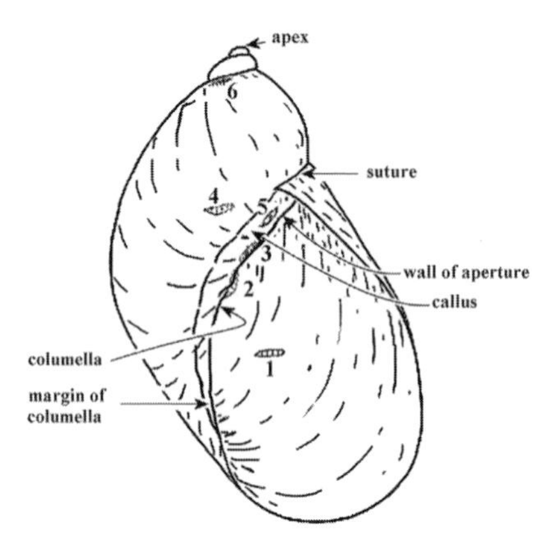

Fig. 12.3 Relative frequencies of variously situated, encapsulated larvae of ***Pherbellia s. schoenherri*** in shells of *Succinea putris*. Position 1 (n = 1), freshly penetrating larva encapsulated at beginning of its movement to its definitive resting place. Position 2 (n = 21), larva around columella. Position 3 (n = 21), larva on inner wall of aperture. Anterior end always directed toward apex of shell. Position 4 (n = 11), various places a short distance from columella. Position 5 (n = 1), around callus (fusion place, which connects outer lip and columella). Position 6 (n = 1), along suture. See Fig. 12.2. Modified from Moor (1980).

collected in nature (Knutson 1970a) indicates a physiological defense mechanism (Fig. 12.4a).

12.2 SCIOMYZID DEFENSE MECHANISMS AGAINST HOSTS/PREY AND NATURAL ENEMIES

Sciomyzids seem relatively defenseless. The paucity of defense mechanisms may be related to the facts that (1) mollusc populations are invariably enormously greater than sciomyzid populations, (2) most adult sciomyzids produce many progeny, and (3) the larvae seem to have little competition among the relatively few other insect natural enemies of molluscs. These features would seem to result in little evolutionary pressure resulting in sciomyzid defense mechanisms. Most morphological features of sciomyzid larvae seem to be adaptations to a predaceous or parasitoid life style, to their feeding site, and to their microhabitat, not to defense. Also, the patchy distribution of the food resource likely has contributed to the evolution of short life cycles, thus decreasing the exposure of larvae in search of food to parasites, predators, and inimical conditions.

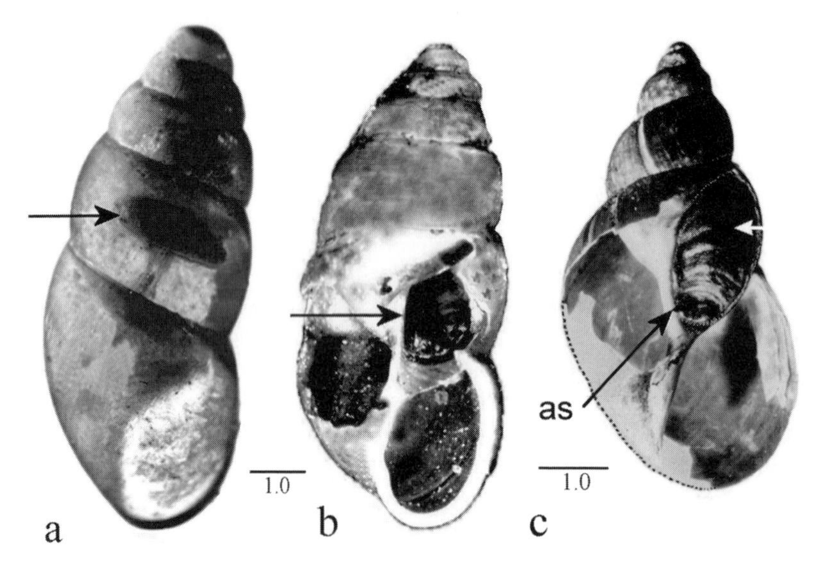

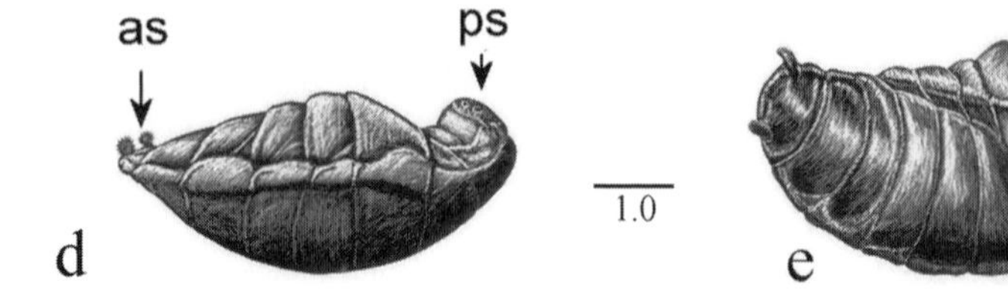

Fig. 12.4 (a) Encysted larva (arrow) of ***Tetanura pallidiventris*** on inner wall of shell of *Cochlicopa lubrica* (from Knutson 1970a). (b) Anterior end of puparium (arrow) of ***T. pallidiventris*** in shell of *C. lubrica* (photo by L. Knutson). (c) Puparium (arrow) of ***C. bifasciella*** at beginning of second whorl in shell of *Stagnicola* sp. (d, e) Puparium of ***Colobaea bifasciella*** removed from shell of *Stagnicola* sp. (c, d, e from Knutson & Bratt unpublished data). as, anterior spiracle; ps, posterior spiracle.

Unlike all of the essentially naked parasitoid and parasitoid/saprophagous, terrestrial to semi-terrestrial, Sciomyzini larvae and those of a few parasitoid, terrestrial to semi-terrestrial Tetanocerini, most of which spend most of their life inside the shell with only the posterior spiracles exposed, the free-living aquatic, predaceous Tetanocerini larvae are replete with warts, tubercles, and integumentary processes that may protect them from predators. Surprisingly, the strongest integumentary armature, consisting of many strong, black spines covering the entire body is found in the closely related *Perilimnia albifacies* and *Shannonia meridionalis*, which "feed as predators, parasitoids, or even saprophagously" and which "both tend strongly to remain in snail shells even when not feeding" (Kaczynski *et al.* 1969). Recent stereoscan studies of the larvae of many genera, both aquatic and terrestrial, show they are replete with diverse cuticular sensory structures, especially on segments I–IV and XII and especially in first-instar larvae (Gasc *et al.* 1984a, Vala & Gasc 1990b).

Only in the case of *Atrichomelina pubera* have other, saprophagous, insects been found feeding in the same snail with a sciomyzid that continues feeding on the decaying tissue after the host/prey dies (Foote *et al.* 1960). Behavioral and/or chemical adaptations to prevent this probably exist, but have not been studied. It might be expected that sciomyzid larvae feeding in the liquified tissues of snails, with only their posterior spiracles exposed, would be protected from parasitic Hymenoptera, but there are many records of these being parasitized by Braconidae, Ichneumonidae, and Pteromalidae.

The unusual swimming behaviors of some aquatic sciomyzid larvae are distinctive, and may be important in avoiding some predators. The large air bubble maintained in the gut of the larva of most surface aquatic predators and some terrestrials but not by subsurface predators such as *Eulimnia philpotti* allows the larva to float back to the surface in the event that the prey snail falls beneath the water surface. It also enhances their ability to swim. The combination of an air bubble in the gut and well-developed, hydrofuge interspiracular processes around the posterior spiracles enables aquatic larvae to swim and to forage efficiently at and just below the surface of open water.

The well-developed mouthhooks and dentate ventral arch characteristic of the cephalopharyngeal skeleton of Sciomyzidae and the accessory teeth of Tetanocerini could well be effective against soft-bodied predators of the larvae, but there are no direct observations. The dark, paired, comma-shaped marks seen on the integument of some *Pherbellia* larvae (Plate 2b, c) that have fed together in decaying snails probably result from the larvae slashing at each other, and may be a mechanism to force dispersal and thus limit intraspecific competition. Such slashing behavior may also repel some predators. J. B. Coupland (personal communication, 2002) has noted similar "scars" on larvae of *Sarcophaga* sp. feeding together in snails. Mellini & Baronio (1971) reported that first-instar planidia of *Macquartia chalconota* (Tachinidae) used their mouthhooks to injure competitors.

Surprisingly, few larvae of *Tetanocera clara* that penetrate the "mouth" of slugs were swallowed, and when they were swallowed both the larva and slug died. First-instar larvae routinely attack large (20-mm long) *Pallifera* sp. and small *Philomycus carolinianus* by entering the "mouth cavity" individually, and feed there with the posterior spiracles exposed for 16–25 days until the third instar, when the slug dies. Then the larvae attack a second slug in the same manner (Trelka & Foote 1970). All instar larvae of *Sepedon umbrosa* also routinely attack by penetrating the "mouth opening" of *Subulina octona* terrestrial snails up to 15 mm long (L. G. Gbedjissi, personal communication, 2003). Of 32 first-instar larvae of this species observed, 18 attacked by the mouth; the rest penetrated the foot. Often two or three larvae can be found in the mouth. The snail may continue to move about with the first- or second-instar larvae in the mouth but often is retracted into its shell. It is very rare to see a snail containing a third-instar larva moving about. If the larvae change their position in the snail's mouth, i.e., larval anterior end directed to the exterior, they are consumed by the snail. Certainly, larvae that live in the mouth of snails or slugs must have a highly specialized adaptation to enervate or otherwise escape the chainsaw-like action of the gastropods' radulae (Fig. 12.1).

Puparium formation within the shell of the host by some *Colobaea* (Fig. 12.4c–e), *Pherbellia* (Plate 4g; Fig. 14.45) *Ditaeniella* (Fig. 14.46), *Pteromicra*, *Sciomyza*, and *Tetanura pallidiventris* (Fig. 12.4b), and often by *Atrichomelina pubera* may offer some protection against larger insect predators. Such larvae generally clean the shell before pupariating, pushing out remains of tissue that might attract disease organisms. For species overwintering in the puparium in cold climates, cleaning the shell before pupariating also may be important in removing material that may serve as ice nucleation sites on the surface of the puparium. The calcareous septum formed across the opening of the host's shell by pupariating larvae of 11 species of *Pherbellia* (Plate 4g; Fig. 14.45), *Ditaeniella trivittata* (Fig. 14.46), and *Colobaea americana* (Knutson *et al.* 1967b) seems to provide protection. The inflated, empty, anterior end of puparia of *Colobaea bifasciella* (Knutson & Bratt unpublished data) (Fig. 12.4c) and *Sciomyza varia* (Barnes 1990) (with the pupa in the posterior part of the puparium, behind a sclerotized septum) may foil a predator or parasitoid. But these highly unusual devices seem to be more or less ineffectual as attested by their rarity and the fact that ichneumonid parasitoids have been obtained from puparia of these species. Terrestrial and semi-terrestrial larvae that pupariate outside the shell form their puparia on or slightly beneath the surface of the substrate litter. Aquatic larvae form puparia in the water and some, e.g., *Elgiva* spp., are highly modified for flotation. However, puparia of a few aquatic species sometimes are formed on vegetation far above the surface, where they may be more protected. Neff & Berg (1966) found 14 puparia of *Sepedon tenuicornis* 2–93 cm above the water surface on bur-reed and one *S. spinipes* puparium was found attached to a grass spikelet found in a sweep net. They also quoted an observation by Knutson of *S. spinipes*: "a clear, colorless, mucilaginous substance (evidently produced by the pupating larva) that hold the puparium in place" on the sides of rearing containers.

13 • Population dynamics

La nature ne fait rien sans objet. [Nature does nothing without aim].

Aristotle, Politique, I, 1, 10.

Quantitative field data on population dynamics is one of the least studied aspects of Sciomyzidae and is a deficiency that has retarded their application in biological control. The multivoltine, aquatic predator *Sepedon fuscipennis*, which overwinters as adults in facultative diapause, has been studied the most extensively, i.e., in a series of six related researches in New York and Connecticut. Most other information on populations is mainly on seasonal abundance from quantitative sweep-net and trap collections of adults, collections of snails bearing eggs for two species, and a few collections of larvae, summarized in Chapter 9 on phenology and noted below. Fortunately, there is a rather large amount of quantitative laboratory data on many aspects of the life cycles of several other species, especially *Ilione albiseta* and *Tetanocera ferruginea*, available to complement further field studies. See Chapters 5–9.

13.1 AQUATIC SPECIES

Sepedon fuscipennis . . . those disgusting black maggots . . . .

Dyar (1902).

13.1.1 *Sepedon fuscipennis*

Biologically, *S. fuscipennis* is one of the best known Sciomyzidae, some life-cycle features and morphology of immature stages having been studied as early as 1901 (Needham & Betten 1901) and later by Dyar (1902), Johannsen (1935), and Peterson (1953). The complete life cycle was first provided by Neff & Berg (1966). Their rearings were based on larvae, pupae, and adults collected at several localities around Ithaca, New York and also from Kentucky, Michigan, and Alberta. The material from New York, Michigan, and Alberta are from within the area in which only the "subspecies" *S. fuscipennis nobilis* is known to occur, but the material from Kentucky, south of the known range of that "subspecies," apparently was *S. f. fuscipennis* "intermediates" between the two "subspecies," recognized on morphological features by Orth (1986). Based on an isozyme comparison study, Manguin (1990) did not recognize subspecies within *S. fuscipennis* but concluded "that this is a single species showing geographic cline in several character states." Neff & Berg (1966) stated "No geographic variations were noted in this material." Elsewhere in this text and in the legends to figures and tables where *S. fuscipennis* is referred to the reader should note the conclusions of Orth (1986) and Manguin (1990) in regard to subspecies.

A multifaceted research project on *S. fuscipennis* and its prey snails in both the field and the laboratory in Ithaca included population studies of three species of aquatic snail prey (Eckblad 1973b), summarized in Section 1.2.1; predator–prey studies (Eckblad 1973a), summarized in Chapter 7; and population dynamics (Eckblad & Berg 1972). In the last study, the population dynamics of adults were followed by recording average catch per sweep over emergent vegetation in a 210 ± 50 m^2 area (2–6 pm) and by a mark–release–recapture estimate on seven occasions, recapturing 3h after release, during 1969 and 1970. The relative population densities were estimated with the equation of Bailey (1951, 1952) (see Eq. 13.1). Larval survivorship in the field was estimated using two separate cohorts in 1969 of 12 floating cylinders, each cylinder containing 25 unfed first-instar larvae. An additional cohort series in 1970 consisted of 25 unfed first instars in each of four cylinders, 50 unfed first instars in each of four cylinders, and 25 which had been given one snail meal in four cylinders. A simple exponential model was used to estimate the mean individual weight of larvae in the cylinders. A plot of mean individual weight against survivorship provided net cohort production. A comparison of net production with pupal crop enabled calculation of the percentage of larval production not resulting in puparia but remaining to be recycled in the study area.

The two methods used to study adults during 1969 showed similar population trends. Unimodal population curves were found in 1969 and 1970, with the 1970 peak about 2/3 the size and almost 2 months later than 1969, probably due to an artificial lowering of the water level (Fig. 13.1). It was estimated that there were 50–200 adults in about 210 m^2 (*c.* one fly per m^2). Using adult population estimates and survival rates for larvae and pupae, it was calculated that 5931 first-instar larvae, resulting in 445 third-instar larvae or a density of less than two per m^2, were required to yield the peak adult population (July 29, 1969). The amount of aggregation for the two dates each year with the largest estimated populations was compared with the negative binomial distribution (Waters 1959; Fig. 13.2). Calculations of the mean sizes of adult clumps indicated that aggregation "was probably due to some heterogeneity in the environment rather than to active aggregation by the flies." Mark–recapture data from adjacent habitats in 1969 indicated the absence of immigration and emigration from local demes. The extended egg-laying periods observed in laboratory studies and the finding of eggs, larvae, and puparia throughout the summer (larvae and puparia, May–September) suggested that several overlapping, non-synchronous generations were present during both summers. The number of generations per year was not determined. The sex ratio was close to 1:1. Maximum longevity of marked flies was at least 49 days.

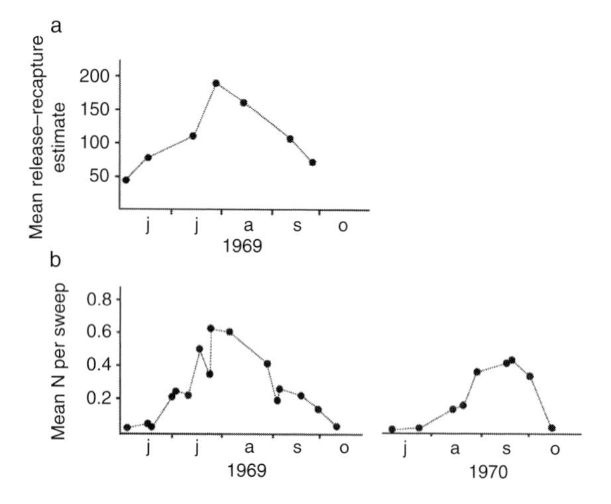

Fig. 13.1 Comparison of methods used to estimate populations of adult *Sepedon fuscipennis* at Bool's Backwater, Ithaca, New York. (a) Mark–release–recapture method, using equation of Bailey (1951, 1952). See Eq. 13.1. (b) Average catch per sweep estimation method (DeLong 1932). Modified from Eckblad & Berg (1972).

Survivorship of four cohorts of larvae in floating cylinders is shown in Fig. 13.3. Survivorship varied during the year, and there were low survival rates by first-instar larvae. Intraspecific competition did not appear to be an important factor influencing larval survival; there were no significant differences due to crowding in cohorts of 25 or 50 larvae. Cohorts of larvae that were provided a snail meal before release had a considerably higher survival rate (Fig. 13.4). Snail density and snail size appeared to be an important factor in total larval survival of all cohorts,

The 17 June to 5 July cohort had a calculated larval production of 64.0 mg, and 32.4 mg (50.6%) of this was recycled, i.e. not realized in puparia. About 24.3% of the 24 August to 11 September cohort production was recycled, and the calculated net production was only 5.2 mg.

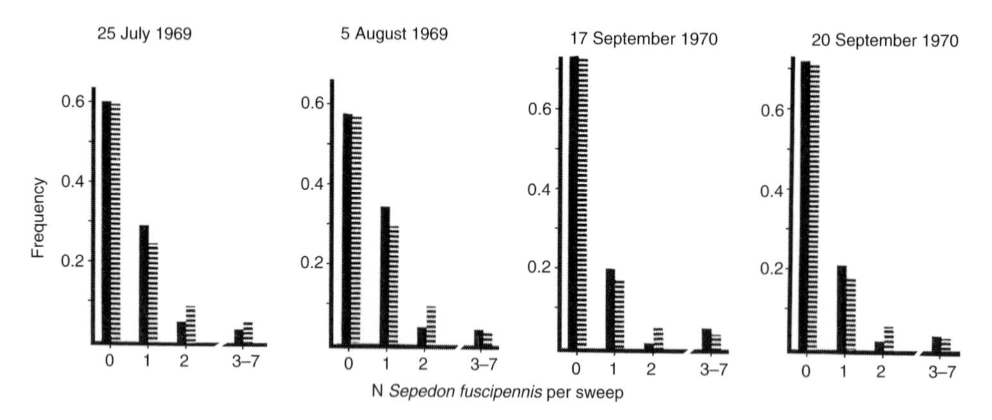

Fig. 13.2 Observed frequency of adult *Sepedon fuscipennis* (solid bar) at Bool's Backwater, Ithaca, New York, and frequency expected assuming a negative binomial distribution (striped bar) for two sampling dates each year when population was most abundant. Differences between observed and expected not highly significant ($P > 0.05$). From Eckblad & Berg (1972).

Table 13.1. Sciomyzidae collected at selected still-water and flowing-water sites in Connecticut, July–October 1972

Habitats	Sciomyzid taxa						
	Dictya sp.	*Elgiva solicita*	*Limnia* sp.	*Sepedon armipes*	*Sepedon fuscipennis*	*Tetanocera* sp.	Total
Still water							
Ponds	5	9	2	19	455	1	491
Backwaters	22	2	0	7	86	0	117
Flowing water							
Riverine	3	2	0	1	86	0	92
Total	30	13	2	27	627	1	700

Note: Modified from Peacock (1973).

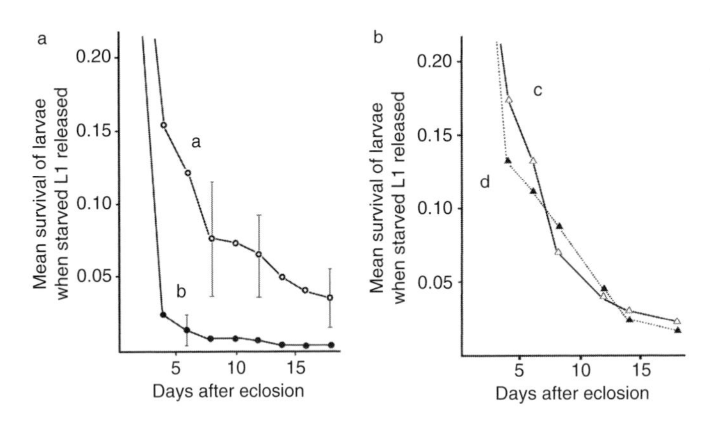

Fig. 13.3 Survivorship of larval cohorts of ***Sepedon fuscipennis*** kept in floating cylinders at Bool's Backwater, Ithaca, New York when starved L1 were released. (a) Cohort a followed from June 17 to July 5, 1969, cohort b followed from August 24 to September 11, 1969; (b) cohort c of 25 and cohort d of 50 followed simultaneously during summer of 1970. Vertical lines are ± typical standard errors of mean. From Eckblad & Berg (1972).

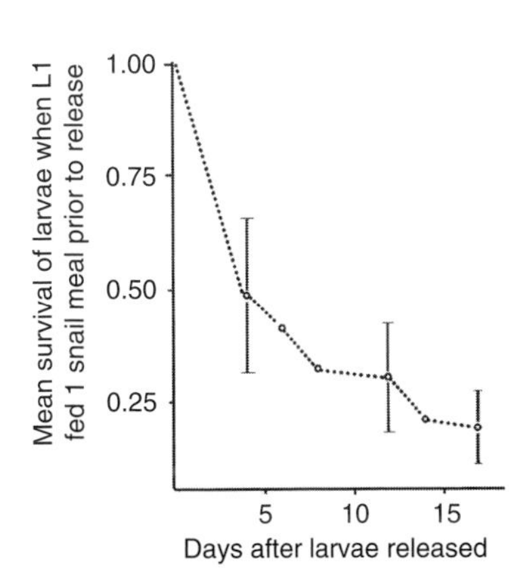

Fig. 13.4 Survivorship of a larval cohort of ***Sepedon fuscipennis*** kept in a floating cylinder at Bool's Backwater, Ithaca, New York when L1 were fed 1 snail meal prior to being released. Vertical lines are ± typical S.E. of mean. From Eckblad & Berg (1972).

Peacock (1973) focused on habitat preferences of *S. fuscipennis*, using data on collection of adults, larvae, and puparia, with limited data from mark–release–recapture trials, to obtain a relative indication of adult survival in different kinds of still-water and flowing-water habitats. His study was carried out along the Willimantic River in northeastern Connecticut with 128 sites sampled October 1971 to October 1972. He concluded that the absence of flies in riverine habitats during May and June was due to high river discharge. His conclusion on the strong ability of flies to emigrate from still-water to flowing-water habitats conflicts with results of others and the general opinion, based on limited data that there is little emigration by Sciomyzidae. Immigration/emigration is a topic well-deserving of further research, especially from population dynamics and biocontrol points of view. He noted "the ability of *Sepedon* species to invade and breed successfully in riverine habitat would greatly increase its value as a biocontrol agent" (Table 13.1) and he suggested that the flies were using "rivers or riverine pathways in locating potential habitat." See Sections 10.1, 16.3, and 18.1.

The above studies were followed by Barnes' (1976) laboratory studies (extensively reviewed in Chapter 9) on development and other aspects of the life cycle of *S. fuscipennis* (parent adults collected at Ithaca) at controlled temperatures and lighting regimes, which are critical in determining the rate of increase and evaluating the climatic limitations of the species. With these studies as background, another multifaceted research program was carried out by Arnold (1978a) who studied natural populations of all stages of *S. fuscipennis* and its snail prey in four adjacent, experimental marshes (20 m^2, 20-cm deep) during 1974–1976 at Ithaca. His research, aimed at developing a general population model usable for any species of Sciomyzidae, included five parts: (1) development of a simulation model for *S. fuscipennis*; (2) preparation of a versatile algorithm for the simulation of temperature-dependent development; (3) study of population dynamics of natural populations of *S. fuscipennis*; (4) study of development of natural populations of *S. fuscipennis*; and (5) development of quantitative sampling techniques for immature aquatic Sciomyzidae (*S. fuscipennis* and *Dictya* spp.). The extensive developmental data of Barnes (1976) were re-analyzed to estimate the parameters of the model. Unfortunately, the only published information on this project is part 3 (Arnold 1978b), and a few comments in Berg & Knutson (1978). Considerable information on population density, survival, emergence, intermarsh migration rates, fertility, and egg mortality and percentage of parasitism was obtained. In part 3 (Arnold 1978b), 1800 adults were captured–marked–released on 28 occasions during May–October, 1974, 1975, and 1976. Over 1–3-day periods, every 1–3 weeks, 50–150 flies were captured by sweep net with a relatively constant effort expended per unit area. Jolly–Seber (Jolly 1965, Seber 1965) estimates of population size peaked during late summer each year at 0.45–1.1 adults per 1.0 m^2 of emergent vegetation (Fig. 13.5). Previously, no one had attempted to make quantitative samples of the immature stages of aquatic Sciomyzidae because of their low densities, patchy distributions, and cryptic microhabitats. Arnold (1978a) developed a stratified random-quadrat sampling method: a 0.25 m^2 quadrat was enclosed by a floating frame, emergent and floating vegetation within was examined for eggs, and then the vegetation was pushed beneath the surface to allow larvae and puparia to float free. This method gave results (densities of larvae and

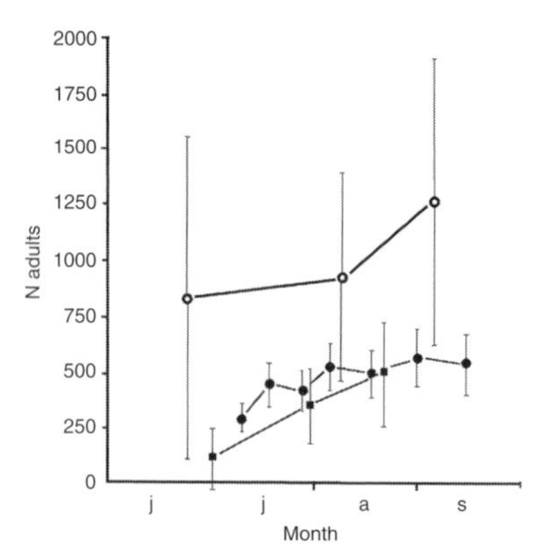

Fig. 13.5 Number of ***Sepedon fuscipennis*** adults in experimental marsh populations at Ithaca, New York in 1974 (black square), 1975 (open circle), and 1976 (black circle), as estimated from data for both sexes and all marshes pooled by Jolly–Seber method. Vertical bars are ± 1 asymptotic S.E. (Jolly 1965). Only sampling dates for which sufficient numbers of recaptures were made are plotted. From Arnold (1978b).

puparia combined of 0.9 and 1.3 individuals, respectively, per 1.0 m^2 during August 1974 and 1975) that correspond well with the result that Eckblad & Berg (1972) obtained by back calculation from adult density and survivorship of larvae.

Juliano (1981) found that near Ithaca mortality of *S. fuscipennis* eggs due to at least three species of *Trichogramma* wasps varied significantly over the summer and reached 43.6% in late July and early August. These *Trichogramma* species also parasitized eggs of *Dictya* spp., *Elgiva solicita*, and *Tetanocera* spp. Juliano (1982) also studied the influence of age of *S. fuscipennis* eggs on host acceptance and suitability for one species of *Trichogramma*. See Section 11.2 on parasitoids.

Berg *et al.* (1982) collected adult *S. fuscipennis* (and other species) between January 18 and March 3, 1973 and 1974, at Ithaca, 10 or more days after temperatures had dropped to –27 °C. Some of these adults held in screen cages outdoors survived 21, 23, and 64 days, when temperatures fluctuated around freezing and dropped to as low as –22 °C. Of 27 adult *S. fuscipennis* collected and placed in a screen cage outdoors on October 27, four were alive on April 21. They also found that winter survival of males seemed necessary in order to carry viable sperm through the winter. A female collected, marked, and released in August 1974 was recaptured on May 29, 1975 and was reproductive when recaptured (Arnold 1978b). See Section 9.1 on phenology and Section 9.2 on reproduction.

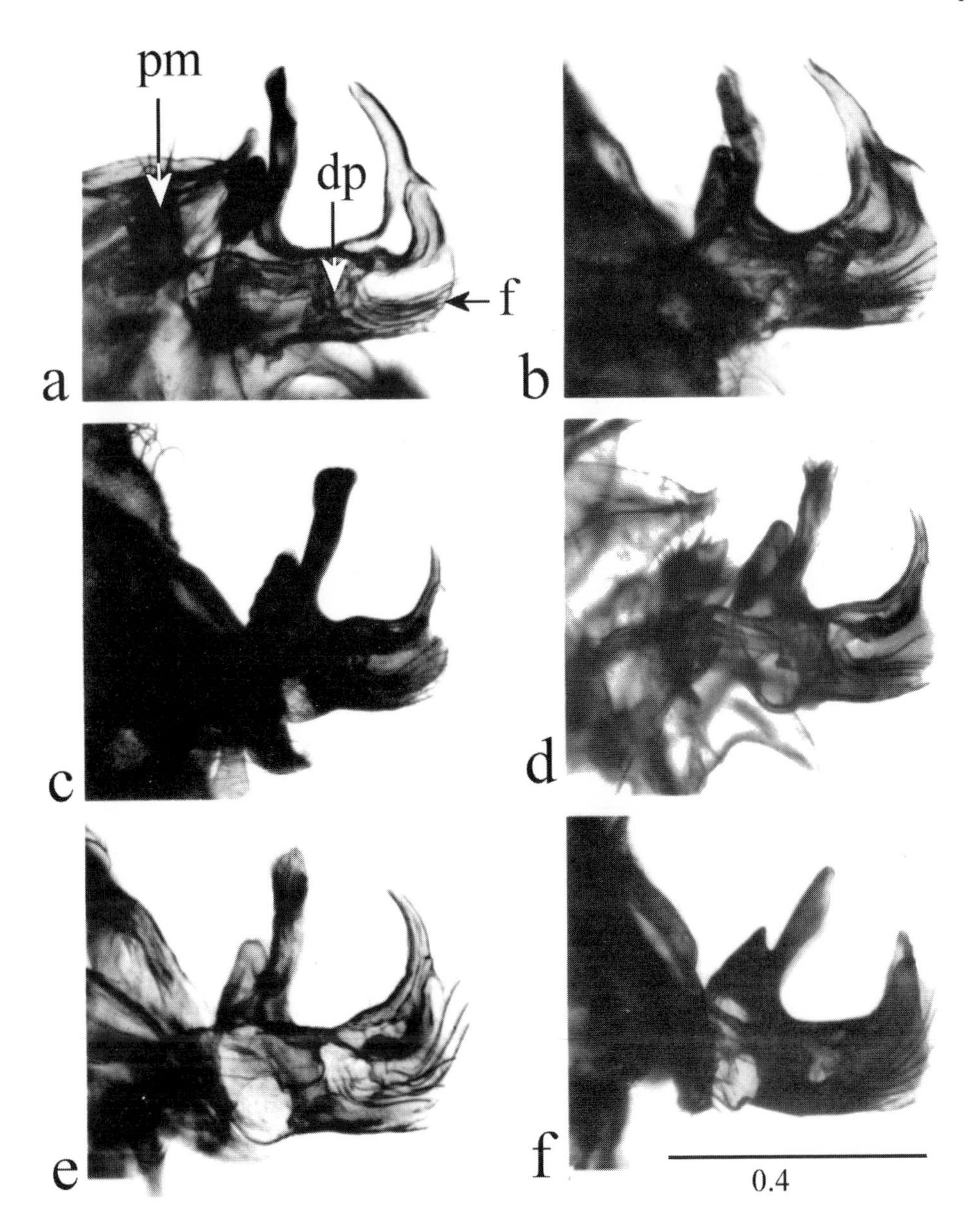

Fig. 13.6 Aedeagi, lateral. (a, b) ***Sepedon f. fuscipennis***; (c, d) ***S. f. nobilis***; (e) ***S. fuscipennis*** (intermediate form); (f) ***S. floridensis***. dp, distiphallus; f, filaments; pm, paramere. Modified from Orth (1986).

The wealth of information briefly summarized above, the most complete field data for any species of Sciomyzidae, gained in central New York and nearby northeastern Connecticut during 1969 to 1978, ostensibly refers to one monomorphic/monotypic species. However, Orth (1986), primarily on the basis of characters of the male aedeagus (Fig. 13.6), separated the species into two subspecies, *Sepedon fuscipennis fuscipennis* ranging from New Jersey across to northern Illinois and south to Texas and Florida, and *S. f. nobilis* ranging north of that area and westward

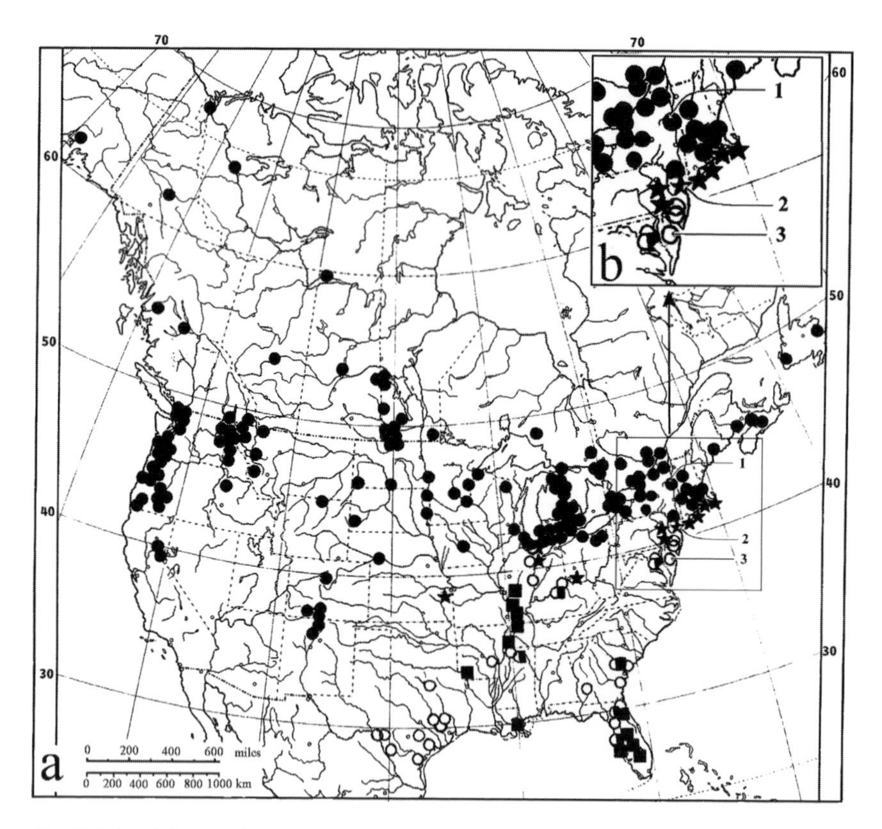

Fig. 13.7 (a) Collection sites for ***Sepedon f. fuscipennis*** (open circle), ***S. f. nobilis*** (black circle), ***S. fuscipennis*** (intermediate form) (star), and ***S. floridensis*** (black square). (b) Details on collection sites for specimens analyzed by Manguin (1990). 1, ***Sepedon f. nobilis***; 2, ***S. fuscipennis*** intermediate; 3, ***S. f. fuscipennis*** (see Table 13.2). Modified from Orth (1986).

transcontinentally, with "intermediate" specimens occurring from Kansas to the northeastern coast (Long Island, New York and Massachusetts) (Fig. 13.7). Only *S. f. nobilis* was recorded in Ithaca where the extensive biological studies had been carried out. Notably, the line separating *S. f. fuscipennis* and *S. f. nobilis* is strikingly similar to the 0 °C isotherm of the coldest month. Subsequently, Manguin (1990) studied the population genetics and biochemical systematics of the *S. fuscipennis* group. She found that in an isozyme comparison of 23 loci among populations of the two subspecies and the intermediate form, six loci were monomorphic in all three populations (Table 13.2). The sampling locations were in close concordance with the ranges of the subspecies and the intermediate form (*S. f. fuscipennis* from Prince Georges Co., Maryland [below the 0 °C isotherm], *S. f. nobilis* from Albany, New York [above the 0 °C isotherm], and the intermediate form from Orange Co., southern New York [along the 0 °C isotherm]) (Fig. 13.7). The F-statistics showed moderate diffentiation among the populations, with high values for only two loci. Nei's index (1978) of genetic identity for the three populations ranged from 0.923–0.990, indicating a high degree of similarity among all populations. Manguin (1990) did not formally synonymize the subspecies. In establishing nine species Groups within *Sepedon sensu strictus* on a worldwide basis, using 20 features of adult morphology and two larval behavioral features, Knutson & Orth (2001) recognized the *S. fuscipennis* Group of Orth (1986) (*S. floridensis*, *S. fuscipennis*, *S. gracilicornis*, and *S. tenuicornis*).

13.1.2 Other aquatic predators

13.1.2.1 LARVAE

Many collections of immature stages, but usually only a few specimens per occasion, have been made throughout the year to determine phenologies and overwintering and to

Table 13.2. Genetic variability of 23 loci in three populations of ***Sepedon fuscipennis*** group

Population	Mean sample size per locus	Mean N alleles per locus	Percentage loci polymorphic[a]	Mean heterozygosity	
				Direct-count	Hardy–Weinberg expected[b]
S. f. fuscipennis (39° N. Lat.)	106.2 (10.6)	2.4 (0.2)	73.9	0.153 (0.035)	0.168 (0.039)
S. f. "intermediate" (41° N. Lat.)	64.6 (5.7)	2.1 (0.2)	69.6	0.201 (0.040)	0.200 (0.039)
S. f. nobilis (42°30′ N. Lat.)	48.7 (3.0)	2.1 (0.2)	65.2	0.135 (0.032)	0.142 (0.035)

Notes: Standard errors in parentheses.
[a]A locus is considered polymorphic if more than one allele was detected.
[b]Unbiased estimate. See Fig. 13.7. Modified from Manguin (1990).

Table 13.3. *Galba truncatula* and ***Hydromya dorsalis*** larvae recapture data for 1974–1976 in Ireland

	June–August 1974	June–August 1975	June–August 1976
N snails released	300	510	588
N snails recaptured	177	256	353
N snails infected with *Fasciola hepatica*	0	0	0
N snails attacked by *Hydromya dorsalis* larvae	52	40	5
N *Hydromya dorsalis* larvae found	13	16	0
% recaptured snails attacked	29.38	15.63	1.4

Note: From Hope Cawdery & Lindsay (1977).

initiate laboratory rearings. Limited routine quantitative or experimental sampling to determine population sizes and fluctuations in populations of larvae have been conducted for only five species – *S. fuscipennis* (above) and the following four species. Lynch (1965) sampled *Austropeplea tomentosa* and larvae of *Dichetophora biroi* with a 10-cm diameter scoop in two nearby habitats at one of the Murray Lakes in South Australia between June and October. The populations of larvae peaked sharply in September, about a month after the snail population peak. An enormous number of fly larvae, up to 23 per scoop, were collected, this being the densest population of sciomyzid larvae ever reported, but they were virtually absent during June, July, and October (Fig. 9.11). Similar numbers, 19 per scoop, were found at a stream near Adelaide. Boray (1969) provided limited data on the abundance of *D. biroi* and *D. hendeli* larvae and puparia collected (methods not given) with *A. tomentosa* fortnightly over a 2-year period in New South Wales, Australia. The numbers of Sciomyzidae were "very much smaller that the numbers of snails and there was no inverse relationship between numbers of live snails and sciomyzids," conclusions quite different from those of Lynch (1965). In Ireland, Hope Cawdery & Lindsay (1977) used the technique of "tracer snails" (uninfested laboratory-reared snails marked and released in groups of about 100 and recovered 96 and 144 hours later) to determine numbers of snails "attacked" (i.e., "soft body tissue partly or completely eaten") and number of first-instar larvae of *Hydromya dorsalis* found during June–August over a 3-year period (Table 13.3). Also in

Table 13.4. Field collections of larvae of *Ilione albiseta* in Ireland

		October	November	December	January	February	March	Total
(a)	1980–81	39	53	29	15	3	4	143
	1981–82	20	15	10	12	4	0	61
		L_1	L_2	L_3				
(b)	1980–81	0	11	132				
	1981–82	2	16	43				

Note: (a) by month; (b) by instar. From Lindsay (1982), Lindsay *et al.* (2011).

Ireland, Lindsay (1982) used "refuge traps" described by Gormally (1987b) as "white plastic cups placed in pools so that the mouth of the trap was flush with the bottom of the pool" to capture overwintering larvae of the univoltine *Ilione albiseta* during October–March 1980–81 and 1981–82 (Table 13.4).

13.1.2.2 ADULTS

Routine, quantitative sampling of adults by standardized sweep-net collections, pan traps, emergence traps, and Malaise traps has been conducted on several occasions, primarily to determine relative abundance, phenology, and habitat range. This has resulted in some information on population sizes (see also Chapter 10). Unfortunately, specifics of collection techniques, e.g., sizes of sweep nets, often are not given. Vala & Manguin (1987) sampled 14 species – seven aquatic, two semi-aquatic, and five terrestrial species – by standardized sweep netting along 400 m of a 1-m wide ravine, which almost completely dried during the summer, once monthly throughout the year during 1982, 1983, and 1984 in southern France. They compared fluctuations in populations with the rainfall records for the 3 years and gastropod populations for 1984 (see Section 9.1). They analyzed specific richness, diversity, and equability (see Section 10.1), and they discussed the results in relation to the type of life cycle and phenology. About 1000 specimens were collected each year. Only three species, the aquatic predators *Tetanocera ferruginea*, *Sepedon spinipes*, and *Ilione albiseta* were collected in large numbers. The effect of numbers of the terrestrial species (35 specimens) and the less common aquatic and semi-aquatic species (141 specimens) on their analysis of population sizes could be discounted. In a global analysis of the pooled numbers, they found three successive population peaks, during April–May, July, and September–October (Fig. 9.7) with considerable variation in the second and third peaks due to variations in rainfall.

Barraclough (1983) provided general comparisons of distinct differences in the sizes of populations of *Sepedon neavei*, *S. pleuritica*, and *S. testacea* at three somewhat different types of still-water sites along the Mpushini River, south of Pietermaritzburg, Natal, South Africa. He found large numbers (at least 50 per 100 sweeps in a 6 × 6 m area) at an impoundment during April, decreasing to none during mid June. But the flies were present "in large numbers" during May–June 30 km south in a different type of impoundment habitat. He attributed the differences primarily to temperature and rainfall.

Appleton *et al.* (1993, citing unpublished data) recorded sampling of adults of seven species of *Sepedon* at three stream sites and three impoundment sites near Pietermaritzburg in 100 sweeps with a 30-cm diameter net per sample during February–June 1991 and 1992. Only 5.4 ± 5.2 males and 3.6 ± 2.8 females were collected per 100 sweeps. There was close correlation between abundance of the flies and snail species richness.

About 20 habitats of *Sepedon sphegea* were sampled by sweep nets in Mazandaran Province, a subtropical area on the Caspian Sea coast in northern Iran, along roads near vast cultivations of rice during the summer of 1974 (S. Tirgari, personal communication, 1983). At one site (Haftan), on July 9, an average of 19.8 individuals (range 6–29) were collected in one stroke of the sweep net and several hundred were captured while dragging the net over emergent vegetation during a walk of 10 m. Eight to ten thousand adults were collected on July 16 during 2–3 hours at two nearby sites. Most of the flies were collected between 5 and 7 am (air temperature 30 °C, water surface temperature 29 °C). By 10 am, only a few flies could be found. During late August, no insect of any kind was seen because of widespread applications of insecticides against the rice-stem borer, but by the end of September, *S. sphegea* began to appear again.

Other, similar studies providing information on relative abundance and population sizes are: (1) *Thecomyia* spp. in Brazil (Knutson & Carvalho 1989); (2) *Sepedon* spp. and related genera in Africa (Verbeke 1963); and (3) 13 species of aquatic and terrestrial species in five kinds of habitats near Avignon, southern France (Vala 1989a). See Sections 9.1 and 10.1.

Quantitative studies of natural populations of assemblages of adults of several species of Sciomyzidae in an ephemeral wetland in Ireland have been conducted by Williams *et al.* (2009). The experimental design, sampling methods, and especially the many modern methods of data analysis used in these studies are of particular interest and will be of value in further research along these lines.

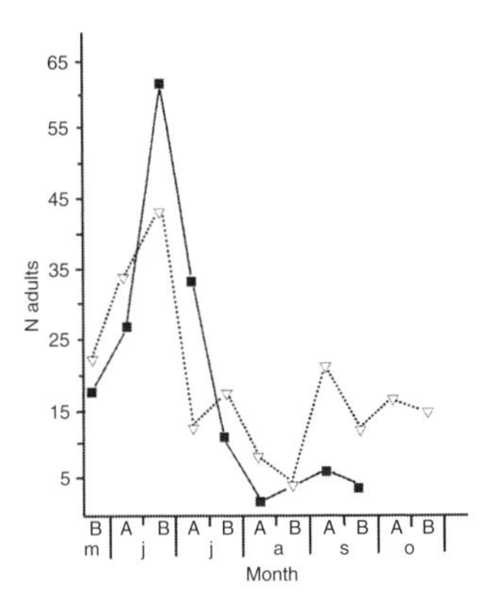

Fig. 13.8 Variation in populations of ***Limnia unguicornis*** collected in two adjacent meadows near Avignon, southern France. Square, vegetation harvested two or three times per year; triangle, vegetation undisturbed for several years. A, 1–15; B, 16–30 of each month. From Vala & Knutson (1990).

13.2 TERRESTRIAL SPECIES

Moor (1980) collected 530 specimens of the hygrophilous, terrestrial snail *Succinea putris* during April–October of one year near Basel, Switzerland and found two population peaks (end of April and mid August) of the parasitoid *Pherbellia s. schoenherri* based on the number of snails bearing eggs (each population peak with 50% of the snails bearing one or two eggs (Fig. 9.30). Vala & Ghamizi (1992) studied this sciomyzid in association with *Oxyloma elegans* during May 1984–June 1985 near Avignon, France. They collected about 1400 snails and also found two population peaks, in April–May and August. Parasitization reached maxima of only about 30% during August. They also collected adults throughout the year, but these were in very small numbers (total 35) (Fig. 9.32).

Vala & Knutson (1990) sampled populations of adults of the terrestrial/hygrophilous, predator/saprophage *Limnia unguicornis* (30-minute sweepings made every 15 days, May–October) at two adjacent meadow localities near Avignon. The stations differed primarily in whether or not the vegetation was harvested annually. They found distinct differences in the sizes of populations, but close approximation in population curves (Fig. 13.8).

Coupland *et al.* (1994) made standardized sweep-net samples of adults of *Salticella fasciata* from May 1991 to January 1993 and random quadrat samples of terrestrial snails during September 1 to December 20, 1991 in a "typical Mediterranean waste ground habitat" (a disused vineyard) near Montpellier, France. Adults were not found

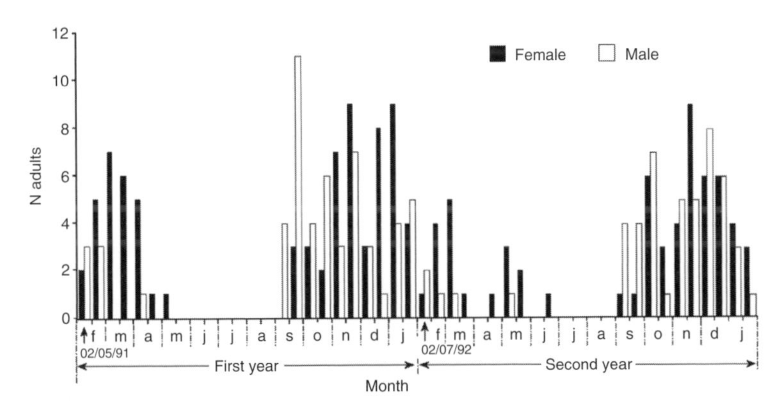

Fig. 13.9 Flight activity of ***Salticella fasciata*** over 2 years near Montpellier, southern France. A period of 2 weeks separates two ticks on abscissa axis. Modified from Coupland *et al.* (1994).

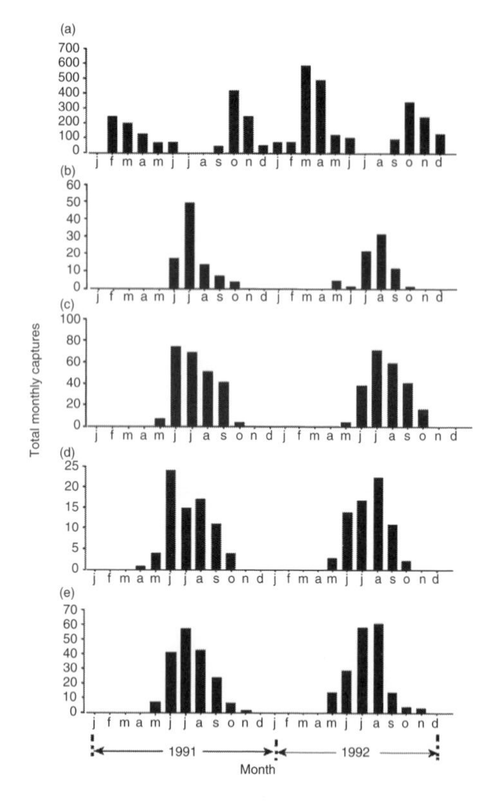

Fig. 13.10 Total monthly captures (1991–1992) of adults of (a) *Pherbellia cinerella*; (b) *Coremacera marginata*; (c) *Euthycera cribrata*; (d) *Dichetophora obliterata*; (e) *Trypetoptera punctulata* near Montpellier, southern France. From Coupland & Baker (1995).

between June and September; the population increased slowly to a peak in late October and November, and then declined (Fig. 13.9). The number of live, adult host snails (*Theba pisana*) decreased as the snail breeding season progressed "due to mortality after breeding," but the proportion of remaining live (but apparently moribund) snails bearing *S. fasciata* eggs increased as the season progressed (Fig. 18.4).

Coupland & Baker (1995) made standardized sweep-net collections of five terrestrial species (*Coremacera marginata*, *Dichetophora obliterata*, *Euthycera cribrata*, *Pherbellia cinerella*, and *Trypetoptera punctulata*) in four distinct types of terrestrial habitats with five stations for each type, during February 1991 to January 1993 near Montpellier. Of 4828 individuals captured, *P. cinerella* comprised over 80%. Whereas *P. cinerella* populations peaked in March and October and were not found during July and August, the populations of the four other species were unimodal, peaking in June, July, or August (Fig. 13.10, Table 10.2).

14 • Morphological, physiological/behavioral, and genetics and related aspects

Animals live for business, not for pleasure;
and all their instincts and their useful structures
are developed for practical purposes.

Snodgrass (1930).

14.1 MORPHOLOGICAL ASPECTS

The truth has such excellence that by praising little
things it makes them noble.

Leonardo da Vinci.

The morphology of all stages is treated in detail from a taxonomic viewpoint in Chapter 15 and from an evolutionary viewpoint in Chapter 17. Here we discuss morphological aspects primarily from functional and eco-physiological viewpoints and we review some recent research on unique and important but incompletely studied features that may be of broad significance, not only to Sciomyzidae but also to other Diptera.

14.1.1 Adults

14.1.1.1 SPECIAL FEATURES

Many special features of adults have been described in numerous taxonomic papers, but few have been described from a functional viewpoint. Color patterns in the eyes have been described for some Nearctic species of *Limnia* (Steyskal *et al.* 1978); *Tetanura pallidiventris* (Knutson 1970a; Fig. 14.1a); *Elgiva* spp. (Steyskal 1954d; Fig. 14.1b, c); and as early as 1902 (Hendel 1902) for *T. pallidiventris* and *Pteromicra glabricula* (Fig. 14.2). Note that the patterns shown for *T. pallidiventris* are not in agreement. The pattern is usually of two, more or less parallel, horizontal, reddish stripes across the middle of the metallic green surface (Plate 1c; Fig. 14.1b, c). We have also observed this pattern for *Pherbina coryleti, Ilione albiseta, I. trifaria, Coremacera marginata, Dichetophora obliterata, Euthycera amoena, E. cribrata*, and *Trypetoptera punctulata*. *Sepedon sphegea* has the anterodorsal angle and lower 1/4 of the eye reddish purple. Many species, e.g., *Salticella fasciata, Euthycera chaerophylli, Tetanocera ferruginea, Hydromya dorsalis*, and *Pherbellia cinerella* lack a pattern. Eye-color patterns are not sexually dimorphic and occur in species frequenting both open and shaded habitats and species with patterned or unpatterned wings. Lunau & Knüttel (1998) noted that the function of colored corneal lenses, known also in Tabanidae, Dolichopodidae, and Tephritidae, is not clear. Possible functions are intraspecific signaling, shifting the rhodopsin-metarhodopsin equilibrium towards more active rhodopsin in the process of photo-reconversion and filtering distinct dominant wavebands in order to enhance the perception of color contrast.

Panov (1994) stated that the retrocerebral endocrine complex of *Sepedon sphegea* differs from that described for any other Brachycera in having paired *corpora allata*, situated above the aorta and connected with an unpaired *corpus cardiacum* by a pair of allatal nerves. The *corpora allata* are paired and separate in the various Nematocera that have been studied but their coalescence with the *corpora cardiaca* and prothoracic glands into a ring gland (= Weismann's gland) in the Cyclorrhapha is considered an autapomorphic state. The condition of these glands, the source of juvenile hormone, in *Sepedon* may have some relationship to the loss of the ptilinum in *Sepedon* and related genera. These, apparently, are the only Schizophora other than Nycteribiidae in which the ptilinum is completely absent.

Some Afrotropical, Oriental, and Australian *Sepedon* (subgenus *Parasepedon*) and *Sepedoninus* species appear to be the only Diptera in which the sperm pump is highly modified into a cochleate vesicle (Fig. 14.3), and ejaculation is effected by compression of a hydrostatic mechanism rather than by muscular action (Steyskal & Knutson 1975a); see also Section 15.1. The cochleate vesicle; the umbrella-like aedeagal apodeme of *Tetanoptera* (Knutson & Vala 1999; Fig. 14.4), *Verbekaria* (Vala *et al.* 2000c), and some *Euthycera* spp.; and the aedeagal apodeme in all genera across the family require further study from a functional viewpoint.

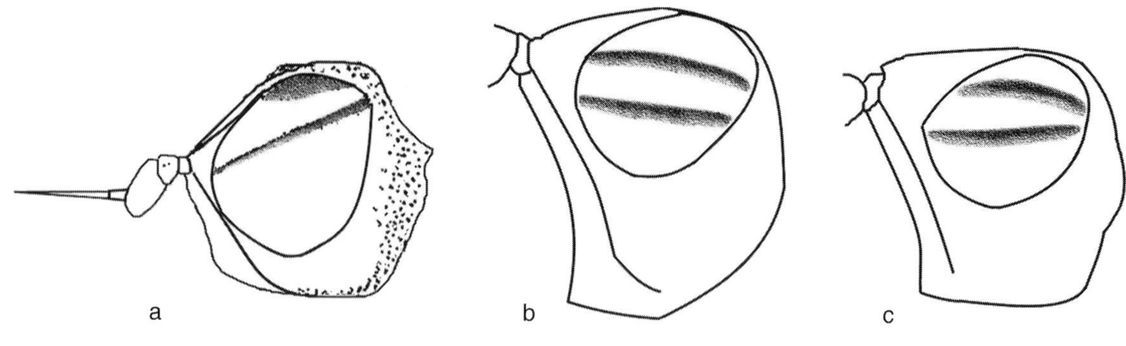

Fig. 14.1 Eye pattern of (a) *Tetanura pallidiventris* (from Knutson 1970a); (b) *Elgiva solicita*; (c) *E. connexa* (modified from Steyskal 1954d).

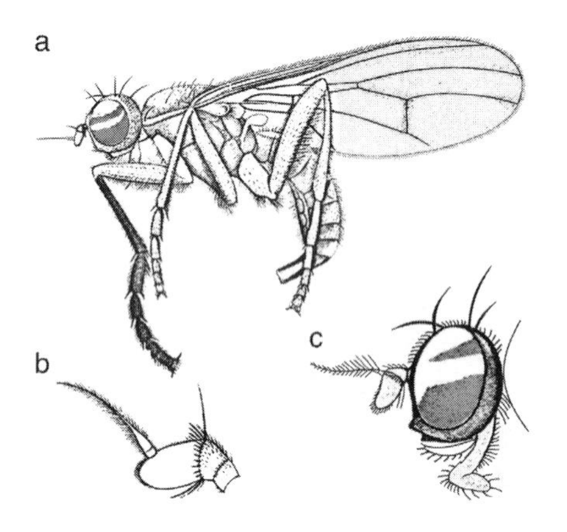

Fig. 14.2 (a), (b) ***Tetanura pallidiventris***; (a) habitus, female. (b) antenna. (c) ***Pteromicra glabricula***, head. From Hendel (1902).

The aedeagus of the male copulatory apparatus has been figured in detail for many species. Especially interesting is the presence of slender, flexible, solid filaments in the distiphallic chamber of the aedeagus only in five of the nine species groups of *Sepedon* (Knutson & Orth 2001). These filaments, when extruded, might function in guiding the male gonopore into contact with the female genital aperture or their apices might in fact be the gonopore(s). They were first figured, for some species of Afrotropical *Sepedon* (subgenus *Parasepedon* = *Trichrooscelis* species group), by Verbeke (1950). He speculated that the long, coiled filaments, which he termed "ejaculateur," were unrolled by pressure of a liquid secreted by the "organe vésiculaire" [= cochleate vesicle]. However, filaments are found in four species groups of *Sepedon* that lack a cochleate vesicle and are not present in the *Lobifera* species group, which has a cochleate vesicle (Knutson & Orth 2001). The filaments also were shown in drawings of three species of the *Armipes* species group (including *S. capellei*) by Fisher & Orth (1969a). They labeled the apices of the filaments as the gonopore. Steyskal & Knutson (1975a) presented drawings of the long, coiled filaments of three species in the *Trichrooscelis* species group and the very short, almost straight, two pairs of filaments of *S. sphegea* (*Sphegea* species group). They described the anatomy of the filaments and noted that they were not able to see a connection between the end of the ejaculatory duct and the base of the filaments. R. E. Orth (personal communication, 2004) has provided a remarkable series of previously unpublished stereomicrographs of the aedeagus of *S. capellei*, which show two pairs of filaments projecting from the apex of the distiphallus and apices of one pair at magnifications up to 6000× (Fig. 14.5).

In the *Sepedon* group of genera, *Thecomyia* lacks palpi, a secondary loss occurring in certain representatives of many groups of Diptera. In another family of Sciomyzoidea, the Sepsidae, palpi are vestigial in all except the genus *Orygma*.

Only two cases of intraspecific variation in body color have been reported in sciomyzids. Zuska & Berg (1974) analyzed variation in five of the seven species of the south temperate Neotropical genus *Tetanoceroides*, along with a detailed review of their distribution and habitats. Exceptionally fine color drawings illustrate key examples. They found clinal progression in color variation opposite to the direction expected according to "Gloger's Rule," i.e., with darker colors in the southern parts of the ranges with lower mean temperatures, rather than in the northern parts with

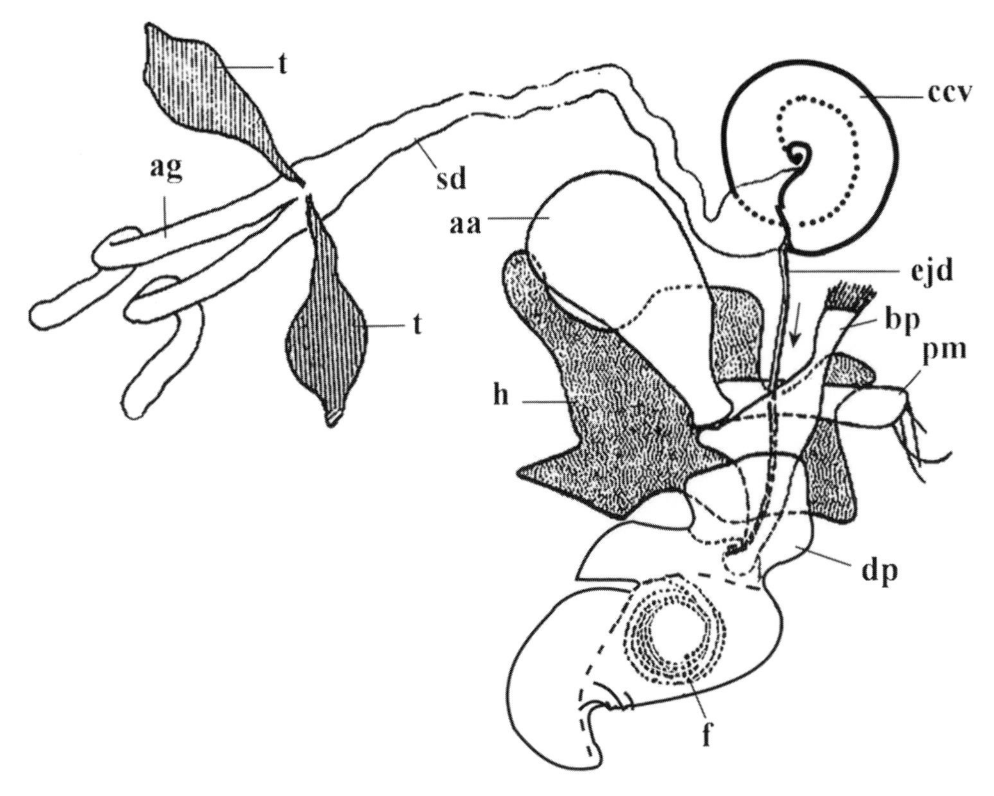

Fig. 14.3 *Sepedon senex*, male genital tract. aa, aedeagal apodeme; ag, accessory gland; bp, basiphallus; ccv, cochleate vesicle; dp, distiphallus; ejd, ejaculatory duct; f, filament; h, hypandrium; pm, paramere; sd, sperm duct; t, testes. From Steyskal & Knutson (1975a).

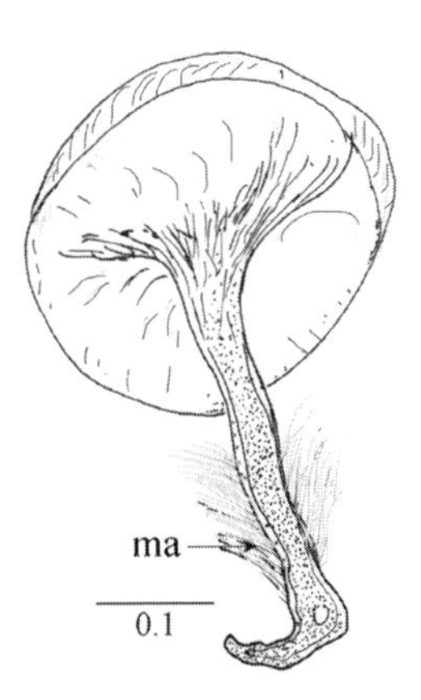

Fig. 14.4 *Tetanoptera leucodactyla*, aedeagal apodeme. ma, muscle attachments. From Knutson & Vala (1999).

higher mean temperature. They speculated that the color differences might be important in thermoregulation. Barnes (1979a) described the extensive clinal color variation in the common and widespread New Zealand species, *Neolimnia tranquilla*. In this species, whereas there is a southward decline in the intensity of color of some features there is a northward decline in the intensity of other features. The author concluded that there was no apparent explanation.

The internal anatomy has hardly been studied except for details of the male postabdomen, widely used in taxonomic studies, and the excellent illustrations of the spermathecae and associated structures for many Palearctic species presented by Rivosecchi (1992), e.g., Fig. 14.6a, b, and more recently by other authors. Dufour (1851) was the first to study some internal organs – brain, digestive system, male and female reproductive system (Fig. 14.6c–e) – of adults of any Sciomyzidae (eight species in four genera). On a historical note, it is of interest that he was also the first author to present information on the biology and morphology of any immature stages (Dufour 1847a, 1847b).

The ptilinum, a fundamental structure of 42 000 species of higher flies, apparently has been lost only in 111 species of Sciomyzidae and in the Nycteribiidae. Strickland (1953) discovered that five Nearctic species of *Sepedon* were the

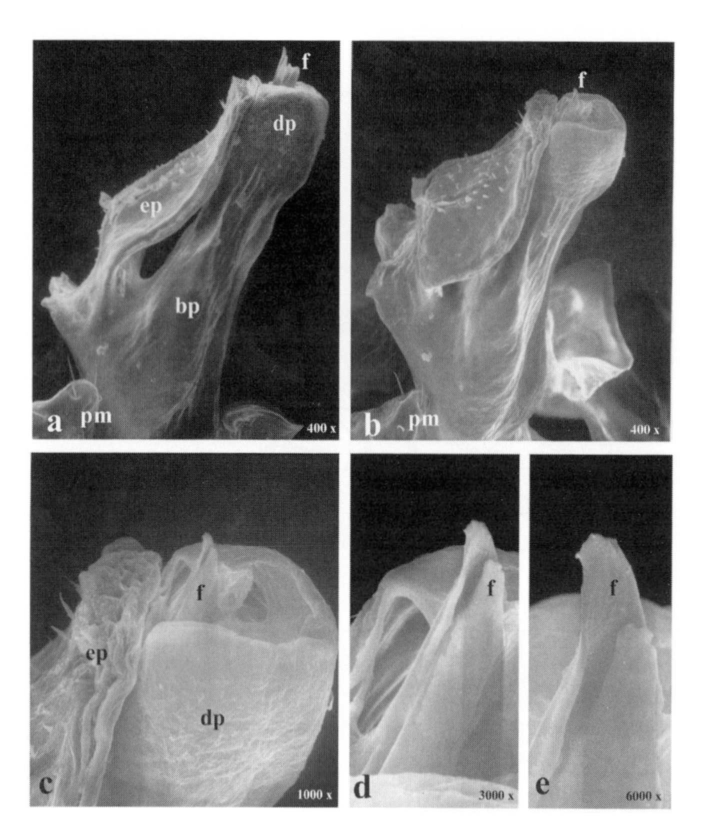

Fig. 14.5 ***Sepedon capellei***, aedeagus. (a) Lateral; (b–e) slightly dorsolateral; (c) apical end of distiphallus, filaments, and epiphallus; (d) pair of filaments; (e) apex of filaments. bp, basiphallus; dp, distiphallus; ep, epiphallus; f, pair of filaments; pm, paramere. Photos by R. E. Orth.

only species of 150 he studied, representing 40 families of Schizophora, that lack a ptilinum (Fig. 14.7), the protrusible bladder above the antennae used for emergence from the puparium and the medium surrounding it. This is an overlooked paper, which we have found cited only by Hennig (1973) and J. F. McAlpine (1981, 1989). Strickland (1953) described the anterior portion of the frons in *S. armipes*, *S. pacifica*, and three unnamed species of *Sepedon* as follows: "The permanently exposed "ptilinal" area is smooth, rigid, and flat. It holds the narrow transverse suture open as a sclerotized area connecting the frons with the transverse frontal lunule." The author examined five other species of Sciomyzidae and found a transformation series from (1) *Pherbellia schoenherri maculata* and *P. n. nana* having a normal ptilinal fissure (his frontal or transverse suture) extending downward to the genal angle and normal scales on the ptilinum to (2) *Tetanocera valida* with a fissure extending only to the level of the antennal foveae and a moderately sized ptilinum to (3) *Limnia* "*saratogensis* (Fitch)" with the ptilinal fissure ending above the antennal foveae and with a ribbon-like ptilinum to (4) *Elgiva solicita* with an even narrower ptilinum and weak scales to (5) the aberrant *Sepedon* spp. This transformation series agrees remarkably well with the relative positions of the genera in the cladograms of Marinoni & Mathis (2000) and Barker *et al.* (2004). Strickland (1953) found that only in a few species of Ephydridae and in the five *Sepedon* spp. are scales absent; in a few species of Ephydridae the fissure is not produced downward; and in Hippoboscidae the ptilinum is ribbon-like, as in most Sciomyzidae, and the fissure bends downward only slightly at the extremities.

Darvas & Fónagy (2000), in describing the emergence of Schizophora, noted that it is "modified" in "Conopidae, which has rudimental ptilinal fissure and in some Sciomyzidae, which has rudimental ptilinal sac," without further detail. Strickland (1953) also noted that in the five species of Conopidae (Schizophora of uncertain relationship, regarded as a separate superfamily), the ptilinal fissure is very short and narrow but the ptilinum is voluminous, strongly developed, and well provided with ptilinal scales, such scales also being found on the proboscis. Darvas & Fónagy (2000) referred only to rhythmic contractions of the abdominal muscles in causing changes in hemolymph pressure resulting in peristalsis of the abdomen, bases of

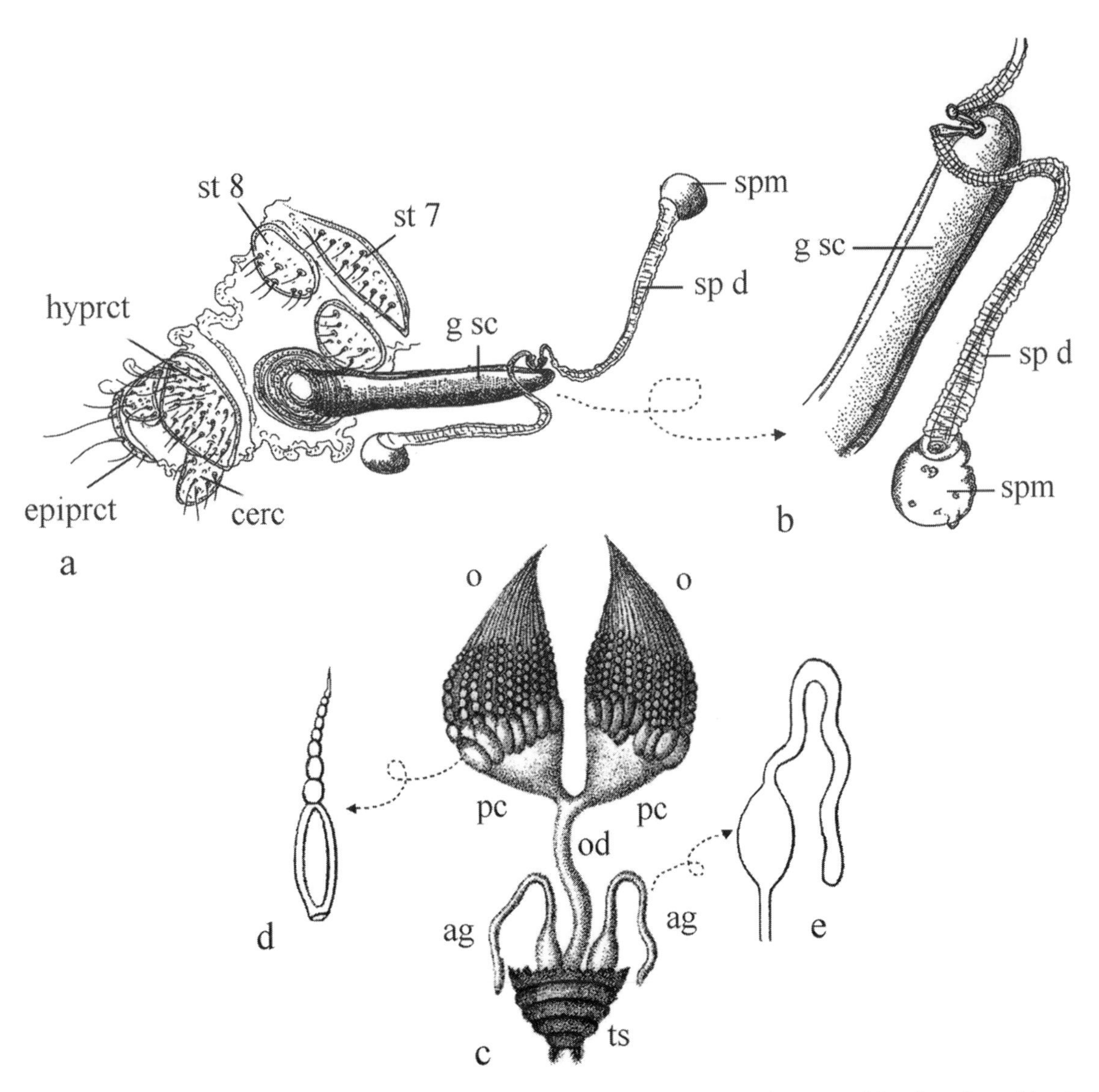

Fig. 14.6 (a) Female terminalia and (b) spermathecae of *Trypetoptera punctulata*. cerc, cercus; epiprct, epiproct; hyprct, hypoproct; g sc, genital sclerite; sp d, spermathecal duct; spm, spermatheca; st 7, 8, sterna 7, 8 (modified from Rivosecchi 1992). (c–e) Female reproductive system of *Sepedon sphegea*; (c) entire system; (d) one ovariole; (e) accessory gland. o, ovaries; pc, posterior calyxes, od, oviduct; ag, accessory glands; ts, terminal segments (from Dufour 1851).

wings, and legs, which apparently are sufficient for non-schizophorans to emerge, and in activating the ptilinum in Schizophora. However, coordinated contraction of all segments of the body are involved in Schizophora (Miyan 1989), first by the abdomen, followed by the head and thorax, which seem to be more important in expansion of the ptilinum. Recent observation of emerging *Sepedon spinipes* (R. J. Mc Donnell, personal communication, 2004) shows that the ptilinal area, although lightly tanned as the rest of the emerging fly, is flat, exposed, without a fissure, and that the proboscis is well extended and functions importantly in emergence.

The loss of a ptilinal fissure was not referred to specifically by Marinoni & Mathis (2000) in their cladistic analysis. However, one of the characters they used in establishing the monophyly of *Sepedon* and related genera was "35. Head

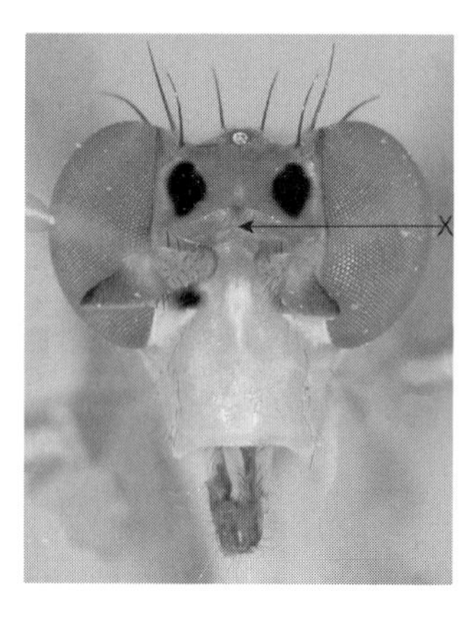

Fig. 14.7 Head of ***Sepedon spinipes***, x shows absence of ptilinal fissure. Arista out of focus. Photo by J. Ebeling.

with sutures: P (0) distinct; A (1) indistinct. The head sutures of *Thecomyia, Sepedoninus, Sepedonella, Sepedon, Sepedomerus*, and *Sepedonea* are indistinct." They also indirectly referred to the ptilinal fissure by their character 25, "Lunula: P(O) not exposed; A(1) exposed; A(2) greatly exposed." J. F. McAlpine (1963) and Barnes (1981) used the extent of exposure of the lunule as a character among families of Lauxanioidea and Sciomyzoidea; mostly concealed being apomorphic in some Phaeomyiidae, Sciomyzidae, and Chamaemyiidae; mostly exposed being plesiomorphic in the other families. Vala *et al.* (2000) noted the presence of a ptilinal fissure as plesiomorphic in *Elgiva, Dichetophora, Euthycera, Verbekaria*, and *Tetanoptera* and as absent (apomorphic) in *Sepedomerus, Sepedon, Sepedonea, Sepedoninus, Sepedonella*, and *Thecomyia*. Mc Donnell & Knutson (unpublished data) examined one or more specimens of 117 species, including in most cases the type species, of 56 of the 61 known genera of Sciomyzidae and one genus in Phaeomyiidae. The ptilinal fissure is absent only in the *Sepedon* lineage, i.e., the six last genera cited above.

As described in Section 14.2, hooks on the labellum have been found only in three genera of the *Sepedon* lineage (*Sepedomerus, Sepedon, Sepedonella*), but are absent in *Thecomyia* of that lineage. Microteeth are scattered over the surface of the labellum in the unrelated genus *Dictya*. These hooks and teeth may aid in emergence of the adult from the puparium and pupariation site, in part replacing the function of the ptilinum.

Other features of adults remaining relatively unexplored include sensilla, musculature of the male genital apparatus, morphology of the testes, and the female abdomen. It might be especially interesting to make a comparative study of the abdomen and its sensilla of females that oviposit onto the flesh of the host (*Tetanura pallidiventris*), deep into the umbilicus of snail shells (*Salticella fasciata*), across or in shell sutures (*Atrichomelina pubera, Colobaea bifasciella, Pherbellia s. schoenherri*, and *Sciomyza varia*), on snail eggs (*Anticheta* spp.), and on vegetation and the microhabitat substrate (all other species).

Possible Batesian mimicry between *Pherbellia cinerella* (widespread in the Palearctic) and *Trixoscelis puncticornis* (Tunisia, Azores, Canary Islands) of the family Trixoscelididae, based on 17 morphological and color characters, was raised by Munari (1996). The striking brown and yellow body color of the 12 species of *Thecomyia* is similar to many predatory wasps. Furthermore, the very exceptionally long haustellum and proboscis of *Thecomyia* spp. might indicate that these species are active nectar feeders on flowers, where they would come into contact with predatory wasps and other predatory insects, and thus would benefit from a mimetic color pattern.

Except for the morphology of the postabdomen and body size there is relatively little sexual dimorphism in adult Sciomyzidae; the instances noted below may be significant in courtship or mating. Sexual dimorphism is common in the extent of short setae in rows on the ventral surface of the hind femur, with males having larger and more numerous setae. The middle of the face is shiny in female *Tetanocera plebeja* and pruinose in the male, and in *Sepedon lobifera* and *S. plumbella* the pruinosity of the face is different between the sexes. In *S. plumbella* and an undescribed Oriental species, tarsomere 1 of the male fore leg is broadened, with a dorsal, diagonal groove (Fig. 14.8d), but simple in the female. In the two species of *Sepedon*, subgenus *Sepedomyia*, tarsomere 5 of the fore leg of the male is much longer and broader than in the female and is silvery pruinose. In one of the species the first three tarsomeres of the fore leg of the male has a comb of erect hairs on the anterior margin, but these are absent in the female. In the eight species of the *Sepedon armipes* group in North America, the males have characteristically ventrally emarginate hind femora with processes (Fig. 14.8a–c), whereas those of the females are simple. We do not know the basis for the statement by J. F. McAlpine (1981) that these processes are used in mating; Neff & Berg (1966) reared four species of the *S. armipes* group but did not comment on the use of the hind femora in mating. The fore femur of the males of some genera of the related family Sepsidae have similar modifications. The hind femur of *Tetanoptera*

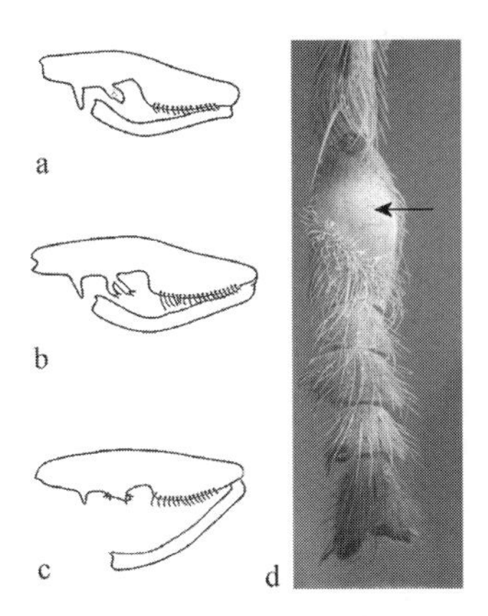

Fig. 14.8 (a–c) Hind femora and tibiae of males. (a) ***Sepedon armipes***; (b) ***S. bifida***; (c) ***S. melanderi*** (from Steyskal 1951a). (d) Apex of fore tibia and tarsomeres, dorsal, of ***S. plumbella***. Arrow, groove across basitarsus (photo by L. Knutson).

leucodactyla is swollen in the male, but simple in the female. In *Hydromya dorsalis* sternum 4 of the male abdomen has a pair of unique, short to elongate cylindrical processes, occasionally greatly reduced, which are absent in the female. No evidence of any possible function during mating was seen in laboratory studies (Knutson & Berg 1963a). In *Protodictya* and *Pherbecta* sterna 4 and 5 of the male abdomen are fused, with sternum 5 modified into a median apical process. In most species of Sciomyzidae, the female is slightly larger than the male.

Many Sciomyzidae have distinctive wing patterns, but in many others the wing is immaculate or with slight shading (e.g., Figs 14.9–14.12). The presence or absence of wing pattern is not correlated with habitat, life style, or sex, and seems not to be associated with mating or other display, but is more or less correlated with taxonomic position. Most genera are characterized by presence or absence of a pattern, except in *Tetanocera* all but two species (*T. clara* and *T. valida*) lack a pattern, one of 11 species of *Colobaea* (*C. bifasciella*) has a pattern, and in *Pherbellia* several species have a pattern. The seven species of *Graphomyzina*, placed by some as a subgenus of *Pherbellia*, have a pattern. The sister group Phaeomyiidae and both the most plesiomorphic genus (*Salticella*) and most apomorphic genera (*Sepedon* and relatives) lack a pattern. The wing pattern in *Trypetoptera punctulata* is very similar to some Tephritidae, and might, in fact, mimic the color pattern of some spiders.

Only in *Anticheta brevipennis* and *A. shatalkini*, which have a short, narrow wing; *Pteromicra angustipennis* which has a somewhat narrowed wing; *Teutoniomyia* spp. in which the wing is almost quadrate; and *Sepedon lobifera* in which the anal lobe is enlarged, are there strong differences in wing shape. The venation and shape of the cells are quite uniform; cell cu*p* is triangularly prolonged in *Salticella* and *Prosalticella*; vein A_1+CuA_2 does not reach the wing margin in *Colobaea* and is evanescent in many species; there are stump veins on vein R_{4+5} in a few *Pherbellia, Tetanocera*, and *Ilione*; vein R_1 may end before or after crossvein r-m; and crossvein dm-cu is sigmoid in a few genera (e.g., *Elgiva, Ilione*). Papp (2004) described the first-known apterous species of Sciomyzidae, from Nepal, as a new genus (Fig. 15.22).

The Sciomyzidae have been included in a number of very comprehensive studies of the morphology of major structural features of Diptera, many having the objective of elucidating relationships between families and superfamilies.

Speight (1969) studied the morphology of the prothorax, especially the shape of the prosternum, of 59 families of Acalyptratae, represented by 800 genera, including 28 genera of Sciomyzidae. He classified prosternal shapes into 31 categories. The Sciomyzidae exhibit primarily shape B (Fig. 14.13), found in 15 genera of Tetanocerini plus *Salticella fasciata* and some species of *Pherbellia*. Shape H is found in the two genera of Huttonininae, four of Sciomyzini, and seven of Tetanocerini. Three genera of Tetanocerini and some species of *Pherbellia* have shape A. One species of *Pherbellia* exhibited shape V (precoxal bridge). *Pelidnoptera* (Phaeomyiidae) was the only genus that exhibited shape R, yet another character that distinguishes that family. Internal views of the basisternum are shown in Fig. 14.13. Speight concluded that no distinctive prothoracic features separate the Sciomyzidae from other acalyptrate families. Prosternal shape is not correlated with Steyskal's (1965a) classification of subfamilies and tribes (as shown above, except for *Pelidnoptera*). Speight (1969) noted that the promotor muscle (one of two anterior sternal muscles) of the fore coxa is present in all of the 2150 species of 59 families of Acalyptratae that he examined, but the adductor muscle is lost in *Sepedon* and a few other species (Fig. 14.13). Ulrich (1991) noted that this muscle also is missing in the Dolichopodidae *s. str.*

Manguin (1989), in laboratory rearings conducted at 20 °C, LD 16:8, showed that there is sexual dimorphism in sizes of adults and puparia of the aquatic predator *Tetanocera ferruginea*. Female adults and puparia are larger (Fig. 14.14) and the shape of female puparia is slightly different (Fig. 14.15).

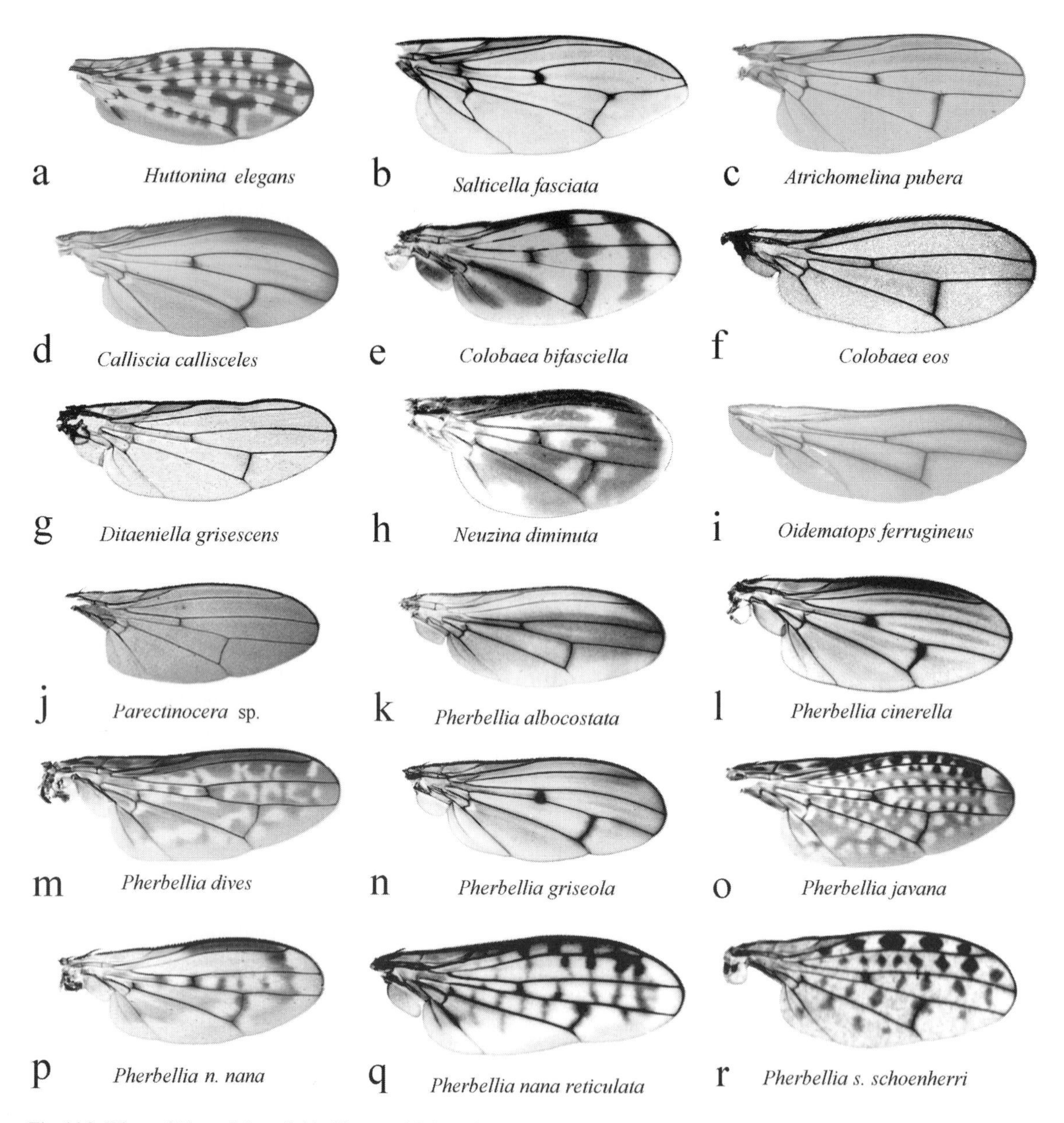

Fig. 14.9 Wings of Huttonininae, Salticellinae, and Sciomyzinae–Sciomyzini. (a, d, g, i, j, l, m, o, p) (original); (b, r) (from Vala 1989a); (c) (photo by J. B. Keiper); (e, k) (Rozkošný 1987a); (f, n, q) (photos by M. Sueyoshi); (h) (photo by L. Marinoni).

14.1.1.2 SETATION PATTERNS

Considering the striking plethora of setae on most flies, their critical importance in taxonomy and cladistics, yet their essentially unknown relationship with major behavioral patterns, some of the primary features of setal patterns of Sciomyzidae are discussed below in regard to recent analytical studies and some of our unpublished observations.

The patterns of bristles, or setae (including the alternative terms macrotrichia, macro- and microchaetae, hairs, and setulae) are very characteristic features of Diptera. Especially those on the mesothoracic dorsum (notum or scutum) are of critical importance in taxonomy, especially of the Cyclorrhapha, including the Sciomyzidae. However, further analysis of the importance of mesothoracic setae in

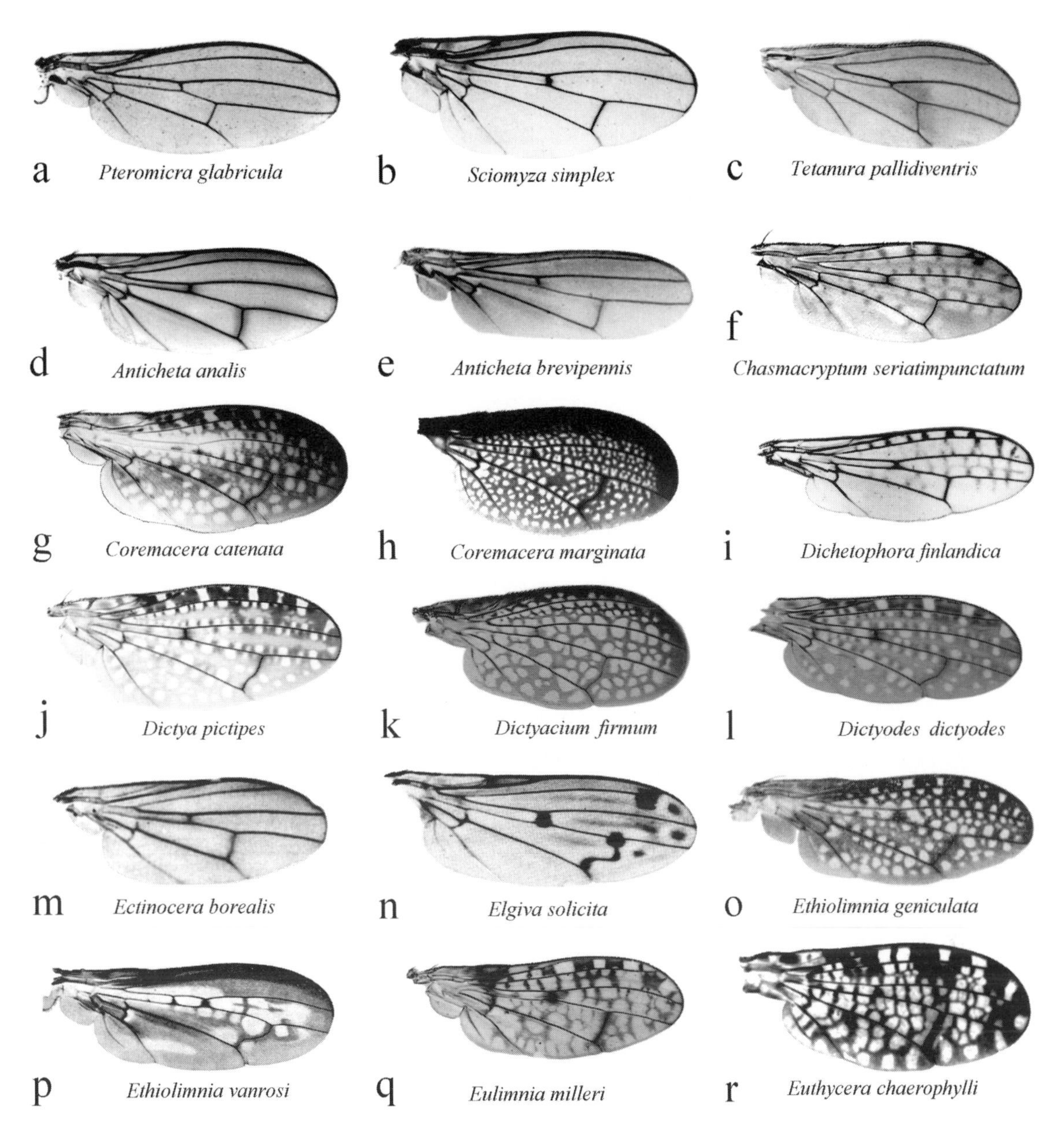

Fig. 14.10 Wings of Sciomyzinae–Sciomyzini and Tetanocerini. (a, g, h, i, m, n, r) (from Vala 1989a); (b, c, d, e) (from Rozkošný 1987a); (f, j, k, l, q) (original); (o, p) (from Verbeke 1962a).

the evolution of Sciomyzidae seems to be needed. While some setae probably function, in part, as mechanical protection from predators, for protection from abrasion of sensilla on the surface of the cuticle, and perhaps in emergence from the puparium and puparial site substrate, and/or mate recognition, their primary function is sensory. Setae have mechanoreceptors at the base and some also serve as chemoreceptors. Superficial extensions of the cuticle, e.g., minute hairs on the wing surface and the pruinosity or pollenosity that dulls the surface of sclerites are not

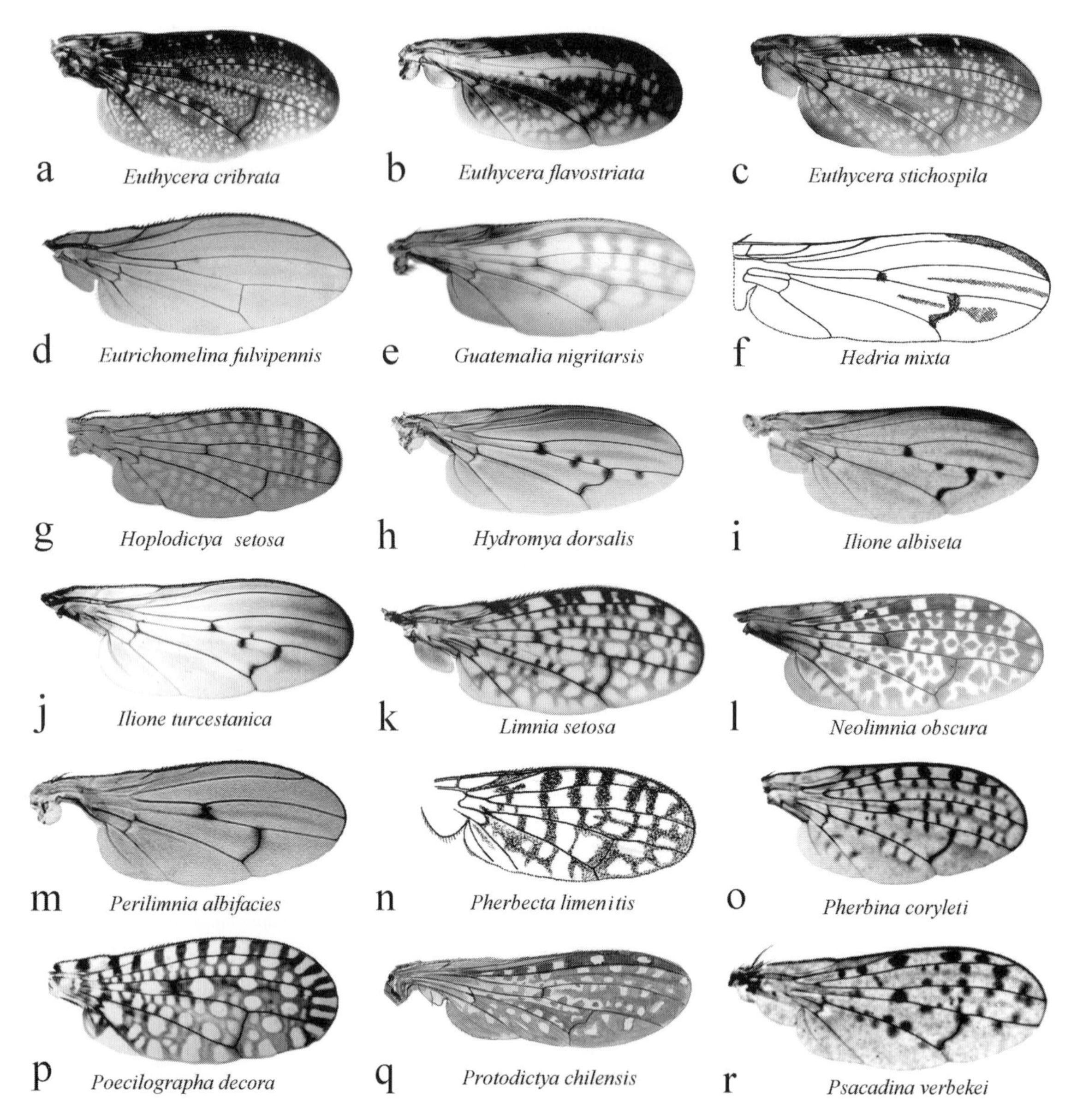

Fig. 14.11 Wings of Sciomyzinae–Tetanocerini. (a, b, c, h, i, o, r) (From Vala 1989a); (d, e, g, j, l, m) (original); (f) (from Steyskal 1954a); (k) (photo by M. Sueyoshi); (n) (from Knutson 1972); (p) (from Foote & Keiper 2004); (q) (drawing by R. Grantsam).

sensory, although the latter are of taxonomic significance among many species of Sciomyzidae. The terminology used herein is that of J. F. McAlpine (1981).

There have been many reviews of the morphology of many aspects of Diptera, especially of the head capsule and (relevant to Sciomyzoidea) its setae in Schizophora (Hennig 1958), mouthparts, antennae, wing venation, structure of the thorax, and most especially the male and female postabdomen. However, setation patterns have been relatively neglected. Thus, considering that we are beginning to relate setation patterns to behavioral patterns in Sciomyzidae, the recent extensive analysis of the development and evolution of setal patterns on the mesothoracic dorsum of Diptera by Simpson *et al.* (1999) is of great

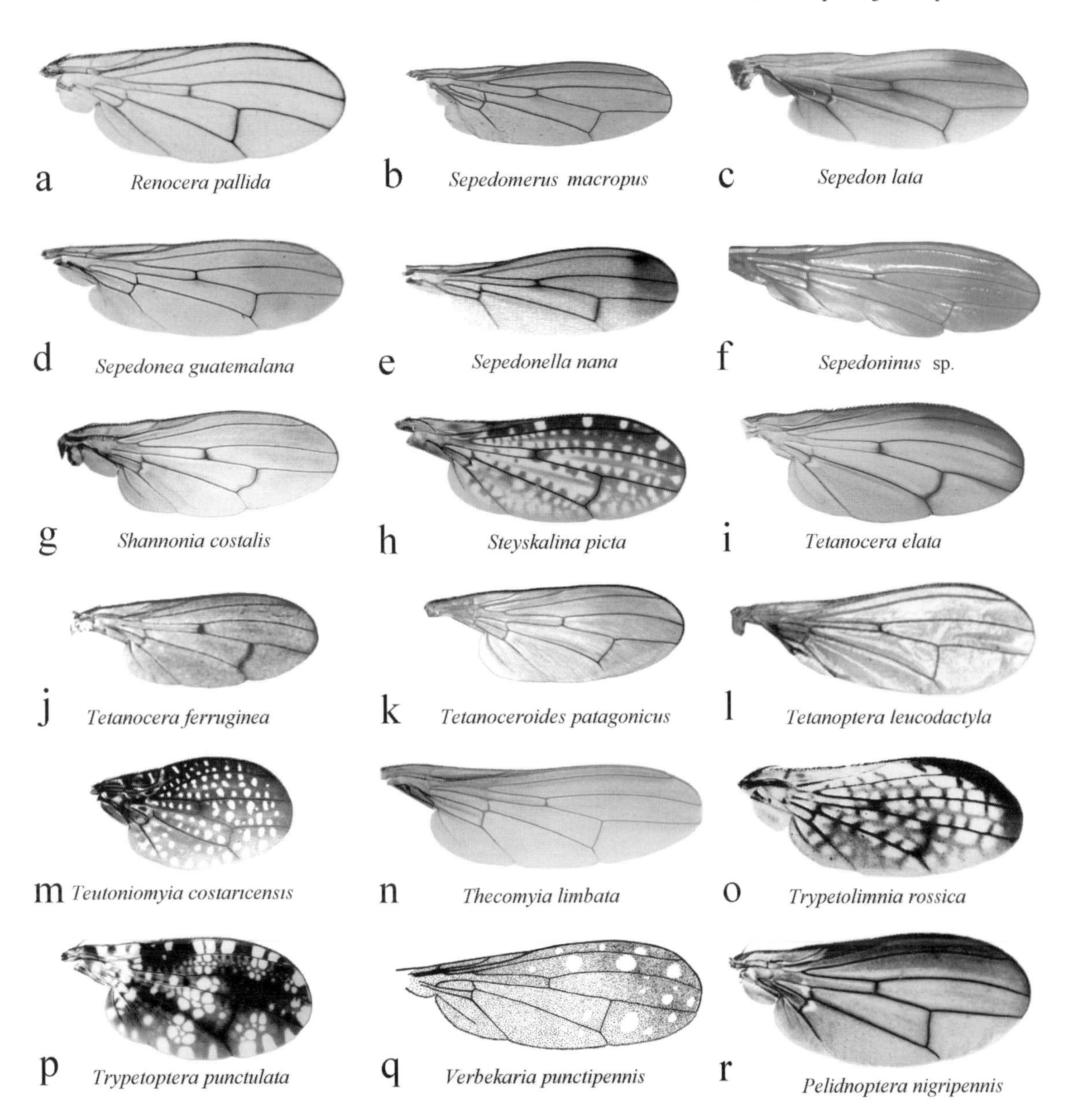

Fig. 14.12 Wings of Sciomyzinae–Tetanocerini. (a, i) (from Rozkošný 1987a); (b, c, d, e, f, g, h, j, l, m, n) (original); (k) (from Zuska & Berg 1974); (o, p, r) (from Vala 1989a); (q) (from Knutson 1968). See also Figs 15.4e, f.

interest. They made special reference to the Drosophilidae, an acalyptrate family not closely related to Sciomyzoidea, but in which family, notably *Drosophila melanogaster*, the genetic control of patterns has been extensively studied. They noted that the remarkably uniform patterns of setae in Muscomorpha is due to a strong system of lateral inhibition, whereby nascent setal precursors prevent neighboring cells from developing into setae (Simpson 1990). However, adventive i.e., duplicate setae are seen fairly often in the small prescutellar acrostichals of the mesonotum of some Sciomyzidae, e.g., in *Pherbecta*, and on other parts of the body (fronto-orbital setae on the head, *Renocera* spp.; notopleural, *Sepedon* spp.; postalar, *Dichetophora* spp.; and other pleural setae, *Pherbellia* spp.)

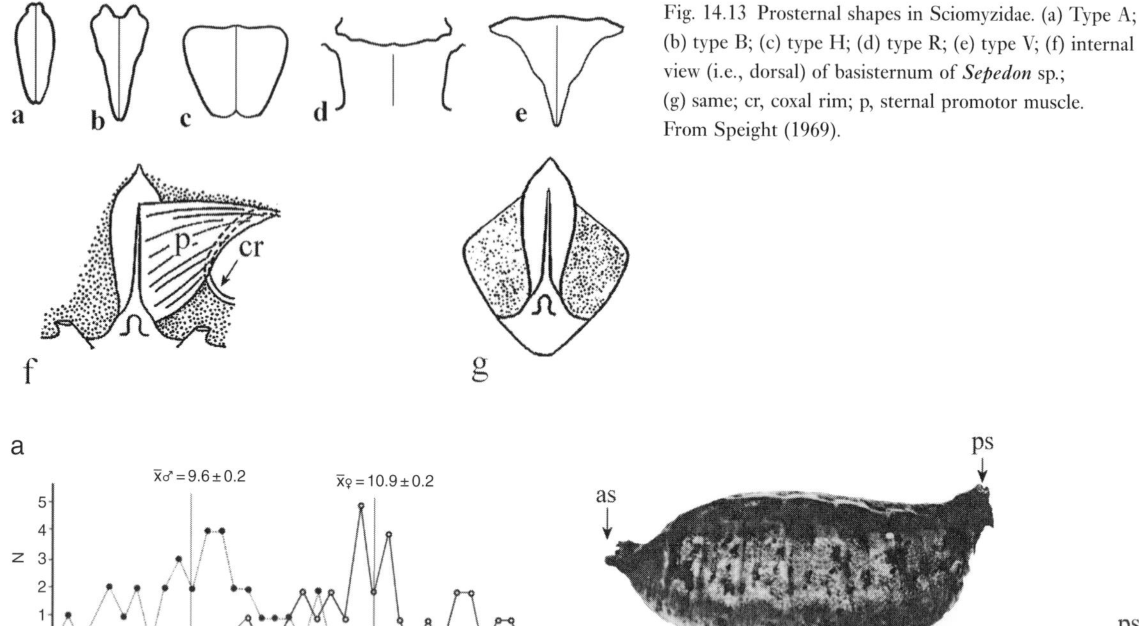

Fig. 14.13 Prosternal shapes in Sciomyzidae. (a) Type A; (b) type B; (c) type H; (d) type R; (e) type V; (f) internal view (i.e., dorsal) of basisternum of *Sepedon* sp.; (g) same; cr, coxal rim; p, sternal promotor muscle. From Speight (1969).

Fig. 14.14 Size distribution of (a) length and (b) width (mm) of adults of *Tetanocera ferruginea*. Males, broken line, solid circle; females, solid line, open circle; N = 30 specimens per sex. From Manguin (1989).

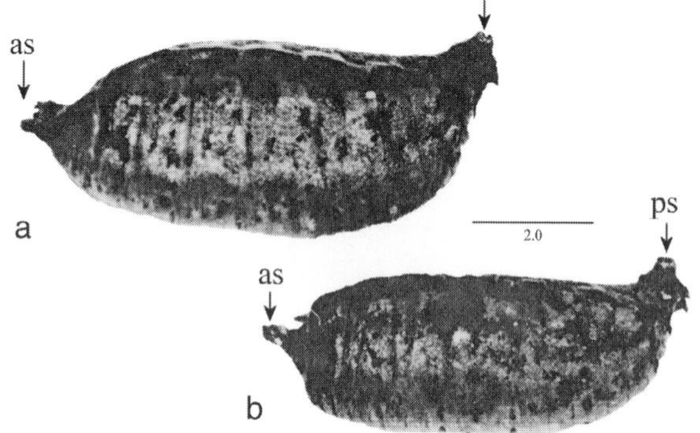

Fig. 14.15 Lateral view of puparia of *Tetanocera ferruginea*. (a) female; (b) male. as, anterior spiracle; ps, posterior spiracle. From Manguin (1989).

and setae on the femora and tibiae, indicating a breakdown of "lateral inhibition." Usually, these aberrations are not symmetrical, i.e., not the same on right and left sides. Occasionally there are instances of real or apparent absence of some setae. Notably, the single female of *Tetanoptera leucodactyla* n. sp. n. gen. described by Verbeke (1950) seemed to lack the important postocellar setae, leading to confusion as to its relationships. However, discovery of swollen spots where the postocellar setae normally arise in this specimen, and well-developed postocellar setae in a subsequently discovered male (Knutson & Vala 1999) and a second female specimen (Knutson unpublished data) clarified the taxonomic position of the genus.

Simpson *et al.* (1999) emphasized that "The number of rows of large bristles on the scutum was probably reduced to four [acrostichal, dorsocentral, intra-alar, and supra-alar] early on in the evolution of cyclorrhaphous Brachyceran flies" (Fig. 15.3a). They referred to Ghysen (1980) in noting that the neuronal specificity of the seta organ is dependent upon the site within the epithelium at which the precursor cell of the seta arises. Thus, they concluded "The positions of bristles are therefore likely to be of importance to the fly's behavior which could explain why the patterns have been maintained over such long periods of time." They also emphasized the row of marginal scutellar setae, which we consider an extension of the supra-alar row. They noted that postalar and notopleural setae are situated on more lateral sclerites, i.e., pleural, not scutal, and in the Acalyptratae setae are most often missing from the anterior part of the scutum. These and other observations of patterns

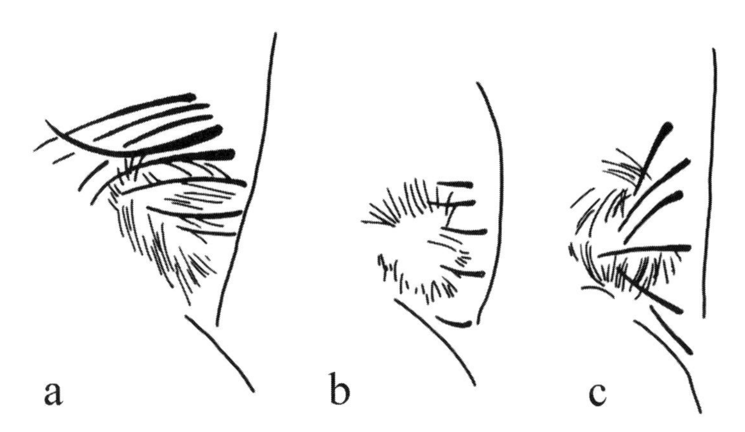

Fig. 14.16 Setulae on posterior margin of posterior thoracic spiracle of *Sepedonea* species. (a) *Sepedonea trichotypa*; (b) *S. lindneri*; (c) *S. guatemalana*. From Freidberg *et al.* (1991).

should give future researchers on cladistics information of value in determining plesiomorphic and apomorphic character states.

The patterns of distinctively large setae (macrochaetae *sensu* Simpson *et al.* 1999) on the thoracic dorsum of Sciomyzidae range from that of the plesiomorphic Salticellinae and Sciomyzini (e.g., *Pherbellia*), which have essentially the full complement of setae characteristic of the ground plan of Schizophora to the greatly reduced pattern of the apomorphic *Sepedon* lineage (Tetanocerini) at the apex of cladograms of Sciomyzidae. Many other character systems distinguish the subfamilies and tribes of Sciomyzidae, as we note in the comparative diagnoses (Section 15.1.2). Major differences in setal patterns on the mesothoracic dorsum of Sciomyzidae primarily involve reductions in the apomorphic *Sepedon* lineage and related genera and in the morphologically and/or biologically aberrant genera *Anticheta* and *Renocera* of the Tetanocerini and *Tetanura* of the Sciomyzini. The "pronotum" was not considered by Simpson *et al.* (1999) but significant differences exist in the setae of this sclerite in the Sciomyzidae.

Patterns of setae on other parts of Diptera – the head, thoracic pleuron, legs, wings, and abdomen – have not been reviewed as extensively in a comparative-evolutionary perspective as have those of the mesothoracic dorsum. However, they show strong differences and similarities of taxonomic, cladistic, and behavioral significance. These setal patterns are also included in the diagnostic description (Section 15.1.2.1). Reduction of the setae of the head and thorax of the apomorphic *Sepedon* lineage is an important feature of Sciomyzidae. Setae of the thoracic pleuron, which show important distinctions between the Acalyptratae and Calyptratae, are also distinct among many genera and species of Sciomyzidae. There is extensive variation among genera in setal patterns of the anepisternum, katepisternum, and anepimeron. In *Sepedonea* there are several distinct setae, presumably protective in function, on the posterior margin of the posterior thoracic spiracle (Fig. 14.16). Meier (1996) noted that this character is a strong synapomorphy for the sister group relationship between the Ropalomeridae and Sepsidae (both in Sciomyzoidea) and is very rare within the remaining Cyclorrhapha.

Setae of the legs have been widely used as family, generic, and specific characters of the Sciomyzidae. Presence of anterodorsal and posterodorsal setae on the mid and hind tibiae, an apomorphic feature of the sister group Phaeomyiidae, are one of the most important features of that family. Rozkošný (1987a) found considerable variation in the numbers of these setae in three described Palearctic species of *Pelidnoptera* (genus of the Phaeomyiidae). In another Palearctic–Oriental *Pelidnoptera* sp., we found that although at least some mid tibial setae were always present, the numbers at each of the four locations ranged from 0 to 5. The strongly bristled legs of Sciomyzidae are considered apomorphic in relation to the ground plan of the Sciomyzoidea (J. F. McAlpine 1981). A comb of a short series of closely spaced setulae on the fore femur is unique to *Pteromicra*. We have seen, in several genera, a feature not previously described for Sciomyzidae: a comb of about ten closely spaced, short setulae mesally between the apical setae of the fore and hind tibiae. This comb might be associated with the excessive grooming of antennae, wings, and legs that is characteristic of Sciomyzidae. Such excessive grooming behavior might be symptomatic of the need for adult Sciomyzidae to be especially receptive to the substrate in order to locate appropriate oviposition sites or mating partners. A row of fine hairs on the inner, posterodorsal margin of the coxa,

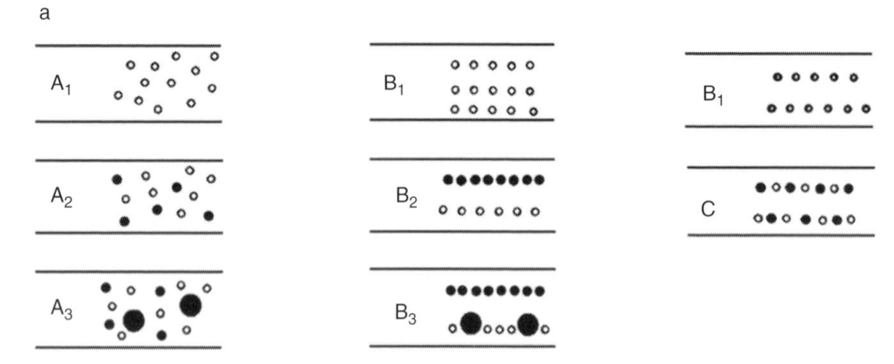

b

Fig. 14.17 (a) Types of costal chaetotaxy in Sciomyzoidea. Open circle, hair or bristle; small dot, spinula; large dot, spine. All Sciomyzidae B_2 except *Tetanura pallidiventris* (AB_1: transitional between A_3 and B_1). (b) Costal chaetotaxy of ***Sciomyza simplex***. See Table 14.1. From Hackman & Väisänen (1985).

although of insignificant structure and thus perhaps of insignificant behavioral value, has however been found to be of significant taxonomic importance at the generic level. The patterns and role of abdominal setae have not been explored. See also the occurrences of setae across some genera of Sciomyzidae in Tables 15.5, 15.7–15.10 and across segregates of Sciomyzoidea and Lauxanioidea in Tables 15.12, 15.13, 15.15.

Hackman & Väisänen (1985) studied an important new character system, the costal chaetotaxy (mechanoreceptors) of the wing, for 875 species in 128 families of Diptera. They examined 12 species of Sciomyzidae in nine genera, *Pelidnoptera fuscipennis*, and other members of the Sciomyzoidea (Fig. 14.17, Table 14.1). They noted that the costal chaetotaxy is "remarkably uniform" in the family (including *P. fuscipennis*), except *Tetanura pallidiventris*, which is unique in the Sciomyzoidea. A regular distribution of non-setate campaniform sensilla (mechanoreceptors) on veins on the dorsal side of the wing of 17 species of Nearctic *Limnia*, *Tetanocera ferruginea*, and *Dictya pictipes* (200× magnification observations) was described by Steyskal *et al.* (1978).

14.1.2 Eggs

Vala *et al.* (1990) described the egg of *Pelidnoptera nigripennis* as having a smooth chorion, an apical micropyle with an arched, transverse dorsal flange, and without aeropyles. The smooth chorion is one of four characters used by those authors to confirm transfer of the (then) three known species of *Pelidnoptera* to the family Phaeomyiidae. The female *P. nigripennis* lays its eggs directly onto the living millipede host in terrestrial, dry habitats.

Eggs of Sciomyzidae have a sculptured chorion and can be characterized as plastron eggs, first described by Brocher (1913) for aquatic Chrysomelidae and other aquatic Coleoptera. Thorpe & Crisp (1947) demonstrated the physiological importance of plastrons in respiration in

Table 14.1. Comparison of costal chaetotaxy in Sciomyzoidea according to seta type shown in Fig. 14.17

	Chaetotaxy								
	A_1	A_2	A_3	AB_1	B_1	B_2	B_3	C	R
Coelopidae 3/4					B_1	B_2			
Dryomyzidae 2/4						B_2	B_3		
Sciomyzidae 10/13				AB_1		B_2			
Sepsidae 6/6					B_1	B_2			
Ropalomeridae 2/3					B_1				
Megamerinidae 2/2					B_1				

Note: Number of genera/species studied given after family name. AB_1, transitional between A_3 and B_1; R, reduced. All Sciomyzidae and ***Pelidnoptera*** are B_2 except ***Tetanura pallidiventris*** is AB_1. From Hackman & Väisänen (1985).

aquatic insects. Hinton (1981) noted, "Sciomyzid eggs have a chorion respiratory system, and many appear to have a plastron, but so far the fine structure of the respiratory system has not been described." According to the classification of plastrons that he proposed, sciomyzid eggs have a plastron spread over the entire surface of the chorion, not confined to the hatching line and the poles, and are placed in group II of plastron eggs. Recent stereoscan studies of surface microstructure, aeropyles at the poles, and ultrastructure of the chorion in transverse section of eggs of 17 Palearctic and one Afrotropical species are summarized here (Gasc *et al.* 1984b, Vala *et al.* 1987, Vala & Gasc 1990a, Vala *et al.* 1999, Vala unpublished data).

The chorion has mainly two types of surface ornamentation. In the ridged type, there are more or less anastomosed longitudinal ridges as in most Tetanocerini (Fig. 14.21a) and a few Sciomyzini. In the other, reticulated type, there are more or less hexagonal polygons, quite regularly arranged with raised edges as in Salticellinae (Fig. 14.20a, e) and most Sciomyzini (Fig. 9.28a). These ridges and polygons correspond to the roof of the chorional galleries system which enables exchange of air. In all examined species the chorion is flared anterodorsally, below which opens the subterminal micropyle (Fig. 14.18b), except in *Salticella fasciata* the micropyle is terminal (Fig. 14.20c). This micropylar flare may have a number of aeropyles permitting air exchange. However, probably most air exchange is from other aeropyles of the same type scattered over the surface of the egg. The posterior end, generally subspherical, has a slight pre-apical constriction (except in *Salticella fasciata*) and always has aeropyles. In addition to the surface ornamentation, in some species of *Sepedon*, *Sepedonea*, and *Sepedomerus* the chorion is raised into two pairs of strong, longitudinal ridges that delimit a dorsal, lateral, and ventral surface (Fig. 14.18); their function is unknown.

The presence or absence of aeropyles at the ends is of ecological significance (Gasc *et al.* 1984b). It seems that the most terrestrial species of Sciomyzidae reduce loss in ovular water by not having aeropyles at the posterior end, but in aquatic species and in some semi-terrestrial species both ends have aeropyles. However, as noted below, there are some exceptions, i.e., *Euthycera chaerophylli*, *Sepedon* (*Parasepedon*) *ornatifrons*, and *Trypetoptera punctulata*. Examination of the eggs of species that have been refractory to rearing attempts, e.g., *Pherbecta limenitis* and *Poecilographa decora*, may be instructive in this regard, i.e., indicating terrestrial or aquatic hosts/prey.

We classify eggs on the basis of chorion structure and distribution of aeropyles as follows (further study of eggs and a reclassification may be needed):

A. *Reticulated chorion:* external surface with polygonal structure.
   - Both ends with aeropyles: aquatic – *Sepedon sphegea*, *S. spinipes* (Gasc *et al.* 1984b), terrestrial – *Sepedon* (*Parasepedon*) *ornatifrons* (Vala & Gbejissi, unpublished data), *S.* (*Parasepedon*) *trichrooscelis* (Vala *et al.* 1995), *Euthycera chaerophylli* (Vala 1989a).
   - Only posterior (abmicropylar) end with aeropyles: *Salticella fasciata*. Unlike other Sciomyzidae, neither end has a distinct tubercle (Vala *et al.* 1999).

B. *Ridged chorion:* external surface with more or less anastomosing longitudinal ridges.

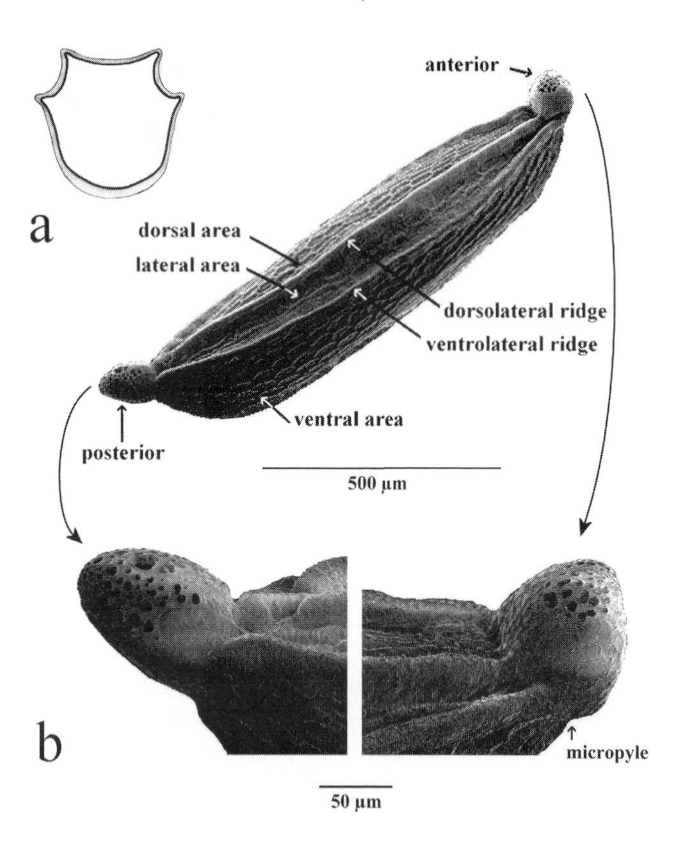

Fig. 14.18 (a) Egg of ***Sepedonea telson***, diagrammatic cross-section (from Freidberg *et al.* 1991); (b) egg of *Sepedon ornatifrons* (from Gbedjissi 2003).

- Both ends with aeropyles: aquatic and semi-aquatic – *Ilione trifaria, Tetanocera ferruginea* (Gasc *et al.* 1984b), *Pherbina mediterranea* (Vala & Gasc 1990a), *Hydromya dorsalis, Ilione albiseta, I. lineata* (Vala 1989a); terrestrial – *Trypetoptera punctulata.*
- Only posterior end with aeropyles: terrestrial – *Dichetophora obliterata* (Vala *et al.* 1987), *Pherbellia cinerella, Euthycera cribrata* (Gasc *et al.* 1984b). *E. stichospila, Coremacera marginata, C. catenata, Limnia unguicornis* (Vala 1989a).

Transverse sections of eggs of the two main types were studied by Gasc *et al.* (1984b) – (*Sepedon sphegea* [Fig. 14.19e, f, h], *Tetanocera ferruginea, Ilione trifaria,* and *Euthycera cribrata*) and Vala *et al.* (1999) – (*Salticella fasciata*) (Fig. 14.20d). The four chorion layers in *S. fasciata* (Fig. 14.20d) from the inside to the outside are: (1) an internal layer (inner endochorion), 2–3 µm thick, spongy and riddled with 0.5 µm diameter pores; (2) a columnar zone, 8–10 µm thick, with 7 µm long cavities between the columns; (3) an alveolar zone, 4–6 µm thick, with 3–5 µm diameter circular alveoli; and (4) an external layer, 3–5 µm thick, consisting of a fine spongy network with 0.5 µm diameter pores. The internal layer disappears at the poles. The layers are slightly different in some species. The chorion surrounds a pellucidous membrane, the ovular envelope (pierced by three micropylar holes at the anterior end). In all species, the longitudinal ridges or the ribs of the exterior polygonal network form the roof of a subchorional galleries system. The confined air clearly increases the mechanical resistance of the egg and provides (partly or totally) oxygen during embryogenesis, even when the egg is submersed.

It would be of interest to examine, by stereoscan microscopy, the eggs of *Tetanura pallidiventris*, the only sciomyzid known to place its eggs on the surface of the host snail's soft parts. Knutson (1970a) described the egg as having a thin chorion, with very weak, reticulate sculpturing. Also, further study of eggs of *Anticheta*, which are the only sciomyzids known to obligatorily lay their eggs on the egg masses of snails (Fisher & Orth 1964, Knutson 1966, Knutson & Abercrombie 1977, Robinson & Foote 1978) (Fig. 14.21d) would be of interest. Eggs of the latter are

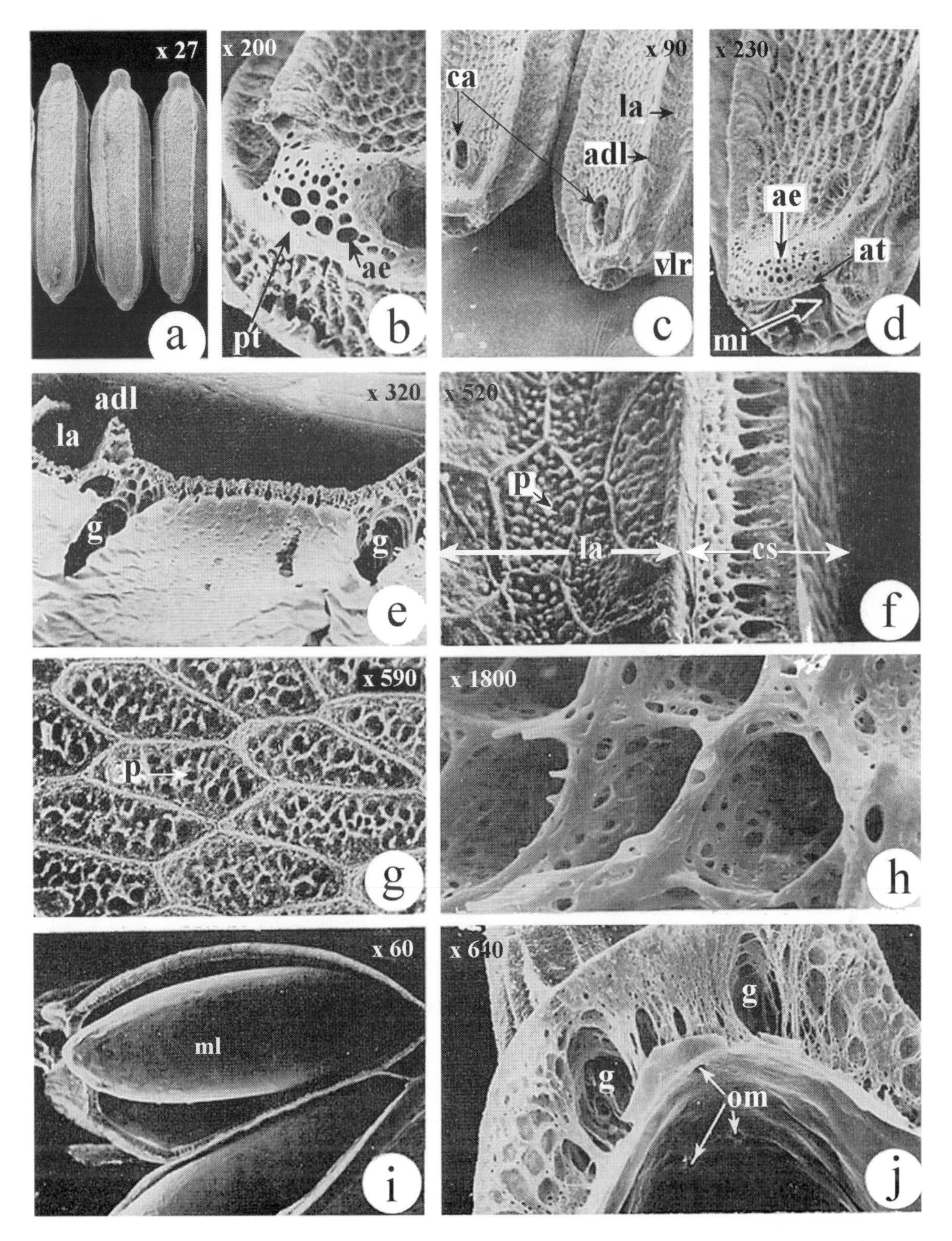

Fig. 14.19 Egg of ***Sepedon sphegea***. (a) Entire eggs; (b) posterior end; (c) anterior end showing chorional aberrations; (d) anterior end; (e) cross-section of dorsal surface; (f) cross-section of chorion and lateral area near emergence line; (g) detail of dorsal surface; (h) alveolar zone; (i) egg without chorion showing pellucidous membrane; (j) cross-section at anterior end showing three openings of micropyles. adl, dorsolateral ridge; ae, aeropyles; at, tubercle of anterior end; ca, chorional aberration; cs, chorional section; g, gallery; la, lateral area; mi, micropyle; ml, pellucidous membrane; om, micropyle openings; p, papilla; pt, tubercle of posterior end; vlr, ventrolateral ridge. Modified from Gasc *et al*. (1984b).

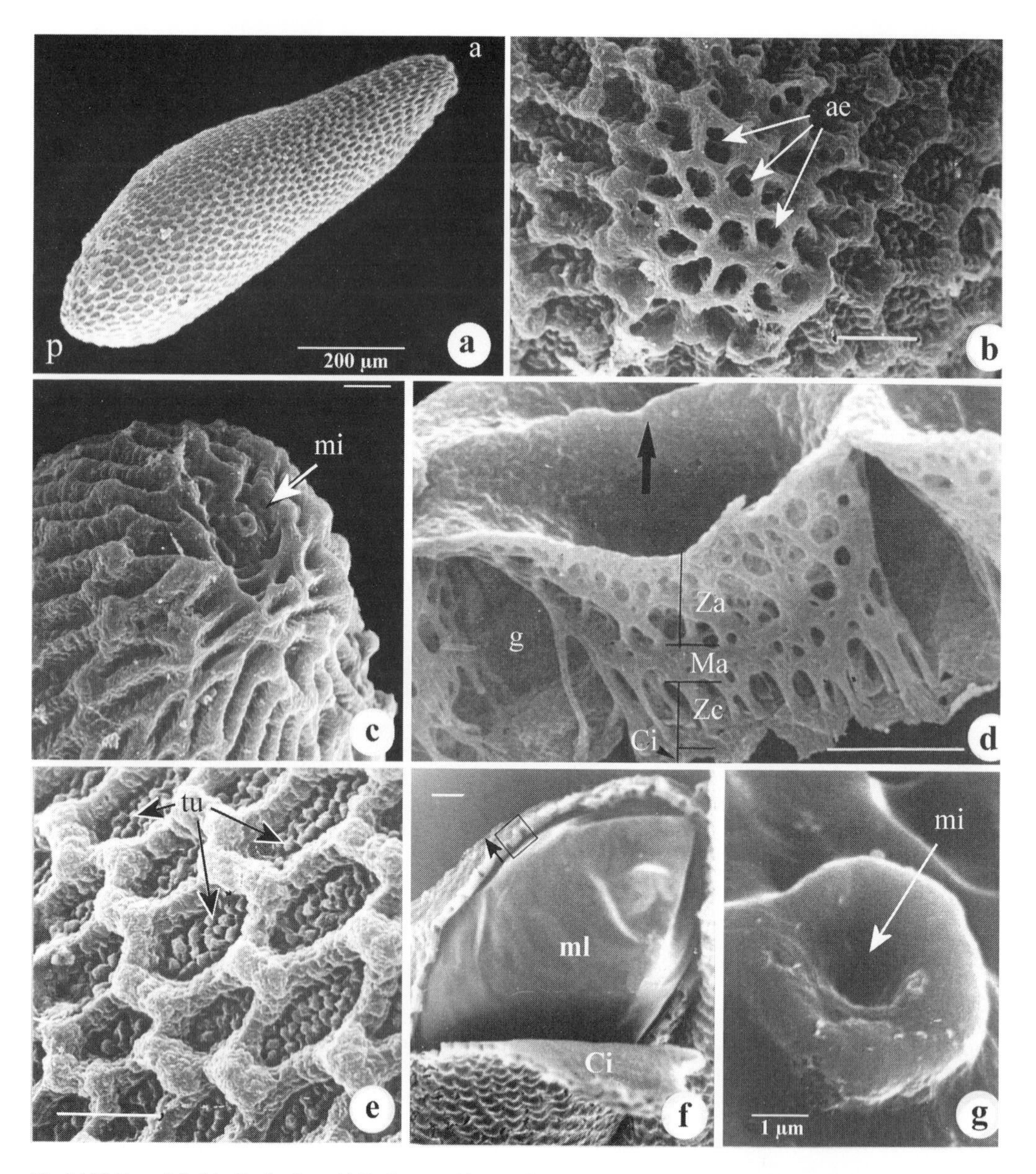

Fig. 14.20 Egg of *Salticella fasciata*. (a) Entire egg; (b) posterior end; (c) anterior end; (d) cross-section of chorion, arrow points from internal to external face of chorion; (e) chorion surface with protuberant hexagons and rounded tubercles; (f) chorion separated to show pellucidous membrane and inner surface of chorion, square part enlarged in d; (g) micropyle. a, anterior end; ae, aeropyles; Ci, inner surface of chorion; g, gallery; Ma, massive area; mi, micropyle; ml, pellucidous membrane; p, posterior end; tu, tubercles; Za, alveolar zone; Zc, columnar zone. Labeled scale lines in µm. Unlabeled scale lines equal to 20 µm. From Vala *et al.* (1999).

mostly surrounded by a clear, hygroscopic, sticky, gelatinous substance, with only the dorsal surface of the micropylar end and the knob-like abmicropylar end exposed, and with two transverse, dorsal patches of high-walled cells (Fig. 14.21b).

14.1.3 Larvae

Detailed studies (primarily by light microscopy) of external features of the larvae of many species in many genera have shown a vast array of features of taxonomic, evolutionary,

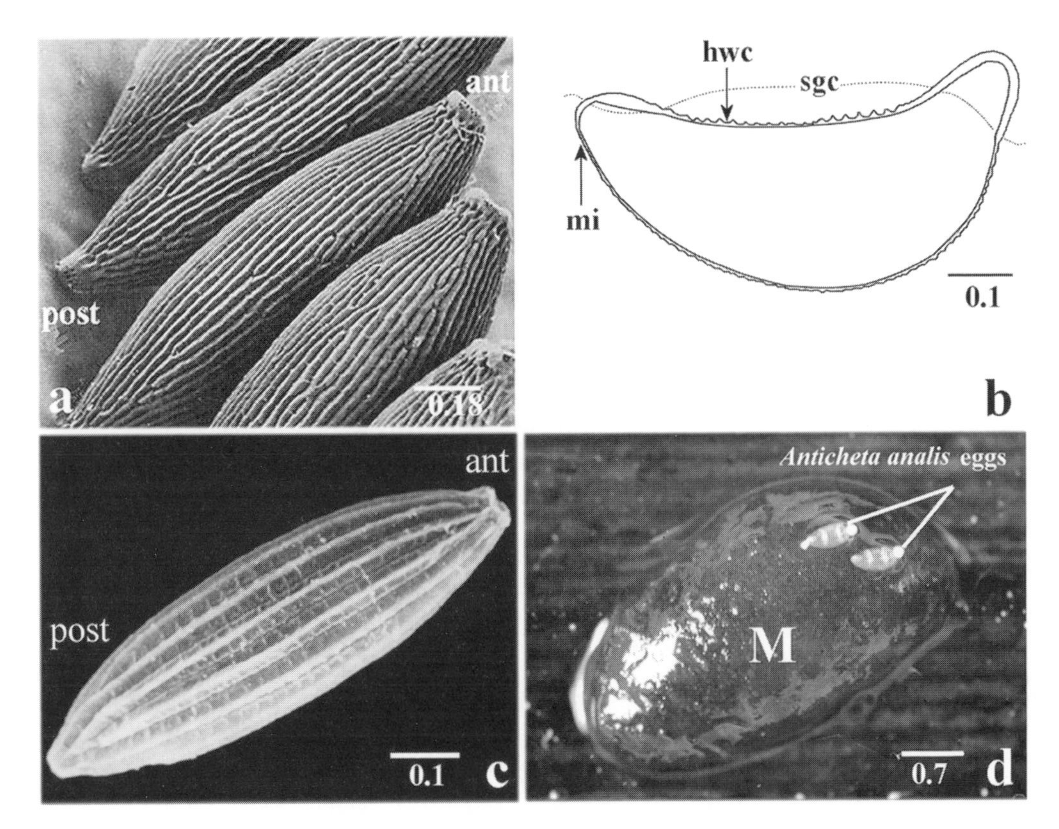

Fig. 14.21 (a) Eggs of *Ilione trifaria* (from Gasc *et al.* 1984b). (b) Lateral longitudinal section of egg of *Anticheta analis* (from Knutson 1966). (c) Egg of *Coremacera catenata*; note unusual chorion sculpturing (from Vala 1989b). (d) Two eggs of *A. analis* on egg mass (M) of *Galba truncatula* (from Knutson 1966). ant, anterior end; hwc, high-walled cells; mi, micropyle; post, posterior end; sgc, surface of gelatinous coat.

and functional–morphological interest. Ferrar (1987) provided a detailed review, with references to 114 species in 28 genera and reproductions of 148 representative figures. Other recent reviews of the morphology of the larvae, by regions, some also including eggs and puparia, plus keys are: western Palearctic (Rozkošný 1998), northern Europe (Rozkošný 1997a), central Europe (Rozkošný 1967, 2002), France (Vala 1989a), Italy (Rivosecchi 1984, 1992), North America (Knutson 1987, Foote 1991), and world (Barker *et al.* 2004).

We describe key features from a taxonomic perspective and provide updated family-level diagnostic descriptions in Chapter 15. Here we focus on recent stereoscan studies and new information on features of diagnostic, behavioral, ontological, and morphological interest. Research is needed on the musculature of the cephalopharyngeal skeleton, especially the mouthhooks and ventral arch, as the latter is a highly derived feature among acalyptrate Diptera and the sole synapomorphic structural feature of Sciomyzidae in the larval stage.

The internal anatomy is essentially unknown except for the unidentified larva reared from a land snail and described in detail by Bhatia & Keilin (1937), which most probably is *Trypetoptera punctulata* (Tetanocerini). The excellence of their light microscopy study is indicated by their description of interspiracular processes which were thought to be absent in Salticellinae, Sciomyzini and terrestrial Tetanocerini. They held living larvae between a slide and cover slip, enabling them to describe sensory papillae, including the nerve supply. Internal features described for this larva by Bhatia & Keilin (1937) that are especially noteworthy, in addition to external features that compare well with the description of *T. punctulata* (Vala 1986), are the smooth pharynx without longitudinal ridges and the conspicuous salivary glands (Fig. 14.22).

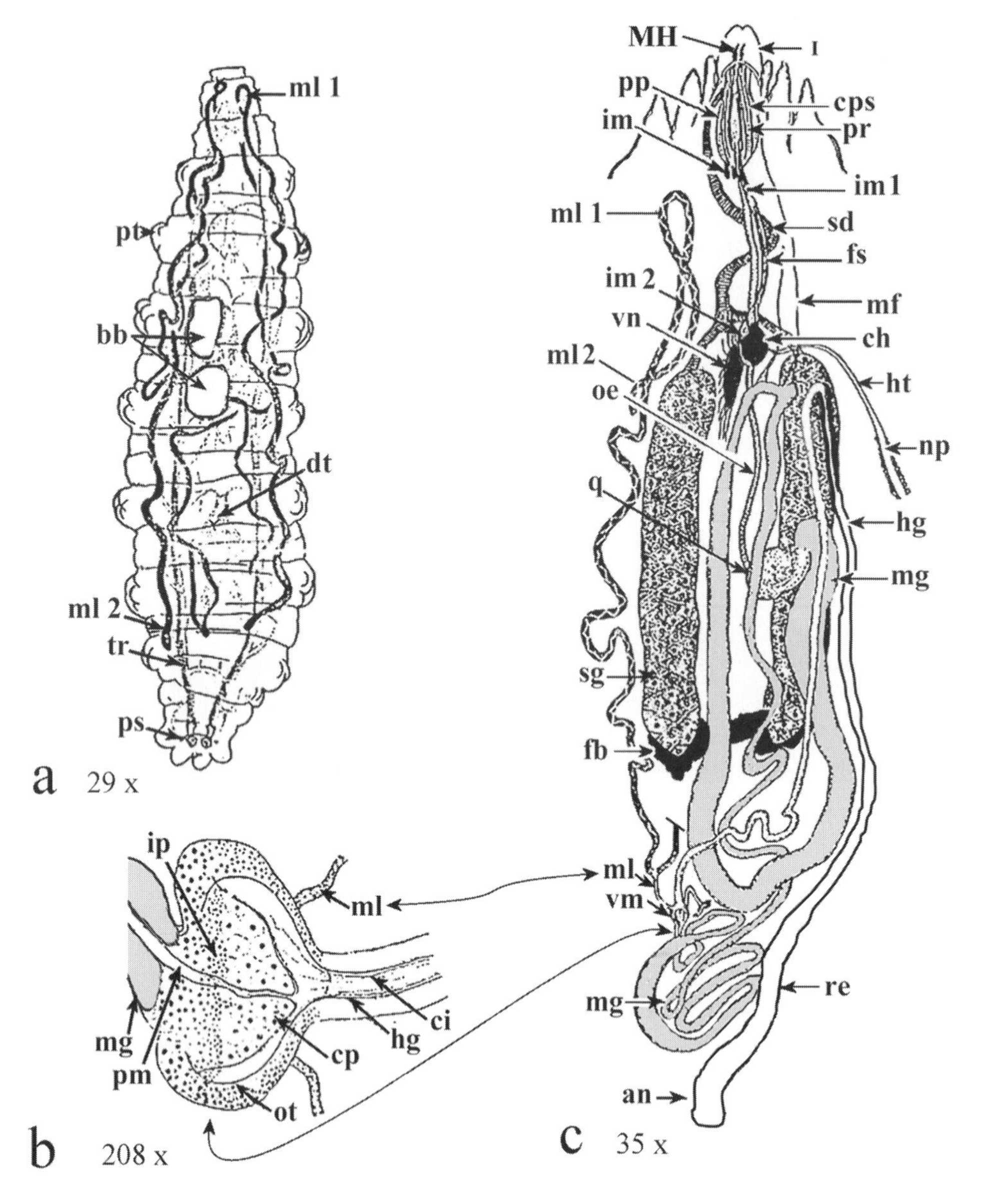

Fig. 14.22 Internal morphology of *Trypetoptera punctulata* (?). (a) Drawing of living L2 in state of rest with segment I retracted. (b) Internal anatomy, from dissection of larva stained with hemalum; membranous frontal sac (fs) and cerebral hemispheres (ch) displaced to right to show course of esophagus. (c) Valve at posterior end of mid gut. I, Segment I; an, anus; bb, air bubbles in alimentary canal; ch, cerebral hemispheres; ci, chitinous intima; cp, conical plug; cps, cephalopharyngeal skeleton; dt, dorsal tracheal commissure; fb, portion of fat body; fs, membranous frontal sac; hg, hind gut; ht, heart; im, im 1, im 2, imaginal buds of labium, pharynx, and compound eyes respectively; ip, basal part of valve containing imaginal tissue; mf, muscle fiber arising from mid gut; mg, mid gut; MH, mouthhooks; ml, ml 1, ml 2, proximal unpigmented, anterior loop, and terminal parts of Malpighian tubules; np, pericardial nephrocytes; oe, esophagus; ot, outer wall of valve containing imaginal tissue; pm, peritrophic membrane; pp, pharyngeal pump; pr, protractor muscles of cephalopharyngeal skeleton; ps, posterior spiracle; pt, lateral protuberance of abdominal segment; q, proventriculus; re, rectum; sd, common salivary duct with spiral of chitinous intima within; sg, salivary gland; tr, tracheal trunk; vm, valve at end of mid gut; vn, ventral nerve mass. Modified from Bhatia & Keilin (1937).

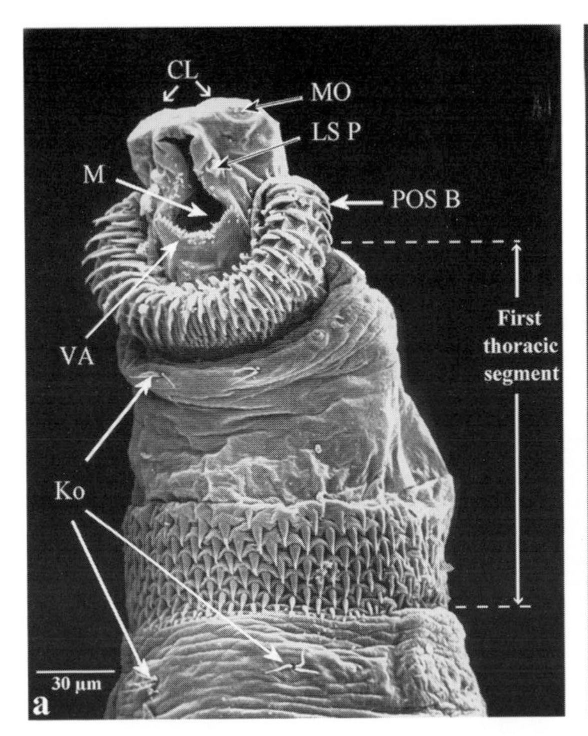

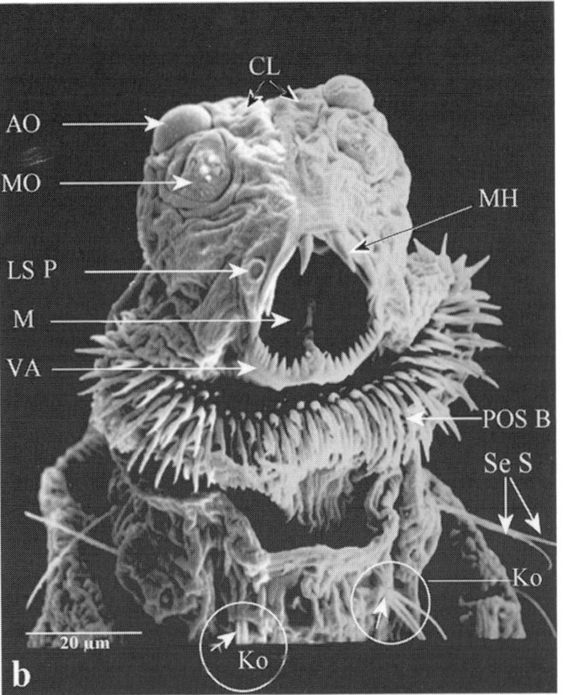

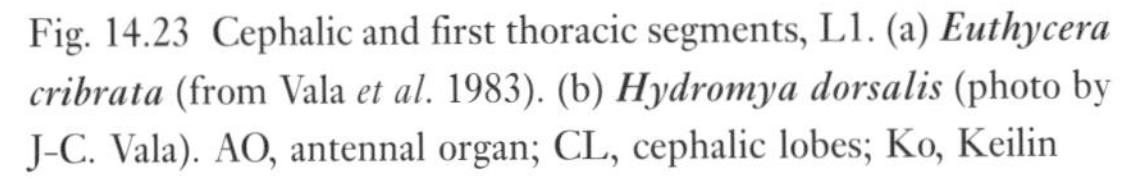

Fig. 14.23 Cephalic and first thoracic segments, L1. (a) ***Euthycera cribrata*** (from Vala *et al.* 1983). (b) ***Hydromya dorsalis*** (photo by J-C. Vala). AO, antennal organ; CL, cephalic lobes; Ko, Keilin organ; LS P, labial sensory papilla; M, mouth; MH, mouthhooks; MO, maxillary organ; POS B, postoral spinule band; Se S, sensile setae; VA, ventral arch.

The first modern, diagnostic descriptions and figures of the larvae of any sciomyzid were those of *Salticella fasciata* by Povolný & Groschaft (1959) and of *Atrichomelina pubera* (plus eggs and puparia) by Foote *et al.* (1960). Hennig (1952b) summarized the earlier publications (1883–1939) on morphology of immature stages. However, even in that masterful review, the very first description (rather detailed) and first figures (rather stylistic), i.e., those of Dufour (1847b) on *Tetanocera ferruginea*, were overlooked (Fig. 14.25).

14.1.3.1 POSTORAL SPINULE BAND AND PATCH

In making a comprehensive review of larval morphology, we discovered that the postoral spinule patch of the Sciomyzini and the postoral spinule band of Tetanocerini (Figs 14.23, 14.24b) seem to be major, distinctive features between the tribes. In Sciomyzini, the postoral spinule patch is very weak to moderately developed with dark, unicuspid to multicuspid spinules arranged in medially converging rows. But, it occupies only the ventral surface of segment I, and lacks a strongly sclerotized posterior margin (Figs 14.24a, 14.34b). In Tetanocerini, there is a similar but much stronger, larger, more rectangular band across the ventral surface that extends laterally to almost encircle segment I, and it has a strongly sclerotized posterior margin (Fig. 14.24b). We retain the common usage of "band" (= complete ring) for this structure although among Sciomyzidae it completely encircles segment I only in *Euthycera chaerophylli*. The vestige of the postoral spinule band is evident on the inner surface of the ventral cephalic cap of Tetanocerini puparia, but not of Sciomyzini.

The spinules of the postoral spinule patch or band are directed anteriorly at rest but project backward when feeding, obviously assisting in feeding movements of active larvae. The unicuspid to multicuspid shape of the spinules (Fig. 14.24c) and their arrangement of medially converging, diagonal rows, forming a mass of minute pincers, likely increases their efficiency.

In the parasitoid/saprophagous *Salticella fasciata* (Salticellinae), there is only a poorly defined patch of dark spinules (Knutson *et al.* 1970) or a spinous strip

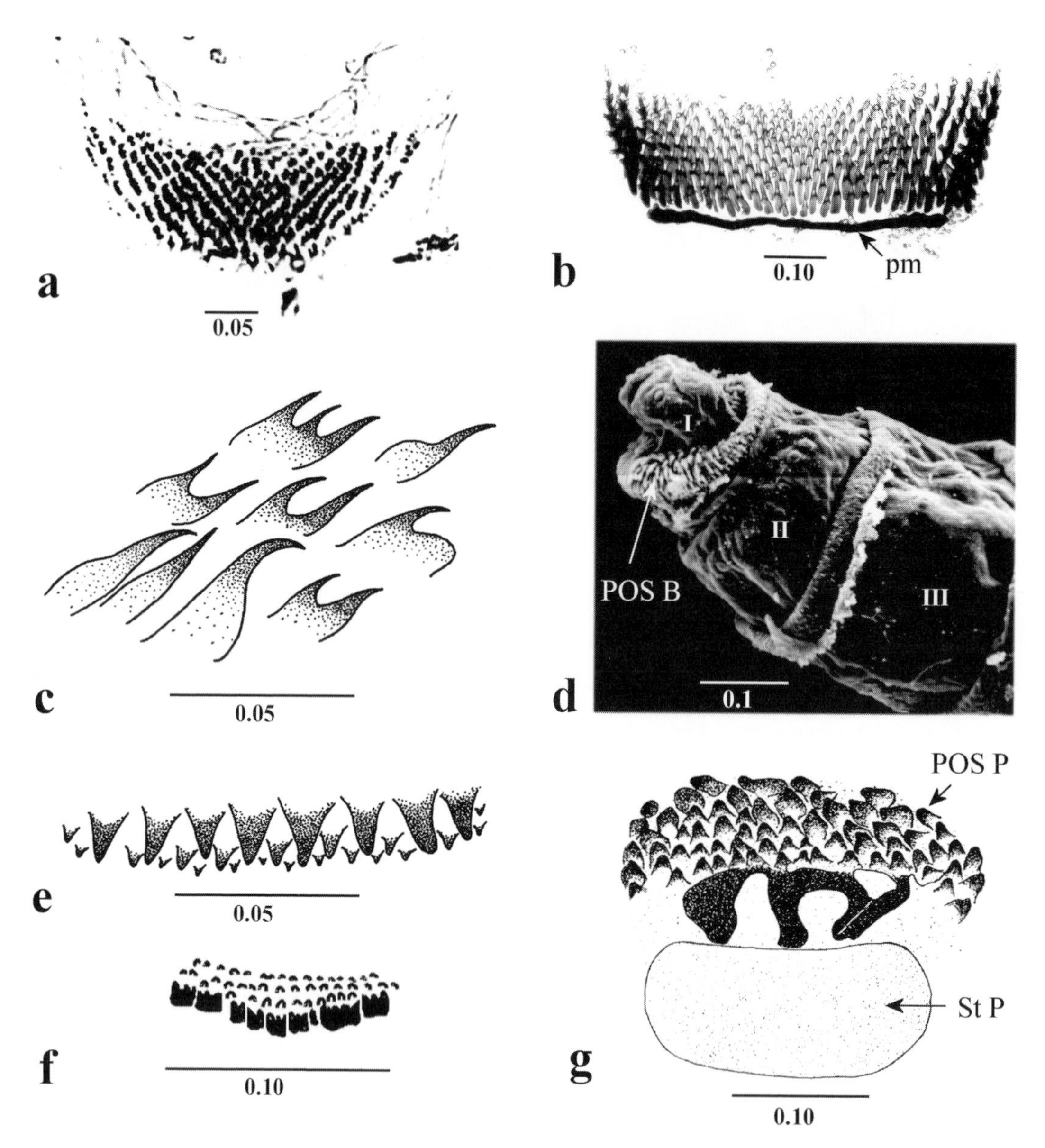

Fig. 14.24 (a) ***Sciomyza varia***, L3, postoral spinule patch (from Barnes 1990). (b) ***Coremacera marginata***, L3, postoral spinule band (from Knutson 1973). (c) ***Pherbellia dorsata***, L3, spinules of postoral patch (from Bratt *et al.* 1969). (d) ***Pelidnoptera nigripennis***, L1, anterior end (from Vala *et al.* 1990). (e) ***Pherbellia dorsata***, L3, spinules of mid ventral patch on segment II (from Bratt *et al.* 1969). (f) ***Pteromicra pectorosa***, L3, spinules of mid ventral patch on segment II (from Rozkošný & Knutson 1970). (g) ***Sciomyza simplex***, L3, postoral spinule patch on segment I and sternal plate on segment II (from Foote 1959a). pm, posterior margin of POS B; POS B, postoral spinule band; POS P, postoral spinule patch; St P, sternal plate on segment II; I–III, segments.

(Vala *et al.* 1999) on the postoral margin of segment I. In first-instar larvae of the internal parasitoid of millipedes, *Pelidnoptera nigripennis* (Phaeomyiidae), there is a completely encircling band of spinules on segments I–V, XI, and XII (Fig. 14.24d) and in second- and third-instar larvae all segments have completely encircling bands (Vala *et al.* 1990).

Several structures that might be confused with the postoral spinule patch or band need to be distinguished:

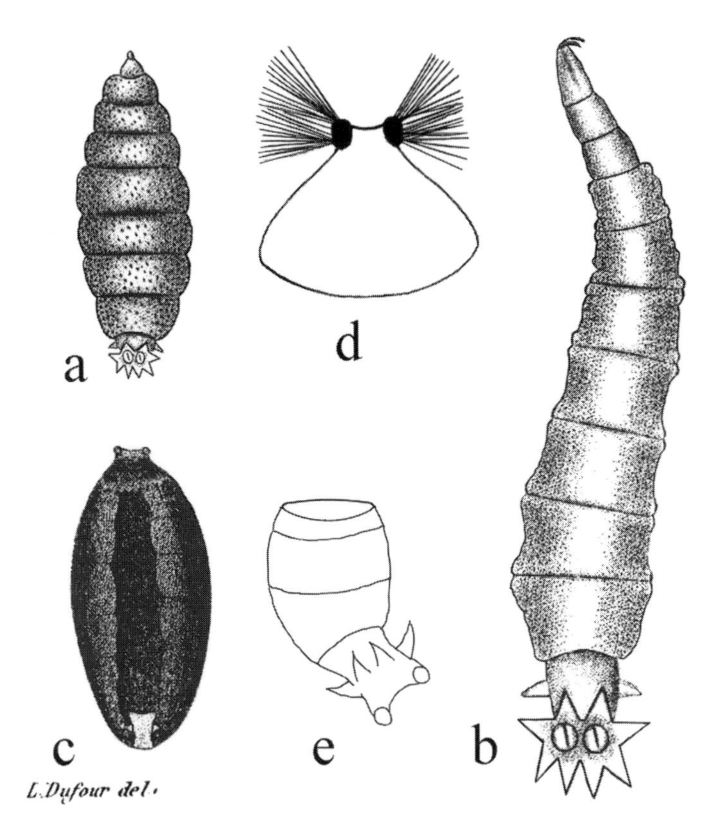

Fig. 14.25. *Tetanocera ferruginea*: (a, b) larva. (a) Body contracted; (b) body extended. (c–e) Puparium. (c) Entire, dorsal; (d) anterior part after emergence of adult; (e) posterior part showing lobes, dorsal. Modified from Dufour (1847b).

(1) a few spinules or a small patch or rows of minute, unicuspid, pre-oral spinules surrounding the mouth opening have been described for *S. fasciata* and many Sciomyzini and Tetanocerini (Fig. 14.35b) (there seems to be some confusion between these and the postoral spinules in some publications); (2) a completely encircling band of spinules is present on the anterior margin of segment II in the succineid parasitoid *Pherbellia schoenherri maculata* (unique in known Sciomyzidae); (3) a small, midventral spinule patch is present on segment II in most *Pherbellia*, *Ditaeniella*, and some *Pteromicra* (Figs 14.24e, f, 14.34b), but is absent in terrestrial, parasitoid *Pherbellia*, *Colobaea*, and Tetanocerini; and (4) a unique patch of curiously shaped, strong spinules is present ventrally on segment I, with a sternal plate on segment II, in *Sciomyza simplex* (Fig. 14.24g).

14.1.3.2 SENSILLA

The distribution of sensilla, potentially important to behavioral and systematic studies and described by light microscopy and stereoscan studies, are summarized below for the species that have been most carefully examined: *Dichetophora obliterata*, Vala *et al.* (1987); *Euthycera cribrata*, Vala *et al.* (1983); *E. stichospila*, Vala & Caillet (1985); *Pherbina mediterranea*, Vala & Gasc (1990a); *Salticella fasciata*, Vala *et al.* (1999); *Sepedon sphegea*, Gasc *et al.* (1984a); *Sepedon trichrooscelis*, Vala *et al.* (1995); and *Trypetoptera punctulata*, Bhatia & Keilin (1937) and Vala (1986) (all Tetanocerini, except *S. fasciata*). A uniform arrangement of sensilla of larvae seems to exist in the family (Fig. 14.26).

(a) Cephalic region. *Pherbina mediterranea* (Fig. 14.27a–c); *Sepedon trichrooscelis* (Fig. 14.28a, b, d). The cephalic region, traditionally called segment I, is divided dorsoventrally into two cephalic lobes. Each lobe has the same sensilla: (1) one more or less long, retractile antennal organ (AO) at the frontal extremity; (2) one more or less elliptical maxillary organ (MO), delimited by a well sclerotized outline, with about ten coeloconical or basiconical sensilla according to the species; (3) one pair of intermediate coeloconical sensilla (SPC) laterally between the antennal and maxillary organs; and (4) two labial coeloconical sensilla (LSP) laterally near the mouth opening.

When compared with other terrestrial sciomyzid larvae, the coeloconical sensilla on the cephalic segment of

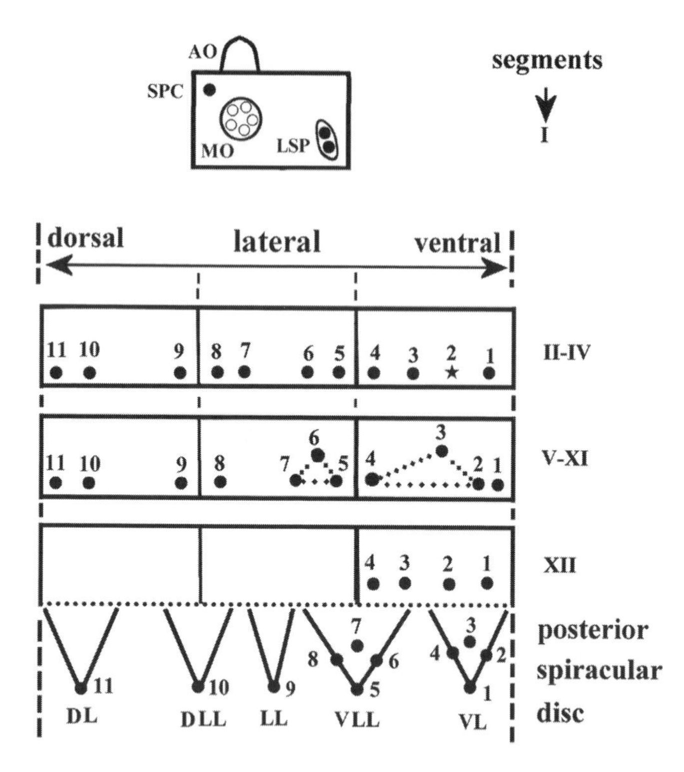

Fig. 14.26 Generalized distribution of sensilla in half of sagittal section of Tetanocerini larva. AO, Antennal organ; DL, dorsal lobe; DLL, dorsolateral lobe; LL, lateral lobe (absent or fused with VLL in some genera); LSP, labial sensillum; MO, maxillary organ; SPC, intermediate cephalic sensillum; VL, ventral lobe; VLL, ventrolateral lobe; I–XII, body segments; black circle, 1–11, sensilla positions; star, Keilin organ. Note: as indicated in Sections 14.3.3 and 15.1.2.3, on the posterior spiracular disc the distinction between dorsolateral, lateral, and ventrolateral often is unclear and may appear different between instars and between larvae with three, four, or five pairs of lobes. Modified from Vala (1989a).

S. fasciata larvae (Fig. 14.29) are more robust and have a short but strong central peg. The structures within the maxillary organ are very different from other Sciomyzidae. The four pits with sensilla, commonly at the center of the maxillary organ in other species, are lacking in *S. fasciata*; the three bulbous sensilla (p) are unusual; the outline of the star-shaped sensillum (s) is weakly indented; and knob sensilla are more robust. Also, the intermediate coeloconical sensilla are very well developed in *S. fasciata*.

(b) Thoracic and abdominal regions (Fig. 14.26); *Sepedon trichrooscelis* (Fig. 14.28c, d); *Pherbina mediterranea* (Fig. 14.30a–g). Eleven sensilla, in some cases with an associated seta, are distributed on each side of the mid sagittal section of each segment as follows: 1, 2, 3, and 4 are ventral; 5, 6, and 7 are lateroventral; 8 and 9 are lateral or sometimes slightly laterodorsal; 10 and 11 are dorsal. Sensile setae (SS) are always long in aquatic larvae (Fig. 14.23b) and short in terrestrial and parasitoid larvae. Thoracic segments (II to IV) have a tuft of three trichoid setae arising together from a single pit in position 2. These formations were first described in Diptera larvae by Keilin (1915) and are designated as Keilin organs (Ko). The sensilla without setae are coeloconical. In terrestrial or parasitoid larvae, their margins are festooned, contrasting with a regular margin in aquatic larvae. The sensilla in positions 5 and 10 are always double. In *S. sphegea* and *Eulimnia philpotti* many sensilla are more or less masked by long and distinctive cuticular setae. The slender, leaf-like sensilla of the abdominal segments of *S. fasciata* are unique among the Sciomyzidae (Vala *et al.* 1999).

On the caudal segment, probably a compound segment (traditionally called segment XII), only ventral sensilla 1, 2, 3, and 4 are easily visible. On the posterior spiracular disc the number of sensilla varies according to the lobe: the ventral lobes have sensilla 1, 2, 3, and 4; the ventrolaterals have sensilla 5, 6, 7, and 8; the laterals have sensillum 9; the dorsolaterals have sensillum 10; and the dorsals have sensillum 11. When the number of posterior spiracular disc lobes is less than five pairs, the sensilla located on the missing lateral lobes can be seen on the composite ventrolateral lobes, as in *Ilione albiseta* and *Tetanocera* spp. i.e., there are two setae on these lobes.

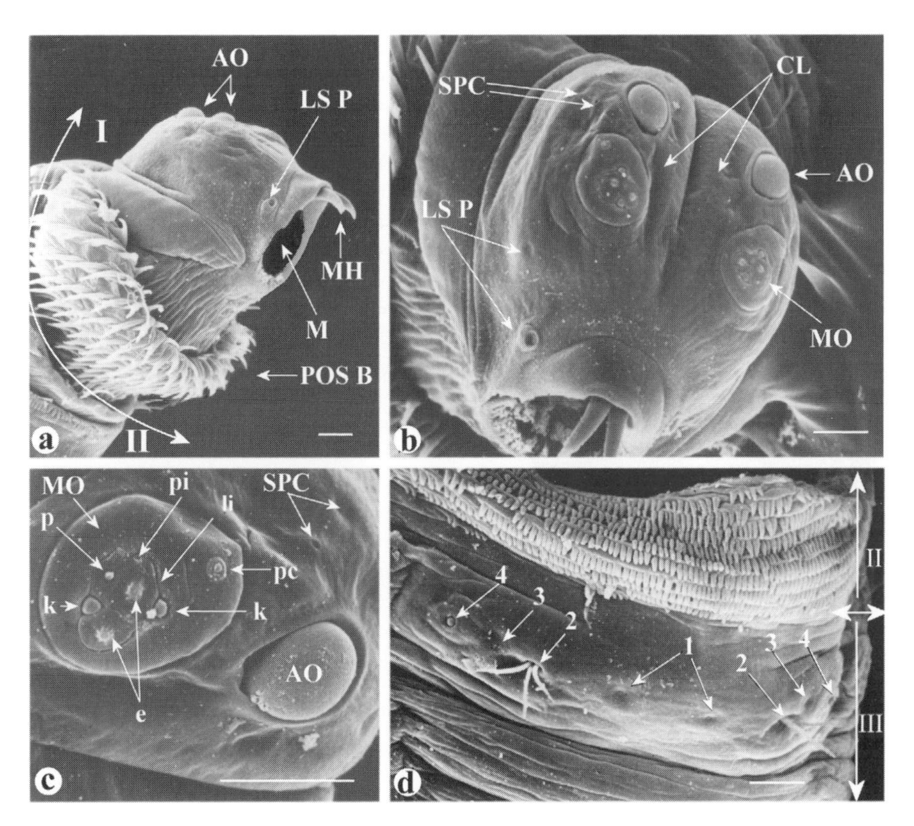

Fig. 14.27 ***Pherbina mediterranea***, L1. (a) Segments I–II, lateral view; (b) segment I, frontolateral view; (c) details of maxillary organ sensillum; (d) segments II and III, showing ventral and lateral sensilla on III. AO, antennal organ; CL, cephalic lobes; e, star-shaped sensilla; k, knob sensillum; li, longitudinal indentation; LS P, labial sensillum; M, mouth; MH, mouthhook; MO, maxillary organ; SPC, intermediate cephalic sensilla; p, bulbous sensillum; pc, coeloconical sensillum; pi, pit sensillum; POS B, postoral spinule band; 1–4, ventral sensilla; I, II, segments I, II. Scale lines equal 10 μm. Modified from Vala & Gasc (1990a).

14.1.3.3 POSTERIOR SPIRACULAR DISC AND OTHER CHARACTERS

A major feature long used to separate subfamilies and tribes within the Sciomyzinae, the interspiracular processes (or "float hairs") between the spiracular slits of the posterior spiracles, has been known to be characteristic of aquatic and semi-aquatic Tetanocerini (Fig. 14.31). They were thought to be absent in the terrestrial Salticellinae and terrestrial and semi-terrestrial Sciomyzini larvae, except first-instar larvae of *Pteromicra* (Rozkošný & Knutson 1970). In fact, they are present but reduced and scale-like in *Salticella fasciata* (Vala *et al.* 1999), three species of *Pherbellia* (Sciomyzini) examined, and the terrestrial Tetanocerini examined: *Coremacera, Tetanocera, Dichetophora*, and *Euthycera* (Vala & Gasc 1990b) (Fig. 14.32), but slightly elongate in *Trypetoptera punctulata*. They are also present but scale-like in *Pelidnoptera nigripennis* (Phaeomyiidae) (Vala *et al.* 1990).

In addition, some surprisingly minute papillae, with a regular distribution along longitudinal lines, are visible on the interspiracular processes of the aquatic predator *Ilione trifaria* (Fig. 14.31e).

An ecomorphological classification of larvae based on stereoscan microscope studies of the posterior spiracular disc morphology, primarily of first-instar larvae, with emphasis on the diverse forms of the interspiracular processes, size of peripheral lobes, behavior, and microhabitat was presented by Vala & Gasc (1990b) and is modified and summarized as follows.

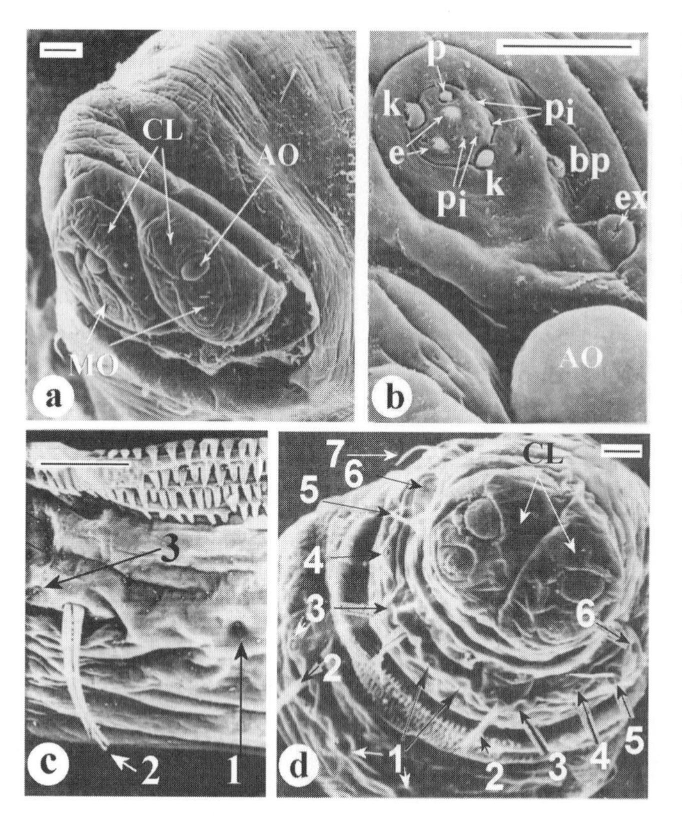

Fig. 14.28 *Sepedon trichrooscelis*, L1. (a) Segment I, fronto-dorsolateral view; (b) details of maxillary organ sensillum; (c) segment II with ventral sensilla 1–3; (d) frontal view showing distribution of sensilla 1–7 on segment II. AO, antennal organ; bp, basiconical sensillum; CL, cephalic lobes; e, star-shaped sensilla; ex, external star-shaped sensillum; k, knob sensilla; MO, maxillary organ; p, bulbous sensillum; pi, pit sensilla; 1–7, sensilla. Scale lines equal 10 μm. See Fig. 14.26 for sensilla numbers. Modified from Vala *et al.* (1995).

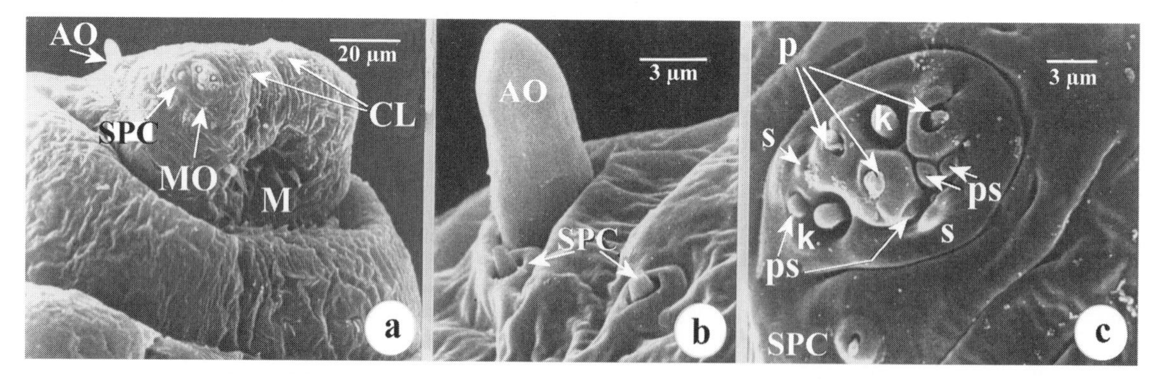

Fig. 14.29 *Salticella fasciata*, L1. (a) Segment I; (b) exposed antennal organ and intermediate cephalic sensilla; (c) maxillary organ. AO, antennal organ; CL cephalic lobes; k, knob sensilla; M, mouth; MO, maxillary organ; p, bulbous sensilla; ps, sensilla without walls; s, star-shaped sensilla; SPC, intermediate cephalic sensilla. From Vala *et al.* (1999).

(a) Strictly aquatic larvae, which live beneath the surface for most of their life, do not swim, and have elongate body tubercles as well as elongate posterior spiracular disc lobes and moderately elongate interspiracular processes. Examples: *Eulimnia philpotti* (Barnes 1980a), *Ilione lineata* (Knutson & Berg 1967), and some *Renocera* spp. Behavior: host-specific parasitoid/predators of fingernail clams.

(b) Strictly aquatic larvae, which live primarily at the surface, can swim well, can stay under water for up to several hours, have some long or conical posterior spiracular disc lobes, and moderately or very elongate

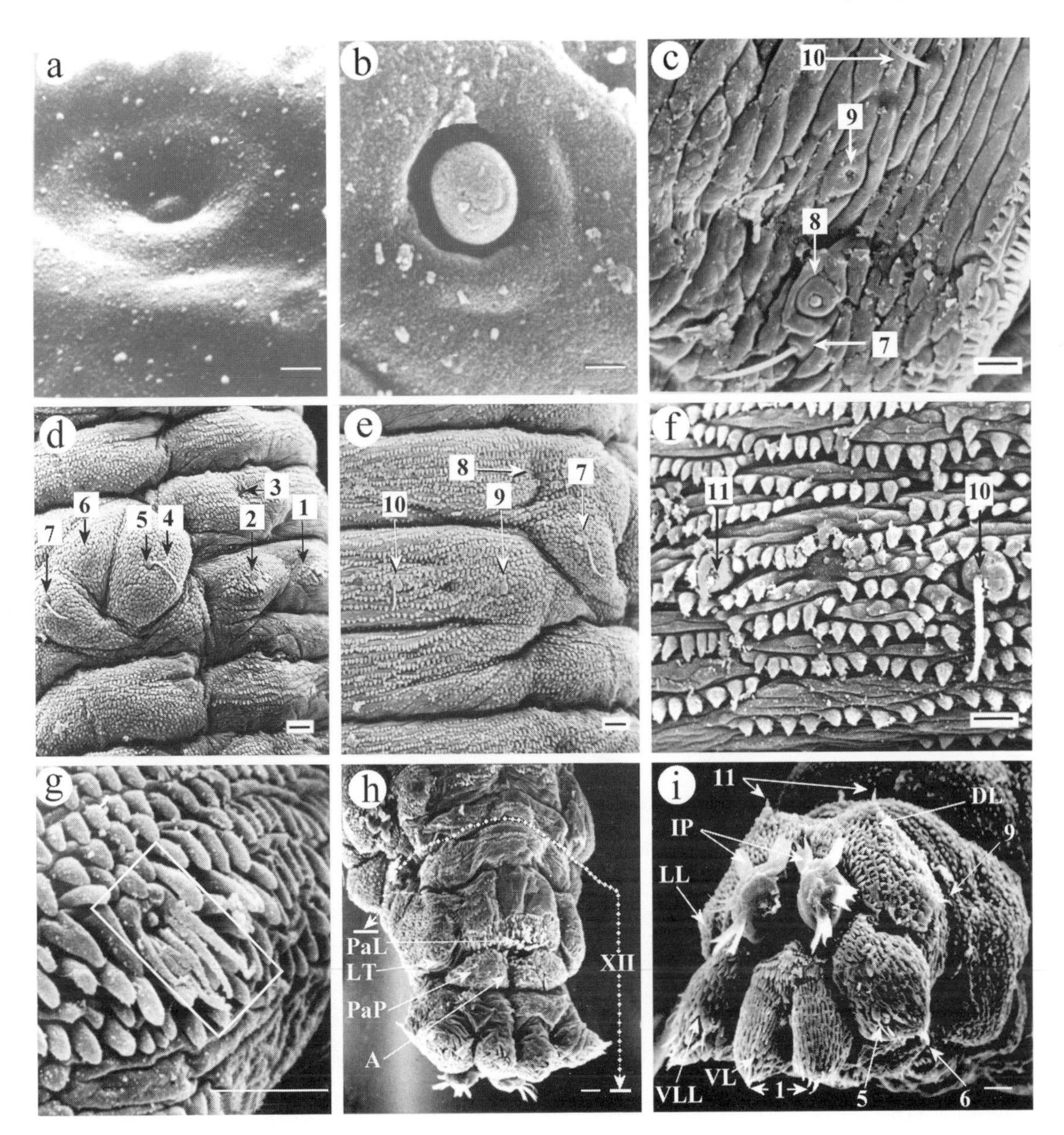

Fig. 14.30 ***Pherbina mediterranea***, L1. (a–c) Details of sensilla on segments II–IV. (a) Sensillum 1; (b) sensillum 4; (c) sensilla 7–10. (d–i) Abdominal sensilla; (d) segment VIII with ventral sensilla 1–3, ventrolateral sensilla 4–7; (e) same segment with sensilla 7–8, and dorsal sensilla 9–10; (f) same segment with dorsal sensilla 10–11; (g) sensillum 2 on ventral tubercle of abdominal segment; (h) segment XII, ventral view; (i) segment XII, posterior view. A, anus; DL, dorsal lobe; IP, interspiracular processes; LL, lateral lobe; LT, lateral tubercle; PaL, pre-anal lobe; PaP, peri-anal pad; VL, ventral lobe; VLL, ventrolateral lobe. XII, segment XII; 1–11, sensilla. Scale lines equal 10 mm except in a and b, where scale lines equal 1 mm. See Fig. 14.26 for sensilla numbers. From Vala & Gasc (1990a).

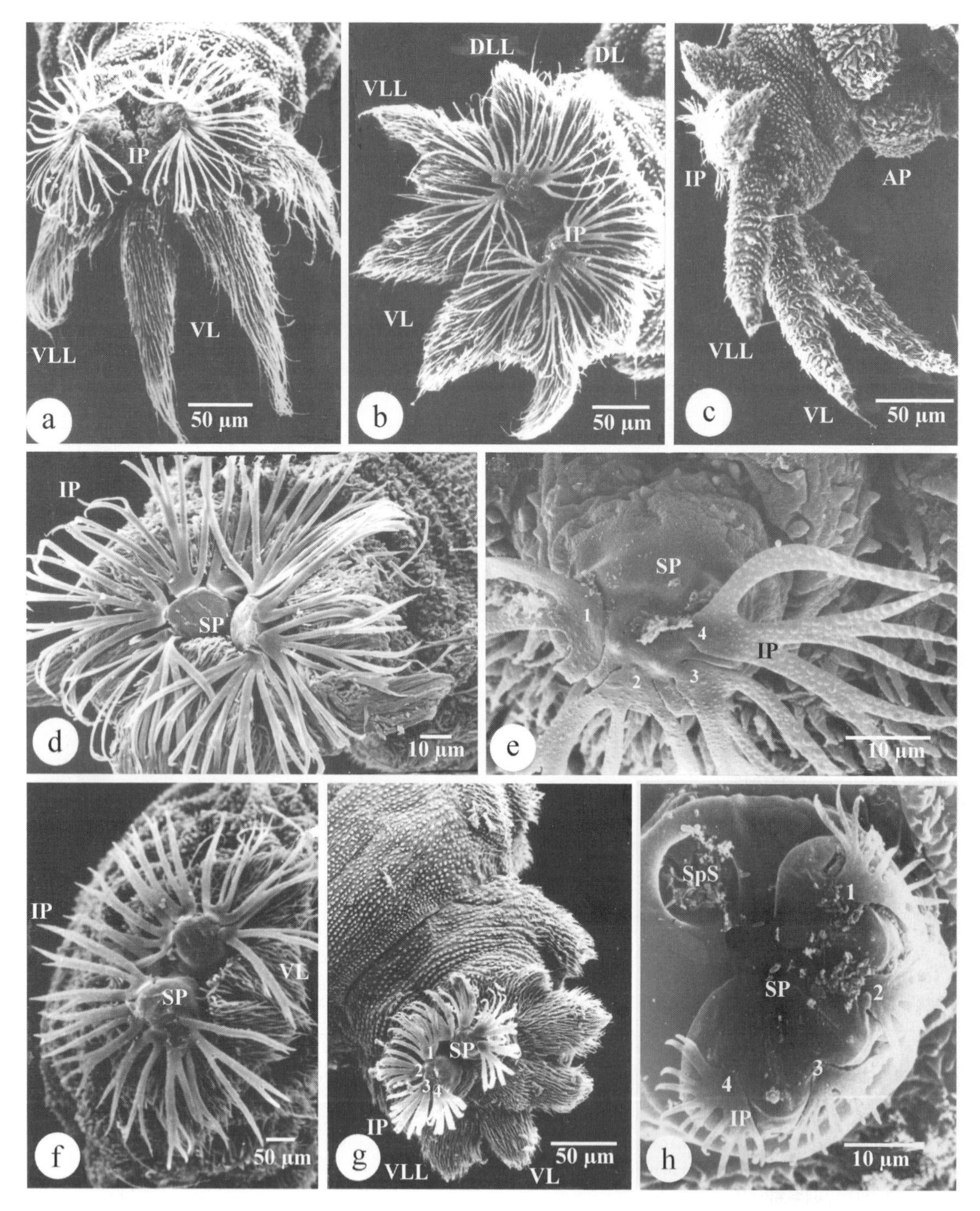

Fig. 14.31 Posterior spiracular discs and spiracular plates of L1 of true aquatic (a–e), semi-aquatic (f–g), and L3 terrestrial larvae (h). (a) *Sepedon sphegea*; (b) *Tetanocera ferruginea*; (c) *Ilione albiseta*; (d) *Sepedon sphegea*; (e) *Ilione trifaria*; (f) *Hydromya dorsalis*; (g) *Psacadina verbekei*; (h) *Trypetoptera punctulata*. AP, anal proleg; DL, dorsal lobe; DLL, dorsolateral lobe; IP, interspiracular processes; SP, spiracular plate; SpS, spiracular scar; VL, ventral lobe; VLL, ventrolateral lobe; 1–4, bases of interspiracular processes. From Vala & Gasc (1990b).

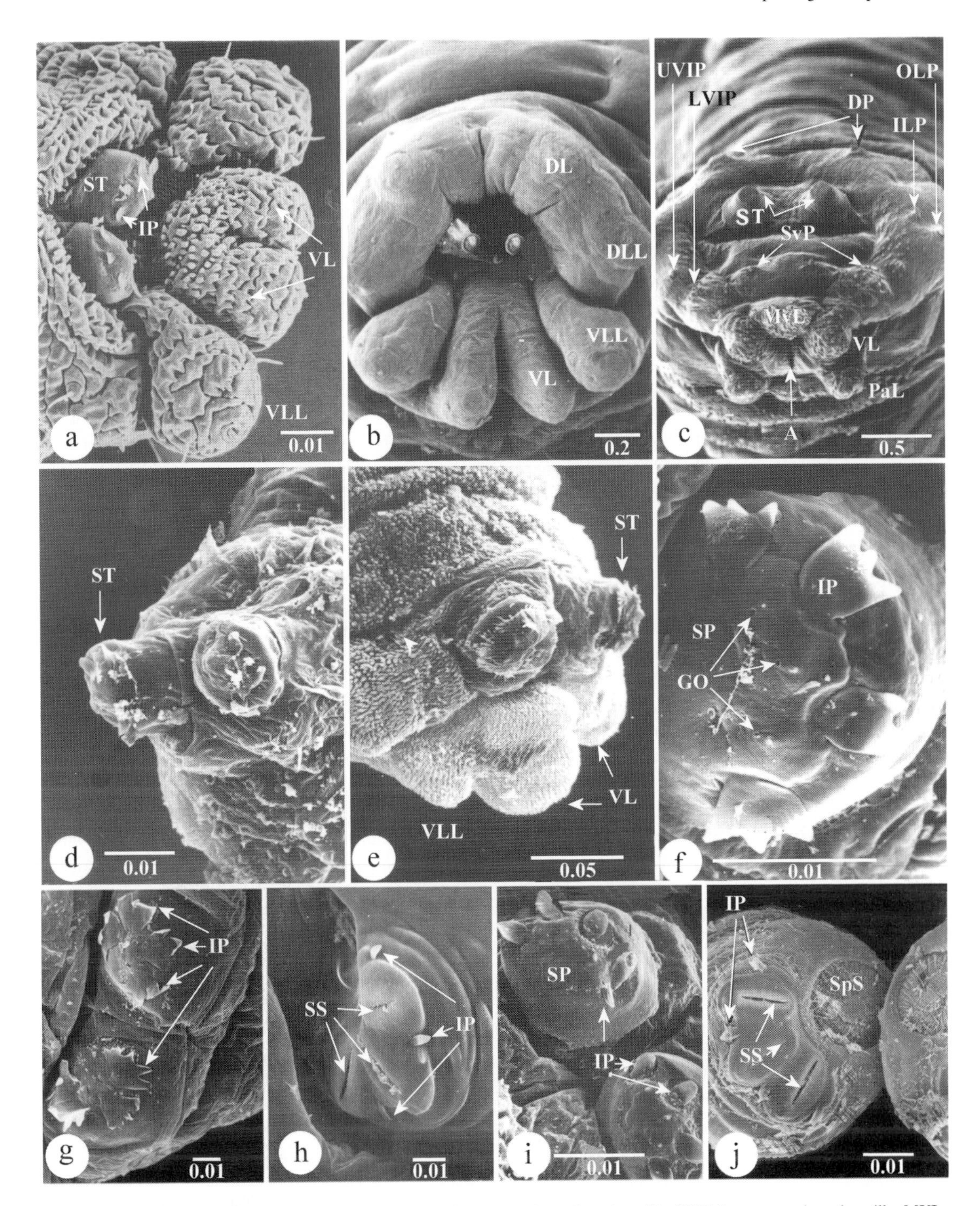

Fig. 14.32 Posterior spiracular discs and spiracular plates of terrestrial larvae. (a) ***Euthycera chaerophylli*** L1; (b) ***E. cribrata*** L3; (c) ***Salticella fasciata*** L1; (d) ***Pherbellia s. schoenherri*** L1; (e) ***Trypetoptera punctulata*** L3; (f) ***E. cribrata*** L3; (g) ***S. fasciata*** L1; (h) ***Dichetophora obliterata*** L1; (i) ***E. chaerophylli*** L1; (j) ***E. cribrata*** L3. DP, dorsal papillae; GO, glandular openings; ILP, inner lateral papilla; LVIP, lower ventrolateral papilla; MVL, midventral lobe; OLP, outer lateral papilla; PaL, pre-anal lobe; SS, spiracular slits; ST, spiracular tube; SvP, supraventral papilla; UVLP, upper ventrolateral papilla (for other abbreviations see Fig. 14.31). From Vala & Gasc (1990b).

interspiracular processes. Examples: *Tetanocera ferruginea*, *Sepedon sphegea*, *S. spinipes*, *Ilione albiseta*, and *I. trifaria* (Vala & Gasc 1990b). Behavior: efficient predators, killing up to 40 snails during larval life, little host specificity.

(c) Semi-aquatic larvae with weak swimming behavior, shorter posterior spiracular disc lobes, and short interspiracular processes. Examples: *Hydromya dorsalis* (Knutson & Berg 1963a), *Psacadina verbekei* (Vala & Gasc 1990b), and *Pherbina coryleti* (Knutson *et al.* 1975). Behavior: predators or parasitoid/predators, killing 10–20 snails during larval life; moderately host specific.

(d) Semi-terrestrial or terrestrial larvae that have short, rounded posterior spiracular disc lobes and moderately developed interspiracular processes. Examples: *Limnia unguicornis* (Vala & Knutson 1990), *Sepedon trichrooscelis* (Vala *et al.* 1995, Knutson 2008), *Sepedon h. hispanica* (Knutson *et al.* 1967a), *Tetanocera arrogans* (Vala 1989a, Vala & Knutson unpublished data), and *Trypetoptera punctulata* (Vala 1986). Behavior: parasitoid/predators of Succineidae or terrestrial snails; moderately host specific.

(e) Strictly terrestrial larvae that have short posterior spiracular disc lobes which are often rounded and interspiracular processes in the shape of minute unicuspid or polycuspid scales. Examples: *S. fasciata* (Vala *et al.* 1999), *Dichetophora obliterata*, *Pherbellia cinerella*, *P. dorsata*, *P. schoenherri*, *Euthycera cribrata*, *E. stichospila*, *E. chaerophylli*, *Coremacera marginata*, and *C. catenata* (Vala & Gasc 1990b). Behavior: parasitoids or parasitoid/predators of terrestrial snails, Succineidae or slugs; moderately to strongly host specific.

Vala & Gasc (1990b) also proposed two major, adaptive types of change in these structures. Assuming the minute, polycuspid interspiracular processes and shape of the posterior spiracular disc lobes in *S. fasciata* to be ancestral they suggested that the scale-like, unicuspid shape of interspiracular processes in first-instar larvae indicates that the species has a high level of host specificity. Also, they recognized a trend in first- to third-instar larvae, in some species, from polycuspid to unicuspid processes (e.g., *E. cribrata*, *E. stichospila*). The second adaptive pathway they recognized led to the aquatic forms with conspicuous interspiracular processes and at least some elongate posterior spiracular disc lobes. Most are not prey specific. The interspiracular processes are reduced when aquatic larvae are prey specific, e.g., *Ilione lineata*, which feed on fingernail clams.

Similar to the ecological approach of Vala & Gasc (1990b) that focused on the posterior spiracular disc, Knutson *et al.* (1975) presented a comparison of "habitat-adaptive" morphological characters of terrestrial Sciomyzini, aquatic Tetanocerini, and behaviorally intermediate *Pherbina* and *Psacadina* (both Tetanocerini). Of the habitat-adaptive morphological features compared in Table 14.2 the following functions are assumed. "Integument color" is related to whether the larva spends its life inside a snail or much time outside the shell and thus benefits from protective body color. However, aquatic Tetanocerini in some genera e.g., *Sepedon*, *Sepedonea*, and *Sepedomerus*, although pigmented, are lightly so. "Integumentary spinule patches" are special adaptations for movement in the viscous tissues of dead and dying molluscs. "Integumentary bristles" are associated with sensory structures needed for active movement outside the prey or host. "Body tubercles" may be an external manifestation of extensive development of muscles for active crawling. Elongate interspiracular processes surrounding the posterior spiracles are adaptations to maintain the spiracles at the surface film in disturbed situations. The development of the ventral spiracular disc lobes and the interspiracular processes are indicated by numerical indices. The "Ventral lobe index" is computed by dividing the length of the ventral lobe (from apex to basimedian juncture with spiracular disc) by the narrowest diameter of the spiracular disc measured transversely just below the paired spiracles, × 100. The "Float hair index" (interspiracular process index) is computed by dividing the greatest length of the longest interspiracular process by the greatest diameter of a posterior spiracle, × 100. The index of "0" given for terrestrial larvae (three species of Sciomyzini, in Knutson *et al.* 1975) is incorrect since it was subsequently shown that interspiracular processes are present in such species. Obviously, the index would be extremely small in terrestrial species, much less than 10 as recorded for *Psacadina* sp., except for *Trypetoptera punctulata*, which has unusually long interspiracular processes for a terrestrial species. In addition, despite the many adaptations for an aquatic existence among aquatic, predaceous sciomyzids, the postanal portion of the caudal segment of the larvae, bearing the posterior spiracles, is very elongate (and strongly upturned and projecting in the puparium) only in *Elgiva* spp. (Knutson & Berg 1964) and in *Renocera striata* (Foote 1976).

An important pair of features previously barely commented upon, i.e., the length of the postanal portion

Table 14.2. Comparison of habitat-adaptive morphological features of terrestrial, semi-aquatic, and aquatic L3

Character	Terrestrial			Semi-aquatic		Aquatic		
	Colobaea bifasciella	*Sciomyza aristalis*	*Pherbellia cinerella*	*Pherbina* spp.	*Psacadina* spp.	*Elgiva* spp.	*Ilione* spp.	*Hedria mixta*
Integument color	transparent	transparent	transparent	yellowish to light tan	yellowish to dark brown	translucent black & tan	yellow & tan	dark brown to black
Integumentary spinule patches	present	present	present	absent	absent	absent	absent	absent
Integumentary bristles	absent	absent	absent	present	present	present	present	present
Body tubercles	reduced	reduced	reduced	developed	developed	developed	developed	developed
Ventral lobe index[a]	20	24	33	34–39	54	35	41–138	150
Interspiracular process index[b]	------------ approaching 0 ----------			21–60	10–14	38	45	48

Notes: [a]Length (apex to basimedian juncture with spiracular disc) divided by narrowest diameter of disc measured just below spiracles × 100.
[b]Greatest length of longest process divided by greatest diameter of one spiracle × 100. Modified from Knutson *et al.* (1975).

of segment XII along with the degree to which it is upturned, are major characters in an analysis of the evolution of larval morphology and behavior of the genus *Tetanocera* (Chapman *et al.* 2006).

14.1.3.4 FUNCTIONAL, ONTOLOGICAL, AND EVOLUTIONARY ASPECTS OF THE MORPHOLOGY OF THE CEPHALOPHARYNGEAL SKELETON

The sclerites of the cephalopharyngeal skeleton of sciomyzid larvae, with associated features of the mouth and integumentary features of the first two apparent body segments, are highly and variously adapted to saprophagous, parasitoid, and predaceous feeding behavior. They show many striking plesiomorphic and apomorphic states. Thus, analyses of the more distinctive features of these structures from functional, ontological, and evolutionary viewpoints are provided here. A research approach that also could be instructive would be to consider the entire cephalopharyngeal apparatus, including not only cuticular structures, but also musculature and enervation. Diagnostic taxonomic comparisons among larvae of the genera of Sciomyzidae are given in Section 15.1.2.

In their overview of the morphology and terminology of Diptera larvae, Courtney *et al.* (2000) discussed the head capsule and feeding apparatus of Nematocera, Orthorrhapha, and Cyclorrhapha. For the last they detailed homologies and previous interpretations of the major sclerites of the cephalopharyngeal skeleton. Other important publications on the mouthparts of higher Diptera include those of Roberts (1971), Sinclair (1992), and the reviews in Teskey (1981) and Ferrar (1987). Equivalent terms used in describing the cephalopharyngeal skeleton (based in part on Courtney *et al.* 2000) are given in Table 14.3.

We note the comment of Meier (1996) in his masterful study of the larvae of Sepsidae that "The latter [cephalopharyngeal skeleton] even became the embodiment of papers on the morphology of 'maggots'." We agree with his admonition to study other character systems of larvae, but in our view the cephalopharyngeal skeleton remains a critical character system for study of Sciomyzidae. Most importantly, as we show below, many of the features of the cephalopharyngeal skeleton reflect the diverse feeding behaviors of the larvae and thus are of special significance in explaining the evolution of life cycles in the family. While it is important to study other features of the larvae, many of those require study of well-preserved specimens by relatively sophisticated techniques and equipment. From a practical viewpoint, however, the cephalopharyngeal skeleton can be studied easily from both the larval stage and from the exuvium in the often more commonly found puparium of species recalcitrant to rearing attempts, as *Pherbecta limenitis* (Knutson 1972) or *Poecilographa decora* (Barnes 1988). Also, it can be studied from the few, usually poorly preserved larvae of rare species, recovered from nature. Furthermore, the cephalopharyngeal skeleton of unfed first instars of species that have been obdurate to further rearing attempts provide material for study, this being a relatively unusual approach.

In Figs 14.33–14.42, all figures of the entire cephalopharyngeal skeleton and of the mouthhooks are lateral views unless indicated otherwise, and all figures of the ventral arch, ligulate sclerite, H-shaped sclerite, and epistomal sclerite are dorsal or ventral views unless indicated otherwise. Also, to save space in the legends, the abbreviations in these figures are given below. All scale lines are in millimeters, except as noted.

ab, abductor muscle apodeme of mouthhook
ad, adductor muscle apodeme of mouthhook
admc, anterior dilator muscles of pharynx
AO, antennal organ
apr, anteroventral process of H-shaped sclerite
AR, accessory rod of pharyngeal sclerite
as, anterior spiracle
AT, accessory teeth
AvP, anteroventral process of pharyngeal sclerite
cl, lumen of ventral channel between pharyngeal ridges
CL, cephalic lobe
crp, carapace over mouthhooks
cs, campaniform sensillum of mouthhook
d, denticle of mouthhook
DB, dorsal bridge between dorsal cornua of pharyngeal sclerite
DC, dorsal cornu of pharyngeal sclerite
de, dorsal extension of H-shaped sclerite
dmc, dorsomedian connection between mouthhooks
DS, dentary sclerite
DW, dorsal window in dorsal cornu
ES, epistomal sclerite
f, floor of pharynx
HS, H-shaped sclerite
Ko, Keilin organ
L, main, upper lumen of pharynx
lap, laterally arched process of mouthhook
ll, lateral lamella of pharyngeal ridge

Lp, longitudinal plate of ventral cibarial ridges
LS, ligulate sclerite
LSP, labial sensory papilla
m, mouth opening
MdL, mid dorsal lobe of ventral cornu of pharyngeal sclerite
MH, mouthhook
MO, maxillary organ
ms, marginal spinules of mouth opening
MSP, midventral spinule patch of segment II
P, arrow showing direction of passage of liquid between lateral lamellae during sieving
PB, parastomal bar
PMP, posteromedian process of mouthook
POS B, postoral spinule band of segment I
POS P, postoral spinule patch of segment II
PR, pharyngeal ridges
PS, pharyngeal sclerite
r, roof of pharynx
RT, rudimentary teeth of ventral arch
SDO, salivary duct opening
t, ventrally directed accessory tooth in *Sepedon knutsoni*
T, teeth on ventral arch
VA, ventral arch
VC, ventral cornu of pharyngeal sclerite
VR, ventral ramus of mouthhook
VW, ventral window of ventral cornu
I, II, III, apparent segments I, II, III

Table 14.3. Terms used in describing larvae and puparia and in keys, with alternative terms

Terms used here	Alternative terms, including terms in German by Rozkošný (R) and in French by Vala (V)
1. **cephalopharyngeal skeleton**	buccal armature, buccopharyngeal apparatus, cephalic skeleton, cephalopharyngeal apparatus or sclerites, head skeleton, mouthparts, pharyngeal skeleton. (R-CP-skelett, Cephalophayngealskelett), (V-squelette céphalopharyngien)
accessory teeth	(R-akzessorischer Zähne), (V-dents accessoires)
dorsal bridge	dorsal thecal arch, anterodorsal bridge. (R-Dorsalbrücke), (V-pont antérodorsal)
dorsal cornu	dorsal wing, dorsal process, dorsal arm, clypeofrontal phragma. (R-Dorsalhorn), (V-corne dorsale)
epistomal sclerite	epipharyngeal sclerite, epistomal plate. (R-Epistomalsklerit), (V-sclérite epistomal)
H-shaped sclerite	hypopharyngeal arch or sclerite, hypostomal sclerite, intermediate sclerite, labio-hypopharyngeal sclerite. (R-Hypostomalsklerit), (V-sclérite hypostomal)
indentation index	pharyngeal index. (R-Pharyngeal-index), (V-indexe pharyngien)
ligulate sclerite	dental sclerite, dentate sclerite, ectosomal sclerite, lingual sclerite. (R-Ligularsklerit), (V-sclérite lingual)
mouthhooks	mandibles, mandibular sclerites, mandibular hooks, maxilla, oral hooks, labial sclerites, lateral hooks. (R-Mundhaken, mandibularsklerit), (V-crochet mandibulaire, mandibule)
parastomal bars	atrial bars, parastomal rods, parastomal sclerites. (R-Parastomalsklerit), (V-barres parastomales)
pharyngeal sclerite	basal piece or plates or sclerite, lateral plates, lateral pharyngeal sclerite, paraclypeal phragma, posterior lateral plates, thecal sclerite, tentopharyngeal sclerite, tentoropharyngeal sclerite. (R-Pharyngealsklerit), (V-sclérite pharyngien)
ventral arch	buccal sclerites, dentary sclerites, ectosomal sclerite. (R-Ektostomalsklerit), (V-arche ventrale)
ventral cibarial ridges	pharyngeal filter, pharyngeal ridges, T or Y ribs
ventral cornu	ventral wing, ventral process, ventral arm, cibarial phragma. (R-Ventralhorn), (V-corne ventrale)
window	(R-Fenster), (V-fenêtre)
2. **cephalic region**	segment I, cephalic segment, head segment, pseuodocephalic segment, pseudocephalon. (R-Kopfsegment), (V-segment céphalique)
antennal organ	antenna, antennal sense organ, dorsal organ. retractile papilla. 2-segmented sensile papilla, (V-organe antennaire)
cephalic papillae	antennomaxillary lobes, cephalic lobes. (V-lobes céphaliques)
intermediate cephalic sensilla	(V-sensilles intermédiaires, isolated sensilla)
labial lobes	mandibular lobes, ventral lobes
labial sensilla	labral sensilla papillae. (V-sensilles labiales)

Table 14.3. (*cont.*)

Terms used here	Alternative terms, including terms in German by Rozkošný (R) and in French by Vala (V)
maxillary organ	circular maxillary organ, circular sensory plate, maxillary sense organ, ventral organ, (V–organe maxillaire, plaque sensorielle circulaire)
maxillary palp	terminal organ
midventral spinule patch	(R-Sternalfläche)
mouth	(R-Atriumöffnung), (V-bouche)
oral grooves	cirri, cuticular food canals, facial combs, facial mask, integumentary grooves, labral rami, genal rami, pseudotracheae, radiating channels
postoral spinule band or patch	(R-Postoralbandes), (V-bande épineuse postorale)
pre-oral cavity	atrium, extra-oral cavity
3. **thoracic segments 1, 2, 3**	segments II, III, IV. (R-Thorakalsegments)
anterior spiracle	(R-Prothorakalstigma, Vorderstigma), (V-stigmate antérieur)
anterior spiracular papillae	(R-Vorderstigmenknopsen, Stigmenpapillen)
Keilin organ	(V–organe de Keilin, groupe de trois sensilles ou soies, récepteur trichoide à 3 soies)
sensilla, 1–11: basiconical, bulbous, coeloconical trichoid sensilla	(V–papilles, récepteurs sensoriels; sensilles 1–11: sensille basiconique, coeloconique, trichoide)
4. **abdominal segments 1–6**	segments V–XI. (R-Abdominalsegmente)
intersegmental welt	(R-Intersegmentalfläche)
lateral tubercle	(R-Lateralwarzen)
sensilla, sensillae 1–11	(V–sensilles, récepteurs sensoriels 1–11)
5. **caudal segment**	segment XII, abdominal segments 8 or 8 + 9, last segment
dorsal lobe	(R-Dorsallappen), (V-lobe dorsal)
dorsolateral lobe	(R-Dorsolaterallappen), (V-lobe dorsolateral)
interspiracular glands	(R-Interspirakulardrüsen)
interspiracular processes	float hairs, hydrofuge setae, interspiracular hairs. (R-Interspirakularborsten), (V-soies hydrofuges)
lateral lobe	(R-Laterallappen), (V-lobe lateral)
peri-anal pad	anal plate. (R-Analplatte), (V-plaque anale, papille anale)
posterior spiracles	(R-Hinterstigmen), (V-spiracle posterieur)
posterior spiracular disc	posterior disc. (R-Stigmenfeld), (V-disque posterieur)
pre-anal welt or anal proleg	(R-Präanalpseudopodium, Anallapen), (V-lobe préanal)
spiracular plate	(R-Stigmenplatte)
spiracular scar	button, ecdysial scar, stigmatic scar. (R-Stigmennarbe), (V-cicatrice stigmatique)
spiracular slits	stigmatic apertures. (R-Stigmenöffnungen, Stigmenknopsen), (V-orifices stigmatiques)
spiracular tube	stigmatic tube. (R-Stigmenträger), (V-tube spiraculaire)
ventral lobe	(R-Ventrallappen), (V-lobe ventral)
ventrolateral lobe	(R-ventrolaterallapen), (V-lobe ventrolateral)

Note: Modified, in part, from Rozkošný (1967) and Courtney *et al.* (2000).

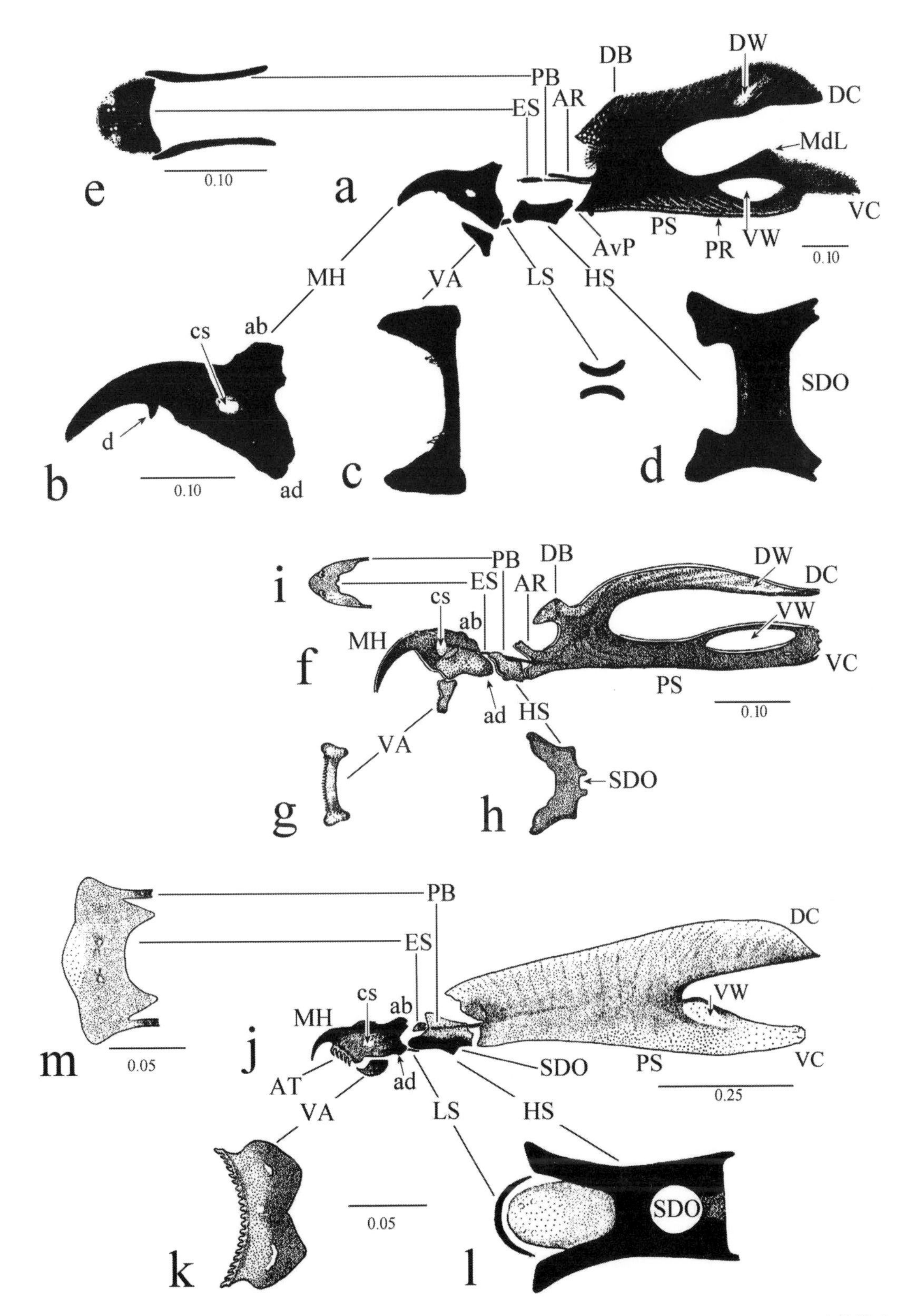

Fig. 14.33 (a–e) ***Salticella fasciata***, L3. (a) Cephalopharyngeal skeleton; (b) mouthhook; (c) ventral arch; (d) ligulate and H-shaped sclerites; (e) epistomal sclerite and parastomal bars (from Knutson *et al*. 1970). (f–i), ***Atrichomelina pubera***, L3; (f) cephalopharyngeal skeleton; g, ventral arch; h, H-shaped sclerite; (i) epistomal sclerite (from Foote *et al*. 1960). (j–m) ***Elgiva solicita***, L3; (j) cephalopharyngeal skeleton; (k) ventral arch; (l) ligulate and H-shaped sclerites; (m) epistomal sclerite (from Knutson & Berg 1964). See text for abbreviations.

14.1.3.4.1 Relationship to cephalic region

The elements of the third-instar cephalopharyngeal skeleton of a Salticellinae (*Salticella fasciata*, saprophagous during third stadium), a typical Sciomyzini (*Atrichomelina pubera*, saprophagous–parasitoid–predatory), and a typical Tetanocerini (*Elgiva solicita*, strictly predaceous) are shown in Fig. 14.33. The relationship of the cephalopharyngeal skeleton to the first two apparent body segments is shown in a ventral view of *Sepedon fuscipennis* (Fig. 14.34a), and in a dorsal view of *Limnia unguicornis* (Fig. 14.34c) (both Tetanocerini), and to the first three apparent segments in a lateral view of *Pherbellia idahoensis* (Sciomyzini) (Fig. 14.34b).

Note the weak development of the postoral spinule patch and presence of a midventral spinule patch on segment II (absent in some species) in the more subtle-feeding Sciomyzini and the massive development of the postoral spinule band, but lack of the small midventral patch, in the more rapacious aquatic and terrestrial Tetanocerini. Many terrestrial parasitoid/predatory Tetanocerini have a well-developed postoral spinule band (Figs 14.23, 14.34a, c, and 14.35a) that almost encircles segment I, except mid dorsally. The slug mesoparasitoid *Euthycera chaerophylli* (Tetanocerini) has completely encircling bands of spinules on the posterior margin of segment I and anterior margin of segment II and III. In the Sciomyzini, the succineid parasitoid *Pherbellia s. maculata* has a completely encircling band of spinules on the anterior margin of segment II in all three instars. Two species of the related, more predaceous genus *Ditaeniella* have the most extensive development of spinule bands, with completely encircling bands on most segments. The apposition of the hook part of the mouthhook and the postoral spinule band is well illustrated in the terrestrial predator, *Dichetophora obliterata* (Tetanocerini) (Fig. 14.35a). Bratt *et al.* (1969) considered the postoral spinule patch of *Pherbellia* spp. (Fig. 14.34b) serving "to anchor the anterior end of the body when the mouthhooks are thrust into the hosts' tissues." In regard to the midventral spinule patch they noted,

> Larvae that have the midventral spinule patch hook and pull downward after sinking their mouthhooks into the snail's flesh. This brings the tips of the mouthhooks into contact with the ventral body wall. The spinules are on large, flat plates. Their function may be to prevent injury and damage to the body wall and also to provide a surface for the mouthhooks to contact, holding the snail's tissues and enabling the ventral arch to cut off chunks of flesh. It seems significant that the spinule patch is missing in the more parasitoid species; their feeding activities are less vigorous and their mouthhooks apparently do not contact the ventral body wall. Observations indicate that they feed more on secretions, hemolymph, and liquefied tissues than the predatory species do.

The fine structure of the integument around the mouth opening has not been studied carefully in most Sciomyzidae. Sciomyzid larvae have been generally characterized as lacking the labral rami leading into the mouth opening, characteristic of saphrophagous Diptera larvae. However, micro-spinules are present in this area in some species. Bratt *et al.* (1969), in characterizing the Sciomyzini, commented, "Rows and patches of spinules usually line the floor of the atrium." They described *Pherbellia* as having a double row of spinules extending back into the oral opening in second-instar larvae of some species and in third-instar larvae of most species. Rozkošný & Knutson (1970) described third-instar larvae of *Pteromicra* as having "Oral opening with 1–2 marginal rows of spinules" (Fig. 14.35b) and such spinules have been described for *Sciomyza* spp. (Foote 1959a). Barnes (1990) described *Sciomyza varia* as having "oral ridges absent," but "Atrium surrounded by small pre-oral patch of minute, unicuspid spinules between mandibular apices and narrow para-oral lines of about 30 unicuspid, posteroventrally directed spinules stretching from mandibular apices to lateral edges of ventral arch."

There have been no studies of the musculature of the cephalopharyngeal skeleton (except *Salticella fasciata*, Fig. 14.41a, b) and head of larval Sciomyzidae. Muscles have not been included in most published figures. However, the location of the apodemes of the abductor and adductor muscles of the mouthhook and apodemes of the ventral arch are obvious. The cornua of the pharyngeal sclerite also are obvious expanses for muscle attachments. Comparisons with the descriptions of the musculature of the cephalopharyngeal skeleton and head of some calyptrate Diptera by Roberts (1971) will be useful in further studies of Sciomyzidae.

14.1.3.4.2 Accessory teeth of mouthhook

The presence of well-developed accessory teeth on the anterior margin of the mouthhook below the hook part in all second- and third-instar Tetanocerini (Fig. 14.33j) except *Renocera* spp. and *Dictya ptyarion*, which generally

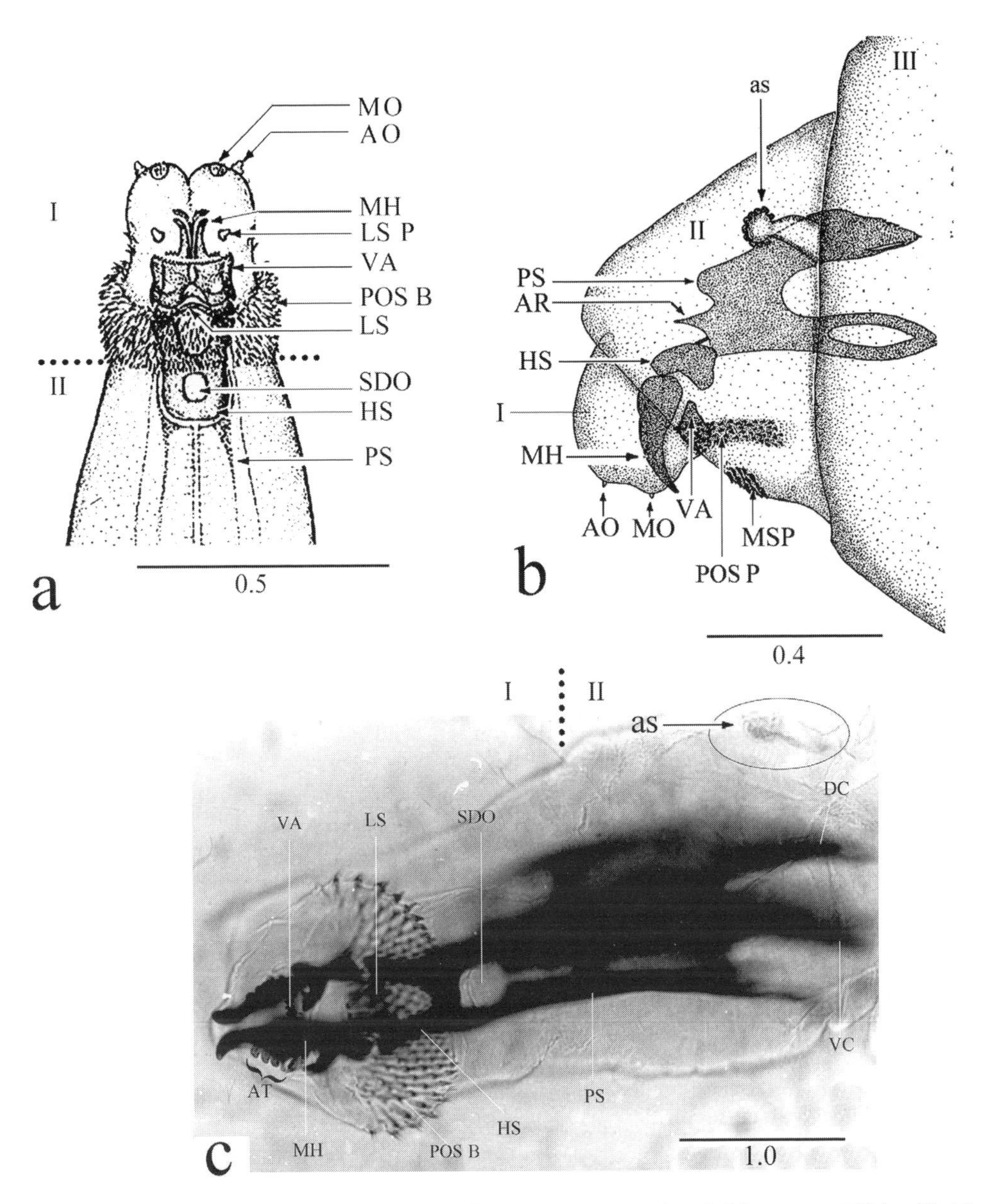

Fig. 14.34 (a) ***Sepedon fuscipennis***, L3, segments I, II, ventral (from Neff & Berg 1966). (b) ***Pherbellia idahoensis***, L3, segments I, II, III, lateral (from Bratt *et al.* 1969). (c) ***Limnia unguicornis***, L2, segments I, II (modified from Vala & Knutson 1990). See text for abbreviations.

are predaceous, is in contrast to the less predaceous, often parasitoid to saprophagous Sciomyzini, which lack accessory teeth (Fig. 14.33f). In the biologically unknown *D. ptyarion*, there are no typical accessory teeth but there are eight or nine minute denticles at the base of the hook part (Fig. 14.38f). The oligochaete-feeding *Sepedon knutsoni* has a normal complement of accessory teeth on the anterior margin of the basal part of the mouthhook, but there is also a strong, downward-directed denticle at the base of the hook part (Fig. 14.39e). *Salticella fasciata* also lacks typical accessory teeth but has a single large denticle at the base of the hook part (Fig. 14.33b).

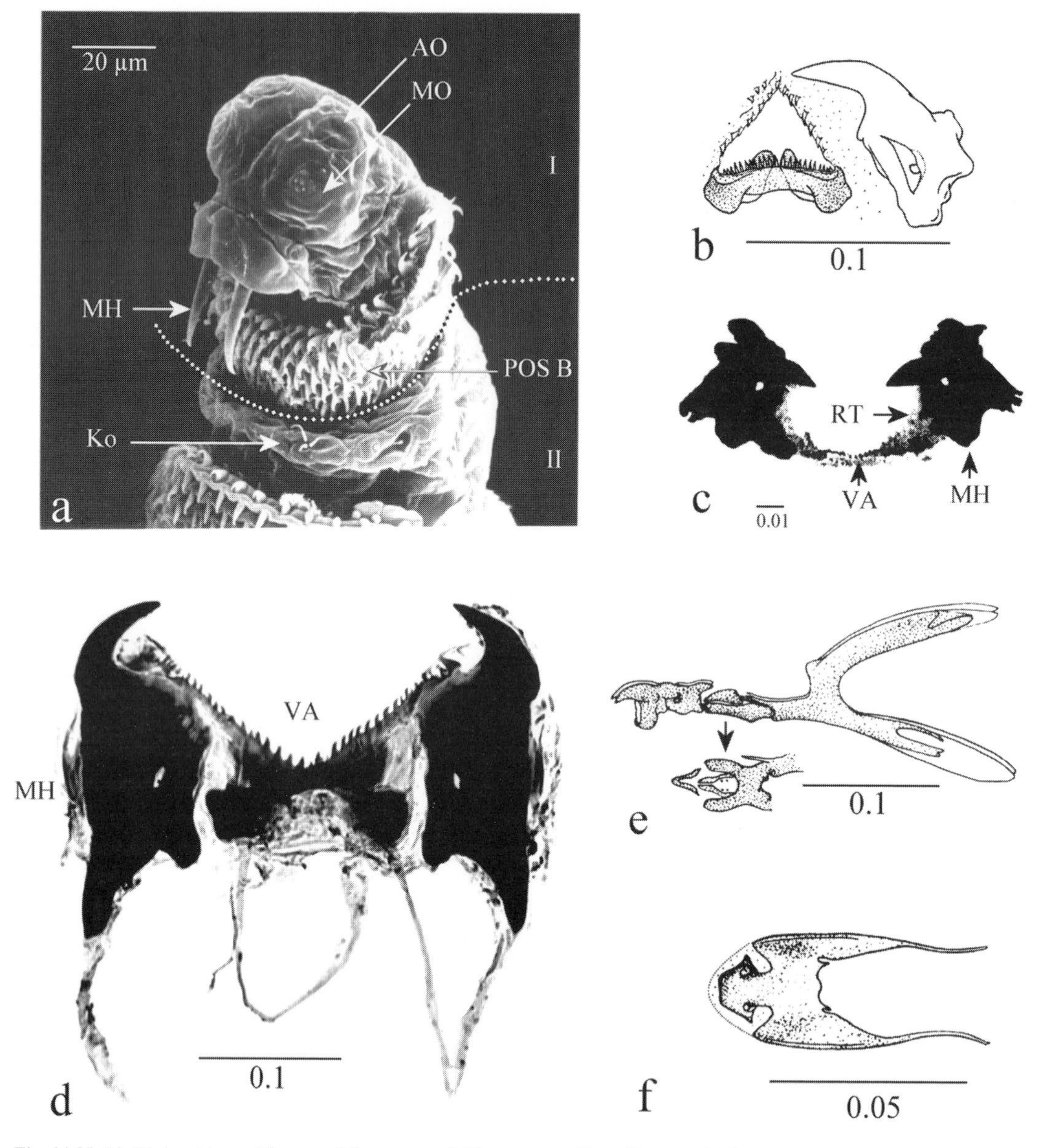

Fig. 14.35 (a) ***Dichetophora obliterata***, L3, segments I, II, anteroventrolateral (from Vala *et al.* 1987). (b) ***Pteromicra glabricula***, L3, ventral arch, left mouthhook, and mouth opening, showing marginal spinules (from Rozkošný & Knutson 1970). (c) ***Tetanura pallidiventris***, L3, mouthhooks and ventral arch (from Knutson 1970a). (d) ***Anticheta brevipennis***, L3, mouthhooks and ventral arch (from Knutson 1966). (e) ***P. glabricula***, L1, cephalopharyngeal skeleton; (f) ***P. pectorosa***, L3, epistomal slerite (from Rozkošný & Knutson 1970). See text for abbreviations.

The correlation between presence/absence of accessory teeth is not as close vis-à-vis differences in feeding behavior as it is vis-à-vis taxonomic (tribal) placement. Some Sciomyzini feeding on aquatic and semi-terrestrial snails exposed in damp situations, e.g., many *Pherbellia* spp., are very similar in feeding behavior to some Tetanocerini, e.g., *Psacadina* spp., attacking the same species of snails in the same microhabitats. Evolutionarily, accessory teeth may have moved gradually onto the mouthhook, in a series of ancestors in which the ventral arch, with its anterior

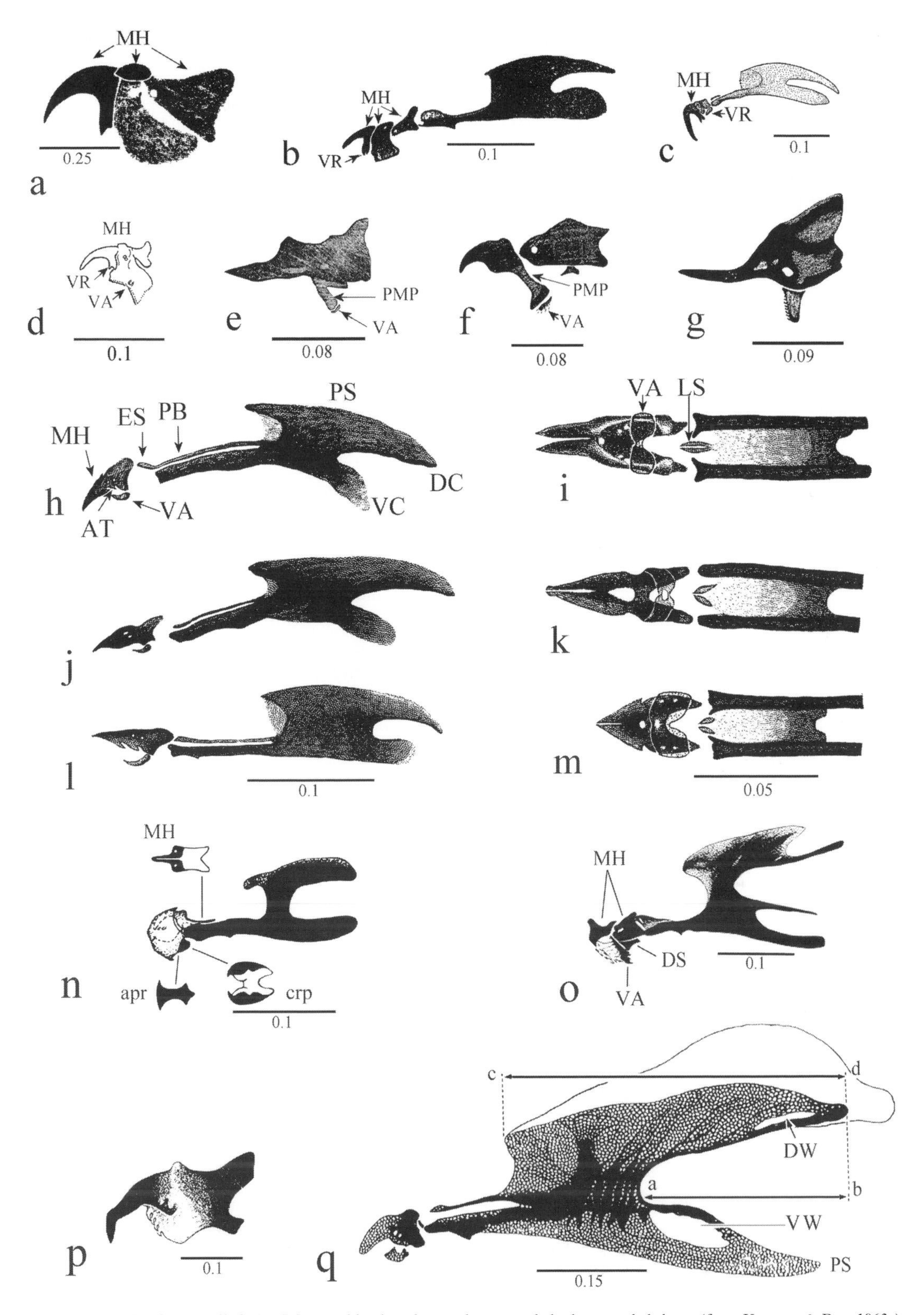

Fig. 14.36 (a) ***Perilimnia albifacies***, L1, mouthhook and ventral arch (from Kaczynski *et al.* 1969). (b) ***Hydromya dorsalis***, L1, cephalopharyngeal skeleton (from Knutson & Berg 1963a). (c) ***Coremacera marginata***, L1, cephalopharyngeal skeleton

margin of similar but smaller teeth, was joined with the anteroventral angle of the basal part of the mouthhook. This condition is present in the unusual terrestrial parasitoid *Tetanura pallidiventris* (Sciomyzini) (Fig. 14.35c) and the snail-egg/snail feeding *Anticheta brevipennis* (Tetanocerini) (Fig. 14.35d), although these species certainly are not closely related to a plesiomorphic ancestor. However, in the Salticellinae (*S. fasciata*), the plesiomorphic sister group of Sciomyzinae, the ventral arch and mouthhooks are separate in the second (Fig. 14.40i) and third instar (Fig. 14.33a) and there are only a few teeth in the anterolateral corners of the ventral arch (Fig. 14.33c).

Ontogenetically, the ventral arch is fused with the anterior or median portion of the mouthhook in most first-instar larvae. It is minute and closely articulated with or fused with a laterally arched process of the mouthhook in many second-instar larvae of Sciomyzini, e.g., *Pteromicra* (Fig. 14.35e) and *Pherbellia* (Fig. 14.36e, f), and Tetanocerini, e.g., *Perilimnia, Hydromya, Coremacera*, and *Pherbina* (Fig. 14.36a–d). The sclerites appear to be separate in first-instar slug-killing *Tetanocera* spp. (Fig. 14.36h, j, l). In most third-instar aquatic and semi-aquatic Tetanocerini predators there are three to seven (usually four or five) strong, hook-like accessory teeth along the entire anterior margin of the basal part of the mouthhook, beginning just below the hook part (Figs 14.37k, 14.38l, m, 14.40a, b).

In the clam-killer *Ilione lineata* (Fig. 14.38n) and the oligochaete-feeding *Sepedonella nana* the large accessory teeth are strongly curved, more or less lightly sclerotized, and closely approximate on a lightly sclerotized projection of the anterior margin of the basal part (Fig. 14.39a). In most Tetanocerini parasitoid/predators of terrestrial and semi-terrestrial snails and slugs (Fig. 14.38g–j), except *Trypetoptera punctulata* (Fig. 14.38l), there are only one to three (usually two) weaker teeth situated well below the hook part, on the ventro-apical angle of the basal part.

As in all instars of Salticellinae and Sciomyzini, accessory teeth below the hook part of the mouthhook are absent in first-instar Tetanocerini (except possibly present in some slug-feeding *Tetanocera* spp., Fig. 14.36h, j, l) but they appear progressively stronger in second- and third-instar Tetanocerini. This obviously is an adaptation to the increasingly predatory nature of larvae of Tetanocerini as they mature. In first-instar Tetanocerini the hook part of the mouthhook and the ventral ramus are more strongly sclerotized than the rest of the cephalopharyngeal skeleton. The ventral ramus is usually acute or bidentate apically in first-instar Tetanocerini and may function in feeding (Fig. 14.36a, b). In third-instar Tetanocerini, the number of accessory teeth varies in aquatic predators from none to three in *Dictya*, to five, e.g., *Elgiva solicita* (Fig. 14.37k) to seven in *Thecomyia limbata*. In several terrestrial, parasitoid/predator Tetanocerini there are only one or two weak accessory teeth, e.g., *Coremacera marginata* (Fig. 14.38g), *Dichetophora obliterata* (Fig. 14.38h), *Euthycera* spp. (Fig. 14.38i, j), and *Hoplodictya spinicornis* (Fig. 14.38k). However, some terrestrial, parasitoid/predators, e.g., *Trypetoptera punctulata* (Fig. 14.38l) have more teeth (five) than some voracious, aquatic predators, e.g., *Ilione unipunctata* (three) (Fig. 14.38m), but the shape and size of the hook parts of the mouthhooks and accessory teeth in these two species are quite different. Among clam-killing Tetanocerini (most of which are subsurface predators), accessory teeth are lacking in *Renocera* (Fig. 14.37b) and are small, lightly pigmented, but present in *Eulimnia philpotti* (Fig. 14.37a) and *Ilione lineata* (Fig. 14.37c).

14.1.3.4.3 Mouthhooks

Griffiths (1994) considered the mouthhooks to be of maxillary origin. However, Courtney *et al.* (2000) concluded that embryology supports a maxillary origin while traditional morphology "supports that the mouthhooks are at least partially of mandibular origin."

The hook part of the mouthhook is variously decurved in most Sciomyzidae, strikingly so in, e.g., the first instar of the terrestrial snail parasitoid/predator *Coremacera marginata* (Tetanocerini) (Fig. 14.36c). However, exceptionally it is awl-shaped in the first and/or second instar in some

Caption for Fig. 14.36 (*cont.*)
(from Knutson 1973). (d) ***Pherbina coryleti***, L1, mouthhook, lateral view, and ventral arch, laterally displaced to show entire sclerite in ventral view (from Knutson *et al.* 1975). (e–g) ***Pherbellia*** spp., L2, mouthhooks and ventral arch; (e) ***P. seticoxa***; (f) ***P. scutellaris***; (g) ***P. obtusa*** (from Bratt *et al.* 1969). (h–m) ***Tetanocera*** spp., L1, cephalopharyngeal skeleton, lateral (h, j, l) and ventral (i, k, m). (h, i) ***T. plebeja***; (j, k) ***T. valida***; (l, m) ***T. clara*** (from Trelka & Foote 1970). (n, o) ***Sciomyza varia***, cephalopharyngeal skeleton. (n) L1, (o) L2 (from Barnes 1990). (p) ***Sepedon sphegea***, L3, mouthhook (from Neff & Berg 1966). (q) ***Renocera pallida***, L3, cephalopharyngeal skeleton (from Horsáková 2003). See text for abbreviations.

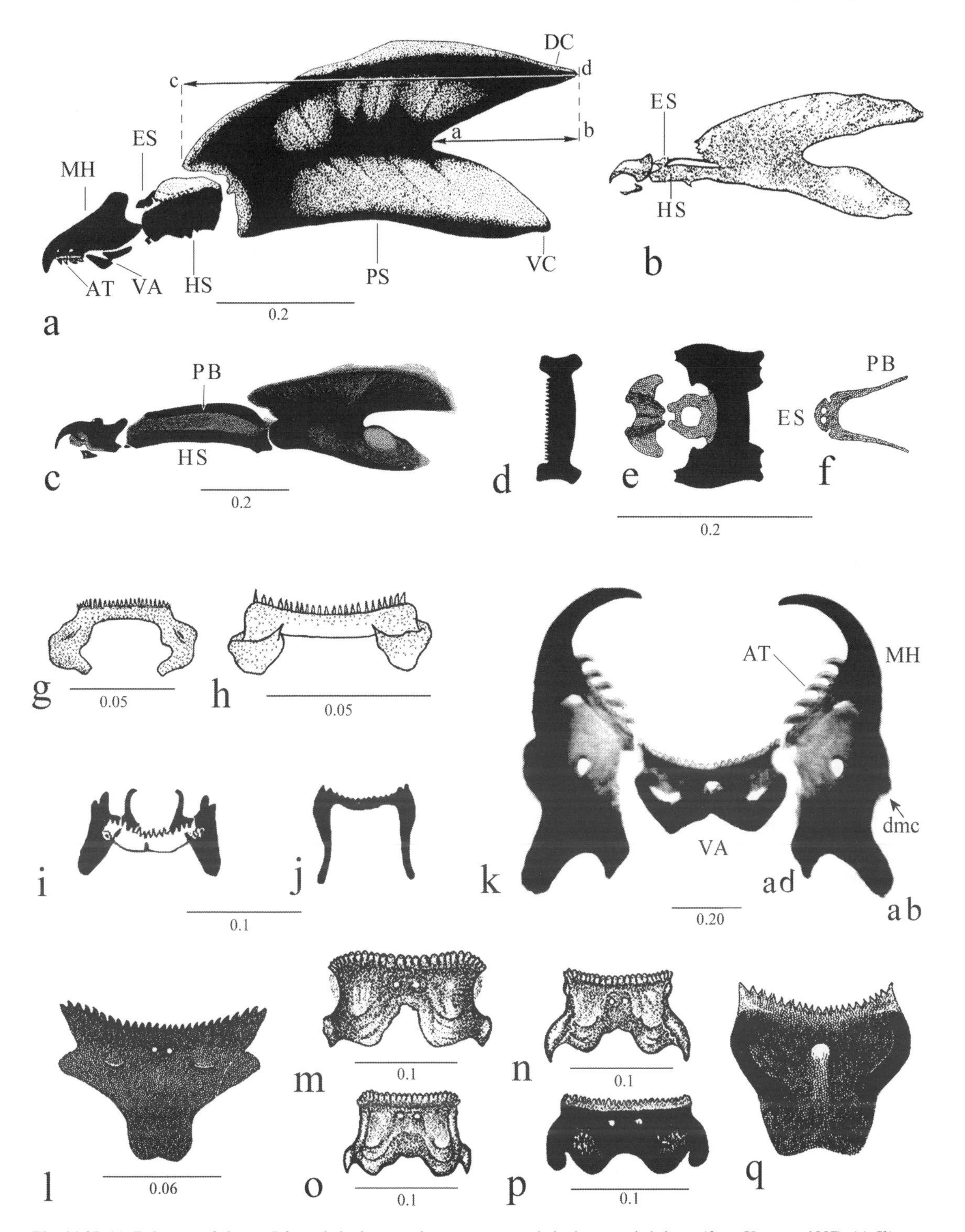

Fig. 14.37 (a) ***Eulimnia philpotti***, L3, cephalopharyngeal skeleton (from Barnes 1980a). (b) ***Renocera*** sp. L3, cephalopharyngeal skeleton (from Knutson 1987). (c) ***Ilione lineata***, L3, cephalopharyngeal skeleton (from Knutson & Berg 1967).

Sciomyzini, e.g., *Pherbellia seticoxa* and *P. obtusa*, parasitoid/predators of snails on moist surfaces (Fig. 14.36e, g) and some Tetanocerini, e.g., first instars of parasitoid/predator slug-killing *Tetanocera* species (Fig. 14.36h, j, l). In the second and third instar of these *Tetanocera* species and third-instar *P. seticoxa* the hook part is decurved. While most of the cephalopharyngeal skeleton of first-instar larvae is lightly sclerotized in Sciomyzini and many Tetanocerini, in the more predaceous Tetanocerini the hook part and the anterior margin of the anteroventral portion is strongly sclerotized. The sclerotization of the latter and the generally bidentate apex of the anterior margin (the ventral ramus, Fig. 14.36a, d) probably are important to the more predaceous feeding of such species. The degree of curvature of the hook part also probably is correlated with the degree of rapacious, predatory feeding. The hook part is curiously fore-shortened and broad with a massive adductor process on the basal part in the aberrant, terrestrial parasitoid *Tetanura pallidiventris* (Fig. 14.35c). The combination of a reduced ventral arch, small hook part, and expanded basal part of the mouthhook of *T. pallidiventris* is similar to some second-instar larvae of *Pherbellia*, e.g., *P. argyra*. The apex of the hook part is angulate, perhaps providing more mechanical strength, in six aquatic, predaceous Tetanocerini, e.g., *Sepedon sphegea* (Fig. 14.36p). The mouthhooks of first- and second-instar larvae of the highly parasitoid Sciomyzini *Sciomyza varia* are especially aberrant (Fig. 14.36n, o), those of the first instar being unpaired, dorsally split, and hollow.

The basal part of the mouthhook is massive and heavily sclerotized in the third instar, generally with two (most Tetanocerini), occasionally with one window (*Salticella fasciata* [Fig. 14.33b] and most Sciomyzini, e.g., *Atrichomelina pubera* [Fig. 14.33f]). These windows, which probably are campaniform sensilla, are variously situated just behind the accessory teeth or closer to the ventral margin or middle of the sclerite. The mouthhooks are joined dorsomedially, which ensures their functioning in unison. The dorsomedian connection is much stronger in the Tetanocerini, e.g., *Elgiva solicita* (Fig. 14.37k) than in the Sciomyzini. Such structural rigidity in the anterior-most portion of the feeding apparatus probably is important in voracious predators. At the same time, the flexibility and strength needed to pursue a snail retracting into its coiled shell may be provided by the hypostomal-pharyngeal sclerite articulation, noted below, and by the massive area of the pharyngeal sclerite cornu for extensive muscle attachment. However, such functional explanation does not entirely account for the structure of the subtle, subsurface feeding of fingernail clam-killers like *Ilione lineata* and *Eulimnia philpotti*, both of which have a cephalopharyngeal apparatus similar to surface-dwelling aquatic predators such as *Tetanocera ferruginea*. Dorsally joined mouthhooks are known elsewhere in the Cyclorrhapha only in a few Ephydridae and Drosophilidae, and they are weakly joined in the Tanypezidae.

Ontogenetically, the mouthhooks appear to be more or less distinctly separated into two or three parts in most first-instar larvae of Sciomyzidae (Fig. 14.36a, b, d) but occasionally are undivided (Fig. 14.36h, j, l). In the first instar of the plesiomorphic *S. fasciata*, which may be parasitoid in feeding behavior, the entire cephalopharyngeal skeleton, including the mouthhooks, is fused into one piece (Fig. 14.40h) but there is the usual separation of sclerites in the second instar (Fig. 14.40i). Most first-instar larvae have one window in the median portion and one in the posterior portion (Fig. 14.36a, b). In second-instar larvae mouthhooks may be undivided in some *Pherbellia* spp. (Fig. 14.36e, g), but in second-instar larvae of many *Pherbellia* spp. each consists of two distinct parts (Fig. 14.36f).

The mouthhooks, especially, and other parts of the cephalopharyngeal skeleton of second-instar larvae were found to be remarkably diverse, i.e., more diverse than in the third instar, among the 28 species of *Pherbellia* and *Ditaeniella* studied by Bratt *et al.* (1969). In five species the mouthhooks are not curved but are fused into a single, anteriorly directed blade. These authors noted, "In

Caption for Fig. 14.37 (*cont.*)
(d–f) ***Pherbellia seticoxa***, L3; (d) ventral arch; (e) ligulate and H-shaped sclerites; (f) epistomal sclerite and parastomal bars (from Bratt *et al.* 1969). (g, h) ***Pteromicra*** spp., L3, ventral arch; (g) ***P. angustipennis***; (h) ***P. pectorosa*** (from Rozkošný & Knutson 1970). (i, j) ***Sciomyza*** spp., L3, ventral arch; (i) ***S. varia*** (from Barnes 1990); (j) ***S. aristalis*** (from Foote 1959a). (k) ***Elgiva solicita***, L3, mouthhooks and ventral arch (from Knutson & Berg 1964). l–q, ventral arch L3; (l) **Dictya expansa** (from Valley & Berg 1977); (m) ***Sepedomerus caeruleus***; (n) ***Sepedon tenuicornis***; (o) ***Sepedonea guatemalana*** (from Neff & Berg 1966); (p) ***Sepedon trichrooscelis*** (from Vala *et al.* 1995); (q) ***Pherbecta limenitis*** (from Knutson 1972). See text for abbreviations.

general, the more insidious feeders have mouthhooks which are less curved and more anteriorly directed than those of predaceous species." They also noted that paired, laterally arched processes extend posteroventrally from the bases of the mouthhooks in all except *Pherbellia annulipes* and *P. beatricis*, predators/parasitoids of terrestrial and exposed aquatic snails, respectively. However, their figure of *P. obtusa* (Fig. 14.36g) shows this process lacking, but a well-developed ventral arch present. Whereas Bratt *et al.* (1969) figured and described a separate, weak, ventral arch below the posteroventral process of the mouthhook for most species, it might be that this process actually is part of the ventral arch, fusion of the lateral portions and separation of the toothed portion resulting in the configuration in most *Pherbellia* species. The carapace is an apparently unique structure described only for the highly specialized parasitoid *Sciomyza varia* (Barnes 1990). It is a moderately sclerotized cover over the unpaired mouthhook and the anteroventral process of the H-shaped sclerite.

14.1.3.4.4 Ventral arch

The ventral arch generally has been considered to have resulted from fusion of the pair of dental (dentary) sclerites present below the mouthhooks in many cyclorrhaphous Diptera (Ferrar 1987). Teskey (1981) noted that in the Sciomyzidae it is possible that the ventral arch does not represent fused dental sclerites but that it is a subhypostomal or similar sclerite that has migrated forward. Ferrar (1987) noted that fusion of dental sclerites usually is associated with a carnivorous habit, but not always so as it occurs in some phytophagous Tanypezidae, Opomyzidae, Psilidae, Chloropidae (*Meromyza*), and Scathophagidae (*Cordilura*). However, the latter families have well-developed ventral cibarial ridges which function as a filter mechanism to concentrate bacteria and yeasts developing in decaying plant material, probably the main food of such "phytophagous" (in fact, saprophagous) larvae. However, a ventral arch with 16–18 sharp teeth on the anterior margin and connected to the paired mouthhooks was described and figured for the succineid parasitoid *Angioneura cyrtoneurina* (Calliphoridae: Rhinophorinae) by Čepelák & Rozkošný (1968). They did not state whether oral grooves and cibarial ridges were present or not. In some Muscidae having a ventral arch there are no teeth on its anterior margin, no accessory teeth on the mouthhooks, and there are accessory oral sclerites beneath the mouthhook. Rudimentary ventral cibarial ridges and oral grooves in the pre-oral cavity are known in Sciomyzidae only in *S. fasciata* (Fig. 14.41a–c).

The ventral arch of sciomyzid larvae is exceedingly diverse in form, which presumably is correlated with diversity in feeding movements. However, considering its uncommon presence among the Cyclorrhapha, its various forms in the Sciomyzidae also likely indicate phylogenetic relationships. The ventral arch in Sciomyzini is strap-like medially and expanded posteriorly on each side where it articulates with the mouthhooks. The lateral apices are weakly expanded in *Atrichomelina* (Fig. 14.33g) and *Pherbellia* (Fig. 14.37d) but more strongly developed in *Pteromicra* (Fig. 14.37g, h) and *Sciomyza* (Fig. 14.37i, j). Bratt *et al.* (1969) noted that of the 25 species of *Pherbellia* and three species of *Ditaeniella* they studied, the ventral arch was absent in *P. schoenherri maculata* and reduced to a lightly pigmented strap in all other second-instar larvae except *P. annulipes* and *P. beatricis*. However, their figures show it also well developed in second-instar *P. obtusa*. In *Sciomyza varia* (Fig. 14.37i) there is an unusual pair of elongate hooks before the anterolateral margins of the ventral arch. In Tetanocerini, the ventral arch usually appears as a bilobed plate (e.g., *Elgiva solicita*, Fig. 14.37k), but in *Dictya* (Fig. 14.37l) and *Pherbecta* (Fig. 14.37q) it is strongly projected posteromesally. In *Hoplodictya spinicornis* (Fig. 14.38e) it is extremely elongated posteromesally and there is a long, narrow strut on each side. In *Sepedomerus*, *Sepedonea*, and *Sepedon*, both in aquatic predaceous species (e.g., *Sepedon tenuicornis*) and terrestrial, parasitoid species (e.g., *Sepedon trichrooscelis*), there is a strong appendage on each side of the bilobed plate (Fig. 14.37m–p). Lobes, appendages, and struts of the ventral arch probably are apodemes for muscle attachments. The ventral arch of third-instar (snail feeding) *Anticheta testacea* (Fig. 14.38d) is more similar to the Sciomyzini than to the Tetanocerini, but in the first instar (snail-egg feeding) it is very weak and the hook part and ventral ramus of the mouthhook are heavily sclerotized and apically acute (Fig. 14.38a, b). In most Tetanocerini there is one or a pair of small windows at the mesal juncture of the lobes and a larger, often less distinct, window in each lobe; these windows are absent in the Salticellinae and Sciomyzini and virtually absent in *Anticheta* and *Renocera*. Also in the subtle-feeding *Anticheta* and *Renocera*, the ventral arch is lightly sclerotized except for the posterolateral lobes, where the major muscles probably are attached, and for the medial connection in *A. brevipennis*. In *Euthycera stichospila* (Fig. 14.38o–q), the

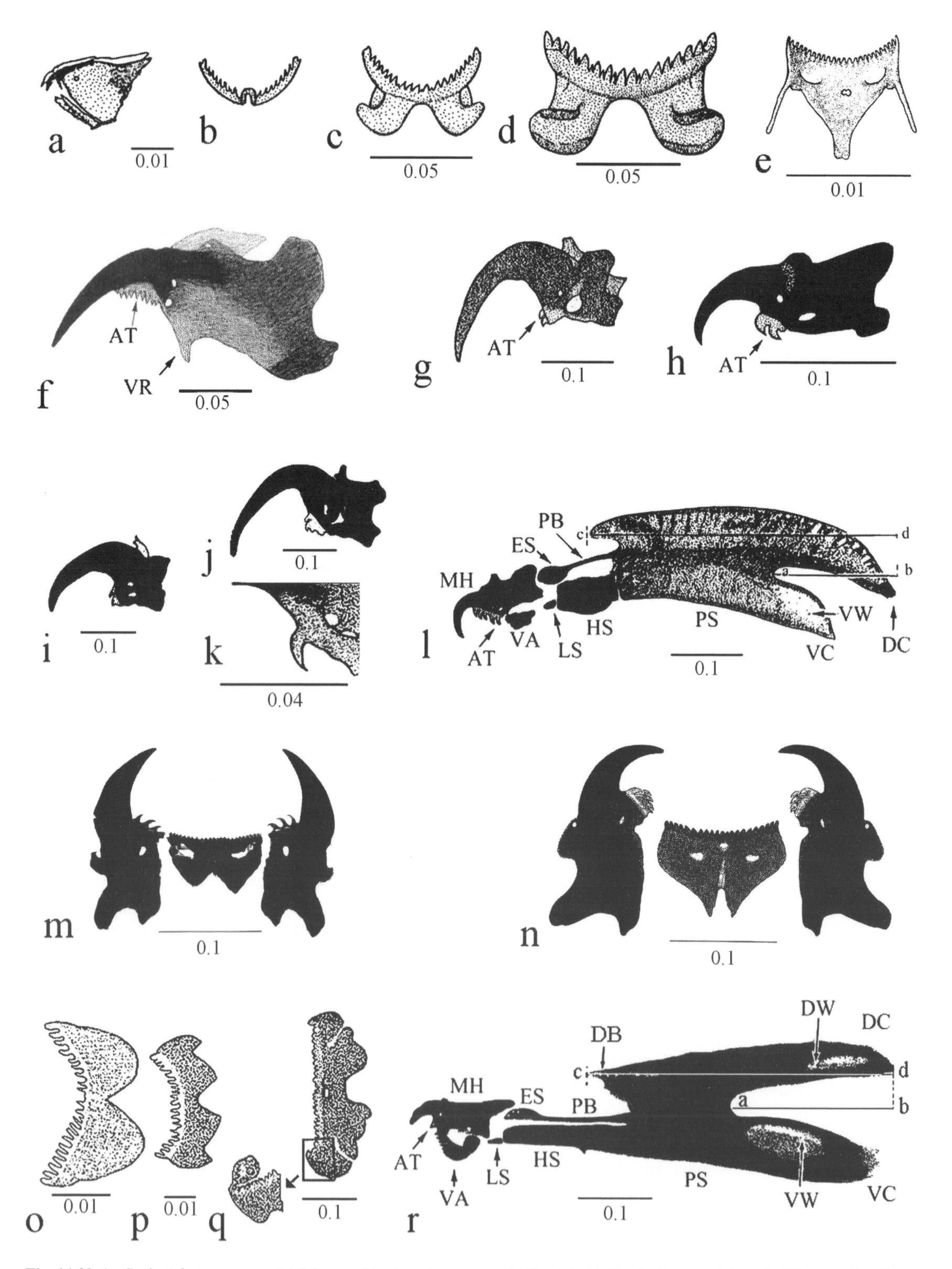

Fig. 14.38 (a–d) *Anticheta testacea*; (a) L1, mouthhook and ventral arch; (b–d) ventral arch, L1, L2, L3 (from Fisher & Orth 1964. *Hilgardia*, 1964, *36*(1). ©Regents, University of California). (e) ***Hoplodictya spinicornis***, L3, ventral arch (from Neff & Berg 1962). (f) ***Dictya ptyarion***, L3, mouthhook (from Valley & Berg 1977). (g) ***Coremacera marginata***, L3,

form of the ventral arch presents a progressive modification from the first- to third-instar larvae with the lateral extremities enveloping the basal parts of mouthhooks in the third instar.

As noted under "Accessory teeth" the ventral arch is clearly articulated or (rarely) joined (Fig. 14.38r) with the mouthhooks anteroventrally just below the accessory teeth. A non-pigmented connection can be seen in this position in photographs of some slide preparations (Fig. 14.35d); this condition might be typical of many genera.

The teeth on the anterior margin of the ventral arch, along with the mouthhooks and accessory teeth, working in conjunction with the postoral spinule patch or band (and mid ventral spinule patch on segment II in a few second-instar and many third-instar larvae of Sciomyzini) make up a formidable feeding apparatus. The ventral arch teeth have not been described in detail, or probably figured accurately except for the photographs of a few species. There are only two or three weak teeth in the anterolateral corners of the ventral arch of *Salticella fasciata* (Fig. 14.33c). The most precise enumerations of variation within a genus or species are those of Valley & Berg (1977) for second- and third-instar larvae of 21 American species of *Dictya*. In second-instar larvae, the number of teeth range from 17 (*D. atlantica*) to 34 (*D. ptyarion*) and the greatest range within a species is 18–26 (*D. texensis*). In third-instar larvae, the number ranges from 20 (*D. floridensis*) to 37 (*D. ptyarion*) and the greatest range within a species is 24–30 (*D. bergi*). In the oligochaete-feeding *Sepedonella nana* (Fig. 14.39b, d) and *Sepedon* (*M.*) *knutsoni* the teeth are strongly reflexed. In *S. nana*, but apparently not in *S.* (*M.*) *knutsoni*, there is a row of widely spaced, unpigmented denticles anterior to the reflexed teeth.

14.1.3.4.5 H-shaped sclerite

The H-shaped sclerite is fused with the pharyngeal sclerite in all first- and second-instar sciomyzid larvae. This probably gives rigidity to the rather weakly sclerotized cephalopharyngeal skeleton of the first and second instars, and an articulation allowing more flexible movement probably is not adaptive in these generally more subtly feeding instars. In most third-instar sciomyzid larvae the H-shaped sclerite abuts and articulates with the truncate anteroventral processes of the pharyngeal sclerite (Fig. 14.33a, f, j). This is quite unlike the condition in related families (except Phaeomyiidae), where the posterior arms of the H-shaped sclerite over lie the sloped anteroventral projections of the pharyngeal sclerite, e.g., in Dryomyzidae (Fig. 14.41d). The abutting type of articulation probably allows for considerable flexibility of dorsal-ventral movement of the mouthhooks, important to predators. However, note that the posterior arms of the H-shaped sclerite and anteroventral processes of the pharyngeal sclerite are both somewhat sloped in the plesiomorphic, parasitoid–saprophagous *Salticella fasciata* (Fig. 14.33a). Also, exceptionally, the H-shaped sclerite is fused with the pharyngeal sclerite in third instars of *Renocera, Anticheta, Perilimnia*, and *Shannonia* (Figs 14.36q, 14.38r, 14.40a, b). Functionally, there is no obvious adaptation to this fusion in these, respectively, clam-killing, snail-egg/snail-feeding, and snail-feeding species. However, as discussed in Section 15.3, these morphologically, and in the case of *Renocera* and *Anticheta* also behaviorally, aberrant genera seem to be related and the fusion may be more a manifestation of their evolutionary relationship than functional adaptations. The H-shaped sclerite is connected to the pharyngeal sclerite by a narrow, lightly pigmented dorsal strap in *Dictya*. The H-shaped sclerite is articulated but remarkably large and elongate in the clam-killing *Ilione lineata* (Fig. 14.37c) but not so in the other, snail-killing members of that genus, and it is of normal length in the clam-killing *Eulimnia philpotti* (Fig. 14.37a). The H-shaped sclerite portion of the fused structure is relatively more elongate in first-instar larvae, especially so in some slug-feeding *Tetanocera* species (Fig. 14.36h, j, l). Posterior rami of the H-shaped sclerite are absent to weak in most Sciomyzini (Figs 14.33h, 14.40k), but elongate in e.g., *Pherbellia dubia*. The rami are elongate in many Tetanocerini genera and in many are

Caption for Fig. 14.38 (*cont.*)
mouthhook (from Knutson 1973). (h) ***Dichetophora obliterata***, L3, mouthhook (from Vala *et al.* 1987). (i) ***Euthycera stichospila***, L3, mouthhook (from Vala & Caillet 1985). (j) ***E. cribrata***, L3, mouthhook (from Vala *et al.* 1983). (k) ***H. spinicornis***, L3, accessory tooth (from Neff & Berg 1962). (l) ***Trypetoptera punctulata***, L3, cephalopharyngeal skeleton (from Vala 1986). (m, n) ***Ilione*** spp., L3, mouthhooks and ventral arch; (m) ***I. unipunctata***; (n) ***I. lineata*** (from Knutson & Berg 1967). (o–q) ***E. stichospila***, ventral arch, L1, L2, L3 (from Vala & Caillet 1985). (r) ***A. brevipennis***, L3, cephalopharyngeal skeleton (from Knutson 1966). See text for abbreviations.

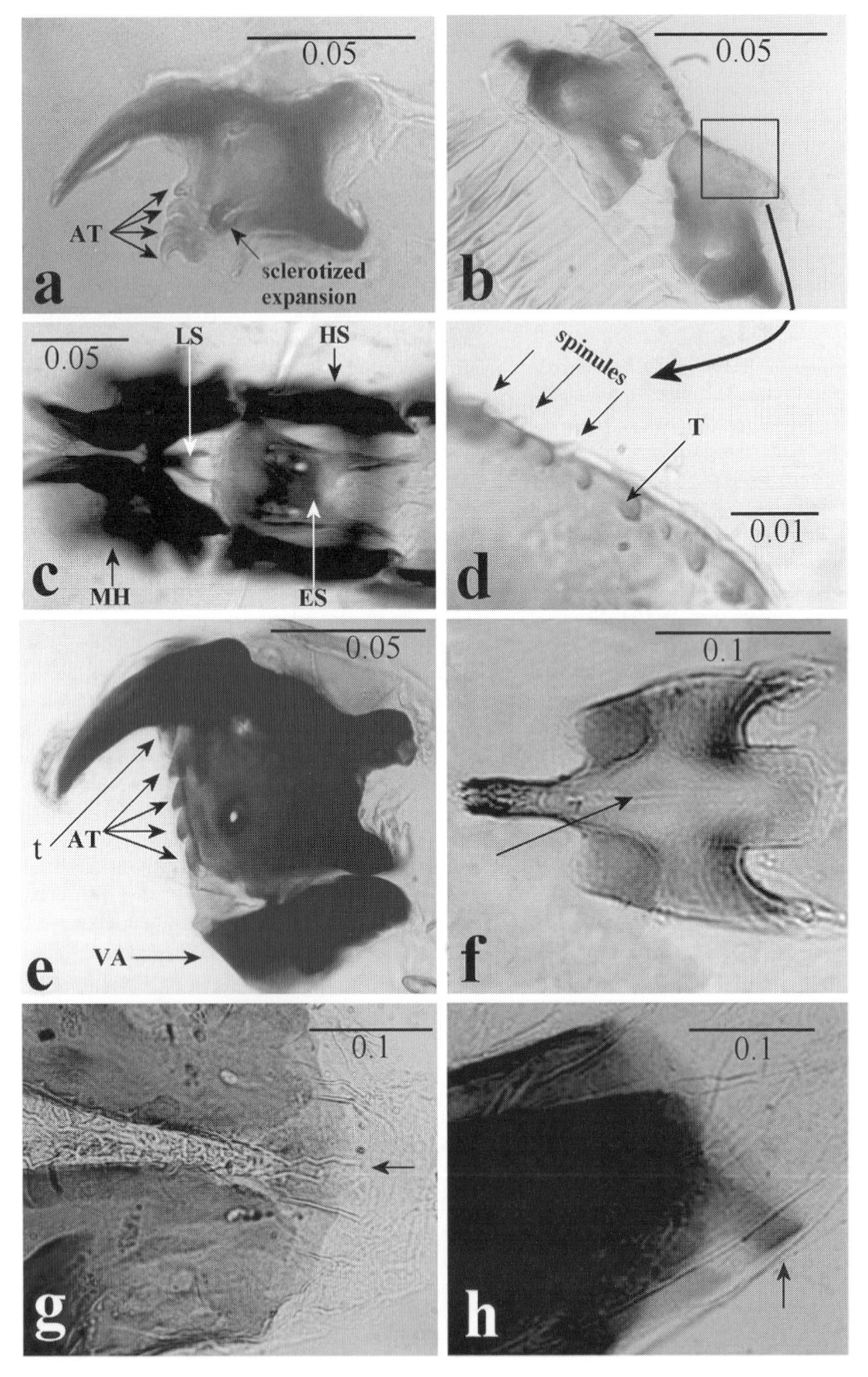

Fig. 14.39 a–d, ***Sepedonella nana***, L3; (a) mouthhook; (b) ventral arch (fractured in middle); (c) anterior part of cephalopharyngeal skeleton, ventral; (d) spinules and teeth on anterior margin of ventral arch. (e–h) ***Sepedon knutsoni***; (e) L3, mouthhook and

connected by a more or less strongly sclerotized transverse bar posterior to the salivary duct passage way (Fig. 14.40e).

The lateral surfaces of the H-shaped sclerite are plate-like extensions, probably providing protection for the food tube (Fig. 14.40d). A more or less lightly sclerotized plate between the anterior rami of the H-shaped sclerite has been figured for most species (Fig. 14.40g). Barnes (1990), in describing the cephalopharyngeal skeleton of *Sciomyza varia*, referred to this plate as a distinct sclerite, i.e., the labial sclerite. The complex pattern of sclerotization of this plate with several windows and the location of the similarly complex, sclerotized epistomal sclerite (Fig. 14.40f) lying just above the plate might indicate a sensory function for these structures. The salivary duct enters the cephalopharyngeal skeleton from below (Fig. 14.33j), and passes between and just posterior to these structures.

14.1.3.4.6 Epistomal sclerite

The epistomal sclerite is connected to the anterior ends of the pharyngeal sclerites by a pair of rod-like parastomal bars (Fig. 14.33j, m) except the bars are not fused to the pharyngeal sclerites in *Dictya* spp., four of five *Protodictya* spp., and *Dictyodes*. In *Ilione lineata*, the epistomal sclerite and the strong parastomal bars are fused to the dorsolateral extensions of the H-shaped sclerite (Fig. 14.37c) The epistomal sclerite is not fused to the anterior apices of the parastomal bars in the plesiomorphic *Salticella fasciata* (Fig. 14.33e), unlike all other Sciomyzidae, but as in Dryomyzidae. In *S. fasciata* and the Sciomyzini genera *Atrichomelina, Ditaeniella, Pteromicra*, and most *Pherbellia* species, but not in *Sciomyza* and *Tetanura pallidiventris*, variously shaped accessory rods project anteriorly from the pharyngeal sclerite, near the attachment of the parastomal bars. These accessory rods are not present in the Tetanocerini.

14.1.3.4.7 Ligulate sclerite

The ligulate sclerite appears as a small C- to V-shaped bar in most Sciomyzidae (e.g., Fig. 14.33l), but is broad and wing-like in at least *Pherbellia seticoxa* (Fig. 14.37e), and *Sepedon* (*M.*) *knutsoni*. It appears as a pair of C-shaped sclerites in third-instar *Salticella fasciata* (Fig. 14.33d). In *T. pallidiventris* there is a pair of hook-like sclerites fused to the plate between the anterior rami of the H-shaped sclerite, which may represent the ligulate sclerite (Fig. 14.40k). J. K. Barnes (personal communication, 2004) has noted that his "dental sclerite" described and figured for *Sciomyza varia* (Barnes 1990) is the same as the ligulate sclerite. The sclerite is absent in *Atrichomelina pubera* and *Dictya* spp., and indistinct in *Pherbecta limenitis*. Its function may be to provide protection to the delicate plate between the anterior rami of the H-shaped sclerite lying just behind it.

14.1.3.4.8 Pharyngeal sclerite

Differences in the pharyngeal sclerite are among the most distinct and important in the Sciomyzidae. *Salticella fasciata* (Fig. 14.33a) and the Sciomyzini (Fig. 14.33f) have an anterodorsal bridge, a deep sinus between the dorsal and ventral cornua, and windows in both cornua in the second and third instar, but these are lacking in the first instar. Almost all Tetanocerini lack an anterodorsal bridge, the cornual sinus is shallow, and there is a window only in the ventral cornu. In addition to their other unique characters or characters shared with the Sciomyzini, the Tetanocerini genera *Anticheta* (Fig. 14.38r) and *Renocera* (Fig. 14.36q) have windows in both dorsal and ventral cornu and the cornual sinus is of intermediate depth; also, the dorsal bridge is present in *Anticheta*. In *Tetanoceroides* (Tetanocerini) there is a diffuse but distinct window in the dorsal cornu. The dorsal bridge and accessory rods may be adaptations to provide rigidity to the weak, paired pharyngeal sclerites of *S. fasciata* and Sciomyzini in lieu of their much more massive structure in the Tetanocerini. In the latter, the shallow cornual sinus results in much more surface area for muscle attachment, consistent with what would be expected of the generally more voraciously feeding Tetanocerini. However, even in those Tetanocerini that are less voracious, e.g., *Trypetoptera punctulata*, the sinus is shallow (Fig. 14.38l). The absence of a window in the dorsal cornu of almost all Tetanocerini also may be correlated with increased area for muscle attachment.

The depth of the cornual sinus is conveniently expressed by the indentation index (proposed by Knutson

Caption for Fig. 14.39 (*cont.*)
ventral arch, t indicates unusual, ventrally directed accessory tooth; (f) L1, cephalopharyngeal skeleton, flattened and spread dorsoventrally; (g) L3, posterior end of pharyngeal sclerite flattened and spread dorsoventrally; (h) L3, posterior end of pharyngeal sclerite, lateral. (f, g, h): arrows show pharyngeal sclerites fused mid ventrally. Photos by L. G. Gbedjissi. See text for abbreviations.

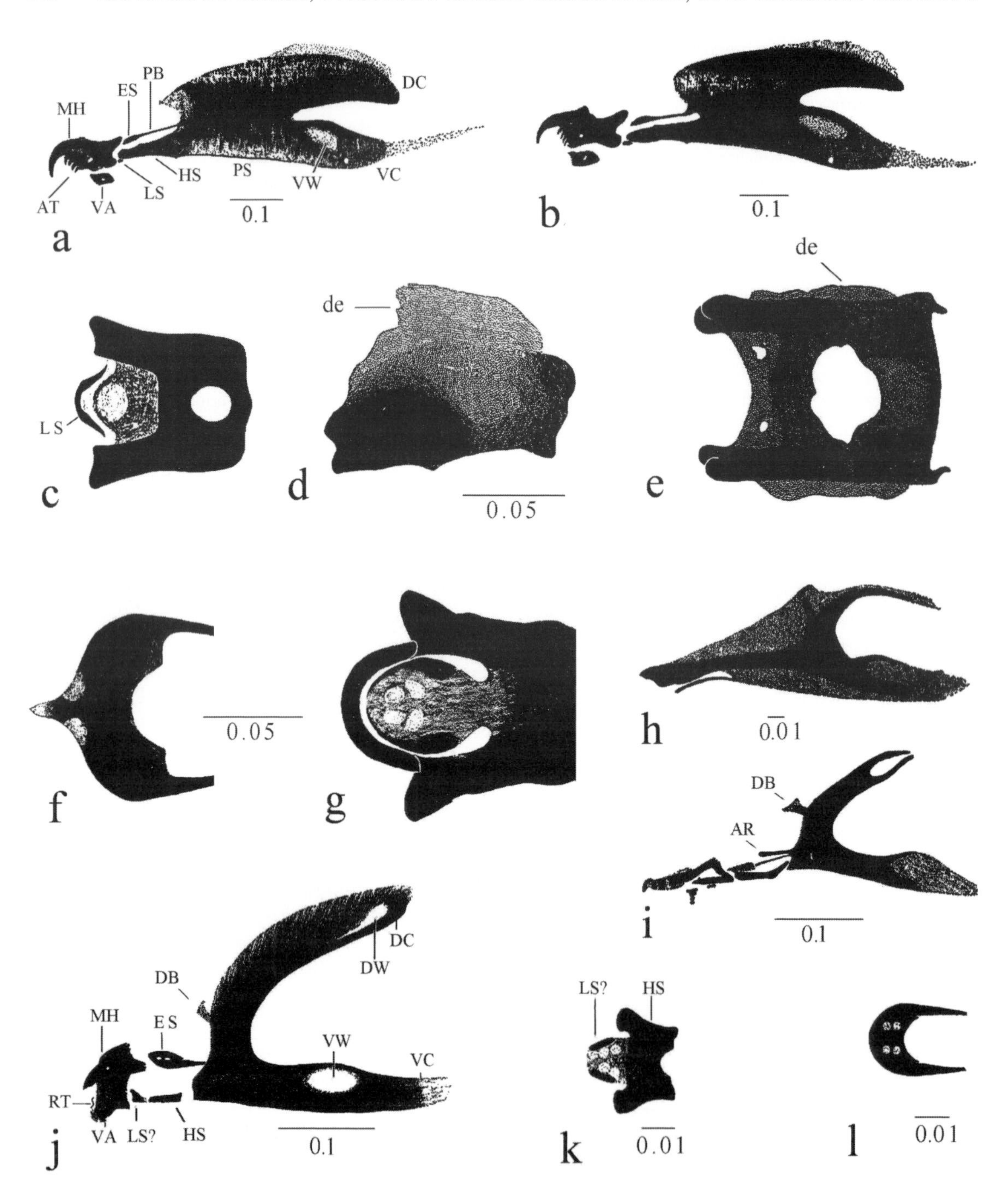

Fig. 14.40 (a) ***Perilimnia albifacies***, L3, cephalopharyngeal skeleton. (b) ***Shannonia meridionalis***, L3, cephalopharyngeal skeleton (from Kaczynski *et al.* 1969). (c) ***Thecomyia limbata***, L3, ligulate and H-shaped sclerites (from Abercrombie & Berg 1975). (d, e) ***Dictya ptyarion***, L3, H-shaped sclerite; (d) lateral, de indicates dorsolateral plate-like extension; (e) ventral (from Valley & Berg 1977). (f, g) *S. meridionalis*, L3; (f) epistomal sclerite; (g) ligulate and H-shaped sclerites (from Kaczynski *et al.* 1969). (h, i) ***Salticella fasciata***, cephalopharyngeal skeleton; (h) L1; (i) L2 (from Knutson *et al.* 1970). (j–l) ***Tetanura pallidiventris***, L3; (j) cephalopharyngeal skeleton; (k) ligulate ? and H-shaped sclerites; (l) epistomal sclerite (from Knutson 1970a). See text for abbreviations.

1966) (ab/cd × 100, as in Fig. 14.36q). The index is 61–67 for *S. fasciata*, 60–90 for the Sciomyzini, 61–62 for *Renocera*, 50–56 for *Anticheta*, and 27–53 for the rest of the Tetanocerini, except the oligochaete-feeding *Sepedonella nana* which has the smallest index known, i.e., 19. While the divergent appearance of the dorsal and ventral cornu in second-instar *S. fasciata* and some Sciomyzini (Fig. 14.40j) might be due to artifacts of slide preparation, their essentially parallel disposition in the Tetanocerini obviously contributes to their rigidity and forceful movement in that lineage of predators.

Salticella fasciata appears to be unique in the family in having ventral cibarial ridges between the ventral cornu, a filter mechanism found in many cyclorrhaphous larvae that feed "saprophagously," i.e., on bacteria and yeasts developing in the nutrient medium (Fig. 14.41a–c). In other species examined, both Sciomyzini and Tetanocerini of all three instars, the pharyngeal sclerites are

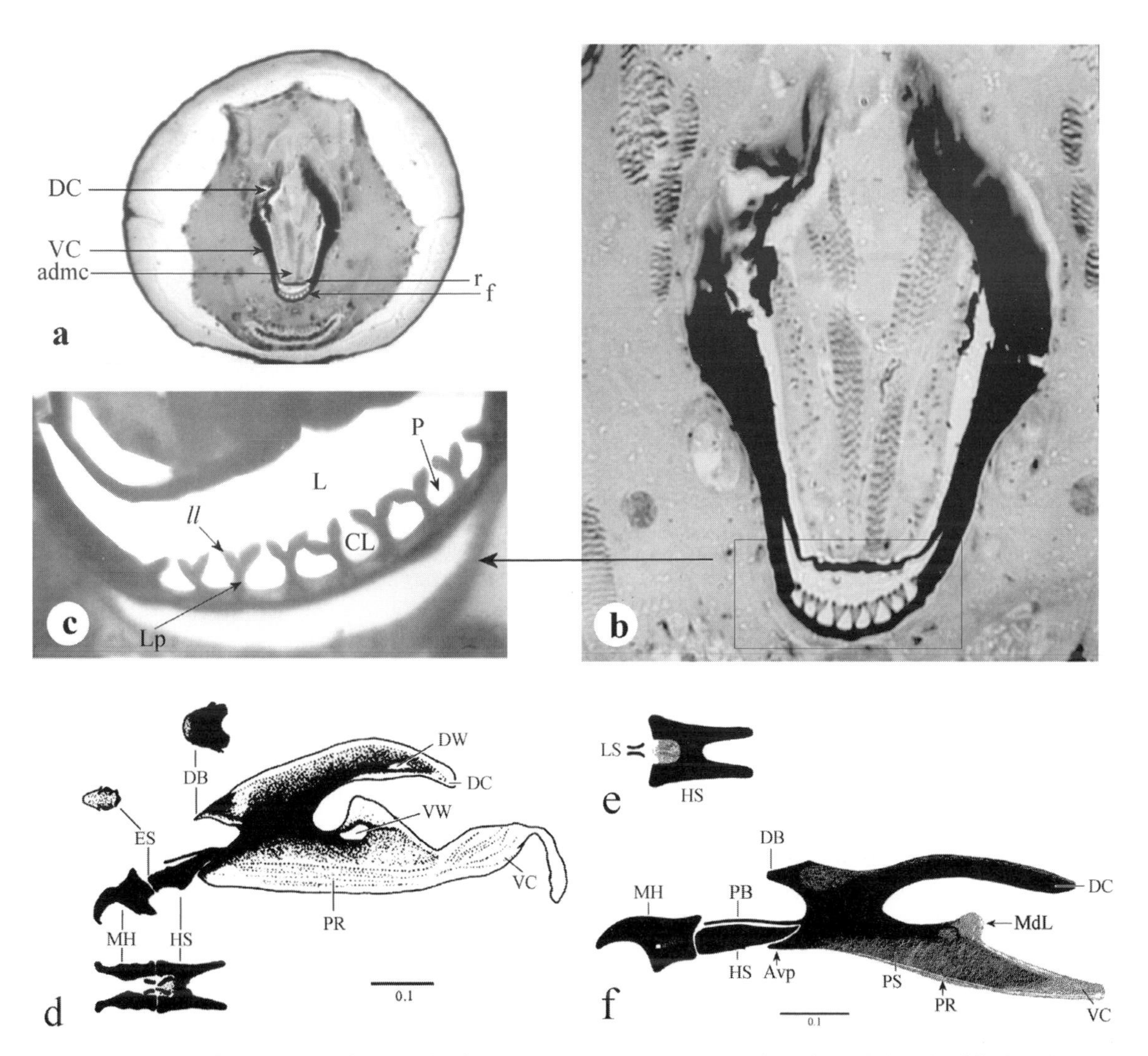

Fig. 14.41 ***Salticella fasciata***, L3. (a) Cross-section through anterior part of pharyngeal sclerite; (b, c) same, at higher magnifications (from Knutson *et al.* 1970). (d) *Dryomyza anilis*, L3, cephalopharyngeal skeleton (from Barnes 1984). (e, f) *Helosciomyza australica*, L3; (e) H-shaped and ligulate sclerites; (f) cephalopharyngeal skeleton (from Steyskal & Knutson 1978). See text for abbreviations.

connected ventrally by a more or less pigmented sheet, or at least a bridge posteriorly (Fig. 14.39f–h). The unsclerotized sheets prolonging the dorsal and ventral cornua posteriorly and dorsally, the clypeal or clypeal-frontal and cibarial phragmata of Roberts (1971), have not been examined after staining and have been only roughly indicated in some figures of the cephalopharyngeal skeleton of Sciomyzidae.

14.1.3.4.9 Aberrant genera

It is important to comment briefly on the cephalopharyngeal skeletons of the aberrant snail-egg/snail killers and clam killers. The snail-egg/snail-feeding *Anticheta* spp. (Figs 14.35d, 14.38a–d, r) and the clam-killing *Renocera* spp. (Figs 14.36q, 14.37b) are placed in the Tetanocerini on adult morphological characters, but each has a peculiar set of morphological features of the larvae, especially of the cephalopharyngeal skeleton. Some are shared with the Sciomyzini and some are shared with the Tetanocerini. These characters are more fully discussed in the taxonomic diagnoses (Section 15.1.2) and are compared in Table 15.6. The morphology of the cephalopharyngeal skeleton of the clam killers (in three unrelated genera of Tetanocerini) shows no special convergence in structure (Figs 14.36q, 14.37a–c). This is surprising, in that these species are the only known members of the Class Insecta that are obligate natural enemies of bivalve molluscs. *Renocera pallida* and *Eulimnia philpotti* have the shortest, least decurved, and bluntest hook parts of the mouthhooks of any Tetanocerini. *Renocera* spp. are the only Tetanocerini lacking accessory teeth in the second and third instar except *Dictya ptyarion*. *Renocera pallida* is the only Tetanocerini in which the hook part is not strongly sclerotized in the third instar. The H-shaped sclerite in third-instar *E. philpotti* is normal, fused to the pharyngeal sclerite in *Renocera* spp., and extremely long, unlike other species of *Ilione*, but separate from the pharyngeal sclerite in *Ilione lineata*.

14.1.3.4.10 Cephalopharyngeal skeleton of Phaeomyiidae

The immature stages, including the cephalopharyngeal skeleton, of *Pelidnoptera nigripennis* (Phaeomyiidae) were described and figured by Vala *et al.* (1990). In general, the cephalopharyngeal skeleton (Fig. 14.42) is more similar to that of the Sciomyzini than the Tetanocerini. Notably, the ventral arch and ligulate sclerite are absent. Accessory teeth are lacking in the first and third instar but there are two very large teeth below the small, blunt hookpart of the mouthhook in the second instar. At least the second instar has a well-developed dorsal bridge (condition not given for first and third instar). In the third instar, the H-shaped sclerite articulates with and abuts the pharyngeal sclerite, the indentation index is 70–74, and both the dorsal and ventral cornua have a window. The posterior margin of segment I in the first and second instar has a completely encircling band of spinules. Oral grooves and ventral cibarial ridges, not mentioned, apparently are absent.

14.1.4 Puparia and pupae

The puparia have not been studied in nearly as much detail as the other immature stages of Sciomyzidae. The pupal stage, per se, has been barely studied, except for comments on the condition of overwintering pupae of a few species in the paper on phenology by Berg *et al.* (1982) and the condition of the tracheal trunks in pupating *Elgiva solicita* by Knutson & Berg (1964) and *Ilione albiseta* by Knutson & Berg (1967). Here, we relate features of the puparia to the microhabitats and behavior of the pupariating larvae and emerging adults and to their evolutionary position. See Figure 17.2 for morphological features of the puparia and pupal stage displayed on the cladogram of Marinoni & Mathis (2000).

In the literature on sciomyzid puparia, particular attention has been given to whether or not they are formed in the shell of the host/prey and, if formed outside, the relative degree of extension and dorsal direction of the last segment, bearing the posterior spiracles. Except for a few species, the condition of the anterior spiracles is relatively less noteworthy. These features have been emphasized because pupariation in the shell has been considered an indication of a close relationship with the host/prey, and as a specialized or derived feature. Also, the degree of extension and dorsal direction of the posterior spiracles, and in some species the anterior spiracles, has been assumed to indicate the microhabitat of the larvae. There are reasons for emphasis on the condition of the posterior, as opposed to the anterior spiracles. As discussed in Section 14.2.3, there is evidence from experiments by Buck & Keister (1953) on *Phormia regina* (Calliphoridae), which pupariates in the open, that the posterior spiracles are far more important than the anterior spiracles for pupal respiration. As shown below, neither assumption, about site of puparium formation or position of the posterior spiracles might be entirely correct when one analyzes these features across the entire family Sciomyzidae.

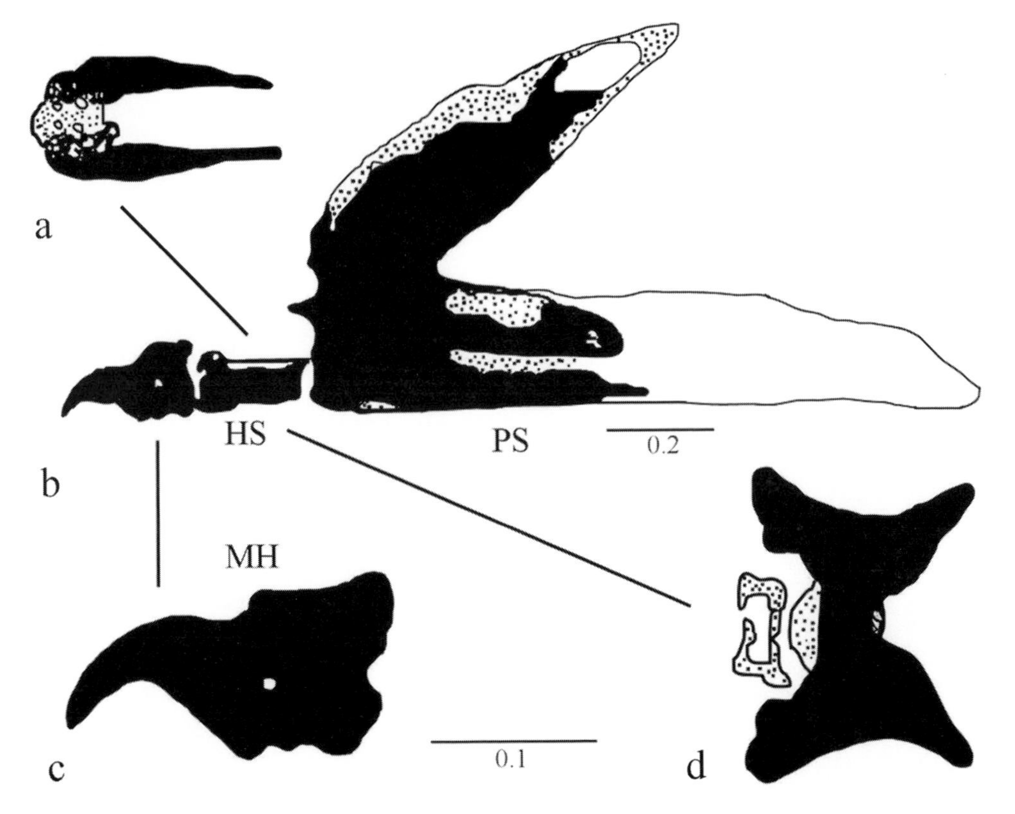

Fig. 14.42 ***Pelidnoptera nigripennis***, L3, cephalopharyngeal skeleton. (a) Epistomal sclerite and parastomal bars; (b) cephalopharyngeal skeleton; (c) mouthhook; (d) H-shaped sclerite. From Vala *et al.* (1990). See text for abbreviations.

The puparium of *Pelidnoptera nigripennis* of the plesiotypic sister group Phaeomyiidae, a parasitoid of millipedes in mesic woodlands in northern Europe to Mediterranean shrubland and heathland, is formed in the cylindrical exoskeleton of its host. The posterior spiracles are moderately elevated above the mid longitudinal axis (Fig. 14.43a).

In the Sciomyzidae, puparia of *Salticella fasciata* of the plesiotypic subfamily Salticellinae, a parasitoid/saprophage of terrestrial snails in western Europe to Mediterranean grasslands, are formed outside the shell, with the unusually elongate posterior spiracular tubes rather strongly directed dorsad (Fig. 14.43c). In the other subfamily, the plesiotypic tribe Sciomyzini of the Tetanocerinae (terrestrial and semi-aquatic parasitoid/predators/saprophages), puparium formation in the shell of the host/prey occurs in all eight reared genera (of a total of 13 genera) and is by far the most common pupariation site. Puparia of 45 of the 50 reared species of Sciomyzini (147 total species) are formed in the shells of at least 43 species in 30 genera of Gastropoda. Puparium formation in the shell is usual for *Tetanura pallidiventris* (Fig. 12.4b), *Oidematops ferrugineus*, *Atrichomelina pubera* (Fig. 14.44, Table 15.1), three of five reared *Pteromicra* spp., three of five reared *Sciomyza* spp. (Fig. 14.43e, f), all five reared *Colobaea* spp. (Fig. 12.4c–e), 15 of 29 reared *Pherbellia* spp. (Plate 4g; Figs 14.43g–j, 14.45), and two of three reared *Ditaeniella* (Fig. 14.46).

When formed in shells that have the body whorl of large diameter and not tightly coiled, as in *Helisoma trivolvis*, the shape of the puparium is not strongly modified, e.g., *Pherbellia seticoxa* (Fig. 14.43g, h), except that the ventral surface, which rests on the inner surface of the outer wall of the shell, is strongly curved and the dorsal surface is almost straight. The puparia of the terrestrial parasitoid/predators *Pherbellia albocostata*, which pupariates in litter under bark of fallen trees, *P. scutellaris*, which pupariates on the substrate, and *P. dubia*, which pupariates in shells of terrestrial snails

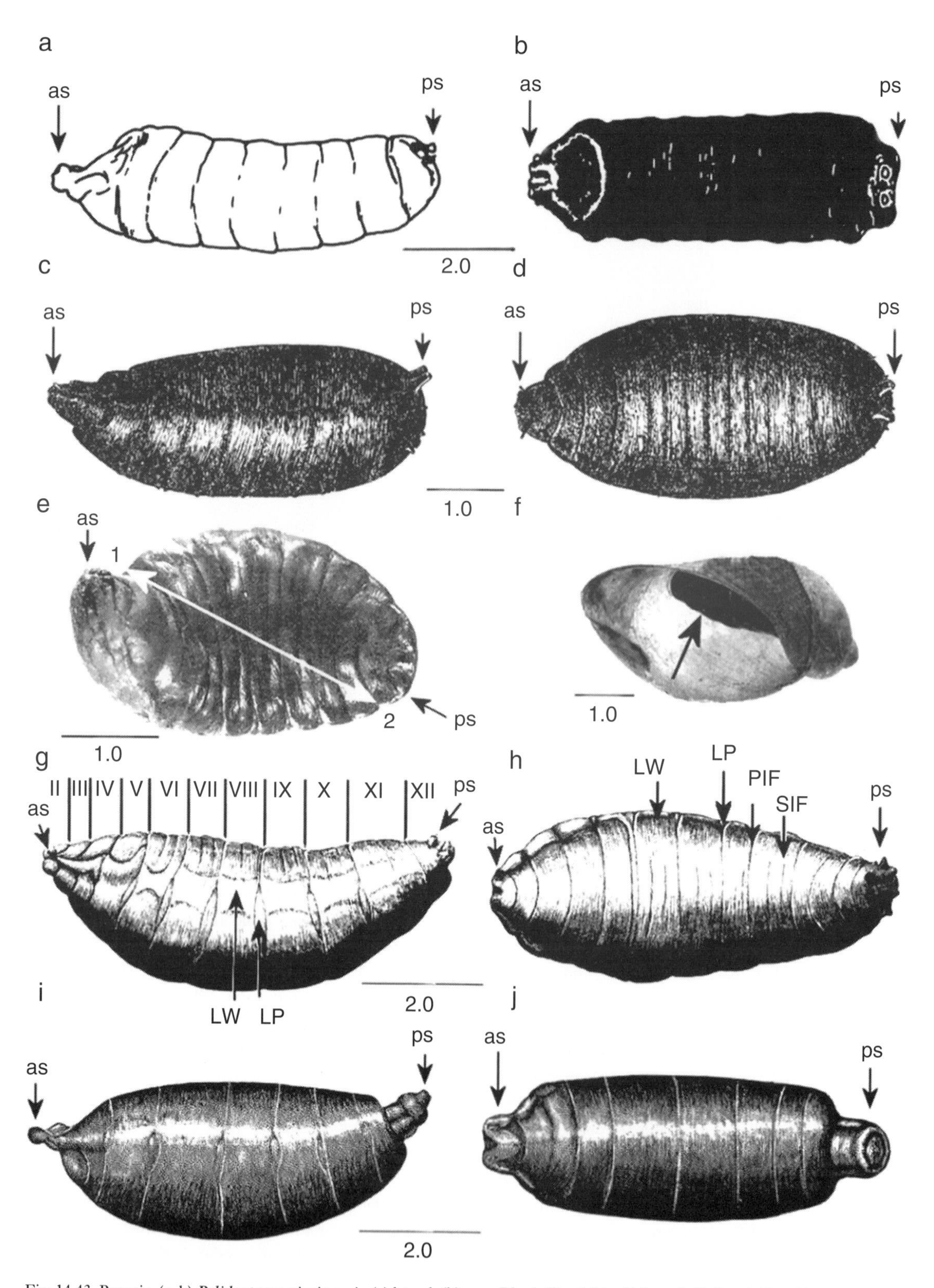

Fig. 14.43 Puparia. (a, b) ***Pelidnoptera nigripennis***; (a) lateral; (b) dorsal (from Vala *et al.* 1990). (c, d) ***Salticella fasciata***; (c) lateral; (d) dorsal (from Knutson *et al.* 1970). (e, f) ***Sciomyza aristalis***; (e) removed from shell; (f) in shell of *Novisuccinea ovalis* (from Foote 1959a). (g, h) ***Pherbellia seticoxa***; (g) lateral; (h) dorsal. (i, j) ***Pherbellia dubia***; (i) lateral; (j) dorsal (from Bratt *et al.* 1969). as, anterior spiracle; ps, posterior spiracle. II–XII, body segments; LP, lateral pad; LW, lateral welt; PIF, primary integumentary fold; SIF, secondary integumentary fold. 1–2, concave surface that rests against columella of shell.

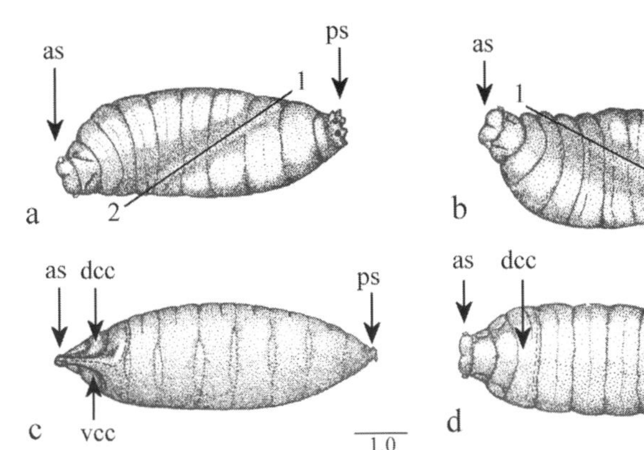

Fig. 14.44 Puparia, ***Atrichomelina pubera***. (a) Formed in shell of *Aplexa hypnorum*, dorsal; (b) formed in shell of *Stagnicola elodes*, dorsal; (c) formed outside shell, lateral; (d) formed outside shell, dorsal. 1–2, concave surface that rests against columella of shell. dcc, dorsal puparial cap; vcc, ventral puparial cap. From Foote *et al.* (1960).

(Fig. 14.43i) are unusual in having the posterior end strongly narrowed and truncate.

Pupariation site (in shells or outside) appears to be labile in two reared species of *Ditaeniella*, three *Pteromicra* spp., eight *Pherbellia* spp., two *Sciomyza* spp., and *Atrichomelina pubera*. However, much of the information on labile pupariation site is based on laboratory rearings and may reflect the rearing conditions, especially the relative density of larvae and snails. Puparia of *A. pubera* formed inside of and outside of shells are shown in Figure 14.44. In a few Sciomyzini, often more than one puparium is formed, in nature, in shells of sufficient size and appropriate shape, e.g., at least two of *A. pubera* have been found in one *Helisoma anceps* and up to five of *Pherbellia seticoxa* in one *H. trivolvis*.

In the Sciomyzini, only the succineid parasitoid/predators *Sciomyza dryomyzina* and *S. simplex*, eight species of terrestrial and semi-aquatic parasitoid or parasitoid/predatory *Pteromicra* and *Pherbellia* spp., and *Ditaeniella grisescens* and *D. parallela* apparently routinely pupariate outside the shell, on or slightly beneath the substrate. In most of these, the puparia are similar in shape to many semi-aquatic Tetanocerini, with the posterior spiracles at or slightly elevated above the mid horizontal axis.

Deposition of calcareous material on the anterior end of the puparium or formation of a calcareous septum produced by larvae of 11 species of *Pherbellia*, *Ditaeniella trivittata* (Plate 4g; Figs 14.45, 14.46) and *Colobaea americana* and *C. punctata* was described by Bratt *et al.* (1969) and Knutson *et al.* (1967b). In *D. trivittata* two separate septa are formed (Fig. 14.46). The origin of septum material in the Malpighian tubules and its chemical composition (32.7–50% calcium plus 11 other elements) were discussed by Knutson *et al.* (1967b). In at least 14 species of *Musca* (Muscidae), *Euleia heraclei* (Tephritidae), *Acletoxenus* spp. (Drosophilidae), seven species of Lauxaniidae, and some *Orygma* and *Saltella* (Sepsidae), the puparium is strengthened by calcification or a thin coat is formed over the puparium (Ferrar 1975, 1987). Gilby & McKeller (1976) made a chemical analysis of the calcified puparium of *Musca fergusoni*. The tough, shagreened surface of larvae of Stratiomyidae (Orthorrhapha) is produced by the deposition of calcium carbonate in the cuticle.

The peculiar "false chamber" present in the anterior segments of puparia of the behavioral equivalents *Sciomyza varia* (Barnes 1990) and *Colobaea bifasciella* (Knutson & Bratt, unpublished data) needs further study from morphological and developmental viewpoints (see also Section 11.2). The puparia of *Ditaeniella* are unusual in having the dorsal cephalic cap moderately to strongly concave.

In the Tetanocerini, of the *c.* 150 biologically known species in 31 genera, pupariation in the shell occurred only in some individuals of the terrestrial parasitoid/predator *Trypetoptera punctulata* in laboratory rearings (Vala 1986).

The likely advantages of pupariation in the shell, (1) protection from natural enemies, (2) insulation from low temperatures, (3) protection from crushing by freezing water during overwintering, and (4) increased buoyancy resulting in dispersal during spring floods and protection from abrasion during this period, indicate its adaptive, evolved nature. Furthermore, the derived nature of pupariation in the shell is also indicated by the occurrence of the very unusual production of calcareous septa (unique in the Diptera: Schizophora) in some *Pherbellia*, *Ditaeniella*, and *Colobaea* species; the unusual false chambers in puparia of *Sciomyza varia* and *C. bifasciella*; and the behavior of pupariating larvae to push out tissue

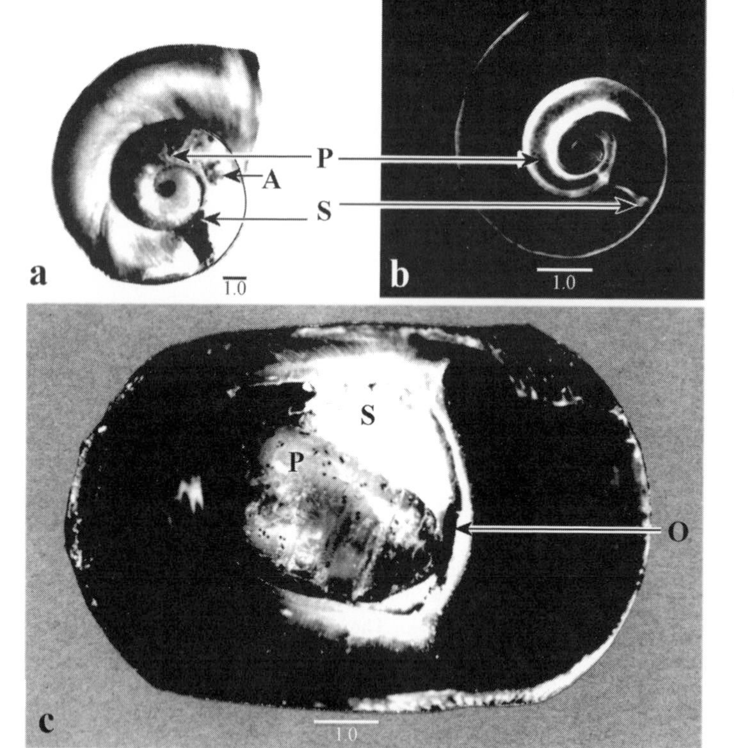

Fig. 14.45 Septa and puparia of ***Pherbellia*** spp. (a) ***Pherbellia seticoxa*** in shell of *Helisoma trivolvis*. (b) ***P. trabeculata*** in shell of *Biomphalaria glabrata*, x-ray photograph. (c) ***P. dorsata*** in shell of *H. trivolvis*. A, anterior end of puparium; O, slit-like opening in septum made by pupariating larva; P, puparium. S, septum. Modified from Knutson *et al.* (1967b).

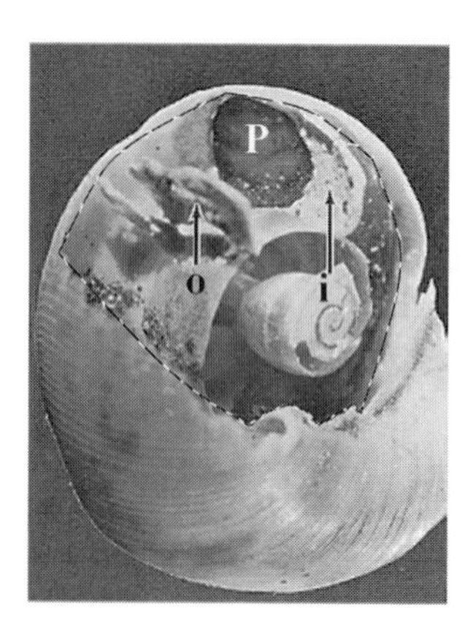

Fig. 14.46 Septa (arrows: i, inner; o, outer) and anterior end of puparium (P) of ***Ditaeniella trivittata*** in shell of *Helisoma* sp. (shell removed along dotted line, part of inner septum removed to show anterior end of puparium). From Bratt *et al.* (1969).

remaining in shells. Pupariation in the host (along with oviposition onto the host and one larva per host–one host per larva, which also occurs only in the Phaeomyiidae, Salticellinae, and Sciomyzini) has been considered a specialized, derived feature. However, the occurrence of these and other so-called derived features is not concordant with the polarity of evolutionary development as indicated on the basis of adult morphology in the cladistic analyses of the genera by Marinoni & Mathis (2000) and Barker *et al.* (2004).

We consider puparia formed outside the shell, with posterior spiracles at or slightly above or below the mid horizontal axis as the plesiomorphic state 0. The apomorphic states are: formed within the host/prey – state 1; formed outside with posterior spiracles not on a postanal extension but above the dorsal surface – state 2; formed outside with posterior spiracles strongly elevated on a postanal extension – state 3; and formed outside with posterior spiracles well below the mid horizontal axis, essentially ventral – state 4. There is little variation in the nature of the anterior spiracles of the puparium; their more common plesiomorphic state is considered as being sessile and their unusual apomorphic state as being strongly extended.

Among Tetanocerini, the condition of the postanal portion of segment XII, along with the relative degree of

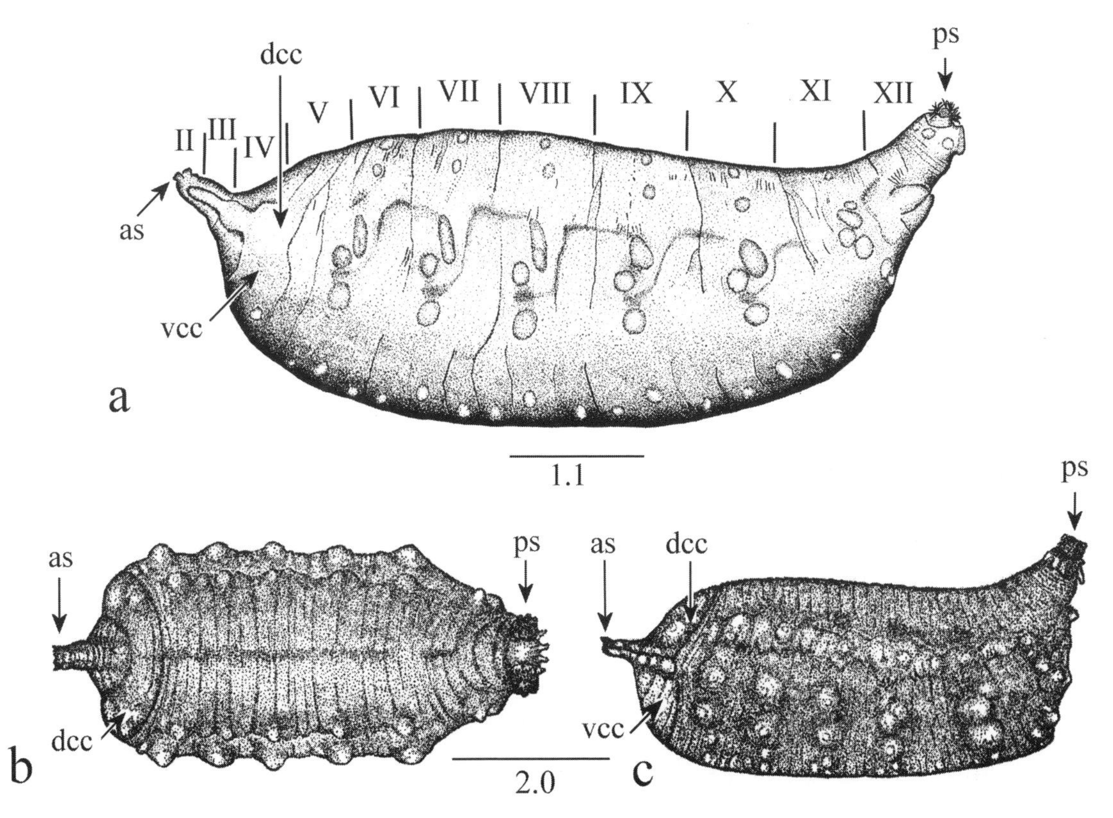

Fig. 14.47 Puparia. (a) ***Dictya texensis***, lateral (from Valley & Berg 1977). (b, c) ***Sepedon ruficeps***; (b) dorsal; (c) lateral (from Knutson *et al.* 1967a). dcc, dorsal puparial cap; vcc, ventral puparial cap.

elevation of the posterior spiracles is the most adaptive feature of puparia. It is varied among and within the genera, and is species-distinctive. Puparia having the posterior spiracles elevated above the dorsal surface of the median segments occur in all those species whose larvae live at the surface of the water: *Dictya* (Fig. 14.47a), *Elgiva* (Fig. 14.49a, b), *Protodictya, Thecomyia, Sepedomerus, Sepedonea, Sepedonella*, most species of *Ilione*, some *Neolimnia* spp., many *Tetanocera* spp. (Fig. 14.15), *Renocera striata*, and almost all *Sepedon* spp. (Figs 14.47b, c, 14.48a–f). Even the puparia of the parasitoid/predators of semi-terrestrial Succineidae, *Sepedon hispanica, S. trichrooscelis*, and *Tetanocera arrogans* have the posterior spiracles elevated (Figs 14.48e, f, 14.51a, b). Among *Sepedon* spp., only *S. borealis* does not have the posterior spiracles elevated (Fig. 14.48g, h). From laboratory rearings *S. borealis* seems to be an aquatic predator but it forms its puparium buried in moss or sand with only the anterior end visible This may indicate that, at least in this species, the anterior spiracles are more important in pupal respiration. In some species of *Sepedon*, e.g., *S. ruficeps* (Fig. 14.47b, c) and *S. sphegea* (Fig. 14.48a, b), lateral tubercles on the body segments are unusually protrudent; these undoubtedly air-filled spaces probably aid in the buoyancy of the puparia.

A very elongate extension of the postanal portion of segment XII occurs only in *Elgiva* spp. (aquatic predators of snails, Fig. 14.49a, b) and *Renocera striata* (aquatic predator of fingernail clams). *Elgiva* spp. and *R. striata* also have the anterior spiracles extended and uplifted; however, there does not seem to be anything unusual about their microhabitat, i.e., the neuston. Interestingly, *Renocera* also includes a species, *R. pallida*, that feeds on fingernail clams that are not in the water but exposed in semi-aquatic situations. Its puparia do not have the posterior spiracles elevated at all (Fig. 14.49c, d).

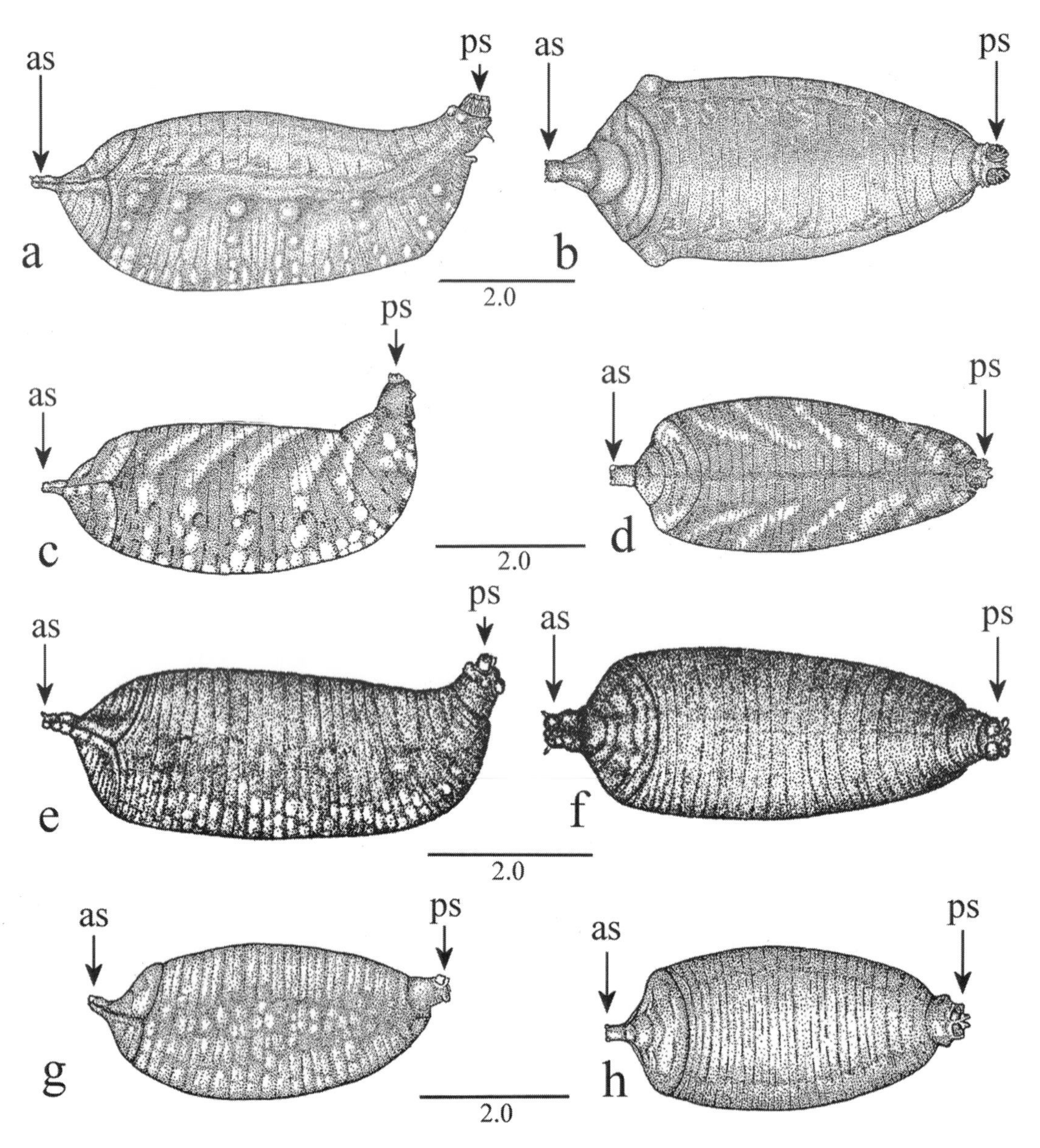

Fig. 14.48 Puparia. (a, b) ***Sepedon sphegea***; (a) lateral; (b) dorsal. (c, d) ***S. anchista***; (c) lateral; (d) dorsal (from Neff & Berg 1966). (e, f) ***S. h. hispanica***; (e) lateral; (f) dorsal (from Knutson *et al.* 1967a). (g, h) ***S. borealis***; (g) lateral; (h) dorsal (from Neff & Berg 1966).

Most other Tetanocerini, both terrestrial and semi-aquatic, have the posterior spiracles more or less above the mid horizontal axis but not above the dorsal surface of the median segments. Surprisingly, puparia of the three species having the most truly aquatic larvae, i.e., larvae that live beneath the surface most of their lives, have posterior spiracles only slightly above or well below the mid horizontal axis, indicating that they are formed out of the water. These are the fingernail clam-killers *Ilione lineata* (Fig. 14.49e, f) and *Eulimnia philpotti* (Fig. 14.50a) and the aquatic snail-killer *Hedria mixta* (Fig. 14.50b, c). Only the dorsal view of *E. philpotti* was figured by Barnes (1980a) but his description states "Segment 12 usually slightly up-turned posteriorly, but not reaching above level of

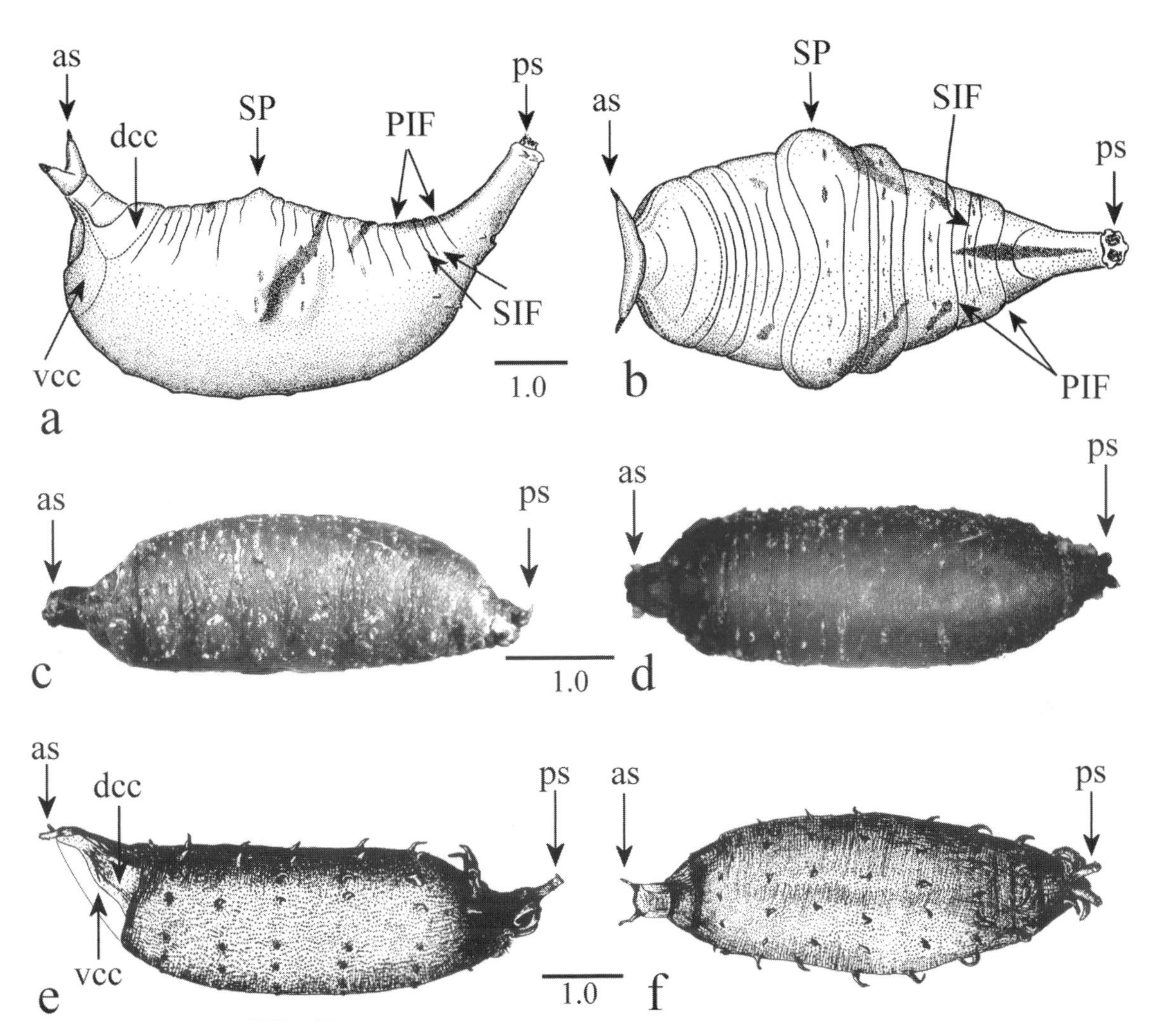

Fig. 14.49 Puparia. (a, b) ***Elgiva solicita***; (a) lateral; (b) dorsal (from Knutson & Berg 1964). (c, d) ***Renocera pallida***; (c) lateral; (d) dorsal (from Horsáková 2003). (e, f) ***Ilione lineata***; (e) lateral; (f) dorsal (from Knutson & Berg 1967). dcc, dorsal puparial cap; PIF, primary integumentary fold; SIF, secondary integumentary fold; SP, spinule patch; vcc, ventral puparial cap.

dorsal surface of middle segments." The posterior spiracles of *H. mixta* are well below the mid horizontal axis, almost completely ventral as in the terrestrial *Dichetophora obliterata* (Fig. 14.50d, e). Puparia of *H. mixta* were not found in nature: "All four puparia obtained during the laboratory rearings were formed above the water line, usually at the juncture of the side and lid of the rearing dish" (Foote 1971). Puparia of *E. philpotti* were found buried under 10–15 cm of *Sphagnum* a few centimeters from the shoreline or firmly wedged among moss and roots at the bases of emergent *Carex* tussocks. In the laboratory they were formed on sand above the water level, in moist *Sphagnum*, or on the lid of the rearing jar (Barnes 1980a). Thus, puparial shape cannot always be used to predict larval microhabitat, as has been assumed in most previous literature.

The puparia show vestiges of larval cuticular structures, segmental protuberances, and primary and secondary segmental folds. Protuberances and folds are particularly well illustrated for *Pherbellia seticoxa* (Sciomyzini) (Fig. 14.43g, h), *Dictya texensis* (Fig. 14.47a), and *Eulimnia philpotti* (Fig. 14.50a) (Tetanocerini). Puparia of most species are tanned light brown to dark

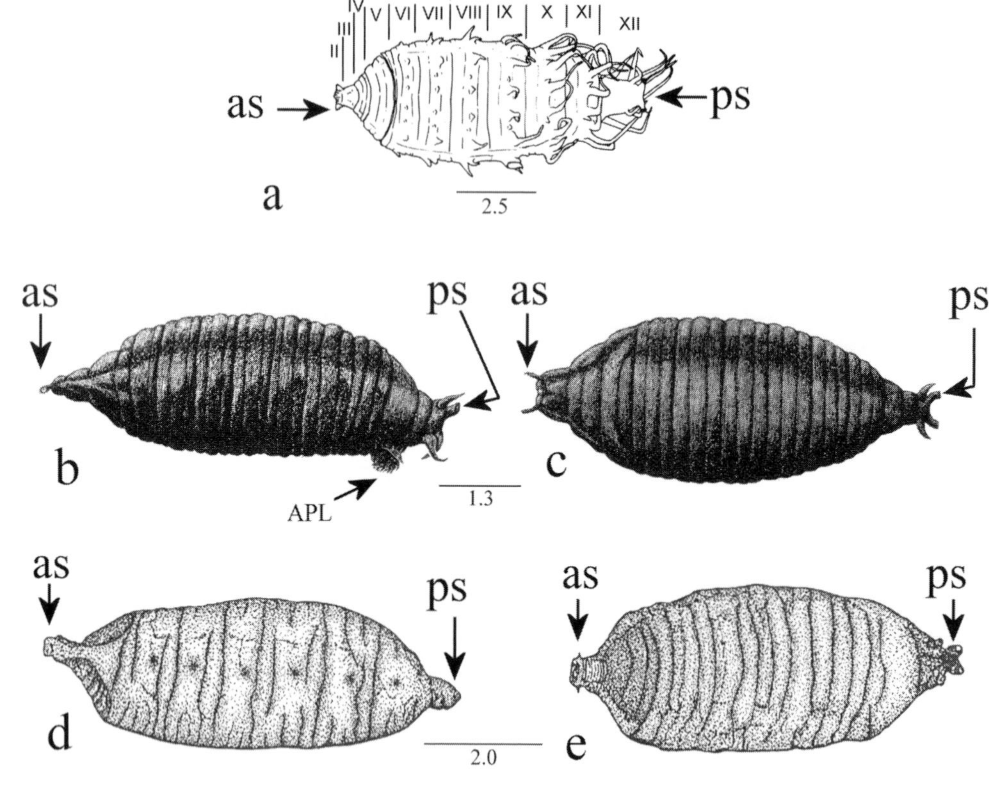

Fig. 14.50 Puparia. (a) ***Eulimnia philpotti***, dorsal (from Barnes 1980a). (b, c) ***Hedria mixta***; (b) lateral; (c) dorsal (from Foote 1971). (d, e) ***Dichetophora obliterata***; (d) lateral; (e) dorsal (from Vala *et al.* 1987). APL, anal proleg.

brown, almost blackish in some species. Puparia of many aquatic species in which the integument of the larvae are light in color and have a pattern are also patterned, e.g., species of *Sepedomerus*, *Sepedon*, and *Sepedonea* (Fig. 14.48c, d) and *Elgiva solicita* (Fig. 14.49a, b). Neff & Berg (1966) noted that the color patterns of *Sepedomerus*, *Sepedon*, and *Sepedonea* are more characteristic of species than genera, sometimes are more obvious in the puparia than the larvae, and exhibit considerable variation, including two color phases in some species. The unusual green color of mature larvae and puparia of *E. solicita* and *E. connexa* (persisting in newly emerged adults in *E. solicita*) but not seen in *E. cucularia* is due to the green color of the fat bodies, visible through the translucent integument.

The dorsal surface (in lateral view) of puparia formed outside the shell may be essentially straight (as in *I. lineata*, Fig. 14.49e), convex (as in *H. mixta*, Fig. 14.50b), but generally is somewhat inflated in the region of segments VI–VIII. In *E. solicita* (Fig. 14.49a, b) a predator of aquatic snails, this region is very strongly inflated. Since it seems not to be related to the form of the developing pupa/pharate adult within, it probably is an adaptation for flotation. The swallowing of air, seen throughout larval life of many aquatic predators and some terrestrial species might be a factor at the time of puparium formation in determining the shape of the puparium and increasing its buoyancy. Buoyancy might be important for semi-terrestrial species, e.g., *Tetanocera arrogans* (Fig. 14.51a, b) as well as for aquatic species that overwinter in the puparium and whose overwintering sites are flooded in early spring. That these puparia are adapted for flotation also is indicated by the dorsal elevation of the posterior spiracles.

Most puparia, even those of many terrestrial species, have an evenly arcuate ventral surface in lateral view, indicating they were formed on a soft, friable substrate or in the water. However, others are flat ventrally for most of their length; this might indicate pupariation on a flat substrate, e.g., stiff leaves, or on similar rigid surfaces.

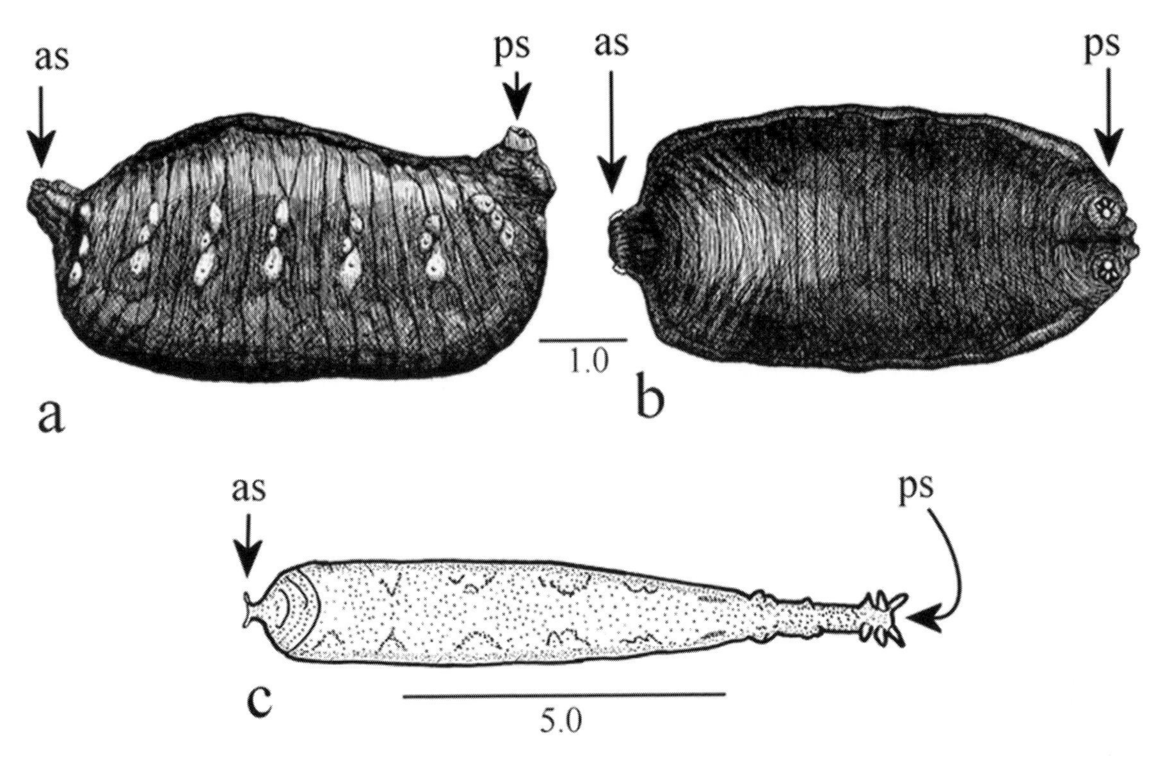

Fig. 14.51 Puparia. (a, b) *Tetanocera arrogans*; (a) lateral; (b) dorsal (Vala & Knutson, unpublished data). (c) ***Hoplodictya setosa***, dorsal (modified from Neff & Berg 1962).

The puparium of *Hoplodictya setosa* (Fig. 14.51c), a predator of *Littorina littorea* in strandline debris on North American Atlantic coasts, is much longer in relation to width than any known sciomyzid puparium. These puparia probably are formed in hollow plant stems, giving them protection and additional buoyancy during tidal flooding. The puparia have not been found in nature, but larvae pupariated in plant stems in laboratory rearings (Neff & Berg 1962).

The exact pupariation sites, in nature, of most Sciomyzidae that pupariate outside the shell is not known and have been more or less identified on the basis of puparial shape. Such conclusions are not always valid. For example, the puparia of *Sepedon tenuicornis* and *S. spinipes* have shapes typical of aquatic *Sepedon* spp. but are the only aquatic Tetanocerini known to pupariate occasionally on vegetation above the water or soil surface (2–93 cm, average 52 cm in *S. tenuicornis*; Neff & Berg 1966). Another clue that the pupariation site might be rather remote from the microsite of the larva is the fact that, in laboratory rearings, many sciomyzid larvae seem to have a non-feeding, prepupariation period of several to many days. The extent and range of movement during this period, in the confined space of laboratory rearing containers, is impossible to determine. In nature it might be a "wandering phase," i.e., search for pupariation site, that is well known for many Diptera (see Section 9.3.5.1). Also, puparia of many typically terrestrial and semi-aquatic sciomyzids may be found floating in spring flood debris, giving an inaccurate impression of the pupariation site. For example, Lundbeck (1923) collected puparia of 26 species in flood debris in Denmark. Of these, we would classify only six species as truly aquatic (neustonic). The techniques of Przhiboro (2001), described in Section 20.1, for collecting surface samples in specifically identified microhabitats and holding them for emergence of adults, and use of emergence traps (e.g., Speight 2001, 2004) should be used more extensively for identifying pupariation sites.

14.2 PHYSIOLOGICAL/BEHAVIORAL ASPECTS

Physiology, per se, is one of the most neglected aspects of study of Sciomyzidae. The most critical questions requiring further research involve food quality and nutrition of adults and larvae. These are important to a more profound understanding of the diverse feeding strategies and use of Sciomyzidae as biocontrol agents. Other matters in regard to larvae include the precise function of ventral cibarial ridges in *Salticella fasciata* (Dowding 1971) and examination of other species for these structures, especially *Atrichomelina pubera*, which might represent the primitive Sciomyzidae. Whether or not the anterior spiracles of larvae, much of the time immersed in fresh or putrid mollusc tissue and thus similar to the condition of many saprophagous Diptera larvae, are functional needs to be studied. Broader studies are needed of: (1) salivary gland toxins (Trelka & Foote 1970, Trelka 1973, Trelka & Berg 1977); (2) the green fat bodies of developing larvae of some *Elgiva* species (Knutson & Berg 1964); (3) the pinkish pigment dispersed in eggs of *Anticheta analis*, which becomes concentrated in the gut of the developing larva before hatching (Knutson 1966); (4) swallowing of air to maintain bubbles in the gut by many aquatic and some terrestrial larvae; and (5) the calcareous septum produced by pupariating larvae of some *Pherbellia*, *Ditaeniella*, and *Colobaea*. Quite a few descriptions of larvae note that the Malpighian tubules of Sciomyzini are white, whereas those of Tetanocerini are brick-red. The Malpighian tubules of what appears to be *Trypetoptera punctulata* (Tetanocerini) were figured by Bhatia & Keilin (1937), and in at least that species there is a very short unpigmented proximal section of the tubules (Fig. 14.22b, c). The nature of the differences in pigmentation need study. Also, since the Malpighian tubules are one of the few organs in which there is cellular carry-over from the larval to the adult stage, examination of Malpighian tubules of adults would be of interest. The nature of the rectal and oral gland secretions produced by adults before or during mating by some Nearctic and Oriental *Sepedon* species (Berg & Valley 1985a, b) requires study. Cold hardiness is characteristic of this predominantly northern hemisphere family, with many species abundant far north of the Arctic Circle in northernmost Alaska, Canada, Fennoscandia, and Siberia, and this aspect also deserves further study. All stages, most commonly the pupae, are known to overwinter in areas with prolonged periods of freezing temperatures, depending on the species. Several special subjects that have received some attention are discussed below.

14.2.1 Adults – nutrition

The food of adults and how adults feed is poorly known. There are many observations of adults applying their mouthparts to dead or living snails in laboratory rearings and a few such reports from nature. As early as 1874, Glover reported that adults of *Sepedon* "frequent putrifying substances." Needham & Betten (1901) observed *Sepedon fuscipennis* sitting "close to the surface of the water, and apparently feeding on the stuff which collects about the bases of the leaves just above the water line." Bratt *et al.* (1969) noted that adults of many species of *Pherbellia* were observed feeding "on living snails placed in the breeding jars. The flies crawled over the exposed tissues and appeared to lap the secretions. When the snails withdrew into their shells, the flies often followed them in, disappearing from sight."

There have been many observations that the availability of crushed and/or living snails in laboratory rearings increases fecundity. Neff & Berg (1966) noted that protein is needed for continued egg production of *Sepedon* spp. Three female *Sepedomerus macropus* fed brewer's yeast and honey and two that were also provided crushed snails all laid roughly the same number of eggs during the first 2 weeks but the latter two females produced 395 and 323 eggs each (average 4.3 and 5.2 per day) and the three without protein produced only 196, 178, and 140 each (average 2.1, 1.8, and 0.9 per day). O. Beaver (1989) showed that protein food for adults increased larval food consumption, conversion efficiency, and growth rate, and shortened the developmental period at every stage (Tables 9.20–9.22).

Neff & Berg (1966) recorded a few instances of three Nearctic species of *Sepedon* collected on apple, willow, *Cardamine*, and *Caltha* flowers, but noted that Sciomyzidae are not commonly collected on flowers. Yano (1978) observed *Sepedon aenescens* (as "*Sepedon sauteri*") applying their mouthparts to a freshly dead earthworm on a leaf of rice plant in Thailand. Johnson & Hays (1973) observed a "*Sepedon* sp." feeding on egg-masses of *Chrysops* sp. (Tabanidae) in Alabama. Barraclough (1983) observed *Sepedon neavei* feeding on *Spirogyra* filaments and freshly crushed and boiled snails about one day after they had been offered, with feeding lasting 20–60 minutes.

S.A. Marshall (2006 and personal communication, 2005) photographed feeding by sciomyzid adults in nature

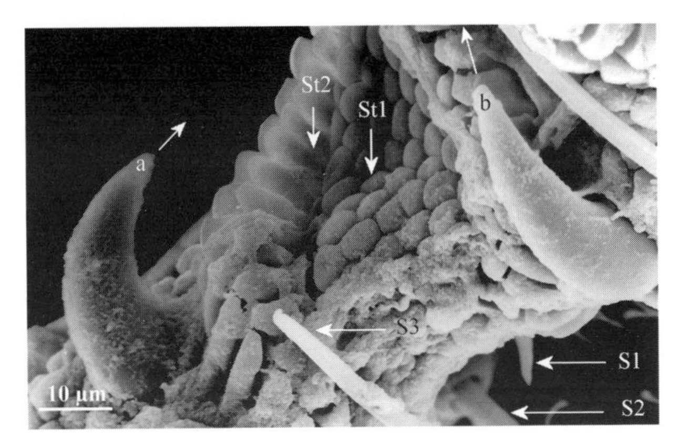

Fig. 14.52 Two hooks on labellum of ***Sepedon neavei*** curving away from side of labellum. Arrows a and b point toward prestomum. Note also presence of basiconic (S1), chaetical (S2), and trichoid (S3) sensilla and hexagonal (St1) and oval (ST2) structures on the labellum surface. Modified from Barraclough (2007).

in Canada. These included a *Sepedon* sp. feeding on nectaries of *Parnassia palustris* (Plate 1e), *Tetanocera valida* feeding on "a bird dropping loaded with insect parts," *Sepedon fuscipennis* "eating deer fly eggs on a cattail leaf," *Limnia loewi* feeding on a "dead earthworm," and many other species applying their mouthparts to leaf surfaces and to "amorphous masses that look like old spider prey." N. Vikhrev (personal communication, 2007) has provided photographs in nature of a male *Tetanocera* sp. feeding on the protective liquid ("spittle") surrounding the larval nest of an Aphrophoridae species, and another male *Tetanocera* sp. feeding on a dead Trichoptera.

Carles-Tolrá (2001) presented data on about 25 000 specimens of acalyptrates (304 species, 116 genera, 31 families) collected in 36 kinds of habitats in Spain. Of the 20 species of Sciomyzidae, most were found in typical habitats but one specimen of the terrestrial predator/parasitoid *Euthycera cribrata* was taken in a trap baited with a dead chicken.

Mouthparts were described and figured for *Trypetoptera punctulata* and "*Limnia magnicornis* Scopoli" (name not recognized, possibly not Sciomyzidae) (Becher 1882); *Pherbellia cinerella* (Wesché 1904); *Tetanocera plumosa* and *Sepedon fuscipennis* (Peterson 1916); *T. elata* and *T. ferruginea* (Frey 1921); *Tetanocera* sp., clearly showing the reduced clypeus (Gouin 1949); and *Ilione albiseta* (Oldroyd 1964). Barraclough (1983) presented a stereoscan micrograph of labellar hooks referred to as "prestomal teeth" of *Sepedon* (*P.*) *neavei* (Fig. 14.52). These are two rows of 7–9 opposing, curved, strong hooks on the inner surface of the labellum. He considered them as rasping–hooking structures that function during feeding, particularly when feeding on *Spirogyra* filaments. Barraclough (2007) presented a more detailed study of the labellar hooks of *S. neavei* and three additional Afrotropical species of *Sepedon* (*Parasepedon*), importantly showing that they are not extensions of pseudotracheal ring tips.

Detail from a stereoscan survey of the labellae of 35 families of Cyclorrhapha (Elzinga & Broce 1986) is of special interest. They found no prestomal teeth or ability to fully evert the labellar lobes in Sciomyzidae or any other acalyptrate Diptera. Five sciomyzid species were examined, one Sciomyzini (*Pherbellia n. nana*) and four Tetanocerini (*Dictya incisa, Limnia* sp., *Tetanocera plumosa*, and *Tetanocera* sp.). The morphology of the labellae of Sciomyzidae and the unrelated families, Psilidae, Diopsidae, and some Tephritidae was similar, but there was often more variability within families than between families. The authors found that in *Pherbellia, Dictya, Limnia* (Fig. 14.53a), and *Tetanocera* (Fig. 14.53b) the labellar lobes are above average to large in size, but are not capable of being everted beyond stage III of six stages. Stage VI is typical of direct feeders like *Calliphora* spp. in which prestomal teeth are rotated out for scraping (Fig. 14.54). Pseudotracheal tips appeared capable of minor abrasive activity, especially in *Dictya*, which have terminal microteeth on pseudotracheal ring tips.

In an extensive analysis of the mouthparts of adult Diptera, Zaitzev (1992) provided numerous stereoscan micrographs of the labellae of Bombyliidae and other Diptera. These included two sciomyzids, *Pherbina coryleti* and *Ilione rossica*, in which prestomal teeth were not seen. Zaitzev commented (translation by R. Rozkošný),

> The structures forming the pseudotracheal closing apparatus of many groups of Acalyptratae cannot be

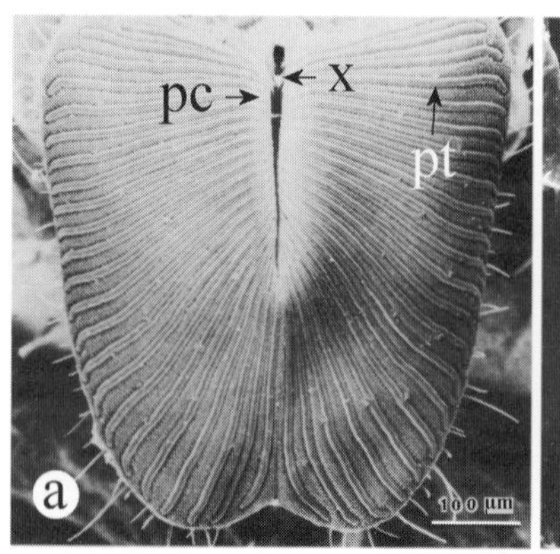

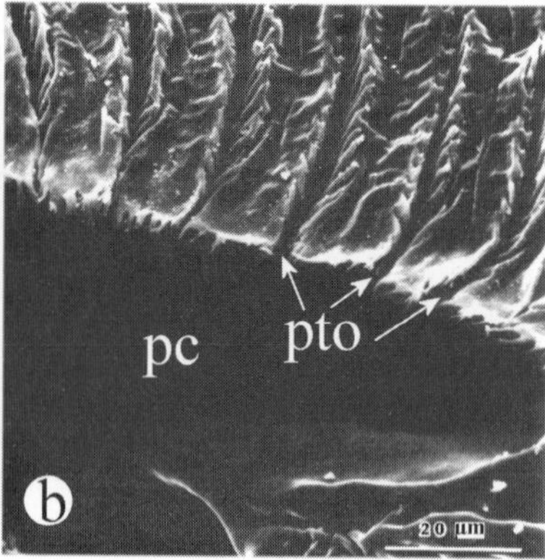

Fig. 14.53 (a) ***Limnia*** sp., terminal view of labellum with lobes in Stage 3, cupping position. (b) ***Tetanocera plumosa***, enlarged view of prestomum with pseudotracheal openings (pto). Note small brush of tentacle-like structures projecting from pseudotrachea (pt) into prestomal cavity (pc). x, location of prestomal teeth, when present. Modified from Elzinga & Broce (1986).

> compared with any of the above mentioned subgroups [lower Muscomorpha]. As a rule, the main difference is in the form of tooth-plates that are strongly sclerotized and thanks to this fact the pseudotracheal teeth are transformed into a grater covering the whole of the inner side of the labellae. In this way the labellae and the pseudotracheae are formed, e.g., in Lonchaeidae, Pallopteridae, Sciomyzidae.

Heads were prepared in potassium hydroxide and examined for 63 species in 31 genera of Salticellinae, Sciomyzini, and Tetanocerini by Mc Donnell & Knutson (unpublished data). Labellar hooks are present in all 19 species of *Sepedomerus, Sepedonella*, and *Sepedon* examined, but are absent in all other genera, including *Thecomyia*, a member of the *Sepedon* lineage. *Dictya* spp. has minute teeth (pseudotracheal ring tip extensions) scattered over the surface of the labellum. It is possible that the labellar hooks also aid in the emergence of the adult from the puparium and pupariation site, replacing, in part, the function of the ptilinum, which along with the ptilinal fissure, is absent in *Sepedon* and related genera (see Section 14.1.1).

14.2.2 Eggs

The only physiological aspects that have been studied are respiration (see Section 14.1), the effect of humidity and wetting on hatching, the effect of temperature on development, and age of the parent female on egg viability (see Section 9.2.2).

14.2.3 Larvae

14.2.3.1 AQUATIC LARVAE

Further critical studies are needed of the ability of larvae to attack and feed on snails well below the water surface especially in view of their use as biocontrol agents. These include re-examination of the behavior of larvae earlier reported to be restricted to feeding at the water surface with their posterior spiracles exposed to ambient air. Berg & Neff (1959) and Neff (1964) studied attack by ten species of *Dictya, Sepedon*, and *Tetanocera* against 14 species of *Schistosoma* host snails on moist sand and in water deeper than the lengths of the larvae. Referring to these studies, Berg (1964) noted that the larvae kill snails almost as effectively in the latter situation: "Despite their open tracheal systems, they evidently have excellent adaptations to live and obtain their food in water." Larvae of all instars of the typical aquatic, predaceous Tetanocerini (our Behavioral Group 11, ecomorphological category *a* of Vala & Gasc 1990b) characteristically feed infraneustically. That is, they hang from the surface and feed on snails just beneath the surface, as well as attacking snails on wet surfaces. Such infraneustic behavior is enabled by the buoyancy provided by the elongate posterior spiracular disc lobes; the elongate, branched, hydrofuge interspiracular processes; and by the swallowed air bubbles maintained in the gut of these larvae.

14.2.3.1.1 Air swallowing

Larvae have been seen swallowing air during rearings of many species of aquatic, predaceous Tetanocerini. Brocher (1913) was the first to mention the presence of an

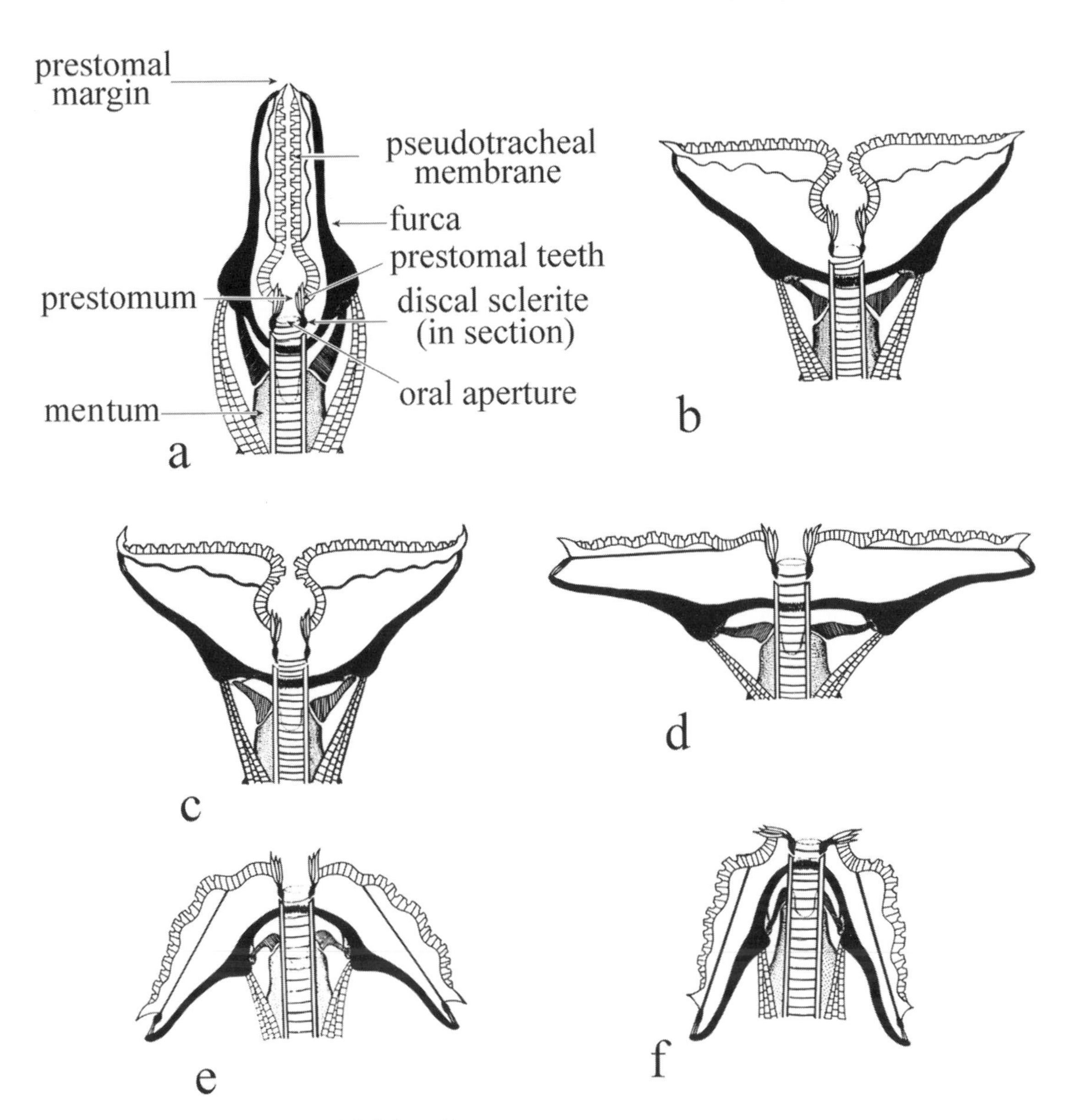

Fig. 14.54 Stages of labellar eversion of *Calliphora vicina*. Positions: (a) resting; (b) filtering; (c) cupping; (d) intermediate; (e) scraping; (f) direct feeding. From Elzinga & Broce (1986).

air bubble in the gut of Sciomyzidae. The air swallowing behavior was excellently described for *Sepedon* spp. by Neff & Berg (1966). It has also been seen during rearings of several terrestrial Tetanocerini. But it has not been seen in any rearings of the few Tetanocerini that live most of the time beneath the surface of the water, nor in any Sciomyzini. Wesenberg-Lund (1943) suggested that the air bubble that develops in the gut also might help to create turgidity needed by the larva to crawl effectively. Neff & Berg (1966) considered that this function explains air swallowing by terrestrial larvae as well.

Air swallowing also might play a part in determining the shape of the puparium of some species whose puparia have unusual, characteristic curvatures and bulges, e.g., *Tetanocera arrogans* (Fig. 14.51a, b). Shortly before larvae of the neustonic, aquatic predator *Elgiva solicita* contracted to form the puparium, they defecated until the gut was empty and swallowed air until a mass of bubbles formed in the gut. The prepupal stage lasted only about 24 hours, and the pupa, seen through the lightly pigmented, translucent puparium, rested on the inner, ventral surface, occupying only about 2/3 of the available space (Knutson & Berg 1964).

14.2.3.1.2 Swimming behavior

Another adaptation to aquatic existence is the strong and characteristic swimming behavior by sciomyzid larvae, described over a century ago by Needham & Betten (1901). They noted that the behavior of *Sepedon fuscipennis* "is most curious. It pulls itself below the surface, turns over on its back, and then progresses by bending and straightening its body, striking the water sharply with the flat face of its caudal disc." Brocher (1913) briefly mentioned swimming by *Sepedon* sp. larvae. The precise description of swimming *Sepedon* larvae by Neff & Berg (1966), who reared 12 species of the genus, is quoted here in full,

> *Sepedon* larvae swim with 2 effective but remarkably different methods. If placed in the center of a large dish or aquarium so they have no contact with floating or emergent objects, most larvae soon use one these techniques. Oriented in the normal floating position with dorsal surface up, a larva extends its anterior body segments forward, then sweeps them downward and backward in a quick stroke which propels the larva forward. This motion is accompanied by an extension, then contraction of the intermediate abdominal segments. The posterior spiracular disc dips into the water during the swimming stroke, but is raised so the spiracular plates are again exposed to the air as the larva returns its extended anterior end to a position parallel to the surface film, prior to the next downward stroke.
>
> Sometimes an equally effective and more remarkable swimming motion is seen. When the larva has lowered its extended anterior end several times and thrust it to both right and left without touching any substrate, it stretches out just below the surface, rolls from side to side to develop momentum, then rolls over completely so that its ventral side is up. Then it bends the posterior 1/3 of the body down and snaps it sharply backward and upward . . . Both swimming motions described above are smoothly coordinated series of rhythmic movements which result in rapid propulsion.

Abercrombie (1970) observed swimming behavior in two species of *Protodictya* and two species of *Sepedonea*. Swimming of larvae was observed in nature as well as in the laboratory for *Thecomyia limbata* by Abercrombie & Berg (1975) and *Dictyodes dictyodes* by Abercrombie & Berg (1978). The behavior of *T. limbata* was like that described by Neff & Berg (1966) as the first method, but *D. dictyodes* displayed both swimming behaviors. The swimming behavior of *Dictya* spp. described by Valley & Berg (1977) is like that of the first method described by Neff & Berg (1966), except the posterior spiracular disc remains in contact with the surface film. Barraclough (1983) noted that first-instar larvae of *Sepedon neavei*: "moved about just below the surface by rhythmical contraction and relaxation of longitudinal muscles," and that in late first-instar larvae, "faster forward movement was achieved by rapid up and down movements of the spiracular disc area in the water . . . Mature L3s often rolled on to their dorsa and swam forward by undulating the spiracular disc area. Thus forward motion could be halted by rolling right side up and holding the spiracular disc curved dorsally as a brake." The author noted that *Sepedon testacea* had "a rather clumsy swimming action, with lateral rolling." Even some of the less aquatic Tetanocerini display swimming behavior. *Hydromya dorsalis* larvae, whose microhabitat seems to be wet surfaces rather than in the water, has been observed swimming by first rolling over in a "clumsy" manner, so that its ventral surface is uppermost, then by peristaltic movements along the body length, creeping below the surface film, without stroking movements of the anterior or posterior ends (Knutson, unpublished data). Similar swimming behavior was described for *Psacadina verbekei* and *P. vittigera* except that *P. vittigera* crept upside down for at most one minute, then rolled over so its dorsal side was up, floated for several minutes, then alternately repeated the behavior (Knutson *et al.* 1975). Foote (1999) noted that larvae of the aquatic predator *Tetanocera annae* swam by rapid down strokes of the anterior half of the body. Barnes (1979b) described most aspects of the behavior of the four species of *Neolimnia*, subgenus *Pseudolimnia*, that he reared in New Zealand as fairly typical of aquatic, predaceous Tetanocerini. However, he wrote, "Larvae are not adept at locomotion in open water. They extend the anterior segments forward and then pull up the posterior segments as if they are crawling across a solid surface. This enables them to move only weakly across the water surface. No other swimming motions were observed."

14.2.3.1.3 Submerged feeding
Larvae of five genera of Palearctic, Nearctic, and Subantarctic Tetanocerini routinely feed completely beneath the surface during only the first stadium or until late into the third stadium, depending on the species. *Hedria mixta* (Foote 1971), *Ilione albiseta* (Knutson & Berg 1967, O. Beaver 1972b, Lindsay 1982, Lindsay *et al.* 2011), *I. trifaria* (Mc Donnell *et al.* 2005a) and four species of *Dictya* (Valley & Berg 1977) feed on submerged snails. Also three species of *Renocera* (Foote 1976), *I. lineata* (Knutson & Berg 1967, Foote & Knutson 1970), and *Eulimnia philpotti* (Barnes 1980a) feed on submerged fingernail clams. These five widely dispersed genera do not seem to be closely related.

Ilione spp., *H. mixta*, and *E. philpotti* share some strong similarities in morphology of the larvae. They have elongate posterior spiracular disc lobes (and elongate lobes on body segments in *I. lineata* and *E. philpotti*); elongate, branched interspiracular processes; and elongate, convoluted tracheoles throughout the body that extend into the posterior spiracular disc lobes and body lobes. Of *Renocera* spp., the immature stages of only the surface-feeding *R. pallida* have been described (Horsáková 2003). Foote (1976) mentioned the elongate anterior spiracles and elongate postanal portion of the puparium of the subsurface feeding *R. striata*. The posterior spiracular disc lobes and interspiracular processes are not strongly developed in *Dictya* spp., but are more similar to those of the surface-dwelling aquatic predators in *Tetanocera* and *Sepedon*.

Submerged feeders differ in their ability to remain beneath the surface, with *E. philpotti* living the longest beneath the surface and the *Dictya* species the shortest. Unfed first-instar larvae of *E. philpotti* lived under the surface 8–15 days; feeding first instars lived under the surface 29–36 days; second instars, 64–67 days; and only late in the third stadium were the posterior spiracles exposed to ambient air. Two second- and two third-instar larvae of *E. philpotti* were collected from substrate mud under 20 cm of water. Larvae of *H. mixta* and *E. philpotti* do not carry an air bubble in the gut and display no swimming behavior. First-instar larvae of *I. lineata* lived beneath the surface without feeding for 16 days, the first and second stadia took place completely submerged, and the entire third stadium was passed with the posterior spiracles exposed to the air. The behaviors of three species of *Renocera* are similar; newly hatched larvae crawl to the water surface and free themselves from the surface by vigorous movements, the first stadium lasts 5–6 days with molting to the second stadium under the surface, then the posterior spiracles are exposed to the air. The older larvae can live only about one hour beneath the surface and at least those of *R. striata* maintain air bubbles in the gut (Foote 1976). The larvae of the other known clam-killer in *Renocera*, *R. pallida*, do not penetrate beneath the water surface but feed on clams exposed on wet substrates (Horsáková 2003). A third-instar larva of *I. trifaria* was found feeding on a *Physella acuta* on a bed of *Fontinalis antipyretica* moss 10 cm below the water surface (Mc Donnell *et al.* 2005a). But during laboratory rearings of this species, many larvae were reared to pupariation on *Helisoma*, *Lymnaea*, *Physa*, and *Planorbis* spp. in dishes containing water to a depth of only a few millimeters (Knutson & Berg 1967). This species also is unusual in that third-instar larvae were able to kill and feed on the operculate snails *Hydrobia* sp. and *Melanopsis praemorsa*.

The behavior of *I. albiseta* is more labile, and the several reports on this species are summarized below; detail is provided also because this species is of interest as a potential biocontrol agent of snail intermediate hosts of *Fasciola* (see Section 18.1). Knutson & Berg (1967) noted that many larvae that hatched from submerged eggs continued to live submerged for 8–11 days. They reared only two larvae, on snails in rearing dishes with only a wet filter paper or wet sand substrate, to pupariation (one of these larvae had been dissected from the egg membrane, the other had hatched naturally). O. Beaver (1972b), who reared the species through the complete life cycle, did not describe her rearing methods but noted, "Between meals, the larvae often remained completely submerged below the filter paper at the bottom of the rearing dish." Lindsay (1982) conducted detailed studies of the biology of *I. albiseta* during 1980–82 in Ireland (204 larvae, primarily third instars, collected in nature and 1065 larvae reared from eggs laid in the laboratory). Most of the larvae he studied "remained submerged until about the middle of the third-instar stage" and then "larvae actively emerged from the water onto the sides of culture dishes. Even though the larvae were routinely replaced in the water at each examination between 50% and 100% were found out of the water at the next inspection." He noted that when the water level in culture dishes fell, both first- and second-instar larvae were found at the water surface, and that larvae of all instars successfully attacked and fed on their prey both in and out of the water. Large numbers of larvae were reared from hatching to pupariation by Gormally (1988b) in Petri dishes lined with filter paper, with water to a depth of 5 mm, covered with parafilm, placed in sealed, weighted containers, and submerged in water baths at controlled temperatures of 14–26 °C and kept at LD 16:8. It was not stated whether or not the larvae kept their posterior spiracles in contact with the air.

Valley & Berg (1977) described in considerable detail the behavior of many species of Nearctic and Neotropical *Dictya* larvae that feed below and at the water surface. *Dictya* species seem to have the least developed ability of the five genera of subsurface-feeding sciomyzids to live beneath the surface. Although larvae of all instars of several *Dictya* species are able to search for and feed on snails while submerged, descent below the surface film seems primarily to attack prey. The larvae then extend their posterior ends to make contact with air, or float to the surface with their prey, or even crawl backwards to the surface film, dragging their prey. Ability to stay under water diminishes as the *Dictya* larvae develop, second-instar larvae remaining submerged for up to 64 minutes, and third instars being able to remain below the surface for no longer than three minutes. Similar to surface dwellers, but unlike most subsurface dwellers, *Dictya* larvae have well-developed swimming behavior and routinely swallow air. Larvae of five *Dictya* species were found in nature, just below the surface among plants and pieces of debris. Barraclough (1983) observed that first-, second-, and third-instar larvae of *Sepedon neavei* could kill and feed on snails while completely submerged for "short periods."

14.2.3.1.4 Respiration

Proof that larvae are capable of using oxygen dissolved in water was obtained for *I. albiseta* by Gormally (1988b). He used cartesian divers in 26 experiments with first-instar larvae and a Warburg apparatus in ten experiments with third-instar larvae to measure oxygen consumption. The rate of oxygen consumption increased as temperatures increased. Mean oxygen consumption (*ul*) per first-instar larva per hour ranged from 13.72×10^{-3} at 7.5 °C to 42.18×10^{-3} at 16.1 °C, and per third-instar larva per hour from 1.63 at 8 °C to 12.98 at 17.5 °C. Consumption per mg per hour for third-instar larvae ranged from 0.04 at 8 °C to 0.3 at 17.5 °C.

The mechanism(s) for intake of dissolved oxygen by sciomyzid larvae are unknown but in some cases seem to be correlated with highly convoluted tracheoles that enter even into the posterior spiracular disc lobes and into the elongate lobes on body segments. Such tracheoles have been described for *I. lineata* and *E. philpotti* but not for other species. The numbers of lobes on the anterior spiracles of second-instar larvae of some of these species are greater than in most other genera (*E. philpotti*: 18, weakly developed; three *Ilione* spp.: 10–25) and third-instar larvae (*H. mixta*: 10–14; *E. philpotti*: 17–22; four *Ilione* spp.: 16–23). The aquatic predator *Ilione turcestanica* has the greatest number of any species in the family (30–35 in the second instar, 33–37 in the third instar). Whether or not this species feeds beneath the surface is unknown; 36 larvae of *I. turcestanica* were reared to pupariation in dishes with water to a depth of a few millimeters, but they were not exposed to deeper water (Knutson & Berg 1967). However, the number of anterior spiracle papillae seems not always correlated with larval life style. Most surprisingly, the second greatest number of papillae (29 in the second instar, 34 in the third) are in *Sciomyza varia*, a parasitoid/saprophage of estivating *Stagnicola elodes*, that has its head immersed in the oxygen-deficient, liquefied, rotting tissues of the prey during later larval life (Barnes 1990).

The functional importance of the anterior spiracles of sciomyzid larvae is not known. As in many other Diptera, the anterior segments of the larvae of many terrestrial and semi-terrestrial Sciomyzidae are immersed for long periods in the tissues of the host/prey and these tissues often are in a liquified, decaying state that must be oxygen deficient. Buck & Keister (1953) studied the relative importance of anterior and posterior spiracles in overall gas exchange in mature larvae "at about the time of emptying the digestive tract preparatory to pupation" of the blowfly *Phormia regina* (Calliphoridae) by reciprocal ligation experiments. They found that the posterior spiracles by themselves can admit enough oxygen from air to supply normal needs and that the anterior spiracles have about half the capacity of the posterior spiracles. According to Darvas & Fónagy (2000) the larvae of most Cyclorrhapha appear to have functional posterior spiracles and non- to low-functioning anterior spiracles in the larval stage but functional anterior thoracic spiracles in the pupal stage. However, the great variety of morphology of the anterior spiracles of the larvae and to some extent the disposition of the posterior spiracles of the puparia of Sciomyzidae and many other Cyclorrhapha indicates strong selective pressure in the evolution of the ontogeny of both anterior and posterior spiracles. Such selection probably reflects functional advantages. See Section 14.2.

14.2.3.2 SEMI-AQUATIC AND TERRESTRIAL LARVAE

14.2.3.2.1 Respiration

The ability to survive submergence can be strikingly different between first-instar larvae that are well known to live habitually at the water surface and usually maintain contact with some support compared with those that live on moist surfaces. First-instar larvae of *Tetanocera annae* (Tetanocerini), which are typical of the many aquatic predators studied by Foote (1999), die if held beneath the surface for 20 minutes. However, newly hatched first-instar larvae of *Pherbellia griseicollis*, a typical Sciomyzini predator of snails on damp surfaces, without morphological adaptations for an aquatic existence, were pushed beneath the water surface and remained there without feeding for 96 hours. When removed, they fed, developed normally, and one formed a puparium (Knutson 1988).

The first- and in some species, early second-instar larvae of the most intimately associated terrestrial parasitoid Sciomyzidae, *Colobaea bifasciella, Pherbellia s. schoenherri, P. dubia, Sciomyza varia*, and *Tetanura pallidiventris*, feed between the mantle and shell of their hosts, completely cut off from the ambient air for up to 12 days (longest, *P. dubia*). These larvae must be respiring through their integument on dissolved oxygen in extrapallial fluid of their hosts. Young larvae of species in several genera that feed in a somewhat parasitoid manner on snails in moist situations but later become predatory and/or saprophagous, e.g., some species of *Pherbellia, Pteromicra*, and *Euthycera*, also might have this ability. It also might be true of young larvae of the terrestrial *Salticella fasciata*.

Bratt *et al.* (1969) noted that in laboratory rearings of *Pherbellia dorsata* (Behavioral Group 2) on *Planorbis planorbis*, when the rearing jars were allowed to become somewhat dry, "a thin crust formed on the surface of the liquefied snail tissues. Air might have been trapped under this crust, because larvae were able to feed beneath it, completely cut off from the outside air for as long as 24 hours. In nature, this phenomenon might occur fairly often when larvae attack snails in open, sunlit areas where drying conditions exist."

Newly hatched larvae of *Euthycera chaerophylli* rapidly penetrate deep into the tissues of their *Deroceras laeve* or *D. reticulatum* slug hosts and apparently feed there, completely cut off from outside air (Knutson, unpublished data). Although the life cycle has not been completed, first-instar and small second-instar larvae were found inside the bodies of hosts, that had died or were dissected, up to 14 days after penetrating the slug. The respiratory requirements of these larvae were indicated by our following simple experiment. Several newly hatched first-instar larvae were pushed beneath the surface of water (depth, 0.5 cm) in 2 × 5 × 5 cm transparent plastic boxes with tightly fitting lids; the water was not changed. Several of these larvae lived for a week or more, up to 20 days, on the bottom.

14.2.3.2.2 Survival without feeding

The ability of first-instar larvae to live for long periods without feeding is pronounced in Sciomyzidae having terrestrial larvae. Vala (1989a) noted that some such larvae can live for 20 days without losing their potential to attack. In general, there appear to be some differences between aquatic and terrestrial larvae in regard to how long first-instar larvae can live without feeding. The terrestrial species, in general, seem to be able to survive longer. Experiments under controlled temperatures on survival of fed and unfed neonate larvae show differences among aquatic species, i.e., between univoltine species with larvae developing during the fall–winter and multivoltine species with larvae developing during the warmer months.

In a study of predation by *Sepedon sphegea*, Ghamizi (1985) compared the results of food available in excess after hatching of larvae with larvae that were starved (unfed) during the first 48 hours. One, 5, 10, and 20 larvae were associated with 25 *Radix balthica*. This provided evaluation not only of the development of individual larvae but also evaluation of the effects of intraspecific competition under precise conditions. As a function of the density of prey, survival of larvae showed a very marked decline of up to 50% at the time of molting from the first to the well-fed second stadium. This decline was accentuated for the larvae that were starved during the first 48 hours. The curves of survival (Fig. 14.55a, b) show slopes which strongly decline when competition for food intensifies. The concave courses of these curves show a type III population development, as indicated by Pearl (1922). The results also were seen in the levels of production of adults of the ensuing generation. This variation is effected by the logistical growth of the population. Likewise it is an element of the numerical response of the individuals, according to Holling (1961) and Beddington *et al.* (1976). In nature, the first cohort of adults thus is regulated, to some extent, by the quantity of food available to the first-instar larvae.

Gormally (1988b) found that survival of unfed first-instar larvae of the univoltine *Ilione albiseta* (most of which hatch during the fall) decreased as temperature increased,

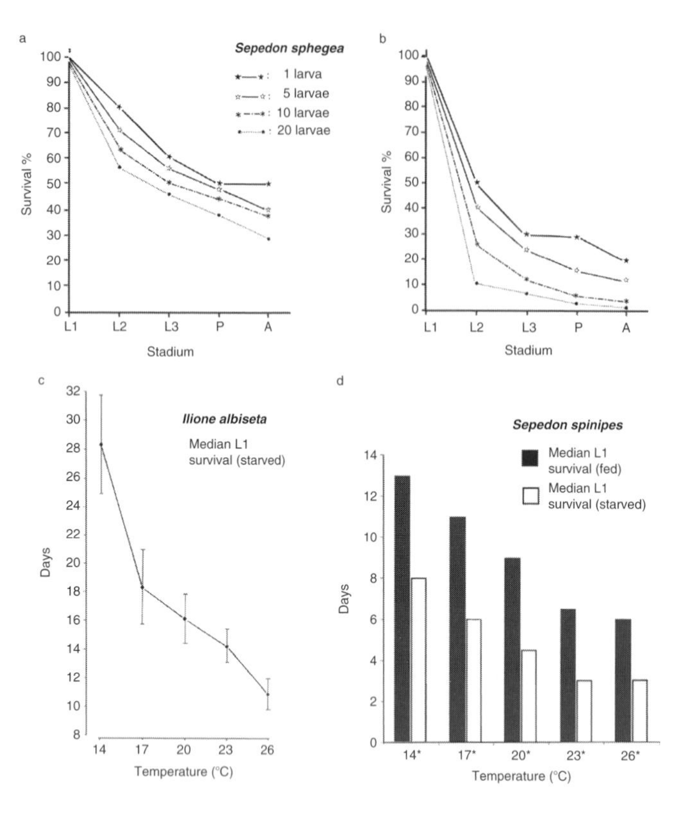

Fig. 14.55 (a, b) ***Sepedon sphegea*** survival from fed or starved neonate larvae; (a) when first-instar larvae fed immediately after hatching; (b) when first-instar larvae were starved for 48 hours after hatching (from Ghamizi 1985). (c) Mean larval survival (days) of starved L1 ***Ilione albiseta*** at five constant temperatures (vertical lines denote 95% confidence limits) (from Gormally 1988b). (d) Comparison of median survival period of starved and fed (one 2.0 mm *Galba truncatula*) L1 ***Sepedon spinipes*** at 14, 17, 20, 23, and 26 °C. Temperatures denoted by * indicate significant difference in median neonate survival between starved and fed larvae (from Mc Donnell 2004).

from a mean of 28.4 days at 14 °C to 11 days at 26 °C, and that first-instar larvae can survive up to one month without food (Fig. 14.55c). R. J. Mc Donnell (2004) reared the multivoltine *Sepedon spinipes* (which breeds during the warmer months) and found the same relationship with temperature but a much shorter survival period (Fig. 14.55d). Notably, the experiments were conducted at the same temperatures and with parent material from populations at the same place (Galway, Ireland), thus differences due to geographical origin can be discounted. See Section 7.3.5.

14.2.4 Puparia and pupae

In regard to physiological factors of puparia and pupae, only aspects of phenology and development (see Section 9.3.5) and respiration have been investigated. As noted in Section 14.2.3.1.4, larvae of most Cyclorrhapha have functional posterior spiracles and non- to low-functioning anterior spiracles, but functional anterior spiracles in the pupal stage. However, we propose that the great variety of morphology of the anterior spiracles of larvae and of the development of segment XII and disposition of the posterior spiracles of Sciomyzidae and many other Cyclorrhapha indicates selective evolutionary pressures that reflect functional advantages to these diverse structures. It might be that the anterior spiracles of Sciomyzidae and many other Cyclorrhapha are only barely, or intermittently, functional in the larval stage but are ontogenetic developments that find their primary function in the pupal stage. Another possibility is that the anterior spiracles, at least in the larvae, function primarily in expelling air from the tracheal system. In the Sciomyzidae, no pupal respiratory processes or horns penetrate through the puparium wall as is known to occur sporadically in some families and genera of Aschiza and Schizophora (Meijere 1902, Ferrar 1987).

Knutson & Berg (1967) noted that pulsations of the posterior tracheal trunks could be seen through the translucent integument of very recently formed puparia of *Ilione albiseta*, collected on March 30 in central Italy. Notably, these puparia contained pre-pupae, not pupae.

The unusually translucent, tan integument of the puparium as well as of the pupariating larva of *Elgiva solicita* allowed Knutson & Berg (1964) to observe the condition of the posterior tracheal trunks of pupariating larvae. They noted,

> Larvae did not feed for 24–48 hours prior to puparium formation. Shortly before the larva contracted to form the puparium, it defecated until the gut was empty and swallowed more air until a mass of bubbles formed in the mid-gut. The posterior tracheal trunks could be seen through the transparent integument of the puparium and the body wall of the prepupa as they extended anteriorly from the posterior spiracles. The prepupal period lasted only about 24 hours. When true pupation occurred, the tracheal trunks detached from the pupa and came to rest on the inner, posterodorsal wall of the puparium, on each side of the midline. The pupa rested on the floor of the puparium and occupied only about two-thirds of the available space.

Also, Knutson & Berg (1967) observed that some of the 98 puparia of *Ilione albiseta* collected on March 30 in central Italy were unpigmented; contained pre-pupae, not pupae; and "pulsations of tracheal trunks could be seen through the integument." Their observations contrast somewhat with the statement of Darvas & Fónagy (2000) that "the larval tracheal epithelium is histolyzed during metamorphosis." Perhaps the air in the vacant space of *E. solicita* puparia is sufficient for the subsequent metabolism of the pupa and pharate adult. Puparia of many aquatic Tetanocerini have characteristic bulges, especially anterodorsally in the region of the pupal respiratory structures. The bulges might function both for flotation and as air reservoirs for development of the animal within. Puparia of Sciomyzini, except *Sciomyza varia* and *Colobaea bifasciella*, and terrestrial Tetanocerini and many other Cyclorrhapha lack such bulges and their vacant space.

Two circumstantial bits of evidence also indicate that both the anterior and posterior spiracles of puparia of Sciomyzidae are functional. First, for those species of *Pherbellia*, *Ditaeniella*, and *Colobaea* whose pupariating larvae excrete a thick calcareous substance from the anus, which is moved forward over the larva and covers the surface or is formed into a septum in the shell aperture, the spiracles remain free of the material. This protection apparently results from the oily film produced by glands around the spiracles. However, we question whether or not the oily secretion persists after pupariation; it is unlikely that it continues to be produced in the tanned, sclerotized integument of the puparium. In any case, the quality of the air, that we presume passes through the spiracles, would seem to be more indicative for parasitoid wasps, as cited below, in determining the viability and suitability of the animal within the puparium than surface oil around the spiracles. Second, parasitoid wasps of the family Diapriidae, which oviposit into puparia of Sciomyzidae, characteristically tap the posterior spiracles and the anterior end with their antennae, apparently being able thus to sense the viability and suitability of the animal within (Knutson & Berg 1963b, O'Neill 1973) (Plate 4h). Furthermore, according to Isidoro *et al.* (2000) bioassays with altered females of the diapriid *Trichopria drosophilae* (apical antennomeres removed) and intact puparia of *Drosophila melanogaster* or with intact *T. drosophilae* and puparia with anterior and posterior spiracles removed or covered with wax showed that spiracular glands elicited recognition response.

Further study of the effects of temperature and light on induction and termination of diapause and ability to survive desiccation in the pupal stage are needed. Considering the pivotal role of the puparium in protecting the animal within from parasitism, predation, and disease, and in protecting it from abrasion during transport by seasonal flooding, which might be important in dispersal, further comparative study of the structure of the puparium is warranted.

14.3 GENETICS AND RELATED ASPECTS

There has been no review of information on genetics and related aspects except for the brief summary by Vala (1989a) of information on karyotypes. Stevens (1908) described the karyotype of *Limnia boscii*. Boyes *et al.* (1969, 1972) provided detailed descriptions, with many photographs of chromosome complements and diagrams of the karyotypes of 71 species in 25 genera. These included *Salticella fasciata* (Salticellinae) and 24 genera of Sciomyzinae (five Sciomyzini, 19 Tetanocerini), from the Nearctic, Palearctic, Australian, Neotropical, and Oriental regions. Slides were prepared from brains of larvae or ovaries or testes of adults. The characteristic arrangement for Diptera of $2n = 12$, with five pairs of autosomes and one pair of sexually heteromorphic chromosomes usually was found in Sciomyzidae (Fig. 14.56a, d). A reduction of $2n = 10$ was seen in *Pteromicra glabricula* (Fig. 14.56c) and *P. similis* of Sciomyzini and *Tetanocera arrogans* of Tetanocerini. *Ditaeniella parallela*, *D. patagonensis* (Sciomyzini) and *Hoplodictya spinicornis* and *Sepedon*

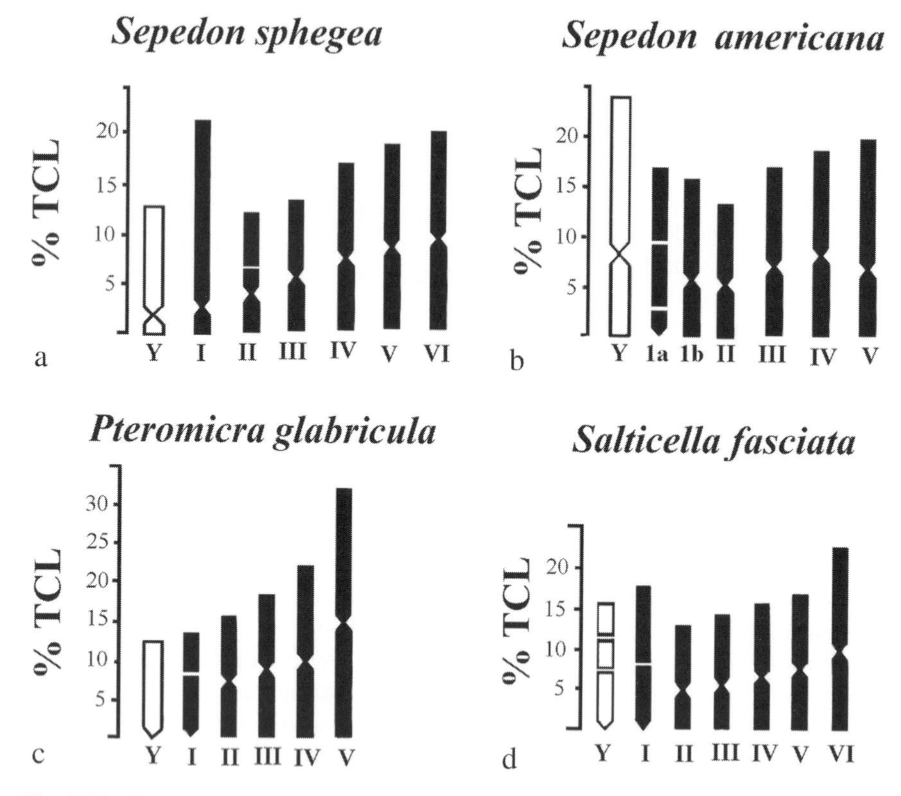

Fig. 14.56 Idiograms of karyotypes of Sciomyzidae. (a) *Sepedon sphegea*; (b) *S. americana* (from Boyes *et al.* 1969); (c) *Pteromicra glabricula*; (d) *Salticella fasciata* (from Boyes *et al.* 1972). TCL, total complement length; Y, sex chromosome; I–VI, autosomes.

armipes (Tetanocerini) lack a Y chromosome in the male, and it may be absent in males of other species. In *Sepedon americana* there is a supernumerary sex chromosome (X1, X2, Y type; Fig. 14.56b); *S. spinipes* was not studied. The karyotype of *S. fasciata* fits with the concept of a primitive karyotype (Fig. 14.56d). There is considerable karyological variation in the Sciomyzini, especially in *Pherbellia*. Variability is conservative in *Dictya*, a genus that is also conservative in most aspects of adult and larval morphology and behavior; moderate in *Tetanocera*; and extensive in *Sepedon*, one of the most apomorphic genera. These data essentially support the major distinctions in the cladistic analyses of adults of the genera of Sciomyzidae, and the general conclusion that conservative karyotypes may be found in relatively primitive and relatively specialized species but modified karyotypes are likely to occur only in relatively specialized species. Karyological re-organizations have occurred without much effect on gross adult morphology, which has apparently changed more in response to genetic mutations.

Manguin & Hung (1991) studied the developmental genetics of larvae, pupae, and adults of *Sepedon f. fuscipennis*. They found 28 isozyme markers, of which few were stage specific. Of the 23 isozyme markers for adults, three were adult specific and of the 25 found for larvae five were larval specific. Pupae had 21 isozyme markers; none was pupal specific, but one of them was mainly active at this stage; 16 were polymorphic and four were monomorphic. Manguin (1990) used these baseline data in comparing 21 enzyme systems and 23 loci of enzymes among two "subspecies" and one "intermediate" population of *Sepedon fuscipennis* described on morphological characters of the adults (male genitalia and parafrontal spot) by Orth (1986). It was concluded that the allelic differences between the three populations studied were intraspecific variation, but the "subspecies" were not formally synonymized (see also

Chapter 13, Table 13.2). The results are particularly instructive as to what might be the relationships among many closely similar Afrotropical species of *Sepedon*, also separated primarily by male genitalia, parafrontal spots, and other color characters.

Genetic compatibility of widely separated (both north–south and east–west) populations has been demonstrated on several occasions by the successful mating of adults from widely separated localities. For example, a virgin female *Thecomyia limbata* laboratory reared from a Rio de Janeiro population mated with a field-collected male from southern Paranà, Brazil (about 1600 km distant) (Abercrombie & Berg 1975). Also, a very small male of *Elgiva solicita* collected in Alberta mated within 1 minute after being placed in a rearing jar with a female, almost twice his size, collected in Belgium (about 1800 km distant). Reared females of this species from North American populations which had been isolated since emergence and were known to be virgin laid viable eggs after mating with males field-collected in Belgium (Knutson & Berg 1964). As noted in Section 5.1, differences in host/prey specificity of proximate populations have been observed in laboratory rearings of a few species.

There have been few attempts to use cross-mating experiments to resolve the identity of species of Sciomyzidae. Fisher & Orth (1969b) attempted cross-mating of *D. fontinalis* with two forms of the closely similar *D. montana* in southern California. No mating was seen and no eggs were laid by either species. In their cross-matings of *D. montana* and the similar *D. texensis* in southern California there was attempted copulation both ways and numerous non-fertile eggs were laid by *D. montana* females mated by *D. texensis* males, but not vice-versa. Fisher & Orth (1983) considered *D. montana* a polytypic species based on differences in size, male and female terminalia, and often by patterns of distribution. They segregated four forms found in Washington, Oregon, California, Nevada, Idaho, Utah, Colorado, and Baja California Norte (Mexico). They found reciprocal cross-mating to be of limited to good success depending on the combinations attempted. Fisher & Orth (1972a) attempted cross-mating of the closely related *Sepedon pacifica* and *S. praemiosa*. The ranges of these species in western USA are disjunct but closely approximate from western Nebraska southwest to northern Arizona. No mating by four pairs of F_1 *S. pacifica* (from northern Arizona) × *S. praemiosa* (from southern California) was seen and no eggs were laid.

Striking abnormalities in adult morphology have been described in regard to six species, but except for the case of gynandromorphism and the aedeagal apodeme they involve only one side of the body, and thus likely are not due to inheritance (Morgan *et al.* 1925). Steyskal (1974a) described and figured a gynandromorphic specimen of a Nearctic species of *Limnia* in the *L. fitchii*–*L. ottawensis* group. The specimen is basically male in the first seven abdominal segments, but segments VIII and following are very abnormal and in part, especially segment IX, are very similar to a normal female. Rozkošný (1964) figured an instance of duplicated aedeagal apodeme in a male of *Pherbellia clathrata* which was explained later as a malformation by Elberg & Rozkošný (1978). Berg (1973b) described and figured a teratological specimen of a male of *Trypetoptera punctulata* collected in Norway and a male of *Tetanocera hyalipennis* collected in Belgium. In each, one antenna is almost perfectly duplicated in the second and third segments, in side-by-side positions, along with abnormalities of the compound eyes, frontal region, other parts of the head capsule, and setae on the same side. O'Neill & Berg (1975) described and figured in detail the morphology and described the behavior of a grossly malformed *Dictya pictipes* female. It had one wing, five legs, and the mesothorax almost completely absent on one side, the longitudinal axis of the body severely displaced, and many thoracic setae absent or deformed. This specimen, reared in the laboratory from a larva collected in nature, lived for 49 days, mated, and laid more than 200 viable eggs. Her offspring were normal morphologically. The F_1 females laid viable eggs after mating with male siblings and field-collected males. It was concluded that the fact that her body, flexed through an angle at about 90 degrees, indicates that female *Dictya*, and perhaps other Sciomyzidae, are recognized and mating is stimulated by olfactory rather than visual cues. Yoneda (1984) described and provided detailed drawings of six malformed specimens of *Sepedon aenescens* obtained during laboratory rearings. The pleural region was especially strongly deformed and reduced on one or both sides, along with loss of thoracic setae and spiracles and wings, and deformed vertical setae, legs, and abdomen. One deformed female laid viable eggs after mating with a normal male. We noted various types of malformed males and females in laboratory rearings of *Tetanocera ferruginea* and, especially, of *Sepedon sphegea*, with duplicated antennae and abnormal wing length, abdominal segments, or legs after three or four generations. The viable adults copulated and females laid eggs.

Occurrence of abnormal specimens was not mentioned in connection with several mass rearing projects: *Sepedomerus macropus* (Chock *et al.* 1961), *Sepedon sphegea* (Tirgari 1986), *Dictya umbrarum* (Willomitzer & Rozkošný 1977), and *D. floridensis* (McLaughlin & Dame 1989). In regard to mass rearing for biocontrol, the phenomenon could obstruct projects if there is a genetic basis, or it could signal inappropriate rearing conditions.

Analyses of the DNA of representative species of genera in families of the Sciomyzoidea are being conducted by R. Meier & B. Wiegmann (personal communication, 2000) to determine relationships within the superfamily. A study of mitochondrial DNA sequences and adult morphological characters of North American species of *Tetanocera* and other genera of Sciomyzidae is being conducted by E. G. Chapman. See also Chapman *et al.* (2006) on evolutionary relationships among species in *Tetanocera* and among genera of Tetanocerini. The objective of the research is to develop phylogenetic hypotheses to determine the direction of ecological shifts in larval feeding strategies and to predict such strategies for taxa with unknown larval hosts/prey. The authors' data on 31 species in the genera *Atrichomelina, Elgiva, Hedria, Limnia, Renocera, Sciomyza, Sepedon, Tetanocera*, and *Trypetoptera* studied have been deposited in GenBank.

There have been no published studies of embryogenesis or the transformations of the pupal to adult stages in Sciomyzidae. Bhatia & Keilin (1937) have provided the only, very limited, information on the location of diploid imaginal cells in a larva (im, im 1, im 2, and ot in Fig. 14.22b, c).

15 • Systematics and related topics

> The society which scorns excellence in plumbing, because plumbing is a humble activity and tolerates shoddiness in philosophy because it is an exalted activity will have neither good plumbing nor good philosophy. Neither its pipes nor its theories will hold water.
>
> Gardner (1961).

15.1 IDENTIFICATION

> *Fó par zòt confon' coco é pi zabrico. Coco tini dlo, zabrico tini grenn.* [Do not take coconut for apricot. Coconut has water, apricot has seed].
>
> Guadeloupean proverb.

15.1.1 Recognition of adults

Adult Sciomyzidae are slender to robust flies, minute (1.7 mm long, *Colobaea americana*) to moderately large (13.0 mm long, *Salticella stuckenbergi*), most about the size of the house fly, *Musca domestica*. They vary in color from shiny black (*Pteromicra* spp.), to dull gray or brown (many *Pherbellia* spp.), to subshiny brownish (*Tetanocera* spp.), or yellowish (some *Pherbellia* spp.). Many species of *Sepedon* are metallic dark blue with yellowish to reddish legs. The wings and abdomen are of normal proportions, with the hind legs somewhat elongate. The body usually is not hairy but major setae are well developed, except variously reduced on the head and thorax in *Sepedon* and related genera.

Adults fly rather low, slowly, and for short distances; have a characteristic, frog-like resting posture, with the head usually directed downward; and pat the surface in front of themselves with the fore tarsi as they walk.

In a collection of Diptera Brachycera, adult Sciomyzidae can be separated from most other groups as follows on some external features:

(1) Clypeus not exposed, thus eliminating most Ephydridae and several Acalyptratae.
(2) Ptilinal fissure present at least dorsally (except in the *Sepedon* group of genera), forming a distinct lunule as in other Cyclorrhapha Schizophora. Thus are eliminated the Aschiza, such as Lonchopteridae, Phoridae, Platypezidae, etc.
(3) Antenna consisting of scape, pedicel, postpedicel, and sub-basal arista (subapical in *Tetanura* and *Prosochaeta*) inserted directly on the front and not on a frontal prominence (as in Conopidae) (Fig. 15.1).
(4) Pedicel without a longitudinal, dorsal seam as in the Calyptratae and some other Muscomorpha.
(5) Oral vibrissae absent, thus eliminating some Calyptratae such as Scathophagidae and certain Anthomyiidae with small calypteres.
(6) Wing with muscoid venation (Fig. 15.4).
(7) Calypteres small, not extending past the halteres, thus eliminating most of the Calyptratae.

Adult Sciomyzidae can be distinguished readily from most other Acalyptratae by the following combination of external characters. Antenna porrect, pedicel often elongated, clypeus (prelabrum) not exposed. Postocellar setae usually strong, and parallel or slightly divergent. No more than two fronto-orbital setae. Female postabdomen not an elongate, strongly sclerotized ovipositor, but flattened and twisted in *Tetanura*. Vein C without elongate setae along margin, unbroken; R_1 not setose above; A_1+CuA_2 reaching wing margin (except *Colobaea* and *Parectinocera*), often weak apically. One or more tibiae with one dorsal pre-apical seta (except absent in *Teutoniomyia*, two on fore tibia in *Sciomyza* [one on fore tibia in *S. sebezhica*], one or two on each tibia in *Oidematops* and two on hind tibia in *Anticheta* and most *Tetanocera*); all tibiae without median setae.

Many families of Acalyptratae include some species with a habitus similar to Sciomyzidae. Most of the more common families, especially in the Palearctic and Nearctic, can be excluded as follows:

(1) Agromyzidae: Costa broken at one place. Vein A_1+CuA_2 not reaching margin. Female with sclerotized ovipositor.

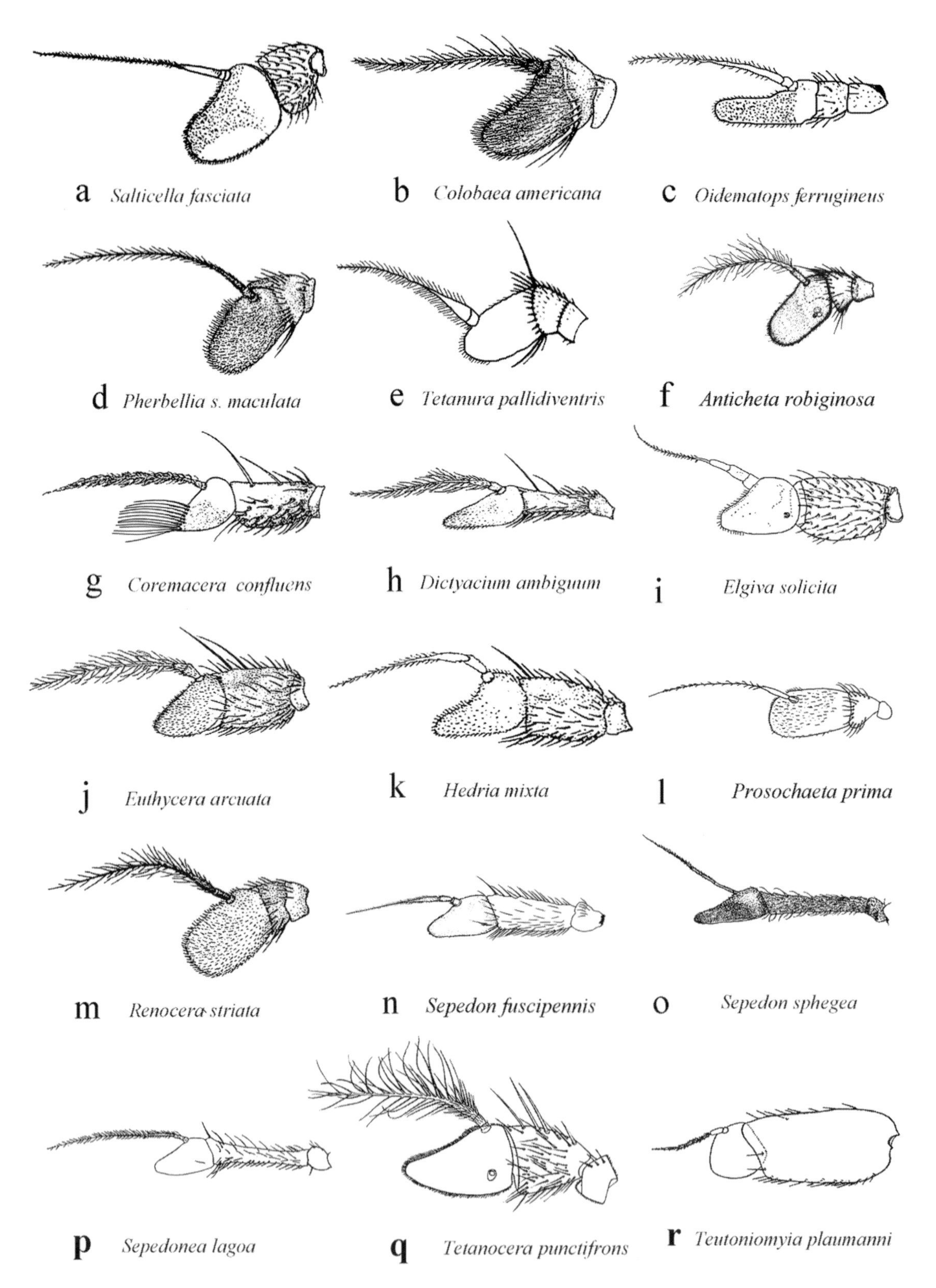

Fig. 15.1 Some different shapes of Sciomyzidae antennae. From various sources.

(2) Coelopidae: Clypeus well developed. Thoracic dorsum strongly flattened. Cell cu*p* elongate apically (also in *Salticella*).
(3) Dryomyzidae: Clypeus conspicuous. Body shiny. Antennae approximate basally. Wing with malleated surface.
(4) Ephydridae: Clypeus exposed or concealed. Costa broken at two places. Subcosta rudimentary. Pre-apical dorsal setae lacking on fore and hind tibiae.
(5) Helcomyzidae: Clypeus conspicuous. Prosternum large, triangular, fused with propleuron. Crossvein bm-cu close to CuA. Mid and hind tibiae with medial setae.
(6) Heleomyzidae: Costa with strong setae.
(7) Helosciomyzidae: Clypeus exposed. Costa with strong setae.
(8) Lauxaniidae: Postocellar setae convergent or cruciate.
(9) Micropezidae: All legs elongate, slender. No pre-apical tibial setae. Wing slender, R_{4+5} and M_{1+2} approximate or fused apically.
(10) Phaeomyiidae: Mid and hind tibiae with anterior and posterior median setae. Vein R_1 setose above in most species.
(11) Tephritidae: Frons with several setae extending to near base of antenna. Sc distinctly turned toward costa at right angle. Costa usually broken in three places. R_1 always setulose above.

15.1.2 Comparative diagnoses

15.1.2.1 ADULTS (SEE ALSO SECTION 14.1.1)

The plesiomorphic Salticellinae are most readily recognized by veins M_{1+2} and R_{4+5} strongly converging at the wing apex, cell cu*p* with a triangular extension (Fig. 14.9b), and a strongly swollen hind femur (Fig. 9.25). In the Sciomyzinae, veins M_{1+2} and R_{4+5} are parallel or only slightly convergent, cell cu*p* is without a triangular extension (Fig. 15.4a), and the hind femur is not strongly swollen. The plesiomorphic Sciomyzini are easily distinguished from the Tetanocerini by the presence of a strong seta on the propleuron above the base of the fore coxa in the Sciomyzini (except the seta is hair-like in *Atrichomelina*, absent in *Pseudomelina*, and short and weak in some *Colobaea*) and its absence in the Tetanocerini (except present in *Eutrichomelina*, *Perilimnia*, and *Shannonia*) (Fig. 15.2a).

Head (Fig. 15.2b) as wide as or wider than thorax, face at least slightly concave in profile. Clypeus small and remote from oral margin. Frontal vitta well developed and shiny in most Tetanocerini except *Sepedon* and related genera, absent or reduced in most Sciomyzini. Occasionally one, usually two fronto-orbital setae, absent in some species of *Dichetophora*, *Sepedon*, *Sepedoninus*, and *Thecomyia*. Ocellar setae absent only in *Hedria*, *Sepedomerus*, *Sepedonea*, *Sepedonella*, *Sepedoninus*, *Thecomyia*, and some species of *Sepedon* and *Dichetophora*. Inner and outer vertical setae well developed. Postocellar setae strong, parallel or slightly divergent; absent in *Sepedomerus*, *Sepedonella*, and *Thecomyia*. Oral vibrissae absent; subvibrissal setae weak. Antenna porrect, short to elongate; pedicel about half to twice length of postpedicel; arista bare to plumose, white or black, sub-basal (subapical in *Tetanura* and *Prosochaeta*). Palpus narrow and elongate, absent in *Thecomyia*. Proboscis moderately short, elongate in *Thecomyia*. Ptilinal fissure extending to about level of antennal bases except fissure, ptilinum, and lunule completely absent in *Sepedon* and related genera. This absence is unique among Schizophora.

The shape of the head is quite different among many genera and groups of genera. Six head shapes were used as a character, consisting of four character states, in the cladistic analysis of Marinoni & Mathis (2000). They used only the lateral view, including the shape of the eye, in outline. A more informative analysis perhaps could be produced by plotting various ratios, e.g., (1) face length : width against frons length : width and (2) eye length : width against eye : gena length, against each other to produce graphic representations (polygons) as shown for some Ephydridae by Drake (2001).

Gaponov *et al.* (2006), using scanning microscopy described several types of sensilla on the head of female *Pherbellia*, *Pteromicra*, *Coremacera*, *Elgiva*, *Limnia*, *Sepedon*, *Tetanocera*, and *Trypetoptera*. His abstract is paraphrased as follows. Chemoreceptors are represented by several types of olfactory sensilla, located mainly on the antennae, especially the postpedicel and arista. The location of sensilla on the postpedicel might be correlated with the behavior of mating and the method of egg laying of certain species. Olfactory cavities with basiconical sensilla are found on the postpedicel. The scape bears several trichoid sensilla, while a group of trichoid olfactory sensilla is located on the pedicel. Basiconical sensilla are usually surrounded by microtrichiae and are distinguished by their shape and thickness. Cylindrical-conic hairs with a narrow or round apex are widespread; in *Pteromicra* they are hook-shaped. The postpedicel has cuticular invaginations, represented by double-walled coeloconical sensilla whose cuticle is multiperforated. Besides basiconical and coeloconical sensilla, there are trichoid and club-shaped sensilla. Trichoid sensilla are thick-walled

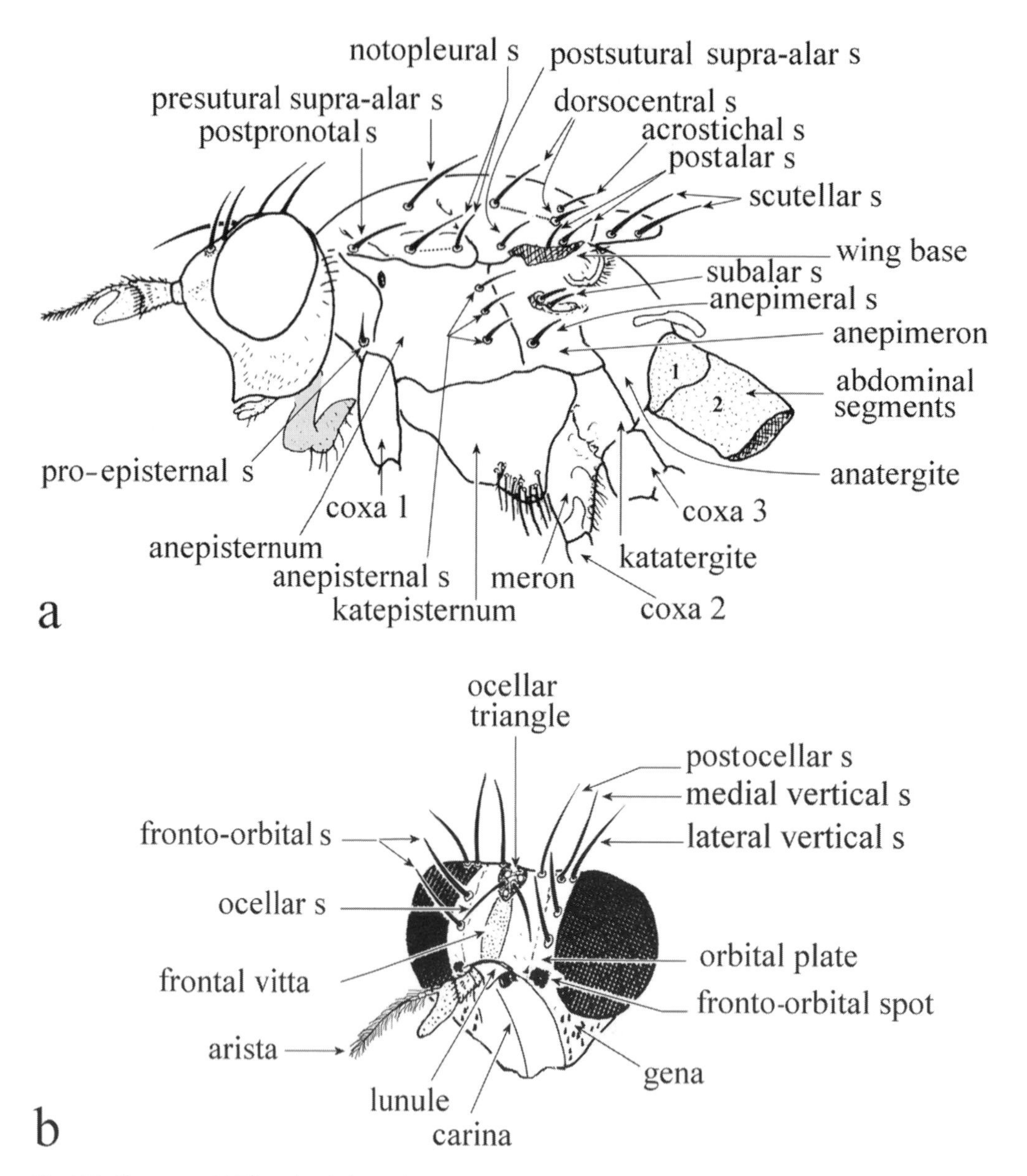

Fig. 15.2 Chaetotaxy. (a) Thoracic sclerites and setae; (b) parts of head and disposition of setae. Modified from Vala (1989a).

receptors with an olfactory sensillum structure, while club-shaped sensilla are typically thin-walled olfactory structures. Basiconical sensilla can be smooth or grooved. The arista is covered with short and sparse hairs in *Pherbellia* but in some species these hairs form compact clusters. Gustatory sensilla are located on the labellum in rows in *Trypetoptera punctulata* and *Sepedon sphegea*, and irregularly in *Tetanocera robusta*. Trichoid, basiconical, and coeloconical sensilla are present on the labellum of all studied species. Basiconical sensilla are by far the most numerous and usually form receptor fields. In some species, basiconical sensory setae are found between the ommatidia. They are mechanoreceptors for receiving information from air streams. Other separate mechanoreceptors are located on the occiput. Paired

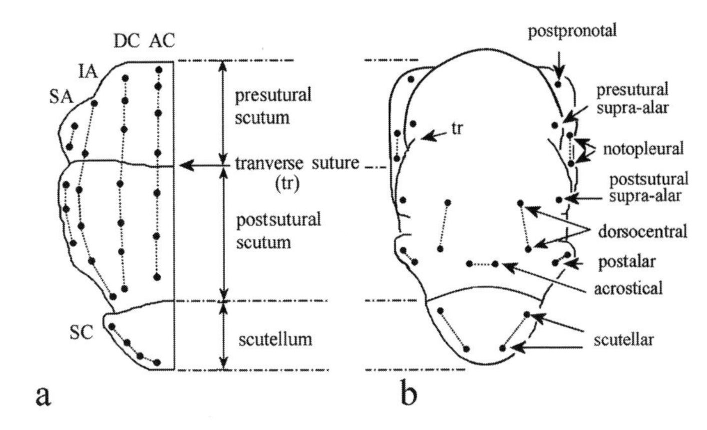

Fig. 15.3 Thoracic dorsal setae. (a) Schematic representation of setae on scutum and scutellum of Muscomorpha labeled with reference to transverse suture; AC, acrostichal setae; DC, dorsocentral setae; IA, intra-alar setae; SA, supra-alar setae; SC, scutellar setae (from Simpson *et al.* 1999). (b) Dorsal thoracic setae with usual names of setae for Sciomyzidae (modified from Vala 1989a).

gravitational receptors are on the lateral sides of the occipital aperture.

Thorax (Figs 15.2a, 15.3) without precoxal bridge. Mesonotal suture complete in *Sepedon* (*Sepedomyia*) and *Sepedoninus*. Postpronotal (= humeral) seta usually present but absent in *Tetanura, Dichetophora, Thecomyia, Sepedomerus, Sepedon, Sepedoninus, Sepedonella, Sepedonea*, and *Renocera johnsoni*. Presutural supra-alar (= posthumeral) seta referred to as intra-alar in Knutson (1987) and as presutural in the majority of publications, e.g., Rozkošný (1987a), Vala (1989a), and Rivosecchi (1992) usually present but absent in some genera. One supra-alar and one or two postalar setae, rarely none. Usually two but sometimes none, one, or three postsutural dorsocentral setae. Presutural dorsocentral seta present only in some *Tetanoceroides* and *Eutrichomelina*. None to two prescutellar acrostichal setae. Usually two pairs of scutellar setae (one basal and one apical), although basal pair absent in *Sepedon*, some *Dichetophora*, some *Anticheta*, and *Hedria;* no scutellar setae in *Sepedon lobifera*. Notopleuron with two setae, except only one in *Sepedoninus, Sepedonella, Thecomyia*, and some species of *Sepedon*. Pro-episternal seta well developed in Sciomyzini except absent or hair-like in *Atrichomelina, Pseudomelina*, and some *Colobaea* spp.; absent in Tetanocerini except present in *Eutrichomelina, Perilimnia*, and *Shannonia*. Anepisternum, anepimeron, and katepisternum bare or with setae or setulae. Subalar (vallar) ridge bare or with setae, never densely pubescent. Sternal-coxal bridge present only in *Thecomyia*.

Wing (Fig. 15.4) usually longer than abdomen, immaculate or infumated along veins, spotted or heavily patterned. Vein C not broken, without elongate spines, extending to end of M_{1+2}. Sc complete, free from R_1 distally, ending in C. Crossvein dm-cu usually straight or slightly arched, strongly S-shaped in *Pherbellia terminalis, Elgiva, Guatemalia, Hedria, Hydromya, Ilione, Neolimnia*, some *Psacadina* spp., *Sepedomerus*, and some *Sepedon* spp. Veins R_{4+5} and M parallel to slightly convergent in apical portion, but strongly converging in *Salticella*. One to three stump veins on terminal 1/3 or 2/3 of M_{1+2} in some *Pherbellia* spp. and *Tetanocera* spp. and in *Hydromya* and *Ilione* spp. $A_1 + CuA_2$ usually reaching wing margin, but often weak apically and abbreviated in *Parectinocera* and *Colobaea*. Cell cu*p* truncate, but with ventro-apical extension in *Salticella*. Halter small to moderately long. Calypteres small. In *Apteromicra*, wing is absent (Fig. 15.22).

Hind coxa often with hair-like setulae on inner posterodorsal margin. Femora well developed, usually strongly setose ventrally in males. One or more tibiae with dorsal pre-apical seta; pair of pre-apical setae on fore tibia in *Oidematops* and *Sciomyza* (one in *S. sebezhica*) and on hind tibia in *Anticheta* and most *Tetanocera* spp.; pre-apical setae barely differentiated in *Teutoniomyia*. Hind tibia with spinous apical process in *Sepedonea* and *Sepedomerus*. Some tarsal segments modified in a few species of *Sepedon*.

Abdomen moderately long, cylindrical. Abdominal spiracles 1–5 in membrane; spiracles 6 and 7 usually in membrane in Sciomyzini, usually in terga in Tetanocerini. Tergum 1 reduced, fused with tergum 2.

Male abdomen with first five segments normal. Postabdomen (Fig. 15.5a, b) ranging from asymmetric (e.g., *Pherbellia*) to symmetric (e.g., *Sepedon*). Sterna 6, 7, and 8 separate and 6 and 7 asymmetrical *or* synsternum 7 + 8 and sternum 6 separate and asymmetrical *or* synsternum 6 + 7 + 8 symmetrical. Small sclerite, possibly a remnant of tergum 8 sometimes present. Epandrium and surstyli almost always symmetric, asymmetric in *Anticheta* and epandrium asymmetric in *Tetanocera robusta*. Epandrium hood-like, free; fused with syntergosternum in *Sepedon*

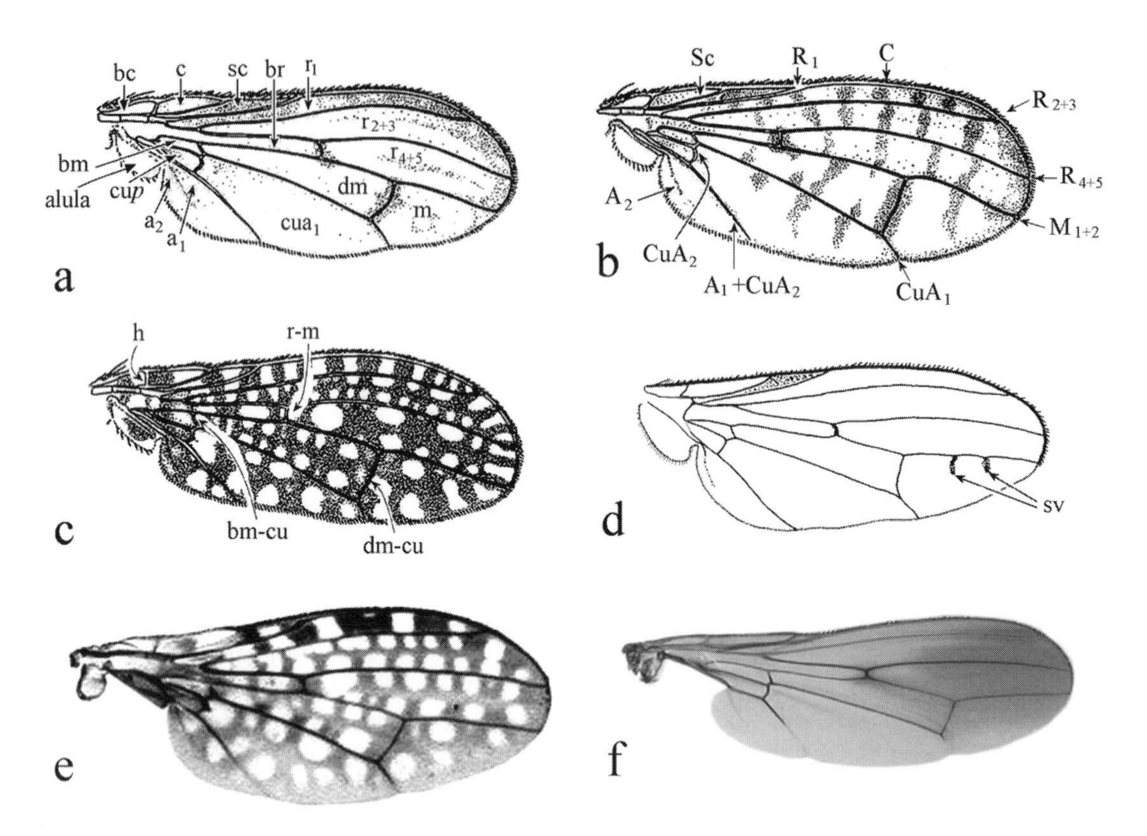

Fig. 15.4 (a–d) wing terminology; a, cells; b, longitudinal veins; c, cross-veins; d, stump veins. (a) ***Tetanocera plebeja***. Cells: a_1, a_2 (anal lobe); bc, basal costal (humeral); bm, basal medial (b_2); br, basal radial (b_1); c, costal; cua_1, anterior cubital (cubital); cu*p*, posterior cubital (anal); dm, discal medial (discal); m, medial (4th posterior); r_1, radial (submarginal); r_{2+3}, radial (1st + 2nd posterior); r_{4+5}, radial (3rd posterior); sc, subcostal. (b) ***T. valida***. Longitudinal veins: A_1+CuA_2 (anal); A_2, branch of anal vein (Ax); C, costal; CuA_1, CuA_2, anterior branches of cubitus; M_{1+2}, posterior branch of media (M); R_1, anterior branch of radius; R_{2+3}, R_{4+5}, posterior (sectoral) branches of radius; Sc, subcosta. (c) ***Poecilographa decora***.Cross-veins: bm-cu, basal medial-cubital; dm-cu, discal medial-cubital (tp, pv, posterior cross-vein); h, humeral; r-m, radial medial (ta, av, anterior cross-vein). (d) ***Pherbellia ditoma***. Stump veins: sv (supernumerary radial cross-veins, recurrent veins). (e) wing, ***Euthycera stichospila***. (f) wing, ***Sepedon plumbella***. Alternative terminology shown in parentheses. a–c, modified from Knutson (1987); d, from Vala (1989a). e, photo by J.-C. Vala; f, photo by L. Knutson.

and related genera, closed below cerci in *Chasmacryptum*, *Guatemalia*, *Neolimnia*, *Steyskalina*, and *Tetanocera* (except *T. chosenica* and *T. stricklandi*), and the *S. neanias* group of *Sepedon*. Both anterior and posterior surstyli well developed in most Sciomyzini; only posterior surstylus present in most Tetanocerini, both well developed in some *Renocera* spp., or posterior and "vestige" (see below) of anterior present in some Tetanocerini; anterior surstylus articulated at anteroventral angle of epandrium; posterior surstylus usually articulated with, rarely (*Huttonina*) fused to posterior margin of epandrium; posterior surstyli fused medially in *Sepedonea* and with small bridge in *S. neanias* group of *Sepedon*. Some authors have referred to a small, setose lobe at the base of the posterior surstylus of some Tetanocerini as a vestige of the anterior surstylus, but it may have no relationship to the anterior surstylus (W. N. Mathis, personal communication, 2004). Cerci well developed, setose, much enlarged and/or of unusual form in several genera. Sternum 10 usually developed as flat sclerite or as pair of sclerites on inner surface of epandrium. Hypandrium U-shaped, well sclerotized, with variably developed hypandrial bridge and anterior apodeme, symmetrical in dorsoventral view except asymmetrical in *Euthycera* and *Pherbina*. Gonopod

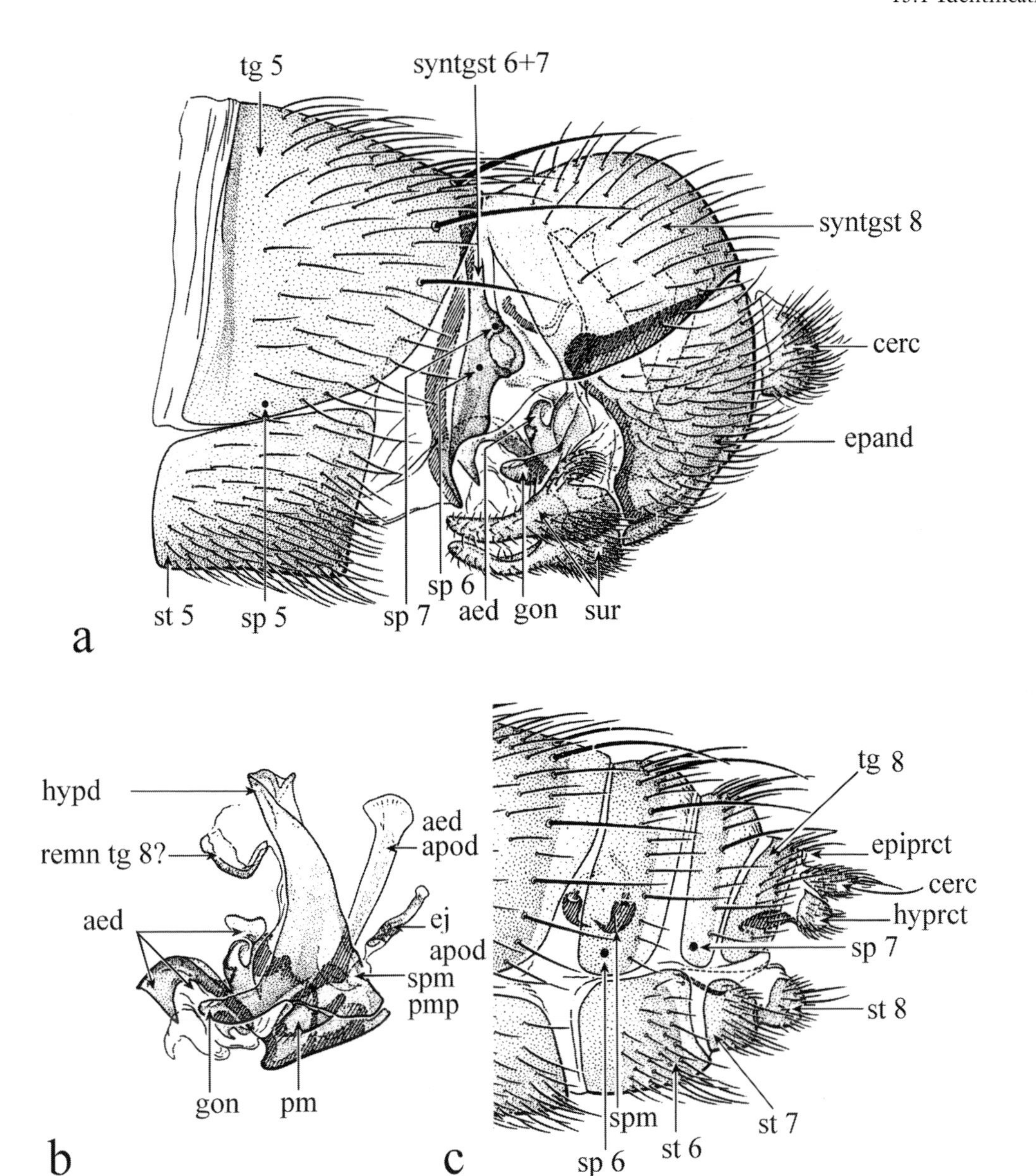

Fig. 15.5 Terminalia of *Tetanocera plebeja*. (a) Ventrolateral, male; (b) lateral, internal structures of male; (c) lateral, female. aed, aedeagus; aed apod, aedeagal apodeme; cerc, cercus; ej apod, ejaculatory apodeme; epand, epandrium; epiprct, epiproct; gon, gonopod; hypd, hypandrium; hyprct, hypoproct; pm, paramere; remn tg 8?, possible remnant of tergum 8; spm pmp, sperm pump; spm, spermatheca; sp, spiracle; st, sternum; sur, surstylus; syntgst, syntergosternum; tg, tergum. Sinistral portion of syntgst 6 + 7 often referred to as sternum 6. From Knutson (1987).

usually well developed, free in Sciomyzini, fused to hypandrium in Tetanocerini. Aedeagus well developed, usually membranous and flexible but frequently well sclerotized and complex, usually asymmetric, symmetric in *Sepedon* and related genera. Aedeagal apodeme free, well developed. Basiphallus well developed to absent. Paramere usually strongly developed, usually with one or more fine hairs apically, but small or absent in *Sepedon* and *Sepedonella*. Sperm pump and ejaculatory apodeme present but sometimes weak; strongly modified into a "cochleate vesicle" in *Sepedon* (*Parasepedon*) and *Sepedoninus* (unique in Schizophora).

J. F. McAlpine (1989) noted that the abdomen of a fly consists of 11 segments, with segment number 11 (proctiger) represented by a pair of cerci; terga and sterna 10 being the epiproct and hypoproct, repectively; and in Muscomorpha tergum 9 is "indistinguishably fused with a composite sclerite sometimes incorrectly called the epiproct or it is absent" and sternum 9 is "indistinguishable or absent."

Female abdomen (Figs 15.5c, 15.6–15.11) usually composed of five normal segments followed by three reduced terga and sterna, with tergum and sternum 9 absent, and terminating in epiproct and hypoproct (= tergum and sternum 10, supra-anal and subanal lamellae, postgenital plates) and cerci. Some authors (e.g., Hippa 1986, Sueyoshi 2001) refer to the epiproct and hypoproct as tergum and sternum 10. Sueyoshi (2005) figured the postabdomen of *T. pallidiventris* as terminating in a fused tergum consisting of 7 + 8, and fused sterna 7 + 8, but latter divided mesially, without cerci. If his numbering is correct, this is the only known instance of the loss of *both* segments 9 and 10 in female Sciomyzidae. Rivosecchi's (1992) figure of the female terminal segments (not numbered) of *T. pallidiventris* showed small cerci as present. The mesially divided apical sterna (not numbered) of *T. pallidiventris* were also figured by Rozkošný (1984b). Curran (1932), Steyskal (1938, 1954b), and Valley & Berg (1977) numbered segments 7 and 8 of *Dictya* spp. the same, considering segments 6 through 10 being present, but without comment on segments 1 through 5. Our examination of the entire abdomens and the posterior part of the thorax of several species of *Dictya* showed nine terga and sterna, *including* the epiproct and hypoproct, to be present. In our figures, the segments are numbered consistently as 1–8 plus the epiproct and hypoproct. Intersegmental membranes only moderately developed but membrane between terga and sterna well developed. Sterna 6–8 separate in the Sciomyzini and in basal (in the cladograms) genera of Tetanocerini, partially to completely fused to form a synsternum in three lineages of Tetanocerini (Fig. 15.7b–d). Sterna 7 and 8 fused in *Pherbellia tricolor* and tergum and sternum 6 also fused. Cerci small to moderately large. Two sclerotized spermathecae, spherical or subspherical, on separate ducts (each bilobed in *Salticella* and *Prosochaeta*). Two accessory glands. The vaginal apodeme (Fig. 15.8) was described for two species of *Pherbellia* (Knutson *et al.* 1990) and noted as present in four other genera of

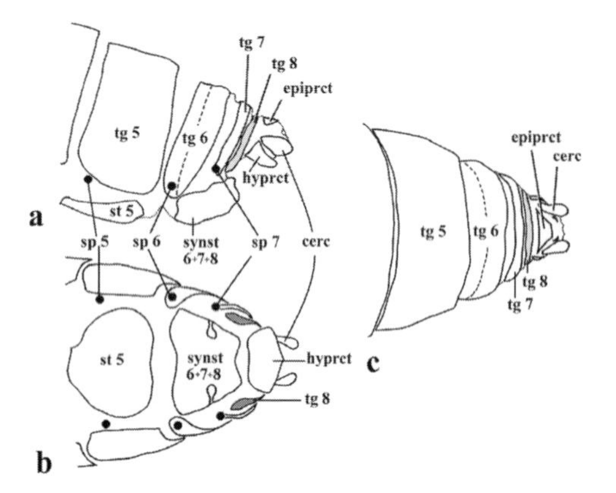

Fig. 15.6 Female terminalia of ***Sepedonea veredae***. (a) Lateral; (b) ventral; (c) dorsal. cerc, cercus; epiprct, epiproct; hyprct, hypoproct; sp. 5–7, spiracles 5–7; st 5, sternum 5; synst 6 + 7 + 8, synsternum 6 + 7 + 8; tg 5–8, terga 5–8. Tergum 8 shaded for clarification. Modified from Freidberg *et al.* (1991).

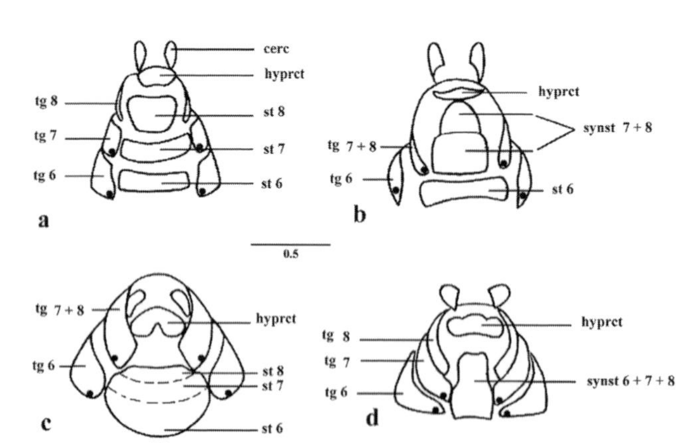

Fig. 15.7 Female terminalia, ventral. (a) ***Perilimnia albifacies***, plesiomorphic character state; (b) ***Elgiva cucularia***, apomorphic character state 1. Note: spiracle 7 in tergum, not in membrane as shown by Marinoni & Mathis (2000); (c) ***Coremacera marginata***, apomorphic character state 2; (d) ***Protodictya chilensis***, apomorphic character state 3. st 6–8, sterna 6–8; synst 6 + 7 + 8, synsternum 6 + 7 + 8. Spiracles indicated by black dots. Modified from Marinoni & Mathis (2000).

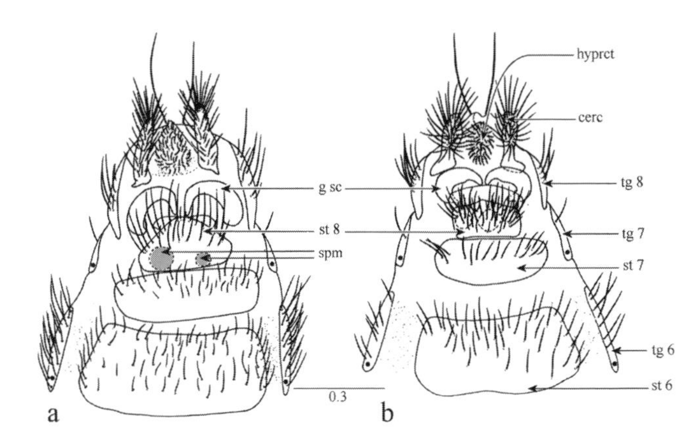

Fig. 15.8 Female terminalia of ***Pherbellia*** spp., ventral. (a) *P. juxtajavana*; (b) *P. javana*. cerc, cercus; g sc, genital sclerite; hyprct, hypoproct; spm, spermathecae; st 6–8, sterna 6–8; tg 6–8, terga 6–8. Spiracles indicated by black dots. Modified from Knutson *et al.* (1990).

Sciomyzini: *Atrichomelina, Colobaea, Pteromicra*, and *Sciomyza*. This structure was referred to as the "scleriti genitali" (genital sclerite) by Rivosecchi (1992).

The female abdomen has only occasionally been used in the classification of Sciomyzidae, e.g., Marinoni & Mathis (2000) used 17 characters of the male postabdomen and only three of the female postabdomen. Further detail is provided here because the female abdomen likely provides very useful characters. Curran (1932) figured sternum 8 for six species of *Dictya*. Steyskal (1954b), in his revision of *Dictya*, relied extensively on the shapes of sterna 6–9, figuring them for 14 species (Fig. 15.9). The more or less gross external morphology of the female abdomen or postabdomen has been figured for *Dictya* –Valley & Berg (1977); *Hoplodictya* – Fisher & Orth (1972b); *Ilione* – Verbeke (1964b), Vala (1989a); *Neolimnia* – Barnes (1979a); *Pherbecta* – Knutson (1972); *Pherbellia* – Knutson *et al.* (1990); *Pherbellia, Dichetophora* – Sueyoshi (2001); *Prosochaeta* – Barnes (1979d); *Protodictya* – Marinoni & Knutson (1992); *Sepedonea* – Freidberg *et al.* (1991); *Sepedoninus* – Verbeke (1950); *Tetanocera* – Verbeke (1964a), Knutson (1987); *Tetanoceroides* – Zuska & Berg (1974); and *Tetanura* – Rozkošný (1984b), Rivosechi (1992; *in copula*), Sueyoshi (2005). Also, Fisher & Orth (1983) figured the female abdomen for 26 species, in six genera, from California.

Also, Rivosecchi (1992) figured the female postabdomen of 46 species, in 18 genera, of Sciomyzidae from Italy, especially emphasizing the spermathecae, associated structures, and ducts (Figs 9.26, 15.10). He provided the most complete descriptions of the female postabdomen. As for his figures of the male genitalia, those of female postabdomens were made from slide mounts and thus display distortion due to the preparation process. However, his figures show details of non-sclerotized structures not seen in most other figures. Hippa (1986) included one sciomyzid, *Tetanocera ferruginea* (Fig. 15.11), in his study of the morphology and taxonomic value of the female postabdomen of Syrphidae by a new methodology, i.e., complete extrusion of the copulatory pocket from segment 8. This method contrasts with the usual examination of the gross external morphology because the details of the complicated internal structures can be compared. The author noted that examination of the female postabdomen in this way yielded characters "as important as those of the male genitalia" in Syrphidae. In comparing the copulatory pocket of *T. ferruginea* with a few other Cyclorrhapha, Hippa (1986) referred to it as of the "membraneous type," i.e., large, long, and difficult to evert. This structure, and the complicated structure of the male distiphallus of many Sciomyzidae may explain, in part, the unusually long copulatory periods of Sciomyzidae, even when they are disturbed. Barnes (1979d) proposed the plesiomorphic/apomorphic state for six important characters of the female abdomen and compared them across *Prosochaeta* and the subfamilies of Sciomyzidae *sensu* Steyskal (1965a) (Table 15.15). Considering the large number of species of Sciomyzidae that have been laboratory reared, with females positively associated with males, there is a wealth of reliably identified material for study of the female abdomen.

Immature stages – introduction

Information on the morphology of the immature stages of 103 species in 28 genera was presented as family-level, diagnostic descriptions by Ferrar (1987) for most of the data published through 1985, with reproductions of 163 published figures. Subsequently, summaries of the morphology of the

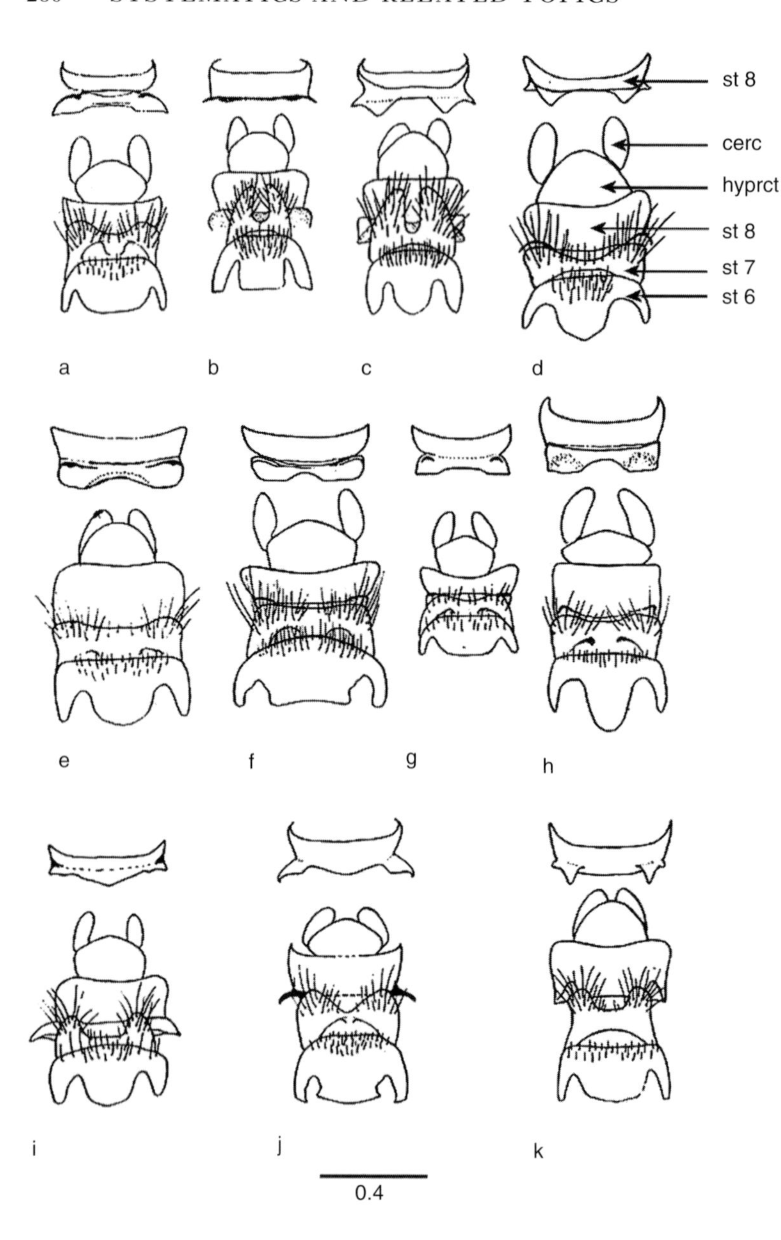

Fig. 15.9 Female terminal sterna of ***Dictya*** spp., ventral, with posteroventral margin of sternum 8 above each (setulae shown on sterna 6 and 7 only). (a) ***D. hudsonica***; (b) ***D. mexicana***; (c) ***D. sabroskyi***; (d) ***D. incisa***; (e) ***D. laurentiana*** (dotted line on sternum 8 indicates extent of emargination); (f) ***D. montana***; (g) ***D. umbroides***; (h) ***D. gaigei***; (i) ***D. expansa***; (j) ***D. stricta***; (k) ***D. pictipes***. From Steyskal (1954b).

immature stages were presented by Knutson (1987), Vala (1989a), Foote (1991), Rivosecchi (1992), Rozkošný (1997a, 1998, 2002) and Barker *et al.* (2004). In the diagnostic descriptions below we expand on Ferrar's (1987) coverage by reference to an additional 49 species in 13 genera, seven genera being included for the first time (Sciomyzini: *Colobaea*; Tetanocerini: *Dichetophora, Limnia, Poecilographa, Renocera, Tetanoceroides*, and *Trypetoptera*). Sources of these new data are as follows:

Colobaea – Knutson & Bratt (unpublished data), Sueyoshi (2005)
Dichetophora – Vala *et al.* (1987)
Limnia – Vala & Knutson (1990), Sueyoshi (2005)
Pherbellia – Narudova-Horsáková & Vala (personal communication, 2008)
Pherbina – Vala & Gasc (1990a)
Poecilographa – Barnes (1988)
Protodictya – Abercrombie (1970)

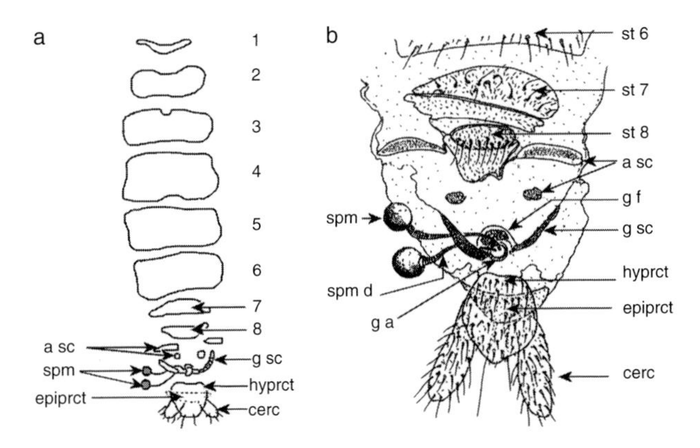

Fig. 15.10 Sterna and postabdomen of ***Pherbellia limbata*** female, ventral. (a) Abdominal sterna and postabdomen. (b) Detail of postabdomen. a sc, accessory sclerites; cerc, cercus; epiprct, epiproct; g a, genital aperture; gf, glandular formations; g sc, genital sclerite; hyprct, hypoproct; spm, spermatheca; spm d, ducts of spermathecae; 1–8, sterna 1–8. From Rivosecchi (1992).

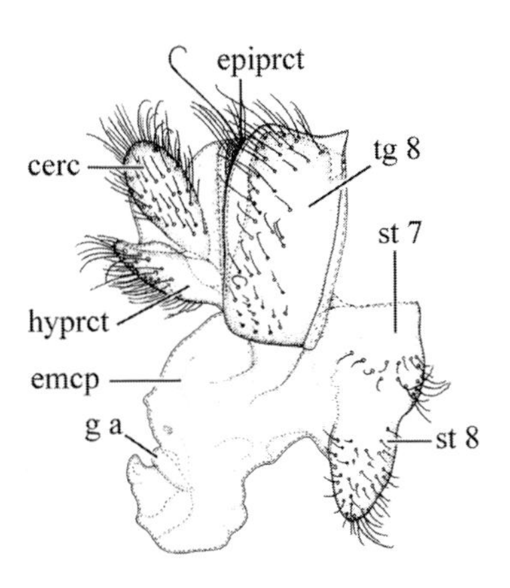

Fig. 15.11 Female postabdomen of ***Tetanocera ferruginea***, lateral. cerc, cercus; emcp, everted membranous copulatory pocket; epiprct, epiproct; g a, genital aperture; hyprct, hypoproct; st, sternum; tg, tergum. From Hippa (1986).

Renocera – Foote, in Knutson (1987), Horsáková (2003)
Sciomyza – Barnes (1990)
Sepedonea – Freidberg *et al.* (1991)
Tetanocera – Foote (1961a), Vala & Knutson (unpublished data)
Tetanoceroides – Abercrombie (1970)
Trypetoptera – Vala (1986)

15.1.2.2 EGGS (SEE ALSO SECTION 14.1.2)

Length, 0.4–1.9 mm; width, 0.1–0.6 mm. Elongate-ovoid, apices usually upturned, boat-shaped in some *Sepedon* spp. White or yellowish when laid, usually unicolorous, rarely spotted (a few species of *Sepedon*). Color change to grayish as eggs mature particularly pronounced in aquatic and semi-aquatic Tetanocerini whose larvae have a pigmented integument. Chorion smooth in Phaeomyiidae. In Sciomyzidae chorion sculpturing of fine hexagonal reticulation. In most Tetanocerini and *Pherbellia cinerella* (Sciomyzini) surface folded into anastomosing ridges and grooves, in some species of *Sepedon* and in *Sepedonea* with a pair each of dorsolateral and ventrolateral ridges. In *Anticheta* spp. (Tetanocerini) eggs laid on snail egg masses, covered by gelatinous material except for exposed dorsal patches of hexagonal reticulation and apices. Anterior and posterior ends of Tetanocerini knob-like, with a lip or one to several tubercules over anterior micropyle and opposite end punctate with aeropyles. Most Sciomyzini eggs gently rounded apically, with terminal micropyle; a few species with subterminal micropyle shielded by arched, transverse lip, and posterior tubercule. In *Salticella fasciata*, terminal micropyle surrounded by thick lip and situated within slight recess, no posterior tubercle, and only posterior pole bears aeropyles.

15.1.2.3 LARVAE (THIRD INSTAR, EXCEPT AS NOTED) (SEE ALSO SECTION 14.1.3)

Rozkošný (1967) presented a comparative table of morphological terms used for the larvae in his own publications and those of Hennig (1952b) and Berg and his students (1953–1960). The terms used here, and alternative terms, are given in Table 14.3.

Elongate, cylindrical, anterior end strongly tapered and retractile (especially so in aquatic predators), posterior end usually truncate, surrounded by two to five pairs of weak to well developed fleshy lobes. First instar 0.5–5.4 mm long, 0.1–1.6 mm wide; second instar 0.6–11.4 long, 0.2–2.2 mm wide; third instar 1.5–18.0 mm long, 0.3–3.5 mm wide.

The most distinctive feature of all Sciomyzidae larvae is the combination of an anteriorly serrate ventral arch sclerite articulated, rarely joined, with the anteroventral margin of the paired mouthhooks and the lack of oral grooves and ventral cibarial ridges, except cibarial ridges are weakly developed in *Salticella* (Fig. 14.41a–c).

While *Salticella* has only two or three weak teeth in the anterolateral corners of the ventral arch (Fig. 14.33c) the entire anterior margin is replete with about 20–30 minute teeth in the Sciomyzinae (Fig. 14.33g, k). These teeth are absent in some first-instar larvae. *Salticella*, as the Sciomyzini, lacks accessory teeth on the anterior margin of the basal part of the mouthhook below the hook part, but *Salticella* has a small denticle at the base of the hook (Fig. 14.33b). One species of *Dictya* also lacks accessory teeth but has several denticles at the base of the hook (Fig. 14.38f).

There are major differences between the cephalopharyngeal skeletons of *Salticella*, Sciomyzini, and Tetanocerini. In addition to lacking accessory teeth, *Salticella* and Sciomyzini have an anterodorsal bridge between the pharyngeal sclerites, both the dorsal and ventral cornua of the pharyngeal sclerite have a window, and the sinus between the cornua is deep (Fig. 14.33a, f). In the Tetanocerini, one to seven accessory teeth are present (Fig. 14.33j), except they are absent in *Renocera* (Fig. 14.36q); there is no anterodorsal bridge between the pharyngeal sclerites, except it is present in *Anticheta*; the dorsal cornu lacks a window, except it is present in *Anticheta* (Fig. 14.38r) and *Renocera* (Fig. 14.36q) and there is a weak window in *Tetanoceroides*. The sinus between the cornua is shallow, but deep in *Renocera* and of intermediate depth in *Anticheta, Perilimnia*, and *Shannonia* (Figs 14.36q, 14.38r, 14.40a, b). The paired mouthhooks are closely apposed anteriorly, joined by a weak to strong dorsal connection, divergent posteriorly. In most species, the hook portion is variously decurved, weakly decurved in *Anticheta*, fore-shortened in *Tetanura*, apically acute (less so and lightly sclerotized in the clam-killing *Renocera pallida*), but straight and fused into a single blade in second-instars of a few species of parasitoid *Pherbellia* and straight and broadly connected dorsally in first-instars of slug-killing *Tetanocera* species. The ventral arch is strong in most species. But it is weak, fused with the basal portion of the mouthhook in most first-instar larvae, fused with but stronger in some second-instar larvae, and reduced in most third-instar *Pherbellia*. The ventral arch is connected to the mouthhook in *Tetanura* (Sciomyzini) and one of the seven reared species of *Anticheta* (Tetanocerini), these being a snail killer and snail-egg feeder, respectively (Fig. 14.35c, d).

The H-shaped sclerite is separate from the pharyngeal sclerite in all third-instar Sciomyzidae, except joined in all first and second instars and in third instars of *Anticheta, Perilimnia, Renocera*, and *Shannonia* (Figs 14.36q, 14.38r, 14.40a, b). The posterior rami of the H-shaped sclerite never overlie the anteroventral projection of the pharyngeal sclerite in Sciomyzidae and Phaeomyiidae as they do in Dryomyzidae (Fig. 14.41d) and many other Sciomyzoidea. The ligulate sclerites are paired in most first- and second-instar larvae and in a few third instars, joined to form a V- or C-shaped strap in most third instars, and absent in *Tetanura* and *Dictya*. The parastomal bars are absent in first-instar larvae, but present and joined to the epistomal sclerite anteriorly and pharyngeal sclerite posteriorly in second and third instars. A pair of accessory rods above the parastomal bars project forward from the pharyngeal sclerite in *Salticella* and some Sciomyzini. The ventral cornu is sometimes slightly expanded above the window, but except for *Salticella* never has a strong dorsobasal lobe as in *Dryomyza* (Dryomyzidae) and some other Sciomyzoidea. The dorsal and ventral cornua are parallel, except divergent in first-instar *Salticella*, Sciomyzini genera, and *Renocera*. The depth of the sinus between the dorsal and ventral cornua is referred to as the indentation index (IIx) in the keys (ab/cd × 100, as in Fig. 14.36q).

In all sciomyzid larvae, segment I has a pair of minute retractile antennae and a pair of maxillary organs with various sensilla. The posterior margin of this segment has a variable number of rows of long and arched spinules, the postoral spinule patch or band. The three thoracic segments (II–IV) are distinguished by the presence, on each, of a ventral pair of "Keilin organs" (= three trichoïd sensilla projecting from a pit). The eight abdominal segments (V–XII) never have Keilin organs. The disposition of sensorial setae is different on thoracic and abdominal segments (Fig. 14.26) and they are longer in aquatic species.

The external morphology of sciomyzid larvae is exceedingly diverse in most features and strongly adaptive to the larva's microhabitat. *Salticella* and the Sciomyzini (Plate 2a, Fig. 15.12e) are muscoid in appearance; colorless to whitish; with various combinations of spinule patches and bands on most body segments, these completely encircling on the anterior-most segments of a few species, and completely encircling on most segments in *Ditaeniella* (Fig. 15.12e). They have reduced body warts or tubercles. There are 7–34 anterior spiracular papillae in the third instar. The spiracle has a short to elongate stem and a transverse or elliptical to rounded head (Fig. 15.18). The posterior spiracular disc lobes are moderately developed to strongly reduced (Figs 15.12, 15.13), the ventrolateral lobe usually somewhat elongate. The posterior interspiracular

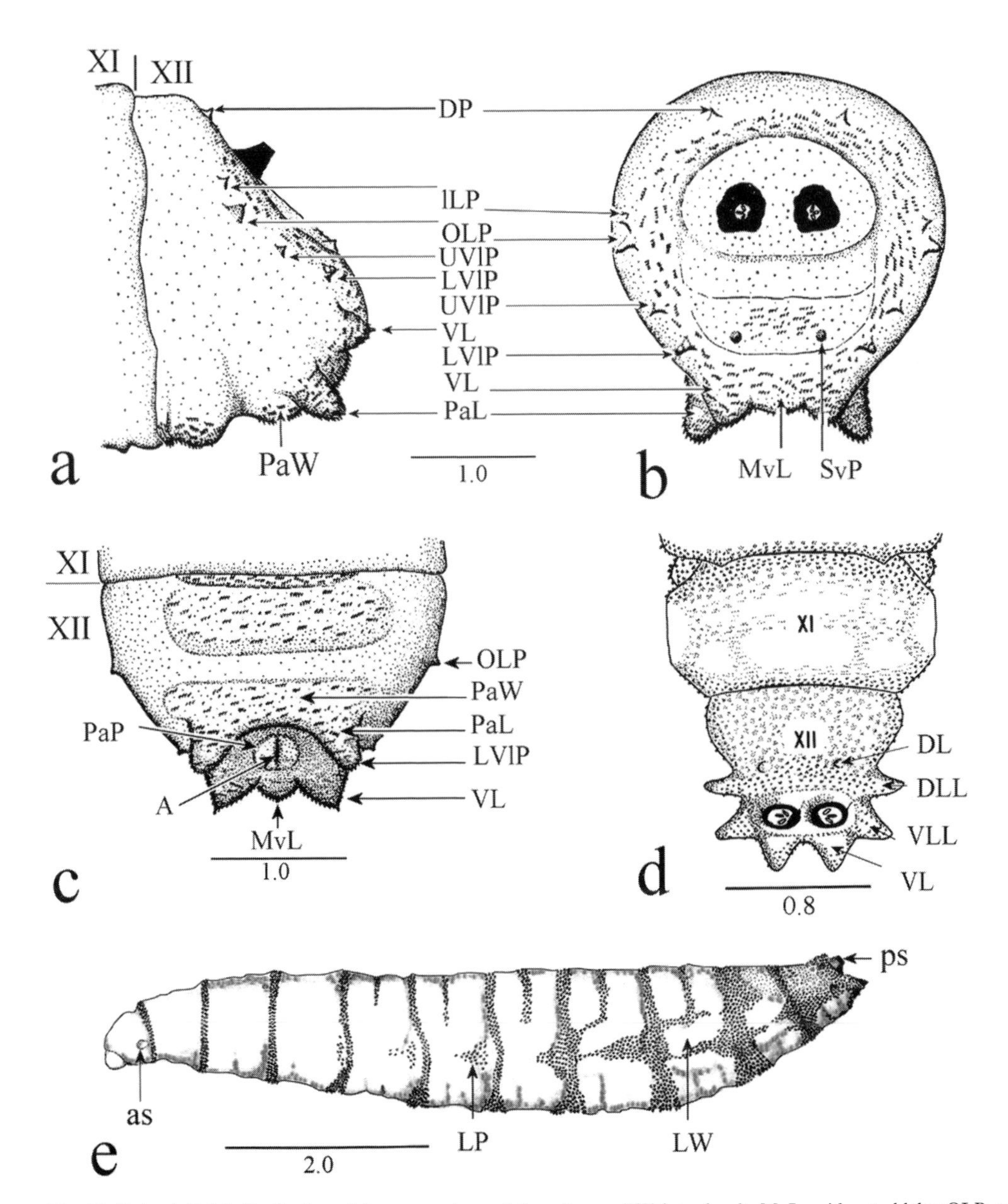

Fig. 15.12 (a–c) ***Salticella fasciata***, L3. a, posterior end, lateral; b, same, caudal; c, same, ventral (from Knutson *et al.* 1970). (d), (e) ***Ditaeniella parallela***, L3; d, posterior end, dorsal; e, lateral (from Bratt *et al.* 1969). A, anus; as, anterior spiracle; DL, dorsal lobe; DLL, dorsolateral lobe; DP, dorsal papilla; ILP, inner lateral papilla; LP, lateral pad; LVlP, lower ventrolateral papilla; LW, lateral welt; MvL, midventral lobe; OLP, outer lateral papilla; PaL, pre-anal lobe; PaP, peri-anal pad; PaW, pre-anal welt; ps, posterior spiracle; SvP, supraventral papilla; UVlP, upper ventrolateral papilla; VL, ventral lobe; VLL, ventrolateral lobe; XI, XII, segments.

processes always are minute, scale-like, and barely visible with a light microscope (Fig. 14.32d, g). Usually there are two slits on the posterior spiracular plate of second instars, three in third instars. The Malpighian tubules are white.

Terrestrial Tetanocerini larvae also are colorless to whitish; muscoid in appearance, but with tubercles, welts, and lobes distributed on segments IV–XII somewhat as in semi-aquatic and aquatic Tetanocerini, although much

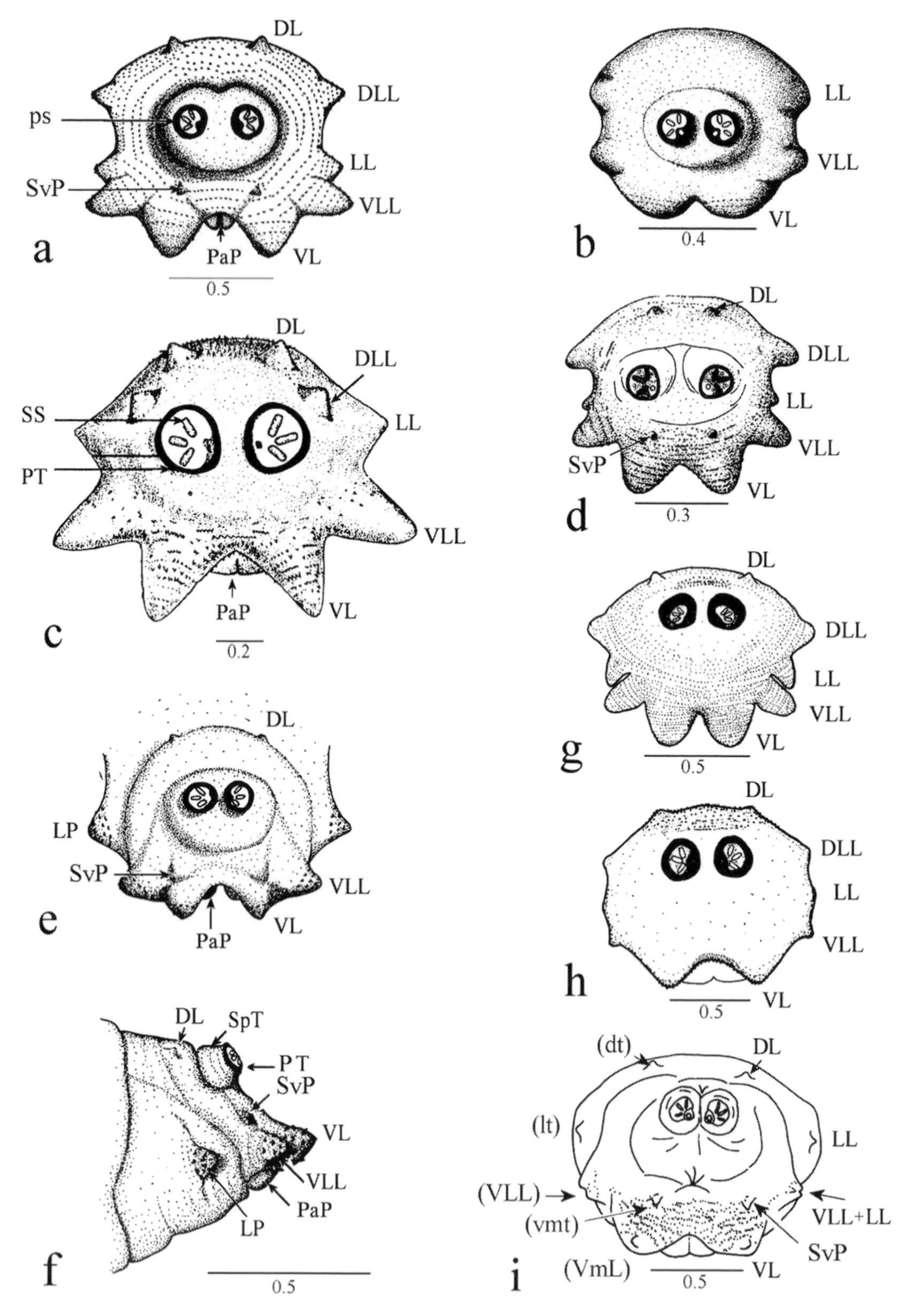

Fig. 15.13 Posterior spiracular disc, L3, caudal except f, lateral. (a) ***Pherbellia quadrata***; (b) ***P. knutsoni*** (from Bratt *et al.* 1969); (c) ***Atrichomelina pubera*** (from Foote *et al.* 1960); (d) ***Pteromicra glabricula*** (from Roskošný & Knutson 1970); (e, f) ***Colobaea***

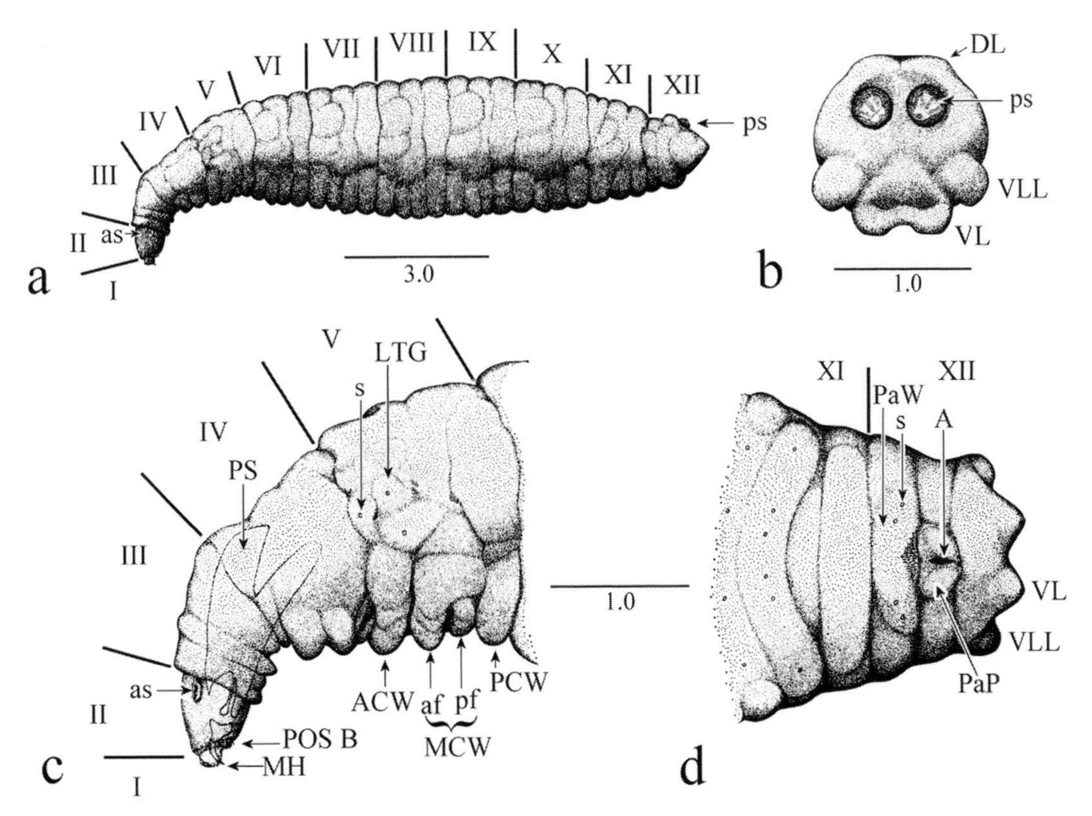

Fig. 15.14 *Tetanocera plebeja*, L3. (a) lateral; (b) posterior spiracular disc, caudal; (c) segments I–V, lateral; (d) segments XI–XII, ventral. ACW, anterior creeping welt; as, anterior spiracle; A, anus; DL, dorsal lobe; LTG, lateral tubercle group; MCW, middle creeping welt (af, anterior fold; pf, posterior fold); MH, mouthhook; PaP, peri-anal pad; PaW, pre-anal welt; PCW, posterior creeping welt; POS B, postoral spinule band; ps, posterior spiracle; PS, pharyngeal sclerite; s, sensillum; VL, ventral lobe; VLL, ventrolateral lobe; I–XII, segments. From Trelka & Foote (1970).

reduced (Fig. 15.14). The anterior spiracles have 7–18 papillae, the stem-part is usually long, and the head-part usually circular (Fig. 15.19). The posterior spiracular disc lobes are reduced, except the ventral and ventrolateral lobes may be somewhat elongate (Figs 15.14b, 15.17a). The posterior interspiracular processes are scale-like, unicuspid, or polycuspid (Fig. 14.32a, f, h–j), except *Trypetoptera punctulata* has rudimentary branched processes (Fig. 14.31h). There are no spinule patches and bands except for a well-developed postoral spinule band.

Aquatic (Fig. 15.16) and semi-aquatic (Fig. 15.15d–f) Tetanocerini larvae, which are the most common forms in the tribe, have segments I–IV translucent and V–XII moderately translucent to densely pigmented. The integument is translucent but with a dense coat of platelets to heavy, black spinules unlike any other Sciomyzidae in *Perilimnia* (Fig. 15.15a–c) and *Shannonia*. It is translucent in

Caption for Fig. 15.13 (*cont.*)
punctata (Knutson & Bratt unpublished data); (g) *Sciomyza simplex*; (h) *S. aristalis* (from Foote 1959a); (i) *Sciomyza varia* (from Barnes 1990 with original terminology in parentheses on left: dt, dorsal tubercle; lt, lateral tubercle; VLL, ventrolateral lobe; VmL, ventro-medial lobe; vmt, ventro-medial tubercle). DL, dorsal lobe; DLL, dorsolateral lobe; LL, lateral lobe; LP, lateral papilla; PaP, peri-anal pad; PT, peritreme; ps, posterior spiracle; SpT, spiracular tube; SS, spiracular slit; SvP, supraventral papilla; VL, ventral lobe; VLL, ventrolateral lobe.

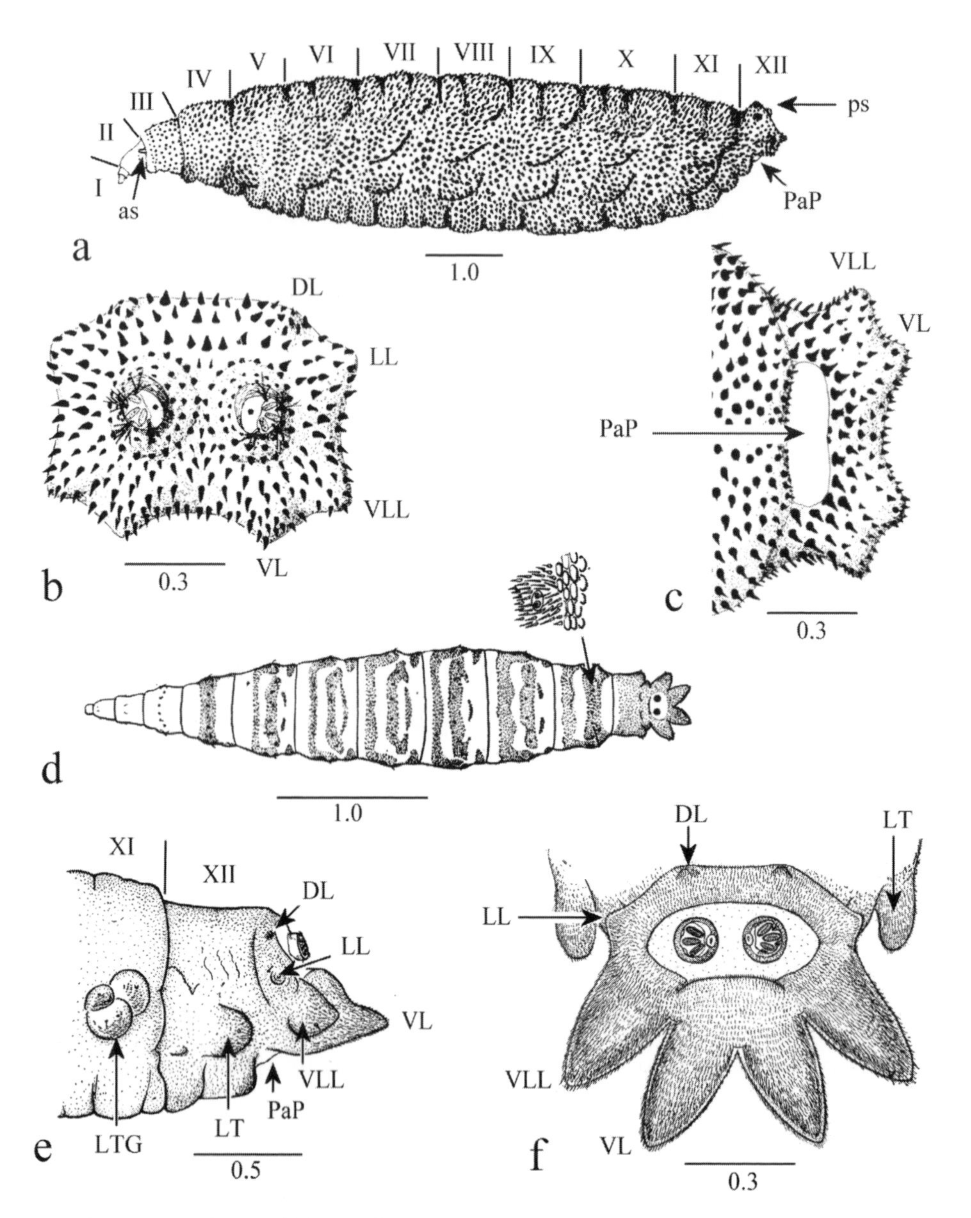

Fig. 15.15 (a–c) ***Perilimnia albifacies***, L3. (a) lateral; (b) posterior spiracular disc, caudal; (c) apex of segment XII, ventral (from Kaczynski *et al.* 1969); (d) ***Psacadina verbekei***, L2, dorsal with expanded view of cuticular structures and sensilla; (e) *P. verbekei*, L3, segments XI, XII, lateral; f, same, posterior spiracular disc, caudal (from Knutson *et al.* 1975). I–XII, segments; as, anterior spiracle; DL, dorsal lobe; LL, lateral lobe; LT, lateral tubercle; LTG, lateral tubercle group; PaP, peri-anal pad; ps, posterior spiracle; VL, ventral lobe; VLL, ventrolateral lobe.

Anticheta and *Renocera* with a dense coat of hyaline to somewhat darkened integumentary scales or minute, hair-like processes, these most strongly developed on and around the sensory papillae. The anterior spiracles have 3–37 papillae; usually with a long stem and rounded to lanceolate head (Figs 15.20, 15.21). Ventral-ventrolateral spinule patches are absent, except present in *Anticheta* and similar but probably not homologous and less darkly pigmented patches present in *Renocera*. Segments IV–XI each are more or less divided into three annuli by primary and secondary integumentary folds which are distinct and constant ventrally, visible dorsally on most segments, but obscured laterally by a group of two or three low tubercles per segment and spindle-shaped, intersegmental welts. Segments IV–XI are ventrally divided transversely into three rather distinct creeping welts. Segment XII is slightly narrowed to the apically truncate apex, except elongated post-anally in *Elgiva* spp. (Fig. 15.16e, f) and *Renocera striata*, with variously developed pre- and postanal welts. The pre-anal welt is developed into an anal proleg and strongly armed with chitinous hooks in the most aquatic species. The posterior spiracular disc is surrounded by three to five pairs of lobes, often less distinct in first and second instars (Fig. 15.17b–f). The dorsal lobe always and dorsolateral lobe frequently are low and rounded. The lateral lobe usually is longer, sometimes indistinct. The ventrolateral lobe is elongate, sometimes with an apical appendage, and sometimes fused with the lateral (indicated by presence of two sensory setae on mid posterior surface). The ventral lobe always is present and usually elongate. The paired posterior spiracular plates each have three slits in both the second- and third-instar larvae. There are four moderately long to elongate, palmately branched interspiracular processes around each spiracular plate (Fig. 14.31a–g). The processes are plate-like with a marginal fringe in *Renocera pallida*. The best-developed lobes and processes are in the most aquatic species. Those of *Anticheta* are similar to Sciomyzini. The Malpighian tubules are reddish. Aquatic Tetanocerini frequently are seen swallowing air, resulting in a large bubble in the gut which aids in flotation. Some terrestrial Tetanocerini also occasionally are seen swallowing air; swallowing of air by non-aquatic larvae perhaps increases body turgor, resulting in more efficient crawling over the substrate.

15.1.2.3.1 Posterior spiracular disc

The lobes of the posterior spiracular disc are distinctive of the species, many genera, the tribes and subfamilies. They are of significance from functional, identification, and phylogenetic viewpoints, thus further detail is provided. Those of *Pelidnoptera* (Phaeomyiidae) were described only as being "very low, peripheral" (Vala *et al.* 1990). Those of *Salticella fasciata* (Salticellinae, Fig. 15.12a–c) are very different from the Sciomyzinae, consisting of six distinct pairs of very small, papilla-like structures, with a pair of supraventral papillae between the posterior spiracular tubes and the bases of the ventral lobes. Notably, a pair of supraventral papillae also is found, sporadically, in several genera of Sciomyzini, i.e., some *Colobaea, Pteromicra*, and *Sciomyza*, and most *Pherbellia* spp. (Fig. 15.13a, d, e, f, i) but not in *Atrichomelina* and *Ditaeniella* (Sciomyzini) or any Tetanocerini. A point of confusion that needs to be clarified is the "lateral protuberances" (Fig. 15.13e, f) which appear as peripheral lobes in caudal view but in fact are situated before the spiracular disc on segment XII; these occur in *Colobaea* spp. and in some *Pherbellia* spp. Also, the "dorsobasal protuberance" (Rozkošný & Knutson 1970) of the ventrolateral lobe of *Pteromicra pectorosa* and *P. glabricula* (Fig. 15.13d), not present in *P. angustipennis*, seems to be equivalent to the lateral lobe of several *Pherbellia* spp. (Fig. 15.13a) and *Atrichomelina pubera* (Fig. 15.13c).

The lobes have been designated in descriptions and figures on the basis of their position and in part orientation; thus it probably is not possible to homologize all of them among all species or all genera. Sometimes there has been inconsistency in the terms applied to the lobes, especially confusion or disagreement with regard to definition of dorsal and dorsolateral lobes. Consequently, we prefer to use the terminology given in Figures 15.12–15.17. It may be possible, to a certain extent, to homologize lobes among many aquatic Tetanocerini, where a relatively strong seta is present postero-basally on "lateral" and "ventrolateral" lobes. These setae probably are homologous with setae on the lateral tubercle groups of the body segments of both terrestrial and aquatic Tetanocerini (Vala & Gasc, unpublished data). The lateral tubercle groups of segments IV–XI generally have been described as consisting of three parts: an anterior tubercle and upper and lower posterior tubercles. Setae usually have been described as absent on the anterior, but one or two present on the two posterior tubercles. However, there has not been enough consistency in description of these to generalize in a definitive manner. Further analysis of the distribution of sensory structures on the posterior spiracular disc in comparison with the fairly well-studied distribution of sensory structures on the body segments probably would result in a more reliable homology of lobes. The distribution of sensory structures on body segments and the posterior spiracular disc have been described best for several semi-terrestrial and terrestrial genera of Tetanocerini by Vala (1989a), Vala & Gasc (1990a), and Vala *et al.* (1995) (Fig. 14.26).

The convergence of lobe structure in many terrestrial to semi-terrestrial Sciomyzini (Fig. 15.13) and Tetanocerini (Figs 15.14b, 15.15b, 15.17a) is quite striking. From a functional point of view, the larger lobes seem to be found

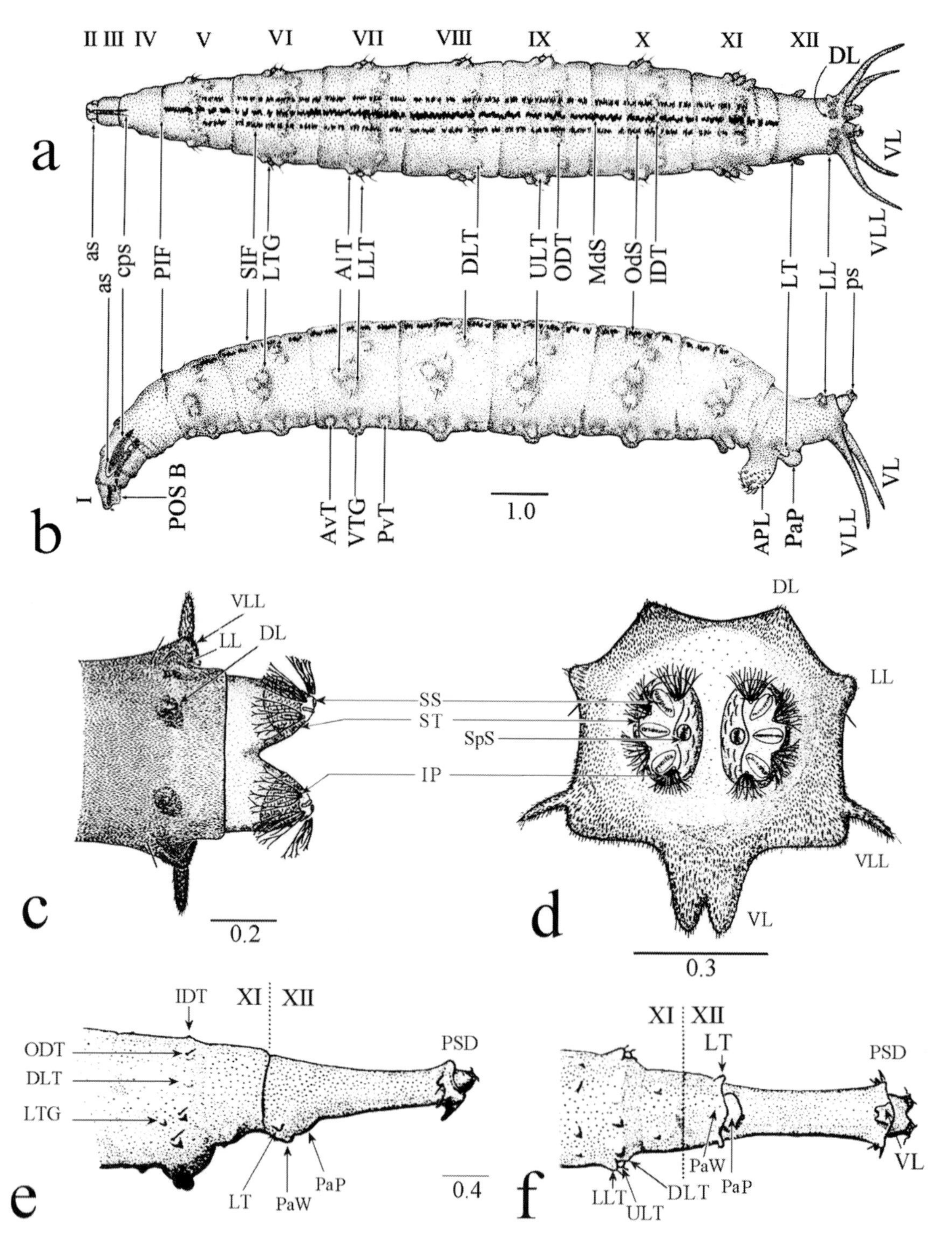

Fig. 15.16 (a, b) ***Ilione turcestanica***, L3; (a) dorsal; (b) lateral (from Knutson & Berg 1967). (c–f) ***Elgiva solicita***, L3; (c) posterior end, dorsal; (d) same, caudal; (e) segments XI–XII, lateral; (f) same, ventral (from Knutson & Berg 1964). AlT, anterolateral tubercle; APL, anal proleg; as, anterior spiracle; AvT, anteroventral tubercle; cps, cephalopharyngeal skeleton; DL, dorsal lobe; DLT, dorsolateral

in species with more mobile behavior, especially among the more aquatic species. A comparison of the extent of tracheation of the lobes, obvious in such truly aquatic species as *Ilione lineata* and *Eulimnia philpotti* (Plate 2d, e), might be useful in characterizing the extent of development of aquatic as opposed to terrestrial/semi-terrestrial microhabitats and behavior of larvae. The "ventral lobe index" proposed by Knutson *et al.* (1975) generally has not been used in descriptions, but it might prove useful after review of its status across the genera.

In the Sciomyzini, the number of posterior spiracular disc lobes ranges from two to five distinct pairs, with the most terrestrial and highly parasitoid species having the fewest pairs (Fig. 15.13). Most Tetanocerini have four distinct pairs, but as noted above, there is extensive disagreement as to which lobes are "absent": either the dorsals, dorsolaterals, or laterals. In the Tetanocerini, *Anticheta* spp. have the fewest number of lobes, i.e., two. There are many genera of Tetanocerini, primarily terrestrial and semi-terrestrial species, in which only the ventrolateral and lateral lobes are distinct, e.g., *Tetanocera plebeja* (Fig. 15.14b). In the Tetanocerini, a distinct set of five pairs of lobes is found only in one of five described *Protodictya* species (*P. hondurana*), one of the two described *Hoplodictya* species (*H. setosa*), and almost all species of the apomorphic lineage including *Sepedomerus, Sepedonea*, and *Sepedon*. Only four pairs were described for the succineid-feeder *Sepedon trichrooscelis*; five pairs were described for the other known succineid-feeding *Sepedon hispanica*, but three of the pairs in that species are barely perceptible (Fig. 15.17a). Surprisingly, *Thecomyia*, a member of the *Sepedon* lineage, has only four pairs of lobes. A two-segmented ventrolateral lobe, usually called "lateral" lobe, is found only in some aquatic genera, both those with four and five pairs of lobes. The two-segmented lobe is found in *Dictya* (Fig. 15.17d), *Dictyodes, Elgiva* (Fig. 15.16d), two species of *Protodictya* (one with four pairs, one with five pairs), *Sepedomerus, Sepedonea* (Fig. 15.17b), *Sepedon* (Fig. 15.17c) (except the two succineid feeders), *Thecomyia*, and most aquatic *Tetanocera* spp. The two-segmented lobe has been interpreted by some authors as being formed by the coalescence of lateral and ventrolateral lobes. But, the presence of a two-segmented ventrolateral lobe *and* the presence of a lateral lobe in *Protodictya hondurana, Sepedomerus, Sepedonea*, and aquatic *Sepedon* spp. argue against this interpretation. The two-segmented lobe very probably is homologous in ontological development across the genera in which it occurs.

Basal coalescence of the ventral and ventrolateral lobes is found in several species of *Tetanoceroides*, and of the ventral lobes in *Elgiva* spp. (Fig. 15.16d) and slightly in some other genera, e.g., *Dictya* (Fig. 15.17d). Two species that are well known to be subsurface feeders, *Hedria mixta* and *Eulimnia philpotti* (Fig. 15.17e, f), have the longest lateral, ventrolateral, and ventral lobes described. A few other species which have similarly elongate lobes, but whose microhabitat is not so precisely known, e.g., several *Ilione* spp., may also be subsurface feeders.

15.1.2.3.2 Anterior spiracles

The anterior spiracles project from a membranous pocket on the posterolateral surface of segment II in second- and third-instar larvae (Figs 15.18a, b, 15.21d, e). They are weakly developed in the second instar, but well developed and usually with a few more papillae on the head-part in the third instar. In the third instar, a scar at the base of the stem-part indicates the point of removal of the second-instar spiracle during ecdysis (Figs 15.18c, 15.19f). Published figures, mostly drawings made from slide mounts, often indicate a continuous membrane covering the papillae (Figs 15.18c, f, 15.19f). Such a continuous membrane appears to be present also in stereomicrographs of second- and third-instar *Sepedonella nana* (Fig. 15.21a, b). Some drawings from slide preparations show one or more pores at the apex of papillae, but some stereomicrographs (Figs 15.18b, 15.21d) show a slit. Prismatic spots on the surface of the head-part of third-instar spiracles also have often been figured (Fig. 15.20d). Their function is unknown but they may indicate points of internal support that keep the fragile structure in an inflated condition. The fine structure of the anterior spiracles has been scarcely examined;

Caption for Fig. 15.16 (*cont.*)
tubercle; IDT, inner dorsal tubercle; IP, interspiracular process; LL, lateral lobe; LLT, lower lateral tubercle; LT, lateral tubercle; LTG, lateral tubercle group; MdS, mid dorsal stripe; OdS, outer dorsal stripe; ODT, outer dorsal tubercle; PaP, peri-anal pad; PaW, pre-anal welt; PIF, primary integumentary fold; POS B, postoral spinule band; ps, posterior spiracle; PSD, posterior spiracular disc; PvT, posteroventral tubercle; SIF, secondary integumentary fold; SS, spiracular slit; ST, spiracular tube; SpS, spiracular scar; ULT, upper lateral tubercle; VL, ventral lobe; VLL, ventrolateral lobe; VTG, ventral tubercle group; I–XII, segments.

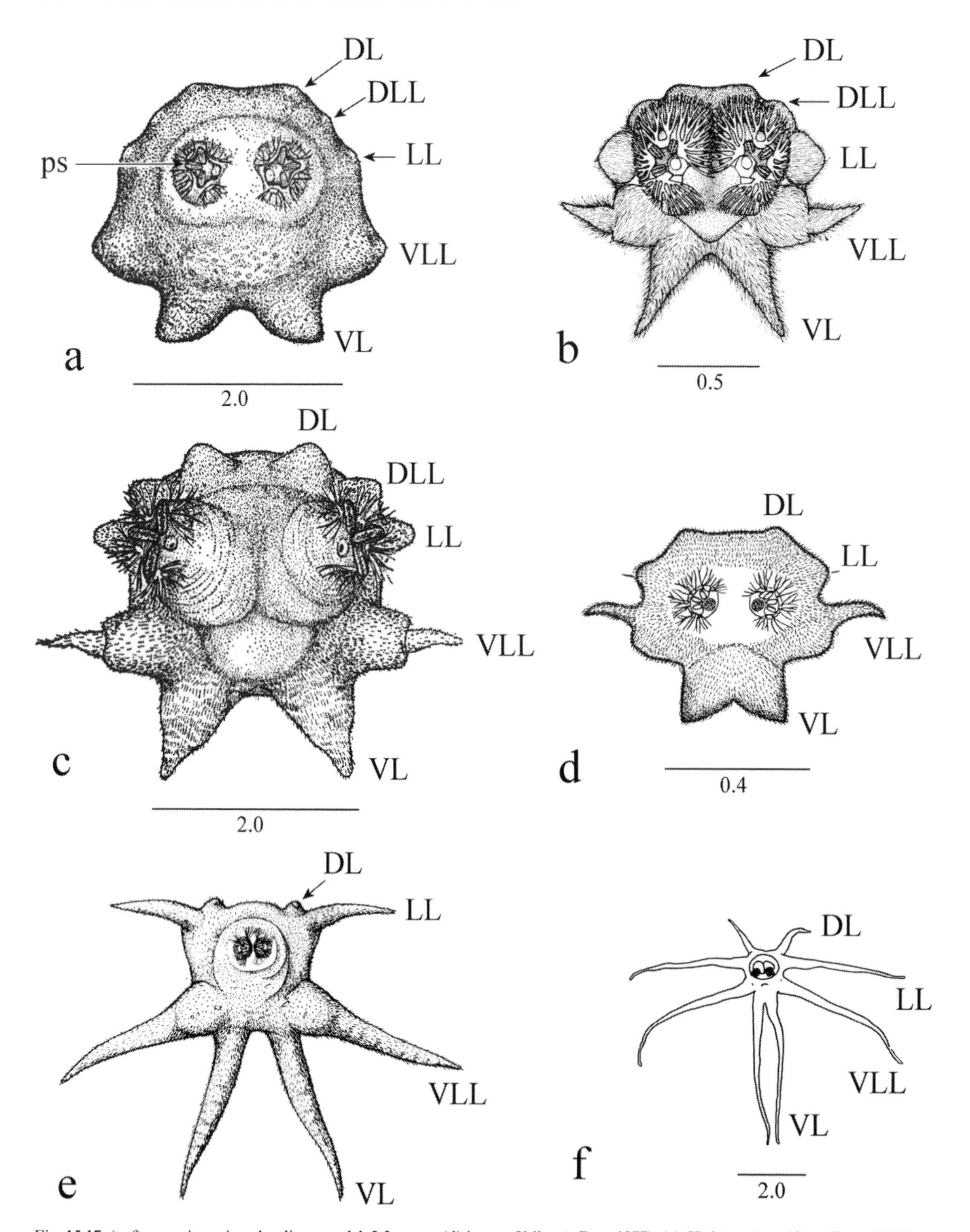

Fig. 15.17 (a–f) posterior spiracular discs, caudal, L3 except (d) is L2; (a) ***Sepedon h. hispanica*** (from Knutson *et al.* 1967); (b) ***Sepedonea isthmi*** (from Knutson & Valley 1978); (c) ***Sepedon ruficeps*** (from Knutson *et al.* 1967a); (d) ***Dictya matthewsi*** (from Valley & Berg 1977); (e) ***Hedria mixta*** (from Foote 1971); (f) ***Eulimnia philpotti*** (from Barnes 1980a). DL, dorsal lobe; DLL, dorsolateral lobe; LL, lateral lobe; ps, posterior spiracle; VL, ventral lobe; VLL, ventrolateral lobe.

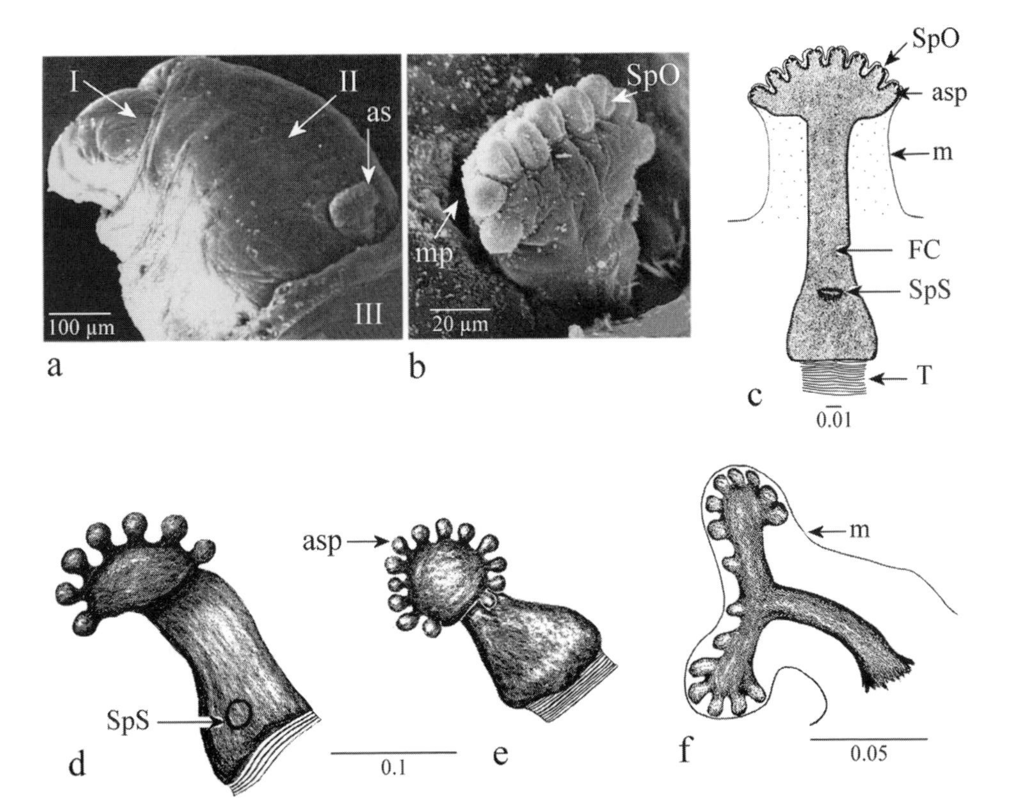

Fig. 15.18 Anterior spiracles, L3, of some terrestrial and semi-terrestrial parasitoid/predators. (a–c) ***Salticella fasciata***; (a) segments I–III; (b) spiracle mostly retracted into membranous pocket (from Vala *et al.* 1999); (c) expanded spiracle (from Knutson *et al.* 1970). (d) ***Pherbellia griseola***. (e) ***P. quadrata*** (from Bratt *et al.* 1969). (f) ***Colobaea americana*** (Knutson & Bratt unpublished data). as, anterior spiracle; asp, anterior spiracular papilla; FC, felt chamber; m, membrane; mp, membranous pocket; SpO, spiracular papilla opening; SpS, spiracular scar; I–III, segments; T, trachea.

there has been published only one stereomicrograph (of *Salticella fasciata*, Fig. 15.18b). Photos of slide preparations of *Sepedon knutsoni* and *Sepedonella nana* and stereomicrographs of *Tetanocera ferruginea* (Fig. 15.21) are presented here.

Ferrar (1987) provided a list of the numbers of papillae described and/or figured for second- and third-instar larvae in most papers published to 1987 (114 species in 30 genera), but without any analysis as to behavioral, microhabitat, phylogenetic, or taxonomic significance. We attempt this in a separate contribution (Knutson, unpublished data), in which we include the data for an additional seven genera and 32 species. Here, we provide a summary and a few observations and figures that should be helpful in identification.

The range of numbers of papillae are similar in the Sciomyzini (second instar, 7–29; third, 7–34) and the Tetanocerini (second, 3–35; third, 2–37). In *Salticella fasciata* (Salticellinae) the ranges are 7–10 and 8–10. Curiously, the largest numbers in second- and third-instar larvae are found in a highly parasitoid, terrestrial member of the Sciomyzini (*Sciomyza varia*, 29 and 29–34) and in an aquatic, predaceous member of the Tetanocerini (*Ilione turcestanica*, 30–35 and 33–37, Fig. 15.20d). Neff & Berg (1966) noted that for the 15 species of *Sepedomerus, Sepedon*, and *Sepedonea* they studied, "Numbers of papillae differ considerably between individuals of a species and even between the right and left spiracles of an individual." Variation in numbers of papillae in second- and third-instar larvae occurs within almost all species. Two is the most common

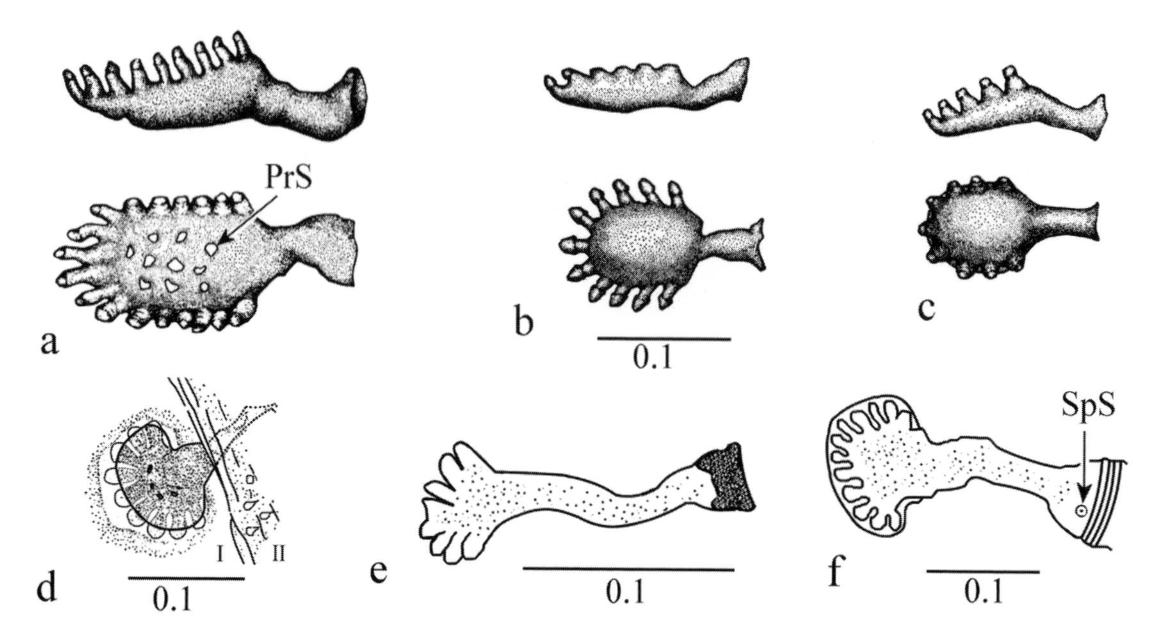

Fig. 15.19 Anterior spiracles, L3, of some terrestrial parasitoid/predators. (a) ***Tetanocera plebeja***. (b) *T. **valida***. (c) *T. **clara*** (from Trelka & Foote 1970). (d) ***Coremacera marginata*** (from Knutson 1973). (e) ***Euthycera stichospila*** (from Vala & Caillet 1985). (f) *E. **cribrata*** (from Vala *et al.* 1983). PrS, prismatic spot; SpS, spiracular scar; I, II, segments.

difference in number in both instars, with a difference of ten in third-instar *Anticheta analis* being the greatest. Considering the use to which numbers of spiracles is being made for identification, even more rigorous analyses of variation in individuals, within populations, and across the geographical and temporal ranges of species would be of interest.

The shapes of the head-part can be described as transverse (Fig. 15.18c), circular (Fig. 15.18e), elliptical (Fig. 15.19b), or longitudinal, either with the papillae closely set (Fig. 15.20a) or broadly spaced (Fig. 15.20b). There are patterns intermediate between some of these. In the Sciomyzini (Fig. 15.18d, e), papillae are arranged in a transverse, circular, or elliptical pattern; none of the Sciomyzini has papillae arranged in an extremely longitudinal pattern. *Colobaea americana* (Sciomyzini, Fig. 15.18f) is unusual in having a strongly transverse, biramous head-part. In the Tetanocerini (Figs 15.19, 15.20) all patterns except biramous are found; many have the papillae arranged along the longitudinal axis. The shape of the head-part and number of papillae in *Anticheta* (Fig. 15.20h) and *Renocera* is more similar to the Sciomyzini than to many Tetanocerini. The stem-part is short in some species and genera (e.g., *Sciomyza, Atrichomelina, Pherbellia*, Fig. 15.18e), its length about equal to the diameter of the head-part, or it may be elongate, up to somewhat more than three times the diameter of the head-part (e.g., *Pherbellia, Euthycera, Sepedon*, Figs 15.18d, 15.19e, f, 15.20e, g, 15.21a).

15.1.2.4 PUPARIA

Length, 2.6–12.0 mm; width, 1.0–4.2 mm. Those of *Salticella* and Sciomyzini (species formed outside the shell) are generally barrel-like in shape (Figs 14.43c, d, 14.44c, d), with the posterior end only slightly upturned in a few species of Sciomyzini (Fig. 14.43i). *Salticella* and many Sciomyzini puparia are formed outside of the shell. Those Sciomyzini whose puparia are formed within shells are adapted to fit within, with the anterior end more or less flattened. A few are strongly modified anteriorly to occlude the shell aperture (Fig. 12.4b–d). Pupariating larvae of a few species of *Pherbellia, Ditaeniella*, and *Colobaea* excrete a calcareous fluid which they mold into one or two septa that close the opening of the shell. Puparia of Tetanocerini are always formed outside of the shell (Figs 14.47–14.51). Those of terrestrial and semi-terrestrial species are barrel-like, with posterior and/or anterior ends usually slightly above or below the mid horizontal axis (Figs 14.49c, 14.50d). Puparia of the more numerous truly aquatic Tetanocerini are well adapted for flotation, with the posterior and usually anterior end elevated from above the mid horizontal axis to well

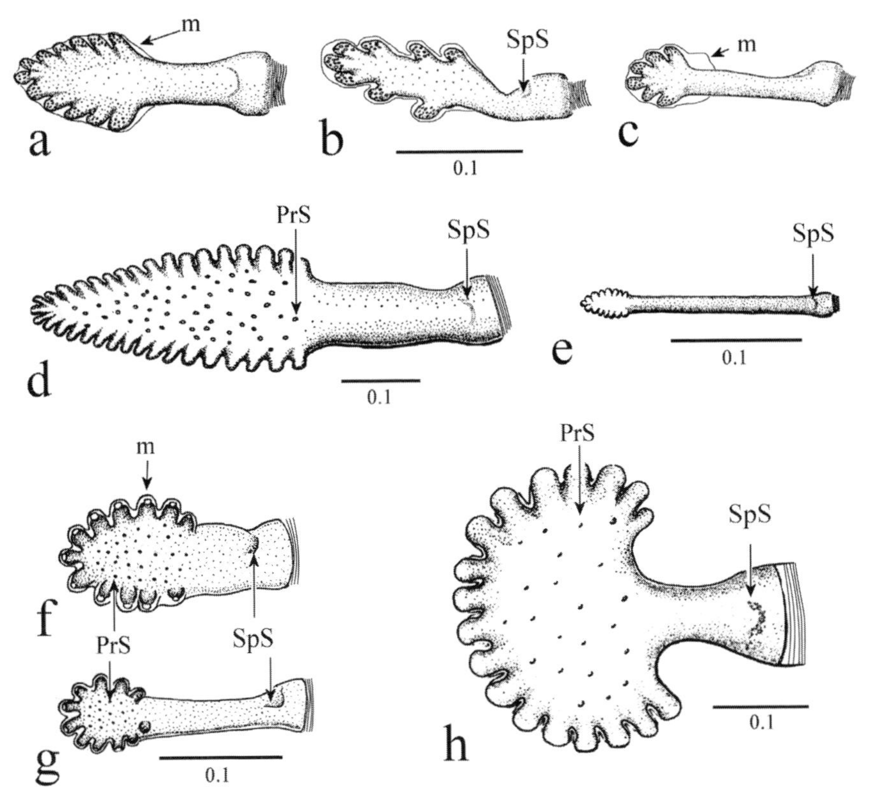

Fig. 15.20 Anterior spiracles, L3 (except e is L2), of some aquatic and semi-aquatic predators. (a) *Elgiva solicita*. (b) *E. cucularia*. (c) *E. connexa* (from Knutson & Berg 1964). (d) *Ilione turcestanica*. (e) *I. albiseta* (L2) (from Knutson & Berg 1967). (f) *Pherbina coryleti*. (g) *P. intermedia* (from Knutson *et al.* 1975). (h) *Anticheta brevipennis* (from Knutson 1966). m, membrane; PrS, prismatic spot; SpS, spiracular scar.

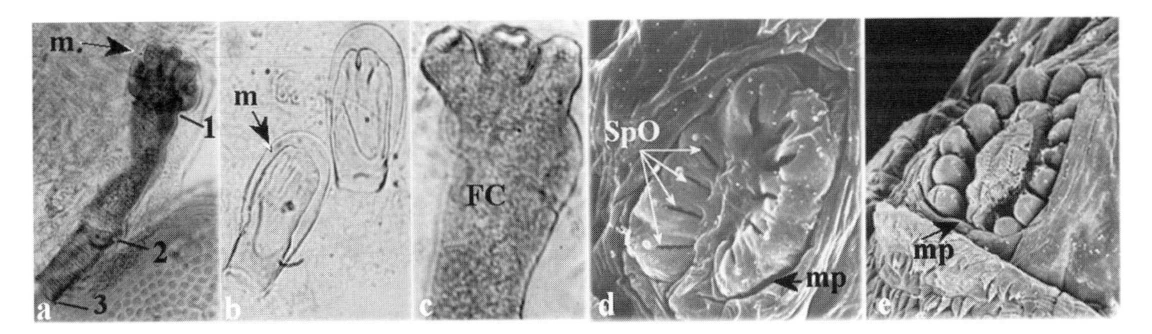

Fig. 15.21 Anterior spiracles of some predators of aquatic oligochaetes (a–c) and aquatic snails (d, e). (a, b) *Sepedonella nana*; (a) L3; (b) L2. (c) *Sepedon knutsoni*, L3. (d, e) *Tetanocera ferruginea*; (d) L2; (e) L3. 1–2, stem; 2–3, tracheal connection; FC, felt chamber; m, membrane; mp, membranous pocket; SpO, spiracular papilla opening. Photos by J-C. Vala.

above the dorsal surface, and with the anterior spiracles projecting more than in Sciomyzini (Figs 14.47, 14.48, 14.49a). In *Elgiva* species and *Renocera striata* the postanal portion of segment XII is strongly elongated and the anterior spiracles are at the end of elongate extensions (Fig. 14.49a, b). Puparia of several aquatic and a few semi-terrestrial Tetanocerini are characteristically inflated, especially anterodorsally and/or laterally (Fig. 14.51a); development of such shapes perhaps is enhanced as a result of air-swallowing by pupariating larvae and are adaptive for flotation.

Vestiges of larval structures, i.e., integument, body and posterior spiracular disc tubercles and lobes, and anterior and posterior spiracles, are useful in identifying puparia, but even more reliable is the third-instar cephalopharyngeal skeleton, which remains attached to the inner surface of the ventral cephalic cap.

15.2 KEYS TO ADULTS AND LARVAE OF GENERA OF SCIOMYZIDAE OF THE WORLD BY REGIONS AND KEYS TO PUPARIA IN THE NEARCTIC AND PALEARCTIC REGIONS

Chyen tini kat pat, I pa ka pwan kat chimen. [Dog has four legs, but only take one way].

Guadeloupean proverb.

Following are the first keys to genera of the adults and larvae of all regions and to genera of puparia of the Nearctic and Palearctic with a uniform format, nomenclature, and terminology. Included are keys to adults in the Nearctic–Neotropical and Oriental–Australian transition areas. For each regional key to larvae and puparia, size ranges, numbers of anterior spiracular papillae, and indentation indices are given only for species occurring in the region, unless otherwise indicated. Space limitations prevent including many illustrations of characters, for which the user is advised to refer to the regional publications and Ferrar (1987). Note: numbers refer to one side, i.e., to number of pairs, e.g., two scutellar setae = two pairs of scutellars.

Because keys to genera of adults and larvae of the well-known Nearctic and Palearctic faunas have been available and used for some decades, we have retained the general formats and characters to which users are accustomed, but with many additions and corrections. Since there have been no or few published keys to adults of the other, more poorly known faunas we have included some additional data, e.g., geographical ranges, to facilitate their use.

15.2.1 Adults

Nearctic (modified from Knutson, 1987)

1. Pro-episternal seta well developed; if hair-like (*Atrichomelina*), then basal tarsomere of fore leg white and other tarsomeres black (SCIOMYZINI)............ 2
- Pro-episternal seta absent, frequently a cluster of several short hairs; basal tarsomere not contrasting strongly with other segments except base of basal and second tarsomere of fore leg pale to white in *Dictyacium* (TETANOCERINI)........................ 8

2. Pro-episternal seta hair-like. Basal tarsomere of fore leg white and other tarsomeres black. Fore coxa without setae. All thoracic pleurites with setulae but without setae. Wing (Fig. 14.9c)............... **Atrichomelina**
- Pro-episternal seta well developed (short in *Colobaea montana*). Basal tarsomere of fore leg at most yellowish and other tarsomeres brownish. Fore coxa with 1 or more setae. At least 1 thoracic pleurite bare or with 1 or more setae.................................... 3

3. Fore tibia with 1 dorsal pre-apical seta............. 4
- Fore tibia with 2 dorsal pre-apical setae............ 7

4. $A_1 + CuA_2$ not reaching margin of wing (Fig. 14.9e, f). Frons dull black or yellow. 2 fronto-orbital setae. Arista usually with a few dorsobasal hairs slightly to much longer than others (Fig. 15.1b)........... **Colobaea**
- $A_1 + CuA_2$ reaching margin of wing, although sometimes faint apically. Frons dull yellow or entirely shiny, never dull black. 1 or 2 fronto-orbital setae. Arista never with elongate dorsobasal hairs (e.g., Fig. 15.1d)...................................... 5

5. Frons entirely shiny. Fore femur usually with a pecten (short series of closely spaced spinules anteromedially). Mesonotum shiny black or brown. Wing (Fig. 14.10a) .... **Pteromicra**
- Frons not shiny. Fore femur without a pecten. Mesonotum pruinose, yellow, brown, or gray........ 6

6. One fronto-orbital seta. Prosternum with a few hairs. Frontal vitta usually reaching anterior margin of frons. Hind coxa with hairs on inner posterodorsal margin. Wing without pattern (Fig. 14.9g). Anterior surstylus vestigial.............................. **Ditaeniella**
- Two fronto-orbital setae. Prosternum bare. Frontal vitta short, triangular, pollinose. Hind coxa without hairs on inner posterodorsal margin. Wing with or without pattern (Fig. 14.9k–r). Anterior surstylus well developed .... **Pherbellia**

7. Arista black. 2 fronto-orbital setae. Presutural supra-alar seta present. 2 postsutural dorsocentral setae. Face not tuberculate. Wing (Fig. 14.10b)....... **Sciomyza**
- Arista white, antenna (Fig. 15.1c). 1 fronto-orbital seta. Presutural supra-alar seta absent. 1 postsutural dorsocentral seta. Face tuberculate. Wing (Fig. 14.9i)......................... **Oidematops**

8. Ocellar setae present and crossvein dm-cu straight to slightly curved. Presutural supra-alar seta usually present......................................9
- Ocellar setae usually absent (hair-like in some specimens of some *Sepedon* spp.), but if well developed (*Elgiva*), crossvein dm-cu bent almost at right angles. Presutural supra-alar seta usually absent......................20
9. Frons strongly convex, glossy. 3 postsutural dorsocentral setae, 1 at suture. Body yellowish with dark spots (Plate 1c). Wing patterned (Fig. 15.4c)..**Poecilographa**
- Frons not strongly convex, occasionally shiny. 1–3 postsutural dorsocentral setae. Body shining black or dull brownish to gray, with or without spots. Wing patterned or not................................10
10. Subalar ridge without setae.......................11
- Subalar ridge with setae..........................18
11. Hind tibia with 2 pre-apical setae, 1 large dorsal and 1 anterodorsal, the latter closer to the apex and slightly shorter. Only katepisternum with hairs. Arista black. Male epandrium and surstyli strongly asymmetrical. Wing (Fig. 14.10d, e)...................... **Anticheta**
- Hind tibia usually with only 1 (dorsal) pre-apical seta; 2 in some *Tetanocera;* in *Teutoniomyia* barely differentiated. 1 or more thoracic pleurites with hairs or setae. Arista black or white. Male epandrium and surstyli symmetrical except *Tetanocera robusta* epandrium asymmetrical... 12
12. Pedicel less than 1/3 length of postpedicel (Fig. 15.1m). Arista black. Wing not patterned (Fig. 14.12a)...**Renocera**
- Pedicel at least 1/2 length of postpedicel (Fig. 15.1i, n). Arista black or white. Wing patterned or not 13
13. Anepisternum and anepimeron with setae. 1 fronto-orbital seta or 3 dorsocentral setae 14
- Anepisternum and anepimeron bare or with hairs. 2 fronto-orbital setae and 1 or 2 dorsocentral setae...15
14. Face white, without central black spot. 2 fronto-orbital setae. 3 dorsocentral setae. All thoracic pleurites setose. Wing (Fig. 14.11g)......................**Hoplodictya**
- Face white, with central black spot (Fig. 11.1b). 1 fronto-orbital seta. 2 dorsocentral setae. Katepisternum with hairs but without setae. Wing (Fig. 11.1b, 14.10j) ..... **Dictya**
15. Arista black (Fig. 15.1q). Wing usually not patterned, at most sparsely patterned with discrete dark spots (Figs 14.12i, j, 15.4b)............. **Tetanocera**
- Arista white. Wing densely patterned with dark network around hyaline spots (e.g., Fig. 14.10r)....16
16. Face flat to slightly concave. Pedicel less than twice as long as wide (Fig. 15.1j). Wing (Figs 14.10r, 14.11a–c, 15.4e) .... **Euthycera**
- Face tuberculate. Pedicel at least twice as long as wide (Fig. 15.1 n–p) 17
17. Pedicel cylindrical, equal in length to elongate, triangular postpedicel; aristal hairs semi-erect (Fig. 15.1h). Head length equal to height; dorsal surface not flattened; lunule normal; lower face vertical; mouth opening normal. Abdomen and legs mainly yellow. Wing (Fig. 14.10k)....................**Dictyacium**
- Pedicel compressed, 3 times length of quadrate postpedicel; aristal hairs appressed (Fig. 15.1r). Head length 1.5 times height; dorsal surface flattened; lunule grossly inflated; lower face strongly retreating; mouth opening strongly reduced. Abdomen and legs mainly black. Wing (Fig. 14.12m).... **Teutoniomyia**
18. Wing cells with several confluent, widely separated spots forming more or less transverse bars. Vein R_{2+3} with several weak, irregular undulations (Fig. 14.11n). Prescutellar acrostichal setae present or absent. Prosternum bare. Arista black.............................**Pherbecta**
- Wing patterned or not. Vein R_{2+3} without undulations. Prescutellar acrostichal setae present. Prosternum bare or haired and arista white, or prosternum haired and arista black.....................................19
19. Arista white. Frontal vitta broad, shiny. Anepisternum and anepimeron with hairs but without setae. Wing (Fig. 14.11k)................................ **Limnia**
- Arista dark brown to black. Frontal vitta narrow, waxy. Anepisternum and anepimeron with hairs and some weak setae. Wing (Fig. 14.12p).............. **Trypetoptera**
20. Ocellar setae well developed. 2 scutellar setae. Hind coxa with hairs on inner posterodorsal margin. Crossvein dm-cu rectangularly bent. Wing (Fig. 14.10i). Antenna (Fig. 15.1l)................................ **Elgiva**
- Ocellar setae absent (hair-like in some specimens of some *Sepedon* spp.). 1 or 2 scutellar setae. Hind coxa usually without hairs on inner posterodorsal margin. Crossvein dm-cu straight to curved...............21
21. Two fronto-orbital setae, anterior 1 smaller. Ptilinal fissure present. Postpronotal seta present. Crossvein dm-cu S-shaped. Wing (Fig. 14.11f). Antenna (Fig. 15.1k)....................................**Hedria**
- One fronto-orbital seta. Ptilinal fissure absent (Fig. 14.7). Postpronotal seta absent. Crossvein dm-cu straight to arched............................22
22. Postocellar setae absent. Mid femur with minute setae, none of which is distinctly longer than the others. Hind femur of male without mid ventral notch. Wing (Fig. 14.12b) .... **Sepedomerus**

- Postocellar setae well developed. Mid femur with 1 or more distinctly longer anterior setae near mid length. Hind femur of male with (Fig. 14.8a–c) or without mid ventral notch. Antenna (Fig. 15.1n, o). Wing (Fig. 14.12c)............................. **Sepedon**

Palearctic (modified from Rozkošný, 1998)

1. Cell cu*p* with triangular extension. Veins R_{4+5} and M_{1+2} strongly converging at apex of wing (Fig. 14.9b) (SALTICELLINAE)......................... **Salticella**
- Cell cu*p* without triangular extension. Veins R_{4+5} and M_{1+2} parallel or only slightly converging (Fig. 15.4a) (SCIOMYZINAE)...............................2
2. Pro-episternum with distinct seta above base of fore coxa. If pro-episternal seta short or indistinct (*Colobaea*), vein $A_1 + CuA_2$ not reaching posterior margin (SCIOMYZINI)..............................3
- Pro-episternum without seta above base of fore coxa, but often with several fine hairs. Vein $A_1 + CuA_2$ reaching posterior margin (TETANOCERINI).....................8
3. Arista subapical (Fig. 15.1e). Female terminalia modified into elongate, flattened ovipositor (Fig. 14.2a). Wing (Fig. 14.10c)....................... **Tetanura**
- Arista basal or sub-basal (Fig. 15.1b, c, f, g, i, j, m, o, q). Female terminalia not modified into elongate, flattened ovipositor (Figs 15.5c, 15.7b, 15.11)................4
4. Arista without long, dorsobasal rays (Fig. 15.1d). $A_1 + CuA_2$ reaching posterior margin of wing (Fig. 14.9n). Pro-episternal seta elongate. Anterior surstylus without thickened subcylindrical setae (except in *Pherbellia knutsoni* and *P. mikiana*)...........................5
- Arista with long, dorsobasal rays (Fig. 15.1b). $A_1 + CuA_2$ not reaching posterior margin (Fig. 14.9e, f). Pro-episternal seta short or indistinct. Anterior surstylus of most species with thickened subcylindrical setae........... **Colobaea**
5. Fore tibia with 1 dorsal pre-apical seta. 6
- Fore tibia with 2 dorsal pre-apical setae (1 in *S. sebezhica*, w. Russia).......................... **Sciomyza**
6. Frons mainly pruinose. Body never shining black. Gena usually broad..............................7
- Frons entirely shining. Body mainly shining black or black and yellow. Gena narrow (Fig. 14.2c). Wing (Fig. 14.10a).......................... **Pteromicra**
7. One fronto-orbital seta. Prosternum with a few hairs. Hind coxa with hairs on inner posterodorsal margin. Wing without pattern (Fig. 14.9g). Anterior surstylus vestigial.............................. **Ditaeniella**
- Two fronto-orbital setae (1 in *P. setosa*, Japan). Prosternum bare. Hind coxa bare on inner posterodorsal margin. Wing with or without pattern (Figs 14.9k, l, n, p–r, 15.4d). Anterior surstylus well developed .... **Pherbellia**
8. Wing with distinct reticulate pattern. 9
- Wing without reticulate pattern, at most with darkened costal margin and infuscated crossveins or with several isolated dark spots...............................20
9. Two fronto-orbital setae.........................10
- One fronto-orbital seta. Wing patterned as in Figs 11.1b, 14.10j.............................. **Dictya**
10. Postpedicel with tuft of long black hairs at tip (Fig. 15.1g). Wing patterned as in Fig. 14.10g,h............................ **Coremacera**
- Postpedicel without tuft of long hairs at tip. Wing pattern different................................11
11. Anepimeron haired and/or with setae............12
- Anepimeron bare. 16
12. Subalar setae present.............................13
- Subalar setae absent. Wing (Fig. 14.11r)... **Psacadina**
13. Arista with short whitish hairs. Mesonotum with conspicuous longitudinal stripes. Wing (Fig. 14.11k) .... **Limnia**
- Arista with long black hairs at least in apical 2/3. Mesonotum with or without conspicuous longitudinal stripes...14
14. Wing dark, with numerous large, pale spots (Fig. 14.12p) .... **Trypetoptera**
- Wing with reticulate pattern......................15
15. Aristal hairs entirely black. Anepimeron with hairs and 1–3 setae. Anepisternum with 1 strong seta. Apex of wing pale, at most with reticulation (Fig. 14.11o)............................ **Pherbina**
- Aristal hairs white basally, black in apical half. Anepimeron and anepisternum with only minute hairs. Apex of wing with dark patch, otherwise wing similar to *Dictya*..................................**Neodictya**
16. Anepisternum with 4–5 hairs. Wing similar to *Limnia*.............................. **Oligolimnia**
- Anepisternum bare...............................17
17. Prosternum setose. Male sterna 4 and 5 densely covered with short setae. Wing (Fig. 14.10f)....**Chasmacryptum**
- Prosternum bare. Male sterna 4 and 5 not densely covered with short setae..........................18
18. Hind coxa with hairs on inner posterodorsal margin. Hind femur with distinct ventral spines. Wing (Fig. 14.10i) with reticulate pattern at least on apical third................................ **Dichetophora**
- Hind coxa without hairs on inner posterodorsal margin. Hind femur without distinct ventral spines. Wing pattern different...........................19

19. One postalar seta. 2–3 short, fine setae at upper posterior margin of katepisternum. Antenna as in *Ectinocera*. Wing pattern (Fig. 14.12o) similar to *Pherbellia limbata*. . . . . **Trypetolimnia**
- Two postalar setae. Katepisternum without setae at upper margin. Antenna (Fig. 15.1j). Wing (Figs 14.10r, 14.11a–c). . . . . **Euthycera**
20. Ocellar setae absent. Ptilinal fissure absent (Fig. 14.7). Scutellum with 1 seta. Pedicel rod-like (Fig. 15.1o). Habitus (Plate 1a, f). . . . . **Sepedon**
- Ocellar setae present. Ptilinal fissure present. Scutellum with 2 setae (except 1 in *Anticheta bisetosa*). Pedicel usually relatively stout and short, at most as long as postpedicel (e.g., Fig. 15.1h). 21
21. Hind tibia with 2 pre-apical setae, 1 dorsal and 1 anterodorsal, the latter closer to apex and slightly shorter. Gena not wider than postpedicel. Wing (Fig. 14.10d, e). . . . . **Anticheta**
- Hind tibia with 1 dorsal pre-apical seta. If rarely 2 (some *Tetanocera*), then gena distinctly wider than postpedicel. 22
22. Anepisternum and anepimeron bare. Hind coxa without hairs on inner posterodorsal margin. 23
- Anepisternum and anepimeron with at least a few minute, scattered hairs. Hind coxa with hairs on inner posterior margin, toward apex. 26
23. Postpedicel elongate, about 3 times as long as broad, pedicel short. 1 postalar seta. Wing (Fig. 14.10m). . . . . **Ectinocera**
- Postpedicel at most twice as long as broad, if rarely longer, then pedicel long. 2 postalar setae. 24
24. Arista only pubescent. Male sternum 4 with pair of cylindrical processes. Wing (Fig. 14.11h). . . **Hydromya**
- Arista moderately short to long haired. Male sternum 4 without cylindrical processes 25
25. Mid femur usually with anterior seta near middle. Frontal vitta indistinct or weak. Pedicel slightly longer than broad (Fig. 15.1q). Male sternum 5 simple. Hind tibia with 1 or 2 pre-apical setae. Wing (Fig. 14.12i, j). . . . . **Tetanocera**
- Mid femur without anterior seta near middle. If present, then dark frontal vitta well developed. Pedicel broader than long (Fig. 15.1m). Male sternum 5 with posterior incision or tubercles. Hind tibia with 1 pre-apical seta. Wing (Fig. 14.12a). . . . . **Renocera**
26. Subalar setae present. Crossvein dm-cu bow-shaped in upper half (Fig. 14.11i, j). Aristal hairs white, partly brown, or entirely brown. . . . . **Ilione**
- Subalar setae absent. Crossvein dm-cu almost rectangularly bent in upper half (Fig. 14.10n). Aristal hairs white . . . . **Elgiva**

Nearctic–Neotropical transition area (modified from Marinoni & Knutson [in press])

1. Pedicel half to less than half length of postpedicel. Pro-episternal seta present; if short, fine, hair-like (*Atrichomelina*) then only 1 well developed fronto-orbital seta. (SCIOMYZINI). 2
- Pedicel subequal to or twice length of postpedicel (shorter in *Renocera*). Pro-episternal seta absent. (TETANOCERINI). 5
2. Pro-episternal seta short, hair-like. Thoracic pleuron with fine hairs, without setae. Basal tarsomere of fore leg white, contrasting with black 4 distal tarsomeres. Wing without pattern (Fig. 14.9c). Mexico . . . . **Atrichomelina**
- Pro-episternal seta well developed. Thoracic pleuron with setae. Basal tarsomere of fore leg not contrasting with 4 distal tarsomeres. Wing with or without pattern. 3
3. One fronto-orbital seta. Frontal vitta subshiny, nearly reaching anterior margin of frons. Wing without pattern (Fig. 14.9g). Mexico, Costa Rica . . . . **Ditaeniella**
- One or 2 fronto-orbital setae. Frontal vitta absent or short, triangular, pruinose. Wing patterned 4
4. Face with median rounded black spot. 1 fronto-orbital seta. Wing strongly patterned (Fig. 14.9h). Costa Rica, Venezuela, Brazil . . . . **Neuzina**
- Face without such spot. 2 fronto-orbital setae. Wing patterned or not (Fig. 14.9k–r). Mexico, Guatemala, Costa Rica . . . . **Pherbellia**
5. Hind tibia with 2 pre-apical setae, 1 dorsal and 1 anterodorsal, the latter closer to the apex and slightly shorter. Wing (Fig. 14.10d). Mexico. . . . . **Anticheta**
- Hind tibia with 1 pre-apical dorsal seta (2 in *Tetanocera plumosa*). 6
6. Face with median rounded black spot. 7
- Face without median rounded black spot. 9
7. Head longer than high. Lunule exposed. 2 fronto-orbital setae. Arista white (Fig. 15.1j). Wing patterned (Figs 14.10r, 14.11a–c). Mexico. . . . . **Euthycera**
- Head higher than long. Lunule exposed or covered. 1 fronto-orbital seta. Arista white or black. Wing patterned or not. 8
8. Lunule exposed. Wing translucid, with weak to strong pattern (Fig. 14.11q). Hind femur slender,

longer than tibia. Fifth abdominal sternum of male with 1 process (Plate 1b, d). Mexico, Guatemala, Honduras, Nicaragua, Costa Rica, Panama, throughout S. America. . . . . **Protodictya**

- Lunule not exposed. Wing strongly patterned (Fig. 14.10j). Hind femur short, almost same length as tibia (Fig. 11.1b). Fifth abdominal sternum of male without process. Mexico to Colombia . . . . **Dictya**

9. Ocellar setae absent. Ptilinal fissure absent (Fig. 14.7). 1 scutellar seta 10

- Ocellar setae present. Ptilinal fissure present. 2 scutellar setae. 13

10. Hind tibia with short spine-like projection ventro-apically. Head not conically developed. With palpi. Thorax and abdomen mostly reddish brown. 11

- Hind tibia without such projection. Head conically developed downwards. Without palpi. Thorax and abdomen dark brown with bright yellow markings. Wing (Fig. 14.12n). Guatemala, Nicaragua, Costa Rica, Panama, throughout n. half of S. America . . . . **Thecomyia**

11. Face with dark brown spot in anteroventral corner. Antenna (Fig. 15.1p). Postocellar setae well developed. Hind femur without markings. Surstyli fused on mid line. Posterior thoracic spiracle with strong setulae on posterior margin (Fig. 14.16). Wing (Fig. 14.12d). Mexico, Guatemala, Honduras, Nicaragua, Costa Rica, Panama, throughout S. America.. . . . . **Sepedonea**

- Face without dark brown spot in anteroventral corner. Postocellar setae present or absent. Hind femur well developed, with or without lateral dark marks. Surstyli separate. Posterior thoracic spiracle without strong setulae on posterior margin 12

12. Postocellar setae present. Antenna (as Fig. 15.1n, o). Hind femur without markings. Mid femur with 1 or more distinctly larger anterior setae near mid length. Wing (Fig. 14.12c). Mexico, s. to Oaxaca . . . **Sepedon**

- Postocellar setae absent. Hind femur with lateral dark marks. Mid femur with minute setae on anterior surface, none of which is distinctly larger than the others. Wing (Fig. 14.12b). Guatemala, Nicaragua, Costa Rica, Panama.. . . . . **Sepedomerus**

13. Very small flies. Head dorsoventrally flattened. Pedicel high and compressed (Fig. 15.1r). Hind tibia arch—shaped. All basitarsi white. Wing almost quadrate (Fig. 14.12m). Mexico, Costa Rica, Brazil. . . . **Teutoniomyia**

- Small to medium-sized flies. Hind tibia not arch-shaped. Head, pedicel, and basitarsi not as above. . . 14

14. Two fronto-orbital setae. Antenna (Fig. 15.1q). Anepisternum and anepimeron without setae or hairs. Wing not patterned (Fig. 14.12j). Mexico, Venezuela.. . . . . **Tetanocera**

- One or 2 fronto-orbital setae. Anepisternum and sometimes anepimeron with setae or hairs. Wing patterned. 15

15. Face not strongly concave, with dark brown vitta below eye. Lunule not exposed. 2 fronto-orbital setae. Frons without spots. Arista with sparse black hairs, not plumose. Wing (Fig. 14.11g). Mexico, Guatemala.. . . . . **Hoplodictya**

- Face strongly concave, without dark brown vitta below eye. Lunule exposed. 1 fronto-orbital seta. Frons with black spots laterad of antenna and ocellar triangle. Arista with black, plumose hairs. Wing (Fig. 14.11e). Mexico, Guatemala, Honduras . . . . **Guatemalia**

Neotropical (modified from Steyskal & Knutson [1975b])

The genera *Anticheta*, *Atrichomelina*, *Euthycera*, and *Tetanocera*, widespread in the Nearctic region and also found near the distinction between the Nearctic and Neotropical or slightly adventive into the Neotropical region, are included.

1. Pro-episternal seta present (Sciomyzini, plus *Eutrichomelina*, *Perilimnia*, and *Shannonia* of Tetanocerini). 2

- Pro-episternal seta absent, only fine hairs on pro-episternum (Tetanocerini, plus *Pseudomelina* of Sciomyzini). 10

2. Pro-episternal seta short, fine, only slightly larger than pro-episternal hairs. Nearly entire anepisternum and katepisternum, center of anepimeron, and inner posterodorsal margin of hind coxa with fine hairs. Fore tarsus with basal tarsomere white, distal 4 tarsomeres black. Wing without pattern (Fig. 14.9c).. . . . . **Atrichomelina**

- Pro-episternal seta long, coarse, if short, fine (*Shannonia meridionalis*) then without above combination of characters. Wing with or without pattern 3

3. Face with median, rounded black spot. 1 fronto-orbital seta. Wing strongly patterned (Fig. 14.9h). . . . . **Neuzina**

- Without above combination of characters. 4

4. $A_1 + CuA_2$ not reaching margin. R_1 not extending beyond crossvein r-m. 2 dorsocentral setae; anterior seta very strong, at suture. Arista short, pubescent. Palpus with 1 strong apical seta. Gena very narrow. Shiny black or brown species. Wing (Fig. 14.9j). . . . . **Parectinocera**

- $A_1 + CuA_2$ reaching margin, although weakly in some species. Without above combination of characters..... 5
5. Four dorsocentral setae, 1 antesutural and 3 postsutural. Frontal vitta very short, tomentose. R_1 extending beyond crossvein r-m. Wing (Fig. 14.11d)........................ **Eutrichomelina**
- Two or 3 dorsocentral setae, none antesutural. Frontal vitta short and tomentose or elongate, subshiny..... 6
6. Three dorsocentral setae. Anepisternal seta present. Katepisternal setae absent. Frontal vitta scarcely extending beyond ocellar triangle. 1 fronto-orbital seta. Wing hyaline (Fig. 14.9d).. .... **Calliscia**
- Two or 3 dorsocentral setae. Both anepisternal and katepisternal setae present or both absent or only katepisternals present. Frontal vitta short or elongate. 1 or 2 fronto-orbital setae............................7
7. Frontal vitta usually reaching anterior margin of frons, subshiny. 1 fronto-orbital seta. Prosternum with a few hairs. Katepisternum without setae. Hind coxa with hairs on inner posterodorsal margin. Wing without pattern (Fig. 14.9g). Anterior surstylus vestigial. .... **Ditaeniella**
- Without above combination of characters.......... 8
8. Frontal vitta short, triangular, pollinose. Postpedicel about as long as wide or slightly longer (Fig. 15.1d). Anepisternum without setae, with or without hairs. Anterior surstylus well-developed. Wing patterned or not (Fig. 14.9k–r). .... **Pherbellia**
- Frontal vitta elongated, parallel-sided, subshiny. Postpedicel 1–5 times longer than wide. Anepisternal setae present or absent, with hairs. Anterior surstylus vestigial. Wing not patterned........................9
9. Anterior fronto-orbital seta smaller than posterior, occasionally absent. Postpedicel elongate, tapered apically. Anepisternum with 1 seta and several small setulae. Male fourth abdominal sternum with short, dense, black spinules on posterior margin. Wing (Fig. 14.11m).......................... **Perilimnia**
- Anterior fronto-orbital seta almost as strong as posterior. Postpedicel rather short, more or less rounded apically. Anepisternum without setae, with small setulae on posterior portion. Male fourth abdominal sternum without short, dense, black spinules on posterior margin. Wing (Fig. 14.12g).............. **Shannonia**
10. Postpedicel broadly rounded apically as in *Pherbellia*. Anepimeron with 1 or 2 bristly hairs. Hind coxa without hairs on inner posterodorsal margin. 2 postsutural dorsocentral setae.................. **Pseudomelina**
- Without above combination of characters..........11
11. Subalar setae present............................12
- Subalar setae absent.............................13
12. Arista with sparse blackish hairs. Lunule more or less covered. Face without black central spot. Male abdominal sternum 5 without medio-apical process. Wing (Fig. 14.10l)............... **Dictyodes**
- Arista densely white pubescent. Lunule broadly exposed. Face with black central spot (Plate 1b, d). Male abdominal sternum 5 with medio-apical process. Wing (Fig. 14.11q).................... **Protodictya**
13. Ocellar setae well developed. Ptilinal fissure present. 2 scutellar setae. 14
- Ocellar setae small and weak or lacking. Ptilinal fissure absent (Fig. 14.7). 1 scutellar seta.. 22
14. Hind tibia with 2 dorsal pre-apical setae. Wing (Fig. 14.10d)................................ **Anticheta**
- Hind tibia with 1 or 0 dorsal pre-apical setae.......15
15. Arista with white pubescence or hairs. Wing with strong pattern..................................16
- Arista with black pubescence or hairs. Wing with or without pattern.................................18
16. Lunule exposed only between antennae. Pedicel about half as long as postpedicel............ **Euthycerina**
- Lunule broadly exposed. Pedicel at least almost as long as postpedicel, sometimes much longer...........17
17. Pedicel about as long as postpedicel (as Fig. 15.1j). Head about as long as high. Wing longer than broad (Figs 14.10r, 14.11a–c).. .... **Euthycera**
- Pedicel much longer than roundish postpedicel (Fig. 15.1r). Head at least 1.5 times as long as high. Wing nearly quadrate, patterned (Fig. 14.12m)........................... **Teutoniomyia**
18. Anepisternum and anepimeron without setae or hairs; if hairs present on posterior part of anepisternum then wing without dense pattern.......................19
- Anepisternum and sometimes anepimeron with setae or hairs. Wing with dense pattern.................20
19. Katepisternum with hairs only. 0 presutural dorsocentral setae, 2 postsuturals. Arista long-haired to plumose (as Fig. 15.1q). Wing (Fig. 14.12i).. .... **Tetanocera**
- Katepisternum with two strong setae. 0 or 1 presutural dorsocentral setae, 3 or 4 postsuturals. Arista with short pubescence. Wing (Fig. 14.12k)..... **Tetanoceroides**
20. Two fronto-orbital setae. 1 strong katepisternal seta. Face without black central spot. Wing (Fig. 14.11g)............................. **Hoplodictya**
- One fronto-orbital seta. 0 katepisternal setae. Face with or without black central spot.....................21

21. Face with black central spot. Body with many dark spots at bases of hairs and otherwise (Fig. 11.1b). Wing (Fig. 14.10j). . . . . **Dictya**
- Face without black central spot. Body largely plain tawny. Wing (Fig. 14.11e). . . . . **Guatemalia**

22. Sternum closed above hind coxae. Lower head drawn out into conical rostrum into which elongate proboscis may be withdrawn. Palpus absent. Postocellar and usually fronto-orbital setae lacking. Pleuron without hairs or setae, except a few hairs on pro-episternum close above fore coxa. Large, brown flies with yellow markings on thorax and abdomen. Wing (Fig. 14.12n). . . . . **Thecomyia**
- Sternum above hind coxae divided by membranous area. Lower head not forming tube into which proboscis may be withdrawn. Palpus well developed. Postocellars present or absent. 1 or 2 fronto-orbital setae. Pleuron frequently with numerous hairs. Body color not brown and yellow. 23

23. Postocellar setae absent. 1 fronto-orbital seta. Mid femur with minute setae or setulae, none of which is distinctly larger than the others. Hind femur almost twice as long as abdomen. Wing (Fig. 14.12b). . . . . **Sepedomerus**
- Postocellar setae well developed. 0, 1, or 2 fronto-orbital setae. Mid femur with 1 or more distinctly larger anterior setae near mid length. Hind femur only about 1/3 longer than abdomen. 24

24. Face with black spot in each lower corner. Antenna (Fig. 15.1p). Fore femur with at least 1 outstanding dorsal seta. Male surstyli fused along median line. Posterior thoracic spiracle with strong setulae on posterior margin (Fig. 14.16). Wing (Fig. 14.12d). . . . . **Sepedonea**
- Face without spots in lower corners. Antenna (as Fig. 15.1n). Fore femur without outstanding dorsal seta. Male surstyli well separated on median line. Posterior thoracic spiracle without strong setulae on posterior margin. Wing (Fig. 14.12c). . . . . **Sepedon**

Afrotropical (modified in part from Miller, 1995)

1. Cell *cup* with triangular extension apically. R_{4+5} and M_{1+2} strongly converging at wing apex (Fig. 14.9b) (SALTICELLINAE). . . . . **Salticella**
- Cell *cup* without triangular extension apically. R_{4+5} and M_{1+2} parallel or slightly converging at wing apex (Fig. 15.4a) (SCIOMYZINAE). 2

2. Pro-episternum with seta above base of coxa, usually strong, may be small or weak. Male with 2 pairs of well-developed surstyli (except 1 pair in *Ditaeniella*) (SCIOMYZINI). 3
- Pro-episternum without strong seta, only fine hairs at most. Male with 1 pair of well-developed posterior surstyli, anterior surstylus sometimes appearing as a vestige. (TETANOCERINI). 6

3. $A_1 + CuA_2$ not reaching posterior margin of wing (Fig. 14.9f). Arista with several dorsobasal hairs black, seta-like, and much stronger than others (as Fig. 15.1b). Frons mostly pruinose. . . . . **Colobaea**
- $A_1 + CuA_2$ reaching posterior margin of wing. Arista without large dorsobasal hairs. Frons shiny or pruinose. 4

4. $A_1 + CuA_2$ reaching margin as a fold (Fig. 14.10a). Gena narrow (Fig. 14.2c). Frons entirely shining . . . . **Pteromicra**
- $A_1 + CuA_2$ distinctly reaching margin of wing. Gena broad. Frons mostly pruinose 5

5. Two fronto-orbital setae. 1 anepimeral seta. Inner posterodorsal margin of hind coxa bare. Wing patterned. . . . . **Pherbellia** (subgenus **Graphomyzina**)
- Only posterior fronto-orbital seta present. 2 anepimeral setae. Inner posterodorsal margin of hind coxa with fine hairs. Wing not patterned (Fig. 14.9g) . . . . . **Ditaeniella**

6. Two scutellar setae. 7
- One scutellar seta. 9

7. Two fronto-orbital setae. Wing with brown costal border; a small, brown spot on M_{1+2} before and after dm-cu crossvein (Fig. 14.11h) . . . . **Hydromya**
- Only posterior fronto-orbital seta present. Wing variously patterned with spots, bands, or reticulation 8

8. Wing brownish, with several large, round transparent spots (Fig. 14.12q). . . . . **Verbekaria**
- Wing either with dense, strongly reticulate pattern or brownish with several, mainly confluent, transparent areas around mid longitudinal axis (Fig. 14.10o, p) . . . . **Ethiolimnia**

9. Ptilinal fissure present. Lunule distinct, broadly exposed. Frons entirely pruinose. Ocellar setae present but weak. 2 notopleural setae. 1 supra-alar seta. Wing (Fig. 14.12l). . . . . **Tetanoptera**
- Ptilinal fissure absent. Lunule not distinct (Fig. 14.7). Frons mostly shiny. Ocellar setae weak or absent. 1 or 2 notopleural setae. 0–1 supra-alar setae. 10

10. Postocellar setae absent. Very small species, length 3.5 to 5.8 mm. . . . . **Sepedonella**
- Postocellar setae present. Mostly larger species, length 5.0 to 9.0 mm. 11
11. Two fronto-orbital setae, anterior proclinate, posterior reclinate, both sometimes greatly reduced. Lunular area very large, inflated. Mesonotal suture complete. Wing (Fig. 14.12f) . . . . **Sepedoninus**
- Usually only posterior fronto-orbital seta present, sometimes with weak reclinate anterior fronto-orbital. Lunular area not greatly inflated. Mesonotal suture incomplete mesially. Wing (Fig. 14.12c) (*Sepedon sensu latum*). . . . 12
12. Scape elongate, at least 2/3 length of postpedicel. Face perpendicular or strongly oblique. Presutural supra-alar seta weak. . . . . **Sepedon** (subgenus **Sepedomyia**)
- Scape short, at most half length of postpedicel (as Fig. 15.1n, o). Face slightly oblique. Presutural supra-alar seta at most weak, usually absent. 13
13. Anterior notopleural seta sometimes present. Usually at most a weak presutural supra-alar seta. Aedeagus usually with spiral filament. Cochleate vesicle usually present (Fig. 14.3). Wing without pattern or darkened apically. . . . . **Sepedon** (subgenus **Parasepedon**)
- Anterior notopleural seta absent. 1 strong presutural supra-alar seta. Aedeagus without spiral filament. Cochleate vesicle usually absent. . . . . . .**Sepedon** (subgenus **Mesosepedon**)

Oriental (modified from Knutson & Ghorpadé, unpublished data)

1. Pro-episternum with 1 strong seta above base of coxa (SCIOMYZINI). 2
- Pro-episternum bare or with a few hairs above base of coxa (TETANOCERINI) 6
2. Apterous. 2 fronto-orbital setae, anterior lateroclinate, posterior reclinate. Dorsal pre-apical tibial and postocellar setae extremely long. Apical pair of scutellar setae dorsally directed, cruciate. (Fig. 15.22) . . . **Apteromicra**
- Winged. 1 or 2 fronto-orbital setae, neither lateroclinate or reclinate. Dorsal pre-apical tibial and postocellar setae of normal length. Apical pair of scutellar setae reclinate, parallel to slightly divergent 3
3. $A_1 + CuA_2$ not reaching wing margin (Fig. 14.9e, f). Arista with several strong dorsobasal hairs (as Fig. 15.1b). Fore coxa usually with 5 setae . . **Colobaea**
- $A_1 + CuA_2$ reaching wing margin, sometimes only as a fold. Arista with normal hairs dorsobasally. Fore coxa with no more than 3 setae. 4
4. $A_1 + CuA_2$ reaching wing margin as a fold (Fig. 14.10a). Body mostly shiny black. Frons black, pruinose or shiny. Gena narrow (Fig. 14.2c). . . . . **Pteromicra**
- Strong $A_1 + CuA_2$ reaching wing margin. Body yellowish, brownish or gray, never shiny black. Frons yellowish, mostly tomentose. Gena broad. 5
5. One fronto-orbital seta. Inner posterodorsal margin of hind coxa with hairs. Wing without pattern (Fig. 14.9g). . . . . **Ditaeniella**
- Two fronto-orbital setae (1 in *P. tricolor*, Japan). Inner posterodorsal margin of hind coxa without hairs. Wing with pattern (Fig. 14.9k–r). . . . . **Pherbellia**
6. Ocellar setae absent or represented only by fine hairs. 7
- Ocellar setae well developed. 9
7. Frons partly tomentose, frontal vitta developed. Eye round. Pedicel less than twice as long as postpedicel. 1 or 2 scutellar setae. At least apical 1/3 of wing patterned (Fig. 14.10i) . . . .**Dichetophora**
- Frons entirely or mostly shiny. Eye oval. Pedicel at least twice as long as postpedicel. 1 scutellar seta (0 in *Sepedon lobifera*). Apical 1/3 of wing clear or darkened, never patterned. 8
8. Two fronto-orbital setae, anterior proclinate, posterior reclinate, both sometimes greatly reduced. Lunular area very large, inflated. Mesonotal suture complete. Wing (Fig. 14.12f). . . . .**Sepedoninus**
- Usually only posterior, reclinate fronto-orbital seta present. Lunular area not greatly inflated. Mesonotal suture incomplete mesially. Wing (Fig. 14.12c). . . . . **Sepedon**
9. Frons with shiny median vitta. Arista with black or white hairs. Epandrium usually open below cerci (closed in *Tetanocera chosenica*). Wing with or without pattern. 10
- Frons entirely tomentose, without shiny median vitta. Arista with long, black hairs. Epandrium closed below cerci. Wing (Fig. 14.12h). . . . . **Steyskalina**
10. Subalar setae present. Anepisternum and anepimeron with hairs or setae. Hind coxa with hairs on inner posterodorsal margin 11
- Subalar setae absent. Anepisternum and anepimeron without hairs or setae. Hind coxa without hairs on inner posterodorsal margin 12
11. Wing with extensive pattern (Fig. 14.11o). Anepisternum with hairs and strong setae. Anepimeron with one

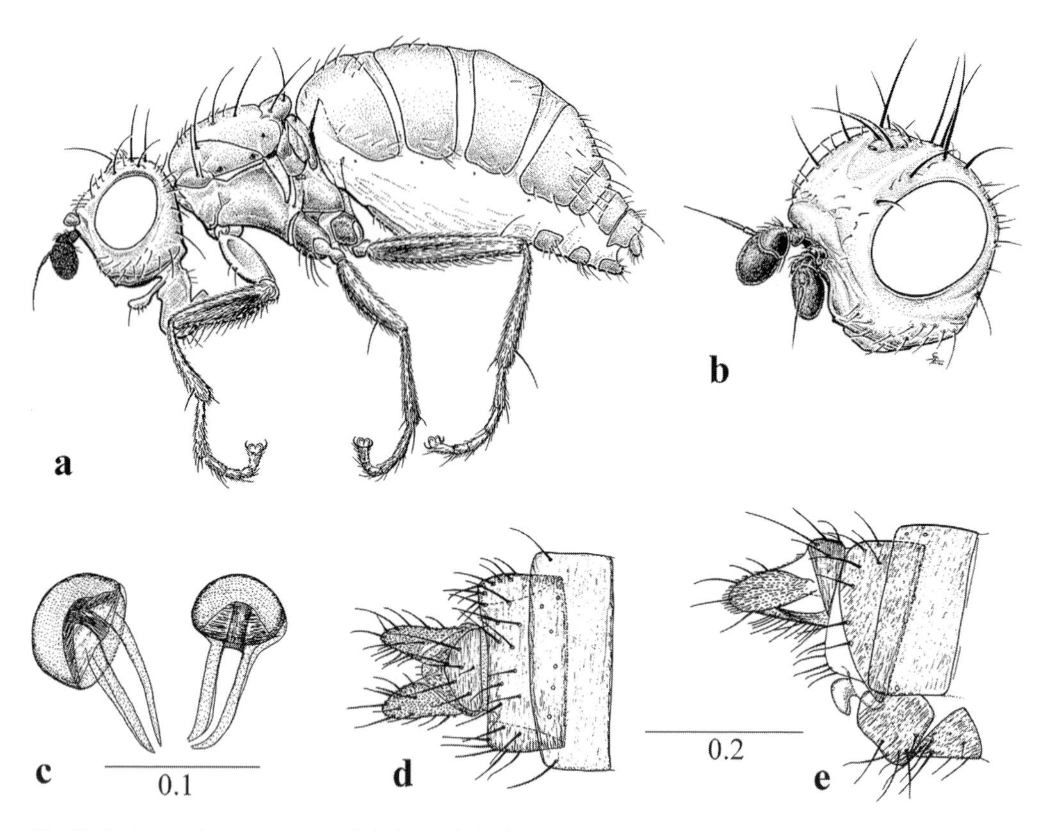

Fig. 15.22 *Apteromicra parva*. a, female holotype; b, head; c, spermathecae; d, postabdomen, dorsal; e, same, lateral. From Papp (2004).

strong seta and no hairs. Arista with long white or black hairs. . . . . **Pherbina**

- Wing without extensive pattern, crossveins infumated, a few dark spots on M_{1+2} between crossveins (Fig. 14.11i, j). Anepisternum and anepimeron with hairs but without strong setae. Arista with white pubescence. . . . . **Ilione**

12. Postpedicel with tuft of long black hairs apically (Fig. 15.1g). Wing extensively patterned. (Fig. 14.10g, h). . . . . **Coremacera**

- Postpedicel without tuft of hairs apically. Wing patterned or not. 13

13. Wing brownish hyaline, without extensive pattern, with or without a few spots on M_{1+2}, crossveins infumated. Arista with black hairs. Lunule not exposed. . 14

- Wing extensively patterned (Figs 14.10r, 14.11a–c). Arista with white or black hairs. Lunule exposed . . . **Euthycera**

14. Wing without spots on M_{1+2} (Fig. 14.12i). Arista with long black hairs (as Fig. 15.1q). Fourth sternum of male without pair of cylindrical processes.. . . . . . . **Tetanocera**

- Wing with spots on M_{1+2} (Fig. 14.11h). Arista only pubescent. Fourth sternum of male with pair of cylindrical processes. . . . . **Hydromya**

Oriental–Australian transition area of Malaysia and surrounding area (modified from Knutson & Ghorpadé [2004])

1. Pro-episternum with 1 strong seta above base of coxa (Sciomyzini). 2

- Pro-episternum bare or with a few short hairs above base of coxa (Tetanocerini). 5

2. A_1 + CuA_2 not reaching wing margin (Fig. 14.9 f). Arista with several strong dorsobasal setulae in addition to aristal hairs (as Fig. 15.1b). Fore coxa usually with 5 setae. . . . . **Colobaea**[1]

[1] Extra-regional genera possibly occurring in the transitional area.

- $A_1 + CuA_2$ reaching wing margin, sometimes only as a fold. Arista with only normal hairs dorsobasally. Fore coxa with no more than 3 setae.................... 3
3. $A_1 + CuA_2$ reaching wing margin as a fold (Fig. 14.10a). Body mostly shiny black. Frons shiny black, at most narrowly yellowish between antennal bases. Gena narrow (Fig. 14.2c) .... **Pteromicra**[1]
- $A_1 + CuA_2$ distinctly reaching wing margin. Body pruinose yellowish, brownish, or gray, never shiny black. Frons yellowish, mostly pruinose. Gena broad 4
4. One pair of fronto-orbital setae. Inner posterodorsal margin of hind coxa with minute hairs. Wing not patterned (Fig. 14.9g)..................... **Ditaeniella**[1]
- Two pairs of fronto-orbital setae. Inner posterodorsal margin of hind coxa bare. Wing patterned or not (Fig. 14.9k–r)................................ **Pherbellia**
5. Frons entirely or mostly shiny. Ptilinal fissure absent. Pedicel at least twice as long as scape. Apical third of wing clear or infumated, never patterned........... 6
- Frons at least partly pruinose. Ptilinal fissure present. Pedicel less than twice as long as scape. Apical third of wing patterned.................................. 7
6. Mesonotal suture complete. Head yellow, with large black spot on upper, inflated face and large, black, anterior frontal spot. Wing (Fig. 14.12f).......... **Sepedoninus**
- Mesonotal suture incomplete medially. Head different. Wing (Fig. 14.12c)......................... **Sepedon**
7. Ocellar setae absent or represented by fine hairs. 1 or 2 fronto-orbital setae. Arista white, pubescent. 1 scutellar seta. Epandrium open below cerci. Wing (Fig. 14.10i)............................. **Dichetophora**
- Ocellar setae well developed. 2 fronto-orbital setae. Arista with long, black hairs. 2 scutellar setae. Epandrium closed below cerci. Wing (Fig. 14.12h)............................... **Steyskalina**

Australian and Oceanic

1. Pro-episternal seta present. Frontal vitta short, extending a little over half distance to anterior margin of frons. 2 fronto-orbital setae. 2 scutellar setae. Hind leg about same length as abdomen. Wing patterned (Fig. 14.9m, o). (SCIOMYZINI)............ **Pherbellia**
- Pro-episternal seta absent. Frontal vitta narrow, shiny, reaching anterior margin of frons or frontal vitta not distinct from entirely shiny frons. 1 fronto-orbital seta. 1 scutellar seta. Hind leg distinctly longer than abdomen. Wing patterned or not. (TETANOCERINI)................................... 2
2. Ocellar setae present. Frontal vitta distinct, frons pruinose laterally. Ptilinal fissure present. Pedicel less than twice as long as scape. Wing patterned (Fig. 14.10i)............................. **Dichetophora**
- Ocellar setae absent. Frontal vitta not distinct from rest of entirely shiny frons. Ptilinal fissure absent. Pedicel at least twice as long as scape. Wing without pattern, slightly infumated apically (Fig. 14.12c)..... **Sepedon**

Subantarctic (modified, in part, from Barnes 1979d)

1. Postocellar setae parallel or convergent. Clypeus large, exposed. CuA_2 forming acute angle (about 45°) with A_1+CuA_2. Suture between abdominal terga 1 and 2 indistinct. Spermathecae 2 + 1 or 2 + 2 (HUTTONININAE)................................ 2
- Postocellar setae divergent. Clypeus small, hidden. CuA_2 forming right angle or only slightly less with A_1+CuA_2. Suture between abdominal terga 1 and 2 distinct. Spermathecae 1 + 1 (SCIOMYZINAE)......... 3
2. Arista not subapical. Pro-episternal seta not well developed. Abdominal spiracles 2 – 5 in membrane. Male tergum 6 fused to protandrium. Surstyli not movable, fused to epandrium or delimited only by a partial suture. Female terga 4 and 5 sclerotized; tergum and sternum 6 separate. Spermathecae 2 + 1. Wing (Fig. 14.9a) (HUTTONININI)............... **Huttonina**
- Arista subapical (Fig. 15.11). Pro-episternal seta well developed. Abdominal spiracles 2–5 in respective terga, at lateral margins. Male tergum 6 not fused to protandrium. Surstyli movable, not fused to epandrium. Female terga 4 and 5 membranous; tergum and sternum 6 fused, forming completely annular somite. Spermathecae 2 + 2. (PROSOCHAETINI)..................... **Prosochaeta**
3. Pro-episternum with 1 strong seta at base of coxa..... ....(SCIOMYZINI) Not known from Subantarctic region
- Pro-episternum bare or with a few hairs above base of coxa (TETANOCERINI)............................. 4
4. Anepimeron with setae. Epandrium fused below cerci. Wing (Fig. 14.11l)...................... **Neolimnia**
- Anepimeron bare. Epandrium not fused below cerci. Wing (Fig. 14.10q)...................... **Eulimnia**

15.2.2. Larvae (third instar)

Notes: (1) Where couplets are weak and/or where the geographical distribution can be especially useful, ranges are included. (2) All freshwater snails refer to non-operculate snails, except *Valvata sincera* which is attacked by *Pherbellia prefixa, Potamopyrgus antipodarum* attacked by *Neolimnia tranquilla*, and several families of operculate snails which probably are attacked by some species of *Dictya*. (3) *Anticheta* and *Renocera* are closely related, aberrant genera of the Tetanocerini; each shares some features with the Sciomyzini and Tetanocerini and each has a unique combination of features (see Section 15.3). (4) asp = number of papillae on anterior spiracular disc; IIx = indentation index (see Fig. 14.36q); I, II, III, etc. = body segments; psd = posterior spiracular disc. (5) Specimens preserved in alcohol often are lighter in color than living specimens. Hosts/prey given in parentheses at end of each couplet part.

Nearctic (modified from Knutson [1987])

Only the biologically unknown Tetanocerini genera *Dictyacium* and *Teutoniomyia* are not included. Questioned characters (couplets 9, 13, 19) are based on Palearctic species.

1. IIx 60–80. Mouthhook without accessory teeth. Dorsal bridge between pharyngeal sclerites. Dorsal and ventral cornua of pharyngeal sclerite with windows. Postoral spines light-colored, in small, sparse patch, primarily ventral; rarely pigmented in larger patch that extends laterally (Figs 14.24a, 14.34b). 7–34 asp. Ventral spinule patches present. No well-developed body tubercles. psd lobe low, rounded, except ventrolateral and ventral may be somewhat longer. Interspiracular processes minute, scale-like. Colorless. Exposed freshwater or terrestrial or semi-terrestrial Succineidae snails. (SCIOMYZINI). 2
- IIx 30–49 (61–62 in *Renocera*, 50–56 in *Anticheta*). Mouthhook with accessory teeth (absent in *Renocera*). Dorsal bridge between pharyngeal sclerites absent (present in *Anticheta*). Dorsal cornu without window (present in *Anticheta, Renocera*). Postoral spines pigmented, in large, dense, well-defined band that extends laterally and nearly or completely meets mid dorsally (Figs 14.24b, 14.34a). 3–23 asp. Ventral spinule patches absent (except weak in *Anticheta*, lightly pigmented in *Renocera*). Body tubercles usually present. At least ventrolateral and ventral psd lobes usually more or less elongate, except low, rounded to conical in most terrestrial species. Interspiracular processes more or less well developed, except minute in terrestrial species. Colorless (terrestrial species) or integument pigmented (aquatic and semi-aquatic species). Freshwater, brackish water, terrestrial snails, Succineidae, snail eggs, slugs, or Sphaeriidae. (TETANOCERINI). 8
2. IIx 67. Encircling spinule bands on segments III–XI (Fig. 15.12e). 14–16 asp. psd with supraventral papilla (Fig. 15.12d). Length 4.5–8.0 mm. Exposed freshwater snails . . . . **Ditaeniella**
- IIx 60–80. Encircling spinule bands entirely absent or absent beyond segments II, V, or VI. 12–34 asp. psd with or without supraventral papilla. Length up to 12.0 mm. Freshwater, terrestrial, or Succineidae snails. 3
3. IIx 80. Mid ventral spinule patch on segment II absent. Anterior spiracle transverse, bifid. 14–16 asp (Fig. 15.18f). psd with dorsolateral lobe (Fig. 15.13e, f). Length 1.5–3.0 mm. Exposed freshwater snails . . . . **Colobaea**
- IIx 60–76. Mid ventral spinule patch on segment II present (Fig. 14.34b) or absent. Anterior spiracle circular, not bifid. 7–34 asp. psd with or without dorsolateral lobe. Length up to 12.0 mm. Freshwater, terrestrial, or Succineidae snails 4
4. IIx 63–68. Ventral arch with elongate, paired, posterior processes extending from lateral apices (except *Sciomyza simplex*) (Fig. 14.37i, j). 12–34 asp. psd flattened in lateral view. Posterior spiracular tubes barely projecting. Length up to 14.5 mm. Exposed Lymnaeidae, terrestrial, or Succineidae snails 5
- IIx 60–76. Ventral arch with lateral apices at most moderately expanded (Fig. 14.37d, g, h). 7–18 asp. psd not completely flattened in lateral view. Posterior spiracular tubes projecting. Length up to 10.0 mm. Freshwater, terrestrial, or Succineidae snails 6
5. IIx 63–68. 24–34 asp. Length 5.3–16.7 mm. Succineidae and exposed Lymnaeidae. . . . . **Sciomyza**
- IIx 76. 12 asp. Length 8.5–14.5 mm. Terrestrial snail, *Stenotrema hirsutum*. **Oidematops**
6. Epistomal sclerite remarkably prominent anteromedially (Fig. 14.35f). IIx 60–65. 12–16 asp. 4 or 5 pairs psd lobes, lateral lobe very small if present; supraventral papilla present (Fig. 15.13d). Length 1.8–7.0 mm. Exposed freshwater, terrestrial, and Succineidae snails. . . . . **Pteromicra**
- Epistomal sclerite strap-like, not prominent anteromedially (Fig. 14.37f), if rarely stronger then IIx 66–76. 7–18 asp. 2–5 pairs psd lobes; supraventral papilla present

or absent. Length 2.5–10.0 mm. Freshwater, terrestrial, and Succineidae snails. 7

7. Accessory rod of pharyngeal sclerite well developed, directed dorsad (Fig. 14.33f). IIx 72. 12–18 asp. All 5 pairs psd lobes distinct, dorsal and dorsolateral smallest, supraventral papilla absent (Fig. 15.13c). Length 4.5–8.5 mm. Exposed freshwater, semi-terrestrial, and Succineidae snails; also large, dead, exposed bivalves. . . . . **Atrichomelina**

\- Accessory rod of pharyngeal sclerite weakly to moderately developed, directed anteriorly. IIx 66–76. 7–18 asp. All 5 pairs psd lobes distinct to only ventral and ventrolateral distinct; supraventral papilla present or absent (Fig. 15.13a, b, Plate 2b, c). Length 2.5–10.0 mm. Exposed freshwater, semi-terrestrial, terrestrial, and Succineidae snails.. . . . . **Pherbellia**

8. Integument colorless, transparent to translucent, micropapillose, covered with minute to moderately long spinules that give surface a pubescent appearance. IIx 61–62 and mouthhooks without accessory teeth (*Renocera*) or IIx 50–56 and accessory teeth present (*Anticheta*). H-shaped and pharyngeal sclerites fused. Dorsal cornu with window. 15–23 asp. 9

\- Integument smooth, transparent (terrestrial species) or slightly to darkly pigmented, often with pattern, with coat of minute integumentary processes (semi-aquatic and aquatic species); tubercles and welts pubescent in *Pherbecta*. IIx 30–49. Accessory teeth present. H-shaped and pharyngeal sclerites separate. Dorsal cornu without window. 3–20 asp. 10

9. Lateral tubercle groups indistinct on body segments. Postanal portion of segment XII not much longer than broad. Posterior spiracular plate without thorn-like dorsal extension. Accessory teeth and dorsal bridge present (Fig. 14.38r). Ventral arch fused with mouthook in some species; hook part acute apically (Fig. 14.35d). IIx 50–56. 16–23 asp (Fig. 15.20h). Length 3.0–8.0 mm. Eggs, then snails, of Lymnaeidae, Physidae, or Succineidae. . . . . **Anticheta**

\- Lateral tubercle groups distinct on body segments. Postanal portion of segment XII much longer than broad, if not then posterior spiracular plate with thorn-like dorsal extension. Accessory teeth and dorsal bridge absent. Ventral arch and mouthook separate. Hook part of mouthhook rounded or acute apically. IIx 61–62 (Figs 14.36q, 14.37b). 15–17 asp? Length 6.2–7.8 mm. Sphaeriidae. . . . . **Renocera**

10. Ventral arch triangular to quadrate, with posteromedial projection. IIx 32–43. 4–17 asp. 11

\- Ventral arch bilobate, strongly emarginate posteriorly. IIx 30–49. 3–20 asp. 13

11. Ventral arch triangular in dorsal outline (Fig. 14.37l). Ligulate sclerite absent. 0–3 accessory teeth; only several denticles at base of hook in *D. ptyarion* (Fig. 14.38f). IIx 32–43. 4–17 asp. Segments V–XII dark golden brown to brownish black (living specimens), white to golden brown (preserved specimens). 4 pairs psd lobes (Fig. 15.17d). Length 5.3–14.2 mm. Freshwater non-operculate or brackish water operculate snails. . . . . **Dictya**

\- Ventral arch quadrate in dorsal outline. Ligulate sclerite present or absent. Segments V–XII dark with rough, pebble-grained texture or white and smooth, or with dense, fine pubescence. 12

12. Integument pubescent, especially on tubercles and welts. 4 accessory teeth on nearly horizontal axis. Ligulate sclerite absent. Ventral arch without posterolateral appendages (Fig. 14.37q). IIx 36. 10 asp. 4 pairs psd lobes. Host/prey unknown. . . . . **Pherbecta**

\- Integument dark with rough pebble-grained texture or white and smooth. 1–3 accessory teeth on nearly vertical axis (Fig. 14.38k). Ligulate sclerite well developed. Ventral arch with elongate posterolateral appendages (Fig. 14.38e). IIx 41. 5–9 asp. 2 or 5 pairs psd lobes. Length 5.8–9.8 mm. Succineidae (*H. spinicornis*) or *Littorina littorea* (*H. setosa*). . . . . **Hoplodictya**

13. Basal part of mouthhook short, about 1/2 length of H-shaped sclerite. 2 accessory teeth (Fig. 14.38i, j). IIx 44–49? 8–17 asp? Integument not pigmented. Length 6.0–13.0 mm? Terrestrial snails and/or slugs. . . . . **Euthycera**

\- Basal part of mouthhook normal, about equal in length to H-shaped sclerite. 2–7 accessory teeth. IIx 30–44. 3–20 asp. Integument pigmented or not. Aquatic or terrestrial snails or Succineidae or slugs. 14

14. Ventral arch with short posterolateral appendages (Fig. 14.37m, n). 5 pairs psd lobes (Fig. 15.17c). IIx 30–36. 3–12 asp. 3 or 4 accessory teeth. 15

\- Ventral arch without posterolateral appendages (Fig. 14.33k). 4 pairs psd lobes (Fig. 15.16d). IIx 32–44. 6–20 asp. 2–7 accessory teeth 16

15. IIx 30–36. 3–12 asp. Length 4.5–13.2 mm. Mainly above 30° N. Lat . . . . **Sepedon**

\- IIx 30. 4–7 asp. Length 7.8–13.5 mm. Southernmost Texas into Neotropical Region. . . . . **Sepedomerus**

16. Five to 7 accessory teeth (Fig. 14.37k). Posterior rami of H-shaped sclerite connected transversely (Fig. 14.33l). IIx about 33. Postanal portion of segment XII 2–4 times longer than wide (Fig. 15.16e, f). 6–13 asp (Fig. 15.20a–c). Green fat bodies visible through integument. Length 7.6–13.5 mm. Freshwater snails. . . . . **Elgiva**
- Two to 7 accessory teeth. Posterior rami of H-shaped sclerite usually not connected. IIx 32–44. Postanal portion of segment XII not much longer than wide. 8–20 asp. Fat bodies not green. Freshwater or terrestrial snails or Succineidae or slugs 17

17. Four accessory teeth. Posterior rami of H-shaped sclerite not connected. IIx 38–40. 10–14 asp. Lateral, ventrolateral, and ventral lobes of psd elongate, subequal (Fig. 15.17e). Lateral tubercle above anal proleg elongate, acute. Interspiracular processes well developed. Length 10.0–14.5 mm. Freshwater snails. . . . . **Hedria**
- Two to 7 accessory teeth. Posterior rami of H-shaped sclerite connected or not. IIx 32–44. 8–20 asp. Lateral lobes of psd always much shorter than ventrolateral and ventral lobes. Lateral tubercle above anal proleg short, rounded. Interspiracular processes well developed or minute. Freshwater or terrestrial snails or Succineidae or slugs. 18

18. Eight to 20 asp (Fig. 15.19a–c). Integument not pigmented. Posterior end posteriorly directed. Postanal portion of segment XII short. 2–4 accessory teeth. IIx 32–42. Posterior rami of H-shaped sclerite connected or not. Interspiracular processes minute. Length 7.3–13.2 mm. (Succineidae, terrestrial snails, or slugs). *Or* 8–16 asp. Integument pigmented. Posterior end directed dorsally. Postanal portion of segment XII relatively long. 3 or 4 accessory teeth. IIx 36–43. Posterior rami of H-shaped sclerite connected or not. Interspiracular processes well developed (Fig. 14.31b, Plate 2g). Length 7.0–23.0 mm. Freshwater snails . . . . **Tetanocera**
- Eight or 9 asp. Integument not pigmented. 4–7 accessory teeth. Posterior rami of H-shaped sclerite not connected. IIx 41–44. Interspiracular processes small. Probably Succineidae and terrestrial snails. 19

19. Integument smooth. 8 or 9 asp? 4–7 accessory teeth? Anterior portion of epistomal sclerite curved, width 3 times length. IIx 41–43? (Fig. 14.34c). Length 6.0–12.0 mm. Hosts/prey unknown, except *L. boscii* attacks Succineidae . . . . **Limnia**
- Integument pruinose. 8 asp. 4 or 5 accessory teeth. Anterior portion of epistomal sclerite rectangular, only slightly wider than long. IIx 44. Hosts/prey unknown. . . . . **Poecilographa**

Palearctic (modified from Rozkošný [1998])

Unknown Sciomyzini genus *Apteromicra* and Tetanocerini genera *Chasmacryptum, Ectinocera, Neodictya, Oligolimnia*, and *Trypetolimnia* are not included.

1. Oral grooves extending into mouth opening, ventral cibarial ridges present (Fig. 14.41a–c). Mouthhook with ventral denticle at base of apical hook (Fig. 14.33b). Ventral arch with a few teeth in anterolateral corners (Fig. 14.33c). IIx 61–67. 8–10 asp (Fig. 15.18a–c). Length 4.6–11.8 mm. (Plate 2a. Figs 14.29, 14.32c, 15.12a–c.) Terrestrial snails. (Salticellinae). . . . . **Salticella**
- Oral grooves and ventral cibarial ridges absent. Mouthhook without ventral denticle at base of apical hook, with or without accessory teeth. Ventral arch with teeth along entire anterior margin (except *Renocera*). IIx 30–90. 5–37 asp. Freshwater or terrestrial snails, snail eggs, Succineidae, slugs, or Sphaeriidae. (Sciomyzinae). 2

2. IIx 60–90. Mouthhook without accessory teeth. Dorsal bridge between pharyngeal sclerites. Dorsal and ventral cornua of pharyngeal sclerite with windows. Postoral spines light-colored, in small, sparse patch, primarily ventral; rarely pigmented and in larger patch that extends laterally (Figs 14.24a, 14.34b). 7–34 asp. Ventral spinule patches present on at least some segments. No well-developed body tubercles. Lobes around psd usually low, rounded except ventrolateral and ventral may be moderately long, conical. Interspiracular processes minute, scale-like. Integument colorless. Exposed freshwater or terrestrial snails or Succineidae. (Sciomyzini). 3
- IIx 30–53, except 61–62 in *Renocera* (Fig. 14.36q), 50–56 in *Anticheta* (Fig. 14.38r). Mouthhook with accessory teeth (absent in *Renocera*). Dorsal bridge between pharyngeal sclerites absent (present in *Anticheta*). Dorsal cornu without window (present in *Anticheta* and *Renocera*). Postoral spines pigmented, in large, dense, well-defined band that extends laterally and nearly or completely meets mid dorsally (Figs 14.24b, 14.34a). 3–37 asp. Ventral spinule patches absent (weak in *Anticheta*, lightly pigmented in *Renocera*). Lobes around psd

usually more or less elongate, except low, rounded to conical in terrestrial and semi-terrestrial species. Interspiracular processes more or less well-developed, except minute in most terrestrial species (moderately long in *Trypetoptera*). Colorless (terrestrial species) or integument pigmented (aquatic and semi-terrestrial species). Freshwater or terrestrial snails, Succineidae, snail eggs, slugs, and Sphaeriidae. (Tetanocerini) 8

3. Ventral arch with elongate paired, posterior processes extending from apices (Fig. 14.37i, j). 24–34 asp. IIx 63. Length 7.5–8.0 mm. Succineidae or exposed freshwater snails. . . . . **Sciomyza**
- Ventral arch with apices only moderately expanded (Fig. 14.37d, g, h). 7–20 asp. IIx 60–90. Length 1.8–10.0 mm. Succineidae, freshwater or terrestrial snails. 4
4. Ten to 14 asp. Mouthhook with short, broad hook part. Ventral arch fused with mouthhook (Fig. 14.35c). IIx 72 (Fig. 14.40j). Terrestrial snails, especially *Cochlicopa lubrica*, *Discus ruderatus*, and *Clausilia* sp. . . . . **Tetanura**
- Seven to 20 asp. Hook part of mouthhook normal, moderately to strongly decurved, except awl-shaped in *Pherbellia obtusa* (Fig. 14.36g). Ventral arch not fused with mouthhook. IIx 60–90. Exposed freshwater or terrestrial snails or Succineidae. 5
5. Mid ventral spinule patch on segment II absent. Cornua of pharyngeal sclerite slender. IIx 75–90. 12–20 asp. Length 2.8–6.5 mm. Exposed small freshwater snails, except *C. bifasciella* attack estivating Lymnaeidae. . . . . **Colobaea**
- Mid ventral spinule patch on segment II present (Fig. 14.34b) or absent. Cornua of pharyngeal sclerite broader. IIx 60–77. 7–20 asp. Length 1.8–10.0 mm. . . 6
6. Epistomal sclerite remarkably prominent anteromedially (Fig. 14.35 f). IIx 60–66. 12–16 asp. Supraventral papilla present. Length 1.8–7.0 mm. Exposed, small freshwater snails. . . . . **Pteromicra**
- Epistomal sclerite usually strap-like, not prominent anteromedially (Fig. 14.37f); if rarely stronger, then IIx 64–77. 7–20 asp. Supraventral papilla present or absent. Length 2.5–10.0 mm. Exposed freshwater or terrestrial snails or Succineidae. 7
7. Segments III–XI with encircling spinule bands (Fig. 15.12e). IIx 72–75. 14–16 asp. Supraventral papilla absent (Fig. 15.12d). Length 4.5–8.0 mm. Exposed freshwater snails. . . . . **Ditaeniella**
- Encircling spinule bands entirely absent or absent beyond segments II, V, or VI (Plate 2b, c). IIx 64–77. 7–20 asp. Supraventral papilla present or absent (Fig. 15.13a, b). Length 2.5–10.0 mm. Exposed freshwater or terrestrial snails or Succineidae. . . . . **Pherbellia**
8. Integument colorless, transparent to translucent, papillose, covered with large spinules that give pubescent appearance or extremely minute papillae. H-shaped and pharyngeal sclerites fused. IIx 50–62. Weak ventral spinule patches present. 15–28 asp. Accessory teeth absent in *Renocera*. Snail eggs, then snails, or Sphaeriidae. 9
- Integument smooth, transparent (terrestrial species) or slightly to darkly pigmented, often with pattern, with coat of minute integumentary processes (semi-aquatic and aquatic species). H-shaped and pharyngeal sclerites separate (fused in *Ilione lineata*). IIx 30–53. 3–37 asp. Ventral spinule patches absent. Accessory teeth present. Freshwater or terrestrial snails, Succineidae, slugs, or Sphaeriidae. 10
9. Lateral tubercle groups indistinct on body segments. Posterior spiracular plate without thorn-like dorsal process. Postanal portion of segment XII not elongate. Accessory teeth and dorsal bridge present (Fig. 14.38r). Hook part of mouthhook acute apically. IIx 50–56. 15–28 asp (Fig. 15.20h). Integument thick, with dense coat of large transparent spinules. Length 3.1–6.9 mm. Eggs, then snails of Succineidae, Lymnaeidae, or Physidae. . . . . **Anticheta**
- Lateral tubercle groups distinct on body segments. Posterior spiracular plate with thorn-like dorsal process or process absent and postanal portion elongate. Accessory teeth and dorsal bridge absent. Hook part of mouthhook rounded or acute apically; IIx 61–62 (Figs 14.36q, 14.37b). 15–17 asp. Integument with moderate coat of large transparent spinules or dense coat of extremely minute papillae. Length 6.2–7.8 mm. Sphaeriidae. . . . . **Renocera**
10. Ventrolateral and ventral psd lobes long, slender (Fig. 15.16a, b). Posterior spiracles relatively small. IIx 35–46. 13–37 asp. Lateral body tubercles developed as elongate, acute appendages in *I. lineata* (Plate 2d). Length 10.4–17.0 mm. Freshwater snails or Sphaeriidae. . . . . **Ilione**
- Ventrolateral and ventral psd lobes shorter. Posterior spiracles relatively large. IIx 30–53. 3–20 asp. Lateral body tubercles not elongate. Length 4.2–20.4 mm. Freshwater or terrestrial snails and/or slugs 11

11. Ventral arch with posteromedian projection (Fig. 14.37l). Ligulate sclerite absent. IIx 35–48? 10 asp. Ventrolateral psd lobe with narrowed apical appendage (Fig. 15.17d). Length 9.0–12.0 mm. Freshwater snails........................... **Dictya**
- Ventral arch without posteromedian projection. Ligulate sclerite present. IIx 30–53. 3–20 asp. Ventrolateral psd lobe with or without apical appendage. Length 4.2–20.4 mm. Freshwater or terrestrial snails, Succineidae, or slugs.. 12
12. Interspiracular processes very short, scale-like. IIx 41–53. 7–17 asp. Colorless. Terrestrial snails and/or slugs.. 13
- Interspiracular processes usually long, conspicuous; if short, then always distinctly branched. IIx 30–48. 3–20 asp. Colorless or integument pigmented. Freshwater or terrestrial snails, Succineidae, or slugs............ 15
13. Mouthhook moderately decurved, basal part of mouthhook longer than hook part (Fig. 14.38h). 7 asp arranged almost transversely. 2 accessory teeth. IIx 41. Length 5.5–8.0 mm. *Lauria cylindracea* and other terrestrial snails (*D. obliterata*)......... **Dichetophora**
- Mouthhook strongly decurved, basal part of mouthhook shorter than hook part (Fig. 14.38g, i, j). 8–17 asp arranged in semi-circle. 1 or 2 accessory teeth, IIx 30–53. Length 6.0–13.0 mm........................ 14
14. One or 2 accessory teeth (Fig. 14.38g). IIx 53. 9 asp (Fig. 15.19d). Length up to 6.7 mm. *Cochlicopa, Discus,* and other terrestrial snails (*C. marginata*)... **Coremacera**
- Two accessory teeth (Fig. 14.38i, j). IIx 44–49. 8–17 asp (Fig. 15.19e, f). Length 6.0–13 mm. *Deroceras* slugs (*E. chaerophylli*) and/or terrestrial snails (*E. cribrata* and *E. stichospila*)....................... **Euthycera**
15. Ventrolateral psd lobe abruptly narrowed in apical half (Fig. 15.16d). IIx 32–37 (Fig. 14.33j). Anterior spiracle elongate, head subtriangular. 6–13 asp (Fig. 15.20a–c). Postanal portion of segment XII 2–5 times longer than wide (Fig. 15.16e, f). Fat bodies of *E. solicita* (but not *E. cucularia*) bright green. Length 7.6–13.2 mm. Freshwater snails........................... **Elgiva**
- Ventrolateral psd lobe simple or with gradually narrowed apical appendage. IIx 30–48. Anterior spiracle rounded or oval. 3–20 asp. Postanal portion of segment XII at most only slightly longer than wide. Fat bodies not green. Length 4.2–20.4 mm. Freshwater or terrestrial snails, Succineidae, or slugs.......... 16
16. Ventrolateral psd lobe simple. Posterior end posteriorly directed. Postanal portion of segment XII not elongate. Ventral arch without lateral appendages (as Fig. 14.7k). IIx 39–48. 5–20 asp.............................. 17
- Ventrolateral psd lobe with narrowed apical appendage (except elongate in *Tetanocera latifibula* and short, rounded in *Sepedon hispanica*). Poserior end uplifted. Postanal portion of segment XII relatively elongate. Ventral arch with (Fig. 14.37n) or without lateral appendages. IIx 30–43. 3–20 asp.................. 22
17. Two to 4 accessory teeth. IIx 39–48. 8–20 asp. Usually colorless. Length 8.9–13.6 mm. Succineidae, terrestrial snails, or slugs................. **Tetanocera** (in part)
- Four to 6 accessory teeth. IIx 30–48. 5–20 asp. Integument usually pigmented. Succineidae, terrestrial, or freshwater snails................................. 18
18. Apex of mouthhook angulate. IIx 44–48. 9–14 asp. (Fig. 15.15d–f). Length 5.1–12.0 mm. Freshwater snails.................................. **Psacadina**
- Apex of mouthhook simple. IIx 40–48. 5–15 asp. Length 4.2–14.0 mm. Freshwater or terrestrial snails, slugs, or Succineidae............................. 19
19. Four accessory teeth. IIx 42–48. 10–12 asp (Fig. 15.20f, g). Ventral psd lobe almost cylindrical. Length 5.6–14.0 mm. Freshwater snails........... **Pherbina**
- Five or 6 accessory teeth. IIx 40–44. 5–15 asp. Ventral psd lobe conical or shorter and rounded. Length 4.2–14.0 mm. Succineidae, slugs, terrestrial or freshwater snails... 20
20. Five asp. 4 or 5 accessory teeth. IIx 40 (Fig. 14.38l). Ventral and ventrolateral psd lobes short, rounded. Length 4.2–8.0 mm. Terrestrial snails, especially *Lauria cylindracea, Vertigo genesii*, small *Candidula unifasciata, Trochulus hispidus*.......... **Trypetoptera**
- Eight to 15 asp. 5 or 6 accessory teeth. IIx 41–44. Ventral and ventrolateral psd lobes short, rounded or moderately long, conical. Freshwater or terrestrial snails, slugs, or Succineidae............................. 21
21. Thirteen to 15 asp. Ventral and ventrolateral psd lobes moderately long, conical. Interspiracular processes extending beyond margin of spiracular plate. Posterior rami of H-shaped sclerite much shorter than anterior rami. IIx 44. Length 10.0–10.5 mm. Freshwater snails.................................. **Hydromya**
- Eight asp. Ventral and ventrolateral psd lobes short, rounded. Interspiracular processes barely extending beyond margin of spiracular plate. Anterior and posterior rami of H-shaped sclerite equal in length. IIx 41–43 (Fig. 14.34c). Length 6.0–12.0 mm. Terrestrial snails, Succineidae, and slugs (*L. unguicornis*)....... **Limnia**

22. Integument light brown. Posterior end uplifted. Postanal portion of segment XII long. (Integument white. Posterior end not uplifted. Postanal portion of segment XII short: *S. hispanica*). 5 pairs of psd lobes (Fig. 15.17a–c). 3–8 asp. Ventral arch with lateral appendages (Fig. 14.37n). 3 or 4 accessory teeth. IIx 30–37. Length 5.5–12.3 mm. Freshwater snails or (*S. hispanica*) Succineidae.......... **Sepedon**
- Integument blackish. Posterior end uplifted. Postanal portion of segment XII relatively long. 4 pairs of psd lobes. 8–20 asp. Ventral arch without lateral appendages. 2–5 accessory teeth. IIx 36–43. Length 8.9–20.4 mm. (Plate 2f, g). Freshwater snails... **Tetanocera** (in part)

Neotropical (including data on South American species from Abercrombie [1970] and on *Tetanocera* species from Foote [1961a])

The biologically unknown Sciomyzini genera *Calliscia, Neuzina, Parectinocera*, and *Pseudomelina* and the biologically unknown Tetanocerini genera *Euthycerina, Eutrichomelina, Guatemalia*, and *Teutoniomyia* are not included. Genera marginally invasive into the Neotropical region (*Atrichomelina, Hoplodictya, Sepedon*, and *Tetanocera*) are included. All of the biologically known Neotropical genera of Sciomyzidae feed on freshwater, non-operculate snails except five genera, as noted in the key.

1. IIx 67–76. Mouthhook without accessory teeth. Dorsal bridge between pharyngeal sclerites. Dorsal and ventral cornua of pharyngeal sclerite with windows. 7–18 asp. Postoral spines light-colored, in small, sparse patch, primarily ventral; rarely pigmented in larger patch that extends laterally (Figs 14.24a, 14.34b). Ventral spinule patches present on at least some segments. No well-developed body tubercles. Lobes around psd low, rounded in most species but ventrolateral and ventral relatively long and conical in some species. Interspiracular processes minute, scale-like. (Sciomyzini).................................... 2
- IIx 28–48. Mouthhook with accessory teeth. Dorsal bridge between pharyngeal sclerites absent. Dorsal cornu without window (except diffuse window in *Tetanoceroides*). 2–17 asp. Postoral spines pigmented, in large, dense, well-defined band that extends laterally and nearly or completely meets mid dorsally (Figs 14.24b, 14.34a). Ventral spinule patches absent. Lobes around psd usually more or less elongate, except in semi-terrestrial species. Interspiracular processes usually moderately to well developed. (Tetanocerini).................................. 4
2. Segments III–XI with encircling spinule bands (Fig. 15.12e). IIx 67. 9–16 asp. Length 4.5–8.00 mm. Exposed freshwater snails. S. Canada to Patagonia............................. **Ditaeniella**
- Encircling spinule bands entirely absent or absent beyond segments II, V, or VI (Plate 2b, c). IIx 67–76. 7–18 asp. Length 2.5–8.5 mm. Exposed freshwater snails and possibly semi-terrestrial snails or Succineidae.................... 3
3. Accessory rod of pharyngeal sclerite well developed, directed dorsad. IIx 72 (Fig. 14.33f). 12–18 asp. All 5 pairs psd lobes distinct, dorsal and dorsolateral smallest (Fig. 15.13c). Supraventral papilla absent. Length 4.5–8.5 mm. Exposed freshwater, semi-terrestrial, and Succineidae snails; also large, dead, exposed bivalves. S. Canada to s. Mexico................ **Atrichomelina**
- Accessory rod of pharyngeal sclerite weakly to moderately developed, directed anteriorly. IIx 67–76. 7–9 asp. All 5 pairs psd lobes distinct to only ventral and ventrolateral distinct (Fig. 15.13a, b). Supraventral papilla present or absent. Length 2.5–10.0 mm. Exposed freshwater, semi-terrestrial, terrestrial, and Succineidae snails. Throughout Neotropical region................................. **Pherbellia**
4. One or 2 accessory teeth. Integument bare, not pigmented, larva appearing white. Ventrolateral and ventral psd lobes short, rounded. Interspiracular processes minute, scale-like, or virtually absent...... 5
- Two to 7 accessory teeth. Integument yellowish white to black, with fine spinules or hairs. Ventrolateral and ventral psd lobes more or less elongate, tapered apically. Interspiracular processes half as long or as long as spiracular tubes, always irregularly branched *or* 3–5 accessory teeth. Integument with dense coat of thick, short, blackish spinules. Ventrolateral and ventral psd lobes short, rounded apically. Interspiracular processes short, branched 6
5. IIx 41. Ventral arch with posteromedian projection and elongate, narrow, posterolateral appendages (Fig. 14.38e) 7–9 asp. Length 5.8–9.8 mm. Succineidae and terrestrial snails. S. Canada to Oaxaca, Mexico... .... *Hoplodictya spinicornis*
- IIx 44–49. Ventral arch incised posteromedially, without posterolateral appendages. 8–17 asp. Length

about 6.0–13.0 mm (based on 3 Palearctic species of *Euthycera*; probably feeding on terrestrial snails). Durango, Mexico. . . . . *Euthycera mira*

6. H-shaped and pharyngeal sclerites fused (Fig. 14.40a, b). Integument with dense coat of short, thick, black spinules (Fig. 15.15a). Ventrolateral and ventral psd lobes short, rounded (Fig. 15.15 b). 7
- H-shaped and pharyngeal sclerites separate. Integument with hairs or spinules, but not as above. Ventrolateral and ventral psd lobes more or less elongate and acute apically.. 8

7. Integumentary spinules dense, overall appearance dark. 13–16 asp. Interspiracular processes extending well beyond margin of spiracular tube. Length 5.0–9.0 mm. Freshwater snails. S. South America to central Chile. . . . . **Shannonia**
- Integumentary spinules more sparse, overall appearance lighter in color. 10 or 11 asp. Interspiracular processes not extending beyond margin of spiracular tube. Length 4.0–9.0 mm. Freshwater snails. S. South America to s. Brazil, Colombia. . . . . **Perilimnia**

8. Four pairs psd appendages (Fig. 15.17d). IIx 30–45. 2–7 accessory teeth. Ventral arch with or without posterolateral appendages, posterior margin incised or with posteromedian process. 5–17 asp. 9
- Five pairs psd appendages (Fig. 15.17b). IIx 28–41. 3–6 accessory teeth. Ventral arch incised on posterior margin, with or without posterolateral lobes. 3–12 asp. 15

9. Ventrolateral and ventral psd appendages fused basally. IIx 33–45. 3–5 accessory teeth. Ventral arch with posterior margin incised, without posterolateral appendages. Both cornua of pharyngeal sclerite with windows. 9–17 asp. Length 7.8–13.0 mm. Freshwater snails. S. South America to central Chile and Uruguay. . . . . **Tetanoceroides**
- Ventrolateral and ventral psd appendages not fused basally. Only ventral cornu of pharyngeal sclerite with window. Without above combination of other characters. 10

10. Ventral arch with posteromedian process, with short, laterally directed anterolateral appendages (Fig. 14.37l). Ligulate sclerite absent. IIx 33–41. 5–8 asp. Color golden brown to black. Length 7.5–13.7 mm. Freshwater and small, operculate, salt marsh snails. North America south to northern Colombia . . . . **Dictya**
- Ventral arch incised posteromesally, with or without posteriorly directed posterolateral appendages. Without above combination of other characters 11

11. Ventrolateral psd lobe triangular or elongate, without apical appendage. 12
- Ventrolateral psd lobe triangular or elongate, with apical appendage. 13

12. Ventrolateral psd lobe moderately elongate, triangular. IIx 40–43. 4 accessory teeth. Parastomal bar fused to pharyngeal sclerite. 11 or 12 asp. Length 12.5–13.5 mm. Exposed freshwater snails and Succineidae. North America south to Caracas, Venezuela. . . . . *Tetanocera plumosa*
- Ventrolateral psd lobe exceptionally elongate, more or less tubular, tapering to apex. IIx 31–41. 3–6 accessory teeth. Parastomal bar not fused to pharyngeal sclerite. 2–12 asp. Length 5.0–11.7 mm. Freshwater snails. S. South America to Bolivia and east-central coast of Brazil (except *P. nubilipennis*, Mexico to Venezuela).. . . . . **Protodictya**

13. Whitish gray, dorsally with pair of median dark lines and several oblique dorsolateral marks; not densely covered with hairs. Dorsal and lateral psd lobes well developed, triangular; ventral lobe elongate, triangular. IIx 40–43. 4 accessory teeth. Posterior rami of H-shaped sclerite not connected by darkly pigmented crossbar. 11 or 12 asp. Length 12.5–13.5 mm. Freshwater snails. Central Mexico. . . . . . . *Tetanocera spreta*
- Dark brown, with or without dorsal pattern; densely covered with hairs. Dorsal psd lobe minute, lateral well developed to barely visible, ventral moderately long or short and conical. IIx 30–34. 4–7 accessory teeth. Posterior rami of H-shaped sclerite connected by darkly pigmented crossbar. 6–8 asp. 14

14. Lateral psd lobe well developed, triangular; ventral lobe elongate, triangular. Anal proleg prominent. With dorsal pattern. IIx 34. 4 or 5 accessory teeth. 6–8 asp. Salivary duct opening near mid length of H-shaped sclerite. Length 6.4–13.0 mm. Freshwater snails. Buenos Aires to central Bolivia . . . . **Dictyodes**
- Lateral psd lobe barely visible; ventral lobe short, conical. Anal proleg inconspicuous. Without dorsal pattern. IIx 30–34. 5–7 accessory teeth. 8 asp. Salivary duct opening near posterior margin of H-shaped sclerite. Length 6.8–13.3 mm. Freshwater snails. S. Brazil to Guatemala. . . . . **Thecomyia**

15. Ventral arch without posteriorly directed posterolateral appendages. IIx 37. 3 or 4 well separated accessory

teeth. 5–7 asp. Upper and lower posterior spiracular slits arcuate. Length 7.1–11.5 mm. Freshwater snails. Mexico to Venezuela.................... *Protodictya nubilipennis*

- Ventral arch with posteriorly directed posterolateral appendages. IIx 30–38. 3–6 more or less overlapping accessory teeth. 3–12 asp. Upper and lower posterior spiracular slits only slightly curved, at most.. 16

16. Three to 6 accessory teeth, upper tooth larger and darker than others. Posterior spiracular tube usually shiny with scalloped basal margin. Length 3.8–10.8 mm. Freshwater snails. Buenos Aires to s. Mexico...... **Sepedonea**

- Three or 4 accessory teeth, subequal in size, all equally lightly pigmented. Posterior spiracular tube usually papillose, without scalloped basal margin.......... 17

17. IIx 30–36. 3 or 4 accessory teeth. Posterolateral appendage of ventral arch elongate, apically acute (Fig. 14.37n). 3–12 asp. Length 4.5–13.2 mm. Freshwater snails. North America s. to Oaxaca, Mexico.................................. **Sepedon**

- IIx 30. 3 accessory teeth. Posterolateral appendage of ventral arch truncate (Fig. 14.37m). 4–8 asp. Length 7.8–13.5 mm. Freshwater snails. Texas s. to Buenos Aires.............................. **Sepedomerus**

Afrotropical

In the Afrotropical region there are data on the morphology of the immature stages of only *Hydromya dorsalis*, *Sepedonella nana*, and a few species of *Sepedon*. The genera *Salticella*, *Colobaea*, *Ditaeniella*, *Pherbellia*, and *Pteromicra* are included in the key below based on species from other regions. The immature stages of the Tetanocerini genera *Ethiolimnia*, *Sepedoninus*, *Tetanoptera*, and *Verbekaria* are unknown.

1. Oral grooves around mouth opening and ventral cibarial ridges present (Fig. 14.41a–c). Mouthhook with a ventral denticle at base of apical hook, normal accessory teeth absent (Fig. 14.33b). Ventral arch with teeth only in anterolateral corners (Fig. 14.33c). IIx 61–67. 8–10 asp (Fig. 15.18a–c) (based on *S. fasciata*). Probably feeding on *Achatina* and other large, terrestrial snails. (Salticellinae). Known only from Lesotho and the Western Cape, South Africa..... **Salticella**[2]

- Oral grooves around mouth opening and ventral cibarial ridges absent. Mouthhook without ventral denticle at base of apical hook, normal accessory teeth present or absent. Ventral arch with teeth along entire anterior margin. IIx 19–80. 3–20 asp (Sciomyzinae) 2

2. IIx 60–80. Mouthhook without accessory teeth. Dorsal bridge between pharyngeal sclerites. Dorsal and ventral cornua of pharyngeal sclerite with windows. Postoral spines light-colored, in small, sparse patch, primarily ventral; rarely pigmented in larger patch that extends laterally (Figs 14.24a, 14.34b). 7–20 asp. Ventral spinule patches present. No well-developed body tubercles. Lobes around psd low, rounded to conical in most species but ventrolateral and ventral conical in some species. Interspiracular processes minute, scale-like (Sciomyzini)................................. 3

- IIx 19–44. Mouthhook with accessory teeth. Dorsal bridge between pharyngeal sclerites absent. Dorsal cornu without window. Postoral spines pigmented, in large, dense, well-defined band that extends laterally and nearly or completely meets mid dorsally (Figs 14.24b, 14.34a). 3–15 asp. Ventral spinule patches absent. Body tubercles more or less well developed. Lobes around psd usually more or less elongate, except in terrestrial species. Interspiracular processes more or less well developed, except minute in terrestrial species (Tetanocerini).................................. 6

3. Midventral spinule patch on segment II absent. Cornua of pharyngeal sclerite slender and divergent. IIx 83. 14–16 asp. Length 2.8–5.0 mm (based on *C. punctata*). Very small freshwater snails. Niger, Nigeria, the Gambia................................ **Colobaea**

- Midventral spinule patch on segment II present (Fig. 14.34b) or absent. Cornua of pharyngeal sclerite broader and almost parallel. IIx 60–77. 7–20 asp. ... 4

4. Epistomal sclerite remarkably prominent anteromedially (Fig. 14.35f). IIx 60–66. 12–16 asp. Length 1.8–7.0 mm (based on the Palearctic species). Exposed, small freshwater snails. Nigeria, Tanzania........... **Pteromicra**

- Epistomal sclerite usually strap-like, not prominent anteromedially. IIx 64–77. 7–20 asp............... 5

5. Encircling spinule bands on segments III–XI (Fig. 15.12e). IIx 72–75. 14–16 asp. Length 4.5–8.0 mm (based on *D. grisescens*). Exposed freshwater snails. South Africa, Namibia..... **Ditaeniella**

- At most only 2 anterior abdominal segments with encircling spinule bands (Plate 2b, c). IIx 76–81. 7–9 asp. Length 3.4–7.2 mm (based on *P.* (*G.*) *trabeculata* and

[2] *S. stuckenbergi*, the only Afrotropical species of this genus, is by far the largest species of Sciomyzidae in the adult stage, so the body length of the third-instar larva probably exceeds 23.0 mm (the length of the largest known sciomyzid larva).

P. (*G.*) *limbata*). Exposed freshwater and terrestrial snails. Nigeria to South Africa. . . . . **Pherbellia** (subgenus **Graphomyzina**)

6. Ventrolateral and ventral psd lobes elongate, conical (Fig. 15.17c). IIx 19–44. 6–15 asp. Aquatic snails or oligochaetes 7
- Ventrolateral and ventral psd lobes short, rounded (Fig. 15.17a). IIx 30–37. 6–8 asp. Length 8.8–11.8 mm. Succineidae (*S. hispanica* and *S. trichrooscelis*) or terrestrial snails (*S. umbrosa*). Throughout Afrotropical region. . . . . **Sepedon** (in part)
7. Five pairs psd lobes, ventrolaterals bipartite. 3 or 4 accessory teeth. Ventral arch with posterolateral appendages (Fig. 14.37p). IIx 19–33. 3–5 asp (Fig. 15.21a–c). Length 4.0–14.2 mm. 8
- Four pairs psd lobes; ventrolaterals not bipartite. 5 or 6 accessory teeth. Ventral arch without posterolateral appendages. IIx 44. 13–15 asp. Length 10.0–10.5 mm. Freshwater snails. Palearctic region south to Ethiopia. . . . . **Hydromya**
8. Mouthhook with broad space between base of hook part and accessory teeth (Fig. 14.39a). IIx 19. Length 4.0–6.0 mm. Freshwater oligochaetes. Nigeria to Lesotho . . . . **Sepedonella**
- Mouthhook without broad space between base of hook part and accessory teeth (Figs 14.36p, 14.39e). IIx 27–33. Length 4.2–14.2 mm. Freshwater snails or oligochaetes (*S. knutsoni*). Throughout Afrotropical region . . . . **Sepedon** (in part)

Oriental

Of the 30 described species in 12 genera occurring in the Oriental region, the immature stages of only four marginally adventive Palearctic species (*Colobaea punctata, Pherbellia cinerella, Hydromya dorsalis*, and *Ilione turcestanica*), and two more intrusively adventive Holarctic and Palearctic species (*Pherbellia nana* and *Ditaeniella grisescens*, respectively) have been described. Thus, the following key is adapted from the key to Palearctic genera. Considering the poor state of knowledge of the immature stages of Oriental sciomyzids, and the large number of undescribed Oriental species known to us, information on distribution and number of species is included in the key to assist the user. The biology and immature stages of *Sepedoninus* (one or two undescribed species from Borneo and Sulawesi), *Apteromicra parva* (Nepal), and *Steyskalina picta* (northeastern Burma) are unknown. Only the freshwater snail predators *Sepedon aenescens, S. ferruginosa, S. plumbella, S. senex*, and *S. spangleri* have been reared in the Oriental region, but their immature stages have not been described. The likely hosts/prey associations for other genera given below are based on rearings in the Palearctic region.

Most species of *Sepedon* and *Tetanocera* are predators of aquatic snails and will arrive at couplet 9. Some Palearctic, Nearctic, and Afrotropical species of these genera are parasitoid/predators of terrestrial snails, slugs, and Succineidae. If species with terrestrial hosts/prey occur in the Oriental region they will key out to couplet 5, but features of the cephalopharyngeal skeleton will distinguish them from *Coremacera, Dichetophora*, and *Euthycera*.

1. IIx 60–83. Mouthhook without accessory teeth. Dorsal bridge between pharyngeal sclerites. Dorsal and ventral cornua of pharyngeal sclerite with windows. Postoral spines light-colored, in small, sparse patch, primarily ventral; rarely pigmented in larger patch that extends laterally (Figs 14.24a, 14.34b). 7–20 asp. Ventral spinule patches present. No well-developed body tubercles. psd lobes low, rounded in most species but conical in some. Interspiracular processes minute, scale-like. Integument colorless. Exposed freshwater or terrestrial snails (SCIOMYZINI). 2
- IIx 27–53. Mouthhook with accessory teeth. Dorsal bridge between pharyngeal sclerites absent. Dorsal cornu without window. Postoral spines pigmented, in large, dense, well-defined band that extends laterally and nearly or completely meets mid dorsally (Figs 14.24b, 14.34a). 3–37 asp. Ventral spinule patches absent. With well-developed body tubercles. psd lobes usually more or less elongate, except low, rounded to conical in terrestrial species. Interspiracular processes more or less well developed, except minute in terrestrial species. Integument colorless (terrestrial species) or pigmented (aquatic species). Freshwater or terrestrial snails, Succineidae. (TETANOCERINI). 5
2. Mid ventral spinule patch on segment II absent. Cornua of pharyngeal sclerite slender and divergent. IIx 83. 14–16 asp. Length 2.8–5.0 mm. Exposed, small freshwater snails; *C. punctata*, se. Pakistan. . . . . **Colobaea**
- Mid ventral spinule patch on segment II present or absent (Fig. 14.34b). Cornua of pharyngeal sclerite broader and almost parallel. IIx 60–77. 7–20 asp. 3
3. Epistomal sclerite remarkably prominent anteromedially (Fig. 14.35f). IIx 60–66. 12–16 asp. Length 1.8–7.0 mm. Exposed, small freshwater snails; *P. leucodactyla*, Taiwan. . . . . **Pteromicra**

- Epistomal sclerite usually strap-like, not prominent anteromedially (Fig. 14.37f), if rarely stronger, then IIx usually more than 70. 7–20 asp. Length 2.5–10.0 mm. 4
4. Segments III–XI with encircling spinule bands (Fig. 15.12e). IIx 72–75. 14–16 asp. Length 4.5–8.0 mm. Exposed freshwater snails; *D. grisescens*, w. Afghanistan to Hong Kong . . . . **Ditaeniella**
- At most only segments V and VI with encircling spinule bands (Plate 2b, c). IIx 64–77. 7–20 asp. Length 2.5–10.0 mm. Exposed freshwater or terrestrial snails, Succineidae. 5 Oriental species. . . . . **Pherbellia**
5. IIx 41–53. Integument not pigmented. psd lobes short, rounded (Fig. 15.17a). Interspiracular processes minute, scale-like. 2 accessory teeth. 7–17 asp. Terrestrial snails, slugs. 6
- IIx 27–48. Integument pigmented. psd with at least ventrolateral and ventral lobes more or less elongate (Fig. 15.17c). Interspiracular processes elongate, branched. 4–8 accessory teeth. 3–37 asp. Freshwater snails. 8
6. Mouthhook moderately decurved. 2 accessory teeth. IIx 41. 7 asp arranged almost transversely. Length 5.5–8.0 mm. Terrestrial snails ?; *D. intermedia*. India, w. Bengal and 3 undescribed spp. from India (Himachal Pradesh, Uttar Pradesh), Nepal, China (Szechwan, Fukien), Myanmar, Taiwan.. . . . . **Dichetophora**
- Mouthhook strongly decurved. 1 or 2 accessory teeth. IIx 44–53. 8–17 asp arranged in semi-circle. Terrestrial snails and/or slugs. 7
7. One or 2 accessory teeth. IIx 53. 9 asp. Length up to 6.7 mm. Terrestrial snails; 1 undescribed sp., China (Chekiang). . . . . **Coremacera**
- Two accessory teeth. IIx 44–49. 8–17 asp. Length 6.0–13.0 mm. Terrestrial snails and/or slugs; 1 undescribed sp., India (Kashmir). . . . .**Euthycera**
8. Ventrolateral psd lobe elongate, with finger-like process (Fig. 15.17c), or short, rounded (Fig. 15.17a). IIx 30–48. 3–20 asp. 9
- Ventrolateral psd lobe elongate, without process (Fig. 15.16a, b). IIx 35–48. 10–37 asp. 10
9. IIx 30–37. 3–5 accessory teeth. Ventral arch with posterolateral appendages (Fig. 14.37p). Anterior spiracle small, 3–8 asp (Fig. 15.21c). 5 pairs psd lobes. Length 5.5–12.3 mm. Aquatic snails; 13 spp . . . . **Sepedon**
- IIx 36–48. 2–5 accessory teeth. Ventral arch without posterolateral appendages. Anterior spiracle larger, 8–20 asp (Fig. 15.21d, e). 4 pairs psd lobes Length 5.5–13.6 mm. Aquatic snails; 2 spp., China. . . . .**Tetanocera**
10. Ventrolateral and ventral psd lobes very elongate, strongly tapered (Fig. 15.16a, b). IIx 35–46. 13–37 asp. Length 10.4–17.0 mm. Freshwater snails in truly aquatic microhabitats; *I. turcestanica*, ne. India. . . . . . **Ilione**
- Ventrolateral and ventral psd lobes moderately elongate, conical, about twice as long as width at base (Fig. 15.15f). IIx 42–48. 10–15 asp. Freshwater and semi-aquatic snails in exposed situations. 11
11. Four accessory teeth. IIx 42–48. 10–12 asp. Length 5.6–14.0 mm. Aquatic snails; *P. coryleti*, nw. Afghanistan and *P. intermedia*, central Szechuan, both marginal to the Oriental region . . . . . . **Pherbina**
- Five or 6 accessory teeth. IIx 44. 13–15 asp. Length 10.0–10.5 mm. Aquatic snails; *H. dorsalis*, ne. Afghanistan and India (Jammu and Kashmir). . . . .**Hydromya**

Australian and Oceanic

None of the immature stages of the Australian and Oceanic Sciomyzidae has been described. The following key is based on a few generic characters of the world *Pherbellia* spp. and *Sepedon* spp. and, for *Dichetophora*, on the terrestrial predator *D. obliterata*. *Sepedon plumbella*, *Dichetophora biroi*, and *D. hendeli* are aquatic predators. Larvae of the Australian and Oceanic species of *Pherbellia* (*Graphomyzina*) probably have features typical of many species of the genus reared from other regions.

1. IIx 64–77. Mouthhook without accessory teeth. Dorsal bridge between pharyngeal sclerites. Dorsal and ventral cornua of pharyngeal sclerite with windows. Postoral spines light-colored, in small, sparse patch, primarily ventral; rarely pigmented in larger patch that extends laterally (Figs 14.24a, 14.34b). 7–20 asp. Ventral spinule patches present. No well-developed body tubercles. psd lobes low, rounded to conical. Interspiracular processes minute, scale-like. Integument colorless. (Sciomyzini). Possibly parasitoid/predators of freshwater snails; 3 species. . . . . **Pherbellia**
- IIx 27–41. Mouthhook with accessory teeth. Dorsal bridge between pharyngeal sclerites absent. Dorsal cornu without window. Postoral spines pigmented, in large, dense, well-defined band that extends laterally and nearly or completely meets mid dorsally (Figs 14.24b, 14.34a). 3–12 asp. Ventral spinule patches absent. With well-developed body tubercles. psd lobes more or

less well developed. Interspiracular processes well developed, branched. Integument pigmented (Tetanocerini). Probably predators of freshwater snails. 2

2. IIx 27–37. Ventral arch with posterolateral appendages. 3–12 asp; 7 species. . . . . **Sepedon**
- IIx 41. Ventral arch without posterolateral appendages. 7 asp; 6 species. . . . . **Dichetophora**

Subantarctic (New Zealand)

None of the immature stages of the genera *Prosochaeta* and *Huttonina* (Huttonininae), which are restricted to New Zealand, is known. Of the other two genera (both endemic Tetanocerini) occurring there the biology and immature stages of one of the two species of *Eulimnia* (Barnes 1980a) and the biology, but not the morphology of the immature stages, of 8 of the 15 species of *Neolimnia* have been published (Barnes 1979b). Since the morphological data are so limited, our key emphasizes behavioral and habitat features.

1. Larvae living beneath the water surface until late third stadium, preying on fingernail clams. IIx 38–46. 4 accessory teeth. Ventral arch strongly emarginate posteromesally. 17–22 asp. psd with 4 pairs of long, filamentous lobes (Plate 2e). Interspiracular processes short, palmately branched. Segments VI to XI with all tubercles becoming less stout and more elongate posteriorly. Length 12.6–16.5 mm. *E. philpotti*, South Island. . . . . **Eulimnia**
- Larvae living at water surface or on moist shores in various permanent, unshaded, freshwater habitats or in terrestrial habitats in diverse forests, river flats, grasslands, and coastal situations. 2
2. Larvae overt predators of several species of small, terrestrial Punctidae snails. Interspiracular processes not visible (probably microscopic, scale-like). No air bubble in gut. Probably univoltine . . . . . **Neolimnia** (subgenus **Neolimnia**)
- Larvae overt predators of non-operculate or operculate freshwater snails. Interspiracular processes well developed. Air bubble in gut. Probably multivoltine. **Neolimnia** (subgenus **Pseudolimnia**). 3
3. Larvae feeding on the operculate snail *Potamopyrgus antipodarum* (Hydrobiidae), probably in exposed situations in a wide variety of freshwater habitats. . . . *N.* (*P.*) *tranquilla*
- Larvae feeding on non-operculate snails of the genera *Gyraulus*, *Lymnaea*, and *Physa*, probably in neustonic situations in a wide variety of freshwater habitats. . . . . *N.* (*P.*) *repo*, *N.* (*P.*) *sigma*, *N.* (*P.*) *ura*

15.2.3 Puparia

Field-collected puparia that seem to be viable should be held individually in vials for emergence of adults. Exposure to low temperatures may help to break diapause. If a puparium does not produce an adult, lift the dorsal cephalic cap to see if it contains a pigmented, pharate adult. Puparia of 44 species in 8 genera of Sciomyzini are formed only or often in snail shells (Tables 15.1–15.3).

The following keys emphasize the shape, length, color, and external features of the puparium which primarily are vestiges of the third-instar larvae. These are the vestiges of posterior spiracular disc lobes, anal proleg, and anterior spiracular papillae which, being shrunken, are not as useful for identification of puparia as they are for larvae. See Section 15.1.2.4 for a general description of puparia.

We focus attention on the cephalopharyngeal skeleton, which can be extracted from the inner surface of the ventral cephalic cap, number of accessory teeth on the mouthhook, and the positions of the anterior and posterior spiracles relative to the mid horizontal axis. The latter character may indicate the relative aquatic to terrestrial nature of the pupariation site. Generally, aquatic species pupariate in the water and have the posterior spiracles uplifted, but some pupariate on the margin of their aquatic habitat. However, the general microhabitat in which puparia are found (except those in shells) is not necessarily a clue as to the aquatic to terrestrial nature of the larvae because puparia often are displaced from the pupariation site by flood waters. For most terrestrial and semi-terrestrial species, the precise locations of pupariation are unknown. As noted, the shape and size of puparia (Fig. 14.15), may differ somewhat between males and females in some species. In a few cases, geographical range is included in the following keys.

Nearctic

Only the biologically unknown genera *Dictyacium* and *Teutoniomyia* are not included.

1. In snail shells. All species of *Atrichomelina*, *Colobaea*, and *Oidematops*, and some species of *Ditaeniella*, *Pherbellia*, *Pteromicra*, and *Sciomyza*. See Tables 15.1–15.3 and use couplets 3–8 below.
- Not in shells. Species and genera usually found in shells are also included in couplets 2–8 so that the generic placement of the species in Tables 15.1–15.3 can be confirmed on morphological features and because some that routinely pupariate in shells may be found outside shells. 2
2. Generally found in semi-terrestrial situations on wet surfaces or strictly terrestrial situations but may be found floating in flood debris. Barrel shaped, without

Table 15.1. Sciomyzini that form puparia within snail shells

MOLLUSCS		SCIOMYZIDAE
Freshwater Snails		
LYMNAEIDAE		
Galba humilis	H	***Pherbellia n. nana***
	N	***Ditaeniella parallela, Pherbellia vitalis***
Galba truncatula	P	***Colobaea bifasciella***
Radix balthica	H	*Sciomyza simplex*
	P	***Colobaea punctata***
Stagnicola elodes	H	***Pherbellia griseola, P. n. nana***
	N	***Atrichomelina pubera, Ditaeniella parallela, Pherbellia seticoxa, P. vitalis, Sciomyza varia***
Stagnicola palustris	H	***Sciomyza simplex***
	P	***Colobaea bifasciella, Pherbellia ventralis***
PHYSIDAE		
Aplexa hypnorum	H	***Pherbellia argyra***
	N	***Atrichomelina pubera***
Physa sp.	H	***Pherbellia n. nana***
	N	***Colobaea americana, Pherbellia vitalis***, *P. beatricis*
Physella gyrina	N	***Atrichomelina pubera***
PLANORBIDAE		
Anisus spirorbis	P	***Colobaea distincta***
Anisus vortex	H	***Pherbellia argyra***
	P	***Colobaea pectoralis***
Bathyomphalus contortus	P	***Colobaea pectoralis***
Gyraulus albus	P	***Colobaea punctata***
Gyraulus parvus	H	***Pherbellia n. nana***
	N	***Colobaea americana***
Helisoma anceps	N	***Atrichomelina pubera***
Helisoma trivolvis	N	***Atrichomelina pubera, Ditaeniella trivittata, Pherbellia seticoxa, P. vitalis***
Planorbarius corneus	P	***Colobaea punctata***
Planorbis planorbis	H	***Pherbellia argyra***
	P	***Colobaea punctata, Pherbellia dorsata***
Planorbula sp.	N	***Pherbellia trabeculata***
Planorbula armigera	N	***Pherbellia vitalis, P. similis***, *Pteromicra similis*
VALVATIDAE		
Valvata sincera	N	*Pherbellia prefixa*
Semi-terrestrial Snails		
SUCCINEIDAE		
Novisuccinea ovalis	N	***Sciomyza aristalis***
Oxyloma sp.	N	*Pherbellia schoenherri maculata*
Oxyloma decampi gouldi	H	***Sciomyza dryomyzina***
Oxyloma elegans	P	***Pherbellia s. schoenherri***
Succinea sp.	H	***Sciomyza dryomyzina***
	N	*Pherbellia schoenherri maculata*

Table 15.1. (*cont.*)

MOLLUSCS	SCIOMYZIDAE	
	P	*Pherbellia s. schoenherri*, ***Sciomyza testacea***
	N	***Pteromicra anopla***
Succinea putris	P	***Pherbellia s. schoenherri***
Terrestrial Snails		
CLAUSILIIDAE		
Clausilia sp.	P	***Tetanura pallidiventris***
COCHLICOPIDAE		
Cochlicopa lubrica	P	***Pherbellia dubia***, ***Tetanura pallidiventris***
CHONDRINIDAE		
Granaria frumentum	P	*Pherbellia limbata*
ENDODONTIDAE		
Anguispira alternata	N	***Pherbellia albovaria***
PATULIDAE		
Discus cronkhitei	N	***Pteromicra steyskali***
Discus patulus	N	***Pherbellia albovaria***
Discus rotundatus	P	*Pherbellia annulipes*, ***P. dubia***, ***Tetanura pallidiventris***
GASTRODONTIDAE		
Zonitoides sp.	N	***Pherbellia inflexa***
Zonitoides arboreus	N	***Pherbellia albovaria***
HYGROMYIIDAE		
Cernuella virgata	P	***Pherbellia knutsoni***
Cochlicella acuta	P	***Pherbellia knutsoni***
Candidula intersecta	P	***Pherbellia knutsoni***
Helicella itala	P	***Pherbellia knutsoni***
Trochulus hispidus	P	***Pherbellia dubia***
OXYCHILIDAE		
Aegopinella nitidula	P	***Pherbellia dubia***
Aegopinella pura	P	***Tetanura pallidiventris***
POLYGYRIDAE		
Stenotrema hirsutum	N	***Oidematops ferrugineus***
Triodopsis tridentata	N	***Pherbellia albovaria***

Notes: Sciomyzini species are listed after the snail host/prey by H, Holarctic; N, Nearctic; and P, Palearctic. For Sciomyzini species in bold, puparia have been found in the shells collected in nature; species in regular italic are known to form puparia in shells only from laboratory rearings. In a few cases, where eggs have been found on shells or larvae in snails in nature, and the resulting puparia were formed in shells in laboratory rearings, the sciomyzid species is assumed to pupariate in shells in nature, and is included above.

adaptations for aquatic existence, posterior spiracles at most reaching dorsal surface (Fig. 14.43g, i). With ventral patches of black spinules on at least a few segments. If in doubt: IIx 60–80. Mouthhook without accessory teeth. Dorsal bridge present between pharyngeal sclerites. Dorsal and ventral cornua of pharyngeal sclerite with windows. 7–34 asp. (Sciomyzini)....................................3

Table 15.2. Species that form puparia outside shells as well as within shells

Sciomyzini	
H	*Pherbellia argyra*
H	*Pherbellia griseicollis**
H	*Pteromicra angustipennis**
H	*Pteromicra pectorosa**
H	*Sciomyza dryomyzina**
H	*Sciomyza simplex**
N	*Atrichomelina pubera**
P	*Ditaeniella grisescens**
N	*Ditaeniella parallela*
N	*Pherbellia anubis**
N	*Pherbellia prefixa**
N	*Pherbellia quadrata*
N	*Pherbellia schoenherri maculata**
P	*Pherbellia knutsoni*
P	*Pherbellia s. schoenherri**
P	*Pherbellia ventralis*
P	*Pteromicra glabricula**
Tetanocerini	
P	*Trypetoptera punctulata**

*, usually formed outside shells.

- Generally found in freshwater, brackish water, semi-terrestrial or strictly terrestrial situations. Most aquatic species not simply barrel shaped, but have the posterior spiracles elevated above dorsal surface (Fig. 14.47a). Semi-terrestrial and terrestrial species barrel shaped and posterior spiracles not elevated above dorsal surface. Without ventral spinule patches (except weak in *Anticheta*). If in doubt: IIx 30–51 (except 61–62 in *Renocera*, 50–56 in *Anticheta*). Mouthhook with accessory teeth (absent in *Renocera*). No dorsal bridge between pharyngeal sclerites (present in *Anticheta*). Dorsal cornu without window (present in *Anticheta* and *Renocera*). 3–23 asp. (Tetanocerini) 9

3. Integument glossy, almost lubricous, dark reddish brown. Body segments without dorsal spinule patches, with intersegmental ventral spinule patches extending to mid lateral surface on segments IV–XI. 12–18 asp. Lateral, ventrolateral, and ventral psd lobes distinct. Length 4.5–7.3 mm.................. **Atrichomelina**
- Integument usually dull, smooth to wrinkled or pebble-grained posteriorly, orange yellow to reddish brown. Spinules on body segments varying from completely encircling bands on most segments to reduced ventral patches only on segments IV or IV and V. 9–34 asp. psd lobes distinct to indistinct.............................. 4
4. Segments XI and XII with shagreened surface due to dense coat of flat spinules. Many body segments with completely encircling spinule bands. Dorsal cephalic cap strongly (*D. parallela*) to weakly (*D. trivittata*) concave. Lateral, ventrolateral, and ventral psd lobes distinct. 9–16 asp. Length 3.7–5.6. mm..... **Ditaeniella**
- Segments XI and XII bare or with sparse coat of spinules. At most, segments III–VI with completely encircling spinule bands. Dorsal cephalic cap flat to convex. Only ventrolateral and ventral psd lobes distinct, at most. 7–34 asp.......................... 5
5. Anterior spiracle strongly projecting, bifurcate. 14–16 asp. Length 3.0–4.0 mm.................. **Colobaea**
- Anterior spiracle appressed to moderately projecting, not bifurcate. 7–34 asp. Length more than 4.0 mm... 6
6. Ventral arch with pair of elongate, posterolateral projections. 12–34 asp............................. 7
- Ventral arch without pair of elongate, posterolateral projections. 7–16 asp.............................. 8
7. 12 asp. Known only from shells of terrestrial *Stenotrema hirsutum*. Maine to Georgia, w. to Kansas; 1 sp., in deciduous woods.................. **Oidematops**
- 24–34 asp. Not in shells or in shells of Succineidae or Lymnaeidae. Northeastern (*S. aristalis*) or transcontinental (3 spp.); in floodplain woods and open marshes................................ **Sciomyza**
8. 7–18 asp. IIx 61–76. Epistomal sclerite usually not prolonged anteriorly. Not in shells or in shells of many genera (Table 15.1). Septum or septum material produced by 9 species (Table 15.3). Length 3.0–8.4 mm..... **Pherbellia**
- 12–16 asp. IIx 60–65. Epistomal sclerite prolonged anteriorly. Not in shells or in shells of *Planorbula*, *Succinea*, or *Discus*. No septum or septum material produced. Length 2.6–4.3 mm.......... **Pteromicra**
9. Torpedo shaped, unusually elongate form (Fig. 14.51c). Probably formed in hollow plant-stems. Integument with pebble-grained appearance. 2 or 3 accessory teeth. 5–7 asp. Length 10.3–12.1 mm. *Littorina littorea* on Atlantic coasts.............................. *Hoplodictya setosa*
- Barrel or boat shaped. Without above combination of features.. 10
10. Postanal portion of segment XII 2–2.5 times longer than width at anus, strongly upturned (Fig. 14.49a, b). Integument translucent, light tan. Green fat bodies of

Table 15.3. Sciomyzini that form a calcareous septum or produce septum material in the aperture of the shell while pupariating

N	*Colobaea americana*	Thin, plate-like encrustation of small amount of calcareous material produced at shell aperture. In *Physa* sp., *Gyraulus parvus*.
P	*Colobaea punctata*	Small amount of calcareous material excreted but not formed into even a rudimentary septum. In *Radix balthica, Gyraulus albus, Planorbarius corneus, Planorbis planorbis*.
N	*Ditaeniella trivittata*	Two septa formed; an outer, often incomplete septum and a complete septum surrounding anterior end of puparium. In *Helisoma anceps*.
N	*Pherbellia idahoensis*	Usually incomplete. Very fragile. In freshwater snails in laboratory rearings.
N	*P. quadrata*	Incomplete, crust of septum material present on most puparia. In freshwater snails in laboratory rearings.
N	*P. seticoxa*	Complete. In *Stagnicola elodes, Helisoma trivolvis*.
N	*P. similis*	Complete. In *Planorbula armigera*.
N	*P. trabeculata*	Complete. In *Planorbula* sp.
N	*P. vitalis*	Complete. In *Galba humilis, Stagnicola elodes, Physa* sp., *Helisoma trivolvis, Planorbula armigera*.
H	*P. argyra*	Complete. In *Aplexa hypnorum, Anisus vortex, Planorbis planorbis*.
H	*P. griseola*	Complete. In *Stagnicola elodes*.
H	*P. nana*	Thin and incomplete in large shells, complete in small, discoidal shells. In *Gyraulus parvus, Galba humilis, Stagnicola palustris, Physa* sp.
P	*P. dorsata*	Complete. In *Planorbis planorbis*.
P	*P. obtusa*	Complete. In freshwater snails in laboratory rearings.

pupa evident through integument or not. Anterior spiracle more or less strongly projecting. 6–16 asp. Length 5.4– 9.0 mm. 11

- Postanal portion of segment XII at most equal to width at anus, strongly upturned to below mid horizontal axis. Integument translucent tan to blackish, opaque. Without green fat bodies. Anterior spiracle strongly projecting to sessile. 3–20 asp. Length 3.8–10.4 mm. 12

11. Integument with moderate coat of strong, sharp spinules in closely set transverse rows on segments V–XI, spinules becoming larger posteriorly; without color pattern. Without green fat bodies. Segments VII–XI not strongly inflated. Without accessory teeth. 16 asp. Length 5.7 mm . . . . *Renocera striata*

- Integument with dense, even coat of very minute spinules or sparse, even coat of minute, rounded projections; with color pattern. With green fat bodies. Segments VII–XI strongly inflated (Fig. 14.49a, b) or not. 5–7 accessory teeth. 6–13 asp. Length 5.4–9.0 mm. . . . . **Elgiva**

12. Anterior spiracle at or slightly above mid horizontal axis, posterior spiracle at or slightly below mid horizontal axis. Integument with dense coat of micro-spinules (0.01–0.09 mm) that give a pubescent appearance. 3 or 4 accessory teeth. 5–23 asp. 13

- Anterior spiracle at or slightly above mid horizontal axis, posterior spiracle well below, at, or well above mid horizontal axis. Integument without pubescent appearance. 0–6 accessory teeth. 3–20 asp.. 14

13. Anterior and posterior spiracles on mid horizontal axis. Pubescence more or less uniform. Lateral tubercle group vestiges indistinct. Interspiracular processes evident, branched. 3 accessory teeth on nearly vertical axis. Ventral arch strongly indented posteriorly. H-shaped and pharyngeal sclerites fused. Dorsal bridge present. 16–23 asp. Length 3.7–5.4 mm. Eggs, then snails of Lymnaeidae, Physidae, Succineidae. . . . . **Anticheta**

- Anterior spiracle above, posterior spiracle below mid horizontal axis. Pubescence uniform but appearing as velvety golden patches on distinct lateral tubercle group vestiges and welts in certain lights. 4 accessory teeth on nearly horizontal axis. Ventral arch strongly projecting mid posteriorly. H-shaped and pharyngeal sclerites separate. Dorsal bridge absent. 10 asp. Length 4.6 mm. Hosts/prey unknown. . . . . **Pherbecta**

14. Posterior spiracle nearly at ventral surface. Upper, outer posterior spiracular slit in thorn-like dorsal process of spiracular plate. Accessory teeth absent. Sphaeriidae. . . . . *Renocera amanda*

- Posterior spiracle nearly at ventral surface to well above dorsal surface. Without thorn-like dorsal process on posterior spiracular plate. Accessory teeth present....... 15

15. Croissant shaped, anterior and posterior ends strongly and evenly arched dorsally. Brownish-black. Surface strongly wrinkled transversely. 5 asp ? Terrestrial snails ? (based on *T. punctulata*, Palearctic)..... **Trypetoptera**

- Barrel or boat shaped, posterior end strongly upturned or not, anterior end at most slightly upturned. Light brown to black, mid dorsal stripes present or absent. Surface smooth to strongly wrinkled transversely. 3–20 asp. Length 3.8–11.2 mm. Freshwater or terrestrial snails, Succineidae, or slugs....................... 16

16. Posterior spiracle uplifted above dorsal surface, anterior spiracle slightly above dorsal surface to level of mid horizontal axis. psd lobe vestiges distinct. 3–20 asp. Length 4.3–10.4 mm. Freshwater snails........... 17

- Posterior spiracle at level of dorsal surface to well below mid horizontal axis. psd lobe vestiges distinct or not. 3–20 asp. Length 3.8–11.2 mm. Freshwater or terrestrial snails, Succineidae, or slugs............ 20

17. Vestiges of lateral tubercle groups well evident, shagreened. In lateral view, more or less truncate posteroventrally or broadly curved. Mid dorsal surface concave or convex. Ventral arch emarginate posteriorly, with posterolateral appendages. 2–4 accessory teeth. 3–12 asp. Length 3.8–10.4 mm. Freshwater snails......... 18

- Vestiges of lateral tubercle groups not very evident. In lateral view, broadly to gently curved posteriorly. Mid dorsal surface essentially straight to convex. Ventral arch emarginate or prolonged posteriorly, without posterolateral appendages. 2–5 accessory teeth. 10–16 asp. Length 6.0–10.4 mm. Freshwater snails........... 19

18. Brown dorsally with faint oblique marking on segments VI–IX, stramineous ventrally or entirely brownish black with irregular light markings ventrally and laterally, posteroventral margins of puparial caps with light dorsoventral stripe. 3 accessory teeth. 4–7 asp. Length 5.6–8.0 mm. Southernmost tip of Texas to Ecuador and Peru.................... **Sepedomerus**

- Color pattern not as above, uniformly light to dark brown, unicolorous or with lateral spots on segments VI–IX or with light oblique stripes dorsolaterally (Fig. 14.48c, d). 2–4 accessory teeth. 3–12 asp. Length 3.8–6.5 mm............................ **Sepedon** (except *S. borealis*, Fig. 14.48g, h)

19. Segments V–XI with strong, transverse wrinkles. Integument shagreened or not. Dorsal surface without bluish-green or coppery iridescence. Ventral arch strongly emarginate posteriorly. 4 or 5 accessory teeth. 10–14 asp. Length 6.0–10.4 mm. Freshwater snails.... .... **Tetanocera** (in part) (*T. annae, T. ferruginea, T. robusta*)

- Segments V–XI without strong, transverse wrinkles. Integument not shagreened. Dorsal surface with bluish-green or coppery iridescence. Ventral arch prolonged mid posteriorly. 8 or 9 denticles below hook part of mouthhook (*D. ptyarion*, North Carolina–Alabama–Florida) *or* 2 or 3 accessory teeth (Distribution: various). 4–17 asp. Length 4.3–8.0 mm. Freshwater or brackish water snails.................................. **Dictya**

20. Posterior spiracle at or slightly below level of dorsal surface. 2–5 accessory teeth. 8–14 asp. Length 5.0–7.5 mm. Freshwater snails..................... **Tetanocera** (in part) (*T. ferruginea, T. fuscinervis, T. silvatica, T. spreta*)

- Posterior spiracle at, slightly above, or well below mid horizontal axis. 1–6 accessory teeth. 5–20 asp. Length 3.8–11.2 mm. Freshwater or terrestrial snails, Succineidae, or slugs................................ 21

21. Posterior spiracle at or slightly above level of mid horizontal axis. 1–4 accessory teeth. 5–20 asp. Length 3.8–11.2 mm................................... 22

- Posterior spiracle well below level of mid horizontal axis. 4–6 accessory teeth. 8–14 asp. Length 4.8–7.0 mm.... 25

22. Light tan, without conspicuous markings, no mid dorsal stripe. Interspiracular processes inconspicuous. Ventral arch emarginate posteromedially, with short posterolateral appendages. 3 accessory teeth. 5–8 asp. Length 3.8–4.8 mm. Freshwater snails........ *Sepedon borealis*

- Reddish brown to black, with or without mid dorsal stripe. Interspiracular processes well developed or inconspicuous. Ventral arch emarginate posteromedially, without posterolateral appendages, or prolonged posteromedially with elongate, strut-like posterolateral appendages. 1–4 accessory teeth. 7–20 asp. Length 4.5–11.2 mm. Freshwater or terrestrial snails, Succineidae, or slugs....... 23

23. Reddish brown, without mid dorsal stripe. Interspiracular processes indistinct. Ventral arch prolonged posteromedially and with elongate, strut-like posterolateral appendages. 1 or 2 accessory teeth. 7–9 asp. Length 4.5–6.2 mm. Succineidae...... *Hoplodictya spinicornis*

- Reddish brown to black, with or without mid dorsal stripe. Interspiracular processes well developed or indistinct. Ventral arch emarginate posteromedially, without posterolateral appendages. 2–4 accessory teeth. 8–20 asp. Length 5.5–11.2 mm. Freshwater or terrestrial snails or slugs. 24

24. Vestiges of psd lobes indistinct. Without mid dorsal stripe. Interspiracular processes minute, scale-like. Ventral arch with angulate lateral margins. Basal part of mouthhook short, equal in length to hook part. 2 accessory teeth. 8–17 asp. Length 6.2–8.0 mm. Terrestrial snails, slugs. (based on Palearctic species, *E. cribrata* and *E. stichospila*). . . . . **Euthycera**
 - Vestiges of psd lobes distinct or not. With or without mid dorsal stripe. Interspiracular processes minute or well developed. Ventral arch evenly curved on lateral margins. Basal part of mouthhook longer than hook part or (*T. plebeja*) triangular. 2–4 accessory teeth. 8–20 asp. Length 5.5–11.2 mm. Freshwater or terrestrial snails, Succineidae, or slugs. . . . **Tetanocera** (in part) (*T. clara, T. loewi, T. montana, T. phyllophora, T. plebeja, T. rotundicornis, T. valida, T. vicina*)

25. Vestiges of psd lobes distinct. Vestige of anal proleg strongly protruding, with strong hooks. Interspiracular processes well developed. Anterior spiracle strongly protruding. 4 accessory teeth. 10–14 asp. Length 5.8–7.0 mm. Freshwater snails. . . . . **Hedria**
 - Vestiges of psd lobes indistinct. Vestige of anal proleg not evident. Interspiracular processes moderately developed or not evident. Anterior spiracle sessile. 4–6 accessory teeth. 8 or 9 asp. Length 4.8–7.0 mm. Succineidae and slugs, or hosts/prey unknown. 26

26. Integument with finely pebbled sculpturing. Dorsal surface strongly convex. Primary and secondary integumentary folds faint. Interspiracular processes indistinct. 4 or 5 accessory teeth. 8 asp. Length 4.8–5.4 mm. Hosts/prey unknown. . . **Poecilographa**
 - Integument with fine asperities. Dorsal surface straight or (*L. boscii*) strongly convex. Primary and secondary integumentary folds distinct. Interspiracular processes of intermediate length, finely branched. 6 accessory teeth. 8 or 9 asp ? Length 5.8–7.0 mm. Succineidae and slugs. (based on *L. boscii* and the Palearctic species, *L. unguicornis*). . . . . **Limnia**

Palearctic

PHAEOMYIIDAE

The only reared species of Phaeomyiidae is *Pelidnoptera nigripennis*, an internal parasitoid of iulid millipedes in mesic woodlands, shrubland, and heathland. The puparium (Fig. 14.43a, b), not yet found in nature, is formed in the host in laboratory rearings. It is 7.3–7.8 mm long, reddish brown to brownish black, subcylindrical, dorsal cephalic cap concave, segment XII truncate in dorsal view, integument covered with minute spinules, strongly wrinkled, anterior spiracle bifurcate with 8 papillae, and the posterior spiracular tubes moderately protruding. The cephalopharyngeal skeleton (Fig. 14.42) lacks the ventral arch and accessory teeth on the mouthhook. Range: England to Azerbaijan, central Finland to Portugal.

SCIOMYZIDAE

Salticellinae

The only described Palearctic species, *Salticella fasciata*, is a parasitoid/saprophage of large, terrestrial snails in open grasslands. The puparium (Fig. 14.43c, d), not yet found in nature, is formed outside the host shell in laboratory rearings. It is 7.2–8.0 mm long, dark reddish brown to black, ovoid, dorsal cephalic cap flat, segment XII rounded in dorsal view, integument pruinose, minutely wrinkled, anterior spiracle quadrate with 8 papillae, and the posterior spiracular tubes strongly protruding. The cephalopharyngeal skeleton (Fig 14.33a–e) has one denticle at the base of the mouthhook and the ventral arch has only three pairs of widely separated teeth. Range: Ireland to Iran, below Scotland and Denmark to North Africa; mainly coastal, inland along river valleys but also in Carpathian Basin (Fig. 16.8).

Sciomyzinae

The biologically unknown Sciomyzini genus *Apteromicra* and Tetanocerini genera *Chasmacryptum, Ectinocera, Neodictya, Oligolimnia*, and *Trypetolimnia* are not included.

1. In snail shells. All species of *Tetanura, Colobaea*, and some species of *Ditaeniella, Pherbellia, Pteromicra*, and *Sciomyza*. See Tables 15.1–15.3 and use couplets 2–7 below.
 - Not in shells. Species and genera generally found in shells are also included in couplets 3–7 so that the generic placement of the species in Tables 15.1–15.3 can be confirmed on morphological features and because some that routinely pupariate in shells may be found to pupariate outside shells. 2

2. Generally found in semi-terrestrial situations on wet surfaces or strictly terrestrial situations but may be found floating in flood debris. Barrel shaped, without adaptations for aquatic existence, posterior spiracles not elevated above dorsal surface. With ventral patches of black spinules on at least a few segments. If in doubt: IIx 60–90. Mouthhook without accessory teeth. Dorsal

bridge between pharyngeal sclerites. Dorsal and ventral cornua of pharyngeal sclerite with windows. 7–32 asp. Length 2.6–6.3 mm. (SCIOMYZINI) 3

- Generally found in freshwater, semi-terrestrial, or strictly terrestrial situations. Most aquatic species not simply barrel shaped, but with posterior spiracles elevated above dorsal surface. Semi-terrestrial and terrestrial species barrel shaped and posterior spiracles not elevated above dorsal surface. Without ventral spinule patches (except weak and lightly pigmented in *Anticheta*). If in doubt: IIx 30–56 (61–62 in *Renocera*, 50–56 in *Anticheta*). Mouthhook with 1–6 accessory teeth (absent in *Renocera*). Dorsal bridge between pharyngeal sclerites absent (present in *Anticheta*). Dorsal cornu without window (present in *Anticheta* and *Renocera*). 3–37 asp. Length 4.3–10.4 mm. (TETANOCERINI) 8

3. In shells of terrestrial *Aegopinella, Clausilia, Cochlicopa*, and *Discus*. Length 2.4–3.5 mm. Anterior end completely occludes whorl of shell like an operculum. Anterior spiracle appressed to dorsal cephalic cap. 10–14 asp. IIx 72. Mouthhook very short, broad. Ends of ventral arch fused to mouthhooks. No septum or septum material. Very restricted habitat: shady, sparse, deciduous woods with low herbaceous vegetation. Ireland to Japan, n. Sweden to Alps and Pyrenees . . . . **Tetanura**

- In shells of above and/or many other snails or outside shells (Tables 15.1, 15.2). Length 2.6–10.0 mm. Anterior end occluding whorl of shell or fitting loosely. Anterior spiracle usually not appressed. 7–34 asp. IIx 60–90. Mouthhook not short, broad. Ventral arch not fused to mouthhook. Wide range of dry, exposed, terrestrial habitats to various woods to open, shoreline situations. Very restricted to throughout the Palearctic 4

4. Usually formed outside of shell. Segments XI and XII with shagreened surface due to dense coat of flat spinules. Many body segments with completely encircling spinule bands. Dorsal cephalic cap strongly concave. Lateral, ventrolateral, and ventral psd lobes distinct. IIx 72–75. 14–16 asp. Length 3.8–4.9 mm . . . . **Ditaeniella**

- Formed free or in shells. Segments XI and XII bare or with sparse coat of spinules; at most, segments III–VI with completely encircling spinule bands. Dorsal cephalic cap flat to convex. Only ventrolateral and ventral psd lobes distinct, at most. IIx 60–90. 7–32 asp. Length 2.6–10.0 mm 5

5. Anterior spiracle strongly projecting. 12–20 asp. IIx 75–90. Length 2.7–6.2 mm. Always in shells of freshwater snails . . . . **Colobaea**

- Anterior spiracle appressed to moderately projecting. 7–32 asp. IIx 60–77. Length 2.7–6.2 mm. In shells of freshwater or terrestrial snails or free 6

6. Not in shells or in shells of *Radix balthica, Stagnicola palustris, Succinea* sp. IIx 63. 32 asp. Ventral arch with elongate, posterolateral appendages. No septa or septum material produced. Length 5.4–6.3 mm . . . . **Sciomyza**

- Not in shells or in shells of above snails and/or many other genera (Table 15.1). IIx 60–77. 7–20 asp. Ventral arch without elongate, posterolateral appendages. Septa or septum material produced or not. Length 2.5–10.0 mm 7

7. 7–20 asp. IIx 64–77. Epistomal sclerite usually not prolonged anteriorly. Not in shells or in shells of many genera (Tables 15.1, 15.2). Septum or septum material produced by 5 species (Table 15.3). Length 2.5–10.0 mm . . . . **Pherbellia**

- 12–16 asp. IIx 60–66. Epistomal sclerite prolonged anteriorly. Not in shells or in shells of freshwater snails (Tables 15.1, 15.2). No septum or septum material produced. Length 2.6–4.2 mm . . . . **Pteromicra**

8. Segments V–XI with vestiges of elongate dorsal, dorsolateral, and ventral lobes (Fig. 14.49e, f). Integument spinulose, not wrinkled. Dorsum flat. Posterior spiracle below, anterior spiracle above dorsal surface. IIx 43. 5 accessory teeth. 16 asp. Length 6.4 mm. Sphaeriidae . . . . *Ilione lineata*

- Segments V–XI without above complement of vestiges of long lobes; without or with vestiges of moderately strong or rounded lateral tubercle groups (Fig. 14.47b, c). Integument smooth to wrinkled, with or without spinules. Dorsum more or less convex. Posterior and anterior spiracle below to above dorsal surface. IIx 30–56. 1–7 accessory teeth. 3–37 asp. Length 4.3–10.4 mm 9

9. Postanal portion of segment XII 2–2.5 times longer than width at anus, strongly upturned (Fig. 14.49a, b). Integument translucent, light tan. Green fat bodies of pupa evident through integument or not. Anterior spiracle more or less strongly projecting. IIx 32–50. 0–7 accessory teeth. 6–16 asp. Length 5.4–7.8 mm 10

- Postanal portion of segment XII at most equal to width at anus, strongly upturned to below mid horizontal axis. Integument translucent tan to blackish, opaque.

Without green fat bodies. Anterior spiracle strongly projecting to sessile. IIx 30–62. 0–6 accessory teeth. 3–37 asp. Length 3.8–10.4 mm. 11

10. Integument with moderate coat of strong, sharp spinules in closely set transverse rows on segments V–XI, spinules becoming larger posteriorly; without color pattern. Without green fat bodies. Segments VII–XI not strongly inflated and without distinct lateral tubercle groups. IIx 50. No accessory teeth. 16 asp. Length 5.7 mm. . . . . *Renocera striata*

\- Integument with dense, even coat of very minute spinules, with or without color pattern. With (*E. solicita*) or without (*E. cucularia*) green fat bodies. Segments VII–XI inflated or not, with or without distinct lateral tubercle groups. IIx 32–37. 5–7 accessory teeth. 5–13 asp. Length 5.4–7.8 mm. . . . . **Elgiva**

11. Posterior and anterior spiracles at or below mid horizontal axis. Integument with large spinules giving pubescent appearance or minute projections giving papillose appearance. H-shaped and pharyngeal sclerites fused (Figs 14.36q, 14.38r). IIx 50–62. 0–3 accessory teeth. 15–23 asp. Length 3.8–5.3 mm 12

\- Posterior and anterior spiracles above to below mid horizontal axis. Integument bare or with spinules, but not as above. H-shaped and pharyngeal sclerites separate (Fig. 14.33j). IIx 30–56. 2–6 accessory teeth. 3–37 asp. Length 4.3–10.4 mm. 13

12. Integument with dense coat of sharp, transparent micro-spinules giving strongly pubescent appearance. Posterior spiracular plate without thorn-like dorsal process. Dorsal bridge present. IIx 50–56. 1–3 accessory teeth. 15–28 asp. Length 3.8–5.3 mm. Eggs, then snails of Lymnaeidae, Physidae, Succineidae. . . . . **Anticheta**

\- Integument with dense coat of minute, rounded micro-processes giving papillose appearance. Upper, outer posterior spiracular slit in thorn-like dorsal process of spiracular plate. Accessory teeth and dorsal bridge absent. IIx 61–62. 15–17 asp. Length 4.1–4.3 mm. Sphaeriidae. *Renocera pallida*

13. Croissant shaped, anterior and posterior ends strongly and evenly arched dorsally. Brownish black. Surface strongly wrinkled transversely. IIx 40. 5 accessory teeth. 5 asp. Length 4.7–5.0 mm. Terrestrial snails (*Lauria* and *Vertigo*). . . . . **Trypetoptera**

\- Barrel or boat shaped, posterior end strongly upturned or not, anterior end at most slightly upturned. Surface smooth to strongly wrinkled transversely. Light brown to black, mid dorsal stripes present or absent. IIx 30–56. 2–6 accessory teeth. 3–37 asp. Length 5.2–10.4 mm. Freshwater or terrestrial snails, Succineidae, or slugs. 14

14. Posterior spiracle above dorsal surface, anterior spiracle slightly above dorsal surface to level of mid horizontal axis. IIx 30–46. 3–6 accessory teeth. 3–37 asp. Length 6.0–10.4 mm. Freshwater snails or Succineidae. 15

\- Posterior spiracle at level of dorsal surface to well below mid horizontal axis. IIx 36–56. 1–6 accessory teeth. 7–20 asp. Length 5.4–8.8 mm. Freshwater or terrestrial snails, Succineidae, or slugs. 18

15. Vestiges of psd lobes long, well evident. Vestige of anal proleg with strong hooks, usually well evident. Postanal portion of segment XII moderately elongate, strongly tapered. Pattern of longitudinal mid dorsal stripes in most species. IIx 35–46. 3–6 accessory teeth. 13–37 asp. Length 6.8–8.5 mm. Freshwater snails. All Mediterranean-Pontic, except *I. albiseta* extends to central Sweden . . . . **Ilione**

\- Vestiges of psd lobes short; if moderately long, without above combination of characters. IIx 30–43. 3–5 accessory teeth. 3–20 asp. Length 6.0–10.4 mm. Freshwater snails or Succineidae. 16

16. Vestiges of lateral tubercle groups well evident, shagreened. In lateral view, more or less truncate posteroventrally or broadly curved. Mid dorsal surface concave or arched. Ventral arch indented posteriorly, with posterolateral appendages. IIx 30–37. 3 or 4 accessory teeth. 3–8 asp. Length 4.3–7.2 mm. Freshwater snails or (*S. hispanica*) Succineidae. . . . . **Sepedon**

\- Vestiges of lateral tubercle groups not very evident. In lateral view, broadly to gently curved posteroventrally. Mid dorsal surface essentially straight to convex. Ventral arch indented or prolonged posteriorly, without posterolateral appendages. IIx 32–43. 3–5 accessory teeth. 10–16 asp. Length 6.0–10.4 mm. Freshwater snails 17

17. Segments V–XI with strong, transverse wrinkles. Integument shagreened or not. Dorsal surface without bluish-green or coppery iridescence. Ventral arch strongly indented mid posteriorly. IIx 40–43. 4 or 5 accessory teeth. 10–16 asp. Length 6.0–10.4 mm. Freshwater snails. . . . . . **Tetanocera** (in part) (*T. ferruginea, T. punctifrons, T. robusta*)

- Segments V–XI without strong, transverse wrinkles. Integument not shagreened. Dorsal surface with bluish-green or coppery iridescence. Ventral arch prolonged mid posteriorly. IIx 32–43? 3 accessory teeth. 10 asp. Length 7.0 mm. Freshwater snails.. **Dictya**

18. Posterior spiracle at level of dorsal surface. IIx 36–48. 2–5 accessory teeth. 8–18 asp. Length 5.2–8.8 mm. Freshwater snails, Succineidae....................19
- Posterior spiracle at, slightly above, or well below mid horizontal axis. IIx 41–56. 1–6 accessory teeth. 7–20 asp. Length 5.4–11.2 mm. Freshwater or terrestrial snails, Succineidae, or slugs.............................20

19. In lateral view, dorsal surface straight to slightly convex; posteroventral surface rounded (Fig. 14.15) or dorsal surface sinuous, strongly inflated medially, and posterior end truncate (*T. arrogans*, Fig. 14.51a). IIx 36–48. 2–5 accessory teeth. 8–18 asp. Length 5.2–8.5 mm. Freshwater snails, Succineidae... **Tetanocera** (in part) (*T. arrogans, T. ferruginea, T. fuscinervis*)
- In lateral view, dorsal surface moderately convex; posteroventral surface rounded. Ventrolateral and ventral lobe vestiges moderately elongate, not bipartite. IIx 42–47. 4 accessory teeth. 10–12 asp. Length 7.8–8.8 mm. Freshwater snails........*Pherbina coryleti*

20. Posterior spiracle at or slightly above level of mid horizontal axis. Anterior spiracle at or slightly above mid horizontal axis. IIx 36–56. 1–6 accessory teeth. 6–20 asp. Length 5.4–11.2 mm... 22
- Posterior spiracle well below level of mid horizontal axis. Anterior spiracle slightly above mid horizontal axis. IIx 41–43. 2–6 accessory teeth. 7 or 8 asp. Length 6.0–7.0 mm... 21

21. Interspiracular processes very small and scale-like, unicuspid. Lobes of ventral arch longer than wide. IIx 41. 2 accessory teeth. 7 asp. Length 6.0–6.5 mm. Terrestrial snails.................... **Dichetophora**
- Interspiracular processes moderately short, finely branched. Lobes of ventral arch with width equal to length. IIx 41–43. 6 accessory teeth. 8 asp. Length 5.8–7.0 mm. Succineidae and slugs............. **Limnia**

22. Anterior spiracle well above mid horizontal axis. IIx 36–53. 1–6 accessory teeth. 9–15 asp. Length 5.4–7.4 mm. Freshwater or terrestrial snails or slugs....23
- Anterior spiracle at or slightly above or below mid horizontal axis. IIx 39–56. 2–6 accessory teeth. 8–20 asp. Length 5.5–8.0 mm. Freshwater or terrestrial snails or slugs...26

23. With elongate vestiges of lateral tubercle groups on segments V–XI. Color: black. psd lobe vestiges distinct. IIx 36–39. 5 accessory teeth. Freshwater snails.........................*Tetanocera hyalipennis*
- Without elongate vestiges of lateral tubercle groups on segments V–XI. Color: white to faint dark pattern. psd lobe vestiges very small to indistinct. IIx 36–56. 1–6 accessory teeth. Freshwater or terrestrial snails...24

24. With psd lobe vestiges. Interspiracular processes well developed, branched. IIx 36–44. 3–6 accessory teeth. 9–15 asp. Length 5.7–7.4 mm. Freshwater snails..25
- Without psd lobe vestiges. Interspiracular processes minute, scale-like. IIx 53. 1 or 2 accessory teeth. 9 asp. Length 5.4–5.8 mm. Terrestrial snails................................ **Coremacera**

25. In dorsal view segments II–V elongate, triangular; in lateral view concave on dorsal surface. Only primary integumentary fold vestiges distinct. IIx 44. 5 or 6 accessory teeth. 13–15 asp. Length 6.0–6.3 mm. Freshwater snails............................ **Hydromya**
- In dorsal view segments II–V not elongate, quadrate; in lateral view almost straight on dorsal surface. Both primary and secondary integumentary fold vestiges distinct. IIx 36–39. 3 or 4 accessory teeth. 9–13 asp. Length 5.7–7.4 mm. Freshwater snails .....*Tetanocera silvatica*

26. With mid dorsal and pair of dorsolateral longitudinal dark stripes. 4 pairs psd lobe vestiges distinct. IIx 46. 5 accessory teeth. 9–12 asp. Length 6.7 mm. Freshwater snails..........................*Pherbina mediterranea*
- Without mid dorsal dark stripes. At most, 2 pairs psd lobe vestiges distinct. IIx 39–56. 2–6 accessory teeth. 8–20 asp. Length 5.5–8.0 mm. Freshwater or terrestrial snails or slugs..................................27

27. Ventral surface almost straight. Ventrolateral and ventral psd lobes distinct. IIx 44–56. 5 or 6 accessory teeth. 9–14 asp. Length 6.1–7.6 mm. Freshwater snails................................ **Psacadina**
- Ventral surface almost straight or strongly curved anteriorly and less so posteriorly. IIx 39–49. 2–4 accessory teeth. 8–20 asp. Length 5.5–8.0 mm. Terrestrial snails or slugs..................................28

28. Strongly or moderately wrinkled. Dorsal and ventral surfaces nearly straight and parallel or more or less equally, slightly convex. Basal part of mouthhook short, about equal to length of hook part. Lateral margin

of ventral arch somewhat lobate (Fig. 14.38p, q). IIx 44–49. 2 accessory teeth. 8–17 asp. Length 5.8–8.0 mm. . . . . **Euthycera**

- At most moderately wrinkled. Dorsal and ventral surfaces more or less equally convex. Basal part of mouthhook longer than hook part or (*T. plebeja*) triangular. Lateral margins of ventral arch more or less evenly curved (as Fig. 14.38m). IIx 39–41. 2–4 accessory teeth. 13–20 asp. Length 5.5–7.1 mm. . . . . **Tetanocera** (in part) (*T. elata, T. phyllophora, T. plebeja*)

15.3 ALPHA-LEVEL TAXONOMY

I found it in a legendary land
all rocks and lavender and tufted grass
where it was settled on some sodden sand
hard by the torrents of a mountain pass.
I found it and I named it, being versed
in taxonomic Latin: thus became godfather
to an insect and its describer . . .
and I want no other fame.

Nabokov (1943).

The state of basic, descriptive and revisionary taxonomy of Sciomyzidae is in fairly good condition, although as indicated under zoogeography (Chapter 16) further collection and research are needed, especially for the Afrotropical, central Asian, and Oriental faunas. A few monotypic genera, as in most groups of insects, are known from only one or a few specimens, and from only one sex. Of the 61 genera of Sciomyzidae, these mostly enigmatic genera are (distribution and known specimens in parentheses): *Apteromicra* (Nepal, one female), *Ellipotaenia* (northeastern Tibet, one female), *Neodictya* (Siberia, one male; Mongolia, one female recently collected, Knutson unpublished, in Philadelphia Academy of Natural Sciences), *Oligolimnia* (Morocco, one female), *Verbekaria* (Tanzania, two females plus one male recently collected, Vala *et al.* 2000), and *Tetanoptera* (Democratic Republic of Congo, one female plus one male recently discovered by Knutson & Vala (1999), and one female recently collected by A. Freidberg, in U. S. National Museum). A few other genera, e.g., *Trypetolimnia* (Ukraine to Amur area) are represented by more material, of one or more species, but are incompletely described and/or illustrated. Study of genera with small and relatively nondescript adults, e.g., *Colobaea, Parectinocera, Pherbellia, Pteromicra*, and of micro-Diptera in museum collections should be productive. Collecting in unlikely habitats and in off-seasons, i.e., during fall to spring in the northern hemisphere, is likely to produce new species and genera. Hill-topping is unknown in Sciomyzidae. Molecular systematics studies are needed, especially for some species groups of *Sepedon* in the Afrotropical region. The immature stages of about one-third of the species have been described, but further stereoscan studies of external morphology are needed. Relatively unexplored character systems of adults need further study, e.g., eye-color patterns, labellae, costal-vein microtrichiae, wing-vein sensilla, aedeagal apodeme, testes, spermathecae, internal anatomy, and external anatomy by stereoscan.

15.4 RELATIONSHIPS AMONG AND WITHIN GENERA

Discovery consists in seeing what everybody has seen and thinking what nobody has thought.

A. von Szent-Györgyi.

Two preliminary cladistic analyses of the type species of 50 of the 61 genera, with the Phaeomyiidae as the outgroup, have been published. Barker *et al.* (2004) re-analyzed the character data of Marinoni & Mathis (2000), omitting the two larval characters, adding seven adult characters of chaetotaxy, wing venation, hind femur shape, and using PAUP* 4.0 software. Higher-level topology generally was consistent with the consensus tree of Marinoni & Mathis (2000), with inclusion of genera within the subfamilies and tribes consistent with the former analysis. However, there were significant differences in placement of a number of genera within the Tetanocerini. Especially, seven genera identified as higher Tetanocerini by Marinoni & Mathis (2000) occur near the base of the Tetanocerini in Barker *et al.* (2004). Notably, most of the *Teutoniomyia* to *Sepedonea* lineage remained exactly the same at the terminus of the cladogram. The original cladograms are shown in Figures 15.23, 15.24.

A "bootstrap analysis" (Felsenstein 2004) is an important procedure for evaluating a cladistic analysis. Considering our use of the cladogram of Marinoni & Mathis (2000) and the slightly modified but similar cladogram of Barker *et al.* (2004) we asked E. G. Chapman to provide a bootstrap analysis. His results (personal communication, 2006) are presented below and in Figure 15.25. Bootstrapping takes a random subsample of characters (= columns) from the original data set to build a tree. The tree provided by E. G. Chapman was built with 1000 bootstrap replicates, each time replacing the characters used in the previous replicate. Then, a consensus tree was built from the 1000

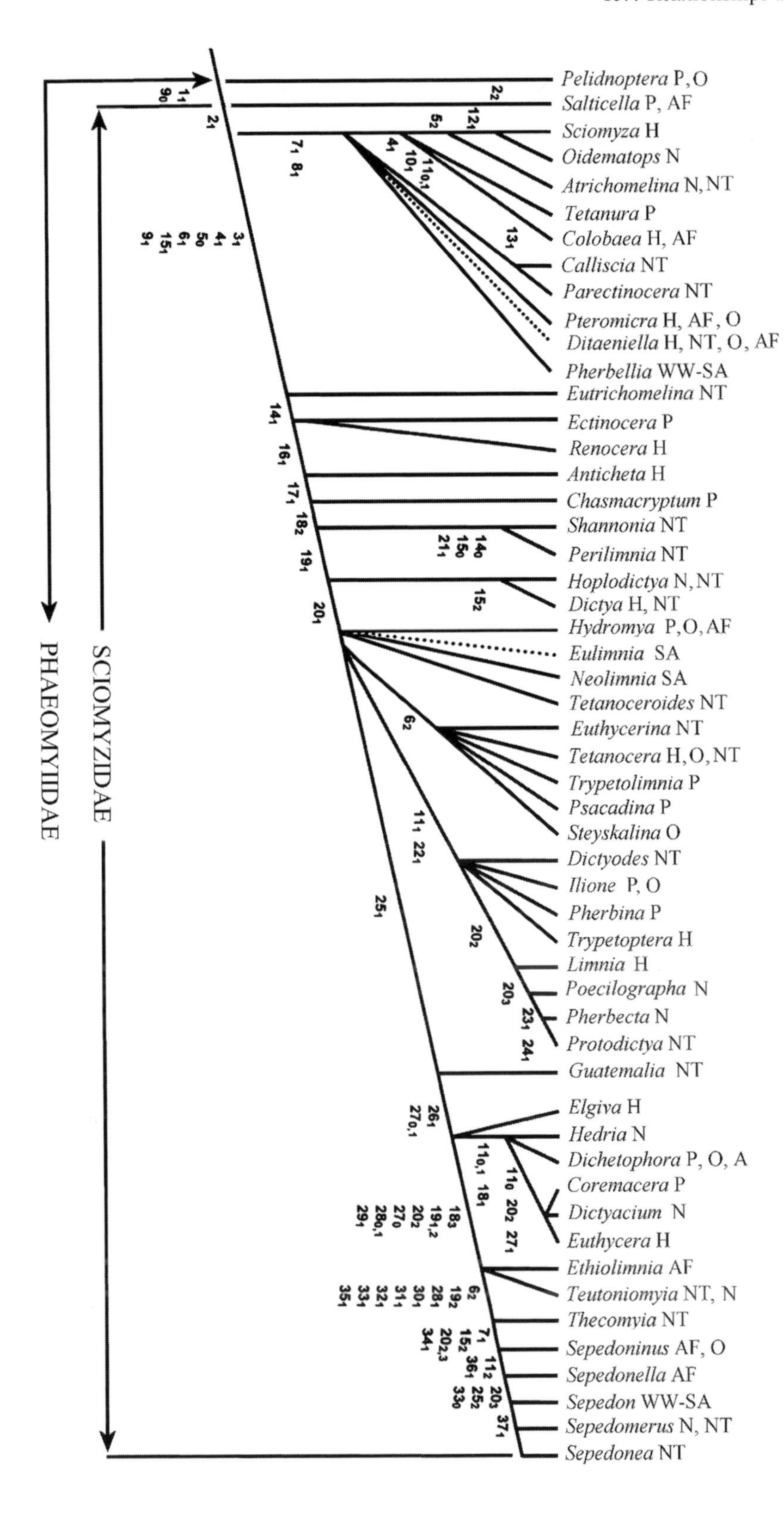

Fig. 15.23 Consensus cladogram of phylogenetic relationships among 50 of the 61 genera of Sciomyzidae based on maximum parsimony analysis (using Hennig86, Version 1.5 software; Farris 1988) of 2 larval and 35 adult morphological characters of the type species of each genus. Strict consensus tree derived from equally parsimonious trees (consistency index 0.60, retention index 0.88). Large numbers at branches are numbers of characters found to establish lineages. Subscript numbers indicate character state (0 = plesiomorphic, 1–3 = apomorphic). Modified from Marinoni & Mathis (2000). See Table 16.1 for abbreviations of zoogeographical regions given after genera.

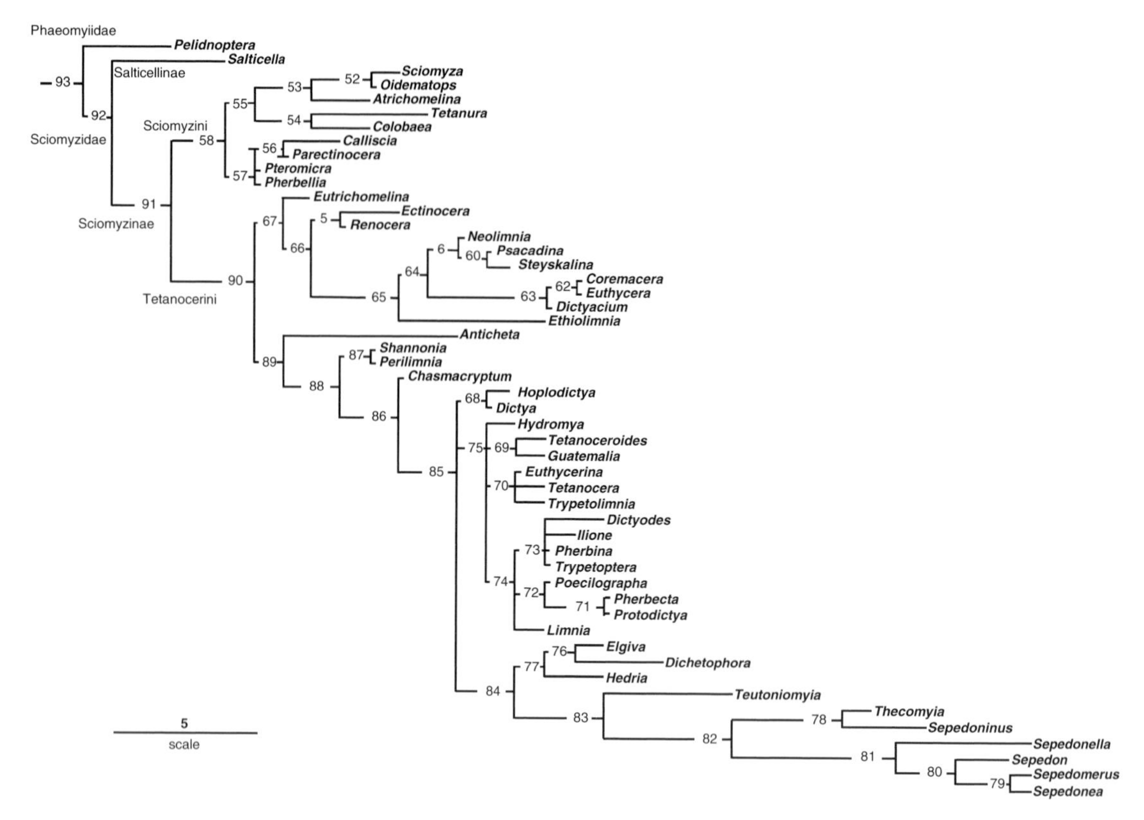

Fig. 15.24 Consensus cladogram of phylogenetic relationships among 50 of the 61 genera of Sciomyzidae based on maximum parsimony analysis (using PAUP 4.0 software; Swofford [1998]) of 43 adult morphological characters. Branch lengths are proportional to numbers of character changes (scale bar = five character state changes). Strict consensus tree derived from four equally parsimonious trees (tree length 750, consistency index 0.42, retention index 0.89, homoplasy index 0.62). Heuristic searching employed, with multiple states of characters within taxa interpreted as polymorphism, preference given to reversals over parallelisms using DELTRAN option, branch swapping by tree bisection-reconnection (TBR) and a random addition sequence with ten replicates. "While analysis without *a priori* assumptions is desirable, preliminary analyses with unordered character states indicated a number of implausible transformations, so characters were variously treated as irreversible or Dollo (transformation polarity generally consistent with analysis of Marinoni & Mathis [2000])." Modified from Barker *et al.* (2004). See Fig. 15.23.

trees produced in the bootstrapping procedure. If, for example, a given clade appears in 800 of the 1000 trees, then a bootstrap value of 80 appears at that node on the bootstrap tree. Generally speaking, the higher the bootstrap value linking a particular group of taxa, the more probable it is that the taxa included are a natural group. That is, all of them having evolved from a common ancestor, exclusive of the other taxa in the analysis.

Chapman's bootstrap analysis supported the primary branches of the two cladograms: *Salticella* as the sister group of the Sciomyzinae, the two major clades within that subfamily, and the genera within these two clades the same as in the cladograms. Chapman noted that "The bootstrap tree says nothing about which genera [of Tetanocerini] are more closely related to the Sciomyzini – any one of the lineages within the Tetanocerini could eventually be found to be closest to the Sciomyzini." He pointed out that the Tetanocerini clade is supported by a moderately robust value of 79. But placement of any of the clades within the Tetanocerini is totally ambiguous since the portion of the tree from *Eutrichomelina* to *Teutoniomyia* contains taxa for which no grouping (of any or all of these taxa) occurred in over 50% of the bootstrap trees. While the value of 98 linking *Sepedon* – *Thecomyia* is a good indication that this clade represents a natural grouping, the

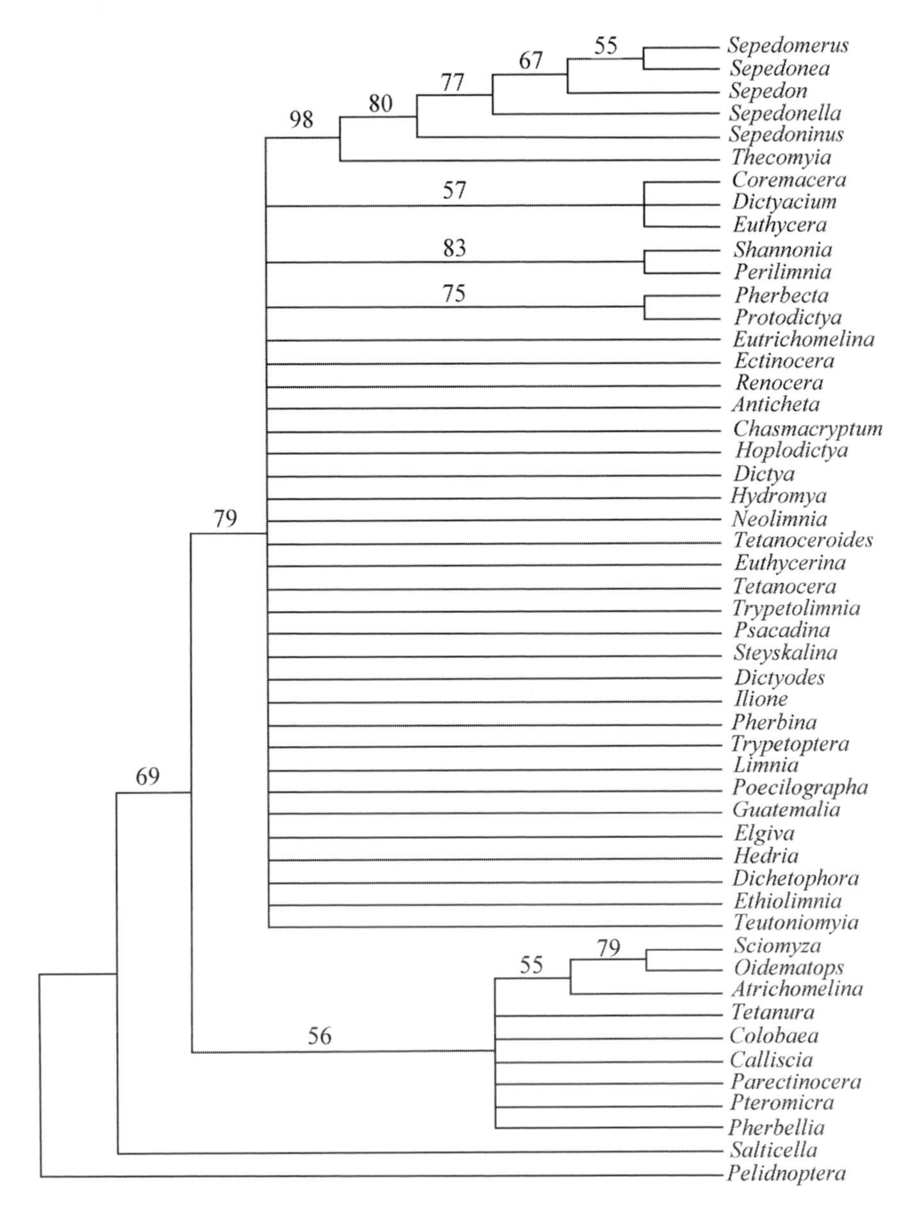

Fig. 15.25 Bootstrap analysis of the cladogram of Marinoni & Mathis (2000). (E. G. Chapman, personal communication, 2006.)

clade containing *Coremacera*, *Dictyarium*, and *Euthycera* (bootstrap = 57) could easily be re-arranged with the addition of just a little more data, according to E. G. Chapman.

15.4.1 Sciomyzini

Several attempts have been made to delimit subgenera or species groups of *Pherbellia*, the largest genus in the family. Sixty-three of the 95 valid species were placed in subgenera when described or by subsequent authors. In several papers, primarily on the 39 Holarctic and Nearctic species, Steyskal (1949, 1961, 1963, 1967) divided the genus into four major groups. Rozkošný (1964) provided a very useful table of the classifications and synonymies of *Pherbellia* and other generic and subgeneric names used by the major authors of these taxa (see Table 15.4), recognizing seven

Table 15.4. Overview of synonymy in ***Pherbellia*** *sensu latum*

<table>
<tr><th>Robineau-Desvoidy (1830)</th><th>Hendel (1902)
Becker (1905)
Séguy (1934)</th><th colspan="2">Hendel (1910)</th><th colspan="2">Cresson (1920)</th><th>Sack (1939)</th></tr>
<tr><td>Pherbellia R-D.</td><td>Ditaenia Hendel</td><td rowspan="2">Melina R-D.</td><td>Subg.
Ditaenia Hendel</td><td rowspan="3">Melina R-D.</td><td>Subg.
Ditaenia Hendel</td><td>Pherbellia R-D.
Ditaenia Hendel
Oxytaenia Sack
Ditaeniella Sack</td></tr>
<tr><td>Dyctia R-D. p.p.
Melina R-D.
Arina R-D.
Chetocera R-D.</td><td>Sciomyza Fallén</td><td></td><td>Subg.
Melina R-D.</td><td>Sciomyza Fallén</td></tr>
<tr><td>—</td><td>Graphomyzina
Macquart</td><td colspan="2">—</td><td>Subg.
Graphomyzina
Macquart</td><td>Graphomyzina
Macquart</td></tr>
</table>

<table>
<tr><th>Enderlein (1939)</th><th colspan="2">Steyskal (1949)</th><th>Steyskal (1961)</th><th colspan="2">Rozkošný (1964)</th></tr>
<tr><td>Dictyomyza Enderlein
Ditaenia Hendel
Heloclon Enderlein
Monopassalus Enderlein</td><td>Pherbellia R-D.</td><td>Subg.
Pherbellia s. str.
Subg.
Ditaenia Hendel
Subg.
Oxytaenia Sack
Subg.
Ditaeniella Sack</td><td rowspan="2">Pherbellia R-D.</td><td rowspan="2">Pherbellia R-D.</td><td>Subg.
Pherbellia s. str.
Subg.
Dictyomyza Enderlein
Subg.
Ditaenia Hendel
Subg.
Oxytaenia Sack
Subg.
Ditaeniella Sack[1]</td></tr>
<tr><td>Callyntropleura Enderlein
Dasypleura Enderlein
Sciomyza Fallén
Dasyclon Enderlein
Gomphoclon Enderlein</td><td colspan="2" rowspan="2">—</td><td>Subg.
Chetocera R-D.</td></tr>
<tr><td>Graphomyzina Macquart</td><td>—</td><td colspan="2">Subg.
Graphomyzina
Macquart[2]</td></tr>
</table>

[1]Generic status recognized by Rozkošný (1987a). [2]Generic status recognized by Miller (1995); p.p., in part. Slightly modified from Rozkošný (1964).

subgenera. Steyskal (1967) considered Rozkošný's (1964) subgeneric classification, "a step toward the solution of our problem, although details of the male postabdomen of most of the Palearctic species are still unknown." Bratt *et al.* (1969) also reviewed the classification of *Pherbellia sensu latum*, proposed eight Groups based on features of the immature stages, and compared their Groups with the subgeneric classification of Rozkošný (1964). Elberg (1978) recognized four subgenera for seven species of *Pherbellia*. In a key to 48 Holarctic and Palearctic species, Rozkošný (1991) noted that a definitive subgeneric classification is still lacking, and he recognized only the genus *Pherbellia*. Over the past two decades, most of the needed details on male genitalia have been provided, primarily by Rozkošný (1984b, 1987a), Vala (1989a), and Rivosecchi (1992) for Palearctic species and by Orth (1982, 1983, 1984a, 1987), Orth & Steyskal (1981), and Orth *et al.* (1980) for Nearctic species. Many other publications on new species of *Pherbellia* discuss their relationships to other species in the groups; see especially Knutson & Freidberg (1983), Orth (1984a, 1987), Rozkošný & Kozánek (1989), Rivosecchi (1989), Knutson *et al.* (1990), and Merz & Rozkošný (1995a).

Ditaeniella, with four species, was re-instated as a valid genus by Rozkošný (1987a), and he re-confirmed this in his (1995) checklist. Miller (1995) proposed generic status for *Graphomyzina*, including the picture-winged *P. cingulata, P. costata, P. kivuana*, and *P. trabeculata*. We recognize *Graphomyzina* as a subgenus, to which we here add *P. limbata, P. javana*, and *P. juxtajavana*. In the cladistic analysis of Marinoni & Mathis (2000), *Pherbellia sensu latum* appears in a polytomy with *Pteromicra*; *Ditaeniella* or other possible segregates of *Pherbellia* were not treated as separate entities. Admittedly, the process of separating off more or less distinctive groups from the suspected polyphyletic genus *Pherbellia* on an intuitive taxonomic basis is suspect. Our excuse is that it is likely a step in the right direction, to be proved or disproved.

Behavioral and morphological features of the adults and immature stages of the unusual terrestrial parasitoid *Tetanura pallidiventris*, based on a complete life-cycle study, were compared with *Anticheta* and other genera by Knutson (1970a). *Tetanura* appeared in a polytomy with *Colobaea* near the middle of the Sciomyzini clade in the cladograms. *Colobaea, Pteromicra, Ditaeniella*, and *Pherbellia* were compared on the basis of 13 characters of adults denoted as plesiomorphic or apomorphic by Knutson *et al.* (1990) (Table 15.5) but they did not speculate on possible relationships. Twenty-eight characters of the eggs, larvae, and adults of *Colobaea* were compared with those of other genera of Sciomyzini, except the biologically unknown Neotropical genera *Calliscia, Eutrichomelina, Neuzina*, and *Pseudomelina*, by Knutson & Bratt (unpublished data).

15.4.2 Tetanocerini

Several genera of uncertain position and relationships, i.e., *Anticheta, Renocera, Shannonia*, and *Perilimnia*, and the apomorphic genera surrounding *Sepedon* at the terminus of the cladograms have been of particular concern. The tribal relationships of the predatory to parasitoid or saprophagous *Perilimnia* (two spp.) and *Shannonia* (two spp.), based on rearings of one species in each genus and a comparison of 15 characters of adult and immature stages, were discussed extensively by Kaczynski *et al.* (1969). They concluded that the two genera are closely related and belong to the Tetanocerini, despite the presence of a pro-episternal seta in both. In the cladogram of Marinoni & Mathis (2000) the two genera appear in a polytomy between *Chasmacryptum* and the *Hoplodictya* + *Dictya* polytomy, five genera remote from the base of the Tetanocerini. In the cladogram of Barker *et al.* (2004) *Perilimnia* and *Shannonia* appear as sister genera between *Anticheta* and *Chasmacryptum* in the second sublineage of the tribe, 11 genera remote from the base of the tribal clade.

The systematic position of the snail-egg/snail-feeding genus *Anticheta* has been discussed by several authors. Verbeke (1950) placed *Anticheta* with *Renocera* in his new subfamily Renocerinae, which was not recognized by later authors (except by Hennig 1965), between the Sciomyzinae and Tetanocerinae. Steyskal (1960b) commented on the phylogeny of the five Nearctic species of *Anticheta* based on postabdominal structures. He placed the genus in the Tetanocerini and considered it as "one of the more highly specialized (apomorphic) of Sciomyzidae." Based on comparisons of 13 features of adult and immature stages with the Sciomyzini and Tetanocerini, Fisher & Orth (1964) noted that *Anticheta testacea*,

> possesses characteristics which seem to place it well toward the Sciomyzinae [= Sciomyzini] in a credible phylogenetic continuum of species within the family .. and .. *Anticheta* may serve as the type genus of a new subfamily.

Table 15.5. Character states of major generic features of ***Colobaea*** and related genera

	Colobaea	*Pteromicra*	*Ditaeniella*	*Pherbellia*
1	1 (0)	0	0	0
2	1 (0)	0	0	0
3	1	0 (1)	0	0
4	1 (0)	0	0	0 (1)
5	1 (0)	0	0	0
6	0	1 (0)	0	0
7	0	1 (0)	0	0
8	0	1	0	0
9	0	1 (0)	1	0 (1)
10	0	0	1 (0)	0
11	0	0	1	0 (1)
12	0	0	1	0
13	0	0	1	0

Notes: 0, plesiomorphic state; 1, apomorphic state; exceptions noted in parentheses.

[1]Arista, dorsobasal and ventrobasal hairs: P (0) same length; A (1) dorsobasal hairs slightly to much longer than ventrobasals (same length in *C. canadensis*, *C. eos*, and *C. flavipleura*).

[2]Ending of anterior branch of R_1: P (0) remote from Sc; A (1) close to Sc (except in *C. canadensis*).

[3]$A_1 + CuA_2$: P (0) continuous to wing margin; A (1) evanescent apically (variable in *Pteromicra leucothrix*).

[4]Anterior surstylus with peg-like processes: P (0) absent; A (1) present (absent in *C. eos*, *C. canadensis*, and *C. flavipleura*; present in *Pherbellia knutsoni* and *P. mikiana*).

[5]Hypandrium with narrow, long posterior process: P (0) absent; A (1) present (absent in *C. bifasciella*).

[6]Frons: P (0) mainly tomentose; A (1) entirely shining (tomentose in an undescribed species of *Pteromicra* from Japan and India).

[7]Fore femur: P (0) without pecten; A (1) with pecten (absent in *Pteromicra anopla*).

[8]Hypandrium with pair of pubescent lobes in posteroventral position: P (0) absent; A (1) present.

[9]Fronto-orbital setae: P (0) 2 pairs; A (1) 1 pair (2 pairs in *Pteromicra angustipennis* and *P. leucopeza*, 1 pair in *Pherbellia tricolor*).

[10]Prosternum: P (0) bare; A (1) haired (bare in *D. trivittata*).

[11]Inner posterior margin of hind coxa: P (0) without hairs; A (1) with hairs (present in *Pherbellia chiloensis*, *P. seticoxa*).

[12]Anterior surstylus: P (0) normal size; A (1) considerably reduced.

[13]Paramere, apical spines: P (0) absent; A (1) present.

Modified from Knutson *et al.* (1990).

Based on complete rearings and descriptions of the immature stages of two additional species of *Anticheta*, Knutson (1966) compared 16 features of adults and immature stages, including all but two of those compared by Fisher & Orth (1964). He maintained the genus in the Tetanocerini. In the cladistic analysis of Marinoni & Mathis (2000), *Anticheta* appears close to the stem of the Tetanocerini, between *Chasmacryptum* and the polytomy *Ectinocera* + *Renocera*. However, in the cladogram of Barker *et al.* (2004) while *Renocera* remains near the base of the Tetanocerini in the first sublineage of the tribe, with *Ectinocera* as the sister genus and eight other genera, *Anticheta* appears considerably beyond *Renocera* as the sister genus of the rest of the Tetanocerini in the second sublineage of the tribe.

A revised comparison of 39 morphological and behavioral characters of the adult and immature stages of *Anticheta*, *Renocera*, *Perilimnia*, *Shannonia*, and *Ectinocera* with the Sciomyzini and Tetanocerini is presented herein (Tables 15.6, 15.7). While the placement of *Renocera* in the Sciomyzini or in the Tetanocerini cannot be determined from this comparison alone, the data further highlight the unusual features of *Renocera* and will be useful in further cladistic and/or phylogenetic analyses. These should include *Eutrichomelina, Ectinocera*, and *Chasmacryptum*, biologically unknown genera whose life styles might be of great importance to further development of an evolutionary scenario of Sciomyzidae.

Twenty-four characters of adults (not denoted as plesiomorphic or apomorphic) among the monotypic Afrotropical *Tetanoptera* and *Verbekaria* and 12 supposedly related genera were compared in a table by Knutson & Vala (1999) (Table 15.8). Eleven additional characters (denoted as plesiomorphic or apomorphic) among *Tetanoptera, Verbekaria*, and eight of the same genera were compared by Vala *et al.* (2000) (Table 15.9).

Twenty-four characters, denoted as plesiomorphic or apomorphic, were compared among nine species groups proposed for *Sepedon* and with six related genera by Knutson & Orth (2001) (Table 15.10). Based on our unpublished morphodiversity and biodiversity data, we conclude that further studies likely will result in taxonomic segregation of *Sepedon* and related genera.

The only cladistic analyses within genera in the family are those for the eight species of *Protodictya*, Marinoni & Carvalho (1993); the 12 species of *Thecomyia*, Marinoni *et al.* (2003); 17 of 39 species of *Tetanocera*, Chapman *et al.* (2006); the six species of *Renocera*, Knutson, Mathis & Vala (unpublished data); and the 13 species of *Sepedonea* by Marinoni & Mathis (2006). Certain natural groupings of species of *Tetanocera* were mentioned by Steyskal (1959) in his revision of the genus. All but seven poorly known Palearctic *Tetanocera* species were placed in four Groups, the *ferruginea* Group including eight Sections, by Boyes *et al.* (1972), primarily on sclerotization of the epandrium and postero-apical bristles on the mid femur. Probable phyletic relationships among the seven species of *Tetanoceroides*, based on a non-cladistic comparison of adult features, were presented by Zuska & Berg (1974). Subgenera and species groups for the eight species of *Ilione* were proposed by Knutson & Berg (1967) on the basis of characters of adults, immature stages, and biology. Taxonomic groups also have been proposed for species of *Dictya* (Steyskal 1954b), and *Limnia* (Steyskal *et al.* 1978).

15.4.3 Huttonininae

Barnes (1979d) redescribed the monotypic New Zealand genus *Prosochaeta*, incorporating morphological characters of the male and female postabdomen, and placed it in his new tribe Prosochaetini of the Huttonininae on the basis of four apomorphic characters (arista pre-apical, female terga 4 and 5 membranous with their spiracles displaced dorsally, female tergum and sternum 6 fused, and spermathecae 2 + 2). He also discussed and compared 16 other apomorphic and plesiomorphic characters of the subfamilies of Sciomyzidae (*sensu* Steyskal 1965a) and redescribed the Huttonininae.

15.5 REGIONAL STUDIES

Comprehensive regional studies are available for Denmark (Knutson & Lyneborg 1965), Fennoscandia (Rozkošný 1984b), West Palearctic and Mediterranean area (Vala 1989a), Central Europe (Rozkošný 2002), Italy (Rivosecchi 1992), Israel (Knutson & Freidberg 1983), the Palearctic (Rozkošný 1987a), Japan (Sueyoshi 2001), California (Fisher & Orth 1983), Idaho (Foote 1961b), Alaska (Steyskal 1954d, Foote *et al.* 1999), Ohio (Foote & Keiper 2004), central Africa (Verbeke 1950, 1956, 1961, 1962a, 1962b, 1963), and Malaysia plus surrounding areas (Knutson & Ghorpadé 2004). Keys to species and genera of adults are included in the above publications.

Keys to genera of adults are available in Steyskal & Knutson (1975b, Americas south of the USA), Knutson (1987, Nearctic), Miller (1995, Afrotropical), Rozkošný (1997a, northern Europe), Rozkošný (2002, central Europe), and Rozkošný (1998, Palearctic). Catalogs are available for all regions: Nearctic (Knutson *et al.* 1986), Neotropical (Knutson *et al.* 1976), Palearctic (Rozkošný & Elberg 1984), Afrotropical (Knutson 1980), Oriental (Knutson 1977) and Australian and Oceanian [including the Subantarctic region] (Barnes & Knutson 1989).

Keys to immature stages are found in Rozkošný (1967), genera and species of larvae and puparia – central Europe; Abercrombie (1970), genera and species of eggs, larvae, and puparia – South America; Rivosecchi (1984), genera and species of larvae and puparia – Italy; Knutson (1987), genera of larvae – Nearctic; Rozkošný (1997a), genera of larvae and puparia – northern Europe; Rozkošný (1998), genera of larvae – Palearctic; Rozkošný (2002), genera and species of larvae and puparia – Central Europe; and Sueyoshi (2005), genera of larvae and puparia – Japan.

Table 15.6 Comparison of behavioral attributes and morphological characters of immature stages of ***Renocera*** with those of the Sciomyzini, the related Tetanocerini genera ***Anticheta***, ***Perilimnia***, and ***Shannonia***, and the rest of Tetanocerini

Morphological Characters and Behavioral Attributes of Immature Stages	Sciomyzini	*Renocera*	*Anticheta*	*Shannonia*	*Perilimnia*	Tetanocerini
Egg						
1. Laid on microhabitat substrate P (0); on host A (1)	0 (1)	0	1	0	0	0
2. Chorion reticulate, without longitudinal ridges and grooves P (0); with longitudinal ridges and grooves A (1); reticulate with a few longitudinal folds A (2)	0 (1)	1	1	1	1	1 (2)
Larva						
3. Hosts/prey: aquatic, semi-terrestrial, and/or terrestrial snails P (0); slugs A (1); snail eggs A (2); fingernail clams A (3); oligochaete worms A (4)	0	3	2	0	0	0 (1, 3, 4)
4. Most segments with transverse ventral spinule patches P (0); some segments with encircling spinule bands A (1); no spinule patches or bands A (2)	0 (1)	2	0	2	2	2
5. Integument of segments III–XII unpigmented, smooth, no strong tubercles P (0); integument pigmented or not, with coat of hairs, spines, etc. and moderate to strong tubercles, some may be elongated A (1); integument unpigmented, smooth, tubercles moderate A (2); entire integument covered with platelets and spinules A (3)	0	1	1	3	3	1 (2)
6. Anterior spiracles round or transverse P (0); oval or elongate A (1)	0	0	0	1	1	0 (1)
7. Posterior spiracular plate without spur-like process P (0); with A (1)	0	1(0)	0	0	0	0
8. Posterior spiracular disc with weak, rounded marginal lobes P (0); lobes weak except ventral and ventrolateral moderately to well developed A (1); elongate lobes A (2)	0 (1)	1	0	0	0	2 (1)
9. Posterior interspiracular processes minute, scale-like P (0); moderately to well developed, branched A (1)	0	1	1	1	1	1 (0)
10. Accessory teeth absent P (0); present A (1)	0	0	1	1	1	1
11. Apices of mouthhooks acute, strongly chitinized P (0); rounded, weakly chitinized A (1)	0	0 (1)	0	0	0	0
12. Ventral arch-mouthhook fusion absent in third instar P (0); present A (1)	0 (1)	0	1	0	0	0 (1)
13. Indentation index 60–90 P (0); 44–62 A (1); 19–53 A (2)	0	1	1	1	1	2
14. Pharyngeal-hypostomal fusion absent in third instar P (0); present A (1)	0 (1)	1	1	1	1	0 (1)
15. Dorsal bridge between pharyngeal sclerites P (0); absent A (1)	0	1	0	1	1	1
16. Window in dorsal cornu P (0); absent A (1)	0	1	1	1	1	1 (0)
Puparium						
17. Formed in/on substrate P (0); in shell of host/prey A (1); in water A (2)	0 (1)	0 (2)	0	0	0	2 (0)
18. Form not adapted for flotation P (0); adapted A (1)	0	0 (1)	0	0	0	1 (0)
19. Anterior spiracles sessile but protruding more or less P (0); appressed, flush with surface of cap A (1); on elongate extensions A (2)	0 (1)	0 (2)	0	1	1	0 (2)

Note: 0, plesiomorphic; 1, 2, 3, 4, apomorphic. Where the taxon includes species with more than one condition, the more common condition is shown first, with the less common condition(s) following in parentheses.

Table 15.7. Comparison of morphological characters of adult stage of ***Renocera*** with those of the Sciomyzini; the related Tetanocerini genera ***Ectinocera***, ***Anticheta***, ***Perilimnia***, and ***Shannonia***; and the rest of Tetanocerini

Morphological Characters of Adults	Sciomyzini	*Ectinocera*	*Renocera*	*Anticheta*	*Shannonia*	*Perilimnia*	Tetanocerini
6. Anterior surstylus well developed P (0); vestigial A (1); absent A (2)	0 (2)	1	1 (0, 2)	1	1	1	1 (2)
7. Aedeagus asymmetrical P (0); symmetrical A (1)	1	0	0	1	0	0	0 (1)
8. Gonopod fused to hypandrium P (0); free A (1)	1	0	0	0	0	0	0 (1)
9. Paramere digitiform and well developed P (0); not digitiform and well developed A (1); absent A (2)	0	1	1	0	1	1	1 (0, 2)
14. Pro-episternal seta present P (0); absent A (1)	0	1	1	1	0	0 (1)	1
15. Male sterna 6, 7, 8 separate, asymmetrical P (0); sternum 6 and synsternum 7 + 8 separate, asymmetrical A (1); synsternum 6 + 7 + 8, symmetrical A (2)	0	1	1	0	0	0	1 (2)
16. Male right spiracle 6 in membrane P (0); in tergum A (1)	0	0	0	1	1	1	1 (0)
17. Male right spiracle 7 in membrane P (0); in tergum A (1)	0 (1)	0	0	0	1	1	1 (0)
18. Head shape group a, P (0); group b, A (1); group c, A (2); group d, A (3)	0	0	0	0	2	2	1 (2, 3)
19. Pedicel ½ length of postpedicel P (0); equal A (1); twice as long A (2)	0	0	0	0	0	0	1 (2)
20. Female sterna 6, 7, 8 separate P (0); 6 separate, 7 + 8 fused A (1); 6, 7, 8 incompletely fused A (2); 6 + 7 + 8 fused A (3)	0	0	0	0	0	0	1 (0, 2, 3)
21. Aedeagus with lobed apex P (0); without A (1)	0	1	0	0	1	1	0 (1)
22. Subalar setae absent P (0); present A (1)	0	0	0	1	0	0	0 (1)
29. Two postalar setae P (0); 0–1 postalar setae A (1)	0	1	0	0	0	0	0 (1)
32. 2 pairs of scutellar setae P (0); 1 pair A (1); absent A (2)	0	0	0	0 (1)	0	0	0 (1, 2)
a. anepisternum with setae P (0); without A (1)	0 (1)	1	1	1	1	0	0 (1)
b. anepimeron with setae P (0); without A (1)	0 (1)	1	1	1	1	1	0 (1)
c. inner posterior margin of hind coxa with hairs P (0); without A (1)	1 (0)	1	1	1	0	0	0 (1)
d. male cerci normal P (0); enlarged and modified A (1)	0 (1)	0	0 (1)	0	0	0	0 (1)
e. mid femur with anteromedian seta P (0); without A (1)	0	0	1 (0)	0	0	0	0 (1)

Note: 0, plesiomorphic; 1, 2, 3, 4, apomorphic. Where the taxon includes species with more than one condition, the more common condition is shown first, with the less common condition(s) following in parentheses. Characters 6–32 are numbered as in Marinoni & Mathis (2000); characters a–e are additional characters important to the comparisons.

Table 15.8. Character matrix of some genera of Tetanocerini presumed to be related to ***Tetanoptera***

Characters	Genera: *Elgiva*	*Hedria*	*Oligolimnia*	*Verbekaria*	*Tetanoptera*	*Dichetophora*	*Neosepedon*(*)	*Sepedon*	*Sepedonea*	*Sepedoninus*	*Sepedomerus*	*Sepedonella*	*Thecomyia*
1. mid frontal vitta	+	+	+	–	–	+	+	–	–	–	–	–	–
2. lunule present	+	+	+	+	+	+	+	–	–	–	–	–	–
3. palpi	+	+	+	+	+	+	+	+	+	+	+	+	–
4. gena/head ratio	1/2	1/2	?	1/4	1/3	1/3	½–1/3	1/1	1/1.5	1/4	1/1.5	1/2	1/2
5. pedicel/postpedicel ratio	1/1	1/1	1/1	1/1.5	1/1	1/1	1/1.5	1/1–2/1	1/1	2/1	1.5/1	1.5/1	4/5
6. aristal plumosity	w-s	w-s	w-s	w-s	w-s	w-s	w-s	w-s	w-s	w-s	w-s	w-s	w-l
7. fronto-orbital setae	2	2	2	1	1	1–2	0–1	0–1	1–2	0–2	1	1	0–1
8. ocellar setae	st	–	+	st	wk	0–wk	0–wk	0–wk	–	–	–	–	–
9. postocellar setae	+	+	+	+	+	+	+	+	+	+	–	–	–
10. h cx3	+	–	–	+	+	wk	0–wk	–/+	+	wk	+	wk	–
11. tb hind coxa	–	–	?	–	–	–	–	–	–	–	–	–	+
12. ab f2	+	+	–	–	–	+	+	+	+	–	–	–	+/–
13. wing spots	wk	wk	+	+	–	+	+	–	–	–	–	–	–
14. dm-cu curved	+	+	–	–	–	–	sl	–	–	–	–	–	–
15. humeral seta	+	+	+	–	–	+	+	–	–	–	–	–	–
16. postpronotal setae	2	2	2	2	2	2	2	1–2	2	1	2	1	1
17. postalar setae	2	2	2	1	1	1–2	1–2	1	1	1	0–1wk	1	1
18. subalar ridge setae	–	–	–	–	–	–	–	–	–	–	–	–	–
19. supra-alar setae	+	0–wk	1	1	1	–	–	+/–	+	–	–	–	–
20. anepimeral callus hairs	–	–	?	+	+	–	–	+/–	–	–	+	–	–
21. no. ps dc setae	1	1–2	1	1	1	1–2	1	0–1	1	1	1	1	1
22. prosternum setose	+/–	–	–	–	+	–	+/–	+/–	+	–	+	–	–
23. cochleate vesicle	–	–	?	–	–	–	–	+/–	–	wk	–	–	–
24. posterior surstyli fused	–	–	?	–	–	–	–	–	+	–	–	–	–
25. distribution	H	N	P	AF	AF	P, O, A	A	H, AF, A, O, NT, OC	NT	AF, O	NT, N	AF	NT

Notes: [1]Characters: ab f2, anterior seta on mid femur; h cx3, inner posterodorsal margin of hind coxa with hairs; ps dc, prescutellar dorsocentral setae; ; tb, thoracic bridge between hind coxae.

[2]Character states: s, short; sl, slight; st, strong; w, white; wk, weak; ?, unknown; + present; – absent.

[3]Distribution: A, Australian; AF, Afrotropical; H, Holarctic; N, Nearctic; NT, Neotropical; O, Oriental; OC, Oceanian; P, Palearctic. (*), ***Neosepedon*** is a [illegible]

Table 15.9. Relationships of ***Verbekaria*** to other genera of Sciomyzidae

	Genera										
Characters	*Elgiva*	*Dichetophora*	*Euthycera*	*Verbekaria*	*Tetanoptera*	*Sepedon*	*Sepedonea*	*Sepedoninus*	*Sepedomerus*	*Sepedonella*	*Thecomyia*
1. Ptilinal fissure present (0), absent (1)	0	0	0	0	0	1	1	1	1	1	1
2. Frontogenal suture present (0), absent (1)	0	0	0	0	0	0	0	0	0	0	1
3. Presutural supra-alar seta strong (0), weak (1), absent (2)	2	2	0, 2	1	2	0, 1, 2	1	1, 2	2	2	2
4. 2 pairs scutellar setae (0), 1 pair (1), absent (2)	0	1	0	0	1	1, 2	1	1	1	1	1
5. Hind tibia without spinous process (0), with (1)	0	0	0	0	0	0	1	0	1	0	0
6. Wing without pattern (0), reticulate (1), discrete dark spots (2), discrete white spots (3)	0	1	1	3	0	0	0	0	0	0	0
7. Male terga & sterna 6, 7, 8 asymmetric (0), symmetric (1)	0	0	0	0	0	1	1	1	1	1	0
8. Syntergosternum & epandrium free (0), fused (1)	0	0	0	0	0	1	1	1	1	1	1
9. Anterior surstylus vestige present (0), absent (1)	1	0	0	0	0	1	1	1	1	1	0, 1
10. Ejaculatory apodeme straight–curved (0), umbrella-like (1), cochleate vesicle (2), absent (3)	0	0	1	1	1	0, 2	0	2	0	3	0
11. Aedeagus dorsoventrally asymmetrical (0), symmetrical (1)	0	0	0	0	0	1	1	1	1	1	0
	a					b					c

Notes: 0, plesiomorphic; 1, 2, 3, apomorphic. a, n°7, ventral process of syntergosternum 6 + 7 simple and pointed. b, n°2, frontogenal suture in upper 2/3 weak and displaced to along eye margin. c, n°8, syntergosternum and epandrium narrowly separated, narrowly fused ventrally on left side. Modified from Vala *et al.* (2000).

Table 15.10. Matrix of characters of adults in groups of ***Sepedon*** *sensu strictu* and related genera

Characters	1	2	3	4	5	6	7	8	9	10	11	12
Taxa	midfacial hairs	anepimeral callus hairs	fronto-orbital setae (1)	postocellar seta	anterior notopleural seta	presutural supra-alar seta	scutellar setae	dorsal setae on fore femur	palpi	fronto-orbital spot	orbito-antennal spot	scape elongate
Groups in *Sepedon s. s.*												
1. *Sphegea* – P, O, NA	±	+	1	+	+	–	+	–	+	–	+	–
2. *Neanias* – P, O	+	+	1	+	+	–	+	–	+	+	+	–
3. *Spinipes* – P, NA	–	+	1	+	+	+	+	–	+	+	+	–
4. *Fuscipennis* – NA, P	±	–	1	+	+	–	+	–	+	±	+	–
5. *Armipes* – NA	+	+	1	+	+	–	+	–	+	±	+	–
6. *Trichrooscelis* – AF, O, AO	–	–	1	+	–	±	+	+	+	±	–	–
7. *Dispersa* – AF	–	–	1	+	–	±	+	+	+	+	–	–
8. *Lobifera* – O	–	–	0	+	–	–	–	–	+	+	–	–
9. *Nasuta* – AF	–	–	1	+	–	+	+	+	+	+	–	+
Related Genera												
1. *Sepedomerus* – NA, NT	–	+	1	–	+	–	+	–	+	–	+	–
2. *Sepedonea* – NT	–	–	1, 2	+	+	+	+	+	+	–	–	–
3. *Thecomyia* – NT	–	–	0, 1	–	–	–	+	–	–	+	–	–
4. *Sepedonella* – AF	–	–	1	–	–	–	+	–	+	–	–	–
5. *Sepedoninus* – AF, O	–	–	0, 2	+	±	±	+	–	+	+	–	–
Plesiomorphic state within Tetanocerini	+	+	2	+	+	+	+	+	+	+	+	–

Table 15.10. (*cont.*)

Characters	13	14	15	16	17	18	19	20	21	22	23	24
Taxa	rostrum extended	hind femur modified	sternal - coxal bridge	complete mesonotal suture (2)	mesonotum angulate anteriorly	cerci fused	epandrium closed below cerci	posterior surstylus & epandrium fused	cochleate vesicle	aedeagus filaments	aquatic predaceous larvae	terrestrial parasitoid larvae
Groups in *Sepedon s. s.*												
1. *Sphegea* – P, O, NA	–	–	–	–	–	±	–	–	–	+	+	–
2. *Neanias* – P, O	–	–	–	–	–	+	+	–	–	–	+	–
3. *Spinipes* – P, NA	–	–	–	–	–	–	–	–	–	+	+	–
4. *Fuscipennis* – NA, P	–	–	–	–	–	–	–	–	–	+	+	–
5. *Armipes* – NA	–	+	–	–	–	–	–	–	–	+	+	–
6. *Trichrooscelis* – AF, O, AO	–	–	–	–	–	–	–	–	±	±	+	+
7. *Dispersa* – AF	–	–	–	–	–	–	–	–	–	–	?	?
8. *Lobifera* – O	–	–	–	–	–	–	–	–	+	–	?	?
9. *Nasuta* – AF	–	–	–	–	+	–	–	–	–	+	?	?
Related Genera												
1. *Sepedomerus* – NA, NT	–	–	–	–	–	–	–	–	–	–	+	–
2. *Sepedonea* – NT	–	–	–	–	–	–	–	–	–	–	+	–
3. *Thecomyia* – NT	+	–	+	–	–	–	–	–	–	–	+	–
4. *Sepedonella* – AF	–	–	–	–	–	–	–	–	–	–	+	–
5. *Sepedoninus* – AF, O	–	–	–	+	–	+	–	–	+ (3)	+	?	?
Plesiomorphic state within Tetanocerini	–	–	–	–	–	–	–	–	–	–	+	–

Notes: [1]Usual number present given first; [2]Distinct and continuous across middle of mesonotum; [3]Rudimentary; + , present; -, absent; P = Palearctic; NA, Nearctic; NT, Neotropical; AF, Afrotropical; O, Oriental; AO, Australian-Oceanian. Modified from Knutson & Orth (2001) with corrections to character states for nos. 5, 16–20, and 22.

15.6 FOSSILS

> The ruins of an older world are visible in the present structure of our planet.
>
> Hutton (1795).

Thirteen fossil species in six genera, all from the Tertiary, have been described in the Sciomyzidae (Evenhuis 1994). Meunier (1904) described one species in his new genus *Palaeoheteromyza* (extinct) from Baltic amber (Eocene). Scudder (1877, 1878a, 1878b, 1878c) described three species in the recent genus *Sciomyza* from Eocene and Oligocene compression fossils from the USA and Canada. Förster (1891) and Theobald (1937) described four species in the recent genus *Tetanocera* from compression fossils from France (Oligocene) and Germany (Miocene). Hennig (1965) stated that he would not dare to identify the "*Sciomyza*" and "*Tetanocera*" species of Scudder and Theobald even as Sciomyzidae on the basis of their descriptions and figures. Hennig (1965, 1969) described species in the extinct genera *Palaeoheteromyza* (Sciomyzinae, three spp.), *Prosalticella* (Salticellinae, one sp.), and *Sepedonites* ("Sepedoninae," one sp.) from Baltic amber; Berg & Knutson (1978) commented on these papers. A magnificently preserved specimen of *Prosalticella succini* has recently been obtained by C. & H. W. Hoffeins (Plate 1e). Hennig (1969) referred "*Palaeoheteromyza*" *curticornis*, described in the Sciomyzinae, questionably to the Phaeomyinae (= Phaeomyiidae), but he retained *P. crassicornis* in the Sciomyzinae.

A fossil sciomyzid puparium was described from the Upper Jurassic/Lower Cretaceous of Spain by Whalley & Jarzembowski (1985); if correctly identified, this is the oldest record for the family. Zherichin (1985) and Gomez Pallerola (1986) recorded undescribed specimens from the Lower Cretaceous of Spain. Poinar & Poinar (1999) presented a photo of a species of "Sciomyzidae" from Dominican amber, but the identification requires confirmation. Several species from Baltic amber are being described (Knutson unpublished data). Also, Hennig (1965) described one species of the extinct genus *Prophaeomyia* in the Phaeomyiinae, now recognized as the family Phaeomyiidae. Hennig's two male specimens have medial setae on the mid and hind tibiae, a diagnostic character for the Phaeomyiidae that is lacking in the Sciomyzidae.

15.7 SUPRAFAMILIAL AND SUPRAGENERIC CLASSIFICATIONS AND PHYLOGENY

The position of the Sciomyzidae in the suprafamilial classification of the Muscomorpha and the names, ranks, and content of the various categories have had tortuous histories. J. F. McAlpine (1989) presented an excellent review of the higher classifications of Diptera and the nomenclature, and a detailed analysis of his own classification. Placement of the Sciomyzoidea and Sciomyzidae in his classification, with authors and synonymies of some of the categories, is shown in Table 15.11. Placement of the Sciomyzidae among a more recent cladogram of the infra-orders and other higher categories of Diptera is shown in Fig. 15.26.

Sciomyzides Fallén (1820b) is the "earliest available family-group name that established its [Sciomyzidae] date for priority" (Sabrosky 1999). The spelling Sciomyzidae dates from Macquart (1846). The status of the synonym, Tétanoceridae Macquart (1844), depends upon ICZN designation of *Tetanocera* Duméril (1800) versus Tetanocerites Newman (1834). A proposal is being prepared by F. C. Thompson (personal communication, 2005).

Several higher-category names, based on Sciomyzides or Tetanocerites, are of interest from the point of view of history of suprafamilial classification, especially possible family relationships, as well as nomenclatorial history. In this regard, the different, major kinds of characters used is notable. Below we trace the major, historical, suprafamilial classifications, then focus on the modern concept of Sciomyzoidea.

"Sciomyzoidea" was proposed by Hendel (1916) as a "Verwandtschaftsgruppen" (relationship group) to include the Sciomyzidae, Ropalomeridae, Dryomyzidae, Coelopidae, Neottiophilidae, and Clusiidae as one of nine categories of Acalyptratae. His classification was based on chaetotaxy of the head, extent of ptilinal fissure, and wing venation. Surprisingly, with the exclusion of a couple of families and addition of a few more, this remains the core content of the modern superfamily. Hendel (1922) proposed Sciomyzomorphae to include Sciomyzoidea (Sciomyzidae, Ropalomeridae, Dryomyzidae, and Neottiophilidae) and Sepsoidea (Sepsidae, Megamerinidae, Diopsidae, Piophilidae, Thyreophoridae, and Psilidae). He placed Coelopidae in his Helomyzoidea and Clusiidae in his Anthomyzoidea. Then Hendel (1924) included under his "1. Superfamilie: Sciomyzoidea" only the Sciomyzidae, Dryomyzidae (including Helcomyzinae), and Neottiophilidae. Of course, during this period Hendel was publishing a huge number of superb alpha-taxonomic papers across the families of Acalyptratae. His knowledge of the broad subject was profound, to be equaled or surpassed by Hennig, Steyskal, etc. of the following generation.

Table 15.11. Placement of Sciomyzoidea and Sciomyzidae in classification of Diptera. Based on Woodley (1989), J. F. McAlpine (1989), and other authors with addition of Natalimyzidae Barraclough and D. K. McAlpine (2006). See Fig. 15.26

Order Diptera
A. 1 – Suborder Nematocera Berthold (1827)[1]
A. 2 – Suborder Brachycera Zetterstedt (1842) = Brachocera Macquart (1834)
  B. 1 – Infraorder Xylophagomorpha *sensu* Woodley (1989)
  B. 2 – Infraorder Stratiomyomorpha *sensu* Woodley (1989)
  B. 3 – Infraorder Tabanomorpha *sensu* Woodley (1989)
  B. 4 – Infraorder Muscomorpha *sensu* J. F. McAlpine (1989). Including cyclorrhaphous Brachycera (= Cyclorrhapha Brauer [1863])
   C. 1 – Section Aschiza Becher (1882)
   C. 2 – Section Schizophora Becher (1882) (= Myodaria Robineau-Desvoidy [1830]; Muscoidea of authors)
    D. 1 – Subsection Calyptratae Robineau-Desvoidy (1830) (= Schizometopa Brauer [1880]; Thecostomata Frey [1921]
    D. 2 – Subsection Acalyptratae Macquart (1835) (= Holometopa Brauer [1880]; Haplostomata Frey [1921])
     E. 1 – Subgroup 1
     E. 2 – Subgroup 2
      F. 1 – Subgroup 2. 1
       G. 1 – Superfamily Lauxanioidea Hendel (1916)
       G. 2 – **Superfamily Sciomyzoidea** Hendel (1916)
        H. 1 – Family Coelopidae
        H. 2 – Family Dryomyzidae (including Helcomyzidae)
        H. 3 – Family Helosciomyzidae
        H. 4 – Family Natalimyzidae
        H. 5 – Family Sciomyzidae (including Phaeomyiidae and Huttoninidae)
        H. 6 – Family Ropalomeridae
        H. 7 – Family Sepsidae

[1]Berthold (1827) is a translation of Latreille (1825) in which Nemocera (without –at – and Latinized) was first proposed by Latreille (1825), but the French version, Némocères, can be found as early as in Latreille (1817) (H. Ulrich, personal communication, 2006).

The Sciomyzaeformes was proposed by Frey (1921) as one of three "Reihe" (= superfamilies according to Sabrosky, 1999) of Schizophora–Haplostomata, including the Sciomyzidae plus 27 other families, and he based it on characters of the mouthparts. Séguy (1937) proposed the superfamily Tétanocéroides. The superfamily Tetanoceratoidea was proposed by Crampton (1944), as one of 12 superfamilies of Acalyptratae, to include the Sciomyzidae, Sepedonidae, and Sepsidae, based only on the male genitalia. Griffiths (1972) proposed the "prefamily" Sciomyzoinea (Sciomyzidae minus *Pelidnoptera*, plus Cremifaniidae and Megamerinidae) as one of seven prefamilies of Acalyptratae, on the basis of wing venation, chaetotaxy of the head, and, especially, male genitalia. We next focus on major aspects of the modern superfamily.

The modern supra- and subfamilial classification and phylogeny of Sciomyzidae and related families has been treated extensively by Hennig (1958, 1965, 1971), Speight (1969), Griffiths (1972), Barnes (1979d, 1981), J. F. McAlpine (1963, 1989), and D. K. McAlpine (1991, 1992). Hennig (1958) included the Coelopidae, Dryomyzidae, Ropalomeridae, Helcomyzidae, Sepsidae, and Sciomyzidae in the superfamily Sciomyzoidea. J. F. McAlpine (1963) discussed the relationships of Sciomyzoidea and Lauxanioidea. Speight (1969) included the Sciomyzoidea and Sciomyzidae in his detailed discussion of the

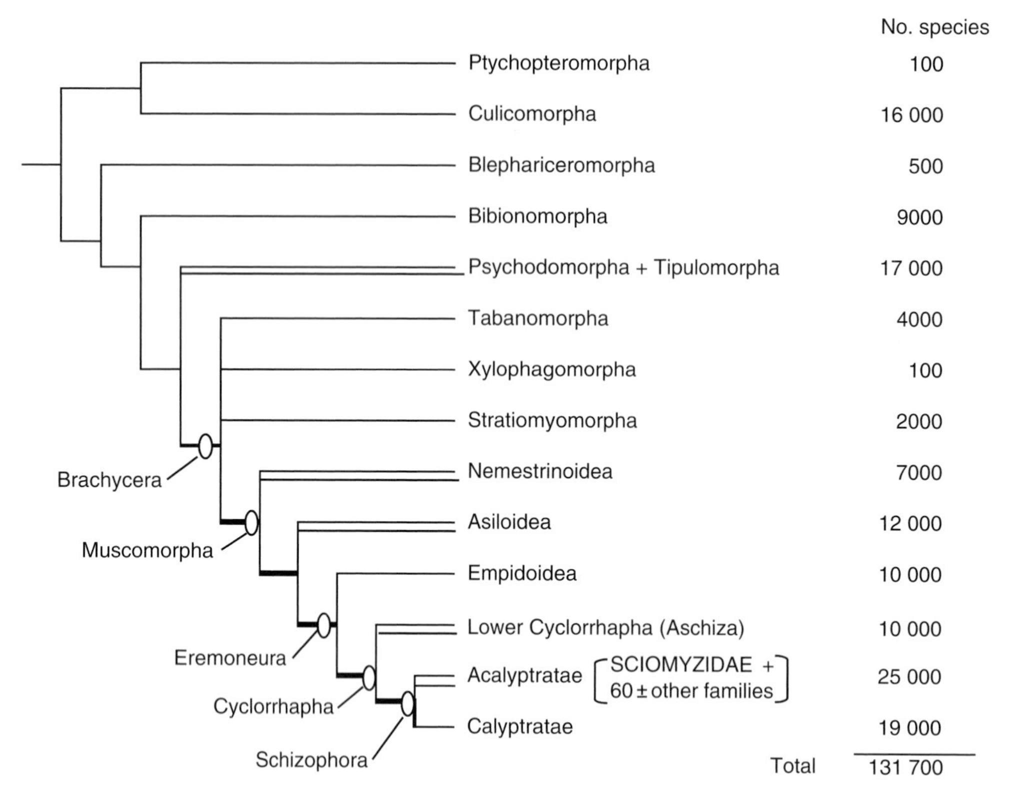

Fig. 15.26 Relationships between infra-orders and other higher categories of Diptera and estimates of number of described species. Bold internodes, those considered to have robust empirical support; parallel branches leading to single terminals, possible or actual paraphyly. Modified from Yeates & Wiegmann (1999).

relationships of 59 families of Acalyptratae based primarily on his comprehensive study of the prothoracic morphology. He especially compared his data and analysis with those of Hennig (1958), noting that "The taxonomic concept 'Sciomyzoidea' seems to have arisen as a result of the disintegration of earlier, " 'ill-defined' family groups." Speight (1969) commented that Hennig's (1958) "justification for grouping together the families he puts in the Sciomyzoidea ... is tortuous," and that the Sciomyzoidea lack unique apomorphic features shared by all members. But he concluded that prosternal shape cannot be used to distinguish families of Sciomyzoidea.

Hennig (1971) commented extensively on Speight's (1969) method of argumentation and his conclusions as follows (translated),

> Until now, no further character has been found which helps with the monophyly of the Sciomyzoidea. Perhaps a certain reduction of abdominal tergite 6 in the male and its articulation with the postabdomen can be seen as such .. In all Sciomyzoidea known to me tergite 6 is smaller than tergite 5, often reduced or fused with tergites 7 + 8 into a tergite complex.

Griffiths (1972) presented a figure showing the distribution of the plesiomorphic and apomorphic states of 68 characters (11 head; eight wing; 45 male postabdomen; four female postabdomen) of adults among the 50 families of Muscoidea that he recognized. In his "prefamily" Sciomyzoinea he included the families that Hennig (1958) included in the Sciomyzoidea, added Cremifaniidae and Megamerinidae, raised Helosciomyzinae to family status, and included Helcomyzidae in Dryomyzidae as a subfamily. He noted, "The reservations expressed by Hennig (1958, 1965) on the validity of this group still remain ... I am not able to give any convincing characterization in terms of autapomorphous conditions." In an analysis of the family relationships of Helosciomyzidae,

Barnes (1981) presented a figure of the distribution of the plesiomorphic and apomorphic states of 25 characters of adults of Sciomyzoidea (the six families included by Hennig [1958]), plus Huttonininae and Phaeomyiinae as subfamilies of Sciomyzidae, Helosciomyzidae as a family, and *Cremifania*. Only eight of his 25 characters were essentially the same as in Griffiths (1972). He noted that the Sciomyzoidea "are difficult to characterize as a monophyletic group on the basis of apomorphous conditions" (Table 15.12). However, J. F. McAlpine (1989) characterized the Sciomyzoidea as "one of the least contentious of all superfamilies of Acalyptratae" in his comprehensive analysis of 82 families of Muscomorpha. His cladogram and comparison of ground-plan characters of Lauxanioidea and Sciomyzoidea are reproduced here (Fig. 15.27 and Table 15.13). He recognized five of the families included by Hennig (1958) in the Sciomyzoidea, Helcomyzinae as a subfamily of Dryomyzidae, and Helosciomyzidae as a family. He excluded *Cremifania* and Megamerinidae, as did Hennig (1958), and agreed with Hennig (1965) that the Sciomyzoidea is the sister group to Lauxanioidea. D. K. McAlpine (1991) included in the Sciomyzoidea the six families included by Hennig (1958). He placed *Heterocheila* in the new family Heterocheilidae, regarded Huttoninidae as a separate family, did not recognize Phaeomyiidae as a family, and included the three families of Lauxanioidea (Lauxaniidae, Chamaemyiidae, and Eurychoromyiidae). Meier (1996) compared adult characters versus larval characters in the Sciomyzoidea, and concluded that larval characters do not necessarily clarify the phylogenetic discussion. He called the Sciomyzoidea a "problematic taxon within the acalyptrate flies." Sueyoshi (2002) made a cladistic analysis of 32 characters of 13 species, in six genera, of Dryomyzidae *s. str.* and one representative species, each, of the Coelopidae, Phaeomyiidae, Helcomyzidae, Heterocheilidae, and Sepsidae. However, the important character, presence or absence of the clypeus, was not included in his study. His cladogram showed *Pelidnoptera fuscipennis* as the sister group of the Dryomyzidae + Helcomyzidae and *Coelopa frigida* as the sister group of the Phaeomyiidae + Helcomyzidae + Dryomyzidae + Heterocheilidae + Sepsidae. R. Meier and B. Wiegmann (personal communication, 2003) are conducting a study of molecular characters of groups within the Sciomyzoidea.

In 2006, Barraclough & D. K. McAlpine described a new family, Natalimyzidae, which they "provisionally treated as *incertae sedis* within the superfamily [Sciomyzoidea]." The authors stated, "A detailed review of all families of Sciomyzoidea suggests no definite sister group relationship with *Natalimyza*." The family is based on a new species, new genus and the authors noted that "numerous additional new species [from Nigeria to South Africa] await description." They also noted that "The larvae appear to be microflora grazers on decaying grass."

The subfamilial and tribal classifications of Sciomyzidae, beginning with that of Verbeke (1950) are shown in an updated and corrected comparison (Table 15.14). Earlier comparisons were presented by Knutson *et al.* (1970), including the classifications of Enderlein (1939) and Sack (1939), and by Berg & Knutson (1978). The paper by Hennig (1965) has been somewhat overlooked. Surprisingly, Hennig's use of the "subfamily group Sciomyzinae," "Groups," and his recognition of Renocerinae, Sepedoninae, and "? Neolimniinae" were not mentioned by subsequent authors. It seems that Hennig (1965) used "Groups" and recognized Verbeke's (1950) Renocerinae and Sepedoninae primarily to clarify discussion and for comparison with Steyskal's (1965a) classification. However, in describing the fossil Baltic amber genus and species *Sepedonites baltica*, Hennig (1969) placed it in the subfamily Sepedoninae! Subsequently, Hennig (1973), in his brief treatment of the Sciomyzidae in the "Handbuch der Zoologie," recognized Huttoninae, Helosciomyzinae, Pelidnopterinae, Salticellinae, and Sciomyzinae in the Sciomyzidae, including *S. baltica* and the *Sepedon* group of genera in Sciomyzinae.

Most of the major historical classifications in which genera were placed in new suprageneric categories or in different combinations of suprageneric categories, mainly within the family Sciomyzidae, up to Verbeke (1950) are shown chronologically below. Steyskal (1965a) listed some of the group names published through 1950, and Sabrosky (1999) provided an exhaustive review of family-group names in all Diptera.

Fallén (1820b)
- Sciomyzides (minus *Tetanura pallidiventris* and *Colobaea bifasciella* described in his Opomyzides 1820a)

Latreille (1825)
- Muscidae [as tribe]
  - Dolichocerae
    - *Lauxania* Latreille
    - *Loxocera* Meigen

Sepedon Latreille
Tetanocera Duméril

Newman (1834)
Tetanocerites (= Chaetomacerini, Cresson 1920)

Macquart (1846)
Sciomyzidae

Schiner (1862)
Tetanura Fallén
Muscidae
Dryomyzinae
Sciomyzinae (minus *Tetanura*, placed in his Tanypezinae)
Tetanocerinae

Loew (1862a)
Sciomyzidae

Brauer (1883)
"Sciomyzinae" and "Tetanocerinae" (as "families" plus 23 other "families" in the Group Holometopa; based on morphology of larvae)

Acloque (1897)
The distribution of genera of Sciomyzidae, recognized at the time, within Acloque's numbered categories is shown to highlight his early, first designation of *Sepedon* and related genera as a suprageneric category, i.e., Sepedonini.
V. Muscidi
8. Muscii
7. Muscini
Melinda Robineau-Desvoidy (synonym of *Pherbellia*)
9. Sepedonini
Sepedon Latreille
Thecomyia Perty
Tetanocera Latreille
11. CordyluLrini
Tetanura Fallén
12. Scatomyzini
Lucina Meigen (synonym of *Salticella*)
Sciomyza Fallén
22. Piophilini
Graphomyzina Macquart (subgenus of *Pherbellia*)

Hendel (1902)
Sciomyziden ("subfamilie")
Sciomyzinae ("subsectio")
Tetanocerinae ("subsectio")

Cresson (1920)
Dryomyzinae
Sciomyzinae
Oidematopsini (new tribe)
Sciomyzini (new tribe)
Euthycerinae (new subfamily)
Chaetomacerini (new tribe)
Euthycerini (new tribe)
Sepedontini

Melander (1920)
Dryomyzinae
Sciomyzinae
Tetanocerinae

Hendel (1924)
Tetanurinae (new subfamily)
Sciomyzinae
Tetanocerinae
Tribe Salticellina (new tribe)
Tribe Sepedonina
Tribe Tetanocerina
Tribe Euthycerina

Malloch (1928)
Tetanurinae
Sciomyzinae
Tetanocerinae
Sepedonini

Séguy (1934)
Salticellinae (new subfamily)
Tetanurinae (new subfamily according to Séguy)
Sciomyzinae
Tetanocerinae

Enderlein (1939)
Tetanoceriden
Sciomyzinae (including *Tetanura* Fallén)
Saltícellinae (new subfamily according to Steyskal 1965a)
Tetanocerinae

Sack (1939)
Sciomyzinae
Ditaeniinae (new subfamily)
Tetanocerinae
Tetanurinae (new subfamily according to Sack)

Crampton (1944)
Superfamily Tetanoceratoidea
Tetanoceratidae or Sciomyzidae
Sepedonidae (new family, "possibly merely a subfamily of the Tetanoceratidae")
Sepsidae

Table 15.12. Distribution of some characters in Helosciomyzidae *sensu strictu* and some other taxa in Sciomyzoidea

Characters	Coelopidae	Helcomyzidae	Helosciomyzidae *sensu strictu*	Dryomyzidae *sensu strictu*	Huttonininae	Phaeomyiinae	Cremifania Czerny	Sciomyzidae *sensu* Griffiths	Sepsidae	Ropalomeridae
1. Lunule mostly exposed (+); mostly concealed (-)	□	□	□	□	□	▥	▥	▥	□	□
2. Oral vibrisse absent (+); present (-)	▥	■	■	■	■	■	■	■	▥	■
3. Clypeus small (+); large (-)	□	□	□	□	□	■	■	■	□	□
4. Postvertical setae long and parallel or divergent (+); otherwise (-)	□	■	■	■	□	■	■	■	■	■
5. Pro-episternal and propleuron fused (+); separate (-)	▥	■	□	□	□	□	□	▥	□	■
6. Propleural seta absent (+); present (-)	□	□	□	□	□	□	■	▥	▥	■
7. Metastigmatal setae present (+); absent (-)	□	□	□	□	□	□	□	□	■	■
8. Number of rows of costal spinules	□	▥ 0-1	■ 1-2	□	□	□	□	□	□	□
9. Cross vein bm-cu close to vein Cu2 (+); apicad to Cu2 (-)	▥	■	□	□	□	□	□	□	□	□
10. Vein A_1+CuA_2 not reaching wing margin (+); reaching margin (-)	□	□	□	□	■	□	■	▥	■	□
11. Middle and hind tibiae with medial setae (+); without (-)	□	□	□	□	□	■	□	□	□	□
12. ♂ fore and hind basal tarsal segments with apical processes (+); without (-)	■	■	□	▥	□	□	□	□	□	□
13. Suture on abdominal terga 1 and 2 indistinct (+); distinct (-)	□	□	□	■	■	□	□	□	■	□
14. ♂ 6th tergum fused to epandrium (+); free (-)	□	□	□	▥	▥	□	▥	▥	▥	□
15. Surstyli fused to epandrium (+); free (-)	□	□	□	□	▥	□	□	□	■	□
16. Aedeagal apodeme expanded and fused to hypandrium (+); otherwise (-)	□	□	□	□	□	□	□	□	■	■
17. Anterior epandrial processes present (+); absent (-)	□	▥	▥	■	□	□	■	□	□	□
18. Basal surstylar processes present (+); absent (-)	□	■	▥	□	□	□	□	□	□	□
19. Aedeagus clothed with hairlike processes (+); otherwise (-)	□	▥	□	■	□	□	▥	□	□	□
20. Epiphallus present (+); absent (-)	□	▥	▥	▥	□	□	■	□	□	□
21. Number of spermathecae	□ 3	■	□ 3	□ 3	□ 3-4	□ 3	□ 3	■ 2-4	■ 2	■ 2
22. Number of spermathecal ducts entering oviduct	■ 2	■ 2	■ 2	■ 2	■ 2	■ 2	□ 3	■ 2	■ 2	■ 2
23. Spiracles 1-5 in terga (+); in membrane (-)	■	□	□	■	□	□	□	□	□	□
24. ♀ spiracles 6-7 in terga (+); in membrane (-)	■	□	▥	■	▥	□	□	▥	■	□
25. Ovipositor short, suddenly tapered (+); elongate, gradually tapered (-)	□	□	□	□	□	□	□	▥	■	□

Note: Black rectangles, characters interpreted as apomorphous groundplan conditions; white rectangles, characters interpreted as plesiomorphous groundplan conditions; crosshatched rectangles, apomorphous character present in some but not in all members of a taxon. From Barnes (1981).

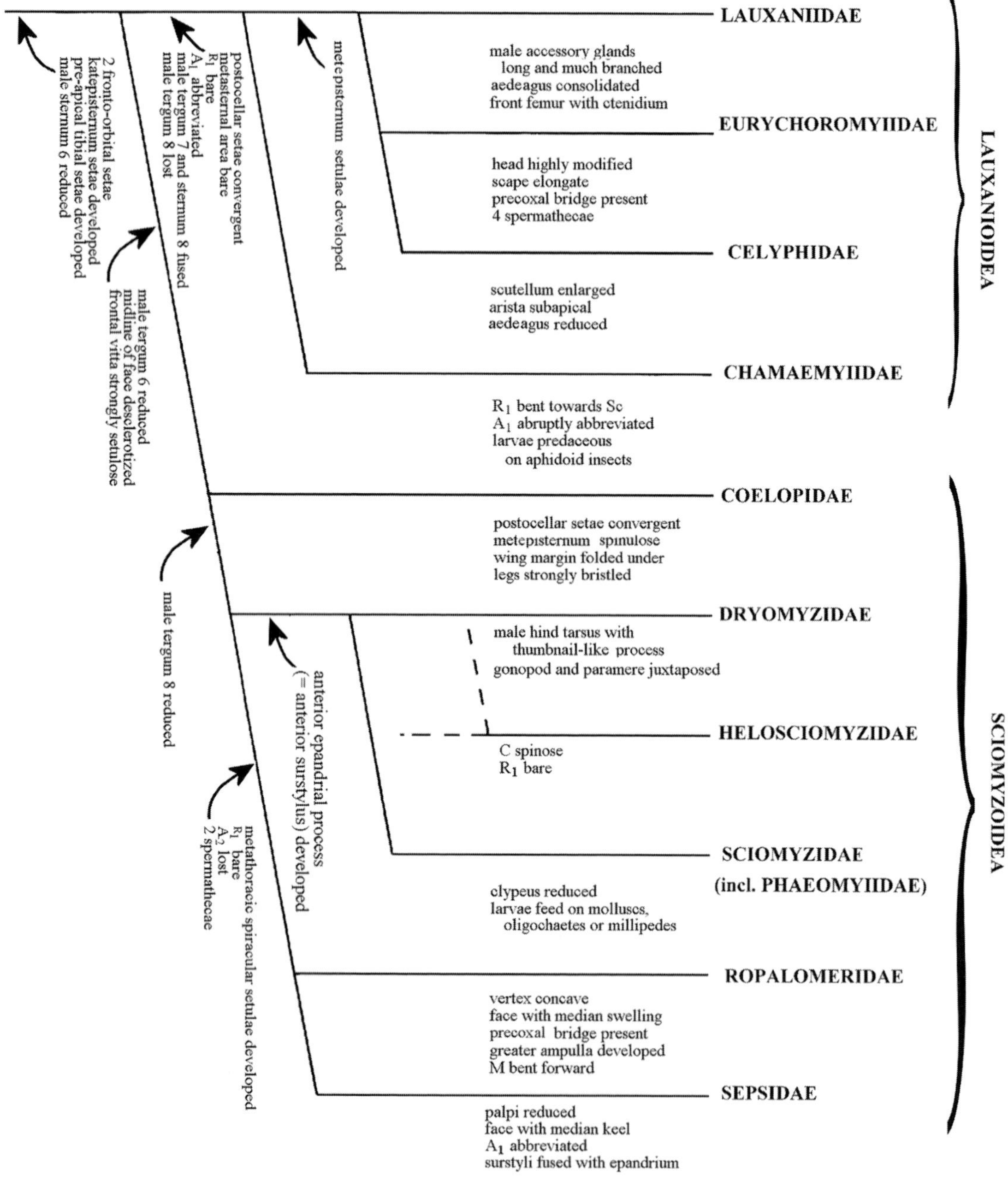

Fig. 15.27 Cladogram of Lauxanioidea and Sciomyzoidea summarizing relationships and apomorphies of subgroups and families. Modified from J. F. McAlpine (1989).

Table 15.13. Comparison of ground-plan characters of Lauxanioidea and Sciomyzoidea. (restricted to those characters that are apomorphic with respect to ground plan of Acalyptratae)

Character	Lauxanioidea	Sciomyzoidea
1. Fronto-orbital plate reduced anteriorly	+	+
2. Reclinate fronto-orbital setae reduced to 2	+	+
3. Inclinate frontal setae absent	+	+
4. Lunule bare	+	+
5. Prosternum haired	+	+
6. Katepisternal seta present	+	+
7. Laterotergum bare	+	+
8. Pre-apical dorsal tibial seta present	+	+
9. Sternum 6 of male reduced	+	+
10. Midfrontal vitta densely and strongly setulose	–	+
11. Face weakly sclerotized along vertical midline	–	+
12. Tergum 6 of male reduced	–	+
13. Postocellar setae convergent	+	–
14. Pro-episternum bare on disc	+	–
15. Metasternal area bare	+	–
16. Meron bare	+	–
17. R_1 bare	+	–
18. CuA_2 recurrent	+	–
19. A_1 abbreviated, not reaching wing margin	+	–
20. Abdominal setae strong	+	–
21. Tergum 7 of male fused with sternum 8	+	–
22. Syntergosterna 7 + 8 of male secondarily nearly symmetric	+	–
23. Tergum 8 of male atrophied	+	–
24. Gonopod and paramere more or less fused at bases	+	–

Note: +, condition present; –, condition not as described. From J. F. McAlpine (1989).

Note that a separate subfamily was newly proposed three separate times for the morphologically apomorphic *Tetanura* – by Hendel (1924), Séguy (1934), and Sack (1939). Surprisingly each time was without reference to the other proposals! Also, as stated by Hackman & Väisänen (1985),

> Tuomikoski (unpublished) even suggested that *Tetanura pallidiventris* should be placed in a [apparently new] subfamily of its own (Tetanurinae) and that it could be the sister group of the other Sciomyzidae.

Also, note that Fallén (1820a) described his new genus and species, *Tetanura pallidiventris* in his group Opomyzides, with *Colobaea bifasciella*, not in his Sciomyzides (1820b). Schiner (1862) placed *Tetanura* in his Tanypezinae, not in his Sciomyzinae. The Tetanurinae has not been recognized by authors after Sack (1939). Knutson (1970a) presented the biology and morphology of immature stages of the terrestrial parasitoid *T. pallidiventris*, commented on its systematic position, and noted "it seems advisable to retain *Tetanura* in the Sciomyzini for the present." In the cladogram of Marinoni & Mathis (2000), *Tetanura* appears in a polytomy with *Colobaea* near the middle of the Sciomyzini clade and essentially in the same position in the cladogram of Barker *et al.* (2004).

It is striking that seven authors have separately proposed a new tribe, or even family, for *Sepedon* and related genera. The proposals were reviewed by Knutson & Orth (2001). This historically is the most concerted agreement on the establishment of a suprageneric category in Sciomyzidae, other than the more recent widely held recognition of Phaeomyiidae, Salticellinae, and Sciomyzinae with two tribes. We believe that the status of *Sepedon* and the related

Table 15.14. Comparison of recent suprageneric classifications of Sciomyzidae and closely related families

Verbeke (1950, 1961, 1963)	Steyskal (1965a)	Hennig (1965)	Griffiths (1972)	Barnes (1979d, 1981)	J. F. McAlpine (1989)	Marinoni & Mathis (2000), Barker *et al.* (2004)
← **SCIOMYZIDAE** →						
Phaeomyiina (P)[1] [new subtribe of Sciomyzini]	Phaeomyiinae n.sf.	Phaeomyiinae	**PHAEOMYIIDAE**	Phaeomyiinae (Sciomyzidae)	**PHAEOMYIIDAE**	**PHAEOMYIIDAE**
not treated	Heloscio- myzinae (A, SA) n.sf.[3]	Helosciomyzinae[4]	**HELOSCIO-** **MYZIDAE**[5] including Huttonininae	**HELOSCIO-** **MYZIDAE**[6]	**HELOSCIO-** **MYZIDAE**	**HELOSCIO-** **MYZIDAE** including Huttonininae
			← **SCIOMYZIDAE** →			
not treated	Huttonininae (SA) n.sf.	Huttonininae		Huttonininae Huttoninini n.t. Prosochaetini n.t.	Huttonininae[8]	
Sciomyzinae (WW)	Sciomyzinae Sciomyzini	Sciomyzinae *sens.lat.*	Sciomyzinae	Sciomyzinae Sciomyzini	Sciomyzinae	
Sciomyzini (WW-SA)[2] Oidematopsini (N)		Subfamily Grp. I. Sciomyzinae	not treated		not treated	Sciomyzinae Sciomyzini
Renocerinae (H) n.sf. Tetanocerinae (WW) *Thecomyia* (NT) *Tetanoptera* (AF) Sepedoninae (WW-SA) n.sf. →	→ Tetanocerini ←	Subfamily Grp. II IIA Tetanocerinae ? Neolimniinae (SA) n.sf. IIB Renocerinae Sepedoninae	not treated	Tetanocerini[7]	not treated	Tetanocerini[9]
Salticellinae (P, AF)	Salticellinae	Salticellinae	Salticellinae	Salticellinae	Salticellinae	Salticellinae

Table 15.14. (*cont.*)

[1]Including ***Pelidnoptera*** Rondani and ***Akebono*** Sueyoshi.
[2]Including subtribes "Phaeomyiina," "Sciomyzina," and "Ctenulina (?)."
[3]Including *Helosciomyzya* Hendel, *Xenosciomyza* Tonnoir & Malloch, and *Polytocus* Lamb.
[4]Including the three genera, above, included by Steyskal (1965a), plus ***Prosochaeta*** Malloch.
[5]Including the four genera included by Steyskal (1965a) and Hennig (1965), plus ***Huttonina*** Tonnoir & Malloch.
[6]Including the genera included by Steyskal (1965a) and Hennig (1965), plus *Sciogriphoneura* Malloch, *Cobergius* Barnes, *Dasysciomyza* Barnes, *Napaeosciomyza* Barnes, *Neosciomyza* Barnes, and *Scordalus* Barnes.
[7]Including ***Neolimnia*** Tonnoir & Malloch and ***Eulimnia*** Tonnoir & Malloch.
[8]Raised to family status by D. K. McAlpine (1991).
[9]Including ***Neolimnia*** but not ***Eulimnia***.

Notes: The abbreviations n.sf., new subfamily and n.t., new tribe in the table are for the first use of the noted taxonomic level. In some cases these are more precisely "new status", as the group name had been used previously. Hennig (1973) included the "Pelidnopterinae (Phaeomyiinae), Huttonininae, Helosciomyzinae, Salticellinae, and Sciomyzinae" in the Sciomyzidae, without reference to categories below the subfamily level.

genera *Sepedoninus, Sepedonella, Sepedomerus, Sepedonea*, and *Thecomyia* should be re-examined. They appear as a monophyletic group at the terminus of the cladograms, based on eight apomorphic features, and are the most highly resolved lineage in those analyses. Also, many other apomorphic characters in all or many species argue for re-study (Knutson & Vala, unpublished data). A few of the more important of these are: (1) lack of ptilinal fissure in all genera, unique among the Schizophora, (2) hooks on labellum of all genera except *Thecomyia*, (3) cochleate vesicle (= hydrostatically activated, not compressive, sperm pump) in many Afrotropical and Oriental species of *Sepedon* and *Sepedoninus*, also unique among the Schizophora, (4) highly unusual *corpora allata*, at least in *Sepedon sphegea*, (5) setae on posterior margin of metathoracic spiracle in *Sepedonea*, (6) loss of palpi in *Thecomyia*, and (7) obligate oligochaete feeding by *Sepedonella nana* and *Sepedon knutsoni*, and facultative feeding on oligochaetes by *Sepedon ruficeps*. Since only character 1 is known to be shared by all of the *Sepedon* group of genera, but some poorly studied features, e.g., *corpora allata* may occur in all, the methods of Hennigian phylogenetic systematics do not allow the non-shared characters to be used to establish a lineage. However, when such an unusual set of highly apomorphic features occurs, although even sporadically, in a lineage that has a large suite of other apomorphic features, these features – while not meeting phylogenetic systematics critera – may indicate an evolutionary lineage that is distinct from the other groups of species in which *none* of these features occurs.

Much of the recent suprageneric work has concerned the rank given to *Pelidnoptera* (Phaeomyiidae), *Huttonina* + *Prosochaeta* (Huttonininae), and the tribes within Sciomyzinae (*sensu* Steyskal 1965a). Griffiths (1972) characterized the Sciomyzidae by four adult characters: (1) clypeus separate from epistoma and not visible in profile when the proboscis is withdrawn, (2) two spermathecae, (3) tergum 6 much reduced in males, and (4) terga 7 and 8 vestiges lost in males, and two larval features "larvae malacophagous, with serrate ventral arch (of labial origin) below mouth hooks." These are apotypic in comparison to the ground plan of the Muscoidea and his prefamily Sciomyzoinea (= Sciomyzoidea plus Cremifaniidae and Megamerinidae). Rozkošný (1984b) pointed out that "some of these characters need further study (e.g. character 1) or are problematic to a certain extent" (e.g., character 4 and malacophagous habits of the larvae). Griffiths's (1972) analysis was based on examination of only eight species: four genera of Sciomyzini and two genera of Tetanocerini, and the descriptions and figures available at that time. He erected the Phaeomyiidae on the basis of five adult characters: (1) mid and hind tibiae with median setae, (2) male tergum 6 asymmetrically reduced, (3) male left spiracles 6 and 7 lying within the sterna, (4) both anterior and posterior surstyli present, the posterior furcate, and (5) aedeagus short, with small pubescent lobes. These characters are apomorphic in comparison to the Muscoidea and Sciomyzoinea. Rozkošný (1984b) noted that only character 1 (mid and hind tibiae with several median setae) is indisputable. Unfortunately, not many of Rozkošný's evaluations of Griffiths's characters of Sciomyzidae and Phaeomyiidae were followed up in the subsequent studies. J. F. McAlpine (1989) noted that the affinities of the "Phaeomyiinae" are uncertain, but that if the larvae were found to be malacophagous that would support a "closer relationship" with Sciomyzidae, and a finding otherwise would exclude them from the Sciomyzidae. He did not refer to the presence or absence of the ventral arch. Subsequently, Bailey (1989) detailed the life cycle of *Pelidnoptera nigripennis*, showing the larvae to be parasitoids of iuliid millipedes, and Vala *et al.* (1990) showed that the serrate ventral arch is absent, described other differences between *Pelidnoptera* and Sciomyzidae in both adult and immature stages, and recognized the family Phaeomyiidae. Marinoni & Mathis (2000) stated that "larval feeding behavior: Plesiomorphic feeding on Diplopoda; Apomorphic feeding .. on .. Mollusca" (without reference to the ventral arch) was the only character in their cladistic analysis that "establishes the monophyly of Salticellinae + Sciomyzinae." As noted above, true but highly apomorphic sciomyzids, *Sepedonella nana* (Vala *et al.* 2000b) and *Sepedon knutsoni* (Vala *et al.* 2002), have been discovered not to feed on Mollusca but to be specific predators of aquatic oligochaete worms. But they have a serrate ventral arch. The lineage to which *S. nana* and *S. knutsoni* belong (*Sepedon* and related genera) was found in the cladistic analyses to be monophyletic and the most corroborated lineage in the analyses. One of our reviewers (R. Meier, personal communication, 2000) noted "there is no obvious reason why feeding on Diplopoda should be plesiomorphic and feeding on molluscs apomorphic. It could be the other way around; after all there is no outgroup that could be used to polarize the character."

Barnes's (1979d, 1981) tabulations and analyses of 25 and 20 characters (respectively), denoted as apomorphic or plesiomorphic, of adults distributed in entities of the Sciomyzoidea included extensive comparison of the subfamilies of Sciomyzidae (Tables 15.12, 15.15). Barnes

Table 15.15. Distribution of some characters in ***Prosochaeta*** and subfamilies of Sciomyzidae *sensu* Steyskal (1965a)

Characters	Prosochaeta	Huttonininae	Helcomyzinae	Phaeomyiinae	Salticellinae	Sciomyzinae
1. Arista pre-apical (+); not pre-apical (-)	■	□	□	□	□	▥
2. Clypeus small (+); large (-)	□	□	□	■	■	■
3. Postverticals parallel or convergent (+); divergent (-)	■	■	□	□	□	□
4. Pro-episternal seta absent (+); present (-)	□	□	□	□	■	▥
5. Costa spinose (+); not spinose (-)	□	□	■	□	□	□
6. Ventro-apical angle anal cell A_1 + CuA_2 distinctly obtuse (+); otherwise (-)	■	■	□	□	□	□
7. Anal vein abbreviated (+); complete (-)	■	■	□	□	□	▥
8. Suture on abdominal terga 1 + 2 indistinct (+); distinct (-)	■	■	□	□	□	□
9. Abdominal spiracles 2-5 in terga (+); in membrane (-)	■	□	□	□	□	□
10. ♂ 6th tergum fused to epandrium (+); free (-)	□	■	□	□	□	▥
11. ♂ 7th left spiracle in 7th sternum (+); in membrane (-)	■	■	■	■	□	▥
12. Anteroventral corner epandrium extended (+); not extended (-)	□	□	▥	□	□	□
13. Surstyli fused to epandrium (+); free (-)	□	■	□	□	□	□
14. Aedeagus bilobed, membranous, scaled (+); otherwise (-)	■	■	□	□	□	□
15. ♂ 4th and 5th terga membranous (+); sclerotized (-)	■	□	□	□	□	□
16. ♂ 6th abdominal tergum and sternum fused (+); separate (-)	■	□	□	□	□	□
17. ♂ 7th abdominal tergum and sternum fused (+); separate (-)	■	■	□	□	□	□
18. ♂ 6th abdominal spiracle in sclerite (+); in membrane (-)	■	□	□	□	□	▥
19. ♀ 7th abdominal spiracle in sclerite (+); in membrane (-)	■	■	□	□	□	▥
20. Number of spermathecae	4 (■)	3	3	3	4 (■)	2 (■)

Note: Black rectangles, characters interpreted as apomorphic groundplan conditions; white rectangles, characters interpreted as plesiomorphic groundplan conditions; crosshatched rectangles, apomorphic character present in some but not all members of a taxon. From Barnes (1979d).

(1979d) transferred *Prosochaeta* from Griffiths's (1972) Helosciomyzidae and he placed it in his new tribe Prosochaetini, with *Huttonina* (Huttoninini), in the Huttonininae of the Sciomyzidae. Barnes (1981) compared 20 morphological features, which he characterized as plesiomorphic or apomorphic, of adults of *Prosochaeta* with the subfamilies of Sciomyzidae as delimited by Steyskal (1965a). As with Steyskal (1965a), he retained *Helosciomyza*, *Xenosciomyza*, and *Polytocus* in the Helosciomyzinae. Barnes (1981) revised the latter three genera along with descriptions of five new genera and transfer of *Sciogriphoneura* from the Dryomyzidae, regarded the group as the family Helosciomyzidae, and redescribed the family. He concluded that the Helosciomyzidae are more closely related to the Helcomyzidae and Dryomyzidae than to the Sciomyzidae or other entities of Sciomyzoidea.

The cladistic analysis of 50 of the 61 genera of Sciomyzidae (not including Huttonininae) by Marinoni & Mathis (2000) utilized 36 morphological features of adults, of which 17 are male postabdomen features. Only the type species as exemplars of each genus were studied. As noted above the data were re-analyzed by Barker *et al.* (2004). Both cladistic studies confirmed the monophyly of the family (Salticellinae + Sciomyzinae) and the monophyly of the two tribes, Sciomyzini and Tetanocerini, in the Sciomyzinae.

16 • Zoogeography

L'univers est une espèce de livre dont on n'a lu que la première page quand on n'a vu que son pays.
[The universe is a kind of book of which, if you read only the first page, it is like knowing only your country].

Fougeret de Monbron (1750).

16.1 DISTRIBUTION

The geographical distribution, by zoogeographical regions, of all species of Sciomyzidae is noted in Table 21.1. Rozkošný (1995) presented a checklist of the 505 described species in the 58 genera (including Huttonininae) known at that time, recorded their distribution by region, and summarized statistics on genera by region.

An updated list of the genera showing the numbers of species in the regions is given in Table 16.1. The current totals are 539 species in 61 genera. This table shows the total number of species per genus ("Total Known Species" column) and the number of species per genus per region. Most species are restricted to one region. Some are slightly adventive from one region to one or two neighboring regions. Some are broadly shared between regions, especially the 28 Holarctic species and some Oriental species. Numbers of adventive species are shown in parentheses. Thus the total number of species per genus per region for each regional box is the sum when two numbers appear in such a box. Notes 1–9 refer to notes in Rozkošný (1995); notes 10–30 are by us on species added, deleted (synonyms), or transferred to different genera.

Rozkošný (1995) commented on extent of endemicity and taxa shared between regions. He distinguished between a species shared between regions, that is, broadly distributed in more than one region, and a species of a region that has penetrated, usually only slightly but in some cases rather deeply, into an adjacent region. That distinction is maintained in the present work. Rozkošný (1995) concluded that the family is a cool-adapted group that probably arose in a north temperate region. About 66% of the species and 82% of the genera are distributed in the Nearctic and Palearctic regions, with 201 species in 23 genera in the Nearctic and 176 species in 27 genera in the Palearctic. The 28 Holarctic species in six genera are included in the totals for both the Nearctic and Palearctic regions. The numbers of species and genera in each region are shown in Tables 16.1, 16.2 and in Figs 16.1, 16.2 (showing numbers shared between regions and numbers of endemics).

Our delimitations of the major zoogeographic regions are generally outlined in Fig. 16.1 and characterized in more detail as follows. Certainly, there are major transitional areas between most regions, as shown by a–f in Fig. 16.3. We follow Griffiths's (1990) distinction of theapproximate boundary between the Nearctic and Neotropical regions (Fig. 16.4) and his delineation of the Nearctic region,

> The whole of Canada, Alaska and the contiguous United States (including all Florida and the Florida keys) are included, but Hawaii and Puerto Rico excluded. The Bahamas and the rest of the West Indies are also excluded, but Bermuda is included. The boundary between the Nearctic and Neotropical Regions in Mexico is constituted near the coast (in northern Mexico) by the boundary between desert or mesquite grassland (considered Nearctic) and tropical (thorn or deciduous) forest, and in the interior by the boundary between pine-oak forest (Nearctic) and tropical (evergreen or deciduous) forest or scrub at lower elevation. All land East and South of the Isthmus of Tehuantepec is considered to belong to the Neotropical Region . . . In the Beringian area the present territorial boundary between the United States and the Soviet Union may be accepted as the boundary between the Nearctic and Palaearctic Regions. Greenland is included in the Nearctic Region, but Iceland, Jan Mayen and Spitzbergen excluded.

Table 16.1. Numbers of species in genera of Sciomyzidae in zoogeographical regions

	Genus	Region									Total Known Species	Note
		H	N	P	NT	AF	O	A	OC	SA		
1.	*Anticheta*		8	7							**15**	10
2.	*Apteromicra*						1				**1**	11
3.	*Atrichomelina*		1		(1)						**1**	1
4.	*Calliscia*				1						**1**	
5.	*Chasmacryptum*			1							**1**	
6.	*Colobaea*		3	10		1	(2)				**14**	12
7.	*Coremacera*			10							**10**	13
8.	*Dichetophora*			5			1	6			**12**	14
9.	*Dictya*		33(4)	1	9(2)						**43**	2, 15
10.	*Dictyacium*		2								**2**	
11.	*Dictyodes*				2						**2**	
12.	*Ditaeniella*		2	1	1(1)	1	(1)				**5**	3, 16, 30
13.	*Ectinocera*			1							**1**	
14.	*Elgiva*	2	3	2							**7**	
15.	*Ethiolimnia*					7					**7**	17
16.	*Eulimnia*									2	**2**	
17.	*Euthycera*		2	19							**21**	18
18.	*Euthycerina*				2						**2**	
19.	*Eutrichomelina*				2						**2**	19
20.	*Guatemalia*				2						**2**	
21.	*Hedria*		1								**1**	
22.	*Hoplodictya*		5		(2)						**5**	4
23.	*Huttonina*									8	**8**	
24.	*Hydromya*			1		(1)	(1)				**1**	
25.	*Ilione*			8			(1)				**8**	5
26.	*Limnia*		17	5							**22**	

Table 16.1. (*cont.*)

	Genus	Region									Total Known Species	Note
		H	N	P	NT	AF	O	A	OC	SA		
27.	*Neodictya*			1							**1**	
28.	*Neolimnia*									14	**14**	
29.	*Neuzina*				1						**1**	20
30.	*Oidematops*		1								**1**	
31.	*Oligolimnia*			1							**1**	
32.	*Parectinocera*				3						**3**	
33.	*Perilimnia*				2						**2**	
34.	*Pherbecta*		1								**1**	
35.	*Pherbellia*	8	29(2)	42(1)	6(1)	3	2(4)	2	1		**94**	6, 21
36.	*Pherbina*			4							**4**	
37.	*Poecilographa*		1								**1**	
38.	*Prosochaeta*									1	**1**	
39.	*Protodictya*				8						**8**	
40.	*Psacadina*			5							**5**	
41.	*Pseudomelina*				1						**1**	
42.	*Pteromicra*	3	11	3		1	1				**19**	22
43.	*Renocera*	1	3	2							**6**	
44.	*Salticella*			1		1					**2**	
45.	*Sciomyza*	2	2	3							**7**	23
46.	*Sepedomerus*		(1)		3						**3**	7
47.	*Sepedon*		22	4(5)	(1)	43	12(1)	2(2)	(3)		**80**	8, 24
48.	*Sepedonea*				13						**13**	25
49.	*Sepedonella*					4					**4**	
50.	*Sepedoninus*					2	1				**3**	26
51.	*Shannonia*				2						**2**	
52.	*Steyskalina*						1				**1**	27

Table 16.1. (*cont.*)

	Genus	Region									Total Known Species	Note
		H	N	P	NT	AF	O	A	OC	SA		
53.	*Tetanocera*	11	18	8(1)	(1)		2				**39**	28
54.	*Tetanoceroides*				7						**7**	
55.	*Tetanoptera*					1					**1**	
56.	*Tetanura*			2							**2**	
57.	*Teutoniomyia*		(1)		2						**2**	
58.	*Thecomyia*				12						**12**	29
59.	*Trypetolimnia*			1							**1**	
60.	*Trypetoptera*		1	1							**2**	
61.	*Verbekaria*					1					**1**	
Total species/Region		28	201	176	89	65	30	12	4	25	539	
Total genera/Region		6	23	27	23	11	12	3	2	4		
Endemic genera/Region		3	6	7	14	4	2	0	0	4		
Endemic species/Region		26	156	153	73	61	13	8	1	23		

H, Holarctic; N, Nearctic; P, Palearctic; NT, Neotropical; AF, Afrotropical; O, Oriental; A, Australian; OC, Oceanic; SA, Subantarctic. Species marginally adventive from an adjacent Region indicated by ( ). Species widespread in one – three adjacent Regions counted for each Region where they occur, but only once for total known species in a genus. Holarctic genera and species also included in Nearctic and Palearctic totals. Modified from Rozkošný (1995).

Notes: 1-9 as in Rozkošný (1995); 10-30 highlight major nomenclatorial changes and additions (see Table 21.1).

10. Added ***Anticheta shatalkini*** (P).
11. Added genus ***Apteromicra*** with one species, ***A. parva*** (O).
12. Added ***Colobaea*** n. sp. 1 (AF), and ***C. acuticerca*** (P).
13. Added ***Coremacera scutellata*** (P).
14. Added ***Dichetophora japonica*** (P) and ***D. kumadori*** (P).
15. Added ***Dictya orthi*** (N).
16. Transferred ***Pherbellia trivittata*** (NA) to ***Ditaeniella*** according to Knutson *et al.* (1990).
17. Deleted ***Ethiolimnia capensis*** Schiner 1868 (AF) = ***E. geniculata*** according to Miller (1995).
18. Deleted ***Euthycera nigrescens*** Becker 1907a (P) = ***E. cribrata*** according to Vala (1989a).
    Added ***Euthycera vockerothi*** (P), ***E. atomaria*** (P), ***E. korneyevi*** (P), and ***E. merzi*** (P). In Rozkošný (1995) ***E. mira*** was listed as a Neotropical species; however, the collection site (2664 m, 28 km southwest of El Salto, Durango, Mexico) is in the Nearctic Region.
19. Transferred ***Eutrichomelina*** (NT) to Tetanocerini according to Marinoni & Mathis (2000).
20. Added genus ***Neuzina*** with one species, ***N. diminuta*** (NT).
21. Added ***Pherbellia brevistriata*** (P), ***P. dentata*** (P), ***P. juxtajavana*** (A), and ***P. tricolor*** (P). Deleted ***P. stylifera*** Rozkošný 1982 (P) = ***P. goberti*** according to Stuke (2005).

Table 16.1. (*cont.*)

22. Added ***Pteromicra*** n. sp. 1 (AF).
23. Added ***Sciomyza pulchra*** (P) and ***S. sebezhica*** (P).
24. Added ***Sepedon floridensis*** (N) as valid species (Orth 1986), ***S. spinipes*** (P), and ***S. americana*** (N) as valid species (Elberg *et al.* 2009), ***S. knutsoni*** (AF), ***S. tuckeri*** (AF), ***S. mcphersoni*** (N), ***S. notei*** and ***S. hecate*** (P). Deleted ***S. oriens*** Steyskal 1980 (O + P) = ***S. noteoi*** according to Rozkošný (personal communication, 2010) and *S. fuscipennis nobilis* (Orth 1986) (N) = *S. fuscipennis* according to Elberg *et al.* (2009).
25. Added ***Sepedonea giovana*** (NT).
26. Added ***Sepedoninus*** n. sp. 1 (O).
27. Added genus ***Steyskalina*** with one species, ***S. picta*** (O).
28. Added ***Tetanocera nigrostriata*** (O). Deleted ***T. gracilior*** Stackelberg 1963 (P) = ***T. spirifera*** (herein).
29. Added nine new species described by Marinoni *et al.* (2003). Deleted ***T. trilineata*** Hendel 1932 (NT) = ***T. lateralis*** according to Marinoni *et al.* (2003).
30. Added *Ditaeniella* n. sp. 1 (AF).

Table 16.2. Matrices of genera (a) and species (b) shared between regions

	NT	N	P	AF	O	A	OC	SA	Total instances
(a) Genera shared. The number of genera in P and N includes Holarctic genera.									
NT	23	9	5	2	4	2	2	0	24
N	9	23	14	4	6	2	2	0	37
P	5	14	27	6	9	3	2	0	39
AF	2	4	6	11	6	2	2	0	22
O	4	6	9	6	12	3	2	0	30
A	2	2	3	2	3	3	2	0	14
OC	2	2	2	2	2	2	2	0	12
SA	0	0	0	0	0	0	0	4	0
(b) Species shared. The number of species in P and N includes Holarctic species.									
NT	89	16	0	0	0	0	0	0	16
N	16	201	28	0	1	0	0	0	45
P	0	28	176	2	12	0	0	0	42
AF	0	0	2	65	1	0	0	0	3
O	0	1	12	1	30	3	2	0	17
A	0	0	0	0	3	12	2	0	5
OC	0	0	0	0	2	2	4	0	4
SA	0	0	0	0	0	0	0	25	0

Note: Shaded diagonal rows, numbers of genera and species per region. Total *instances* of sharing for all regions i.e., extent of sharing with other regions, shown in right-hand, vertical column for each region.

We follow K. D. Ghorpadé (personal communication, 2005) in delimiting most of the Oriental region. In the west, this includes the Baluchistan and Afghanistan subregions, the Hindu-Kush-Karakoram, southern Tadjikistan, and southeastern Tibet. The limits extend across India–Nepal–Bhutan–Burma, and continue across the southern part of the People's Republic of China in an arc extending from the northwestern-most

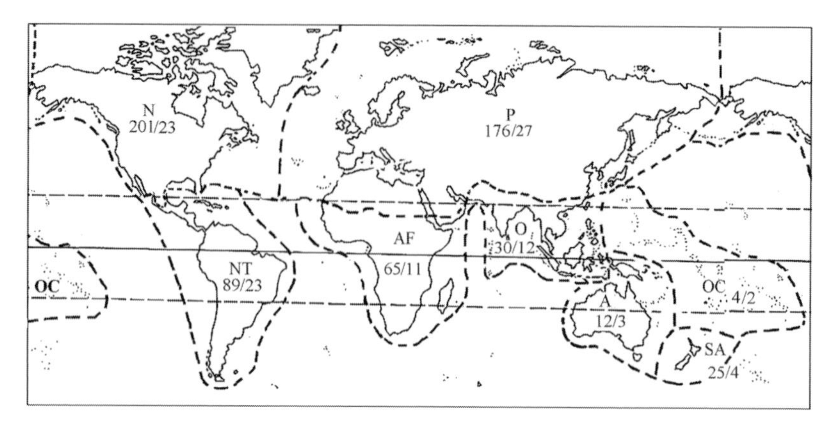

Fig. 16.1 Numbers of species and genera of Sciomyzidae in main zoogeographical regions of the world. A, Australian, AF, Afrotropical, N, Nearctic, NT, Neotropical, O, Oriental, OC, Oceanic, P, Palearctic, SA, Subantarctic. Shared species are included in each region where they occur.

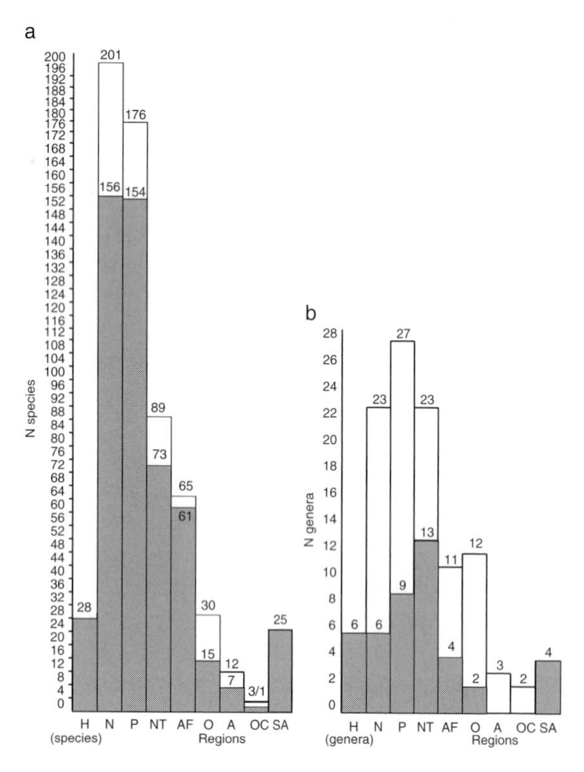

Fig. 16.2 Numbers of species (a) and genera (b) of Sciomyzidae by zoogeographical regions. Numbers above bars are total number of species and genera; numbers above solid portion of bars are endemic species and genera.

corner of Yunnan, along the valley of the lower half of the Yangtze River to north of Shanghai. Included are the Ryukyus and islands to the south enclosed by Weber's Line (Fig. 16.6). This is essentially equivalent to the regional limits as defined in the catalogs of Palearctic and Oriental Diptera, except that it is more inclusive in the northwestern part of the region than are those catalogs. The vast, funnel-shaped area of north-to-south and northeast-to-southwest ranging mountains and valleys crossing the western part of the Palearctic/Oriental contact zone and the broad plain in the eastern part account for extensive interdigitation of the Palearctic and Oriental faunas.

The division between the Oriental and Australian regions has been of particular interest. We recognize the line of Weber (1902) (Fig. 16.6). While some authors, e.g., Müller (1974), regard the Australian region in a broad sense, we consider New Zealand and nearby islands as the Subantarctic region, and the oceanic islands as the Oceanic region. Ghorpadé (personal communication, 2005) does not recognize the Oceanic region, but considers the Oriental region to extend east of Weber's Line, including the Moluccas, New Guinea, and all tropical Pacific Ocean islands.

The demarcation between the Palearctic and Afrotropical regions also has been contested (Fig. 16.7). There are essentially no records of Sciomyzidae between the southernmost and northernmost of the ±18 divisions that have been recognized. We consider the southernmost line as pertinent to the Sciomyzidae.

The distributional patterns of Palearctic Sciomyzidae have been analyzed the most extensively. Verbeke (1964a) recognized five major patterns in the western Palearctic: northcentral, southcentral, western and eastern Mediterranean, and boreal. The distributions of 73 aquatic and semi-aquatic species in 16 genera in 25 areas of the western Palearctic were tabulated by Verbeke & Knutson (1967), and Knutson (1978). Rozkošný (1984b) placed the 83 species known to occur in Fennoscandia and Denmark

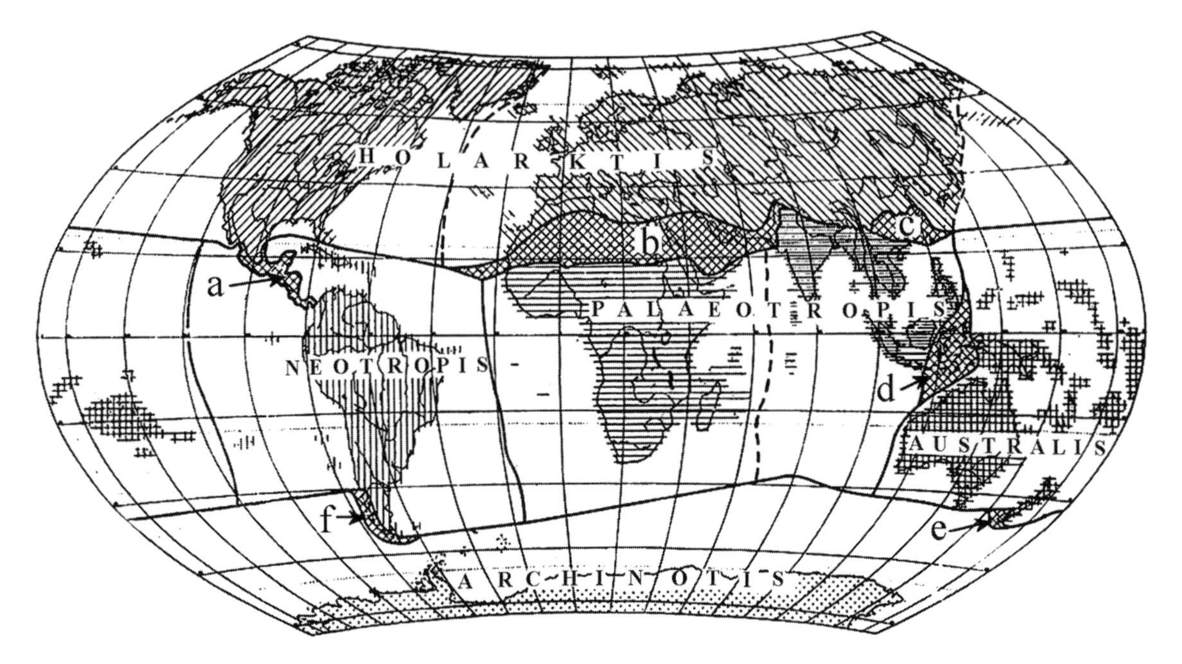

Fig. 16.3 Zoogeographical regions, with transition zones (diagonally crosshatched areas a–f). From Müller (1974).

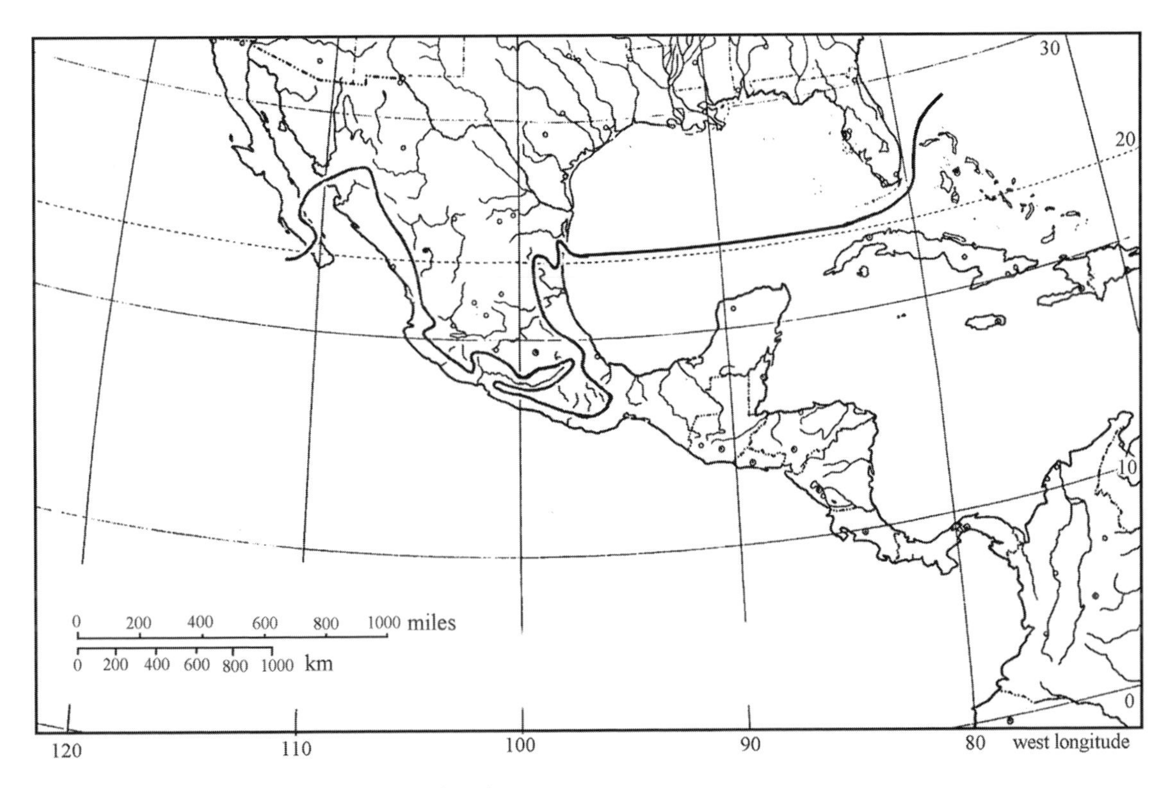

Fig. 16.4 Approximate boundary between Nearctic and Neotropical regions. Modified from Griffiths (1990).

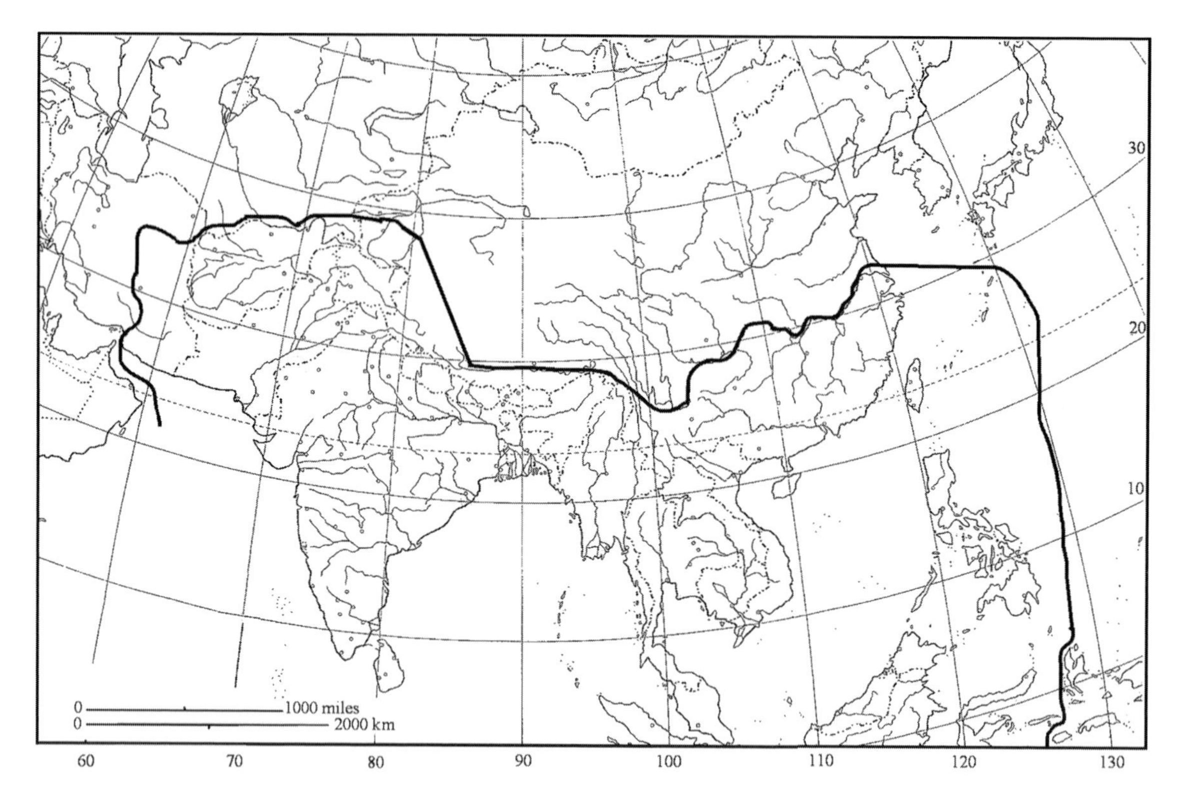

Fig. 16.5 Limits of the Oriental/Palearctic regions. (K. D. Ghorpadé, personal communication, 2005).

in eight zoogeographical groups (Holarctic, Palearctic-Oriental, Palearctic, West Palearctic, Eurasian, Eurosiberian, European, and Boreal) and he commented on their distributions. The 142 Palearctic species occurring west of the Ural Mts. treated by Vala (1989a) were placed in the eight zoogeographical groups recognized by Rozkošný (1984b) plus eight additional groups. The distributions of all European species were tabulated, by country, by Rozkošný & Knutson (2005a, 2005b). The unusual distribution of the plesiomorphic subfamily Salticellinae is of particular interest, with *Salticella stuckenbergi* known from southern Africa, *S. fasciata* from the eastern and southeastern Palearctic, and *Prosalticella succinei* from Baltic amber. *Salticella fasciata* has an essentially Lusitanian distribution in the eastern half of its range, occurring mainly along coasts but inland as far as about 150 km to Zaragoza, Spain on the Ebro River, on the Basento River and Po River in Italy, and on the Rhone River and Seine River in France. It occurs far inland in the eastern part of its range, into the Carpathian Basin, Black Sea and Caspian Sea coasts, and central Turkey (Fig. 16.8). Its distribution in the eastern part of its range might be explained, as for certain Curculionidae with similar distributional patterns, as relict populations representing its former distribution on the margins of the Tethys Ocean (E. Colonnelli, personal communication, 2004). The only records of *S. fasciata* from moderately high elevations are 1000 m at Riffredo, southern Italy (Rivosecchi 1992) and 700 m near Budapest (Knutson, unpublished data).

Only a few species have a distinctly maritime distribution. These are three species of *Dictya* that are predators of small, operculate brackish water snails; *Hoplodictya setosa*, a predator of *Littorina littorea* in seaside situations along the eastern Nearctic coast; and the biologically unknown *Pherbellia mikiana* in marine shoreline situations on eastern Mediterranean coasts. Wagner *et al.* (2008) presented an overview of the biology and diversity of aquatic and semi-aquatic species of continental waters of the world, with a tabulation of the number of such

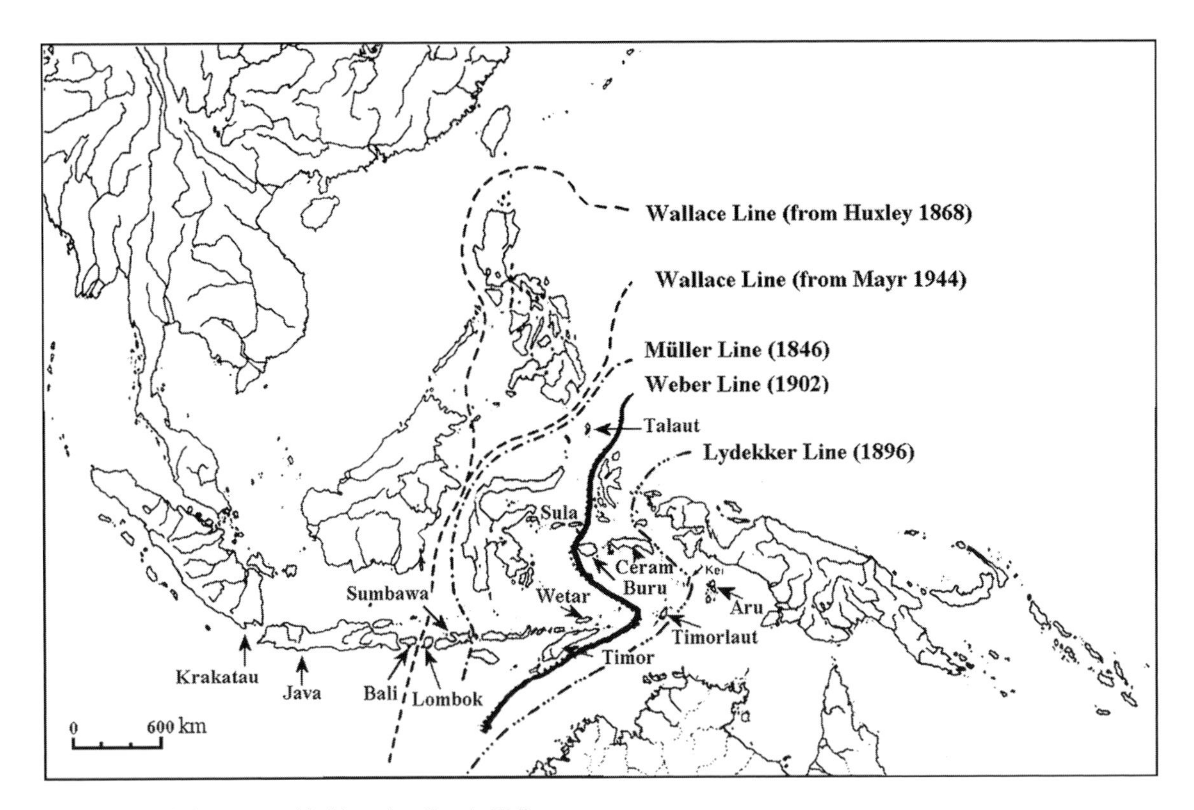

Fig. 16.6 Classical zoogeographical boundary lines in Wallacea. From Müller (1974).

species (154 of 539) by genus (31 of 61) by major zoogeographical regions.

Most Neotropical Sciomyzidae, even those species found in lands bordering both the Atlantic and Pacific Oceans, have a more or less coastal distribution, ranging only up to about 800 km inland (South America being about 2400–48 000 km broad for about half its length). The only relatively broad pathway across the continent (which stretches from about 10° to 65° S. Lat.) appears to be the relatively narrow temperate–tropical transition zone at about 25° to 35° S. Lat., where a few species (e.g., *Perilimnia albifacies, Protodictya guttularis, P. lilloana, Sepedonea guianica, S. lindneri, Tetanoceroides bisetosus*, and *T. mendicus*) are widely distributed, coast-to-coast. Kaczynski *et al.* (1969) commented on the geographical distribution of *Perilimnia* and *Shannonia*, particularly in relation to habitats. Zuska & Berg (1974) analyzed the distribution patterns of *Tetanoceroides*, which is confined to the South Temperate zone of South America. Recently, Ciprandi Pires *et al.* (2008) analyzed the distribution patterns of the 13 species of *Sepedonea* (all Neotropical) using the panbiogeographic method of tracking analysis and compared their results with the previously proposed hypotheses for the evolution of the region, as well as the cladistic phylogeny of the species (Marinoni & Mathis 2006).

Lists of species, some of which include information on habitats and/or distribution maps, are available for Belgium: Denis & Leclercq (1985, including maps of all 67 species); Great Britain: Ball & McLean (1986, including maps and histograms of seasonal distribution of all 64 species); the British Isles and Ireland: Knutson & Stephenson (1970), McLean (1998); Czechoslovakia: Rozkošný (1987b); Czech Republic and Slovakia: Rozkošný (1997b); Switzerland: Merz (1998); Germany: Rozkošný (1999a); and Spain, Portugal, and Andorra: Carles-Tolrá (2003) and other papers cited therein. Important, detailed distributional information is included in publications by Chandler (1972) for Ireland; Kassebeer (2001b) for Iceland; Kasseebeer (2000) for Schleswig-Holstein and Hamburg; Kassebeer (2001a) for the Berlin area; Leclercq & Schacht

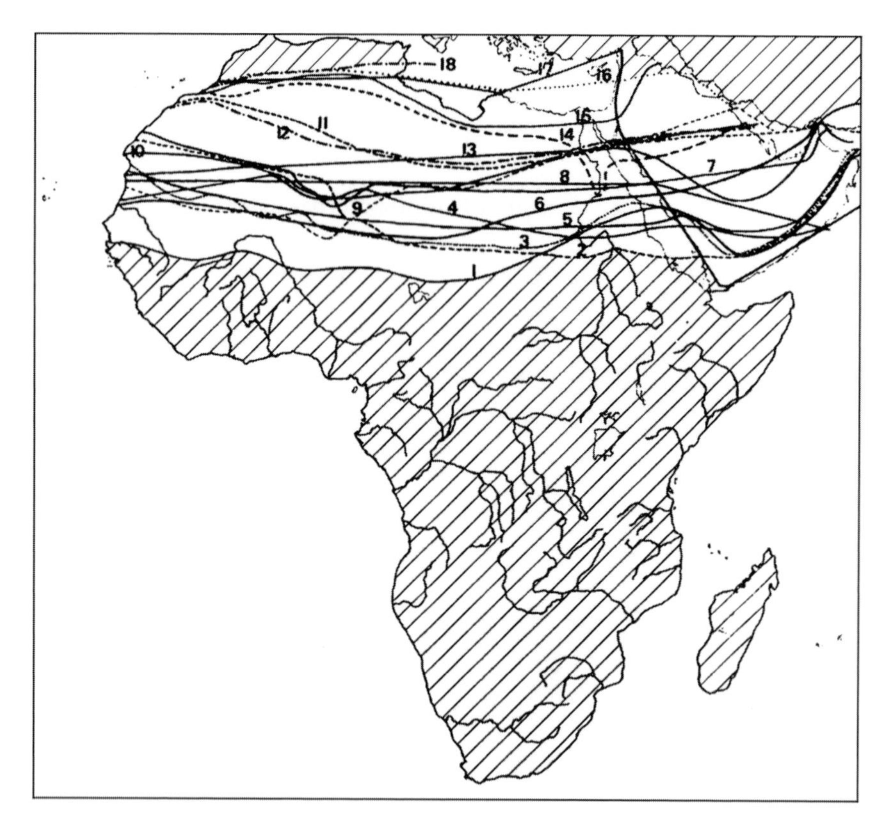

Fig. 16.7 Boundaries between Palearctic and Afrotropical regions as proposed by various authors. From Müller (1974).

(1986, 1987a) for Turkey; Kassebeer (1999a) for North Africa; Vala & Ghamizi (1991) for Morocco; Merz & Rozkošný (1995b) and Rozkošný & Knutson (2006) for Central Asia; Rozkošný (1985) for Cyprus; Rozkošný (1969) and Rozkošný & Knutson (1972) for Afghanistan; Elberg (1978) for Mongolia; Rozkošný & Kozánek (1989) for Korea; Fisher & Orth (1975) for Oregon; and Mello & Bredt (1978a) for D. F. Brazil. Knutson & Ghorpadé (2004) tabulated the distribution of 19 species in five genera known from Burma to northern Australia (Table 16.6).

The distributions of many species have been mapped and/or recorded in many publications on biology and in regional taxonomic studies. These include: Palearctic (Rozkošný 1987a), Fennoscandia (Rozkošný 1984b), Mediterranean Europe (Vala 1989a, including maps of the Palearctic distribution of 25 species), Italy (Rivosecchi 1992, including maps of all species in Italy), Japan (Sueyoshi 2001, including maps of 22 species), Ohio (Foote & Keiper 2004, including maps of all 72 species), California (Fisher & Orth 1983, including maps of all 49 species), Idaho (Foote 1961b), Alaska (Foote *et al.* 1999, including maps of 58 species), and Central America (Marinoni & Knutson in press).

Collection and alpha-level taxonomy still require work, as indicated by recent discoveries, e.g., of a new species of *Colobaea* (Carles-Tolrá, 2008) in the well-known western Palearctic, and a new species of *Dictya* in the mid-Atlantic states of the USA (Mathis *et al.* 2010). The quality of the database is also indicated by the recent discovery of new species of three Holarctic genera (*Colobaea, Ditaeniella* and *Pteromicra*) from sub-Saharan Africa (Knutson, Kirk-Spriggs & Deeming, unpublished data), and major range extensions, e.g., *Tetanura pallidiventris* in Japan (Sueyoshi 2001), are being reported.

Since there is a relatively low level of host/prey specificity in most species, with no evidence of co-evolution, the zoogeography of the food resources sheds little light on the zoogeography of Sciomyzidae.

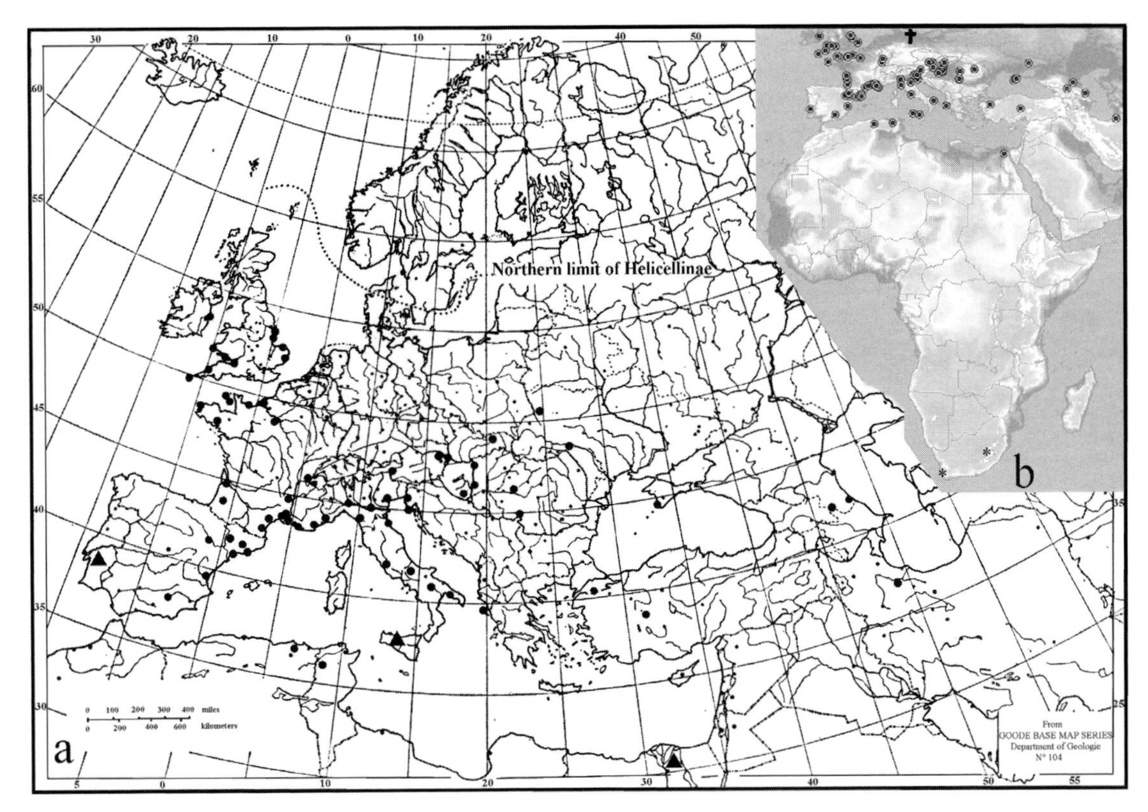

Fig. 16.8 (a) Distribution of *Salticella fasciata*: black dot, precise locality; black triangle, general locality (modified from Knutson *et al.* 1970). (b) Distribution of Salticellinae: dot, *Salticella fasciata*; cross, *Prosalticella succini*; asterisk, *S. stuckenbergi* (modified from Knutson *et al.* 1970 by A. Kirk-Spriggs, personal communication, 2010).

16.2 FAUNAL RELATIONSHIPS AND DISPERSAL ROUTES

If the basic idea is too complicated to fit on a T-shirt it's probably wrong.

L. Lederman (quoted in Ferris 1997).

The plesiotypic subfamily Salticellinae is known from only three species, one each from southern Africa, the south-western Palearctic, and Baltic amber. The Huttonininae (nine species in *Huttonina* and *Prosochaeta*) are restricted to the Subantarctic region. Both tribes of Sciomyzinae, Sciomyzini and Tetanocerini, are found in all regions, except the Sciomyzini are absent from the Subantarctic. The dominant genera *Pherbellia* (Sciomyzini, 95 species) and *Sepedon* (Tetanocerini, 80 species) are found in all regions except the Subantarctic. *Sepedon* is essentially absent from the Neotropical region, but the closely related genera *Sepedonea* and *Sepedomerus* are present there.

Although no species is cosmopolitan, there are common, eurytopic, widespread species and rare, stenotopic, geographically restricted species. Both types include species of aquatic and terrestrial life styles. The 28 Holarctic species are primarily common and abundant species, including both aquatic predators (e.g., *Tetanocera ferruginea*) and terrestrial parasitoids (e.g., *Pherbellia schoenherri*). Post-Pleistocene speciation and radiation may explain some unusual patterns of distribution of, for example, the biologically and morphologically very similar species of the aquatic predator *Dictya* (one Palearctic, seven Neotropical, and 35 Nearctic species) and the morphologically more diverse, but primarily aquatic, predaceous *Sepedon* (42 of 80 species in the Afrotropical region). Other than the five genera mentioned in Section 15.4.2, there are no cladistic analyses of the genera of Sciomyzidae on which vicariant zoogeographical considerations could be based. Thus, dispersal relationships predominate in zoogeographic analyses. Some genera and many species are known from only one or a few specimens and localities.

Relationships between major zoogeographical regions and suspected major dispersal routes are shown in

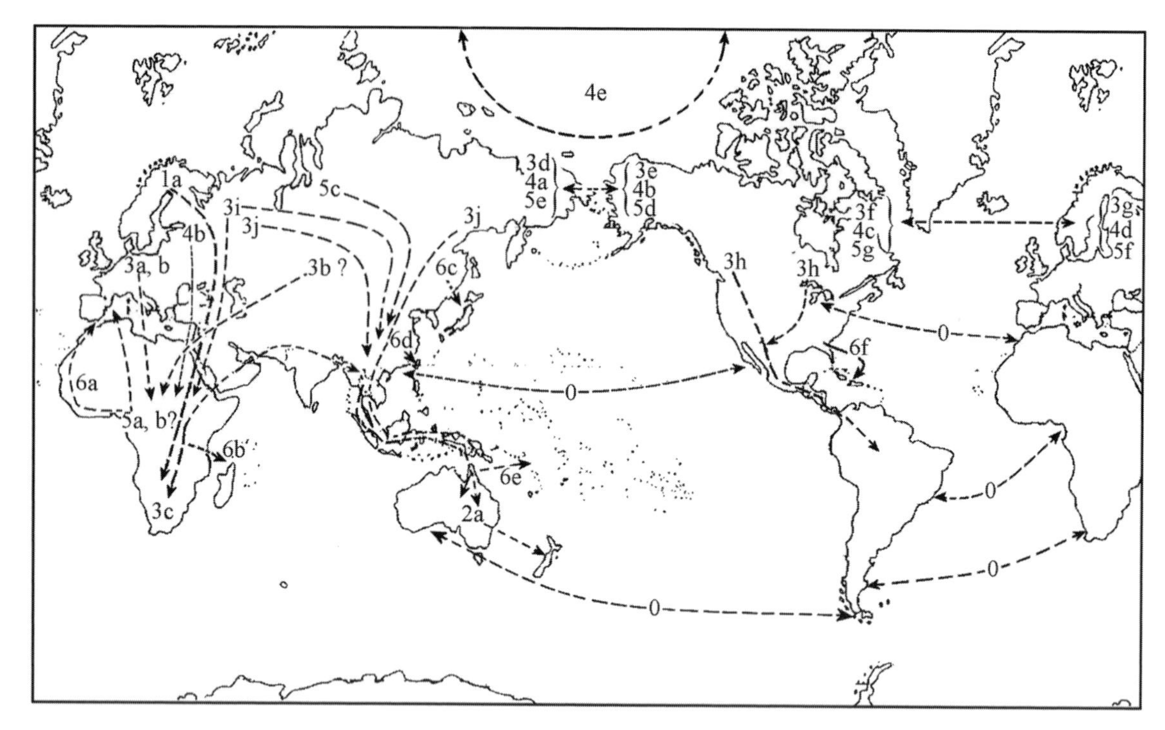

Fig. 16.9 Postulated major dispersal routes of Sciomyzidae.

Fig. 16.9. The relationships and routes, by taxonomic level, are discussed below and indicated by the numbers and letters in Fig. 16.9. Arrows indicate the directions of dispersal. Limited penetrations from one region into another are not included in this portrayal.

There are no indications of dispersal between the Afrotropical and Nearctic, Afrotropical and Neotropical (except proliferation of genera of the *Sepedon* lineage in both), Nearctic and Oriental regions, or between the south temperate areas of Africa, South America, and Australia/New Zealand.

1. *Subfamily Level.* The subfamily Sciomyzinae is found in all Regions, the Huttonininae is restricted to the Subantarctic.
   a. Salticellinae. There is a major disjunction between the southwestern Palearctic *Salticella fasciata* plus the Baltic amber *Prosalticella succinei* and *S. stuckenbergi*, the latter known only from southern Africa.
2. *Tribe Level.* The Sciomyzini is found in all regions except the Subantarctic. Connection between the Subantarctic and Australian regions is indicated by the presence in New Zealand of the typically Tetanocerini, endemic genus *Neolimnia*; otherwise, the New Zealand fauna is restricted to the endemic Huttonininae.
3. *Genus Level.*
   a. The presence of two undescribed species of the Holarctic genus *Pteromicra* in Nigeria and Tanzania may indicate dispersal from the Palearctic across North Africa during pluvial periods (3a, b in Fig. 16.9).
   b. The occurrence of an undescribed species of the Holarctic genus *Colobaea* in Nigeria, the Gambia, and Ethiopia, closely related to *C. punctata* which ranges across Europe to southwestern Iran and southern Pakistan, also may indicate a dispersal route as in Fig. 16.9 (3b?).
   c. The occurrence of many species of *Sepedon* subgenus *Parasepedon* throughout the Afrotropical and Oriental regions, with a few species extending to Australia and Oceania, indicates this dispersal route. The relatedness of these species is essentially confirmed by the presence in *Parasepedon* of the

highly apomorphic cochleate vesicle type of sperm pump. This structure is not found in *Sepedon* species (or in any other Diptera) in other regions. This faunal connection is also supported by the presence of an undescribed species of *Sepedoninus* in Borneo and Sulawesi, a genus otherwise known from two species from central Africa. There are no indications of dispersal through the Rift Valley except the presence of the widespread Afrotropical *Sepedon ruficeps* in Egypt and possibly the eastern Palearctic to southern African distribution of *Salticella*. This is strongly supported by the absence of Afrotropical species in Israel, whose sciomyzid fauna is well known.

d/e, f/g. The most extensive connections are between the Nearctic and Palearctic, which share 28 species in six genera that range across both continents and probably dispersed eastward and/or westward across either or both the Atlantic and Beringia (Table 16.7).

h. Two genera that are very speciose in the Nearctic, *Dictya* with 35 Nearctic species and *Pherbellia* with 31 Nearctic species, are represented by 13 and seven species, respectively, in the Neotropical region. One species of the Holarctic + Afrotropical *Ditaeniella*; the Nearctic *Atrichomelina pubera, Tetanocera plumosa*, and *Sepedon praemiosa*; and two of the Nearctic *Hoplodictya* also occur in the Neotropics. The rest of the 16 genera in the Neotropical region are endemic to that region except *Teutoniomyia* extends to southernmost Texas.

i. *Pherbellia* and *Pteromicra*, genera with numerous species across the Palearctic, are represented by a few species in the Afrotropical and Oriental regions.

j. The Tetanocerini genus *Dichetophora*, with two eastern and central Palearctic and two Japanese species, is represented by at least one Oriental species and at least five endemic Australian species.

4. *Species Level.*

a/b, c/d.The occurrence of the 28 Holarctic species in eastern and western portions of the Nearctic and Palearctic regions is shown in Table 16.7. The data are based on Alaska (Foote *et al.* 1999), northeastern Nearctic (Knutson *et al.* 1986), eastern Palearctic (Rozkošný 1987a) and far western Palearctic (Rozkošný & Knutson 2005a), supplemented by a few other publications. The distributions of most species are well known, but for some there are a lack of data, especially from the eastern Palearctic. Twelve species are so broadly distributed across the Holarctic that it is not possible to speculate simply from distributional data whether their current ranges resulted from movement across Beringia (Fig. 16.9, 3d, e; 4a, b; 5e, d) or across Atlantic areas (Fig. 16.9, 3f, g; 4c, d; 5f, g). Nine species (*Elgiva divisa, Pherbellia griseicollis, P. hackmani, Pteromicra angustipennis, Tetanocera brevisetosa, T. freyi, T. latifibula, T. plebeja*, and *T. spirifera*) probably dispersed across Beringia. Seven species (*Elgiva solicita, Pherbellia obscura, Pteromicra leucopeza, P. pectorosa, Renocera striata, Sciomyza dryomyzina*, and *S. simplex*) probably dispersed across Atlantic areas. Most species currently have rather extensive north–south ranges, with only seven being considered boreal. Mitochondrial DNA analyses (sequence divergence) of specimens of the 28 Holarctic species from their easternmost and westernmost ranges in the Palearctic and Nearctic might be employed to determine the direction of dispersal of Holarctic species. That is, low divergence between specimens from Alaska/Far Eastern Russia compared with high divergence between specimens from the western Palearctic and eastern Nearctic could indicate Beringian dispersal, and vice-versa, Atlantic dispersal.

5. *Subspecific Level.* Subspecies are recognized within only eight species in six genera of Sciomyzidae. Those represented by different subspecies in two regions are:

a,b? The widespread Afrotropical *Sepedon hispanica ruhengeriensis* and the Palearctic subspecies *S. h. hispanica* known only from the southeastern corner of Spain.

c. The extremely widespread *Pherbellia nana nana* in the Holarctic and *P. nana reticulata* in the eastern Palearctic and adjacent Oriental.

d/e, f/g. *Pherbellia schoenherri schoenherri* (Palearctic) and *P. s. maculata* (Nearctic) might have had dispersal histories as proposed for some Holarctic genera and species.

6. *On islands.* The genera *Huttonina, Prosochaeta*, and *Neolimnia* are endemic to New Zealand. There are a few widely dispersed continental species occurring on islands up to about 1500 km distant. There are only a

few endemic island species, mostly of genera that are widespread on the closest continental areas, as follows.

a. *Pherbellia argyrotarsis* and *Euthycera guanchica* – Canary Islands.
b. *Sepedon* (*Parasepedon*) *madecassa*, *S.* (*P.*) *stuckenbergi*, *S.* (*Sepedomyia*) *alaotra* – Madagascar.
c. *Pherbellia tricolor*, *Coremacera scutellata*, *Dichetophora japonica*, *D. kumadori*, *Limnia japonica*, *L. setosa* – Japan.
d. *Pteromicra leucodactyla* – Taiwan.
e. *Pherbellia dives* – Fiji, Tonga; *Sepedon batjanensis* – Maluku, Sulawesi; *S. costalis* – Maluku, Papua New Guinea; *S. lata* – Fiji, Papua New Guinea; *Sepedoninus* n. sp. – Borneo, Sulawesi.
f. *Hoplodictya kincaidi* – Bermuda.

In the northern hemisphere, many Sciomyzidae have very deep north–south ranges as well as broad east–west ranges. The most striking of such widely distributed species in the Palearctic is the common, abundant, semi-terrestrial, predator/parasitoid *Ditaeniella grisescens* which ranges from the High Arctic (Novya Zemlya, about 75° N. Lat., the most northern record for any sciomyzid) and Iceland to Morocco, Egypt, Israel, southwestern Iran, Kyrgyzstan, and Afghanistan, and from Ireland to Japan and marginally into the Oriental region. Detailed examination of the male genitalia and other features of adult morphology of specimens from throughout its range has not given any indication that there are any subspecific differences attributed to *D. grisescens*.

In the Nearctic, some species range from the southern USA (including Mediterranean areas of California) to the Subarctic and analogous northern alpine areas. Several of these broad-ranging species occur in the Low Arctic and a few strictly northern species are restricted to Subarctic, Low Arctic, and analogous alpine areas. The northern-most distributed species in the Nearctic are *Pherbellia griseicollis* and *P. hackmani*, known from about 70 °N. Lat. in Alaska and northwestern Canada. Several other species of *Anticheta*, *Elgiva*, *Pherbellia*, *Sciomyza*, and *Tetanocera* closely approach that latitude in the Nearctic. The southern-most records of any Sciomyzidae are for *Eutrichomelina fulvipennis* and *Shannonia costalis*, recorded from Tierra Del Fuego, Argentina at about 55°S. Lat. *Perilimnia albifacies* and three species of *Tetanoceroides* closely approach that latitude in the Neotropical region.

Further study of adaptations, especially of the north–south broadly ranging species, in regard to ability to survive under strikingly different climatological conditions would be of interest. Are the genetic bases for temperature tolerances, developmental rates, and phenological attributes of northern populations of such species different from the southern and low-altitude populations?

Comparison of the sciomyzid faunas of the southern Mediterranean area (Morocco and Israel, Tables 16.3, 16.4) and the northwestern Palearctic also is interesting in this regard. While northern species might be expected to be found in the cold, higher elevations of Morocco, the occurrence of these species in Israel, at low elevations, is more surprising.

The Sciomyzidae of North Africa (Table 16.3) and the Near East (Table 16.4) show the greatest affinity with the Palearctic, the faunas of Morocco, Israel, and Turkey being well known. The closer proximity of northern and southern land masses to the western as compared with the eastern end of the Mediterranean Sea/Tethys Ocean since the Tertiary, the greater extent and diversity of habitats in the western part, and a comparison of the distribution of gastropods (Table 16.5) indicate the western Mediterranean as having been the more important dispersal route between Europe and Africa. While Pleistocene glaciation of the Pyrenees likely was a major barrier in the west, the more recent arid conditions in the east, despite the relatively rich sciomyzid fauna of the Turkey to Israel area, are now a more important barrier. The gastropod fauna of Morocco is only slightly more diverse than that of Egypt (Table 16.5), while its sciomyzid fauna is significantly richer (Table 16.3). There are 11 genera of gastropods in both countries, but only six genera of the total of 15, and five of 27 species, are shared. The depauperate fauna of Egypt probably is due to extensive disturbance or destruction of habitats, rather than to a lack of suitable kinds of hosts/prey.

Verbeke (1967a) attempted to explain the current eastern Palearctic distribution of two widespread central and northern Palearctic species of *Pherbellia* (*P. argyra* and *P. obtusa*) by the location of the terminal moraines of three glaciations. Knutson & Berg (1967) speculated on the evolutionary and distributional history of the eight European and Middle Eastern species of *Ilione* as related to post-Pleistocene climatic changes. Munari & Cerretti (2006) emphasized that some disjunct distributions, with

Table 16.3. Distribution of Sciomyzidae in North Africa and Israel

Taxa	Morocco	Algeria	Tunisia	Libya	Egypt	Israel	Distribution
Salticellinae							
1. *Salticella fasciata*		x	x				P
Sciomyzinae, Sciomyzini							
2. *Colobaea pectoralis*					?		P
3. *Colobaea punctata*						x	P
4. *Ditaeniella grisescens*	x				x	x	P (O)
5. *Pherbellia albicarpa*						x	P
6. *Pherbellia cinerella*	x	x	x			x	P (O)
7. *Pherbellia dorsata*	x					x	P
8. *Pherbellia griseola*	x		x				H
9. *Pherbellia hermonensis*	x					x	P
10. *Pherbellia kugleri*						x	P
11. *Pherbellia n. nana*	x	x					H
12. *Pherbellia pilosa*						x	P
13. *Pherbellia priscillae*						x	P
Sciomyzinae, Tetanocerini							
14. *Coremacera amoena*						x	P
15. *Coremacera catenata*						x	P
16. *Coremacera obscuripennis*						x	P
17. *Dichetophora obliterata*	x					x	P
18. *Elgiva cucularia*	x	x					P
19. *Elgiva solicita*		x					H
20. *Euthycera alaris*	x	x	x				P
21. *Euthycera algira*	x	x	x				P
22. *Euthycera cribrata*		x	x				P
23. *Euthycera flavostriata*						x	P
24. *Euthycera formosa*						x	P
25. *Euthycera stichospila*		x					P
26. *Euthycera nigrescens*			x				P
27. *Euthycera zelleri*	x	x	x				P
28. *Hydromya dorsalis*	x	x	x			x	P (AF, O)
29. *Ilione albiseta*	x	x	x			x	P
30. *Ilione trifaria*	x	x	x				P
31. *Ilione truquii*						x	P
32. *Ilione turcestanica*						x	P (O)
33. *Ilione unipunctata*	x	x	x				P
34. *Oligolimnia zernyi*	x						P
35. *Pherbina mediterranea*	x	x					P
36. *Psacadina disjecta*	x						P
37. *Psacadina verbekei*	x		x				P
38. *Sepedon h. hispanica*	x						P
39. *Sepedon h. ruhengeriensis*					x		
40. *Sepedon ruficeps*					x		P
41. *Sepedon sphegea*	x					x	P (O)

Table 16.3. (*cont.*)

Taxa	Morocco	Algeria	Tunisia	Libya	Egypt	Israel	Distribution
42. *Sepedon spinipes*	x						P
43. *Tetanocera ferruginea*		x					H
44. *Trypetoptera punctulata*	x						P
Total	23	16	13	0	4 (?)	20	

Note: Data from Steyskal & El-Bialy (1968) (their data from Sack [1939] and Verbeke [1964a]), Knutson & Freidberg (1983), Leclercq & Schacht (1987b), Vala (1989a), Vala & Ghamizi (1991), Kassebeer (1999a), and our new records. ?, Becker's record (1903) of ***Colobaea pectoralis*** from Cairo probably pertains to ***C. punctata***, which was not recognized as a separate species until 1923.

Table 16.4. Distribution of Sciomyzidae in the Near East

Taxa	Egypt	Israel	Jordan	Lebanon	Syria	Cyprus	Turkey	Distribution
Salticellinae								
1. *Salticella fasciata*	x						x	P
Sciomyzinae, Sciomyzini								
2. *Colobaea punctata*	?	x					x	P
3. *Ditaeniella grisescens*	x	x				x	x	P (O)
4. *Pherbellia albicarpa*		x						P
5. *Pherbellia cinerella*		x		x	x	x	x	P (O)
6. *Pherbellia dorsata*		x					x	P
7. *Pherbellia griseola*							x	H
8. *Pherbellia hermonensis*		x						P
9. *Pherbellia pilosa*		x			x			P
10. *Pherbellia priscillae*		x						P
11. *Pherbellia s. schoenherri*							x	P
12. *Pherbellia ventralis*							x	P
13. *Sciomyza testacea*							x	P
Sciomyzinae, Tetanocerini								
14. *Coremacera amoena*		x					x	P
15. *Coremacera catenata*		x		x	x		x	P
16. *Coremacera marginata*							x	P
17. *Coremacera obscuripennis*		x				x	x	P
18. *Coremacera trivittata*							x	P
19. *Dichetophora obliterata*		x				x	x	P
20. *Euthycera chaerophylli*							x	P
21. *Euthycera flavostriata*		x		x			x	P
22. *Euthycera formosa*		x				x		P
23. *Euthycera stictica*				x		x	x	P
24. *Euthycera sticticaria*		x			x			P
25. *Euthycera zelleri*						x		P
26. *Hydromya dorsalis*		x		x		x	x	P (AF, O)

Table 16.4. (*cont.*)

Taxa	Egypt	Israel	Jordan	Lebanon	Syria	Cyprus	Turkey	Distribution
27. *Ilione albiseta*		x					x	P
28. *Ilione truquii*					x		x	P
29. *Ilione turcestanica*		x			x		x	P (O)
30. *Ilione unipunctata*							x	P
31. *Limnia unguicornis*							x	P
32. *Pherbina coryleti*					x	x	x	P
33. *Pherbina mediterranea*							x	P
34. *Psacadina verbekei*							x	P
35. *Sepedon hispanica ruhengeriensis*	x							AF
36. *Sepedon ruficeps*	x							AF
37. *Sepedon sphegea*		x			x		x	P (O)
38. *Sepedon spinipes*		x			x		x	P
39. *Tetanocera arrogans*							x	P
40. *Tetanocera elata*							x	P
41. *Tetanocera ferruginea*							x	H
42. *Tetanocera montana*							x	H
43. *Tetanocera punctifrons*							x	P
44. *Trypetoptera punctulata*							x	P
Total	5 (?)	22	0	5	9	9	35	

Note: Data from Knutson & Freidberg (1983), Knutson (1985), Leclercq & Schacht (1986, 1987a), Rozkošný (1985). ?, Becker's record (1903) of ***Colobaea pectoralis*** from Cairo probably pertains to ***C. punctata***, which was not recognized as a separate species until 1923.

Table 16.5. Freshwater, non-operculate gastropods of Mediterranean Africa

Species/Genus	Palearctic	Morocco	Northwestern Africa	Egypt
Acroloxus lacustris		x	–	–
Afrogyrus oasiensis		–	–	x
Ancylus strictum		x	x	–
A. fluviatilis	x	x	x	–
Anisus spirobus		x	x	–
Anisus sp.	x	x	x	–
Biomphalaria alexandrina		–	–	x
Bulinus forskali		–	–	x
B. truncatus		x	x	x
Ferrissia sp.		–	–	x
Galba truncatula	x	x	x	x
Gyraulus crista	x	x	x	–
G. ehrenbergi	x	–	–	x

Table 16.5. (*cont.*)

Species/Genus	Palearctic	Morocco	Northwestern Africa	Egypt
G. sp. *(? laevis)*	x	x	x	–
Helisoma sp.		–	–	x*
Hippeutis complanatus		x	x	–
Lymnaea maroccana		x	–	–
Lymnaea stagnalis	x	x	–	x
Physella acuta		x	x	x
Planorbarius metidjensis	x	x	x	–
Planorbis planorbis	x	x	x	x
Pseudosuccinea columella		–	–	x
Radix labiata	x	x	x	–
R. natalensis		–	–	x
Segmentina nitida	x	x	–	–
Segmentorbis eussoensis		–	–	x
Stagnicola palustris	x	x	x	–
Total species		18	14	14
Total genera		14	11	13

Note: The list for Morocco is from Ghamizi (1998). The lists for northwestern Africa and Egypt are from D. S. Brown (1980), who noted "The list for north-west Africa is provisional and based on nominal species recorded for Algeria (Bourguignat 1864), Morocco (Morelet 1880), and Tunisia (Germain 1908)." *Introduced from North America.

the submontane *Pherbellia dentata* as an example, are most likely the result of knowledge gaps whereas others, e.g., the Mediterranean *Euthycera vockerothi*, are due to climatic changes that led to rarefaction of humid zones and extinction of intermediate populations. Foote *et al.* (1999) commented briefly on the Bering Land Bridge. Fisher & Orth (1972a) analyzed the distributions of the closely related *Sepedon f. fuscipennis*, *S. pacifica*, and *S. praemiosa* in the USA based on landscape and river basins. This might be a useful rationale in analyzing the distributions of other Sciomyzidae. (See also Tables 16.6, 16.7.)

16.3 DISPERSAL MECHANISMS

> *...j'ai appliqué aux insectes les lois de la résistance de l'air, et je suis arrivé avec M. Sainte Lague à cette conclusion que leur vol est impossible.* [I applied to insects the laws of air resistance and I arrived at the same conclusion as M. Sainte Lague, that their flight is impossible].
>
> Magnan (1934).

Both saltational dispersal (short-distance and long-distance or jump) and neighborhood diffusion dispersal likely account for the post-Tertiary distribution of Sciomyzidae. Adult sciomyzids are slow flying, do not actively aggregate, and are more or less limited to the general area of the microhabitat of the larvae. Short-distance saltation and neighborhood diffusion seem to be the main explanations of dispersal for five reasons: (1) the continuing, gradual, shifting pattern of the mosaic of terrestrial and aquatic mollusc habitats due to climatic, river system, water table, and landscape changes; (2) major, area-wide, seasonal fires; (3) ecological succession; (4) the ability of adults to fly; and (5) the ability of the larvae and, especially, puparia of many species to float, some notably adapted for flotation (many Tetanocerini) and those of many species pupariating in buoyant snail shells (many semi-aquatic Sciomyzini), coupled with annual and large-scale periodic flooding. The fifth reason would seem to be especially important in the dispersal of the many species that thrive in the backwaters, oxbows, and other still-water and humid to wet habitats associated with streams. At least short-distance

Table 16.6. Sciomyzidae known from Malaysia, surrounding countries, and major islands

	ORIENTAL REGION												AUSTRALIAN REGION			
	MYA	THAI	LAOS	CAM	VIET	PMA	IMA	SUM	JAV	BOR	PHIL	SUL	MOL	PNG	NAU	BIOLOGY
Pherbellia																
1. *javana*								x	x						x	
2. *juxtajavana*															x	
3. *terminalis*	x	x			x						x					
Dichetophora																
4. *biroi*														x	x	x
5. sp. nov.	x															
Sepedon																
6. *aenescens*		x						x			x					x
7. *costalis*													x	x		
8. *crishna*															x	
9. *ferruginosa*	x	x	x								x					x
10. *lata*														x		
11. *lobifera*		x														
12. *neanias*			x								x					
13. *plumbella*	x	x	x		x	x	x	x	x	x	x	x	x	x	x	x
14. *senex*	x	x	x			x		x	x		x					x
15. *spangleri*		x														x
16. sp. nov. 1														x	x	
17. sp. nov. 2									x							
Sepedoninus																
18. sp. nov. (1, 2)										x		x				
Steyskalina																
19. *picta*	x															
N SPECIES	6	7	4	0	2	2	1	4	4	2	6	2	2	5	6	6

Notes: MYA, Myanmar; THAI, Thailand; CAM, Cambodia; VIET, Vietnam; PMA, peninsular Malaysia; IMA, insular Malaysia; SUM, Sumatra; JAV, Java; BOR, Borneo; PHIL, Philippines; SUL, Sulawesi; MOL, Moluccas; PNG, Papua New Guinea; NAU, northern Australia. An x in the column "Biology" indicates that the life cycle is known. From Knutson & Ghorpadé (2004).

Table 16.7. Occurrence of Holarctic species in eastern and western portions of the Nearctic and Palearctic regions

	Taxa	Dispersal	Alaska	Eastern P	Northeastern N	Far western P
Sciomyzinae, Sciomyzini						
1.	*Pherbellia albocostata*	?	+	+	+	+
2.	*P. argyra*	?	+	(+)	+	+
3.	*P. griseicollis*	B?	+	(+)	–	+
4.	*P. griseola*	?	+	+	+	+
5.	*P. hackmani*	B?	+	(+)	–	+
6.	*P. n. nana*	?	+	+	+	+
7.	*P. obscura*	A	+	–	+	+
8.	*P. schoenherri*	?	+	+	+	+
9.	*Pteromicra angustipennis*	B	+	+	–	+
10.	*P. leucopeza*	A	–	–	?[a]	+
11.	*P. pectorosa*	A	+	–	+	+
12.	*Sciomyza dryomyzina*	A?	(+)	(+)	+	+
13.	*S. simplex*	A?	+	(+)	+	+
Sciomyzinae, Tetanocerini						
14.	*Elgiva divisa*	B	+	+	–	+
15.	*E. solicita*	A	–	(+)	+	+
16.	*Renocera striata*	A?	(+)	+	+	+
17.	*Tetanocera brevisetosa*	B	+	+	–	–
18.	*T. ferruginea*	?	+	+	+	+
19.	*T. freyi*	B?	(+)	–	–	+
20.	*T. fuscinervis*	?	+	+	+	+
21.	*T. kerteszi*	?	+	(+)	+	+
22.	*T. latifibula*	B	+	+	–	+
23.	*T. montana*	?	+	+	+	+
24.	*T. phyllophora*	?	+	+	+	+
25.	*T. plebeja*	B	+	+	+	–
26.	*T. robusta*	?	+	+	+	+
27.	*T. silvatica*	?	+	+	+	+
28.	*T. spirifera*	B?	+	–	–	+

Notes: A, dispersal across Atlantic. B, dispersal across Beringia. [a] In North America, reported from Massachusetts, New Hampshire, and South Dakota (all females, Steyskal 1954c) and Washington State (two females) and British Colombia (one male) (det. LK). Eastern P indicates species occurring eastward at least as far as Siberia (those in parentheses), with the other species known from the eastern-most Palearctic. Alaskan species not occurring west of Anchorage or north of Fairbanks, i.e., restricted to southeastern Alaska, are also shown in parentheses.

saltation by adults is proved by the records of two exotic species of *Sepedon* released in steep-sided, isolated valleys in Hawaii and now widely established on other Hawaiian islands (Davis 1974).

Population studies using capture–recapture methods (Eckblad & Berg 1972, Arnold 1978b) did not include information on dispersal per se. The only experimental study of dispersal, by Peacock (1973) of *Sepedon fuscipennis*, an aquatic predator, is cited in some detail below as it is an unpublished thesis. The basic life cycle of that species was presented by Neff & Berg (1966), who noted that it over-winters as an adult and breeds continuously from early spring to fall, with several overlapping generations. The species has been studied intensively: dynamics of natural populations (Eckblad & Berg 1972, Arnold 1978a), experimental predation (Eckblad 1973a), biomass and energy transfer in larval feeding (Eckblad 1976), development under controlled conditions (Barnes 1976), phenology (Berg *et al.* 1982), and population and developmental genetics and biochemical systematics (Manguin 1990, Manguin & Hung 1991). The species was intensively sampled by Peacock (1973) from emergent vegetation in eight lentic (marshes, ponds, backwaters of streams) and nine lotic (stream) sites in the Willimantic and Fenton River watersheds in northeastern Connecticut during June–October 1972. Peacock also surveyed the amounts and distributions of the 18 genera of emergent vegetation. He considered the monthly mean discharge of streams; surface water velocity – determined as the single most important physical factor; fluctuations in water level; amount of sunlight reaching sites; and the relative abundance of pulmonate snails in the sites. Although the number of flies captured and released was small (612), the percent recapture was limited, and some sampling and measurement methods might be questioned, the approach in this unique study is interesting. The partly subjective results definitely show adult short-distance dispersal ability. Peacock concluded that, overall, use of lotic habitats appeared to be at least equal to lentic habitats and that the species can invade riverine habitats about as well as lentic habitats. He did not find *S. fuscipennis* in swamps or in wooded areas around aquatic sites. He noted,

> Relatively rapid invasion of seasonably favorable riverine sites and low recapture rates in well established populations is probably greater than formerly thought and about equal to that of Diptera in other studies reviewed by Johnson (1969)... The rapid build-up of flies in seasonally suitable lotic habitats indicates that dispersing flies from drying backwaters or from other lotic habitats, were using rivers or riverine valleys.

The recapture rates were 0.02% to 0.90% in the seven studies reviewed by Johnson (1969) and 0.19% in Peacock's studies. This type of study needs to be repeated, and comparison of studies on snail migration in the same sites could be illuminating.

In studying the distribution and diversity of aquatic Coleoptera in Europe, Ribera & Vogler (2000) noted that rivers and streams persist over long periods and are semi-permanent features of the landscape. In contrast, small to medium-sized stagnant water bodies are short-lived, and discontinuous in space and time. They also noted that in general the dispersal ability of insects is inversely correlated with permanence of the habitat. Whiles & Goldowitz (2001) conducted a detailed, field-experimental study of emergence abundance and production of insects, including Sciomyzidae, in two ephemeral, two intermittently wet, and one permanent "slough" (wet meadow) in the midwestern USA. These are prime habitats for many Sciomyzidae; similar habitats are found throughout much of the northern hemisphere. They concluded that,

> ... a landscape containing a mosaic of hydrologically distinct wetlands will maximize aquatic insect diversity and productivity at larger spatial and temporal scales ... temporal patterns of insect emergence production show that sites with different hydrology generate peaks of adult insect biomass at different times of the year. A landscape of wetlands with different annual hydroperiods also facilitates the process of cyclic colonization, whereby colonists from wetter sites colonize more temporary systems when they become inundated.

Further detailed comparison of both Sciomyzidae and aquatic snails in ponds, small lakes, and marshes, the primary habitats of aquatic Sciomyzidae, with stagnant water habitats associated with streams, as approached by Peacock (1973), likely would be of interest in regard to dispersal as well as to habitat characterization. These comparisons should be conducted throughout the year to identify also the effects of seasonal and aseasonal changes in water levels, flow, and quality.

Speight (2001, 2004) studied the occurrences of 23 species of Sciomyzidae in 13 kinds of terrestrial and aquatic habitats in a 41-ha farm in Co. Cork, Ireland over a 4-year period by Malaise and emergence traps. He found that "the great majority of the species collected would seem to be extremely sedentary within this farmed landscape," and that "there is little evidence for movement of species through inappropriate habitat on the farm." He concluded that "study of permeability of landscape to flighted insects probably requires to be conducted over a long period of time in order to yield meaningful results," to judge by the data on the Sciomyzidae.

As most adults are relatively large, about the size of a housefly, long-distance saltational dispersal as aerial plankton probably has not been significant except in the case of some minute (1 mm long) species of *Colobaea*. However, Yano (1978) noted the collection of two specimens of *Sepedon aenescens* (about 6.0–7.0 mm long) on a ship in the East China Sea, about 160 km from the coast and about 32 °N. Lat. during July. In a study of distribution of insects in the air by airplane traps, 26 adult sciomyzids (unidentified beyond family except one *Tetanocera* sp. collected at 305 m) were collected at about 31 m (one specimen), 61 m (16), 300 m (4), 600 m (3), and 900 m (2) (Glick 1939). Limited long-distance saltational dispersal is also indicated by the paucity of species on Oceanic islands. There are only three species in the Oceanian region: the endemic *Pherbellia dives* – Fiji, Tonga; *Sepedon lata* – New Guinea, Fiji; and the widespread Oriental and Australian *S. plumbella* which reaches to New Caledonia. There are no native species in Hawaii. Within 10 years of the formation of Surtsey Island in 1963, 35 km southeast of Iceland, five of the six species known from Iceland were found on the island (Lindroth *et al.* 1973). *Shannonia meridionalis* has been collected on Juan Fernandez Is., 644 km off the coast of Chile (Kaczynski *et al.* 1969).

17 • Evolutionary considerations

> Indeed, the varying conditions of life in the same individual or species; the remarkable metamorphoses; the rapid development; the phenomena of dimorphism and heteromorphism; of phytophagic and sexual variation; the ready adaptation to changed conditions, and consequent rapid modification; the great prolificacy and immense number of individuals; the three distinctive stages of larva, pupa, and imago, susceptible to modification, as well as other characteristics in insects – render them particularly attractive and useful to the evolutionist.
>
> Riley (1882).

17.1 SCIOMYZIDAE – A PARADIGM FOR EVOLUTION OF SAPROPHAGOUS–PREDATORY–PARASITOID BEHAVIOR AND MORPHOLOGICAL ADAPTATIONS TO HABITATS?

> In scientific work, those who refuse to go beyond fact rarely get as far as fact; and anyone who has studied the history of science knows that almost every step therein has been made by . . . the invention of a hypothesis which . . . often had little foundation to start with . . .
>
> T. H. Huxley (1868) (cited in Horvitz 2000).

Although evolutionary scenarios have been proposed for many phytophagous, mycophagous, mutualistic, and ectoparasitic insects, there have been very few scenarios detailed for parasitoid and/or predaceous insects, other than for parasitoid wasps. Feener & Brown (1997) pointed out the several unique features of Hymenopteran parasitoids that are not essential to the success of parasitoids in general, thus making this group "of questionable value in understanding the evolution and general significance of the parasitoid lifestyle." First, parasitoid Hymenoptera represent only one evolutionary lineage, that arose from mycophagous ancestors (Whitfield 1998), in comparison to the estimated hundreds of lineages in Diptera. Thus the heuristic value of Hymenoptera with regard to the general phenomena leading to the origin of parasitoids is low. Second, the important capability of direct access to the hosts provided by the lepismatid form of ovipositor and the use of venom that subdues and manipulates hosts are special adaptations of Hymenoptera not found in other parasitoids. Finally, Hymenoptera are uniquely haplodiploid, giving the females unique control over the sex ratios of their progeny and thus manipulation of host populations. Also, as pointed out by Feener & Brown (1997), the Diptera attack a broader spectrum of hosts/prey than any other group of parasitoids, utilizing 22 orders in five phyla, while Hymenoptera utilize only 19 orders, all Arthropoda.

The Hymenoptera account for nearly 80% of parasitoid insects. The estimated 16 000 described parasitoid Diptera account for nearly 20%. A few other parasitoids are found in the Coleoptera, Lepidoptera, and Neuroptera. Eggleton & Belshaw (1992) recognized parasitoids in 21 families of Diptera. These Diptera are exceedingly diverse in behavior, habitats, distribution, and morphology as well as evolutionary origins and are an under-utilized but highly suitable group for analyses of evolution of behavior. Among parasitoid Diptera, the behavior of the Bombyliidae, Calliphoridae, Phoridae, Rhinophoridae, Sarcophagidae, Sciomyzidae, and Tachinidae have been the subject of recent specialized reviews, but few have been the subject of evolutionary scenarios. Evolutionary scenarios of parasitoid and predatory Diptera, based on cladistic frameworks, are available for the Keroplatidae (Matile 1990, 1997), some Syrphidae (Gilbert *et al.* 1994), Bombyliidae (Yeates & Greathead 1997), and certain Phoridae (Brown 1992, 1993). A preliminary version of the scenario given here for Sciomyzidae was presented by Knutson & Vala (2002). Another approach was presented by Barker *et al.* (2004). That study was based on (1) a revision of the cladistic analysis of Marinoni & Mathis (2000), (2) a linking of the phylogeny of Sciomyzidae with the phylogeny of molluscan clades that have representatives in sciomyzid habitats, and (3) Principal Axis Correlation analyses of a new classification of ecomorphological groups based on the egg and larval stages. The main conclusions of Barker *et al.* (2004) are given in Section 17.9.

With the extensive information on the diverse and almost unique feeding behavior of the larvae and on the co-adapted behavioral and morphological features in all stages of the life cycles, the Sciomyzidae should be a rich resource for development of a paradigm of the evolution of predaceous/parasitoid feeding behavior in Diptera. There are about 138 200 described species of Diptera (Schumann *et al.* 1999), projected as a quarter of a million species (Yeates & Wiegmann 1999) in 130 ± families. Also, the fact that 41 of the 61 genera are known biologically seems to be a strong point for development of such a paradigm. However, there is need for more information on an even greater percentage of species in some genera, as well as on the 25 biologically unknown genera.

Reflection on the unpredictable course of the history of research on sciomyzid life cycles gives little reassurance in our ability to predict the behaviors of biologically unknown species. An hypothesis of strict malacophagy for Sciomyzidae developed beginning with the discovery of aquatic and semi-terrestrial predators by Berg (1953). Then, many authors accumulated a mass of data on various aspects of life styles. Berg (1964) and Barnes (1990) reported on terrestrial parasitoids (e.g., *Sciomyza varia*); Knutson *et al.* (1965) and Trelka & Foote (1970) on slug killers; Fisher & Orth (1964), Knutson (1966), Knutson & Abercrombie (1977), and Robinson & Foote (1978) on snail-egg feeders; and Foote & Knutson (1970), Foote (1976), Barnes (1980a), and Horsáková (2003) on clam killers. Discovery of parasitoid feeding on millipedes by *Pelidnoptera nigripennis* by Baker (1985) and Bailey (1989) did not particularly challenge the malacophagy hypothesis, because Griffiths (1972) had removed *Pelidnoptera* recently to a separate family (Phaeomyiidae) on morphological characters of the adults. Later, Vala *et al.* (1990) confirmed this exclusion on the larval characters.

However, the supposition of exclusive malacophagy for Sciomyzidae has been destroyed by the discovery of obligatory predation on freshwater oligochaete worms by *Sepedonella nana* (Vala *et al.* 2000b), and *Sepedon knutsoni* (Vala *et al.* 2002), which are true but highly derived sciomyzids based on morphology of the adults. Although malacophagy remains as the most parsimonious working hypothesis in discovering sciomyzid life cycles, we should expect the unexpected.

A fundamental deficiency in the use of knowledge of Sciomyzidae as a model of evolution is that we know practically nothing of the extent of extinction or the rates of extinction of species and higher categories. Some of these probably had life styles key to development of an historical narrative of evolution. Extinction of Sciomyzidae may have been very extensive. The phylum of their hosts/prey, the Mollusca, experienced major extinctions at the end of the Cretaceous and end of the Oligocene as well as several earlier extinctions, and the molluscs' natural enemies may have experienced them as well. Particularly for groups of organisms such as the Sciomyzidae, in which complex and finely adapted life cycles have evolved, one might expect extensive extinction. But on the other hand, since Sciomyzidae have captured a trophic niche barely exploited by other Insecta, many phyletic lines within the family might have been protected from extinction.

The Sciomyzidae still can, indeed, be a useful resource for evolutionary studies. Unlike the biologically diverse, taxonomically complex, and species-rich Tachinidae – parasitoids of arthropods, about 8000 described species with a small percentage studied behaviorally – the Sciomyzidae have a modest number of extant species (539 described), with about one-third behaviorally known in considerable detail. The taxonomy is relatively well worked out. There is a wide range of feeding behaviors, habitats utilized, and correlated adaptations throughout the life cycles available for analysis. As with the diversely phytophagous Tephritidae – 4257 valid species and 471 valid genera, of which the biologies of only 291 species (6.8%) and 91 genera (19.3%) are known (Norrbom *et al.* 1998) – and a few other families of Diptera whose life cycles are comparatively well known, there has been a deficiency in the study of Sciomyzidae that now is being met. That is, the need for comprehensive phylogenetic or cladistic analyses on which to array and test evolutionary scenarios. The cladistic analyses of the genera within the family by Marinoni & Mathis (2000) and Barker *et al.* (2004) are steps in this direction, but greater resolution of genera within lineages and analyses of species within individual genera are needed.

As our manuscript was being completed an analysis that could well serve as a model for future studies was published by Chapman *et al.* (2006). This study of parallel evolution of larval morphology and habitat in the behaviorally diverse genus *Tetanocera*, using phylogenetic comparative methods and molecular and morphological features, is the first study of this nature in the

Sciomyzidae. In fact, it is one of the first modern studies to explore morphological adaptations in both aquatic and terrestrial habitats in a dipteran lineage.

17.2 EVOLUTION OF MALACOPHAGY IN THE DIPTERA (SEE ALSO CHAPTER 3)

The present contains nothing more than the past, and what in the effect was already in the cause.

Bergson (1907).

The biologies of other families of Acalyptratae, none mostly malacophagous, and the relatively rare occurrence of true malacophagy in a few other Diptera are not very instructive in developing scenarios of the evolution of malacophagy in the Sciomyzidae. True malacophagy has arisen in just a few genera in four other, unrelated families of Diptera. These represent some lineages that are much more basal, Culicomorpha and Aschiza, and some that are more derived, Calyptratae, than the Sciomyzidae. All four of the other families including malacophages are more successful in terms of total numbers of species and genera and sizes of populations than the Sciomyzidae. We do not include in this overview the facultative predators or facultative parasitoids in Rhinophoridae (one sp.), Muscidae (six spp.) or Fanniidae (two spp.). See Coupland & Barker (2004) for the most recent review of Diptera, other than Sciomyzidae and Chironomidae, that feed on molluscs, especially terrestrial forms.

In the Chironomidae (Culicomorpha), which are generally microphagous on detritus and small plants and invertebrates, there are a few carnivorous forms and two malacophages: the Palearctic *Parachironomus varus varus* which is ectoparasitic on *Physa* spp. and *P. v. limnaei* which is an hematophagous endoparasite of "*Lymnaea*" spp. (Guibé 1942). In the nutritionally diverse Phoridae (Aschiza), three species of *Megaselia* are predators of slug eggs. This is a niche not known to be exploited by any sciomyzid. *Megaselia fuscinervis* is a parasitoid of terrestrial *Vitrea* snails in the Palearctic (Disney 1979, 1982) and Afrotropical *Wandolleckia* spp. are commensals of terrestrial snails (Disney 1994). Other species of *Spinophora* and *Megaselia* have been implicated as natural enemies of snails and slugs. In the Sarcophagidae (Calyptratae), which includes parasitoids of many arthropods and many saprozoic forms, about 70 species of several genera are known or likely parasitoids or predators of several genera of terrestrial snails (Coupland & Baker 1994, Coupland 1996a, Coupland & Barker 2004). Many of the sarcophagid snail killers seem to be more or less host specific. In the Calliphoridae (Calyptratae), which includes species parasitoid on earthworms, feeding on fresh tissues of vertebrates, and many saprozoic species, several Australian, Nearctic, and Palearctic species are parasitoids of terrestrial snails (Keilin 1919, Čepelák & Rozkošný 1968). Perhaps the most instructive point taken from this brief survey (also see Chapter 3) is that evolutionary experiments in malacophagy in these families did not lead to as extensive speciation as it did in the Sciomyzidae. But, it is interesting that almost the entire range of feeding habits of Sciomyzidae are displayed by the few malacophagous species in these four families.

17.3 BEHAVIOR OF FAMILIES RELATED TO SCIOMYZIDAE

The supposed sister group of the Sciomyzoidea, the Lauxanioidea (J. F. McAlpine 1989), includes aphidoid predators (Chamaemyiidae, Sluss & Foote 1971, 1973) and saprophytic forms (Lauxaniidae, Miller & Foote 1975, 1976 and Miller 1977; and Celyphidae, Sen 1921). The Eurychoromyiidae, a family of the Lauxanioidea from Bolivia, are biologically unknown. J. F. McAlpine (1989) noted that "The ancestry of the Lauxanioidea + Sciomyzoidea probably traces back to the earliest stem of the Acalyptratae . . . but no specific outgroup has yet been established."

No species in other families of the Sciomyzoidea are known to be truly malacophagous or to attack oligochaete worms. The Phaeomyiidae (*sensu* Griffiths 1972) are parasitoids of millipedes (Baker 1985, Bailey 1989). Other dipterous parasitoids/predators of Diplopoda are known only in the families Phoridae, Eginiidae, Muscidae, and Sarcophagidae. The Dryomyzidae include predators of intertidal barnacles (*Oedoparena* spp., Burger *et al.* 1980). No other Diptera larvae are known to attack barnacles. Most Dryomyzidae are saprozoic/coprophagous/fungivorous (*Dryomyza* spp., Barnes 1984). The Helcomyzidae, Coelopidae (Egglishaw 1960a), and Heterocheilidae (Egglishaw 1960b), all in the family Coelopidae *sensu* J. F. McAlpine (1989), live on maritime beaches where the larvae feed on stranded seaweed. The Helosciomyzidae include species that attack ant larvae and are saprozoic (*Helosciomyza* spp., Barnes 1980b; *Polytocus* spp., Barnes 1980b, 1980c), and possibly fungivorous (*Eurytocus* spp., Steyskal & Knutson 1978). The Ropalomeridae are saprophytic or feed on slime fluxes (Prado 1965). Larvae of Sepsidae are exclusively saprophagous in excrement, carrion, sewage sludge, and decaying vegetation, including one species in decaying seaweed on maritime beaches (Meier 1996). The biology of the Huttoninini (Huttoninidae *sensu* D. K. McAlpine 1991, Huttonininae in Sciomyzidae *sensu* Barnes 1979d) is unknown. The Natalimyzidae "appear to be microflora grazers on decaying grass" (Barraclough & D. K. McAlpine 2006).

A dryomyzid-like fly may be the closest relative of the ancestral Sciomyzidae considering the feeding habits of related families and some features of extant Dryomyzidae. These include the northern hemisphere distribution, saprophagous behavior of one of the two genera, presence of ventral cibarial ridges in the larvae of some species, and adult morphology. This is also indicated in the cladogram of Lauxanioidea and Sciomyzoidea presented by J. F. McAlpine (1989) (Fig. 15.27).

17.4 DIVERSITY OF SCIOMYZIDAE

Before considering the evolution of Sciomyzidae, it is useful to note the diversity of species and life cycles. The sciomyzids have exploited an adaptive zone – feeding on molluscs – that has been utilized by relatively few other insects. The only known exceptions to malacophagy in the lineage are *Pelidnoptera* (Phaeomyiidae), an obligate parasitoid of terrestrial Diplopoda, and one species of *Sepedonella* and one species of *Sepedon*, obligate predators of freshwater Oligochaeta. Most non-sciomyzid malacophagous insects, primarily in the Hemiptera and Coleoptera, are opportunistic predators of gastropods and other invertebrates. Only a few Coleoptera (in the Carabidae, Drilidae, Lampyridae, Silphidae, and Staphylinidae) are obligate or preferential predators of gastropods (see Chapter 2). The Sciomyzidae are the most diverse of all malacophagous insects in terms of number of species, kinds of hosts/prey, types of feeding behavior, phenology, microhabitat, and in other behavioral and ecological features. They also are more diverse in morphology of larvae and in geographical distribution in comparison to the other, unrelated, known or suspected parasitoid or predatory, obligatory or facultative malacophagous Diptera. "Malacophagous" flies are found in only seven families: 2 of 15 000 species of Chironomidae; 20 of 3000 Phoridae; 2 of 265 Fanniidae; 11 of 1100 Calliphoridae; 6 of about 4000 Muscidae; 1 of 80 Rhinophoridae and perhaps as many as 70 of 2750 Sarcophagidae (Coupland & Barker 2004).

In comparison to other families of acalyptrate Diptera, the Sciomyzidae are not as species rich as one would expect according to some hypotheses of diversity of numbers of species in a clade. The superfamily classification and families used in our comparison are according to J. F. McAlpine (1989), except we recognize Huttonininae in the Sciomyzidae; the Phaeomyiidae, Helcomyzidae, and Heterocheilidae as families in the Sciomyzoidea plus the recently described family Natalimyzidae; Marginidae in the Opomyzoidea; Rhinotoridae and Trixoscelididae in the Sphaeroceroidea; and we include Risidae in the Ephydridae. Of the 66 families of Acalyptratae only 12 include more than 500 species. The numbers of species per family we considered are primarily from Schumann *et al.* (1999); a few totals are from subsequent publications. The numbers are of known species, not estimates of possible total numbers, and do not include fossil species. In the 11 families of Acalyptratae more diverse, in terms of number of species, than Sciomyzidae, almost all of the species are phytophagous or saprophagous. Exceptions include the 800 species of Conopidae, of which all biologically known species are endoparasitoids of aculeate Hymenoptera, cockroaches, and calyptrate Diptera. Also, some of the 3300 species of Drosophilidae, which are ectoparasitoids of Cercopidae, predators of Pseudococcidae, or predators of frog eggs. And, a few of the 2000 species of Chloropidae that are predatory on aphids or eggs of spiders and insects, and a few of the 1800 Ephydridae, that are predators of Chironomidae or parasitoids in eggs of spiders. Most of these hosts/prey are fairly immobile and defenseless.

Wiegmann *et al.* (1993) used the method of multiple sister group comparisons – primitively parasitic lineages compared with their predaceous or saprophagous sister groups – to test some major hypotheses that have been invoked to account for diversification of insect parasitoids and parasites ("carnivorous parasites" in their terminology). That paper and a similar study by Mitter *et al.* (1988), in which greater diversification was found in clades attacking higher plants than in their predaceous or saprophagous sister groups, are the only statistically replicated of such tests. Wiegmann *et al.* (1993) analyzed 15 clades of Dermaptera, Phthiraptera, Hemiptera, Neuroptera, Coleoptera, Strepsiptera, Siphonaptera, and, in the Diptera, the endoparasitoid Nemestrinoidea, Bombyliidae, Pipunculidae, Conopidae, *Blaesoxipha* (Sarcophagidae), Tachinidae, and the ectoparasitic Hippoboscoidea, along with the sister groups of each. Since most species in the phaeomyiid-sciomyzid lineage are predatory or parasitoid or have a mixture of behaviors, not "parasitic" in the sense of Wiegmann *et al.* (1993), their conclusions cannot be applied directly to the Sciomyzidae, but might provide some insight. They found that six clades of "carnivorous parasites" were more diverse than their predaceous or saprophagous sister groups and nine less diverse. Of the parasitoid Diptera, only *Blaesoxipha* (in part) and Tachinidae were more diverse than their sister groups. The phaeomyiid-sciomyzid lineage (545 species) is much more diverse than its putative sister group, the Dryomyzidae, with ten species in *Dryomyza* (saprophagous) and two species in *Oedoparena* (predators of barnacles).

Wiegmann *et al.* (1993) concluded that it is unlikely that the hypothesis that parasitism "in the broad sense" in and of itself is a primary explanation of the rate of diversification, while the hypothesis that parasitism "in the strict sense" is an evolutionary cul-de-sac is plausible. They found no evidence for ascribing the differences in rate of diversification to age of the parasite clade. The ages are 25–150 million years for the clades they studied; perhaps 65 + million years for phaeomyiids-sciomyzids. There was no evidence for mode of "parasitism" (phytophagous or carnivorous). Nor was there evidence for diversity of host clade; the terrestrial and freshwater Gastropoda are much more diverse than the Sciomyzidae. Finally, there was no evidence for host specificity. One to 89 host species were used by species in the carnivorous parasite clades that they studied. For sciomyzids, the number of host/prey species utilized per species ranges from one to perhaps 50. Fifty is the estimated greatest number of species of suitable freshwater, non-operculate snails within the range of the most common and widely distributed, polyphagous, aquatic, predaceous sciomyzid, *Tetanocera ferruginea*.

There are interesting cases of diversification in behavior and number of species within some genera of Sciomyzidae. The aquatic, predaceous tetanocerine genus *Dictya* is represented only by *D. umbrarum* in the Palearctic, ranging from Iceland, the Faroes, and Ireland across Eurasia to far eastern Russia and from northern Fennoscandia to southern France, northern Italy, and southern Greece. But in the Nearctic there are 37 described species of *Dictya*, ranging across North America, and from above the Arctic Circle into Mexico, many with broad distributions. There are 13 Caribbean and northern Neotropical species of *Dictya*, ranging from the Bahamas to Colombia (Orth 1991). Most adults, including *D. umbrarum*, can be identified only by minute differences in the male genitalia, especially in the surstylus and hypandrial process. The biology and morphology of 22 of the Nearctic and Neotropical species (Valley & Berg 1977, McLaughlin & Dame 1989, Mc Donnell *et al.* 2007b) and the biology of the Palearctic *D. umbrarum* (Willomitzer & Rozkošný 1977) have been studied intensively. Notably, they are very similar in behavior and ecology, except for three Nearctic coastal salt marsh species that seem to prefer operculate snails, and in the morphology of the immature stages, except *D. ptyarion*, southeastern USA. Other than the aberrant *D. ptyarion*, whose larval food preference has not been well determined and which is distinct also on adult characters, the immature stages of most *Dictya* spp. are more similar to each other than are those in any other genus of the family.

Sepedon in the Afrotropical region and *Dictya* in the Nearctic and Neotropical regions perhaps can be considered as examples of explosive speciation, with *Sepedon* radiating into diverse behavioral niches and *Dictya* remaining more or less behaviorally homogeneous. However, considering the relatively basal position of *Dictya* and apical position of *Sepedon* in the cladograms, one might expect *Dictya* to have become more behaviorally diverse than *Sepedon*.

In the Afrotropical region, the subgenus *Parasepedon* includes 33 of the total 80 species worldwide in the genus *Sepedon*. Few species of *Parasepedon* have been studied biologically, but behaviorally the subgenus is much more diverse than *Dictya*. Four are predators of freshwater snails: *S.* (*P.*) *ruficeps* (Knutson *et al.* 1967a, Gbedjissi *et al.* 2003), *S.* (*P.*) *neavei* and *S.* (*P.*) *testacea* (Barraclough 1983), and *S.* (*P.*) *scapularis* (Maharaj *et al.* 1992). Two are parasitoid/predators of semi-terrestrial Succineidae: *S.* (*P.*) *hispanica* (Knutson *et al.* 1967a) and *S.* (*P.*) *trichrooscelis* (Vala *et al.* 1995, Knutson 2008). Two are parasitoid/predators of strictly terrestrial snails: *S.* (*P.*) *ornatifrons* and *S.* (*P.*) *umbrosa* (L. G. Gbedjissi, personal communication, 2003). *Sepedon* (*Mesosepedon*) *knutsoni* (Vala *et al.* 2002) is an obligate predator of freshwater oligochaetes. All other reared species of *Sepedon* – 11 Nearctic, one Palearctic, one Holarctic, and five Oriental – are typical freshwater predators. Also, the adults of *Parasepedon* are much more structurally diverse than in *Dictya*.

Except for *Pherbellia* (total, 95 species – Sciomyzini) and *Dictya* (43), *Sepedon* (80), and *Tetanocera* (39), the latter three genera in the Tetanocerini, species diversity in genera of the phaeomyiid-sciomyzid lineage is low (Fig. 17.1). The four above-mentioned genera have the broadest habitat range in the family, from aquatic (except *Pherbellia*) to

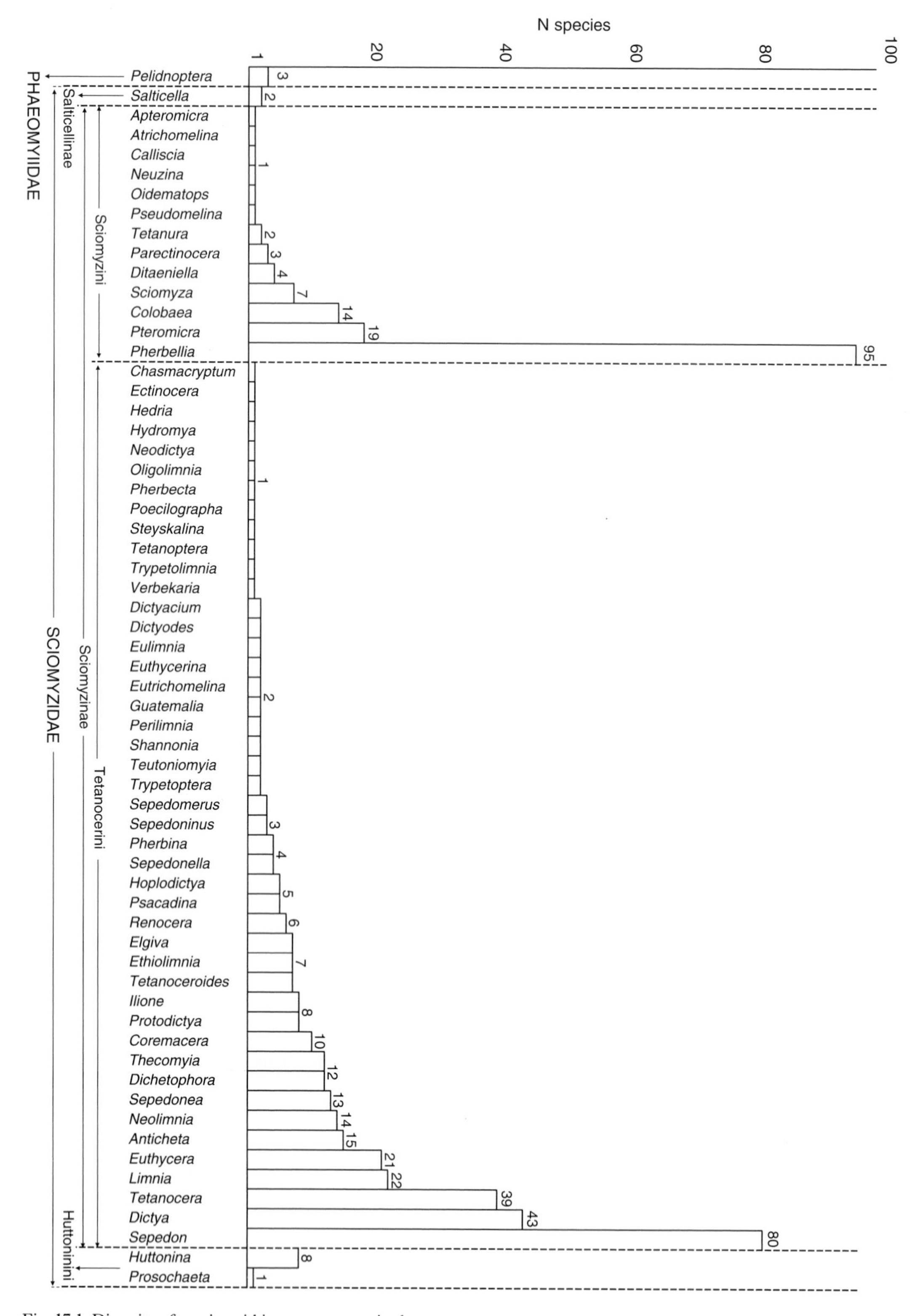

Fig. 17.1 Diversity of species within extant genera in the phaeomyiid-sciomyzid lineage. Note: Phaeomyiidae also includes the genus *Akebono* and an additional species of *Pelidnoptera*.

semi-terrestrial to terrestrial. They also have the broadest range of behaviors – parasitoid, parasitoid/predatory, predatory, and partly saprophagous. Notably, 22 of 45 genera of Tetanocerini have only one or two species. Only a few of the small genera are biologically diverse. *Sciomyza* with seven species and *Colobaea* with 13 species both include a true parasitoid and parasitoid/predators. *Dichetophora* with 12 species includes a terrestrial parasitoid/predator and aquatic predators. *Ilione* with eight species includes a clam killer and aquatic snail killers. *Hoplodictya* with five species includes a predator of operculate Littorinidae and a parasitoid/predator of semi-terrestrial Succineidae.

There is considerable reliability in these numbers as about 25% of the species and 20% of the genera were described by Steyskal (1938 to 1980) and Verbeke (1948 to 1967). Their specific and generic discriminations are in close agreement. Today, only 12 of 109 species (some co-authored) and none of the eight genera described by Steyskal are in synonymy. Seven of 56 species and one of five genera described by Verbeke are in synonymy.

17.5 ORIGIN OF SCIOMYZIDAE

> The difference between what the most and the least learned person knows is inexpressibly trivial in relation to that which is unknown.
>
> A. Einstein.

Oldroyd (1964) proposed that "the pursuit of still more nutritious food" has been an especially important selective pressure in the evolution of the Diptera. This general rationale may apply to some extent in postulating the evolutionary steps leading to the origin and development of malacophagous behavior in the Sciomyzidae. Oldroyd's assessment of data published on the biology of Sciomyzidae was that members of this family are,

> in process of adapting themselves from a saprophagous diet, feeding on decaying animal remains, first to specialising in dead snails, like many Phoridae; then to killing a snail for their own use; and finally to preying actively on a succession of living snails.

Whether living molluscs are more "nutritious" than dead animal remains can be questioned. However, it is likely that non-degraded proteins and other substances in living organisms may be important to parasitoids and predators.

With a broad brush, Fisher & Orth (1983) painted a general impression of the origin of Sciomyzidae, which we expand on below. The fossil record has not revealed when the first connection between primitive sciomyzids and molluscs occurred, but the stage was set over a very long time span for this relationship to evolve. The first land snails are considered to have appeared in the Upper Carboniferous, about 300 million years ago (mya) (Solem & Yochelson 1979). Then, certain pulmonate gastropods entered freshwater habitats and became the ancestral stock from which modern freshwater pulmonate snails evolved. The earliest evidence of Diptera appears in the Triassic (250–206 mya), with the Jurassic (206–144 mya) as the major period of innovation for Diptera. The origin of Cyclorrhapha is thought to be in the Lower Cretaceous (about 142 mya), the Period of the greatest dispersion of the continental shields. The Schizophora apparently diverged during the Upper Cretaceous (98–74 mya) (Yeates & Wiegmann 1999, Wiegmann *et al.* 2003). The time between the Cenozoic Paleocene (65–55 mya) through the Pliocene (1.8 mya) was a period of increasingly cooler and drier climates, with many wetlands beginning to dry or change from permanent to temporary. This timing fits well with the likely early radiation of some Sciomyzidae into terrestrial situations and the major radiation of the terrestrial stylommatophoran snails (Wade *et al.* 2000). Considering the time span involved, it seems reasonable to presume that malacophagous habits of Sciomyzidae were in place by the early Tertiary (about 65 mya).

We speculate that the Sciomyzidae arose during the early Upper Cretaceous from a rather primitive stock of acalyptrate Diptera whose saprophagous larvae developed a proclivity for a diet high in non-degraded protein. We propose a generally adapted, saprozoic, acalyptrate ancestor that fed on all sorts of dead animals and did not need specialized features to find and kill its food. An earlier stage of evolution may have involved generalized feeders on organisms (fungi, bacteria, yeasts) associated with plant decay in wet habitats also supporting gastropods, but we do not propose a simple, direct ecological shift from feeding saprophagously on plant material to snail killing. The presence of ventral cibarial ridges in the plesiomorphic *Salticella fasciata* supports our supposition. These ridges are characteristic of saprophagous larvae but not present in predaceous or parasitoid larvae.

Because nearly all biologically known sciomyzid larvae are restricted in nature to feeding on snails, we speculate that some of the earliest ancestral forms were those whose saprophagous larvae had a proclivity for feeding on dead molluscs in general. A subsequent restriction to dead snails might be taken as one of the major features of the ancestral Sciomyzidae. Dead snails probably were abundant, widely dispersed, not as susceptible to rapid decay as dead slugs, and more of them die in more exposed situations than do most bivalves. After the lineage became restricted to dead snails and concomitant microhabitats and seasonal aspects, opportunities for speciation lay in an adaptive radiation and filling of niches which, almost exclusively, are not occupied by other insects, even to the present.

It seems unlikely that the family evolved from a specially adapted lineage whose genetic make-up already limited it to a narrow range of parasitoid or predaceous behavior; this would argue against the inclusion of the parasitoid Phaeomyiidae in the Sciomyzidae. Also, many important characters of adult and immature Phaeomyiidae are distinctly more plesiomorphic than those of the monophyletic Salticellinae + Sciomyzinae lineage.

Restriction to feeding on dead snails raises the question as to the sort of habitat in which the ancestral forms arose. It seems probable that a malacophagous diet among acalyptrates would have developed in those species living in common habitats widely dispersed over broad areas where dead snails were most abundant and exposed in sheltered situations for rather long periods of time. Some of the most common of such ideal situations are the retreating margins of smaller bodies of water and temporary aquatic habitats with gradual, seasonal changes in water levels where snails are rather continuously being exposed. However, J. F. McAlpine (1963) postulated a seashore habitat for the progenitors of the Sciomyzidae and related families (in the Sciomyzoidea and Lauxanioidea). From a cladistic analysis of 32 characters in 20 taxa of Dryomyzidae, Coelopidae, Phaeomyiidae, Helcomyzidae, Heterocheilidae, and Sepsidae, Sueyoshi (2002) proposed that the marine habitat has arisen separately as parallel evolutionary trends in the Dryomyzidae, Coelopidae, and Heterocheilidae. We note that also one Sepsidae (*Orygma luctuosum* [Meier 1996]) and five Sciomyzidae mentioned below have maritime habitats. Sueyoshi (2002) proposed that, the "ancestral habitat of the Dryomyzidae is inland." The dominance of well-protected operculate snails in marine and brackish water situations, and the fact that only five extant sciomyzids are known to live in such habitats (the specialized *Hoplodictya setosa*, three species of *Dictya*, and *Pherbellia mikiana*), plus perhaps *Calliscia callisceles*, argues against a maritime habitat for ancestral Sciomyzidae.

17.6 PRIMITIVE SCIOMYZIDAE

> A great truth is a truth whose opposite is also a great truth.
>
> N. Bohr.

Behavioral features of the primitive Sciomyzidae may have been similar to the extant *Atrichomelina pubera* which Berg *et al.* (1959), Foote *et al.* (1960), and Berg (1966) suggested as a model for the common ancestor of the Sciomyzidae. This hypothesis is supported by the presence of *Atrichomelina* very close to the base of the cladograms of Sciomyzidae. *Atrichomelina* is a monotypic genus most closely related to the polytomy that includes *Sciomyza* (semi-terrestrial parasitoids) and *Oidematops* (parasitoid/saprophage on a terrestrial snail). The biology of *A. pubera* (Foote *et al.* 1960) is similar to what might be postulated for primitive Sciomyzidae. Its seemingly close relationship to the behaviorally derived *Sciomyza*, including *S. varia*, which is one of the most specialized and intimately associated parasitoids in the family, and to the terrestrial *Oidematops* may indicate extensive extinction of genera with life histories bridging these forms. An alternative explanation, which we find more likely and which we see in other genera, could be a surprisingly rapid rate of evolution leading to quite different life styles in forms that are found to be closely related as based on adult morphology.

Atrichomelina pubera, distributed across North America and from southern Mexico to southern Canada, is an abundant, multivoltine, non-host-specific, saprophage/predator/parasitoid of exposed semi-aquatic and aquatic, non-operculate snails in diverse moist habitats. Females oviposit on the shell or other substrates. The larvae have been reared from five genera of snails collected in nature, and they feed on other genera, including terrestrial snails, in the laboratory. As noted in Chapter 1, the species also is able to develop on dead bivalves. Foote *et al.* 1960) stated,

> Apparently the larvae are associated with an ecological assemblage of snails and not with any particular taxonomic group . . . The manner in which *pubera* larvae feed on snails is variable and dependant on circumstances. They have parasitoid, predatory, and saprophagous capabilities, and the feeding of a given larva may even change day to day depending on the food available and the intensity of intraspecific competition.

This labile type of feeding behavior is seen in laboratory rearings of many other Sciomyzini and semi-terrestrial Tetanocerini, especially during the first part of larval life.

Pupariation by *A. pubera* is usually within the shell, but may occur outside. Overwintering is in the puparium as a quiescent pupa, without diapause. Morphologically, the immature stages are typical of the Sciomyzini. The large number of parasitoid Hymenoptera, six species in four families, reared from puparia of *A. pubera* in nature also might indicate that this is an ancient species relative to other Sciomyzidae, sufficient time having elapsed to allow parasites to exploit it. Unlike any other sciomyzids with somewhat similar behavior, habitats, and hosts/prey, many saprophagous Diptera (Ephydridae, Phoridae, Piophilidae, and Sarcophagidae) have been reared from snails also containing *A. pubera*. Such communal feeding might have been typical of the situation of primitive sciomyzids before behavioral mechanisms to prevent competition evolved.

17.7 ECOLOGICAL SPECIALIZATION AND GENERALIZATION

> If a man will begin with certainties, he shall end in doubts; but if he will be content to begin with doubts, he shall end in certainties.
>
> Bacon (1605).

An overview of specialized and generalized behavioral attributes is given in Section 4.2. Based on their unusual larval diets and diverse modes of feeding, the entire phaeomyiid + sciomyzid lineage may be considered specialized. However, there are many specialized sub-lineages within the higher lineage. In comparison to many other acalyptrate Diptera, the primary niche width of Sciomyzidae in terms of diet, mainly non-operculate freshwater and terrestrial snails, can be considered narrow. But within the phaeomyiid-sciomyzid lineage the niche widths of diet (diplopods, non-operculate and operculate snails, snail eggs, slugs, fingernail clams, and oligochaetes), behavior (saprophagous, parasitoid, predatory, and mixed), and microhabitat (aquatic to terrestrial) are broad. The direction of changes in niche widths have been postulated as from mixed saprophagous/predatory/parasitoid feeding on a broad ecological assemblage of semi-terrestrial and freshwater snails in shoreline or exposed, moist situations to various terrestrial-parasitoid and aquatic-predaceous restricted types. In fact, that is what is seen, not only in behavioral and ecological features, but in associated morphological features of the larvae (see classification of behavioral groups in Section 4.3). This polarity is largely supported by the cladistic analyses based on adult morphological features. Although there are no cladistic phylogenies available for the three largest genera (*Pherbellia* in the Sciomyzini and *Dictya* and *Sepedon* in the Tetanocerini), specialized species and groups of species within each of these genera are manifestly obvious. Chapman *et al.* (2006) detailed a phylogenetic comparative study of larval morphology and habitat which shows at least three independent transitions from aquatic to terrestrial in the aquatic and terrestrial snail- and slug-killing genus *Tetanocera*, with one reversal.

How did the phaeomyiid + sciomyzid lineage come to be so specialized in such different ways? There are several micro- and macroevolutionary generalizations, some of which are conflicting, that can be used to approach this question. We refer, below, to a few that seem applicable to Sciomyzidae. One classical view, perhaps applicable to the Sciomyzidae, is that when a lineage enters a new, unsaturated adaptive zone (in the case of Sciomyzidae, primarily malacophagy) radiation is promoted by temporary escape from competition and/or predation. A related view is that rapid diversification is a population-genetic consequence of ecological specialization per se. That is, ecological specialization with population fragmentation and diversifying selection promotes speciation by reducing gene flow. Futuyma & Moreno (1988) contended that the origins of specialization must be sought at the level of variation within populations, and that "trade-offs between adaptations in two environments . . . are central . . . to arguments for the evolution of specialization." This idea is of interest with regard to the Sciomyzidae where alternative responses – but sometimes more than just two alternative trade-offs – are found in some behavioral attributes, especially larval feeding behavior, of many species. However, Futuyma & Moreno (1988) concluded that,

> In many instances the successive evolution of numerous adaptations to a special resource or habitat constitutes an increasing commitment that makes reversion to a generalized habit, or shift to a very different specialization, increasingly unlikely . . .

This contrasts with their conclusion, more appropriate to the Sciomyzidae, that "Many, perhaps most, specialists arise from other specialists with either the same or different habits."

That is, for example: (1) repeated evolution of specialized terrestrial/predatory behavior seen among some genera in the specialized, aquatic/predatory Tetanocerini lineage (e.g., *Dichetophora*); (2) repeated evolution of specialized, terrestrial/parasitoid species within some specialized aquatic/predatory Tetanocerini genera (e.g., *Tetanocera*, Chapman *et al.* 2006); (3) evolution of the aquatic oligochaete-feeding *Sepedonella* from the lineage of snail-feeding *Sepedon* and related genera; and (4) evolution of the aquatic oligochaete-feeding *Sepedon* (*M.*) *knutsoni* within the snail-feeding genus *Sepedon*. The fact that some Naididae oligochaetes are commensals on freshwater snails suggests an evolutionary pathway for the development of oligochaete feeding by some Sciomyzidae.

Chapman *et al.* (2006) began their studies of the biologically diverse species of the Holarctic genus *Tetanocera* with the questions (1) do the aquatic (13) and terrestrial (14) species each comprise distinct lineages or were there multiple habitat transitions, and (2) what was the ancestral larval habitat? They concluded that "the ancestral larval habitat was aquatic" and "phylogenetic niche conservation has been responsible for the maintenance of aquatic associated larval morphological features in *Tetanocera*." They characterized phylogenetic niche conservation as involving the action of stabilizing selection on phenotypic traits when ancestral ecological conditions are maintained within a lineage. They also concluded that "concerted convergence and/or gene linkage was responsible for parallel morphological changes that were derived in conjunction with habitat transition."

In a few isolated cases, generalists may have evolved from specialized ancestors, for example, perhaps the "shoreline" predatory/parasitoid *Tetanocera* species, *T. fuscipennis* and *T. silvatica*, similar in behavior to *Atrichomelina pubera*, from the specialized aquatic predatory *Tetanocera* lineage. On a macro-evolutionary level the common presumption that specialists evolve from generalists supports the argument for placing the specialist Phaeomyiidae and Sciomyzidae in the same lineage, with the sister group of that lineage to be found in the generalist part of the dryomyzid lineage.

17.8 CLADISTIC-PHYLOGENETIC BASIS FOR EVOLUTIONARY CONSIDERATIONS

If you come to a fork in the road, take it.

Yogi Berra.

It is now generally believed that consideration of evolutionary patterns should have a basis in cladistic phylogeny. A cladistic phylogeny of the generic and suprageneric clades of the Sciomyzidae was produced by Marinoni & Mathis (2000), and re-analyzed by Barker *et al.* (2004). Although 33 of the 50 genera that Marinoni & Mathis (2000) analyzed are in 13 unresolved polytomies "due to the great number of homoplasies and low number of synapomorphies" at least the major clades within the family were proposed. This seems to be a reasonable delineation of primitive and derived lineages. As their study is used as a framework on which we display and discuss our behavioral groups (Fig. 17.3), we note their major conclusions and how the cladogram of Barker *et al.* (2004) differs. Marinoni & Mathis (2000) utilized one behavioral attribute of larvae (feeding on Diplopoda or Mollusca) associated with one morphological character of larvae (presence or absence of a serrate ventral arch below the mouthhooks) and 36 morphological characters of adults, which included 17 characters of the male postabdomen. Their analysis was based on the type species of 50 of the 61 genera of Sciomyzidae – they excluded the Huttonininae from the Sciomyzidae – and on the (then) sole genus (*Pelidnoptera*) in the Phaeomyiidae. We have interpolated *Ditaeniella* and *Eulimnia* into their cladogram as the life histories of both of these genera are known. Of the seven remaining genera not analyzed by Marinoni & Mathis (2000) or Barker *et al.* (2004), including the two genera of Huttonininae, none is biologically known. Thus, in the modified cladogram biologies of 38 of the 53 genera are known.

The cladistic phylogenies agree remarkably well with the classification of genera and the suprageneric categories proposed in the many papers by the dominant taxonomist of the family, G. C. Steyskal (published 1938 to 1980), including descriptions of eight of the 61 valid genera. As described in Chapter 19 on the history of research, Steyskal did not propose his classifications in terms of Hennigian cladistics, but clearly he was well versed in this approach to systematics. Thus, we suggest that in those parts of the Marinoni & Mathis (2000) phylogeny that are not resolved, considerable reliance be placed on Steyskal's classifications. Also, to a lesser extent the classifications of Verbeke (1948 to 1967), primarily with reference to the Afrotropical and Palearctic faunas, especially Afrotropical *Sepedon* and related genera, probably were based to some extent on Hennigian cladistic rationale.

Marinoni & Mathis (2000) used *Pelidnoptera*, represented by *P. fuscipennis*, as the outgroup and considered it as the potential sister group, based on the non-malacophagous habits of its larvae and lack of a ventral arch. Barker *et al.* (2004) omitted those two characters in their re-analysis because they did not believe that immature and adult stage characters should be combined in cladistic analyses, but they included seven additional adult characters. With the discovery (Vala *et al.* 2000b, 2002) of non-malacophagous behavior – obligate feeding on freshwater oligochaetes – in the true but highly derived sciomyzids *Sepedonella nana* and *Sepedon* (*M.*) *knutsoni* and possibly facultative oligochaete feeding in *Sepedon* (*P.*) *ruficeps*, selection of *Pelidnoptera* on the basis of non-malacophagous habits could be challenged. Also, the absence of a ventral arch in the cephalopharyngeal skeleton of larvae of *Pelidnoptera nigripennis* (Fig. 14.42) (Vala *et al.* 1990) must be considered a plesiomorphy and its presence in *S. nana*, *S.* (*M.*) *knutsoni*, and *S.* (*P.*) *ruficeps* an apomorphy. It is essential to note that this character is not found in any known Acalyptratae except Sciomyzidae and a few unrelated families. Thus, it cannot be used to support the designation of the Phaeomyiidae as the sister group of Sciomyzidae as originally proposed by Griffiths (1972) and supported by J. F. McAlpine (1989). The family-level rank given to *Pelidnoptera* was questioned by D. K. McAlpine (1991 and personal communication, 1999). He considered the clypeus being reduced and remote from the face as the only autapomorphy for the Sciomyzidae. He included *Pelidnoptera* and excluded the Huttonininae in which the clypeus is well developed. He considered the phylogenetic significance of millipede feeding as doubtful. For the purpose of discussing the evolution of behavior and ecology we think it is more appropriate to consider a parasitoid-predatory, phaeomyiid-sciomyzid lineage with its sister group in the saprophagous part of the dryomyzid lineage. Based on extant organisms, that part of the Dryomyzidae that has a large clypeus is a plausible sister group. But it is even more likely that the true sister group of the phaeomyiid-sciomyzid lineage is an extinct group. While we do not encourage the creation of phantom sister groups, it seems reasonable in this case.

Marinoni & Mathis (2000) referred to their study as a "preliminary" cladistic analysis. They and Barker *et al.* (2004) confirmed the Sciomyzidae and the subfamilies Salticellinae and Sciomyzinae to be monophyletic. *Salticella* is at the base of both cladograms. Its closer relationship to the Sciomyzini than to the Tetanocerini as based on the morphology of larval characters and behavior (Knutson *et al.* 1970, Coupland *et al.* 1994) was corroborated. The monophyly of the two main clades in the Sciomyzinae, the Sciomyzini and the Tetanocerini, was confirmed. *Eutrichomelina* was transferred from the Sciomyzini to the Tetanocerini. In the Sciomyzini, the group *Atrichomelina* + *Sciomyza* + *Oidematops* at the base of that clade have well-defined relationships in both cladistic analyses. In the Tetanocerini, it is notable that in the cladogram of Marinoni & Mathis (2000) the clam-killer *Renocera* and snail-egg feeder *Anticheta*, both of which share a different set of larval morphological characters with both the Sciomyzini and Tetanocerini, are at the base of the clade (with *Chasmacryptum*, *Ectinocera*, and *Eutrichomelina*; all three biologically unknown). Whereas, in the cladogram of Barker *et al.* (2004) *Renocera* and *Anticheta* are remote from each other. However, the only other clam-killing Sciomyzidae, *Eulimnia philpotti* and *Ilione lineata*, are in genera typical of the Tetanocerini, near the middle and upper end of that clade, and do not share any characters, except at the subfamily level, with the Sciomyzini. In the cladogram of Marinoni & Mathis (2000), after the base of the Tetanocerini clade, *Shannonia* + *Perilimnia*, which are predators/saprophages of non-operculate, primarily freshwater snails on moist surfaces, appear as the sister group of the rest of the genera of Tetanocerini. In the cladogram of Barker *et al.* (2004) *Shannonia* + *Perilimnia* also are found next to *Anticheta* but are separated from *Renocera* by seven genera. The bulk of the genera in the rest of the Tetanocerini clade of Marinoni & Mathis (2000) form five groups. Most genera are in polytomies, but two groups, *Hoplodictya* + *Dictya* and *Limnia* + *Poecilographa* + *Pherbecta* + *Protodictya*, and one subgroup, *Thecomyia* to *Sepedonea*, are monophyletic. The latter subgroup, at the end of both cladograms, is the most corroborated lineage in the analyses, established by eight characters. Thus it is notable that the most plesiomorphic and most apomorphic genera of the Tetanocerini are well established in the cladogram.

Chapman *et al.* (2006) used 14 species in two genera of Sciomyzini and six genera of Tetanocerini as the outgroup in their cladistic analysis of parallel evolution of larval morphology and habitat in *Tetanocera*. They sequenced one nuclear and three mitochondrial DNA loci to construct a phylogenetic framework for their application of behavioral and morphological traits. They also conducted maximum likelihood analyses of character state transformations of four larval morphological characters of 17 species of *Tetanocera* and the outgroup species. In their cladogram, *Atrichomelina pubera* and *Sciomyza simplex* appeared in the Sciomyzini and the rest of the genera appeared in the Tetanocerini, as expected. Surprisingly, the three species of *Sepedon* appeared near the base of their cladogram, as the sister group of the two Sciomyzini genera, above.

Elsewhere in this book, the *Sepedon* lineage is more or less convincingly, we believe, shown as the most derived of all Sciomyzidae. The lineage appears at the derived end of the cladograms of Marinoni & Mathis (2000) and Barker *et al.* (2004). In the cladogram of Chapman *et al.* (2006) the two species of *Renocera* appear as the sister group of *Tetanocera*, as surmised by Steyskal (1959). But *Renocera* was remote from *Tetanocera* in the cladograms of Marinoni & Mathis (2000) and Barker *et al.* (2004). Of course, only a few genera of Tetanocerini were included in the study of Chapman *et al.* (2006), and they did not discuss the relationships of the genera in the outgroup.

We show the cladogram of Marinoni & Mathis (2000) with some of the most important morphological characters of the immature stages (significant at the generic and suprageneric levels) mapped onto it (Fig. 17.3). The characters and character states (0, plesiomorphic; 1–4, apomorphic) are given below for the egg, third-instar larva, and puparium. Highly apomorphic features unique to a species within a genus are not included, except character 8. Other important morphological characters of the immature stages are discussed in Sections 14.1 and 15.1.2.

Egg

1. Egg chorion entirely smooth = 0, with hexagonally reticulate sculpturing but without longitudinal ridges and grooves = 1, with surface sculpturing and more or less anastomosing longitudinal ridges and grooves = 2, laid on snail egg masses and covered with gelatinous coat except reticulate patches exposed dorsally = 3, with reticulate sculpturing, ridges and grooves, and pairs of dorsolateral and ventrolateral ridges = 4.

Larva (third instar)

2. Oral grooves radiating from pre-oral cavity = 0, oral grooves absent = 1.
3. Ventral cibarial ridges present = 0, absent = 1.
4. Postoral spinule patch weak to moderately developed, restricted to ventral surface of segment I, without sclerotized posterior margin = 0, postoral spinule band well developed, almost encircling segment I, with strong posterior margin = 1, postoral spinule band very strong, completely encircling cephalic segment = 2.
5. Cephalopharyngeal skeleton without dental sclerites or a ventral arch = 0, ventral arch present = 1.
6. Ventral arch not fused with mouthhooks in third instar = 0, fused = 1.
7. Anterior margin of ventral arch with a few teeth laterally = 0, with many teeth along entire margin = 1.
8. Mouthhook without accessory teeth on anterior margin just below hook part = 0, with 1 (*Salticella fasciata*) or several (*Dictya ptyarion*) denticles ventrobasally = 1, with 1 to 7 teeth on anterior margin = 2.
9. Ligulate sclerite absent = 0, present, paired or tripartite (*Eulimnia*) = 1, fused into a single structure = 2.
10. H-shaped and pharyngeal sclerites separate or (*Ilione lineata*, *Dictya* spp.) faintly connected = 0, completely fused = 1.
11. Without accessory rods above parastomal bars = 0, accessory rods present = 1.
12. Bridge anterodorsally between pharyngeal sclerites present = 0, absent = 1.
13. Indentation index of pharyngeal sclerite 60–90 = 0, 19–56 = 1.
14. Dorsal cornu of pharyngeal sclerite with window = 0, without = 1.
15. Anterior spiracle rounded or transverse apically = 0, oval or elongate = 1.
16. Posterior interspiracular processes minute, scale-like = 0, branched, moderately to well developed = 1.
17. Posterior spiracular disc with at most weak, rounded marginal lobes = 0, lobes weak except ventral and/or ventrolateral moderately developed to elongate = 1, lateral and/or ventrolateral lobe moderately elongate and ventral lobe very elongate = 2.
18. Integument of segments V–XII unpigmented = 0, pigmented = 1.
19. Integument of segments V–XII bare, except at least some segments with ventral spinule patches = 0, bare or with ultramicroscopic spinules and with encircling macroscopic spinule bands on most segments = 1, apparently completely bare, but ultra-microscopic integumentary processes may be present, without spinule patches or bands = 2, with coat of small to large, transparent processes and at least a few ventral spinules patches = 3, with coat of minute, scale-like or hair-like processes and/or patches or rows of seta-like processes but without ventral spinule patches = 4, entire body covered with heavy, black spinules = 5.
20. Body segments weakly subdivided, without strong lobes or tubercles = 0, each segment more or less strongly subdivided into rings, with encircling or partly encircling set of lobes or tubercles = 1.

Puparium

21. Puparium with posterior spiracles sessile, at or slightly above or below mid horizontal axis = 0, posterior spiracles elevated above dorsal surface = 1, on elongate extension of segment XII = 2, well below mid horizontal axis = 3, formed in snail shells (Sciomyzini), inside grass stems (*Hoplodictya setosa*), or inside remains of millipede host (Phaeomyiidae, *Pelidnoptera*) = 4.

Note: Character states 0 to 3 are for puparia never formed within shells.

The distribution of characters of the immature stages agrees well with the family, subfamily, and two tribal clades identified on the basis of adults in the cladistic analyses and with the classification of Steyskal (1965a), except *Anticheta* and *Renocera* (see Section 15.4). Interspiracular processes were previously heavily relied upon to characterize the Salticellinae and Sciomyzini by their presumed absence in those clades and presence in all except terrestrial Tetanocerini. Recent stereoscan studies have shown their presence, although reduced, in *Salticella fasciata* and in those Sciomyzini examined, terrestrial Tetanocerini, and in the Phaeomyiidae. We found four apomorphic characters of the immature stages to separate the Sciomyzidae clade from the outgroup Phaeomyiidae, four to separate the Sciomyzinae from the Salticellinae, and five to separate the Tetanocerini clade (except *Renocera* and *Anticheta*) from the Sciomyzini clade. On the basis of adults, Marinoni & Mathis (2000) found one apomorphic character separating Salticellinae, one separating Sciomyzinae, and five separating the Sciomyzini.

The immature stages of all but 13 of the 53 genera in the cladogram have been described, but only the puparium is known for *Pherbecta* and only the egg, first-instar larva, and puparium are known for *Poecilographa*. However, for the latter two genera important characters of the third-instar larva, especially of the cephalopharyngeal skeleton, can be seen as vestiges in the puparium. The Phaeomyiidae (based on *Pelidnoptera nigripennis*) is separated from the Sciomyzidae primarily by the plesiomorphic lack of a ventral arch, completely encircling postoral spinule band, and a smooth egg chorion. Salticellinae (based on *Salticella fasciata*) has a ventral arch but it is quite different from all other Sciomyzidae in having just a few teeth restricted to the anterolateral corners rather than the entire anterior margin with teeth. The presence of oral grooves and ventral cibarial ridges in Salticellinae are ground-plan characters of the Muscomorpha, lacking in all other Sciomyzidae and Phaeomyiidae. Lack of accessory teeth just below the mouthhook in Salticellinae is a plesiomorphic character state shared with the Sciomyzini, but the presence of a basal denticle on the mouthhook is an apomorphic feature unique to Salticellinae. The aberrant, unrelated Tetanocerini *Dictya ptyarion* also lacks typical accessory teeth, but has a row of eight denticles at the base of the mouthhook (Fig. 14.38f); other *Dictya* spp. have typical accessory teeth. Most other characters of Salticellinae are shared, in the plesiomorphic state, with the Sciomyzini.

Within the Sciomyzinae, the Sciomyzini have a preponderant uniformity of plesiomorphic character states. Especially significant characters separating them from the Tetanocerini are the lack of anastomosing ridges and grooves on the egg chorion (present in *Ditaeniella*), postoral spinules reduced, lack of accessory teeth, presence of a dorsal bridge, presence of a window in the dorsal cornu of the pharyngeal sclerite, an indentation index of at least 60, presence of at least ventral spinule patches, and the muscoid appearance of the larvae. Within the Sciomyzini, *Tetanura* has several apomorphic character states in the immature and adult stages, the latter so striking that a separate subfamily has been proposed, independently, for the genus four times (see Section 15.4). Notably, recognition of *Ditaeniella* as a distinct genus based on adult characters is supported by at least two character states of the egg and larva.

The preponderance of important apomorphic character states in the Tetanocerini supports their position, as a monophyletic lineage, in the upper part of the cladogram, except for the enigmatic genera *Renocera* and *Anticheta* near the base of the Tetanocerini lineage. Both of these genera share unique combinations of character states with the Sciomyzini and Tetanocerini (see Section 15.4). Unfortunately, a few genera on each side of *Renocera* and *Anticheta* in the cladograms are biologically unknown, these being perhaps the most important targets for further research on the biology and morphology of the immature stages. While the likely large extent of homoplasy in the Sciomyzinae, especially in the large genera *Pherbellia*, *Sepedon*, and *Tetanocera*, will be difficult problems to resolve, further cladistic and phylogenetic studies based solely on the morphology of the immature stages, or on the immature and adult stages combined, should be rewarding.

Key questions for further cladistic studies are (1) which monophyletic taxa to include in the Sciomyzoidea; we agree with the conclusion of Yeates & Wiegmann (1999) that "the relationships between acalyptrate superfamilies remain obscure," (2) whether the Phaeomyiidae should be included in the Sciomyzidae or considered as the sister group, and (3) as a result of such studies, re-analysis of the rank to be given to the putative tribes and subfamilies of various authors, as shown in Table 15.14. Comparison of behavioral evolution based on phylogenetic tests in the large and behaviorally and morphologically diverse genera *Pherbellia*, *Sepedon*, and *Tetanocera*, that include both parasitoids and predators, will be especially interesting. The "*Sepedon* Group" of genera (*Sepedon*, *Sepedonea*, *Sepedonella*, *Sepedoninus*, *Sepedomerus*, and *Thecomyia*, plus possibly *Ethiolimnia* and *Teutoniomyia*), the most corroborated lineage in the cladistic analyses, and which includes the freshwater oligochaete-feeding *Sepedonella nana* and *Sepedon* (*M.*) *knutsoni*, is a critical group.

Recognition of the *Thecomyia* to *Sepedon* group as a lineage, tribe, or subfamily is discussed by Knutson & Vala (unpublished data) in their analysis of about 50 primarily apomorphic characters known for one to all species of the group. Other key relationships to define more precisely are the apparently unrelated Tetanocerini clam-killers *Renocera*, *Eulimnia* (not included in the cladistic analyses), and *Ilione lineata* and snail-egg feeders (*Anticheta*).

17.9 SUBSEQUENT ADAPTIVE RADIATION

> A mystery solved is a mystery filed and at least half forgotten. A mystery unsolved, subject to endless speculation, has the power to haunt us as long as memory persists, as long as the human animal retains its power to sip Chardonnay and contemplate life's enigmas.
>
> Schickel (2005).

The earlier scenarios of evolution of behavior within the Sciomyzidae presented by Berg *et al.* (1959), Foote *et al.* (1960), and Berg (1966) assumed that the suprageneric classification of Steyskal (1965a), with most of the species in the Sciomyzini and Tetanocerini, represented true evolutionary lineages. The evolutionary veracity of this classification was supported by the subsequent phylogenetic and cladistic analyses of Hennig (1965), Barnes (1979d, 1981), Marinoni & Mathis (2000), and Barker *et al.* (2004). The evolutionary scenario presented by Knutson & Vala (2002) used the cladistic analysis of Marinoni & Mathis (2000) as a framework for discussion. Here, we build upon that discussion and incorporate results and ideas from the cladistic analysis, linkage of sciomyzid-molluscan clades, and ecomorphological classification and analysis of Barker *et al.* (2004).

The earlier comparisons of the biology and morphology of the immature stages of Sciomyzidae resulted in characterizations of the Sciomyzini primarily as terrestrial-parasitoids/predators/saprophages and Tetanocerini primarily as aquatic predators. But even in those early proposals both tribes were recognized as having representatives more or less typical of each of these major behavioral groups, as well as including species that are intermediate or mixed in behavior. In general, subsequent publications on the biology and morphology of immature stages of additional species have tended to support the main thesis that terrestrial-parasitoid behavior is most highly developed in the Sciomyzini (e.g., *Sciomyza varia*, Barnes 1990) and aquatic-predaceous behavior is most highly developed in the Tetanocerini (e.g., *Hedria mixta*, Foote 1971). The morphological distinctions between the larvae of the two tribes, with adaptations for an aquatic existence only in the Tetanocerini but some Tetanocerini modified like the Sciomyzini for a terrestrial existence, also have been maintained (e.g., Vala & Gasc 1990b). However, the distinction between terrestrial parasitoid/predaceous/saprophagous Sciomyzini and aquatic predaceous Tetanocerini has been modified further by discovery of fairly intimately associated terrestrial parasitoids/saprophages in the primarily aquatic predaceous Tetanocerini genera *Tetanocera* (e.g., Foote 1996b) and *Sepedon* (e.g., Vala *et al.* 1995). The distinction also was eroded by the discovery of terrestrial parasitoid/predaceous/saprophagous behavior with correlated morphological characters in several other genera of Tetanocerini (*Dichetophora*, Vala *et al.* 1987; *Euthycera*, Vala & Caillet 1985; *Limnia*, Vala & Knutson 1990; and *Trypetoptera*, Vala 1986).

Even the supposition of strict malacophagy of Sciomyzidae no longer is valid, with the discovery that the apomorphic Tetanocerini *Sepedonella nana* and *Sepedon knutsoni* (Vala *et al.* 2000a, 2000b, 2002) are restricted to freshwater oligochaete worms. The lack of malacophagy and a ventral arch below the mouthhooks in the larvae (Vala *et al.* 1990) and some characters of the adults (Griffiths 1972) of *Pelidnoptera nigripennis*, a parasitoid of millipedes, probably should be considered plesiomorphies.

Our present evolutionary scenario of Sciomyzidae does not conflict drastically with the main aspects of the scenario of Berg *et al.* (1959), Foote *et al.* (1960), and Berg (1966). We discuss below some key points and display the distribution of behavioral attributes and a more dissected resolution of behavioral groups on the cladistic framework of Marinoni & Mathis (2000). We also discuss these points in relation to the cladistic analysis of Barker *et al.* (2004) to the extent that it does not agree with that of Marinoni & Mathis (2000). The families, subfamilies, and tribes are indicated below the cladograms in Figs 17.2, 17.3. The numbers of synapomorphic characters used to establish dichotomies are indicated on the cladogram. The geographical distribution and behavioral attributes or behavioral groups are indicated above each genus name. The biology of *Salticella fasciata*, (Salticellinae) (Knutson *et al.* 1970, Coupland *et al.* 1994), the assumed plesiomorphic sister group of the Sciomyzinae, discussed in more detail in Section 18.2, remains unclear. Its combination of oviposition on the terrestrial host, host specificity, and mixed parasitoid and saprophagous larval feeding habits does not fit well in the evolutionary scenarios projected to date. The feeding behavior of the larvae deserve further study. Unfortunately, the clam-killer *Eulimnia philpotti* was not included in the cladistic analyses but obviously this behavior has arisen separately at least three times (*Renocera*, *Ilione*, *Eulimnia*) in three widely separated areas (Palearctic + Nearctic, Palearctic, and Subantarctic, respectively).

The display of Behavioral Groups on the cladogram (Fig. 17.2) shows the Sciomyzini clumped together primarily in Groups 1–4 and 6 but the Tetanocerini scattered in all Groups, except those dominated by the Sciomyzini (1, 3, and 12b). The distribution of the Tetanocerini in Behavioral Groups is highly diffuse in comparison to the positions of the genera in the tribal cladogram. Figure 17.2 shows that a large number of genera have "experimented" with different life styles. Of the largest genera, *Pherbellia* (Sciomyzini) occurs in six Groups and *Tetanocera*, *Dictya*, and *Sepedon* (Tetanocerini) occur in six, three, and four Groups, respectively. Group 11 (typical aquatic predators) is represented by the greatest number of genera (14), all in the Tetanocerini. Only two Groups are broadly shared by the Sciomyzini and Tetanocerini, with four genera of Sciomyzini and six of Tetanocerini in Group 2 and with three genera of each in Group 4. Within both tribes there are few concordances between the distribution of genera in Behavioral Groups and their position in the cladogram. Notable concordances are: (1) *Atrichomelina* (with *Salticella*) in Group 1 and at the base of the Sciomyzini clade; (2) the clam-killing *Renocera* and snail-egg feeding *Anticheta*, both of which share larval morphological characters with the Sciomyzini and Tetanocerini, near the base of the Tetanocerini clade; (3) *Shannonia* and *Perilimnia*, predators/saprophages of non-operculate snails on wet/damp surfaces, as the sister group to the rest of the Tetanocerini; and (4) at the other end of the Tetanocerini clade, the apical position of the oligochaete-feeding *Sepedonella*.

The display of generalized and specialized behavioral attributes in all life stages also shows a very reticulate pattern of distribution across the Phaeomyiidae and Sciomyzidae. There are different sets of specialized attributes found in almost all of the genera. Many specialized attributes obviously have evolved separately in both families, in both subfamilies of Sciomyzidae, in both tribes of Sciomyzinae, and within many genera. The larval stage displays the greatest number of specialized attributes and their more frequent occurrence. The adult stage displays the least number and their most infrequent occurrence. There is little concordance in the number of specialized attributes, but rather strong concordance in the kind of specialized attributes between the Sciomyzini and Tetanocerini and between many genera and their position in the cladogram.

Many of the diverse life styles obviously have evolved separately in many genera of both tribes and within several genera. This is evident especially in three of the largest genera, *Pherbellia* in the Sciomyzini and *Tetanocera* and *Sepedon* in the Tetanocerini. The reticulate and diffuse pattern of evolution of host/prey selection among genera of Sciomyzidae also is illustrated by the number of occasions on which unusual host/prey selection have arisen in mostly unrelated genera (Table 17.1).

It seems likely that species with unusual hosts/prey and feeding behavior would be less common than the more generalist aquatic, semi-terrestrial, and terrestrial predators. However, the limited quantitative data from broad geographical areas, from four publications as analyzed below, clearly indicate that many specialized species are outstandingly successful. Notably, these include several slug killers, which are among the very few members of the Class Insecta restricted to slugs. **(1)** Chandler (1972) recorded Sciomyzidae from 40 of the 51 vice counties in his review of the 40 species of Sciomyzidae then known from Ireland (51°–55°N. Lat.). The terrestrial parasitoid-predator *Pherbellia dubia* was the most widespread (15 vice counties), the clam-killer *Ilione lineata*, with the aquatic snail-killers *Hydromya dorsalis* and *Tetanocera hyalipennis*, ranked number sixth from the top, but the slug-killer *Tetanocera elata* (with *T. robusta*) ranked ninth. **(2)** Fisher & Orth (1983) collected about 20 000 specimens of the 49 species occurring in California (32°–42°N. Lat.) during

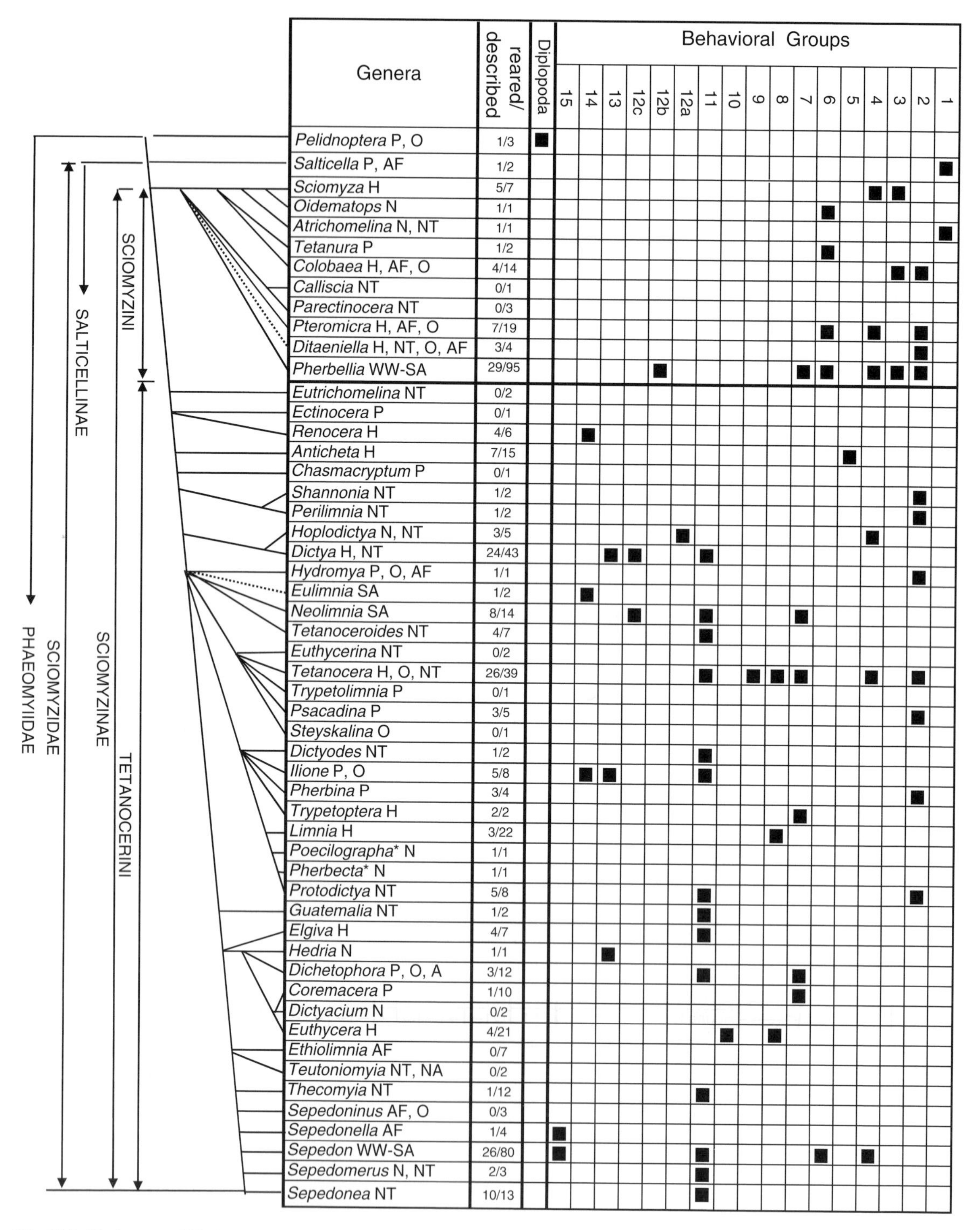

Fig. 17.2 Cladogram of Marinoni & Mathis (2000), with Behavioral Groups. Behavioral Groups are given in Section 4.3. *, host/prey unknown, reared from puparium. Modified from Knutson & Vala (2002). See Table 16.1 for abbreviations of zoogeographical regions given after genera. For Phaeomyiidae, see Note in Fig. 17.1.

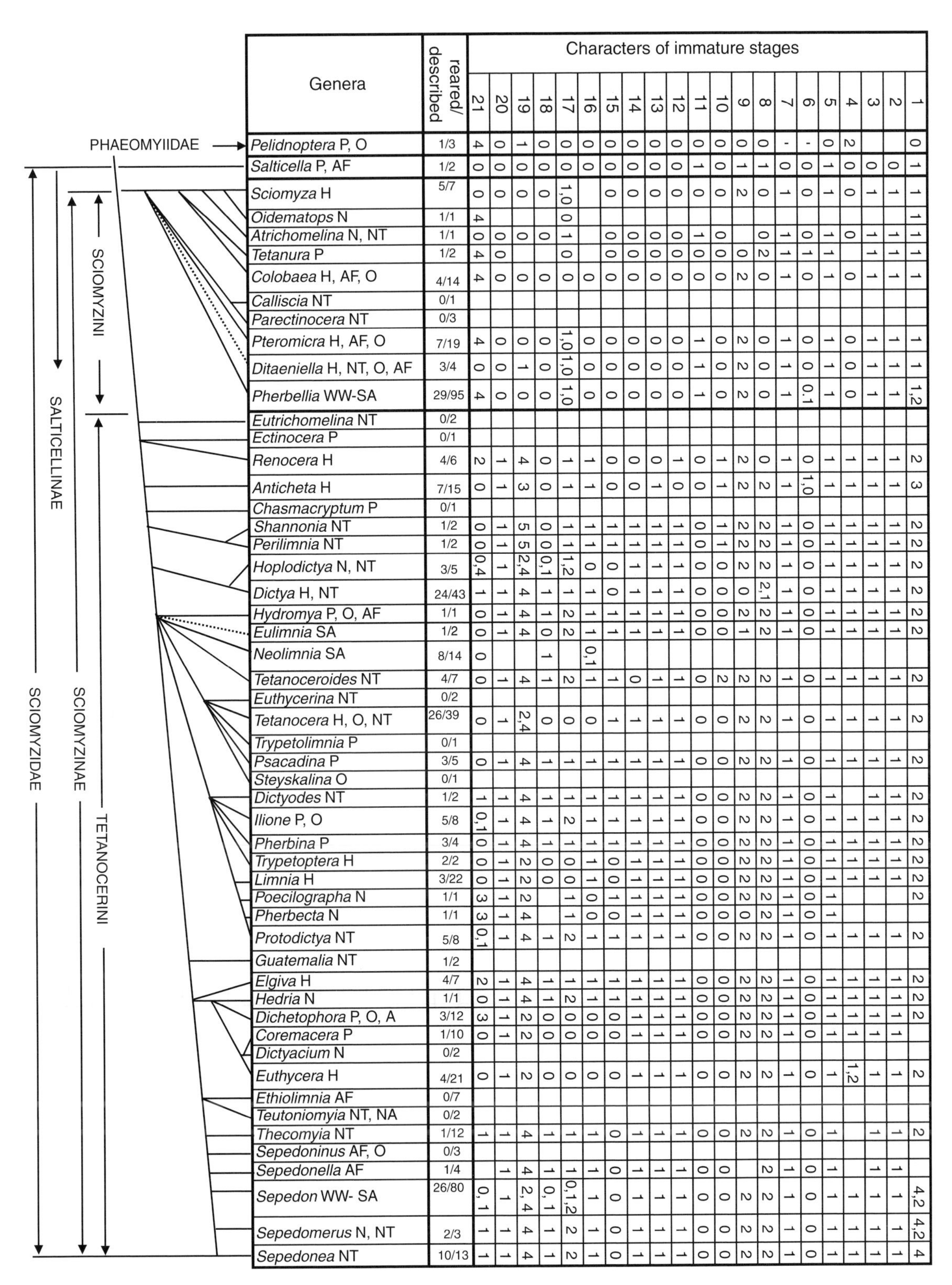

Genera	reared/ described	Characters of immature stages																				
		21	20	19	18	17	16	15	14	13	12	11	10	9	8	7	6	5	4	3	2	1
Pelidnoptera P, O	1/3	4	0	1	0	0	0	0	0	0	0	0	0	0	0	-	-	0	2			0
Salticella P, AF	1/2	0	0	0	0	0	0	0	0	0	0	1	0	1	1	0	0	1	0	0	0	1
Sciomyza H	5/7	0	0	0	0	1,0		0	0	0	0	0	0	2	0	1	0	1	0	1	1	1
Oidematops N	1/1	4				0																1
Atrichomelina N, NT	1/1	0	0	0	0	1		0	0	0	0	1	0		0	1	0	1	0	1	1	1
Tetanura P	1/2	4	0			0		0	0	0	0	0	0	0	2	1	1	1		1	1	1
Colobaea H, AF, O	4/14	4	0	0	0	0	0	0	0	0	0	0	0	2	0	1	0	1	0	1	1	1
Calliscia NT	0/1																					
Parectinocera NT	0/3																					
Pteromicra H, AF, O	7/19	4	0	0	0	1,0	0	0	0	0	0	1	0	2	0	1	0	1	0	1	1	1
Ditaeniella H, NT, O, AF	3/4	0	0	1	0	1,0	0	0	0	0	0	1	0	2	0	1	0	1	0	1	1	1
Pherbellia WW-SA	29/95	4	0	0	0	1,0	0	0	0	0	0	1	0	2	0	1	0,1	1	0	1	1	1,2
Eutrichomelina NT	0/2																					
Ectinocera P	0/1																					
Renocera H	4/6	2	1	4	0	1	1	0	0	0	1	0	1	2	0	1	0	1	1	1	1	2
Anticheta H	7/15	0	1	3	0	1	1	0	0	1	0	0	1	2	2	1	1,0	1	1	1	1	3
Chasmacryptum P	0/1																					
Shannonia NT	1/2	0	1	5	0	1	1	1	1	1	1	0	1	2	2	1	0	1	1	1	1	2
Perilimnia NT	1/2	0	1	5	0	1	1	1	1	1	1	0	1	2	2	1	0	1	1	1	1	2
Hoplodictya N, NT	3/5	0,4	1	2,4	0,1	1,2	0	0	1	1	1	0	0	2	2	1	0	1	1	1	1	2
Dictya H, NT	24/43	1	1	4	1	1	1	0	1	1	1	0	0	0	2,1	1	0	1	1	1	1	2
Hydromya P, O, AF	1/1	0	1	4	1	2	1	1	1	1	1	0	0	2	2	1	0	1	1	1	1	2
Eulimnia SA	1/2	0	1	4	0	2	1	1	1	1	1	0	0	1	2	1	0	1	1	1	1	2
Neolimnia SA	8/14	0			1		0,1															
Tetanoceroides NT	4/7	0	1	4	1	2	1	1	0	1	1	0	2	2	2	1	0	1	1	1	1	2
Euthycerina NT	0/2																					
Tetanocera H, O, NT	26/39	0	1	2,4	0	0	0	1	1	1	1	0	0	2	2	1	0	1	1	1	1	2
Trypetolimnia P	0/1																					
Psacadina P	3/5	0	1	4	1	1	1	1	1	1	1	0	0	2	2	1	0	1	1	1	1	2
Steyskalina O	0/1																					
Dictyodes NT	1/2	1	1	4	1	1	1	1	1	1	1	0	0	2	2	1	0	1		1	1	2
Ilione P, O	5/8	0,1	1	4	1	2	1	1	1	1	1	0	0	2	2	1	0	1	1	1	1	2
Pherbina P	3/4	0	1	4	1	1	1	1	1	1	1	0	0	2	2	1	0	1	1	1	1	2
Trypetoptera H	2/2	0	1	2	0	0	1	0	1	1	1	0	0	2	2	1	0	1	1	1	1	2
Limnia H	3/22	0	1	2	0	0	1	0	1	1	1	0	0	2	2	1	0	1	1	1	1	2
Poecilographa N	1/1	3	1	2		1	0	1	1	1	1	0	0	2	2	1	0	1				2
Pherbecta N	1/1	3	1	4		1	0	0	1	1	1	0	0	0	2	1	0	1				
Protodictya NT	5/8	0,1	1	4	1	2	1	1	1	1	1	0	0	2	2	1	0	1	1	1	1	2
Guatemalia NT	1/2																					
Elgiva H	4/7	2	1	4	1	1	1	1	1	1	1	0	0	2	2	1	0	1	1	1	1	2
Hedria N	1/1	0	1	4	1	2	1	1	1	1	1	0	0	2	2	1	0	1	1	1	1	2
Dichetophora P, O, A	3/12	3	1	2	0	0	0	0	1	1	1	0	0	2	2	1	0	1	1	1	1	2
Coremacera P	1/10	0	1	2	0	0	0	0	1	1	1	0	0	2	2	1	0	1	1	1	1	
Dictyacium N	0/2																					
Euthycera H	4/21	0	1	2	0	0	0	0	1	1	1	0	0	2	2	1	0	1	1,2	1	1	2
Ethiolimnia AF	0/7																					
Teutoniomyia NT, NA	0/2																					
Thecomyia NT	1/12	1	1	4	1	1	1	0	1	1	1	0	0	2	2	1	0	1		1	1	2
Sepedoninus AF, O	0/3																					
Sepedonella AF	1/4		1	4	1	1	1	0	1	1	1	0	0		2	1	0	1		1	1	
Sepedon WW- SA	26/80	0,1	1	2,4	0,1	0,1,2	1	0	1	1	1	0	0	2	2	1	0	1	1	1	1	4,2
Sepedomerus N, NT	2/3	1	1	4	1	2	1	0	1	1	1	0	0	2	2	1	0	1	1	1	1	4,2
Sepedonea NT	10/13	1	1	4	1	2	1	0	1	1	1	0	0	2	2	1	0	1	1	1	1	4

Fig. 17.3 Morphological character states of immature stages applied to genera in the cladogram of Marinoni & Mathis (2000). 0, plesiomorphic; 1–4, apomorphic; -, character not present; blank, no data. Where two character states are present, the more common state is shown first. Questionable character states indicated by ?. See Table 16.1 for abbreviations of zoogeographical regions given after genera. For Phaeomyiidae, see Note in Fig. 17.1.

Table 17.1. Number of occasions on which unusual host/prey selection and feeding behavior (i.e., other than predation on aquatic, shoreline, or terrestrial, non-operculate snails) has arisen in the Sciomyzidae

*Number of occasions	Hosts/prey	Genera/species
Once	a. Eggs of Lymnaeidae, Physidae, and Succineidae	*Anticheta* (H)
		Facultative in *Tetanocera ferruginea* (H) and *Hydromya dorsalis* (P)
	b. Primarily saprophagous on snails	<u>*Salticella fasciata*</u> (P)
		Facultative in ***Atrichomelina pubera*** (N) (also on large bivalves)
	c. Operculate snails living above the water-line in coastal habitats	*Hoplodictya setosa* (N)
Twice	a. Estivating *Lymnaea*	***Colobaea bifasciella*** (P)
		Sciomyza varia (N)
	b. Slugs	*4 *Tetanocera* spp. (H)
		Euthycera chaerophylli (P)
		Facultative in some other *Euthycera* spp. (P, N)
	c. Freshwater oligochaetes	*Sepedonella nana* (AF)
		Sepedon knutsoni (AF)
		Facultative in *Sepedon ruficeps* (AF)
Three times	a. Fingernail clams	4 *Renocera* spp. (P, N)
		Ilione lineata (P)
		Eulimnia philpotti (SA)
	b. Freshwater and brackish water operculate snails	***Pherbellia prefixa*** (N)
		Neolimnia tranquilla (SA)
		*4 *Dictya* spp. (N)
Four times	a. Parasitoids of terrestrial snails	***Oidematops ferrugineus*** (N)
		*Several ***Pherbellia*** spp. (P, N)
		A few ***Pteromicra*** spp. (N)
		Tetanura pallidiventris (P)
Seven (+) times	a. Semi-terrestrial Succineidae	***Sciomyza aristalis*** (N)
		S. testacea (P) ?
		S. dryomyzina (H)
		Pteromicra anopla (H) ?
		Pherbellia schoenherri (H)
		Hoplodictya spinicornis (N, NT)
		* *Tetanocera* spp. (P, N, H)
		Sepedon hispanica (AF, P)
		S. trichrooscelis (AF)
		Limnia unguicornis (P) ?
		L. boscii (N)

Note: *Genera in which the host/prey selection probably arose separately in different lineages within the genus. Sciomyzini in bold italics, Tetanocerini in italics, ***Salticella fasciata*** [Salticellinae] underlined.

1962 to 1976. Ten of these species have unusual biologies (four egg/snail killers, three succineid killers, one parasitoid of estivating *Lymnaea*, one slug killer, and one fingernail clam killer). Of the 58 counties, one egg/snail killer (*Anticheta testacea*) was found in 20 counties and one succineid killer (*Hoplodictya acuticornis*) was found in 28 counties, both throughout the state. The most common species were the predator of exposed aquatic snails, *Pherbellia n. nana* (48 counties) and the aquatic predator, *Sepedon pacifica* (47 counties). (3) Ball & McLean (1986) summarized 6703 records of the 64 species of Sciomyzidae occurring in Great Britain (50°–56°N. Lat.) by 10 km^2 sample plots and vice counties. They noted that "the rarer species have been investigated very much more thoroughly than the commoner ones and therefore appear more frequent than is really justified." Nevertheless, the slug-killer *T. elata* was the most common species (347 records) and most widespread (in 240 of 937 of the 10 km^2 sample plots; 74 of 105 vice counties). The clam-killer *Renocera pallida* was the fifth most common species (241 records) and the highly specialized, terrestrial parasitoid *Tetanura pallidiventris* was the 27th, with the clam-killer *I. lineata* 28th. (4) Most of the 72 species known from Ohio (about 38°–42°N. Lat.) have been recorded from the northeastern part of the state, from less than ten of the 88 counties (Foote & Keiper 2004). Surprisingly, the specialized, Holarctic, slug-killer *Tetanocera plebeja* is equal to the most widely dispersed species, occurring in 18 counties, followed closely by the specialized parasitoid/predator of slugs and snails, *Euthycera arcuata*, from 16 counties and the slug parasitoid/predator *Tetanocera clara*, from 10 counties. The generalist aquatic predators *Tetanocera plumosa*, *T. vicina*, *Sepedon armipes*, and *S. fuscipennis* were recorded from 18, 13, 17, and 17 counties, respectively.

The spotty, dispersed occurrence of species, often closely related species, of the same genera or closely related genera in the broad spectrum of the Behavioral Groups supports criticisms (e.g., Hawkins & MacMahon 1989) of the assumption (e.g., Wiggins & Mackay 1978) that genera or families represent major ecological themes and that species are minor variations. *Tetanocera* is clearly monophyletic, although it appears in a polytomy with four other genera in the middle of the cladogram. *Sepedon* and especially *Pherbellia* are probably polyphyletic and appear at opposite ends of the cladogram. Genera of the two tribes are distributed broadly among the 15 Behavioral Groups: Sciomyzini in 1, 2, 3, 4, 6, 7, and 12 b, and Tetanocerini in 2, 4, 5, 6, 7, 8, 9, 10, 11, 12 a, c, 13, 14, and 15. This distribution tends to support the original distinction between terrestrial-parasitoid/predaceous/saprophagous Sciomyzini and aquatic-predaceous Tetanocerini. It also supports the scenario that terrestrial behavior and morphology in the Tetanocerini are apomorphic features in that tribe.

DeQuieroz & Wimberger (1993) found, among many organisms, a general level of concordance in the extent of homoplasy in behavioral as compared to morphological and molecular characters. Just as Marinoni & Mathis (2000) found a great number of homoplasies in morphological characters of adult Sciomyzidae, we see, in these overall comparisons to their cladogram, a large number of homoplasies in behavioral features.

The study by Barker *et al.* (2004) led to partly different conclusions about the evolution of Sciomyzidae and to new insights. Their results were derived from a revised cladogram of the genera (see Section 15.4), ordination analysis of ecomorphological features of eggs and larvae (see Section 4.3), and linkage of sciomyzid and molluscan cladograms (see Section 5.1). As noted in Section 15.4, their cladogram (Fig. 15.24) of the genera is quite similar to that of Marinoni & Mathis (2000) except for significant differences in the place of a number of genera within the Tetanocerini. Note especially the much broader separation of *Anticheta* and *Renocera*, with seven genera, identified by Marinoni & Mathis (2000) as higher Tetanocerini, separating them. The linkage of Sciomyzidae and Mollusca cladograms (Fig. 5.3) is especially instructive. Stylommatophoran pulmonate gastropods in dry land situations are the hosts/prey of *Salticella* and also are widely utilized by the Sciomyzini and Tetanocerini. Thus it was suggested that these, rather than the non-operculate aquatic and semi-aquatic snails on moist surfaces, may be the plesiomorphic hosts/prey of Sciomyzidae as a whole. Barker *et al.* (2004) noted that stylommatophorans,

> comprise a highly diverse assemblage of 71–92 families and represent the dominant mollusc radiation on land . . . In this context it is interesting to note that the Phaeomyiidae, the likely sister taxon to Sciomyzidae, is comprised of the genus *Pelidnoptera* that parasitizes terrestrial millipedes in the same types of habitats as the stylommatophoran prey of Salticellinae and many Sciomyzinae.

They concluded that "it remains uncertain whether terrestrial or freshwater pulmonates represent the primary [original] prey of Sciomyzidae." They also concluded that "the considerable number of unutilized molluscan clades indicates the ecological conservation of the Sciomyzidae," but noted that "conservatism is relative" and the Sciomyzidae remain the most diverse of all malacophagous insects by many measures. Barker *et al.* (2004) linked a cladogram of 35 biologically known genera, based on adult morphological characters, with a dendrogram of a hierarchical grouping of genera based on an ordination analysis of 30 ecomorphological features of the eggs and larvae (Fig. 17.4). This generally supports the intuitive analysis of the distribution of behavioral groups presented by Knutson & Vala (2002) and here, but resulted in fewer (ten) groups and highlighted the aberrant nature of *Anticheta*.

17.10 SPECIATION

> In determining whether a form should be ranked as a species or a variety, the opinion of naturalists having sound judgement and wide experience seems the only guide to follow.
>
> Darwin (1859).

> Arthur Cronquist quoted the old "joke" that a species is what a competent taxonomist calls a species. For me, this is no joke, but a statement of fact.
>
> Munroe (1977).

Most of our conclusions and speculations regarding speciation of Sciomyzidae are found in other parts of this chapter. In addition we note that some micro-evolutionary rationales for the origins of specialization seem to have no application to the Sciomyzidae. There is no evidence for co-evolution of species of Sciomyzidae and their hosts/prey, obviously due to lack of species-level host specificity among almost all Sciomyzidae. However, the appearance in the fossil record of the greatest number of genera of gastropods during the time that Sciomyzidae probably were originating and rapidly evolving may indicate some relationship between the evolution of Sciomyzidae and Gastropoda. A total of 271 genera of gastropods appeared in the Cretaceous (145–65 mya) and 716 in the Tertiary (65–3 mya), with 109 and 444 genera, respectively, surviving until today. This compares to 129 genera appearing in the Triassic (250–205 mya) with only seven surviving (Schindewolf 1950). Also see Section 17.5.

There is no indication that the somewhat patchy local distribution of most sciomyzids and their hosts/prey is intimately enough related or exists for a long enough duration to promote divergence and speciation. However, one of the most patchily distributed sciomyzids that has been studied biologically, *Tetanura pallidiventris*, is restricted not only in local distribution, but also in diet, habitat, and duration of adult activity. Notably, this boreo-alpine, trans-Palearctic species is in a genus distinct enough from other genera that a separate subfamily has been proposed for it on four occasions on the basis of adult morphology.

Danks (2002), in an analysis of the range of insect dormancy responses, stated that the evolution of a species' life history has been to habitat conditions in the past rather than to present conditions (citing Pritchard *et al.* 1996) or to past rather than present competition (citing Ayal *et al.* 1999). But he also noted, "Adaptations respond to past environments *as well as to current ones . . .*" (our italics). We agree with the latter, broader idea. Danks concluded that "It is then difficult to interpret how particular adaptations [of a species] developed if only their correlation with current environments is considered." Danks's publications and the often-used apt phrase, "the ghost of competition past," raise the spectre of even other kinds of past adaptations that might limit or favor the success of the species today. Thus we assume that in the evolution of a species there have been adaptations to somewhat different hosts/prey, some of which probably are extinct or at very low population levels or have much smaller distributional patterns, but which might have had certain life-cycle aspects different from the populations or species of current hosts/prey. Also, we assume past adaptations to natural enemies, some different from current natural enemies. Further, we assume past adaptations to annual climate regimes different to some degree from present regimes. While the influence of community structures on the evolution of a species may be debated, instances of past adaptations to some aspects of community structure, especially to competition, may have occurred. Also, with modification of the genome, pleiotropic effects that once might have been adaptive and contributed to the success of a species possibly persist and, like life-cycle adaptations, limit or favor the success of a species today. However, we surmise that adaptations of the most critical aspects of the life cycle in most species are primarily to recent or current habitats, hosts/prey, climate regimes, etc. The balance of "residual" adaptations to current conditions might account, to some extent, for the relative success and commonness/rarity of some species, again assuming that most of the recent and current adaptations generally are more effective than relatively ancient adaptations.

The greatest diversity of species and genera of Sciomyzidae and the greatest number of genera endemic to major

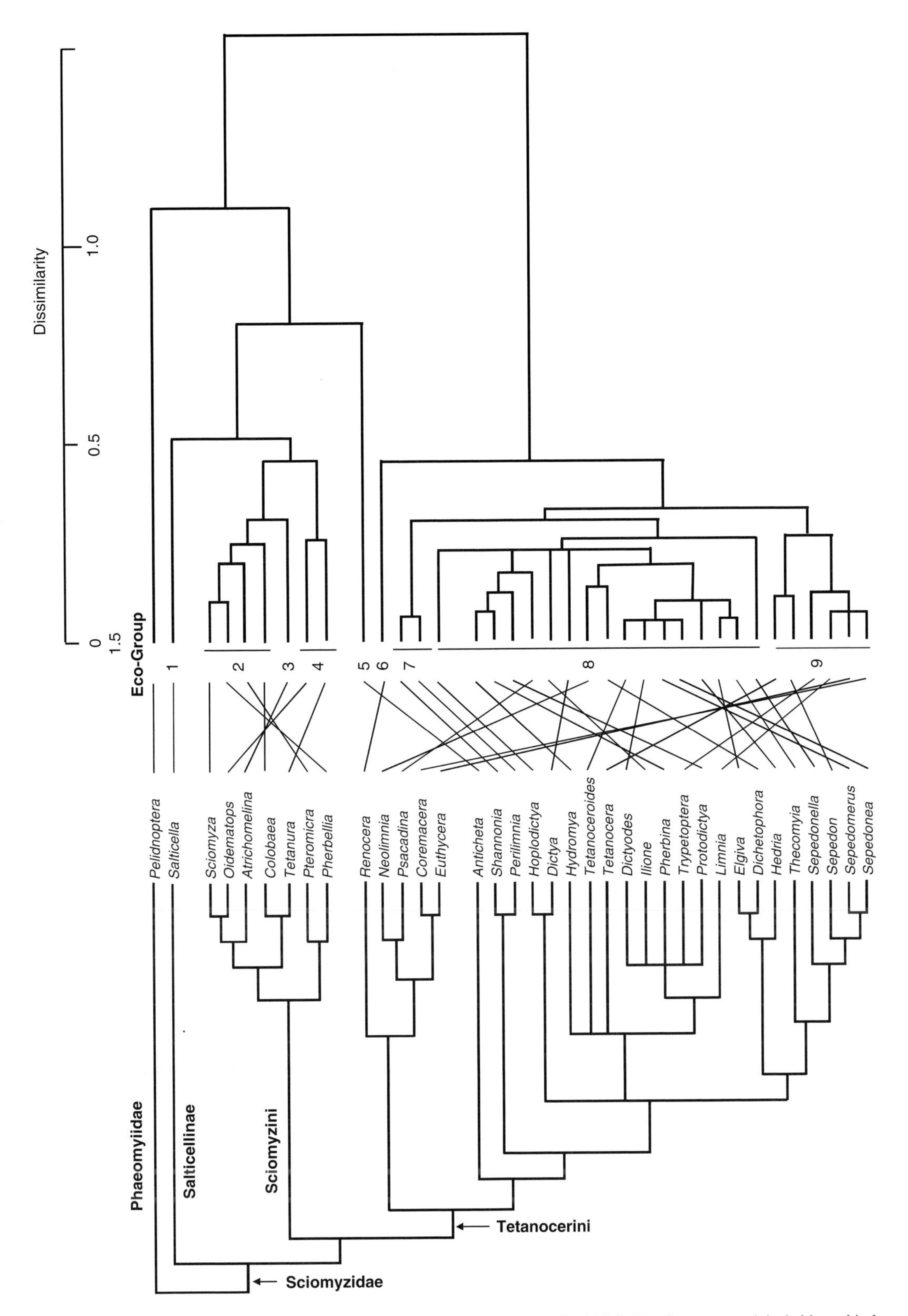

Fig. 17.4 Dispersion of biologically known genera of Sciomyzidae (except ***Ditaeniella*** and ***Eulimnia***) by Eco-Groups identified by ordination analysis of morphological, behavioral, and habitat features of immature stages mapped onto the phylogeny. Dendrogram on left is revision of cladogram of Marinoni & Mathis (2000) by analysis of morphological characters only of adult flies (see Fig. 15.24). Dendrogram on right is hierarchical grouping of genera from ordination analysis of morphological, behavioral, and habitat features in immature stages (see Fig. 4.3). Scale bar indicates dissimilarity based on Bray & Curtis (1957) metric. Modified from Barker *et al.* (2004).

zoogeographic regions is among forms having aquatic and semi-aquatic larvae. Speciation among these forms probably is largely a result of the stochastic and isolating nature of their freshwater habitats, especially still waters, which are often ephemeral, subject to strong seasonal and aseasonal changes, especially in water level or drying completely over longer periods, and subject to spatial subdivision, including altitudinal aspects. These habitat characteristics, combined with the fast, multivoltine life cycles of most aquatic and semi-aquatic forms likely resulted in extensive phenotypic lability and population differentiation.

17.11 CONCLUDING COMMENTS

> If man was to think beyond what the senses had directly given him, he must first throw some wild guess-work into the air, and then, by comparing it bit by bit with nature, improve and shape it into a truth.
>
> Smith (1817).

The "intuitive" suprageneric classification of Sciomyzidae by Steyskal (1965a) based on morphological characters of adults, most having no obvious or proven adaptive value, has been supported by subsequent analyses. The cladistic analyses of Marinoni & Mathis (2000) and Barker *et al.* (2004) were based on most of the same characters, plus some other characters, most of which also are apparently non-adaptive. The value of such non-adaptive characters in determining relationships of species and genera of Sciomyzidae might be questionable. Their phyletic linkages, if at all, with evolutionarily important adaptive features of the larvae, the far more important trophic stage in the Sciomyzidae, are unknown. It is true that some of the structures that have been relied upon strongly in phylogenetic analyses of adult flies and which might appear to be non-adaptive are in fact functional, for example, the macro- and microsetae which are sensory receptors. However, the adaptive significance, if any, of the differences in the presence/absence of certain setae and patterns of setae to the success or survival of the adults is unknown except for a few features of a few Drosophilidae. In the Sciomyzidae there is relatively weak correlation of the distribution of setal patterns, and of most other adult characters, across the entire lineage of the family with the distribution of major life-history attributes of the larvae. For example, the seven genera that include semi-terrestrial succineid snail feeders, noted below, range from those having nearly the full complement of head and thoracic setae in the ground plan of Acalyptratae (*Pherbellia*) to those having lost about half of the major setae (*Sepedon*).

Some such behavioral and morphological characters perhaps are controlled by pleiotropic genetic mechanisms. In the context of their study on intraspecific variation in *Physa* sp. as related to snail-entry and snail-crushing predators, DeWitt *et al.* (2000) noted that,

> Recent theory suggests that trait correlations are important in determining the nature and magnitude of natural selection on individual traits. If a trait conferring positive fitness effects is correlated with a trait conferring negative effects, it may be hard to determine the ecological consequences of one trait studied without reference to (or control over) other traits.

Paraphrasing and modifying one of their conclusions, that since organisms perhaps always face multiple (often conflicting) agents of selection concurrently, it is likely that there exist networks of multiple traits and multiple selective factors that are correlated phenotypically and genetically. While DeWitt *et al.* (2000) make the above comments in regard to intraspecific variation at population levels, the cumulative results pertain to species as well as to that handy construct of the human mind, the genus. Old-fashioned natural history observation might allow us to tease out likely correlates at all levels; modern experimentation or analysis might confirm or deny these; and relating the results to cladistics/phylogenetic approaches provide insights on their evolution.

The weak concordance between phylogeny as based on adult characters with observed behavioral differences in life cycles in many genera of Sciomyzidae led Knutson (2008) to speculate about two points with special reference to the seven genera that include succineid feeders,

> (1) the diversity of host/prey and larval feeding behavior in some small but distinct genera of Sciomyzidae (e.g., *Sciomyza* and *Hoplodictya*) and (2) the lack of concordance between life-cycle strategy and degree of relatedness to other species in some large genera (e.g., *Sepedon*, *Pherbellia*, and *Tetanocera*) as based on adult morphology . . . may indicate that a limited extent of evolutionary pressure (as based on the similarity of adults) has resulted in strikingly different life cycles. This may cast doubt on our ability to recognize the degree of relatedness based on comparative morphology (in the case of Sciomyzidae) of the less-adaptive adult life stage and indicates (as is

> generally applicable) that strong evolutionary pressure on the part of the genome controlling the structure and behavior of one life stage may not result in significant differences in another life stage.

In reviewing the present manuscript, R. Rozkošný (personal communication, 2004) pointed out that analysis of the "more adaptive larval stage" is beset with another difficulty, i.e., the huge extent of convergences and other homoplasies.

A conclusion similar to the above, in part, was inferred by Marinoni & Mathis (2000) in their brief comments on the relationship of larval behavior to their cladistic analysis of sciomyzid genera: "More studies related to the evolution of larval behavior are necessary to *confirm* that the larval habit is *responsible* for relationships among the genera within the family" (our italics).

Although not specifically referring to a conflict in use of adaptive (functional) or non-adaptive characters in determining relationships, Ferrar (1987) concluded,

> I still believe that to some extent one can draw conclusions on such [phylogenetic] relationships from immature stages, but I now consider that larval morphology is predominantly *functional* and that larvae show a number of interesting examples of parallel evolution.

With reference to this statement, Meier (1996) felt that, on the basis of his study of many character systems of larvae of Sepsidae, the problem lies with the relatively low number of morphological characters of larvae used, especially in phylogenetic studies of Cyclorrhapha. Chapman *et al.* (2006) also quoted the above sentence by Ferrar (1987) and concluded that their study of parallel evolution of larval morphology and habitat in the aquatic-terrestrial genus *Tetanocera* supported Ferrar's generalization.

We note Rohdendorf's (1960) conclusion (translated) that "features of adult insects change only slowly and they are thus mostly primitive. Characters of larval and adult stages have therefore a different value in phylogenetic considerations. It seems clear that any phylogenetic evaluation must be based on the primitive (plesiomorphic) characters of adults and on a critical evaluation of features of immature stages." Of course, a guiding principle of Hennigian cladistics is that plesiomorphic characters are useless in establishing relationships, that only apomorphic characters are useful.

A related perspective on some of these questions is afforded by regarding the larval stadia and the adult stage as behaving like two different animals, responding to different ecological factors, as a result of their complete metamorphosis. Krivosheina (1991) attributed "the hypothesis of the independent evolution of adults and larvae," particularly in Diptera, Lepidoptera, and Hymenoptera to Hendel (1936–1937) in her article on "Larval Morphology and the Classification of Diptera." She pointed out that another precursor of these ideas, that the different stages of Diptera have evolved at varying rates, was proposed by Monchadsky (1937), with mosquitoes as an example. She also noted Rohdendorfs' (1960, 1964) hypothesis "that ecological specialization of the larvae has played a leading role in the formation of several groups of Diptera." Krivosheina (1991) emphasized that the greater diversity of larval forms of Diptera compared with that of adults is determined by the differences in microhabitats – in a substrate or aerial – and by the fact that the nutritive stage is the larva, while the adults have only reproductive and dispersal functions. We add the primary function of the adult in microhabitat selection and resultant egg deposition site, and the adults' complementary role with the larva and pupa in phenological sensitivity and adaptation. These features are more critical to the more voracious and less mobile larva being at the "right" place at the "right" time than to the generally more lightly feeding and more vagile adult. Some would consider these part of the "dispersal function." Furthermore, we distinguish the "reproductive" function from both a population (fecundity) point of view and a genetic recombination (selection) point of view. Truman & Riddiford (2002) noted that the evolution of complete metamorphosis led not only to resource partitioning between larva and adult, thus essentially eliminating the potential for competition between them, but also to the evolution of rapid life cycles. They concluded that "These two changes have made holometabolous insects the consummate exploiters of ephemeral food sources."

Emden (1957) in his "The Taxonomic Significance of the Characters of Immature Insects" traced the main idea of "independent evolution" back to several authors, especially Börner (1909), Borradaile *et al.* (1951), and Hennig (1948–1952), without mention of Hendel (1936–1937) or Monchadsky (1937). The essence of the idea, at least in relation to adaptations of larvae as energy transformers, was traced back over 100 years by Price (1997) to Darwin's (1859) "principle of divergence . . . operating only within species, so that more individuals per species can coexist per unit area." Armitage *et al.* (1995) provided an instructive summary of the conclusions from research by Roback & Moss (1978), Cranston (1990), and others on the problem of incongruence of classifications based on different life stages of Chironomidae.

Lim & Meier (2006) asked whether data on larvae or adults have more homoplasy and which provide more information for phylogenetics. They used Partitioned Bremer Support analysis of 20 datasets on Diptera with an average of 33 characters of larvae and 74 of adults, including DNA characters. They found that "adult characters are less homoplasious as measured by RI and CI [Retention & Consistency Indexes] and provide more information than larval characters, with the average adult character providing 74% more support than the average larval character."

Nijhout (2000) agreed that, "Insect evolution is characterized by progressive divergence in the morphology of larval and adult stages," that larval and adult forms have different sets of adaptations, responding to the specific stimuli they receive. He reminded us that the endocrine system (juvenile hormones, ecdysteroids, and neurohormones) provides the transduction mechanism between the stimuli and the development pathways. How endocrine signaling, and especially juvenile hormone secretion and the way that tissues respond to it, have been important forces in the evolution of insect life histories has been reviewed by Truman & Riddiford (2002). The physiological–biochemical–genetic factors of these sensory–regulatory–developmental mechanisms are well known for certain aspects in certain fly species, e.g., diapause in *Sarcophaga* spp. However, the selective pressures that drive the evolution of the magnificent morphodiversity and biodiversity of fly larvae and adults, and the mechanisms that coordinate these resulting in single species comprised of adult and immature stages – which, despite the appearances, are one species – remain largely a mystery. That is, the central mystery of metamorphosis, which has intrigued biologists at least from the writings of S. Aristotle (384–322 BC) (Thompson 1910). We quote from Price (1997), in his reference to ideas unrelated to the above, "However, these ideas stimulate our imagination, draw us back in time through the millennia, and create a broader perspective in time and space. It is "the sensation of the mystical," after all, that is our source for good science in the future."

Theoretical analyses of these questions combined with more profound and extensive information on biologies, along with refined cladistic analyses using, separately and jointly, characters of both adult and immature stages, including all genera, are needed for further development of an evolutionary scenario of Sciomyzidae. In developing the most parsimonious evolutionary scenario of feeding behavior of Sciomyzidae, key points will be the origins of malacophagous behavior, the direction(s) of evolution of behavior, and the occurrences of homoplasy, i.e., conflict of taxonomic status and morphology due to convergence, parallelism, and reversal (loss of derived features). A major conclusion from several studies on other groups (Grandcolas 1997) was that homoplasy is found even in supposedly stable and complex traits – probably including various types of malacophagy. Largely for this reason, phylogenetic tests often do not corroborate former evolutionary hypotheses.

Finally, while generally guided by the principle of parsimony in trying to account for the mechanisms and events that have led to the broad range of morphodiversity and biodiversity of Sciomyzidae, we have been reminded by our colleague E. Colonnelli that evolution might not have been, necessarily, an entirely parsimonious process. Complete parsimony in evolution probably never would have resulted in the numerous evolutionary products that are so rare in occurrence, so unusual in form and function, but so successful, at least at low population levels, as we see in the Sciomyzidae. For example, the three unrelated genera of unique clam killers in Eurasia, North America, and New Zealand. Parsimony and complete order are, of course, constructs of the human mind, and probably neither is true of all of nature – despite the admonition of Evans – see p. xiii.

18 • Biological control

Green streams, blue hills – but all to what avail?
This tiny germ left even Hua To powerless;
Weeds choked hundreds of villages, men wasted away;
Thousands of households dwindled, phantoms sang with glee.

Farewell to the God of Plague.
Mao Tse-Tung.
Cited in Berg (1971a)

Since C. O. Berg's 1953 publication, sciomyzids have been of interest as potential biological control agents of undesirable gastropods, especially of freshwater snails that are intermediate hosts of trematode worm parasites of man and domestic livestock. There have been a few attempts to use native, predaceous species in augmentation programs and to introduce exotic, predaceous species against freshwater snails. More recently, there has been research on the introduction of exotic species for control of terrestrial snails. Also, Williams *et al.* (2009), in a study of populations of a natural assemblage of Sciomyzidae in Ireland emphasized that they likely provide an ecosystem service in their effect on some populations of trematode-bearing snails. The authors' sampling and data analysis methods will be useful to further research on conservation of such natural control agents.

With the increasing incidence of resistance to antihelminthic drugs (Fairweather & Boray 1999) and renewed activity in construction of dams in many countries there likely will be renewed interest in the use and conservation of Sciomyzidae and other potential biocontrol agents of snail intermediate hosts of *Fasciola* spp. and *Schistosoma* spp. For example, the World Bank made a US$ 270 million grant in 2005 to construct a 38-meter high dam on a tributary of the Mekong River in Laos (Fountain 2005). Over 35 years ago Bardach (1972) reviewed the ecological implications of the comprehensive water resources development projects in the Lower Mekong River Basin. He noted that among many other impacts, "there can easily be a sharp rise in schistosomiasis" and at least five other diseases, with snails and mosquitoes probably of the greatest concern as vectors. Schelle *et al.* (2004) focused on the environmental impact of dam construction, especially its impact on biotic diversity of natural ecosystems. They referred to major, recently published sources of information on dam construction and plans, pointing out that more than 45 000 large dams are operational in 150 countries, with an estimated 1500 under construction, nearly 400 of which are over 60-m high. Finkelstein *et al.* (2008), in the first analysis of the global burden of schistosomiasis since the prior study of 10 years ago, found that the impact of symptoms associated with *Schistosoma japonicum* is 7–46 times greater than current estimates.

There have been many reviews of natural enemies of gastropods (see Chapter 2) as well as many that are more specifically on biological control of snails, especially aquatic snails. Those which include more or less detailed reference to Sciomyzidae were presented by Michelson (1957), Gorokhov (1971), Berg (1973a), Thomas (1973), Baronio (1974), Hairston *et al.* (1975), Bay *et al.* (1976), Ferguson (1977), Godan (1979), and Jordan *et al.* (1980). McCullough (1981) considered that, of the predators of snail intermediate hosts of *Schistosoma* spp., only fish and insects "have been studied in enough detail to merit serious consideration." He stated, "further investigations [of Sciomyzidae] deserve encouragement." He proposed a preliminary scheme for screening and evaluating the efficacy and safety of biological agents for control of disease vectors. Madsen (1990) concluded that in regard to schistosomiasis, "snail control operations are still important as a support to chemotherapy, and to reduce the risk of reinfection." He noted that "many predators have been suggested as biological control agents against *Biomphalaria* and *Bulinus* spp. . . . but only [a] few groups, e.g., the Sciomyzidae, certain leeches and some fish are specific snail predators" and "little information is known about the effects of invertebrate predators on snail densities." Berg (1964), Petitjean (1966), Berg & Knutson (1978), Rozkošný (1979b), Greathead (1981), Munari (1983), and Barker *et al.* (2004) provided reviews primarily concerning

Sciomyzidae as biocontrol agents. Many of the basic research papers on the biology of Sciomyzidae and regional studies (e.g., Vala 1989a) also discuss biological control attempts and prospects. Below, we discuss Sciomyzidae as biological control agents of aquatic and terrestrial snails and slugs, and the potential risks of introductions. We include references to some basic research papers overlooked in previous major reviews and offer some new considerations.

18.1 AQUATIC SNAILS

> You can be the most beautiful fish that ever swam. You can be perfectly equipped to survive. Then, one day the pond you live in dries up, and that's it, you die, no matter how fit you are.
>
> S. J. Gould (cited in Horvitz 2000).

As background, the life cycle of *Sepedon*, most species of which are typical aquatic predators, and of *Schistosoma* parasites of man are shown in Fig. 18.1.

18.1.1 Previous reviews

We discuss here the information on biological control and related aspects presented in two of the most important reviews. The first, by Greathead (1981), who had extensive experience in biocontrol of insect pests of agriculture, focused on sciomyzid biologies and identified some needed research. The second, by Barker *et al.* (2004), the first author having had extensive experience in gastropod biology, classification, and pest status, relates information on Sciomyzidae (aquatic and terrestrial species) to modern biological control and ecological theory more extensively than any other review. It is only fair to note that neither Greathead nor Barker has had direct experience in research on sciomyzid biology,

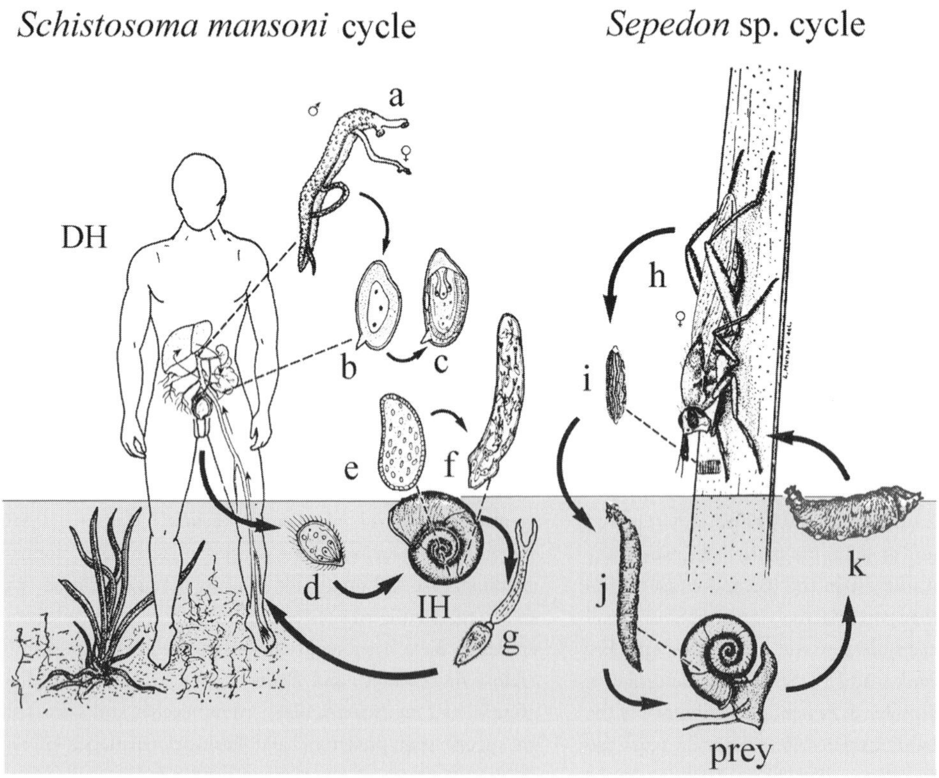

Fig. 18.1 Life cycle of *Schistosoma mansoni* (a–g), parasite of man and life cycle of ***Sepedon*** sp. (h–k), a typical aquatic predator. DH, definitive host; IH, intermediate host; a, pair of adult *S. mansoni*; b, c, eggs; d, aquatic miracidium; e, type I sporocyst; f, type II (daughter) sporocyst; g, furcocercarium; h, adult ***Sepedon***; i, egg; j, larva; k, puparium. Modified from Munari (1983).

although Barker's co-authors have had such delightful experience.

Greathead's (1981) review was focused on consideration of introduction of exotic species for control of snail hosts of *Schistosoma* parasites of man. He covered much but not nearly all of the literature on sciomyzid biology published from 1959 to 1979. Shortly after his review the results of research being accomplished at about that time began to appear, in part answering Greathead's questions. He concluded that biocontrol by introduction of exotic Sciomyzidae should be investigated along with other methods and that detailed studies are required to determine the impact of natural enemies on snail intermediate hosts. His emphasis on integration of biocontrol with other snail control methods reflects a conclusion that has been reached in biological control in general over the past decades, and that also was emphasized by Appleton *et al.* (1993) as well as having been recognized by Berg (1964) and Berg & Knutson (1978).

Specifically with reference to Sciomyzidae, Greathead (1981) stated that there were not enough data on population density, host/prey preferences in the field, and impact on host/prey populations. He noted that Eckblad's (1973a) and Eckblad & Berg's (1972) *Sepedon fuscipennis* studies were the only quantitative ecological studies of a sciomyzid. They measured numbers of adults by the mark–recapture technique, but were unable to develop a method to measure the larval population. However, they did measure larval survival in field cages. Also, Eckblad & Berg (1972) calculated the density of larvae required to produce the adult population of *S. fuscipennis* they studied. The first direct estimates of the absolute densities of immature sciomyzids were obtained by Arnold (1978a) in a 3-year field study at the Cornell University experimental ponds. We quote from his abstract,

> Emergent vegetation was searched for eggs, and the surface film was searched for floating larvae and puparia, of *Sepedon fuscipennis* Loew and *Dictya* spp. in four shallow marshes. Densities were estimated from counts in randomly selected 0.25 m^2 quadrats stratified by presence/absence of emergent vegetation. Peak densities of larvae plus pupae were 0.9–1.3 m for *S. fuscipennis* and 0.1–1.7 m for *Dictya*. Stratified sampling was found to be superior to simple random sampling.

Greathead (1981) also overlooked the brief paper by Lynch (1965), who estimated the relative densities of larvae of the aquatic predator *Dichetophora biroi* by dipping with a 10-cm sieve. He did not comment on the experimental studies of O. Beaver (1974b, 1974c) on food conversion ratios and Eckblad (1976) on biomass and energy transfer that provided evidence that some aquatic predators are "wasteful predators," an important, desirable feature of biocontrol agents; these aspects are discussed in Chapter 7.

Subsequent to Greathead's (1981) review, many publications have appeared on quantitative, experimental studies, both in the field and in the laboratory. These are reviewed primarily in Chapter 7 on feeding behavior and Chapter 13 on population dynamics. Other needs noted in Greathead's review have been met to a significant extent in recent years. He stated that most data on number of prey killed were anecdotal. He referred to the quantitative, experimental studies of Geckler (1971), Eckblad (1973a), and Eckblad & Berg (1972) but overlooked the paper by Neff (1964) and some other papers cited below (see Section 5.2). During the past 2 + decades, extensive quantitative, experimental data from field and laboratory studies have been gained on aquatic predators. These are: *Sepedon fuscipennis* (Barnes 1976, Arnold 1978b), *Ilione albiseta* (Lindsay 1982; Gormally 1985b, 1987a, 1988a, 1988b), *Sepedon sphegea* (Haab 1984, Ghamizi 1985), *S. senex* (Bhuangprakone & Areekul 1973, O. Beaver 1989), *Tetanocera ferruginea* (Manguin *et al.* 1986, 1988a, 1988b; Manguin & Vala 1989), *Sepedon neavei* and *S. scapularis* (Barraclough 1983, Maharaj 1991, Maharaj *et al.* 1992, Appleton *et al.* 1993), and *Sepedon spinipes* (Mc Donnell 2004, Mc Donnell *et al.* 2005b). Records of the natural prey of ten species of aquatic predators have been brought together by Mc Donnell *et al.* (2005a). Greathead (1981) also noted that most information is on temperate, northern hemisphere species. The above-mentioned papers on Oriental and Afrotropical *Sepedon* spp., and the papers by Vala *et al.* (1995, 2000b), Gbedjissi *et al.* (2003), and Knutson (2008) are steps in filling the need for more experimental data on larval feeding by tropical species.

Barker *et al.* (2004) presented the most detailed, recent analysis of Sciomyzidae as biocontrol agents in the context of modern theory and practice. They focused on the role of Sciomyzidae in gastropod population dynamics, with emphasis on terrestrial gastropods. The Sciomyzidae were discussed in relation to seven requirements for successful biological control agents: (1) laboratory rearing, (2) long-distance transport, (3) favorable recipient environment, (4) pest suppression, (5) stability in permanent crops versus opportunism in temporary crops, (6) freedom from natural enemies, and (7) minimal adverse impacts on biodiversity. While information on Sciomyzidae meets requirements 1–3, there is insufficient information concerning pest suppression and the impact of natural enemies. Requirement 7 is discussed in Section 18.4. Barker *et al.* (2004) emphasized the need for six types of information in estimating levels of gastropod mortality caused by the larvae, (1) the average levels of predation or parasitism per host/prey generation, (2) its variability from generation to generation and whether or not this source of host/prey mortality is a key factor, (3) the extent to which predation or parasitism tends to act as a density-dependent factor, (4) other prey mortalities that combine with that caused by the sciomyzid larvae to counter the host/prey's potential rate of increase, (5) any important mortalities suffered by the sciomyzids that reduce their effectiveness, and (6) the density of searching sciomyzid adults.

18.1.2 Pilot field and laboratory studies

There have been three extensive field trials, two inoculative and one augmentative, only the first figuring in Berg's and Greathead's reviews, and a recent, controlled field-experimental trial. Information on releases and recoveries is given in Tables 18.1, 18.2.

The first and best-documented inoculative release was that of adults of the aquatic predator *Sepedomerus macropus* from Nicaragua into Hawaii (where there are no native Sciomyzidae) against *Austropeplea viridis*, host of liver fluke, *Fasciola gigantica* (Chock *et al.* 1961). Only a few puparia were shipped, yet rearings of large numbers of flies were made successfully by persons inexperienced with Sciomyzidae. There were no adequate pre- or postintroductory surveys of snail populations. However, the attempt showed that sciomyzids could be shipped successfully, reared in large numbers with local expertise and materials, released effectively, and that the flies could establish and disperse to other locations. Larvae have been found feeding on the target snails in nature. Subsequently, the species was redistributed to several Pacific Islands, but again without adequate field surveys, and results remain inconclusive. Several other species subsequently were released in Hawaii. Of these, only *Sepedon aenescens* from Japan apparently became established. It was released on five major islands in 1966 and was reported established as early as the same year. It was also recovered in 2003 on Maui (Englund *et al.* 2003).

In previous reviews of the biology of Sciomyzidae or on their biocontrol prospects, there has been no reference to an interesting paper by Hope Cawdery & Lindsay (1977) that initiated a series of studies that continue to the present, on biology and possibilities for biocontrol in Ireland. They studied the decline of populations of the liver fluke, *F. hepatica*, and the intermediate host/vector snail, *Galba truncatula*, over a 5-year period in sheep pastures in Co. Mayo, Ireland. They evaluated weather, pasture contamination, soil fertility, and the presence of the semi-aquatic predator, *Hydromya dorsalis* as limiting factors. During the last 3 years of the study (1974–1976), they released marked, uninfected, laboratory-reared snails in groups of about of 100 during June–August, recovering them 96 and 144 hours later. The snail-marking technique is of interest: the apex of the shell was dipped in white paint, then in reflective glass beads, and the snails were searched for during the night by flashlight. Of the factors studied, predation by *H. dorsalis* was the only one that the authors could demonstrate as being responsible for the decline in snail populations. They found that during June–August of the 3 years larvae had attacked 1.4%, 15.63%, and 29.38% respectively of recaptured snails. They calculated the mean daily attack rate, assuming that predation followed a simple experimental survival curve, as 0 to 9.349, and concluded that an attack rate of the order observed could result in a substantial reduction in the snail population. They then included changes of temperature in their equation and determined that at a mean daily temperature of 13–14 °C, a change of 1 °C would, at an attack rate of 0.04, halve or double the relative survival, depending upon direction of change. Thus, for example, at an attack rate of 0.04, 15 times more snails will survive long enough for flukes to complete development at 18 °C than at 13 °C. In a subsequent paper, also not previously referred to in reviews of Sciomyzidae, Hope Cawdery (1981) presented mathematical models based on temperature data for predicting development times for the extra-mammalian stages of *F. hepatica*. He further considered the relative effect on survival of infected snails by predators, without reference to specific predators, at various temperatures and attack rates. He concluded that predation as a limiting factor on disease transmission is most effective at low temperatures; likely to be confined to areas with cool, wet summers; and that cold weather predators such as *Ilione albiseta* are of particular interest.

Table 18.1. Releases and establishments of Sciomyzidae for biocontrol of aquatic snail intermediate hosts of cattle liver fluke

Species released	Origin	Released at, number, stage	Date of first release (reference)	Recoveries (reference)
1. *Atrichomelina pubera*	Ithaca, New York	Oahu, Hawaii, Kauai – 100 A	September 21, 1961 (Davis & Krauss 1962)	Not recovered (Steyskal 1980)
2. *Pherbellia dorsata*	Sjaelland, Denmark	Oahu – 530 A	December 8, 1960 (Davis 1961b)	Not recovered (Steyskal 1980)
		Subsequently, many thousands throughout the major islands	(Davis *et al.* 1961) (Berg 1964b)	
3. *Ditaeniella parallela*	Riverside, California	Oahu, Hawaii, Kauai – 75 A	September 1961 (Davis & Krauss 1962)	Not recovered (Steyskal 1980)
4. *Dictya abnormis*	Ithaca, New York	Oahu – 15 A, 89 P	June 30, 1958 (Davis 1960)	Not recovered (Steyskal 1980)
5a. *Sepedomerus macropus*	Managua, Nicaragua	Oahu – 8 A	December 5, 1958 (Davis 1959)	Established on all major Hawaiian Islands by 1961 (Chock *et al.* 1961, Davis 1961a, 1974, Berg 1964)
		←	(see Table 18.2)	→
5b. *Sepedomerus macropus*	?	Guam	?	Established (Beaver 1989)
6a. *Sepedon aenescens*	Fukuoka, Kyushu, Japan	Oahu – 75 A	November 15, 1966 (Davis & Krauss 1967)	Established August 29, 1967 (Davis & Chong 1968)
		Kauai – 200 A	November 22, 1966 (Davis & Krauss 1967)	Established May 16, 1969 (Davis & Chong 1969)
		Hawaii	?	?
		Molokai	?	
		Maui – 100 A	March 29, 1967	Puparia, August 29, 1967 (Chong 1968) Adult, January 23, 2003 (Englund *et al.* 2003)
6b. *Sepedon aenescens*	Fukuoka, Kyushu, Japan (from eggs received from Hawaii)	Riverside, California		Not recovered (Knutson & Orth 1984)
		300 A	August 12, 1975	
		300 A	August 22, 1975	
		400 A	August 26, 1975 (Knutson & Orth 1984)	

Table 18.1. (*cont.*)

Species released	Origin	Released at, number, stage	Date of first release (reference)	Recoveries (reference)
7. *Sepedon oriens*	Ajime, Japan	Kauai, Oahu, Mauhi, Molokai, Hawaii	July 1972 (Davis 1972)	Not recovered (Steyskal 1980)
8. *Sepedon pacifica*	Riverside, California	Maui – 20 P, 24 A	November 1970 (Davis 1971)	Not recovered (Steyskal 1980)
9. *Sepedon plumbella*	?	Oahu, Kauai	August 1971 (Davis 1972)	Not recovered (Steyskal 1980)
10. *Sepedon praemiosa*	Riverside, California	Maui – 60 P	September 1970 (Davis 1971)	Not recovered (Steyskal 1980)
		Oahu, Hawaii, Kauai "substantial nos."	October 1961 (Davis & Krauss 1962)	
11. *Sepedon senex*	?	Hawaii	December 1971 (Davis 1962)	Not recovered (Steyskal 1980)

Note: All releases in Hawaii except 5b and 6b. A, adults; P, puparia. From various sources.

Table 18.2. Initial releases (a) and recoveries (b) of ***Sepedomerus macropus*** in Hawaii

Island	Locality	Date released	N Released	Total released to December 31, 1959
a. Initial releases				
Ohau	Kahaluu	December 5, 1958	8	3738
Hawaii	Honolua	January 20, 1959	35	1280
Kauai	Hanapepe	March 19, 1959	100	1684
Molokai	Keopukaloa	May 20, 1959	315	615
Maui	Lahaina	May 21, 1959	300	316
Total				7633

Island	Locality	Date of initial release (1959)	N Recovered	Date (1959) recovered
b. Recoveries				
Kauai	Waimea Ditch above Kekaha	April 29	2	June 10
Oahu	Waiahole	January 6	33	September 23
Oahu	Waimalu	August 26	12	December 15
Oahu	Luluku	November 3	6	December 29

Note: Total number of adults trapped in McPhail traps as of December 31, 1959 was 507. Modified from Chock *et al.* (1961).

In Iran, Tirgari (1977, 1986) and Tirgari & Massoud (1981) conducted trials of mass rearing and release of the native aquatic predator *S. sphegea* against the *Schistosoma* host snail, *Bulinus truncatus*, in irrigated rice fields. They concluded that mass releases of larvae should be made during spring and fall and that relatively small numbers should be released at 10–15 day intervals. They recorded, within 15 days of release of the larvae, a 76–94% decrease in snail populations when the water depth was 4.0 cm and a 48–85% decrease when the water depth was 30 cm. Whether or not these decreases were due to predation by larvae was not proven.

A series of papers (1–5, as follows) from South Africa present conflicting evaluations.

(1) On the basis of his 1-year thesis project Maharaj (1991) concluded that *Sepedon neavei* and *S. scapularis* "are sufficiently efficient predators of medically important snails to be considered as candidates for field trials in the search for suitable biological control agents."
(2) Appleton *et al.* (1993) noted that previous general reviews of biocontrol of snails concluded that sciomyzid larvae might be useful but that they have not been evaluated adequately. They carried out a 3-year study in six sites (three streams and three impoundments) in South Africa to determine the value of using the endemic *S. neavei* and *S. scapularis* in an augmentative, mass-rearing/release approach, run at the rural community level, against the *Schistosoma* host snails *Bulinus africanus* and *Biomphalaria pfeifferi.* Relying primarily on their own several unpublished and published studies, they considered mainly laboratory results on aspects of predation, prey specificity, prey range, competition, etc. Of the many other published experimental studies, they referred only to Manguin & Vala (1989). They projected costs of setting up and running a mass-rearing facility that could treat several foci of transmission at only about US$ 700 per year (1992 US$), not including the salary of one trained employee and transportation, in a rural community. Their study reached stage 3 b (controlled field trials – effectiveness under local conditions) of the Plan for Development of Biological Vector Control Agents of WHO (1984), which is the stage of final laboratory tests for indigenous control agents. They concluded that *S. neavei* is the agent of choice that satisfies most of

the recognized criteria proposed by Samways (1981) and WHO (1984) for an effective, predaceous biocontrol agent. They further concluded, "*Sepedon* spp. used in large numbers (i.e., biological control by predator augmentation) are likely to be effective only in shallow transmission foci and as part of an integrated control programme." They noted the disadvantage of *S. neavei* larvae in being confined to the surface and suggested that Nearctic subsurface predators in the genus *Dictya* might be more effective. However, they felt that implementing such exotic agents would be more expensive and require extensive safety tests, and that the use of imported species was unlikely to be more effective than using indigenous species. But, we note that exotic species might have the important advantage of being less susceptible to host-specific parasitoid wasps than indigenous species. Appleton *et al.* (1993) concluded that polyphagy probably is common among Sciomyzidae. This is true in most laboratory rearings but it is not known if it is true in nature. They did not refer to possibilities of managing water levels to increase the effectiveness of sciomyzid larvae, as proposed by Eckblad (1973b) and Peacock (1973).

(3) Maharaj *et al.* (2002) noted: "Experimental success on snail prey and ease of laboratory rearings indicate that *S. scapularis* is important and has a future in the control of bilharzia."

(4) Tscheuschner *et al.* (1993, 1994, 2002) published the same abstract, word for word, and concluded that due to ". . . the limited active depth of larvae it appears unlikely that augmentation of *Sepedon* [*S. neavei* and *S. scapularis*] larvae will reduce populations of medically important snails in South Africa."

(5) Miller & Appleton (2002) stated, "Both fly species were highly fecund, with *S. neavei* more so, as well as killing more snails than *S. scapularis*, particularly at high prey densities." They speculated, in regard to biocontrol prospects in southern Africa that ". . . interference competition [by other Sciomyzidae] might inhibit the effectiveness of any one species used in large numbers (i.e. biological control by predator augmentation. . .)," and that ". . . in much of southern Africa a major limiting factor for adult survival and species establishment is the requisite for emergent and marginal aquatic vegetation for resting and refuge."

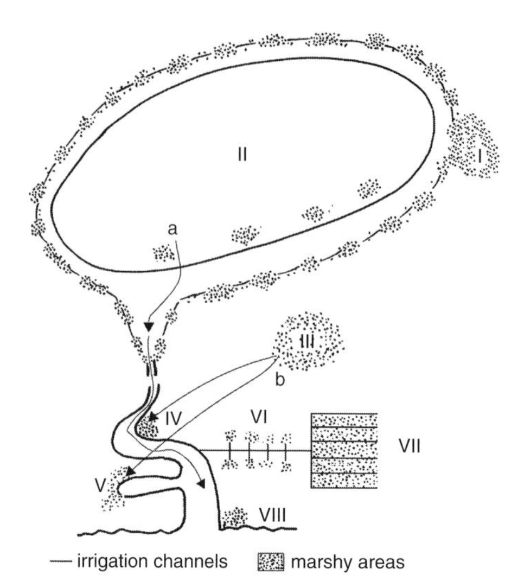

Fig. 18.2 Composite diagram of eight general types of aquatic and semi-aquatic habitats where Sciomyzidae and *Schistosoma* and/or *Fasciola* intermediate host snails live. I, Remote marshland headwater source. II, Remote lake headwater source. III, Wet pasture. IV, Marshy area where water moves slowly along shallow margin of stream. V, Backwater of stream. VI, Irrigation channels and marshy spots. VII, Irrigated cropland. VIII, Coastal marshland. It seems likely that puparia are dispersed, especially downstream, during flood periods (e.g., arrow a), and that adults disperse from drying habitats to permanent aquatic habitats and residual, wet foci during dry periods (e.g., arrow b).

18.1.3 Considerations for biological control

18.1.3.1 FUNCTIONAL AND NUMERICAL RESPONSE (SEE SECTION 7.3.3)

The predation experiments to date, in terms of the functional and numerical response models of Holling (1959) and Watt (1959), show that several aquatic predators are at least theoretically able to regulate prey populations. Eckblad (1973a) concluded that *Sepedon fuscipennis* studied in 5.0 mm water exhibited the type of functional and numerical responses leading to population regulation, but in experiments conducted in increased water depth neither response was apparent. Haab (1984) showed that *S. sphegea* displays density-dependent feeding behavior of the type II functional response of Holling (1959). However, O. Beaver (1989) found that the functional response of first- and second-instar larvae of *S. senex* was of type I and that of third-instar larvae was of type III. The most detailed experiments on functional and numerical response, of

S. sphegea, are those of Ghamizi (1985, unpublished). Manguin & Vala (1989) concluded that *T. ferruginea* chose a prey that gave the greatest return per unit of energy expended, fitting the foraging theories of Emlen (1966), Hassell & Southwood (1978), and Luck (1984). Also, their conclusion that the ability of the larvae of *T. ferruginea* to select the larger prey permit them to kill sexually mature snails, thus increasing their effect on the snail population, is of interest from a biocontrol perspective.

18.1.3.2 HABITAT

In Fig. 18.2 we present a composite diagram of eight general types of aquatic habitats where *Schistosoma* and/or *Fasciola* intermediate host snails live. Following are suggestions on how Sciomyzidae might be implemented in these areas. The diagram and suggestions pertain to tropical, subtropical, and temperate regions, but note that only aquatic/predaceous, and succineid and terrestrial parasitoid/predator life cycles are known for tropical areas, while all life styles are known for temperate areas. Perhaps some species and/or populations from southern parts of the northern hemisphere could be acclimatized for use in tropical areas. Special emphasis might be put on implementing species with broad north–south ranges, such ranges indicating the adaptability of these species to diverse climatic regimes. There are no studies on acclimatization of Sciomyzidae.

Habitats I and II: Remote headwater sources (I, marshland; II, ponds and lakes). When snail populations downstream are dependent, to some extent, on the build-up of populations in headwaters and subsequent drift downstream, introduction of exotic species of natural enemies into or augmenting indigenous natural enemies on a routine basis in such remote areas (that are difficult to reach and costly to treat with molluscicides) might be of value. It is more likely that larger populations of Sciomyzidae could build up in such relatively undisturbed areas than in disturbed areas and that dispersal downstream by floating puparia during seasonal floods would increase downstream populations of Sciomyzidae at certain times.

Habitats III, IV, and V: Wet pastures, marshy areas along slowly moving margins of streams, and backwaters. Many species of semi-terrestrial Sciomyzidae, both Sciomyzini (e.g., some *Pherbellia*) and Tetanocerini (e.g., some *Tetanocera*), could be considered for use in such areas.

Habitats VI and VII: Irrigation channels and irrigated cropland. In these situations, manipulation of water levels would enable use of a broad range of aquatic and semi-aquatic species.

Habitat VIII: Coastal marshland. Only a few Nearctic species of *Dictya* would be of use in these habitats.

Attempts at biocontrol of aquatic snails have been primarily with aquatic, predaceous Sciomyzidae in more or less open water. Another promising approach might be the use of some Sciomyzini and Tetanocerini whose larvae attack aquatic snails that often are not in the water but stranded, foraging, estivating, or for some other reason on damp or drying margins of the habitats (Behavioral Groups 2 and 3 in Chapter 3). Killing snails in these situations, where molluscicides are not effective, would impinge little on the short-term level of on-going disease transmission but might be significant in reducing snail populations over time. As discussed in Section 10.1 aquatic and semi-aquatic Sciomyzidae can breed in somewhat disturbed situations but they cannot breed in locations where the substrate is highly disturbed, for example, wet pastures and margins of water bodies disturbed by domestic livestock and margins of streams and irrigation channels disturbed by human activities. As Miller & Appleton (2002) noted, a limiting factor for adult survival and establishment is the need for emergent and marginal vegetation for resting and refuge.

18.1.3.3 MASS REARING

Ability to mass rear Sciomyzidae could be important to biocontrol efforts. Records from the mass release projects indicate the ease of rearing large numbers of individuals. Tirgari & Massoud (1981) stated that 1000–2000 eggs of the indigenous aquatic predator *S. sphegea* were obtained daily from adults established in insectaries in natural situations in Iran. Tirgari (1986) stated that using this technique, 17 000+ young larvae were released in 1979 and 50 000+ in 1985 in southwestern Iran against the *Schistosoma* host, *Bulinus truncatus*.

From a shipment of 58 larvae of the aquatic predator *Sepedomerus macropus*, received in October 1958, a culture was started with ten pairs in Hawaii (Chock *et al.* 1961). Mass rearing of larvae in wooden troughs outdoors produced 4500 flies per month. Between December 1958 and December 1959, 7633 flies were released against *Austropeplea viridis*, host of *F. gigantica*. The species became established (Tables 18.1, 18.2). From a shipment of 80 puparia of the semi-terrestrial predator *Pherbellia dorsata* received in

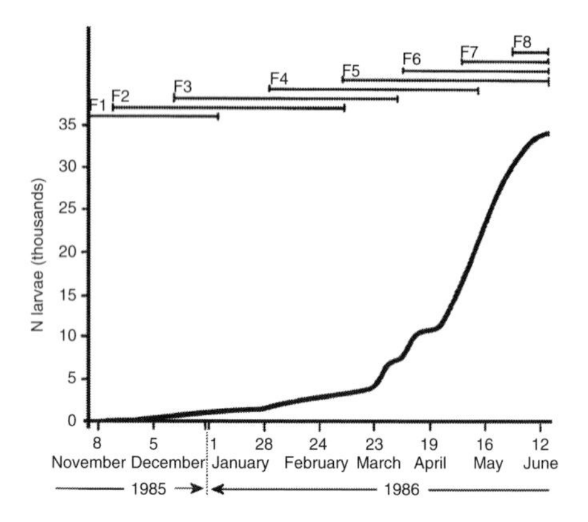

Fig. 18.3 Cumulative production of ***Dictya floridensis*** larvae and duration of each generation during development of rearing methods for large-scale, continuous propagation. F1–F8, generations. Modified from McLaughlin & Dame (1989).

October 1960, 12 592 third-generation flies were obtained (Davis *et al.* 1961). Many thousands were released on the major islands beginning in December 1960; apparently the species did not become established.

Based on their small-scale trials, Appleton *et al.* (1993) projected that about 30 female adults of the endemic aquatic predator *S. neavei* would be needed to produce 1000 once-fed first-instar larvae per day to inoculate several foci of transmission of *Schistosoma* in South Africa.

Willomitzer (1970) attempted to mass rear, in the laboratory, four species of semi-terrestrial and aquatic predators collected near Brno, Czech Republic. Rearings were successful only with the aquatic predator *Dictya umbrarum*. Based on the high mortality of the young larvae he suggested release of second- and third-instar larvae. However, there have been conflicting conclusions on survival of first-instar larvae from experimental studies (see Eckblad 1973a, O. Beaver 1989). Willomitzer & Rozkošný (1977) subsequently were more successful rearing *D. umbrarum* under semi-controlled conditions. From four laboratory trials of 1492 first-instar larvae against *Austropeplea tomentosa* and *Stagnicola palustris* at larva/snail ratios of 1:1, 1:2, 5:2, and 5:6, they concluded that a ratio of three first-instar larvae per two snails would give effective control. They also provided data on survival of eggs and puparia held at 7 °C.

Methods for large-scale, continuous propagation of *Dictya floridensis* on laboratory-reared snails were developed by McLaughlin & Dame (1989). Larvae of this species consume 20–22 snails during the development of an individual (Valley & Berg 1977). During the F_7 generation, colony production was in excess of 1000 neonate larvae per day. At that time the colony was considered to be continuously producing, its size limited only by the availability of food and rearing facilities. The cumulative production through eight generations over a period of about 7 months reached nearly 35 000 individuals (Fig. 18.3). Larvae were reared on living, freshly crushed, and frozen *Physella h. hendersoni* and *Pseudosuccinea columella*. Larval development and survival rate to pupariation were the same whether reared on freshly crushed or live snails. Larvae reared on frozen snail tissue, compared with those reared on live snails, were lighter in weight and had extended development time to pupariation and lowered survival rates. Considerable detail on rearing methods and materials, colony growth, development, and survival were provided. Notably, colonies were initiated with egg masses of the two snail species, and the diet was changed to live or freshly crushed snails at the end of the fourth day, when larvae were in the third stadium.

Ability to rear large numbers of larvae successfully on crushed snails can expedite mass rearings, as shown by McLaughlin & Dame (1989). Detailed studies of growth, duration of all life stages, and survival rates of the aquatic predator *Sepedon plumbella* reared from hatching to pupariation on living and crushed non-operculate snails (*Gyraulus*, *Indoplanorbis*, and *Lymnaea*) and operculate snails (*Filopaludina*, *Melanoides*, *Pila*, and *Bithynia*) were carried out in Thailand by Bhuangprakone & Areekal (1973). Development on crushed snails could not be completed only on *Bithynia* and *Pila*. Survival rates, from larvae to adults, reared on six species of crushed snails were 25% to 55%, with no marked difference in longevity of adults reared from these snails except the longer life of adults and higher rate of oviposition observed with those feeding on the operculate *Melanoides*. However, O. Beaver (1989) found that larvae of all instars of the aquatic predator *S. senex* that fed on live *Gyraulus convexiusculus* had higher conversion efficiencies (growth rate : food consumption rate), higher growth rates, and consumed more food when fed freshly crushed snails. O. Beaver *et al.* (1977) found that larvae of the aquatic predators *Sepedon ferruginosa*, *S. plumbella*, *S. senex*, *S. spangleri*, and *S. aenescens* would feed on crushed operculate snails (*Hydrobioides nassa*, *Idiopoma ingallsiana*, and *I. pilosa*), and *S. ferruginosa* was reared from hatching to pupariation on crushed snails. Surprisingly, the survival rate of newly hatched *S. ferruginosa* appeared to be greater if they were fed crushed snails, even when the alternative was living, non-operculate snails of suitable sizes. Considering the interest in the use of *Ilione albiseta* as a biocontrol agent of snail hosts of *F. hepatica*, it would be useful to see if this sciomyzid could be reared on frozen snails.

18.1.3.4 OTHER CONSIDERATIONS

Ability to store Sciomyzidae at low temperatures would be useful in some biocontrol efforts. There have been no studies of cryopreservation of Sciomyzidae. All stages, depending upon the species, are known to overwinter in cold climates, some species in diapause and others simply quiescent. Extensive information on collection of all stages during low temperatures and exposure outdoors of all stages during laboratory rearings, primarily to break diapause, was included in the review of adaptive differences in phenology by Berg *et al.* (1982). Adults of some species, e.g., most *Sepedon* and *Pherbellia cinerella*, are active in all except the coldest weather in northern latitudes and some *Sepedon* spp. collected during winter become reproductively active when brought into the laboratory. Eggs of several species, especially those that overwinter within the egg membranes, have been kept in a viable condition at low temperatures for long periods. Eggs of the univoltine *Tetanocera latifibula* laid during late July and August were placed at 5–8 °C on September 11, removed to room temperature on February 24, and some hatched February 24–March 2 (Foote 1999). Eggs of the univoltine *Hedria mixta* hatched after 213 days at 5–7 °C when warmed to 20 °C (Foote 1971). Barnes (1980a) noted that eggs of the clam-killer *Eulimnia philpotti* ". . . stored in a refrigerator at 4 °C for 4 days survived a malfunction that forced the temperature below freezing during the last night. Many eggs hatched soon after they were warmed to room temperatures." Newly hatched larvae of *Ilione albiseta* survived for up to 1 month without food at 14 °C (Gormally 1988b). Several European parasitoid/predators of terrestrial snails (*Euthycera cribrata*, *Dichetophora obliterata*, *Limnia unguicornis*, and *Trypetoptera punctulata*) were found to survive for about 20 days after hatching, without feeding (Vala 1989a). Bratt *et al.* (1969) held puparia of species of *Pherbellia* and *Ditaeniella* at various temperatures for different periods of time. Fourteen puparia of *Pherbellia similis*, a species that overwinters in puparia in diapause, were held for 60 days at room temperatures, then stored at 5 °C for 407 days; 13 adults emerged 8–10 days after being returned to room temperatures. They found that species overwintering in puparia in quiescence also survive prolonged periods of storage at 5 °C. Puparia of *Ditaeniella parallela* formed during September 1961 were placed at 5 °C, then removed to room temperatures on December 9, 1962 (15 months later). Four adults emerged on December 21, mated, and the females laid viable eggs.

Table 18.3. Sex ratio of ***Ilione albiseta*** pupae reared at constant and ambient temperatures (± S.E.)

Temperature (°C)	N pupae	Adult sex ratio (male : female)
14	10	1:1
17	12	1:1.4
20	5	1:0.3
23	5	1:0
Ambient (8.2–24)	239	1:1.1

From Gormally (1988a).

Yoneda (1981) found that at temperatures of 15 and 19 °C, the male : female sex ratio of emerging adults of *Sepedon aenescens* was 1.7:1 and 1.4:1, respectively, while the sex ratio in two separate experiments at 30 °C was 1 : 2 and 1 : 2.6. The proportion of females was greater at higher temperatures whereas the proportion of males was greater at lower temperatures. Gormally (1988b) found the opposite in rearings of *I. albiseta* under controlled conditions. The ratio of male:female pupal emergence was greater for males at 20 and 23 °C than at 14 and 17 °C (Table 18.3). A possible association between temperature and sex ratio was tested by comparing the sex ratio at one temperature with each of the other three temperatures in turn. A significant difference in male : female ratio was found between 17 and 23 °C at the 5% level ($P = 0.044$). At the higher temperatures, male emergent adults predominated whereas at the lower temperatures females predominated slightly or the sex ratio was more even. At ambient temperatures emergent adults were found to have a male : female sex ratio of 1 : 1.1 ($n = 239$). We suggest, for biocontrol purposes and other studies, it would be possible to control the proportion of males and females produced by conducting rearings at certain temperatures.

Manguin (1989) indicated the usefulness of being able to predict the sex of flies, based on size of puparium prior to emergence (with 80% accuracy) for population studies, biological control programs, and study of hymenopterous parasites. Although she noted that sexual dimorphism rarely appears among Diptera larvae, she failed to make the inference that differences in sizes of *Tetanocera ferruginea* puparia (formed by the pupariating larva *before* the

Table 18.4. Dimensions of male and female puparia (a) and maximum weight of male and female L3 (b) of *Ilione albiseta*

	a. Puparia			b. Larvae	
	Length Mean ± S.D.	Width Mean ± S.D.	N	Max. Wt. (g) Mean ± S.D.	N
Males	8.57 ± 0.55	3.42 ± 0.14	24	0.057 ± 0.005	16
Females	8.99 ± 0.51	3.73 ± 0.14	27	0.068 ± 0.007	24
Total	8.79 ± 0.56	3.58 ± 0.21	51	0.064 ± 0.009	40
	t = 5.4, d.f. 49	t = 37.7, d.f. 49			
	P < 0.01	P < 0.01			

Note: t, Student t. Modified from Lindsay (1982).

pupa is formed) most likely are due to sexual dimorphism in growth of the larvae. Lindsay (1982) showed that there are significant differences in weights of male and female larvae and in both length and width between male and female puparia of the aquatic predator *I. albiseta* (Table 18.4). He also suggested the usefulness of being able to determine the sex of larvae for experimental studies. Sexual differentiation also permits easy manipulation of inert puparia rather than active flies and eliminates the need to hold them for emergence if only sampling is needed. Mc Donnell *et al.* (2005b) found that adult sex could be predicted with 95% certainty from puparial weights when *Sepedon spinipes* were reared over a temperature range of 14–26 °C. The authors proposed use of an automated (air-stream separator) method, as used in mass-rearing parasitoid wasps in puparia of Tephritidae.

Fontana (1972) conducted experiments in Australia that proved that larvae of the aquatic predator *Dichetophora biroi* do not show any obvious preference when given the simultaneous choice of *Austropeplea tomentosa* infected or not infected with *Fasciola hepatica*, and suggested that "There does not seem to be any a priori reason to believe that this is a special characteristic of only a few species." Lindsay (1982) conducted experiments in Ireland that showed that third-instar larvae of the aquatic predator *I. albiseta* had no preference between *Galba truncatula* infected with *F. hepatica* or not. Gbedjissi *et al.* (2003) showed that first- and third-instar larvae of *Sepedon ruficeps* indifferently killed and ate *Bulinus forskali* parasitized or not by *Schistosoma intercalatum*.

Rondeleau *et al.* (2002) reported that the presence of predators modifies larval development of *Fasciola hepatica* in surviving, experimentally infested *Galba truncatula* in central France. They conducted controlled laboratory experiments using field-collected "first and third-instar larvae" of "*Tetanocera arrogans*" and the predatory snail *Zonitoides nitidus*. They found that for *G. truncatula* "the survival rate at day 30 post-exposure, the duration of cercarial shedding, and the number of cercariae shed by surviving snails were significantly lower when predators were present in snail breeding boxes, whatever the type of predator used." But "the prevalences of *Fasciola* infections in snails, and the length of time between exposure and the onset of cercarial shedding showed no significant variation." They concluded that "The progressive development of a stress reaction in surviving snails against predators during the first 30 days of experimental exposure to *F. hepatica* would influence snail survival during the cercarial shedding period and, consequently, the number of cercariae shed by the snails." Although identification of the predator *Z. nitidus* probably is correct, there is a question concerning the *T. arrogans* larvae. After examination of some specimens and many of the authors' photos by J-C. Vala in 2005, the sciomyzid referred to is *T. ferruginea*, not *T. arrogans*.

Barnes (1976) presented the only original, detailed analysis and discussion of rates of population increase of Sciomyzidae. He used data from his controlled rearings of the multivoltine, aquatic predator *Sepedon fuscipennis* (New York stock) to determine the capacity for increase (r_c) and the intrinsic rate of increase (r_m). His analyses show that *S. fuscipennis* has the intrinsic capacity to reproduce faster than two of its likely snail prey, *Lymnaea* sp. and *Physella* sp. Barnes (1976) found that the intrinsic rate of increase of *S. fuscipennis* peaks at 30 °C, where $r_m = 0.137$ per day. At 15 °C $r_m = 0.013$ per day, at 21 °C $r_m = 0.076$, and at 33 °C it was 0.057. DeWitt (1954) estimated that the intrinsic rate of increase of *Physella gyrina* (Michigan stock; rearings presumably at room temperature) ranged from 0.014–0.058 per day. Of course, mortality factors in the field probably alter the r_m rates of both predators and prey. Since many aquatic snails have a broader microhabitat range than their sciomyzid enemies, it would be useful to determine r_m for a snail species reared in the range of its microhabitats and the r_m for the several species of sciomyzid natural enemies found across that range.

Tirgari & Massoud (1981) stated that the highest average rates of increase, "(RI)", for *Sepedon sphegea* in Khuzestan, Iran were 3.29 (April 1975) and 4.5 (October 1975) and the highest local RI was 9.6 (April 1975). For "snails," apparently species of *Biomphalaria*, *Bulinus*, *Gyraulus*, *Lymnaea*, and *Physa* spp. combined, the highest average RIs were 3.25 (March 1974) and 2.29 (April 1975); the highest local RI was 6.17 (June 1975).

Larvae that penetrate the egg masses of snails and consume the embryonic snails might be of special interest as biocontrol agents (Table 5.1). Only one genus, *Anticheta*, with 14 species, is known to have larvae that are restricted to snail eggs during at least early larval life. The natural prey of the seven species that have been reared include *Aplexa*, *Quickella*, *Lymnaea*, *Oxyloma*, and *Succinea*, with *Physa* and *Radix* being fed upon in laboratory rearings (Fisher & Orth 1964, Knutson 1966, Knutson & Abercrombie 1977, Robinson & Foote 1978). Eggs and larvae of *A. analis* were found in egg masses of the *Fasciola* host, *Galba truncatula* (Knutson 1966). However, larvae of two typical aquatic predators of snails, *T. ferruginea* and *Hydromya dorsalis* (Mc Donnell *et al.* 2005a), have been found feeding on embryonic snails in egg masses in nature. In laboratory rearings other aquatic predators have fed on eggs, e.g., *Dictya brimleyi* on *Physella integra*, *D. hudsonica* on *Physella* sp. (Valley & Berg 1977), and first-instar *S. neavei* on *Bulinus* sp. (Barraclough 1983). Knutson & Berg (1963a) reared *H. dorsalis* through to pupariation on egg masses of *Lymnaea* spp. Egg masses of *Physella h. hendersoni* and *Pseudosuccinea columella* were used exclusively in mass rearing *Dictya floridensis* larva to the third instar, when the diet was changed to living or freshly crushed snails, by McLaughlin & Dame (1989). Bhuangprakone & Areekul (1973) routinely reared newly hatched larvae of the aquatic predator *Sepedon plumbella* on egg masses of *Radix auricularia rubiginosa* and *Indoplanorbis exustus*, changing the diet to living or freshly crushed snails for later instars. Some of the most interesting experimental data on aquatic snail predators developing on eggs unfortunately have not been published. K. Durga Prasad (personal communication, 1972) reared *S. aenescens* larvae from hatching to pupariation and emergence solely on eggs of *Radix luteola* in Bangalore, India, and counted the number of eggs consumed per larva per day. Of the four larvae that pupariated and produced adults (eggs hatched 16 August, pupariated 28 August, emerged 2–3 September) 2–17 eggs were consumed during the 3–4 days of the first stadium, 49–83 during the 3–4 days of the second stadium, and 302–352 during the 5–6 days of the third stadium. Totals of 378, 402, 418, and 437 eggs were consumed during larval life. Further experiments to identify species having strong proclivities to attack and develop normally on snail egg masses are needed.

The detailed study of various aspects of the biology of the aquatic predator *I. albiseta* in Ireland by Lindsay (1982) is of interest to biocontrol. The experimental results on feeding time (Section 7.1) showed that only a small proportion of time, during at least the third stadium, is occupied with actual feeding on snails. This implies that in the field larvae can spend large amounts of time searching for prey and so could survive at fairly low snail densities.

The effect of various temperatures on the duration of the egg, larval, and pupal stages on adult longevity and on the rate of predation and biomass consumption by larvae of *I. albiseta* were studied in detail under controlled conditions by Gormally (1985b, 1987a, 1988a, 1988b). The effect of temperature and prey density on snail predation and the duration of the larval stage of *S. spinipes* were studied by Mc Donnell (2004) and the effect of temperature on the pupal and adult stages by Mc Donnell *et al.* (2005b), with particular reference to biological control (see Section 18.3). From his quantitative studies under a range of controlled conditions, Gormally (1985a) advised on the potential use of *I. albiseta* as a biocontrol agent. He provided specific information on many aspects of the biology that could be important in its use, optimal laboratory rearing conditions needed for mass production, and many aspects that could be manipulated in developing large-scale colonies. Particularly important among these key features are:

(1) The most rapid developmental rates: eggs (23 °C + treatment with 0.1% ascorbic acid); larvae (23 °C for first and second stadia, 17 °C for third stadium); and pupae (26 °C). Using this combination of temperatures, laboratory cultures could be completed in about half the time taken by individuals in the field, thus producing two generations per year in the laboratory compared to one in the field.
(2) Survival: especially the ability of first-instar larvae to live for up to one month without food at 14 °C.

(3) Prey range: inclusion of snails other than the target, *Galba truncatula*; thus fly populations might be maintained by increasing the populations of non-vector snails in release sites.
(4) Number of larvae needed to control a known snail population over a given period of time: this probably can be determined to some extent by laboratory experiments on effects of temperature on predation rate and biomass consumption.

Considering that any snail-control program involving Sciomyzidae likely will be an integrated program including molluscicides, that researchers differ in their recommendations vis-à-vis the release of once-fed or starved neonate larvae, and the limited data on survival of such larvae, the unique study by McCoy & Joy (1977) on the tolerance of larvae to molluscicides is of special interest. The authors tested once-fed and starved neonate larvae of *Sepedon fuscipennis* and *Dictya* sp. against Bayer 73 and sodium pentachlorophenate. They made trials of 10 larvae in 10 ml of aqueous solution in 8.5 cm high × 5.0 cm diameter vials, held for 96 h (*S. fuscipennis*) or 48 h (*Dictya* sp.). They found that in two trials with starved *S. fuscipennis*, the TLm^{96} (concentration of molluscicide required to kill half of the test population in 96 h, abbreviated T below) values were 43 (soft water) and 83 (hard water) ppm for Bayer 73. *Sepedon fuscipennis* allowed to feed before testing were considerably less resistant to Bayer 73; two tests (in soft water) resulting in T values of 6.0 and 1.6 ppm. Starved *Dictya* sp. in various concentrations of Bayer 73 in hard water resulted in a T value of 35.5 ppm. The authors concluded that "Bayer 73, applied in concentrations effective against the snails [*Biomphalaria glabrata*], or applied in the recommended field dosage of 1 ppm (WHO 1961), should not have any immediate detrimental effects on *S. fuscipennis* or *Dictya* sp. 1st-instar larvae." They also noted that "harmful effects could manifest themselves in adult marsh flies whose larvae had been exposed to sublethal concentrations of Bayer 73 or other molluscicides." They found that sodium pentachlorophenate was more toxic to starved larvae than Bayer 73, with T values of 30 (soft water) and 33 (hard water) ppm, and a T value of only 2.4 ppm for larvae that fed before testing. They concluded that starved larvae could tolerate the WHO (1961) recommended field dosages of 10 ppm (flowing water) and 5 ppm (still water) but fed larvae could not. However, larvae would have, of course, to be feeding to effectively control snails. They had no definitive explanation for the low T value in experiments involving larvae that had been fed prior to testing, but that contamination from decaying snail tissues may have contributed to larval mortalities. Such contamination likely was the result of the crowded test conditions. While this study clearly shows the compatibility of Sciomyzidae and molluscicides, further studies that avoid the contamination problem, which is not likely to occur in nature, are needed.

Mc Donnell *et al.* (2005b) noted that puparia formed up on vegetation (e.g., *Sepedon spinipes*) might be subject to grazing by cattle and that the food plants of adult sciomyzids should be identified so that appropriate species could be planted in release sites.

18.2 TERRESTRIAL SNAILS

Barker (2004) presented the most detailed and recent review of all groups of natural enemies of terrestrial snails. Barker *et al.* (2004) placed some emphasis on terrestrial species in their discussion of Sciomyzidae as biocontrol agents. Several species of terrestrial Sciomyzidae have been studied in southern France as potential biocontrol agents of four introduced species of Mediterranean snails (*Theba pisana*, *Cernuella virgata*, *Cochlicella acuta*, and *C. barbara*) that have become serious pests in legume-based pastures and of cereal crops in Australia (Coupland *et al.* 1994, Coupland & Baker 1995, Coupland 1996a). *Salticella fasciata* was studied in the field and laboratory but rejected as a biocontrol agent because, according to Coupland *et al.* (1994), it seems to be inefficient at killing snails, being more of a saprophage than a parasitoid. Females oviposit only onto large, adult snails, thus selecting those that are spent and moribund.

Salticella fasciata was the first Sciomyzidae reported to be associated with molluscs. Perris (1850) reared adults from larvae found in the shells of *Theba pisana* in southwestern France. He did not state if the snails that he collected were found living or dead, but remarked "Cette larve, comme la précédente ["*Sarcophaga* [*Eurychaeta*] *muscaria* Meigen"], dévore l'animal de l'*Helix pisana* probablement après qu'il est mort" [This larva, as the preceding, devours the *Helix pisana* probably after it has died]. Mercier (1921) obtained adults by rearing larvae found in *T. pisana* "récolté vivants [collected alive]" in southwestern France. He stated "mais on peut se demander, avec H. Schmitz (1917), si ces larves sont parasites, ou simplement nécrophages [but we can ask, with H. Schmitz (1917), if these larvae are parasites or simply necrophagous]." Blair (1933) reared adults from larvae found in *T. pisana* in Wales but did not state whether the snails were living or dead when they were collected. In Slovakia, Povolný & Groschaft (1959) found 8% of 130 apparently living *Xerolenta obvia* infested with one larva each. Although they did not discuss the details of feeding behavior, these authors referred to *S. fasciata* as "Parasitische" and remarked that "vernichtet die Made von *Salticella fasciata* ihren Wirt erst kurz vor der eigenen Verpuppung [the larva of *Salticella fasciata* kills its host only shortly before its pupariation]." From 100 *Theba pisana* collected on October 21, 1986 in Veleron, southern France, Vala (1989a) found 83 with one to four eggs of *S. fasciata* on each.

The biology of *S. fasciata* is presented here in some detail, considering its questionable status as a potential biocontrol agent and the fact that it is the only biologically known member of the plesiomorphic subfamily Salticellinae. Also, the techniques might be useful in studies of other species. Based on their laboratory studies and collection of living, infested *Cernuella virgata* during September in Wales, Knutson *et al.* (1970) suggested that the species might be useful as a biocontrol agent of terrestrial snails. They considered its large size, preference for oviposition on terrestrial snails, short period of development, continuous breeding under laboratory conditions, and preference for dry habitats as significant features. The authors concluded that the larvae are primarily parasitoid in behavior, but with labile features. They found it capable of developing as a predator or in a saprophagous manner on dead snails, and even on dead sow bugs (*Oniscus asellus*, Isopoda) – hatching to pupariation – and dead slugs and earthworms – third instar to pupariation. This is the only sciomyzid known to feed on non-molluscan food other than the obligate, freshwater oligochaete-feeding *Sepedonella nana* and *Sepedon* (*M.*) *knutsoni* and the facultative *Sepedon* (*P.*) *ruficeps*. They noted that during the first stadium the larva was completely within the living snail, its posterior spiracles not visible. Only after several days could the posterior spiracles of the second-instar larva be seen protruding between the mantle and the shell. Infested snails (*Cernuella virgata*, *Helicella itala*, *T. pisana*, and *Cepaea nemoralis*) remained alive throughout the first and the second stadium and much of the third stadium, with almost all of the decaying tissues eaten during the few days between the time the snail died and when the larva left the shell to pupariate. When living snails were placed with larvae that had been reared to the third instar on dead snails, many left the dead snail and killed the new prey in a few hours. Six larvae killed a second snail and three larvae each killed three snails. It was clear from their observations that larvae killed healthy snails.

Coupland *et al.* (1994) studied the flight period, seasonal incidence of parasitism based on collections of snails bearing eggs, host specificity based on collections of snails with eggs and pair-wise oviposition tests in the laboratory, longevity and fecundity of adults, and infection rate by larvae on living and dead snails. They did not present observations on the movements or feeding behavior of larvae. They conducted an experiment consisting of four replicates each of 50 living and 50 dead (killed by a scalpel jab) *T. pisana* bearing eggs, with four replicates of 50 living *T. pisana* without eggs as a control in boxes with sufficient food for the living snails. Of the living snails with eggs, an average of only about five snails per box were dead at the end of 1 month and only three contained larvae. This mortality was close to that of the control. However, 77% of the 200 snails killed manually contained larvae (mostly third instar) feeding on decaying tissues. The results of this experiment are quite clear but do not fit with the observations of larval movements and feeding behavior by Povolný & Groschaft (1959) and Knutson *et al.* (1970). The field study by Coupland *et al.* (1994) (September 19 to December 18) of snail populations and percent of the populations bearing eggs led them to conclude that *S. fasciata* females select moribund snails for oviposition. The results showed a striking increase in oviposition as snail populations sharply declined (Fig. 18.4). In a collection of several hundred *T. pisana* on October 18 only the larger snails (more than 9.0 mm diameter) were oviposited onto. This was confirmed with laboratory trials in which *T. pisana* of more than 12.0 mm diameter were preferred to individuals less than 8.0 mm diameter (Fig. 18.5). However, 12 larvae developed on a smaller range of sizes and numbers of *C. virgata*, from one of 7.5 mm diameter to three of 7.5–9.5 mm diameter, with none offered being over 10.0 mm diameter during laboratory rearings by Knutson *et al.* (1970).

Also, Coupland *et al.* (1994) collected adults from September through May but none during June, July, and August, with populations peaking during late October and November. Our records, from our field studies, museum collections, and the literature, of 73 collections show 13 collections in June, three in July, and five in August in Wales, England, France, Spain, Hungary, and Tunisia. The greatest number collected during one month

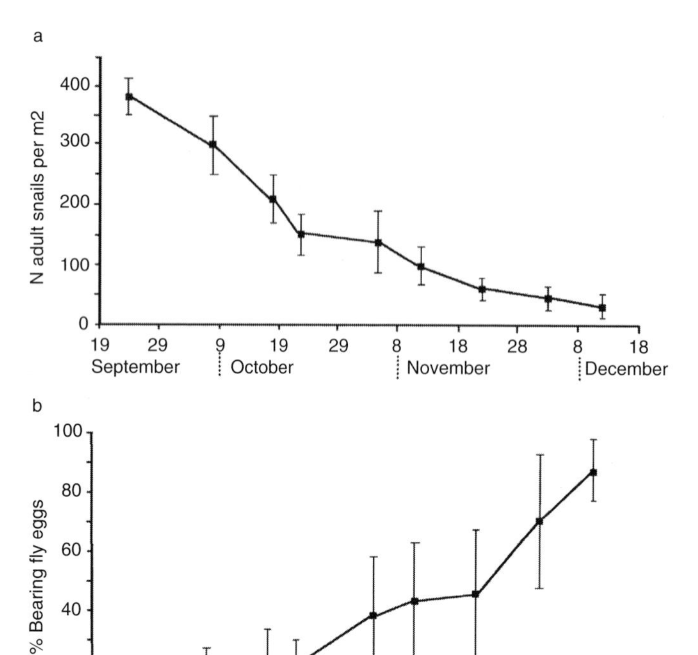

Fig. 18.4 (a) Change in mean density (± S.E.) of adult *Theba pisana* showing gradual population drop due to mortality after breeding. (b) Temporal change in mean percentage (± S.E.) of adult *T. pisana* population bearing eggs of ***Salticella fasciata***. Modified from Coupland *et al.* (1994).

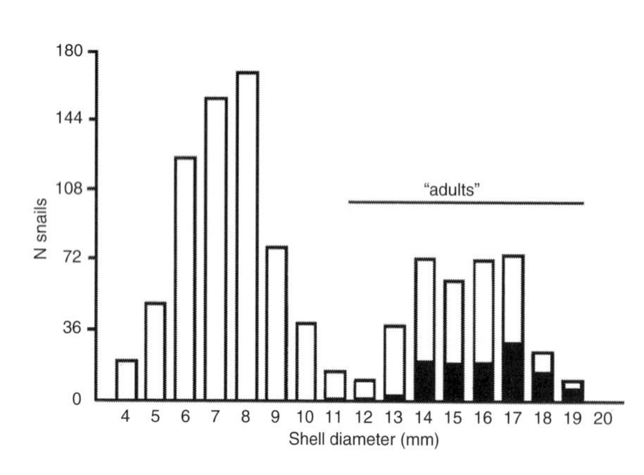

Fig. 18.5 Size frequency distribution of *Theba pisana* in a field near Montpellier, southern France on October 18, 1991. The bimodality indicates a predominately biannual life cycle with larger second-year snails (breeding "adults") having pronounced growth break (due to estivation between two growing seasons, i.e., fall–winter). Black areas indicate snails bearing ***Salticella fasciata*** eggs in their umbilici. Modified from Coupland *et al.* (1994).

was 14, in September. Coupland *et al.* (1994) found that females that emerged in the laboratory had a median life expectancy of 80 days. The mean lifetime fecundity was 205 ± 14 eggs, and the highest number of eggs laid by one female was 389 eggs over 151 days. Knutson *et al.* (1970) found that adults collected on September 7 and 14 lived 3–81 days, laboratory-reared adults lived 17–167 days, and a laboratory-reared female laid 331 eggs, January 9–April 7. In single- and two-host oviposition preference tests, Coupland *et al.* (1994) found that *T. pisana* and *C. virgata* were preferred over the other two species. There was no preference between *T. pisana* and *C. virgata*. Among 1444 living specimens of 18 species of terrestrial snails collected in Wales on September 7, each of five *C. virgata* was infested with one larva, and one third-instar larva was found in a dead *Cepaea hortensis* (Knutson *et al.* 1970). Coupland *et al.* (1994) stated that no eggs were laid on dead snails, whereas Knutson *et al.* (1970) observed eggs occasionally laid within the aperture of living and dead snails and on pieces of crushed snail.

J. B. Coupland (personal communication, 2000) noted that during laboratory rearings in Australia some of the key

Table 18.5. Percentages of nine species of terrestrial snails killed by L1 of five species of Sciomyzidae in southern France

Snail species	*Coremacera marginata*	*Euthycera cribrata*	*Pherbellia cinerella*	*Trypetoptera punctulata*	*Dichetophora obliterata*
Cernuella virgata	80	77	90	0	0
Theba pisana	87	73	77	0	0
Cochlicella acuta	90	77	93	37	23
Cochlicella barbara	83	67	93	20	20
Trochoidea elegans	77	0	90	10	43
Eobania vermiculata	77	37	10	0	0
Pomatias elegans	47	20	3	0	0
Lauria cylindracea	23	10	43	90	80
Discus rotundatus	73	43	67	3	93

Note: From Coupland (1996a).

features that he observed, especially that the snails had to be killed for larvae to feed, also were observed there. But he concluded that the species is worthy of further study. Such studies might include repetition of oviposition rate tests, examination of positions of larvae and anatomical parts fed upon after *x* hours and *x* days of feeding, search for infested snails in nature during the spring and summer, analyses of the condition of the ovaries and testes of adult flies throughout the year, and other aspects to clarify seasonal behavior and killing efficiency.

It would be especially interesting, of course, to find natural enemies of the widespread, invasive *Achatina fulica*, which is native to Africa and now widespread in the Pacific area. Study of immature *A. fulica* collected in the field is strongly suggested. Considering the few species of Afrotropical Sciomyzidae that have been reared, there is some possibility that such natural enemies exist. Most Afrotropical Sciomyzidae (about 60 species) belong to *Sepedon* and related genera, most reared species of which are aquatic predators. However, the highly unusual habit of *Sepedonella nana* and *Sepedon* (*M.*) *knutsoni* feeding obligatorily on freshwater oligochaete worms (Vala *et al.* 2000a, 2000b, 2002) and the discovery of the terrestrial, parasitoid/predatory habitats of *Sepedon* (*P.*) *umbrosa* in Africa (L. G. Gbedjissi, personal communication, 2003) indicate that unexpected trophic relationships have arisen in sub-Saharan Sciomyzidae. In addition, nothing is known about the biology of three Afrotropical genera (*Ethiolimnia*, *Tetanoptera*, and *Verbekaria*). Miller (1995) noted that the type series of *Salticella stuckenbergi*, collected in Lesotho during November "were found scavenging on large agate snails, ?*Archachatina* sp. (Achatinidae) (Stuckenberg: Kilburn personal communication). The larval habits of *Ethiolimnia* are unknown; however, unidentified puparia probably belonging to this genus were discovered in the empty shells of the invasive Grey Snail, *Theba pisana*, along the Cape coast." Examination of the cast third-instar cephalopharyngeal skeleton, found on the ventral cephalic cap of the puparium, for the presence of a ventral arch in these and other unidentified puparia could prove their identity, or not, as Sciomyzidae. Also, see Coupland & Barker (2004) for two species of Muscidae, three species of Phoridae, and one of Sarcophagidae that have been reared from *A. fulica*.

Five Palearctic species of terrestrial sciomyzids were studied by Coupland & Baker (1995) and Coupland (1996a) in southern France during 1990–1995 as potential biocontrol agents of pest snails in Australia. Those are, with references to detailed papers on their biologies: *Coremacera marginata* (Knutson 1973), *Euthycera cribrata* (Vala *et al.* 1983), *Trypetoptera punctulata* (Vala 1986), *Dichetophora obliterata* (Vala *et al.* 1987), and *Pherbellia cinerella* (Bratt *et al.* 1969). The studies in southern France were on phenology, species abundance in four kinds of habitats, pre-oviposition periods, duration of egg to adult development, and prey choice by first-instar larvae (Table 18.5). As *C. marginata*, *D. obliterata*, *T. punctulata*, and *E. cribrata* are univoltine and most abundant in riverine forest or low scrubland habitats, they were not considered as likely biocontrol agents in the target habitats in Australia. *Pherbellia cinerella*, being multivoltine, very abundant in open pasture habitats, and having a native habitat range including areas very similar climatically to South Australia, was considered as a potential biocontrol agent. Work on *P. cinerella* in Australia has stopped (Coupland 1996a) because of its wide prey range and because during quarantine studies in Australia in no-choice tests it was found to kill endemic Australian snails. Coupland & Baker (1995) noted that the strong habitat preference of *P. cinerella* "may reduce the likelihood of it overlapping with the habitats of endemic Australian snails which do not occur in pastures. It should therefore be considered a potential agent as long as snails are an agricultural problem." Bratt *et al.* (1969) noted features of *P. cinerella* that would support its use as a biocontrol agent. It is very successful, being one of the most common and abundant species in the genus. The species is much more common in open habitats. It breeds continuously during warmer weather with 5–7 generations over a 7 ½ month breeding period. Females lay as many as 560 eggs, and longevity of adults is as long as 220 days. However, they noted,

> Not only did [larvae] feed for a long time on the decaying tissues of hosts they had killed, but many fed on dead snails after they had left the first or second host and were in search of food. When dead and living snails were offered to larvae feeding in dead snails, these larvae consistently moved from dead snails to dead snails, rather than attacking living ones.

In addition to the prey recorded by Coupland & Baker (1995), Bratt *et al.* (1969) noted that larvae of *P. cinerella* killed and ate *Candidula intersecta*, *Cepaea nemoralis*, *Trochulus hispida*, *Succinea* sp., and the aquatic *Biomphalaria glabrata*, *Gyraulus* sp., *Helisoma trivolvis*, *Galba truncatula*, and *Physella* sp., and also fed readily on freshly crushed snails. Vala (1989a) also found this species to be most abundant in open habitats in southern France (Fig. 18.6).

Subulina octona, an Afrotropical terrestrial snail, widespread as a pest in gardens and greenhouses in many parts of the world, recently has been discovered to be the host/prey of the parasitoid/predator *Sepedon* (*P.*) *umbrosa* (L. G. Gbedjissi, personal communication, 2003) and possibly could be a biocontrol target.

Also, the fact that the terrestrial, gastropod-killing *Salticella fasciata*, *Pherbellia cinerella*, and *Limnia unguicornis* have been found in disturbed or recently disturbed situations, such as wheat-field stubble, golf links, seaside dune slacks frequented by vacationers, grazed pastures, semi-natural and harvested meadows, and disused vineyards indicates they might have biocontrol potential in such areas (see Section 10.1.2).

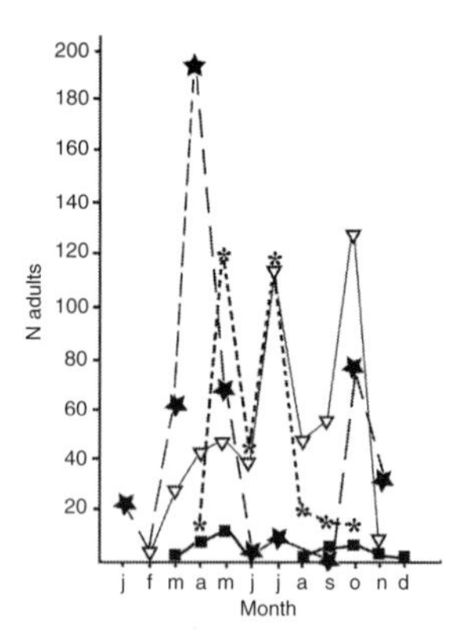

Fig. 18.6 Seasonal variation of adult ***Pherbellia cinerella*** in different habitats in southern France: asterisk, native grass pasture; square, forest; star, abandoned pasture; triangle, cultivated pasture. Modified from Vala (1989a).

18.3 SLUGS

Godan (1979) and Barker (2004) provided the most detailed and recent reviews of natural enemies of slugs. The potential of slug-killing Sciomyzidae as biocontrol agents of crop pests was raised by Reidenbach *et al.* (1989) in a paper primarily on a quantitative study of predation by *Euthycera cribata* and in which they briefly compared the biologies of the seven known slug killers in the genera *Tetanocera* and *Euthycera*. Barker *et al.* (2004) discussed slug killers as an example of adaptive specialization in the Sciomyzidae. From our experience, we suggest that slug-killing sciomyzids have biocontrol potential in semi-natural and greenhouse situations.

The basic life cycles of slug-killing sciomyzids were described for *T. elata* (Knutson *et al.* 1965), *T. plebeja*, *T. clara*, and *T. valida* (Trelka & Foote 1970), *E. cribrata* (Vala *et al.* 1983), *E. stichospila* (Vala & Caillet 1985), and *Limnia unguicornis* (Vala & Knutson 1990). The limited information on *E. chaerophylli* and *E. arcuata* has not been published; Trelka & Foote (1970) mentioned only the unpublished observations of Knutson cited below. Trelka (1973) studied the behavior of larvae of *T. plebeja* and toxicological and neurological aspects of the toxic salivary secretion it uses to immobilize slugs. Trelka & Berg (1977) made a detailed comparison of attack behavior by third-instar larvae of *T. plebeja* and *T. elata* and showed that both species immobilize slugs with a neurotoxic salivary gland secretion. A molecular/morphologically based study of the evolution of slug parasitoidism in *Tetanocera* and a cladistic analysis of the genus was presented by Chapman *et al.* (2006).

Many aspects of the biology of these species support their potential as biocontrol agents in certain situations. Notably, *T. elata* in Europe and *T. plebeja* in North America are widespread and common, occurring in a wide variety of habitats, and are among the most successful species of Sciomyzidae. As an indication of their success, for *T. elata* we have about 500 capture records as compared to about 1000 records for *Pherbellia cinerella*, the most common and abundant sciomyzid in the Palearctic. *Tetanocera valida* also is widespread in North America, and it and the North American *T. clara* prefer mesic woodlands and low-lying forests. Like many other species of *Euthycera*, *E. cribrata* and *E. stichospila* have a predominantly Mediterranean distribution and are adapted to climates where there is a long dry season during which most gastropods estivate. *Euthycera chaerophylli* is quite common in mesic woods throughout much of Europe.

Whereas *E. chaerophylli*, *T. elata*, and *T. valida*, and perhaps *T. clara* appear to be restricted to slugs, the other species also attack terrestrial snails in the laboratory, especially during the third stadium, but do not require snails for development. Of these, only *T. plebeja* has been found in a semi-terrestrial snail (*Oxyloma* sp.) in nature. The *Tetanocera* larvae and those of *E. chaerophylli* appear to be host specific during the first stadium. During laboratory rearings, *T. elata*, *T. plebeja*, *T. valida*, and *E. chaerophylli* attacked only *Deroceras* spp. and *T. clara* attacked only *Philomycus* spp. and *Pallifera* spp. *Euthycera cribrata* and *E. stichospila* attacked *Deroceras reticulatum* and *Deroceras* sp., and terrestrial snails of the genera *Abida*, *Clausila*, *Cochlostoma*, *Helicella*, *Helix*, *Lauria*, *Trochulus*, and *Phenacolimax*. First-instar larvae of *L. unguicornis* attacked *D. reticulatum*, the semi-terrestrial snail *Oxyloma sarsii*, and the terrestrial snail *Lauria cylindracea* but with 95–98% mortality. With only one larva pupariating in laboratory rearings, the natural prey of *L. unguicornis* is questionable.

Larvae of the *Tetanocera* slug killers are parasitoid during early life, living under the mantle, in the eye tentacle, or in the "mouth" of the slug, becoming predatory during the third stadium. They consume 2–9 slugs during the 20–48 days of larval life. Larvae of *E. cribrata* and *E. stichospila* are parasitoid during early life and predaceous/saprophagous in later larval life. They consume 15–25 slugs during the 60–90 days of development. Larvae of *E. chaerophylli* are completely internal endoparasitoids of *Deroceras* spp. in which they live at least several weeks, developing to the second instar; they have not been reared beyond this point. Rozkošný (1967) found in the Czech Republic a puparium on April 14 from which a female emerged on April 25; he described and figured the puparium. Vala (1989a) provided stereoscan photos of the egg and anterior and posterior end of the first-instar larva. First-instar larvae of the North American *E. arcuata* have been observed penetrating into the flesh of living *Pallifera* and *Philomycus*. Larvae of *L. unguicornis* fed for up to 30 days on one or two prey individuals, remaining immersed in the decaying tissues. The *Tetanocera* species are multivoltine; *E. cribrata*, *E. stichospila*, and *L. unguicornis* are univoltine; all overwinter in the puparia, as apparently also does *E. chaerophylli*.

Although host/prey choice tests of slug-killing Sciomyzidae are needed to evaluate further their potential as biocontrol agents, the above biological features, especially the restriction to or predilection for *Deroceras* spp. and the use of a salivary gland neurotoxin to immobilize host/prey, indicate their possibilities. Other salient features include the high fecundity of females, the "ambush" behavior of some newly hatched larvae against their hosts, and the ease of collecting adults and rearing immatures. We suggest that *T. elata* and *T. plebeja* might be useful in biocontrol of *Deroceras* spp. especially in greenhouse situations. It seems likely that means of mass producing eggs of these species, and of storing them at low temperatures to be dispersed as needed can be developed, which could enable commercially successful applications. Notably, although studies of survival of *T. elata* and *T. plebeja* under various temperature/humidity regimes have not been conducted, it is likely that these broadly ranging species could survive in greenhouses. Although *T. elata* and *T. plebeja* are abundant in a broad array of habitats, populations are not known to develop and persist in highly and continuously distributed situations. However, it might be feasible to release adults or overwintered puparia in cultivated fields where cereal strips are employed, after the adjacent production areas are no longer disturbed until harvest and when the developing crop affords protection for slugs and sciomyzids. Speight (2001, 2004) found *T. elata* to be the most abundant of 23 species of Sciomyzidae in emergence traps in production, infrastructure, and disused sectors of a farm in Ireland, including pasture, silage, and cereal strips (see Section 10.3). Knutson *et al.* (1965) found an infestation rate by *T. elata* of 14% in a sample of *D. reticulatum* taken from a disused horticultural plot in southern England. Life-cycle information on these oligophagous, non-cannibalistic species of *Tetanocera* indicates that they could be mass-reared much more economically than polyphagous, cannibalistic Carabidae such as *Pterostichus cupreus*. The retail price in Europe in 2004 for laboratory-reared individuals of the latter was more than £2 each (Symondson 2004). R. J. Mc Donnell initiated a 3-year European Union "Curie" project on the use of slug-killing Sciomyzidae as biological control agents in 2006. C. D. Williams (personal communication, 2006) made the novel suggestion of combining Sciomyzidae that attack adult slugs with Phoridae that attack eggs of slugs in biocontrol attempts. Also, the nematode *Phasmarhabditis hermaphrodita*, which carries symbiotic bacteria pathogenetic to slugs and which has been developed as a commercial biological control agent of slugs, might be included in a combined approach.

18.4 POTENTIAL RISKS OF INTRODUCTIONS

> Nor should dangers of possible consequences of foreign introductions crowd out consideration of inevitable consequences of taking no action. With snail-borne diseases increasing at their present rate, we can ill afford to eschew control measures because some people unfamiliar with a candidate agent of biological control, or with the method itself, can imagine all sorts of dire consequences of an introduction. The situation demands earnest efforts to assess the real probabilities of various allegedly "probable" consequences. And we must ask not only whether we dare to act but also whether we dare not to act.
>
> Berg (1973a).

All of the 203 biologically known species of Sciomyzidae kill and feed solely on molluscs, except *Sepedonella nana* and *Sepedon* (*M.*) *knutsoni* which are restricted to feeding on freshwater oligochaetes, *Sepedon* (*P.*) *ruficeps* which feeds facultatively on freshwater snails and freshwater oligochaetes, and *Salticella fasciata* which developed on various dead invertebrates in laboratory rearings. There are two erroneous reports of Sciomyzidae as agricultural pests. Grist & Lever (1969) stated that *Sepedon* "*senegalensis*" is a minor pest of rice in Senegal and Mali; recent taxonomic research (Knutson, unpublished data) indicates that *S. senegalensis* Macquart is the same species as *S. ruficeps*. Ayatollahi (1971) recorded *Psacadina zernyi* from stems and roots of rice in Iran, but this species is a typical obligate predator of aquatic snails (Knutson *et al.* 1975).

Berg (1964) carefully considered three supposed dangers of introducing exotic Sciomyzidae into new areas: adults might carry disease, larvae might kill desirable molluscs, and the food habits of larvae might change. Berg's 1964 evaluation, reinforced by the vast amount of information obtained over the subsequent 40 years and analyzed herein, shows that adult Sciomyzidae pose no health risk. It is inconceivable that the feeding habits of the larvae will change, except perhaps as do other organisms over geological time. That larvae introduced as biocontrol agents might kill desirable molluscs or, unlikely, have such an impact on snail populations as to alter an entire ecosystem remains a significant concern, especially in light of the increased interest over the past few decades in protecting native biodiversity.

Schalie (1969) questioned the impact of the two exotic aquatic predators established in Hawaii. He asked if the flies would remain confined to the relatively isolated, steep-sided valleys where they were released; if the snail populations were studied before the flies were introduced; and especially, would they attack such lymnaeids as *Pseudisidora producta* (the only known normal sinistral lymnaeid), *E. newcombi*, and other endemic species. To answer his questions: (1) the two sciomyzid species have dispersed to other islands; (2) snail populations and the prey range preferences of the larvae of these two species were not studied before release; and (3) it is likely that the larvae will attack all species of "*Lymnaea*" but there are no records of them attacking the species of concern in nature. Of the 11 species of Sciomyzidae released in Hawaii, only the semi-aquatic *Atrichomelina pubera*, *Ditaeniella parallela*, and *Pherbellia dorsata* were likely to attack the endemic, aesthetically desirable tree snail *Achatinella stewarti* var. *producta* or the introduced terrestrial predators *Gonaxis* spp. and *Euglandina rosea*. And, only individuals of these terrestrial snails that strayed out of their typical habitat onto moist surfaces were likely to be attacked. Host range tests under simulated natural conditions were conducted on *P. dorsata*, whose larval behavior and host range are similar to those of *A. pubera* and *D. parallela* (Davis *et al.* 1961). *Pherbellia dorsata* did not attack the tree snail or *Gonaxis kibweziensis* but *G. quadrilateralis* and *E. rosea* "sustained minor predation." Concern about the impact of these three species is a moot point, since they did not become established.

Barker *et al.* (2004) discussed the potential impact of Sciomyzidae on non-target gastropods in relation to the four scenarios proposed by Hopper (1995) in his analysis of the possible impact of biocontrol agents on endangered arthropod species. They concluded that no sciomyzid could be guaranteed to have no direct non-target effects (scenario 1) and that many might have considerable impact on biodiversity (scenario 4). However, they noted, "with careful selection of sciomyzid species for introduction based on niche requirements and prey preferences, scenario 3 [some small but generally unpredictable mortality if the natural enemy also utilizes one or more co-occurring species in addition to its normal hosts/prey] or even scenario 2 [no impact if the natural enemy utilizes other hosts/prey but is temporally or spatially isolated from them] may be the expected outcome." Generally, in biocontrol, the methods, protocols, and facilities for testing the safety of introduced agents have improved greatly over the past few decades, and if these are followed and used we can be rather strongly assured of the safety of introduced agents (Knutson & Coulson 1997).

The most recent, serious considerations of the introduction of Sciomyzidae, *Salticella fasciata*, and *Pherbellia cinerella* for control of terrestrial snails in Australia (discussed above), are instructive. Plans for introduction of *P. cinerella* were curtailed when the research data indicated it might attack endemic, non-pest snails, and *S. fasciata* was considered to be an inefficient natural enemy.

Boray (1969) noted that the oligochaete *Chaetogaster limnaei* might be a very important predator of miracidia and cercaria of *Fasciola* spp. In this regard, oligochaete predation by Sciomyzidae, known only for *Sepedonella nana*, *Sepedon* (*M.*) *knutsoni*, and *S.* (*P.*) *ruficeps* might be of interest.

19 • History of research on Sciomyzidae

> There is something awe-inspiringly symbolic about the stroke that destroyed his [Linnaeus's] mental competence. It savors of the divine nemesis of which he had once written and long feared. He who in youth had beheld the beautiful radiating lines gleam for an instant like a spider web on a dew-hung morning glimpsed a truth which, as is true of so much human knowledge, was also an illusion. The rainbow bridge to the city of the gods had vanished, leaving an old, memoryless man. The passionate cataloguer of the Systema Naturae no longer knew his book. Finally, and most dreadful fate of all, there passed away from that proud, world-famous man the knowledge even of his own name. There remained in his garden only the dried husk of an old plant among new flowers reaching for the sun.
>
> Eiseley (1958).

As the Sciomyzidae is one of the best known among all flies, it is useful to see how this development of knowledge – the history, the *raison d'être*, the driving forces, the personalities, the opportunities, and the benefits – came about. Little has been published on the history of research on Sciomyzidae. Nearly 3000 papers (solely or in part on Sciomyzidae) including information on all aspects of Sciomyzidae have been published by over 200 authors, worldwide, during the past two and a half centuries. The first was the description of *Musca umbrarum* by Linnaeus (1758) (transferred to *Dictya* by Meigen, 1803).

19.1 SYSTEMATICS

Primarily with regard to systematics, Rozkošný (1984b) reviewed research on the Fennoscandian fauna, and Rivosecchi & Santagata (1978) and Rivosecchi (1992) reviewed research on the Italian fauna. Berg & Knutson (1978) presented a table comparing the suprageneric classifications from Verbeke (1950) to Griffiths (1972). Here, we present an updated table (Table 15.14). Vala (1984a, 1989a) presented a broad summary of the suprageneric classification, and we tabulated in Section 15.7 the classifications from that of Fallén (1820b) to Crampton (1944). It is interesting that Schiner's (1862) division of the family into Sciomyzinae and Tetanocerinae is essentially the major division recognized today (both as tribes of the Sciomyzinae), plus the removal of *Pelidnoptera* to Phaeomyiidae and recognition of the Salticellinae and Huttonininae. The first designation of family-level status was by Fallén (1820b), who included *Sciomyza*, *Tetanocera*, *Dryomyza*, and *Sepedon* in his "Sciomyzides."

The history of research on the systematics of Sciomyzidae can divided into four partly overlapping periods. (1) The classical period of descriptions by the early authors of genera and species and initial suprageneric classifications based on external morphological features of adults, from Linnaeus (1758) until the beginning of the twentieth century. (2) The period of the first revisionary studies, including emphasis on suprageneric classifications, of Hendel (1902) to Sack (1939) and Enderlein (1939). Sack, working in Frankfurt, and Enderlein, working about 400 km away in Berlin, apparently were unaware of each other's studies. Strangely, two revisions of the Nearctic fauna also appeared during the same year: Cresson (March, 1920) and Melander (September, 1920). Melander, working in Pullman, Washington, revised his manuscript in press when he saw the revision of Cresson, who was working across the continent in Philadelphia, Pennsylvania. The first extensive illustrations and use of surstyli in characterizing species was by Melander (1920) for 15 species of North American *Tetanocera*, not, as mentioned (Rozkošný 1984b) by Frey (1924) for 12 species of European *Tetanocera*. However, as early as 1902, Hendel referred to the "Parameren" [surstyli], "Afterglieder" [postabdomen], and abdominal segments of the male, especially the fifth, in his description of the family, genera, and species, and in his keys. Also he figured the undissected postabdomen of three closely related species of *Pherbellia*. (3) The period of extensive

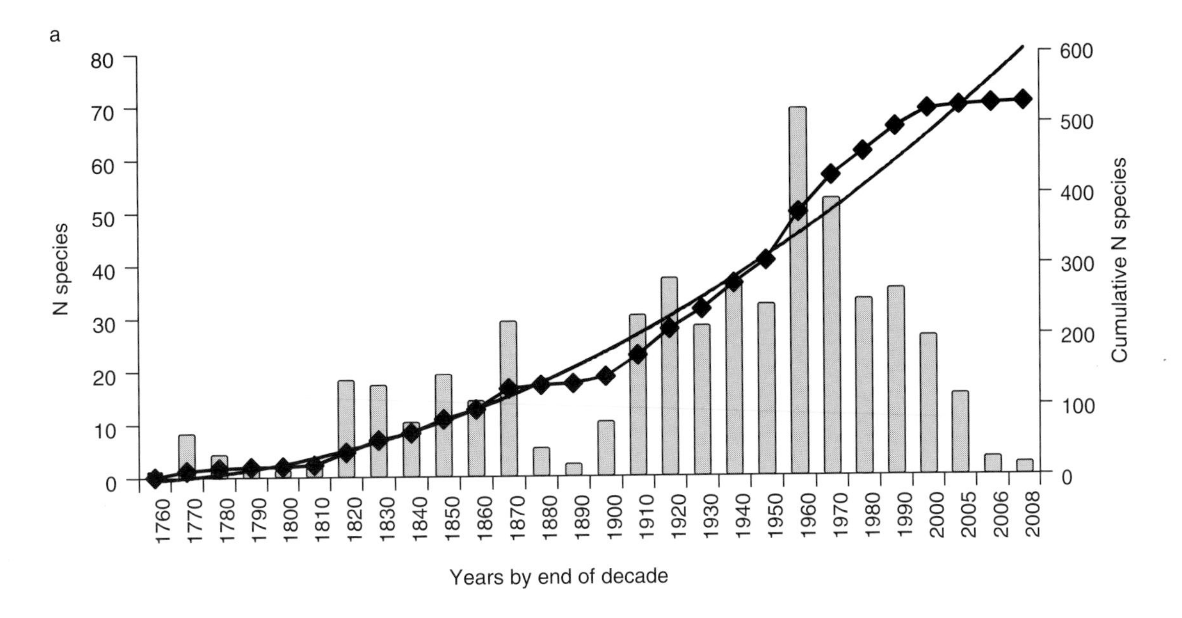

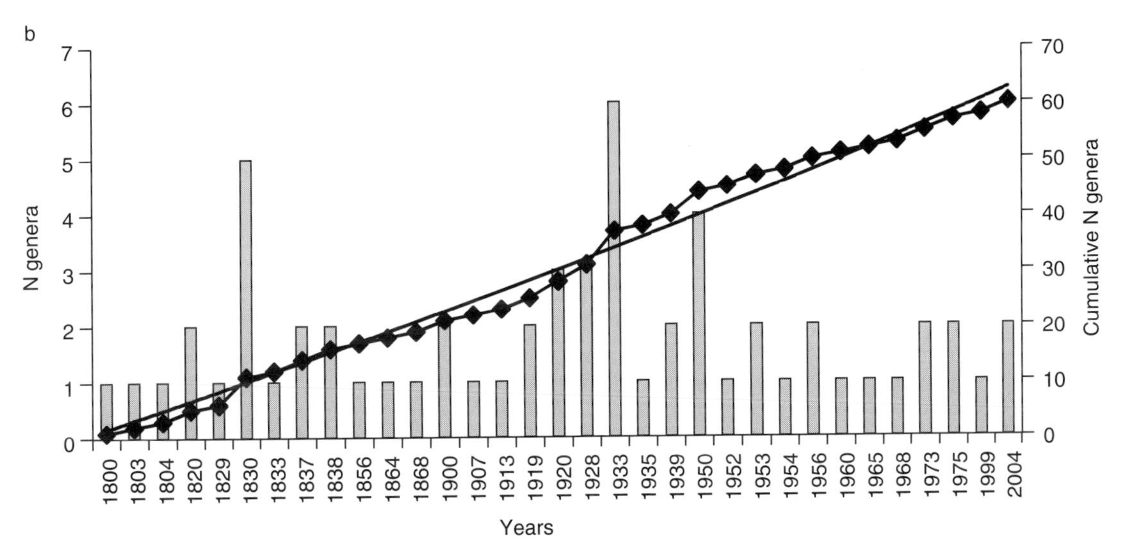

Fig. 19.1 (a) Numbers of valid species of Sciomyzidae described during decades (histogram); curve of cumulative number of species described (row of diamonds); and trend curve (solid line). (b) Numbers of valid genera in years described (histogram); curve of cumulative number of genera described (row of diamonds); and trend curve (solid line). See Eq. 19.1a, b.

description and revisionary studies that include emphasis on use of male genitalia characters, represented primarily by the publications of Steyskal (1938–1980), Verbeke (1948–1967), and Rozkošný (1959–present), and the first theoretical cladistic studies on suprageneric classifications (Hennig 1958, 1965). (4) The period, after the main descriptive work had been accomplished but with some work at this level continuing, of generic revisionary studies, major regional syntheses (e.g., Rozkošný 1984b, 1987a, Vala 1989a, Rivosecchi 1992), and catalogs of all regions (beginning with that of America North of Mexico, Steyskal 1965b). More recently there has been further development of suprageneric classifications using cladistic reasoning or methods (e.g., Griffiths 1972, Barnes 1979d,

Table 19.1. Authors who have described valid genera of Sciomyzidae, not including fossil genera

Becker – 3 (1907, 1919)
Cresson – 3 (1920)
Duméril – 1 (1800)
Elberg – 1 (1965)
Enderlein – 1 (1939)
Fallén – 2 (1820)
Haliday – 2 (1837, 1838)
Hendel – 2 (1900)
Hennig – 1 (1952)
Knutson – 2 (1968, 1999)
Latreille – 2 (1804, 1829)
Lioy – 1 (1864)
Malloch – 7 (1933, 1935)
Marinoni & Knutson – 1 (2004)
Mayer – 2 (1953)
Meigen – 2 (1803, 1838)
Melander – 1 (1913)
Papp – 1 (2004)
Perty – 1 (1833)
Robineau-Desvoidy – 5 (1830)
Rondani – 2 (1856, 1868)
Sack – 1 (1939)
Steyskal – 8 (1954, 1956, 1960, 1973, 1975)
Tonnoir & Malloch – 3 (1928)
Verbeke – 4 (1950)
Zetterstedt – 2 (1837, 1838)

1981, J. F. McAlpine 1989, Marinoni & Mathis 2000, Barker *et al.* 2004).

Concurrent with the latter part of the third period and continuing until today we see an increasing influence of information on the biology and morphology of the immature stages on classification. This has been at the generic level, e.g., *Pherbellia*, Bratt *et al.* (1969); *Pteromicra*, Rozkošný & Knutson (1970), and suprageneric levels, e.g., distinction between Sciomyzini and Tetanocerini (Foote *et al.* 1960, Neff & Berg 1966), Salticellinae as a subfamily (Knutson *et al.* 1970, Vala *et al.* 1999), and Phaeomyiidae as a family (Vala *et al.* 1990).

Diagrams of the numbers of valid species and genera described over time and curves of accumulated numbers are shown in Fig. 19.1. Lists of authors of valid genera and species are shown in Tables 19.1, 19.2. The curves

Table 19.2. Authors who have described valid species of Sciomyzidae, not including fossil species

Adams – 3
Barnes – 4
Barraclough – 1
Beaver – 1
Becker – 8
Bezzi – 2
Carles-Tolrá – 1
Coquillett – 5
Cresson – 15
Curran – 6
Czerny – 1
Day – 2
Elberg – 8
Elberg, Rozkošný & Knutson – 1
Enderlein – 2
Fabricius – 7
Fallén – 16
Fisher & Orth – 9
Foote – 2
Freidberg, Knutson & Abercrombie – 4
Frey – 6
Ghorpadé & Marinoni – 1
Giglio-Tos – 1
Harris – 1
Harrison – 2
Hendel – 23
Hennig – 1
Hutton – 4
Johnson – 1
Karl – 1
Kertész – 3
Knutson – 3
Knutson & Bredt – 2
Knutson & Freidberg – 2
Knutson, Manguin & Orth – 1
Knutson & Orth – 4
Knutson & Zuska – 2
Linnaeus – 2
Loew – 36
Lundbeck – 1
Macquart – 7
Malloch – 18
Marinoni – 1
Marinoni & Knutson – 1
Marinoni & Mathis – 1

Table 19.2. (*cont.*)

Marinoni & Steyskal – 9
Marinoni & Zumbado –1
Mathis, Knutson & Murphy – 1
Mayer – 7
Meigen –11
Meijere – 2
Melander – 13
Merz & Rozkošný – 1
Orth – 22
Orth & Fisher – 2
Orth & Knutson – 2
Orth & Steyskal – 2
Pandellé – 1
Papp – 1
Perty – 1
Przhiboro – 1
Ringdahl – 1
Rivosecchi – 2
Robineau-Desvoidy – 1
Roller – 1
Rondani – 7
Roser – 2
Rozkošný – 12
Rozkošný & Elberg – 2
Rozkošný & Knutson – 4
Rozkošný & Kozánek – 1
Rozkošný & Zuska – 1
Sack – 1
Schiner – 2
Scopoli – 6
Speiser – 2
Stackelberg – 2
Staeger – 1
Steyskal – 80
Sueyoshi – 3
Thomson – 4
Tonnoir & Malloch – 13
Vala – 2
Vala, Gbedjissi & Dossou – 1
Valley – 4
Verbeke – 49
Vikhrev – 1
Villeneuve – 1
Walker – 10
Wiedemann – 7
Withers– 1
Wollaston – 1
Wulp – 5
Yano – 2
Zetterstedt – 7
Zuska – 5

of cumulative numbers of species and genera of Sciomyzidae approach those of poorly known groups, e.g., Culicidae (Fig. 19.2c), as compared with those of well-known groups of organisms, e.g., lower primates and Rhopalocera (Fig. 19.2a, b) presented by Steyskal (1965) in his study of trend curves of the rates of species descriptions of animals. Comparison of the peaks of years of descriptions of species of all Nearctic Diptera with those of the world Sciomyzidae shows interesting, strong differences, no doubt a result of the biological studies initiated during the mid 1950s (Fig. 19.3).

George C. Steyskal, who worked on the world fauna, is the dominant figure of the 27 authors of valid genera and the 91 authors of valid species of Sciomyzidae. During the first period, Robineau-Desvoidy described five valid genera, with 11 species by Meigen and ten by Walker. During the second period, Malloch and Malloch & Tonnoir described ten genera, with 36 valid species described by Loew, 31 by Malloch and Malloch and Tonnoir, 23 by Hendel, 15 by Cresson, and 13 by Melander. During the third and fourth periods, eight valid genera and 80 valid species (worldwide) were described by Steyskal, four genera and 49 species (Afrotropical and Palearctic) by Verbeke, two genera and 25 species (worldwide) by Knutson and Knutson and eight co-authors, 31 species (mainly Palearctic) by Rozkošný and Rozkošný and five co-authors, and 38 species (mainly Nearctic) by Orth and Orth and four co-authors. There are at least 55 names regarded as *species dubia* (doubtful species), the type specimens of which likely are lost or destroyed. Twenty of these were named by Robineau-Desvoidy from specimens from Europe. Only one sciomyzid described by Robineau-Desvoidy remains recognized as a valid species today, i.e., *Limnia boscii*, 1830, but the existence of the holotype of this North American species has not been determined.

The suprageneric classifications proposed from 1950 (Verbeke) to 2004 (Barker *et al.*) are shown in Table 15.14. The earlier classifications of Enderlein (1939) and Sack (1939) are included in a similar table by Knutson *et al.* (1970) and are tabulated here in Section 15.7. History

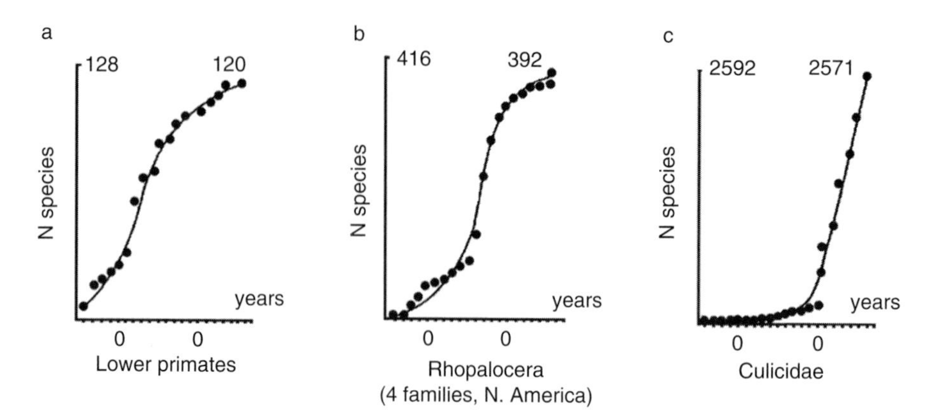

Fig. 19.2 Trend curves (manually fitted) of valid species of (a) lower primates; (b) Rhopalocera; (c) Culicidae. Number of species at upper end of each curve; number of species scaled on ordinates at upper left edge of each image. From Steyskal (1965).

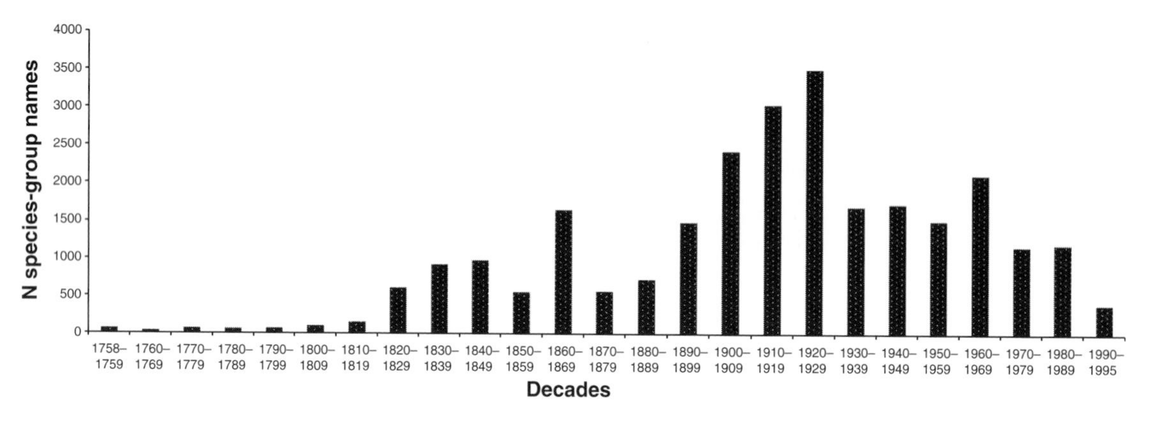

Fig. 19.3 Species-group names (valid and synonymous) of Nearctic Diptera described by decade. Modified from Poole (1996).

of research sometimes can provide a certain level of confidence in published classifications. It is notable that the "intuitive" suprageneric classification proposed by Steyskal (1965a) was supported by the cladistic analyses of Marinoni & Mathis (2000) and Barker *et al.* (2004). However, Steyskal's Phaeomyiinae appeared as a family, the possible sister group to Sciomyzidae, in the cladistic analyses, and the above authors did not include Huttonininae. This is of interest in regard to the great number of species and genera that Steyskal described. Steyskal never explicitly stated in his publications that he used Hennigian methods of cladistic analysis, but he occasionally referred to characters as apomorphic or plesiomorphic (e.g., as early as 1960a in his revision of *Anticheta*). He published (1952) a review of Hennig's 1950 book on phylogenetic systematics, a translation of a passage from that book (1953), and a review of Sokal & Sneath's 1963 book on numerical taxonomy (1964). Steyskal was in extensive communication with Hennig about many families of flies over many years and Hennig met with Steyskal on at least one occasion, as a guest in Steyskal's home near Washington, DC. We consider it likely that Steyskal's classifications were based on a kind of undocumented cladistic approach.

Cooperation and communication have been hallmarks in the development of research on Sciomyzidae, especially in systematics and life-cycle studies, and especially during the third and fourth periods when many researchers were, and some still are, in close contact. This was due, in part, to the moderate but continual monetary support of the program led by Berg at Cornell, and the nature of the professional positions of some others, which allowed considerable international travel, but largely to the attitudes of the persons involved. This exchange of information, ideas, literature, and specimens resulted in many co-authored publications and created a strong sense of identity and enthusiasm. Particularly important for systematics research, the opportunities to study type material in European museums (especially by Rozkošný and Verbeke) and North American museums (especially by Steyskal and Knutson) abetted revisionary studies and resulted in catalogs of all regions being published between 1965 (Steyskal) and 1989 (Barnes & Knutson). In North America, the concentration of research material in museums in rather close proximity (Washington, DC; Philadelphia; Ithaca, New York; and Boston, USA and Ottawa, Canada) was important, as was the cooperation of the curators, S. S. Roback, L. L. Pechuman, P. Wygodzinsky, C. D. Darlington, G. E. Shewell, J. F. McAlpine, and others.

19.2 EVOLUTIONARY STUDIES

> (E)ven mistaken hypotheses and theories are of use in leading to discoveries. This remark is true in all the sciences.
>
> Bernard (1865).

Evolutionary considerations were included to a limited extent in some of the earlier publications on the suprageneric classifications based on adults, notably in the generally overlooked paper of Hennig (1965). As information on the diversity of sciomyzid life cycles began accumulating in 1953, an hypothesis of strict malacophagy for the family developed, but there were predictions of feeding behavior on Mollusca other than snails in the first papers on life cycles. Broader malacophagy was, in fact, a working hypothesis among the earlier researchers on sciomyzid behavior, at least among C. O. Berg's students, and this led to the series of discoveries of diverse malacophagy. An evolutionary scenario of terrestrial, parasitoid Sciomyzini and aquatic, predaceous Tetanocerini was developed by Berg *et al.* (1959), Foote *et al.* (1960), and Berg (1966) with *Atrichomelina pubera* suggested as a model for the common ancestor of the family. This early scenario was elaborated upon in regard to adaptive radiation, especially by Munari (1988). It is important to note that the evolutionary scenario developed in the three early papers was, in fact, not preliminary and not based only on the published information available at the time because Berg and his students had access to a wealth of information being obtained then but in many cases not published until many years later. This explains, in part, why their scenario has not changed drastically. With the publication of the first cladistic analysis of the genera of Sciomyzidae and Phaeomyiidae (based on adults) (Marinoni & Mathis, 2000), a modern framework for evaluation of behavioral attributes finally was available. Knutson & Vala (2002) displayed and discussed the occurrence of the 17 behavioral groups they recognized across the genera of the families as they appear in that cladogram, and presented a revised evolutionary scenario, which was followed by a revised cladogram and additional analyses of Barker *et al.* (2004).

19.3 BIOLOGY

> *Dans les champs de l'observation le hasard ne favorise que les esprits préparés.* [In the field of observation, chance favors only the prepared minds.]
>
> Louis Pasteur (formulated in 1854 and cited by Kubinyi [1999]).

19.3.1 Life cycles

The report by Perris (1850) and the few other reports on the food of larvae before that of Berg (1953) were fragmentary and inconclusive, based primarily on rearing adults from puparia found in snail shells. Surprisingly, three of these reports – by Perris (1850), Mercier (1921), and Blair (1933) – concern the enigmatic *Salticella fasciata*, whose snail-killing ability remains somewhat in doubt to this day even after extensive research by Knutson *et al.* (1970) and Coupland *et al.* (1994). Perris (1850) reared *S. fasciata* from a larva found in *Theba pisana* in southwestern France. Oldham (1912) reared adults of *Pherbellia dubia* in April from shells of *Vitrea* sp. and *Pyramidula* sp. collected during February in southern England. Schmitz (1917) reared *Pherbellia cinerella* from a puparium in an unidentified snail shell. Mercier (1921) reared *S. fasciata* during February from larvae found in living *T. pisana* on the coast in northwestern France during September. Lundbeck (1923) reared *Colobaea bifasciella*, *C. pectoralis*, and *C. punctata* from shells of small aquatic snails found floating in flood debris during the spring in Denmark. Vimmer (1925), with regard to *Sciomyza ventralis* (= *Pherbellia ventralis* ?) stated (translated): "Parasites in the body of the snail *Hecotus auritus* in Vitrin's group," without further detail. He also reared *Tetanocera ferruginea*, *Sepedon spinipes*, and *S. sphegea* from larvae he collected. With regard to the latter, he stated (translated): "Possibly by means of it [the "two-membered head segment"] the larva sucks up the slime which has settled on the plants [*Lemna* sp.?]." He briefly described the larvae and puparia, but captured many important characters of all four species. Blair (1933) reared *S. fasciata* from larvae found in *T. pisana* but did not say whether the snails were living or dead when collected. The early reports on the food of larvae were reviewed by Neff & Berg (1966, primarily in regard to *Sepedon*), Knutson *et al.* (1970, in regard to *Salticella fasciata*), and Foote (1996a, in regard to *Tetanocera*). Most authors before Berg (1953) considered sciomyzid larvae to be phytophagous (e.g., Séguy 1934) or generally saprophagous (e.g., Bertrand 1954).

Berg's discovery led to the development of a major, long-term research program by him and his 16 graduate students on Sciomyzidae at Cornell University, 1953–1978, covering many aspects but focusing on life cycles, natural history, biological control prospects, and descriptions of immature stages. The interesting biologies and possibilities for biocontrol helped to instigate considerable research on biologies of Sciomyzidae in other locations, especially by J-C. Vala and his six graduate students and other associates at Avignon and Orléans (1982–present). Some of Berg's students, especially B. A. Foote at Kent State University, continued very active research programs on life cycles. Notably, the leading systematists on Sciomyzidae, especially R. Rozkošný (Brno), G. C. Steyskal (Washington), and J. Verbeke (Brussels) all were in close contact with the Cornell group and each other. Their extensive taxonomic work was a result, in part, of the new material collected during the extensive field work conducted throughout much of the world or in support of the biological studies.

19.3.2 Other biological studies

The history of research on various other aspects of the biology of Sciomyzidae can be seen best in the reviews incorporated in the special topics in Chapters 5–14. Greathead (1981), in a review of Sciomyzidae as potential biocontrol agents, decried the lack of quantitative studies, detailed ecological investigations, and information on population dynamics and host preferences in nature. We noted in Section 18.1 that Greathead overlooked some publications of this nature, but it is more important to note that somewhat before and after his review some of the research at Cornell shifted to such studies (Eckblad 1971–1976, Barnes 1975–1976, Arnold 1972–1977, and Trelka 1973–1977) as the life cycles of North American species became known more completely. Also, a group led by Vala with emphasis on such research was developing at Avignon (Haab 1981–1984, Ghamizi 1985–1991, Manguin 1985–1988, and Caillet 1980–1986), and more recently in Bénin (Gbedjissi). Furthermore, other researchers were pursuing experimental and quantitative studies, e.g., publications by Fisher & Orth (1964), Bhuangprakone & Areekul (1973), Moor (1980), Lindsay (1982), Gormally (1985–1988), O. Beaver (1973–1989), McLaughlin & Dame (1989), and more recently in Ireland by Gormally, Maher, Mc Donnell, Mulkeen, and Williams. Although this broad area of research is not now as bleak as described by Greathead (1981) it still requires much further effort, not only for biocontrol but also for evolutionary studies.

19.4 MORPHOLOGY OF IMMATURE STAGES

The history of research on the morphology of the immature stages of Sciomyzidae began about 150 years ago with the publication by Dufour (1847a, 1847b) on the larva and puparium of *Tetanocera ferruginea* (Fig. 14.25). The subsequent publications to 1939 treating 13 species in eight genera were summarized in Hennig's (1948–1952) classical review of the literature on the immature stages of Diptera. Rozkošný (1967) reviewed the literature published through 1965, including his new information (on biology as well as morphology) on 25 species in 13 genera, for a total of 32 species in 16 genera for the Palearctic. Rozkošný (1997a, 1998, 2002) incorporated data from later publications in several subsequent reviews of the Palearctic fauna. The world review by Ferrar (1987) summarized information on about 114 species in 30 genera. To date, over 65 publications include descriptions and/or figures of all or some of the immature stages of 172 species in 41 genera of Sciomyzidae.

The first modern descriptions and figures of all immature stages of one or more species in a genus of Sciomyzidae were by Foote (1959a) on species of *Sciomyza* in the Nearctic. That publication and the paper by Foote *et al.* (1960) on *Atrichomelina pubera* established a diagnosis of the immature stages of the family and distinctions between the Sciomyzini and Tetanocerini. Studies of many other Nearctic and Palearctic genera by Foote, Neff, and Knutson at that time contributed to the overall diagnoses and distinctions. Those early papers also established a format and a terminology for descriptions and figures that subsequently were followed by most workers and which enhanced comparisons between species and genera. There followed a long series of papers by the Cornell group on the biology and immature stages of Nearctic, Palearctic, Neotropical, Subantarctic, and Afrotropical species, continuing by some of Berg's students after his retirement in 1978.

The condition of two important characters, the ventral arch of the cephalopharyngeal skeleton and the interspiracular processes on the posterior spiracular plates, were clarified as research developed. The transverse ventral arch with a serrate anterior margin, a distinctive pair of joined sclerites (of labial origin?) below the mouthhooks, was considered to be a unique feature of Sciomyzidae. However, it also is present in a few Muscidae, Chloropidae, Opomyzidae, Psilidae, and Scathophagidae (Ferrar 1987). It is present in *Salticella fasciata* (Salticellinae) (Knutson *et al.* 1970) but absent in *Pelidnoptera nigripennis* (Phaeomyiidae) (Vala *et al.* 1990). By light microscopy studies, it was believed that interspiracular processes were not present in *Salticella*, Sciomyzini (except first-instar larvae of *Pteromicra*, Rozkošný & Knutson 1970), and terrestrial Tetanocerini. Subsequent stereoscan micrography studies (Vala & Gasc 1990b, Vala *et al.* 1999) have shown that small, scale-like processes are present in *S. fasciata* and the Sciomyzini and terrestrial Tetanocerini examined, as well as in *P. nigripennis*. Study of the immature stages of the saprophagous/predatory/parasitoid *S. fasciata* by Knutson *et al.* (1970) confirmed its placement, based on adult characters, in the plesiomorphic Salticellinae, and included description of larval cibarial ridges which are characteristic of saprophagous larvae (Dowding 1971). Detailed studies of cross-sections of the cibarium of other Sciomyzidae, especially the relatively plesiomorphic *A. pubera*, are still needed.

The immature stages of the millipede-feeding *P. nigripennis* were described and figured by Vala *et al.* (1990), supporting the placement of this genus in the new family Phaeomyiidae established by Griffiths (1972) on the basis of characters of the adults.

In 1983, Vala and his students and colleagues began publishing a series of papers on the biology and immature stages of Mediterranean species, with emphasis on terrestrial species. Especially noteworthy is the fact that most of these papers included stereoscan micrographs of eggs and larvae and presented the first detail on cuticular structures and sensilla. A classification of life styles based on stereoscan studies of the posterior spiracular disc and interspiracular processes was proposed by Vala & Gasc (1990b). In a detailed, comparative study of the eggs, Gasc *et al.* (1984b) characterized those of terrestrial and aquatic species morphologically.

A review of sciomyzid biology and morphology of immature stages, including detailed diagnoses and discussion, was presented by Ferrar (1987). Diagnostic overviews of the characters of the immature stages were provided for Nearctic species (Knutson 1987) and Palearctic species (Rozkošný 1967, 1998) and were included in the regional Palearctic studies by Rozkošný (1984b, Fennoscandia and Denmark; 1997a, North Palearctic aquatic and semi-aquatic species; 2002, central Europe), Vala (1989a, West Palearctic and Mediterranean Europe), Rivosecchi (1992, Italy), and Sueyoshi (2005, Japan). Immature stages of a few Afrotropical species were described by Knutson *et al.* (1967a), Barraclough (1983), and Vala *et al.* (1995). Immature stages of some Subantarctic species were described by Barnes (1980a) and of many Neotropical species by Neff & Berg (1961, 1966), Kaczynski *et al.* (1969), Abercrombie & Berg (1975, 1978), Valley & Berg (1977), Knutson & Valley (1978) and Freidberg *et al.* (1991). The immature stages of many Neotropical species remain unpublished in the Ph.D. thesis of J. Abercrombie (1970) and of many Nearctic species of *Tetanocera* in the unpublished part of B. A. Foote's Ph.D. thesis (1961a).

Keys to immature stages were published by Rozkošný (1967, 1997a, 1998, 2002) (Palearctic larvae and puparia), Knutson (1987) (Nearctic larvae), Rivosecchi (1984) (aquatic larvae and puparia of Italian species), and Sueyoshi (2005) (larvae and puparia of Japanese species). Bhatia & Keilin (1937) provided the only detailed information on the internal morphology of a larva, for an unidentified parasitoid of terrestrial snails in Ireland. It was suggested as possibly being a species of *Tetanocera* (Foote 1996a) but subsequently more positively identified as *Trypetoptera punculata* (Knutson & Vala, unpublished data). A comparison of the morphology of the immature stages of the families of Sciomyzoidea led Meier (1996) to conclude that these features contribute little to clarifying relationships within the superfamily.

19.5 EXPERIMENTAL, QUANTITATIVE, AND BIOCONTROL RESEARCH

The development of research on quantitative and experimental aspects can be appreciated to some extent by perusal of those subjects in the foregoing chapters. In summary, pursuit of these aspects began at Cornell University about a decade after the first papers on basic life cycles. These included population dynamics (Eckblad & Berg 1972, Arnold 1978b on *Sepedon fuscipennis*), toxic salivary gland secretions of slug killers (Trelka & Berg 1977 on *Tetanocera* spp.), predation (Eckblad 1973a, 1976 on *S. fuscipennis*), controlled studies on development and diapause (Barnes 1976 on *S. fuscipennis*), experimental studies on hymenopterous parasites (O'Neill 1973, Juliano 1981, 1982), and phenology (Berg *et al.* 1982). Some experimental, laboratory studies also were being carried out at other locations: predation (Geckler 1971 on *S. fuscipennis*), competition (O. Beaver 1974b), development and predation (Bhuangprakone & Areekul 1973 on *Sepedon plumbella*; Lindsay 1982 on *Ilione albiseta*; Yoneda 1981 on *Sepedon aenescens*; Gormally 1985b, 1987a, 1988a, 1988b on *I. albiseta*; O. Beaver 1989 on *Sepedon senex*), oviposition (O. Beaver 1973), and oviposition and development of the parasitoid *Pherbellia s. schoenherri* (Moor 1980, Vala & Ghamizi 1992). Also there were a few quantitative studies on seasonal abundance of adults (O. Beaver 1970 in Wales and Vala & Brunel 1987, Vala 1984b, and Vala & Manguin 1987 in France) and larvae (Hope Cawdery & Lindsay 1977 on *Hydromya dorsalis* in Ireland). Beginning in 1980, Vala and students in Avignon, France, began a series of detailed laboratory studies of development (Vala & Haab 1984 on *Tetanocera ferruginea*), predation (Haab 1984 on *Sepedon sphegea*; Manguin *et al.* 1988a, 1988b, Manguin & Vala 1989 on *T. ferruginea*), and predation and larval survival (Ghamizi 1985 on *Sepedon sphegea*). Some aspects of predation, survival, and development have been studied experimentally for *Sepedon scapularis*, an aquatic predator (Maharaj *et al.* 1992), *S. scapularis* and the aquatic predator *S. neavei* (Appleton *et al.* 1993), the aquatic predator *S. spinipes* (Mc Donnell 2004, Mc Donnell *et al.* 2005b), and the terrestrial *Salticella fasciata* (Coupland *et al.* 1994, Coupland & Baker 1995), and adult seasonal abundance, habitat preference, and prey choice for several terrestrial Mediterranean species (Coupland & Baker 1995, Coupland 1996a).

Particularly in the studies of predation, quite different aspects have been emphasized in the several publications: number of hosts/prey killed, biomass consumed, host/prey preference, effect of parental diet, survival, rate of development, etc., and the experimental methods have varied. Major research needs remain, especially in population biology, physiological/behavioral aspects of feeding by and nutrition of larvae, and diapause/quiescence.

The early research on biocontrol was detailed by Berg (1973a) and the subsequent history on that subject is rather fully described in Chapter 18. Following are a few somewhat subjective points that have played an important role in the history of biocontrol research on Sciomyzidae. The considerable support for the Cornell program provided by the US National Institutes of Health, is recognized. However, a communications gap between the agencies and personnel primarily responsible for public and animal health – parasitologists, veterinarians, public health workers – on the one hand and on the other hand sciomyzid specialists who have been primarily entomologists, has worked against development of biocontrol by Sciomyzidae. Clifford Berg recognized this gap and was partially successful in bridging it. Until relatively recently, the strong reliance on chemical control of disease-transmitting snails was another factor. Biocontrol methods, especially augmentation and periodic release approaches require long-term commitment of effort. The historical difficulty of national, and especially international agencies, in maintaining efforts over the long term also works against biocontrol attempts. Further, few biocontrol specialists have been involved. While there has emerged, over the past couple of decades, a strong emphasis on integrated pest management with biocontrol as a cornerstone, there has also emerged more recently and very appropriately a concern about non-target impact on native faunas. Particularly in this regard further research is needed on the food preferences, especially of aquatic predators under natural conditions.

19.6 BIOGRAPHICAL NOTES

We have had the pleasure and honor of working with or having known most of the researchers on Sciomyzidae active over the past half-century, and we present below biographical sketches of a few. Unfortunately, space limits the number of persons noted.

Clifford Osburn Berg

Born on August 9, 1912 at Stoughton, Wisconsin, Clifford O. Berg spent his childhood on a farm in that area. He died on April 6, 1987, at Ithaca, New York (obituary: Clarke 1987). He was a graduate of Luther College, Decorah, Iowa (B.A., 1934) and the University of Michigan, Ann Arbor, (M.S., 1939; Ph.D., 1949). He received an Honorary D.Sc. from Luther College in 1970. His Ph.D. thesis under

Prof. P. S. Welch was on the biology of insects associated with the aquatic plant *Potamogeton*. He served as a Malaria Control Officer, US Navy, in the South Pacific (1943–1946); Associate Professor, Ohio Wesleyan University, Delaware, Ohio (1947–1953); Consultant Entomologist, Arctic Health Research Center, Anchorage, Alaska (summers of 1950–1952) where he conducted his first research on Sciomyzidae; Professor of Limnology, Department of Entomology, Cornell University, Ithaca, New York (1953–1978); and Resident Ecologist, Office of Environmental Science, Smithsonian Institution, Washington, DC (1970–1971).

Berg moved to Cornell in 1953, assuming the position left by limnologist D. C. Chandler. He established a teaching and research program in limnology and freshwater entomology that flourished for 25 years, during which he supervised 19 Ph.D. and nine M.S. theses. His first Ph.D. students on Sciomyzidae were S. E. Neff and B. A. Foote; the latter had completed a B.A. degree specializing in biology under Berg at Ohio Wesleyan University. Berg's 16 graduate students on Sciomyzidae, with their main thesis subject and year of completion, are listed below. All were on biology and immature stages, except as noted. S. E. Neff: *Sepedon*, *Hoplodictya*, *Protodictya* (also studied *Atrichomelina pubera*), Ph.D., 1960. B. A. Foote: *Tetanocera* (also studied *A. pubera*, *Sciomyza*, *Pherbellia*, *Renocera*, *Hedria*, *Oidematops*), Ph.D., 1961. L. V. Knutson: European Sciomyzidae (*Sciomyza*, *Pherbellia*, *Colobaea*, *Pteromicra*, *Tetanocera*, *Hydromya*, *Pherbina*, *Psacadina*, *Elgiva*, *Ilione*), Ph.D., 1963. A. D. Bratt: *Pherbellia* (also studied *Colobaea*), Ph.D., 1964. J. L. Bath: *Dictya*, M.S., 1964. J. L. Gower Teece: *Pteromicra*, graduate study, 1963–64. V. W. Kaczynski: *Perilimnia*, *Shannonia*, M.S., 1967. J. Zuska: taxonomy of *Perilimnia*, *Shannonia* and *Tetanoceroides*, M.S., 1967. J. Abercrombie: South American Sciomyzidae (*Dictyodes*, *Protodictya*, *Sepedonea*, *Tetanoceroides*, *Thecomyia*), Ph.D., 1970. J. W. Eckblad: population biology of *Sepedon fuscipennis* and gastropods, Ph.D., 1971. W. L. O'Neill: *Trichopria* (Diapriidae) parasitoids, M.S., 1972. D. G. Trelka: behavior of *Tetanocera plebeja* and its toxic salivary gland secretions, Ph.D., 1972. K. R. Valley: *Dictya*, Ph.D., 1974. J. K. Barnes: effect of temperature on development, etc. of *Sepedon fuscipennis*, M.S., 1976; biology and taxonomy of New Zealand Sciomyzidae (*Neolimnia*, *Eulimnia*, *Prosochaeta*) and Helosciomyzidae, Ph.D., 1979 (research primarily carried out at Lincoln University, Canterbury, New Zealand). S. L. Arnold: population dynamics of *Sepedon fuscipennis*, Ph.D., 1978. S. A. Juliano: Trichogrammatidae parasitoids, M.S., 1979.

Berg had 12 other graduate students working in chemical and physical limnology (B. C. Cowell, H. H. Howard, N. J. Lamb, and J. A. Maciolek), biology and taxonomy of Trichoptera (O. S. Flint, Jr.), biology of mayflies and stoneflies (S. B. Fiance), biology and immature stages of Dytiscidae (E. H. Barman), larvae and pupae of Chironomidae (R. E. Bode), biology of aquatic mites (C. A. Lanciani), biology and immature stages of Ephydridae (K. W. Simpson) and taxonomy of the freshwater mussels (A. H. Clarke, Jr.) and gastropods (W. N. Harman) of New York; Berg was the junior author of the publications on the latter two subjects. Not only is the range of subjects impressive, Berg followed the research carefully and, without doubt, read and considered every word of every thesis.

Berg was involved in all aspects of the research program on Sciomyzidae at Cornell and was essentially solely responsible for communication and cooperation on biocontrol applications. The Cornell program was possible, in part, because of his success in obtaining modest but continuing grant support, primarily from the US National Science Foundation and the US National Institutes of Health. He received Guggenheim, Fulbright, and Cornell awards for research travel. He conducted extensive field work with his students around Ithaca, and on many foreign field trips of several months duration: Central America – 1958; Europe – 1959, 1960, 1970; Australia – 1961; Afghanistan – 1964; South America – 1964, 1966, 1967; Ghana – 1970; and Thailand and Indonesia – 1970.

Berg published 17 technical papers and several popular articles on Sciomyzidae as sole or senior author, including the major reviews in *Advances in Parasitology* (1964) and *Annual Review of Entomology* (1978). He was a junior author on many publications with his graduate students. Although he was intimately involved with key aspects of the field and laboratory studies, descriptions of immature stages, and the presentation of results, he left the majority of the research to his students. Berg had a passion for clear writing: lengthy manuscript review sessions with his students were de rigueur. He had little interest in theoretical issues and was adverse to speculation until late in his career, notably in a paper on adaptive differences in phenology (Berg *et al.* 1982). Clifford Berg was a conservative, thoughtful person, professionally "street-wise" in a beguiling manner, very helpful and cooperative, given to understatement (even humorous self-deprecation), and a colorful and enthusiastic public speaker on his professional passion: Sciomyzidae.

George Constance Steyskal

Born on March 30, 1909 in Detroit, Michigan, George Steyskal died on May 30, 1996 in Gainesville, Florida (obituary: Sabrosky 1997). George Steyskal was the leading taxonomist on Sciomyzidae of the twentieth century (Fig. 19.4). His formal education beyond High School (= Lyceum) was only a few years at the Henry Ford Trade School in Detroit, Michigan. He spent much of his working life as a tool-and-die maker, eventually becoming superintendent of a factory. Of impressive intellect and industry, he was self taught in entomology, botany, geography, linguistics, and languages. He published in linguistic journals, translated almost all European languages, spoke and wrote in most, was expert in classical Latin and Greek, and studied languages such as Arabic, Farsi, Turkish, and Japanese, i.e., he read grammars as easily as we read novels. Although lacking the formal educational qualifications for a regular US Government Research Scientist position, in 1962 he was specially appointed to a research position with the US Department of Agriculture's Systematic Entomology Laboratory (SEL) at the Smithsonian Institution, Washington, DC based on his impressive publication record. By that time he had published 80 papers, primarily on Diptera. He was responsible for many acalyptrate and some other families at SEL, especially Tephritidae and Agromyzidae, and published on many of them. He officially retired in 1979 but continued research at the Smithsonian for several years, and spent his last few years as a Research Associate with the Florida State Collection of Arthropods, in Gainesville.

Steyskal published 446 papers in entomology and as noted in the obituary by Sabrosky (1997) the extent of his knowledge of Diptera is evident from the following statistics. He described four new subfamilies in four different families; 24 new genera and two new subgenera in ten families (eight genera in Sciomyzidae); 347 new species (including six *nomina nova* and eight subspecies) in 32 families, as presently recognized, chiefly in 22 of Acalyptratae, with species scattered in ten other families. All species were described by Steyskal alone except for 34 coauthored with nine other persons. All of this was in addition to publications on suprageneric classifications, designations of type species and lectotypes, new synonymies, corrections of authorship or dates of publication, elevations in rank of family-group names, a number of pertinent comments on applications in the *Bulletin of Zoological Nomenclature*, and book reviews.

Fig. 19.4 George C. Steyskal (second from left) and colleagues (from left, L. V. Knutson, K. G. V. Smith, J. C. Deeming) on roof of British Museum (NH), August 1964. (Photo by J. W. Stephenson).

Steyskal worked on the World Diptera but concentrated on the New World Fauna. Most of his papers were singly authored; Knutson had the pleasure of publishing five with him. Steyskal was an excellent collector and cataloger (e.g., he wrote 15 family sections in the Catalog of North American Diptera, 15 in the Catalog of South American Diptera, 16 in the Oriental Catalog, and five in the Afrotropical Catalog). He was not greatly interested in curation. His forté was short, succinct papers on one to many species including fine keys, resolution of nomenclatural problems, and exceptionally detailed and precise drawings of genitalia based on careful dissections. Many of Steyskal's short papers would have amounted to much longer publications in the hands of others. Although he published many generic revisions, it was not his nature to produce comprehensive, area-wide revisions of entire families; he did, however, produce 13 family chapters (plus Tephritidae with R. H. Foote) in volume 2 of the *Manual of Nearctic Diptera*. He was an uncontrollable enthusiast. At least once-daily he took his most recent discovery of a synonymy, resolution of a nomenclatural problem, drawing of genitalia, etc., on a circuit of exposition to his Diptera colleagues at the Museum (R. J. Gagné, L. V. Knutson, W. N. Mathis, C. W. Sabrosky, A. Stone, F. C. Thompson, W. W. Wirth). Knutson's office was next to his, so Knutson usually got the fresh, complete explication, for which he remains grateful, as it was a wonderful learning experience.

Steyskal's first entomological publication was in 1938 at the age of 29. That was 24 years before he gained a scientific position. The paper was on new species of

Stratiomyidae and Sciomyzidae (three new species of *Tetanocera* and four of *Dictya*, including figures of the surstyli). His contributions to research on the systematics of Sciomyzidae also are treated in Section 19.1. Notably, he made contributions at all taxonomic levels and in all areas of systematics except cladistic phylogeny. He described 80 species individually, two with R. E. Orth, and he described eight genera. He revised the suprageneric classification that remains, more or less, the classification accepted to date; resolved many nomenclatural, literature, and distributional questions; catalogued the New World Fauna; examined and documented type material; and provided the first detailed figures of the genitalia of many species and genera. The research programs on biology at Cornell and elsewhere could not have developed as they did without his taxonomic support.

Jean Verbeke

Born on September 24, 1921 in Ghent, Belgium, Jean Verbeke died on September 29, 1973 in Brussels, Belgium. Jean Verbeke was the leading taxonomist on Afrotropical Sciomyzidae and from his first paper (1948) more or less to his last (1967) on the Palearctic Sciomyzidae he was the pre-eminent researcher on that fauna. His taxonomic support, as that of Steyskal's, during the 1960s was critically important to the development of the research on biology.

Verbeke received a Degree in Agronomy from the Rijks Faculteit, Ghent in 1947 and a Doctor of Sciences degree from the University of Lille, France in 1958. From 1948 to 1963 he was a research scientist with the Belgian Institute of National Parks in the (then) Belgian Congo. In 1963, with the collapse of the political situation in the Congo, he joined the Entomology Section of the Royal Institute of Natural Sciences in Brussels. He published 33 papers (1944–1973) on the taxonomy of several families of Diptera, especially on Tachinidae, besides those on Sciomyzidae, and on the hydrobiology of Lakes Kivu, Edward, and Albert.

Verbeke's 1950 paper on Afrotropical Sciomyzidae, in which he described five new genera and 32 new species, established the basis for the taxonomy of the Afrotropical fauna. His 1950 paper also included an analysis of the suprageneric classification of the Sciomyzidae and a detailed study of the male postabdomen, especially of his new subfamily Sepedoninae. He published six more papers on the Afrotropical Sciomyzidae (one with Steyskal, 1956), plus seven on the Palearctic fauna, describing a total of four genera and 49 species. He studied type material in the museums of London, Florence, Rome, Madrid, Vienna, St. Petersburg, and Moscow. Knutson had the pleasure of working with him several times in Brussels and in Harpenden, England, and collecting with him in Almeria and Valencia, Spain.

Rudolf Rozkošný

Born on September 1, 1938 at Kroměříž, Czech Republic, Rudolf Rozkošný has been over the past several decades, and continues, as the dominant figure in the taxonomy of Palearctic Sciomyzidae. His early papers overlapped in time with Verbeke's last papers. Rozkošný has contributed to most aspects of research on Sciomyzidae, predominantly to taxonomy, morphology of immature stages, and faunistics, and especially in broad, area-wide revisions and syntheses. Rozkošný received the Master of Sciences degree (= Ph.D.) in 1965 and the Doctor of Science degree in 1982 from the Czechoslovak Academy of Sciences in Prague. In 1984 he was appointed Professor of Entomology at Masaryk University in Brno. In addition to a wide-ranging research and teaching program, he has held several key administrative and research leadership positions, including Chairman of the Department of Environmental Studies, Vice-Rector of Masaryk University, and member of the Council for the International Congresses of Dipterology. He retired, as Head of the Department of Zoology and Ecology, Faculty of Science of the same university, during 2004 but continues as an active researcher. His taxonomic support and cooperation with the biological research programs at Cornell, Avignon, Orleans, and Galway have been critically important to those programs.

Besides Sciomyzidae, Rozkošný has published extensively on the systematics of Stratiomyidae and Muscidae. He has published 68 papers on Sciomyzidae, most of the major papers alone but others with 11 different co-authors. His most comprehensive revisions (1984b, 1987a) of Palearctic Sciomyzidae, based on study of type material, and including extensive illustrations, distributional data, and overviews of morphology of adults and immature stages, biology, etc., are especially important as the first revisions since that of Sack (1939). They especially expedited further work by other researchers on alpha-level and revisionary studies. He has described 31 species, 17 of them with five co-authors.

Rozkošný also provided the first comprehensive review (1967) of the immature stages and biology of Palearctic Sciomyzidae, including extensive new information and subsequently (1997a) the most comprehensive keys to immature stages of Palearctic genera. He has studied type material from the museums in Copenhagen, Berlin, Budapest, Lund, Stockholm, Helsinki, Vienna, St. Petersburg, Moscow, etc. His very helpful and extensive communication over many years has been of great importance to other researchers, not least to Knutson and Vala.

It is especially appropriate to an understanding of the history of research on Sciomyzidae to highlight the seminal accomplishments of Stuart E. Neff and Benjamin A. Foote, Berg's first two graduate students. They and Berg set the stage on which we continue to play a scenario of eighteenth-century biology, occasionally interrupted by quantitative, experimental, and theoretical concerns.

Stuart Edmund Neff

Born on October 3, 1926 at Louisville, Kentucky, Stuart Neff has a more theoretical nature than Berg, Foote, Knutson, and others, but also is one who enjoyed the sweet smells of the eutrophic habitats of many Sciomyzidae. Neff was Berg's first graduate student on Sciomyzidae. Although he did not continue research on Sciomyzidae after leaving Cornell, the rigor and extent of his work on several genera contributed greatly to establishing the quality of the Cornell program.

Neff received a B.S. degree in 1954 from the University of Louisville. He served in the U.S. Navy from 1945 to 1949 and 1950 to 1952, in the Pacific. After completing his Ph.D. degree at Cornell in 1960 he was Professor of Biology at Virginia Polytechnic University, Virginia; University of Louisville, Kentucky; and, since 1985, at Temple University, Pennsylvania. He retired in 2006. He has published papers on the biology of several families of Diptera, freshwater biology, and limnology.

Neff contributed outstanding papers on the biology and immature stages of *Protodictya* (1961), *Hoplodictya* (1962), and *Sepedon* (1966), with Berg as junior author; on *Atrichomelina pubera* (1960) with Foote and Berg; and on Afrotropical *Sepedon* (1967) with Knutson and Berg. Notably, Neff was the lead researcher on the early studies of the morphology of immature stages of Sciomyzidae and established the basis for this aspect of study and for predation studies as well as relating the early research on life cycles to systematics and ecology (Neff & Berg, 1966). His 1964 publication on trials of ten species of aquatic predators against 13 species of snails of medical importance (Table 5.5) was the first extensive quantitative information on prey consumption and prey preference of Sciomyzidae.

Benjamin Archer Foote

Benjamin Foote was born on October 25, 1928 at Delaware, Ohio. Foote and Neff began studies at Cornell in 1954 a few months after Berg assumed his position there. It is a little difficult to separate out the initial research contributions of this trio, but Foote is acknowledged as the pre-eminent elucidator of life cycles of Sciomyzidae, along with other acalyptrate Diptera in North America, over the past five decades and continuing to the present day.

Foote received a B.A. degree in 1950 from Ohio Wesleyan University, when C. O. Berg was a professor there; an M.S. degree from Ohio State University in 1952; then spent two years driving a tank for the US Army in Germany. He received a Ph.D. from Cornell in 1961, having spent one year in the meantime on the University of Idaho faculty. He joined Kent State University, Ohio, in 1961, retired as Professor of Biology in 1996, and continues as Emeritus Professor.

In addition to research on Sciomyzidae, Foote has produced major publications on the biology of Ephydridae, Tephritidae, and other acalyptrates and was the coordinator for the chapter on Diptera in the manual on North American immature insects (Stehr 1991) as well as the author of many family sections. Over the years, he developed at Kent a group of M.S. students in Diptera biology, D. G. Trelka and K. R. Valley being two, completing their studies in 1968. They then produced outstanding Ph.D. theses on Sciomyzidae under Berg's guidance at Cornell.

Foote has published 29 papers on Sciomyzidae, primarily on life cycles and descriptions of immature stages (34 species in nine genera) and on faunistics, and continues an active research, teaching, and publishing program to date. Foote was the first to show that the Sciomyzidae attack submersed operculates, slugs, and fingernail clams. He began his research in New York but also had opportunities for extensive field work in the western USA (Idaho, Montana), then in the Midwest (Ohio), which resulted in many new life cycles and comparative data from various locations and habitats.

20 • Methods

> There is no unique method for studying things.
>
> Aristotle (384–322 BC).

20.1 COLLECTING (SEE ALSO SECTION 9.1 AND CHAPTER 10)

> The beauty and brilliancy of this insect are indescribable, and none but a naturalist can understand the intense excitement I experienced when I at length captured it. On taking it out of my net and opening the glorious wings, my heart began to beat violently, the blood rushed to my head, and I felt much more like fainting than have done when in apprehension of immediate death. I had a headache the rest of day, so great was the excitement.
>
> Wallace (1869).

In general, areas of alkaline soils supporting dense populations of gastropods are the best collecting sites, but certain species, e.g., some *Renocera* spp. that feed on fingernail clams, can be found in somewhat acidic situations. Populations are seldom dense and are usually patchy, even across stretches of suitable habitat that seem more or less uniform. Sciomyzids cannot tolerate highly disturbed areas where the substrate is frequently upset or trodden upon, such as plowed and heavily grazed fields; semi-natural watering areas for cattle; margins of fast-flowing streams; and margins of irrigation ditches and ponds that are used intensively for human activities. However, they can be found in surprisingly small patches of somewhat disturbed, semi-natural, aquatic and terrestrial situations where the substrate is more or less undisturbed. There is essentially no information on their occurrence in specific plant associations, and adults generally are not attracted to flowers. There is little information on diel periodicity of adults, although the general impression is that sweeping is least productive during the warmest part of the day. Vala (1984b) made sweep-net collections (recording and releasing the captures) of six terrestrial species, hourly between 5 am and 8 pm, from May 28 to June 9 in an oak grove in southern France when temperatures were between 18 and 36 °C. He collected the greatest number of specimens between 18 and 25 °C during early morning and in the evening (Fig. 20.1). Hill-topping and congregating behaviors are unknown. However, the large numbers of *Tetanura pallidiventris* recorded at several small woodland situations during a period of a few weeks in the summer might be an example of congregating behavior or a manifestation of more or less synchronous emergence from puparia, restricted habitat preference, limited adult dispersal, and a short adult lifespan. Collecting in habitats that seem atypical for sciomyzids, e.g., rocky, exposed areas in the Mediterranean (*Euthycera* spp.); inselbergs in northern Nigeria (*Colobaea* sp.); maritime shorelines in the Mediterranean (*Pherbellia mikiana*); thermal springs in Iceland (*Tetanocera robusta*); and during "off seasons," e.g., during the winter in warmer areas of the northern hemisphere, may produce rare or new species and a better understanding of habitats and phenology.

Eggs have been collected in nature only for some species of *Sepedon* that place several eggs together in close groups; those of *Anticheta* that oviposit onto egg masses of *Aplexa*, "*Lymnaea*" spp., *Physa* spp., *Succinea* spp., and *Oxyloma* spp.; and the few species that oviposit onto the shells of host snails. The latter species are *Sciomyza varia* and *Colobaea bifasciella* on estivating *Stagnicola palustris*; *S. aristalis* and *Pherbellia s. schoenherri* on *Succinea* and *Oxyloma* spp.; *Salticella fasciata* on *Theba pisana* and *Cernuella virgata*; and *Atrichomelina pubera* which occasionally oviposits on various aquatic and semi-terrestrial snails.

Interesting species have been collected in Malaise traps, e.g., *Pherbellia goberti* in the Netherlands (Lammertsa 1996), *Steyskalina picta* in northeastern Burma, and *Thecomyia* spp. in Brazil (Knutson & Carvalho 1989). Speight & Chandler (1995) mentioned the difficulty of collecting the secretive genus *Pteromicra* with a sweep net but reported that Speight captured hundreds of individuals of *P. leucopeza* in Ireland using Malaise traps. Speight

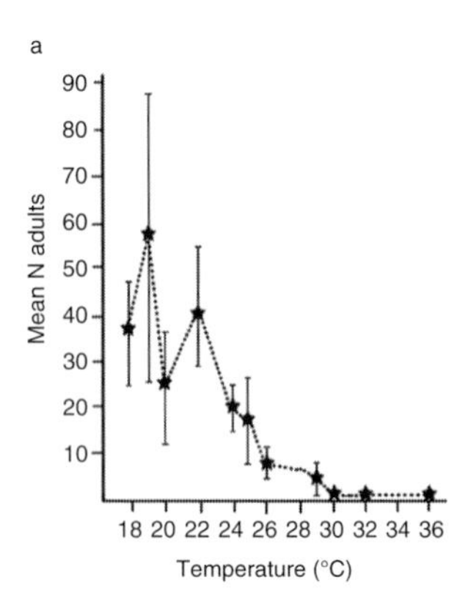

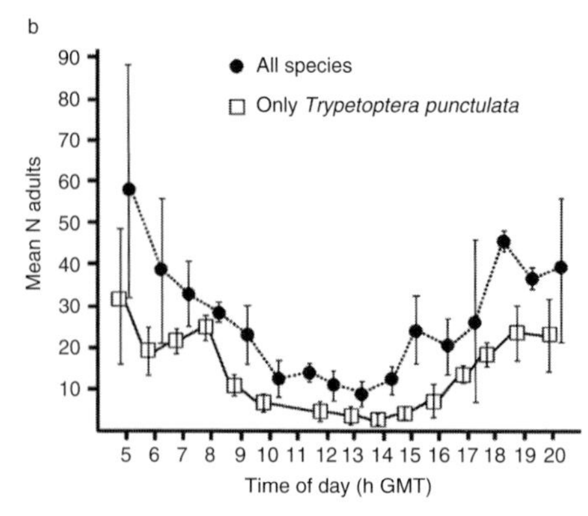

Fig. 20.1 (a) Temperature; (b) time of day and numbers of six terrestrial species of Sciomyzidae collected in an oak grove near Avignon, southern France. From Vala (1984b).

(2001, 2004) presented a detailed analysis and comparison of Sciomyzidae and Syrphidae collected in Malaise traps on a 41-ha farm in Co. Cork, Ireland. He found 182 specimens of 17 species in eight genera, both aquatic and terrestrial species, in traps that were in operation for a series of 20-day periods between April and September 2000. A total of 23 species in ten genera are known from the farm, primarily from the 5-ha disused area. Roller (1995) collected 760 specimens of 30 species in 14 genera in terrestrial situations, with aquatic habitats nearby for some situations, in Slovakia with Malaise traps. Of 4096 specimens of Sciomyzidae collected in the High and Middle Atlas Mountains, Morocco during 1994–1997, only 111 were obtained with Malaise traps, the rest by sweep net (Kassebeer 1999a). In contrast, Malaise traps placed in a terrestrial and an aquatic situation in a bog in Schleswig-Holstein from April to October during 1993 and 1994 yielded 280 specimens of 32 species in 14 genera (of 41 species known from this State), whereas sweep-net collecting yielded only 96 specimens of 15 species in nine genera (Kassebeer 1999b).

No species is known to be collected routinely at light traps, although a few species occasionally have been collected by such a method, e.g., both known specimens of *Pherbellia dentata* (Merz & Rozkošný 1995a) and some specimens of *Limnia unguicornis*, *Teutoniomyia costaricensis*, and *Renocera stroblii* (Knutson & Vala unpublished data). Suction machines were used very successfully by Fisher & Orth (1983) in collecting many species in California. Some minute species, e.g., of *Colobaea*, *Pteromicra*, and *Pherbellia*, which may be difficult to recognize in a sweep net, may be found among micro-Diptera collected with a suction tube. Snail-baited McPhail traps have been used successfully to capture *Sepedomerus macropus* adults in Hawaii, but attempts with such kinds of traps elsewhere have been unsuccessful. A seldom-used kind of trap (2% vinegar in a McPhail-type container) was used by Carles-Tolrá *et al.* (1993) during August–May, 1988–1989 in Spain. About 2500 specimens of 14 families of Acalyptratae representing 37 species in 24 genera were collected. Included were a few specimens of the terrestrial predator/parasitoids, *Dichetophora obliterata*, *Euthycera cribrata*, *E. seguyi*, and surprisingly five males and ten females of *Salticella fasciata*. The large number of the latter species may reflect the supposed predilection of its larvae for moribund snail tissue.

Pan traps have been used successfully on several occasions. Vala & Brunel (1987) collected 295 individuals of 18 species of 12 genera in yellow pan traps across aquatic to terrestrial habitats at a location in northern France. Sixteen of 23 species (in nine of 11 genera) of Sciomyzidae collected in Siberia by S. A. Marshall (determined by L. Knutson) were taken in standard yellow pan traps on the substrate or nestled in surface debris (algae, driftwood, etc.). About 100 specimens of 20 species were collected during a yellow pan trap survey of bogs and fens in Ontario, Canada by Blades & Marshall (1994). However, other major collections by pan-trapping produced very few Sciomyzidae. Of over 100 000 specimens of 27 families of acalyptrate flies collected in yellow and white pan traps on two offshore islands on the Atlantic coast of Germany during 1985, there were only 13 specimens of seven species of Sciomyzidae, whereas 50 specimens of nine species of Sciomyzidae were captured with sweep nets (Schneider 1988). Of over 10 000 Diptera collected in yellow pan traps between June 12 and October 2 in a range of five terrestrial to aquatic habitats in Pas-de-Calais, northern France, only 41 Sciomyzidae of eight species were collected (Denis, 1983). Unfortunately, information as to whether pan traps were placed on the surface of the substrate or were elevated has not been included in most reports.

The few records of adult Sciomyzidae collected in pitfall traps include three individuals of *Pteromicra angustipennis*, one of *Ilione albiseta*, five of *Limnia unguicornis*, and one of *Pherbina coryleti* found in western Ireland (R. J. Mc Donnell, personal communication, 2005). Sciomyzidae and Phaeomyiidae were very uncommon in a pitfall trap survey during 1992 and 1993 in the Palava Biosphere Reserve, southern Moravia (Rozkošný & Vaňhara 1995). Only a single specimen each of *Coremacera marginata*, *Pelidnoptera fuscipennis*, and *Pherbina coryleti* were among the 2715 specimens of 191 taxa in 44 families collected. Barnes (1979b) noted that many specimens of the terrestrial predator *Neolimnia castanea* were collected in pitfall traps in New Zealand. Vala *et al.* (2006 and unpublished) collected during May–August, 2002–2004 two species of sciomyzids around Orleans, France using pitfall traps containing a mixture of 5% formaldehyde and drops of soap, beer, and a little salt. Many *Limnia unguicornis* (23 specimens) in a more or less wet–dry place and two *Pherbellia cinerella* in a dry habitat were collected. In addition, eight *Trypetoptera punctulata* and two *C. marginata* were collected in traps situated in a forest and along the edge of the forest. The terrestrial parasitoid/predators *Euthycera arcuata* and *Limnia boscii* have been collected in interceptor traps in North America and *L. boscii* (one male) was found in a pitfall trap that was in the ground between 1–15 June in a grassy area in Ohio (B. A. Foote, personal communication, 2006). Papp (2004) described *Apteromicra parva*, the first apterous sciomyzid, and placed in a new genus, from a specimen collected under a stone in the mountains of Nepal. This discovery, from an unusual microhabitat for sciomyzids, signals the possibility of finding other apterous Sciomyzidae in pitfall trap collections from other regions of the world.

There has been limited use of emergence traps but their increased use would provide precise information on microhabitats and seasonality. Speight (2001, 2004) presented a detailed analysis and comparison of Sciomyzidae collected in emergence and Malaise traps on a farm in Co. Cork, Ireland. A total of 23 species were collected in Malaise traps. Of these, 18 species in all ten genera except *Coremacera* also were collected in emergence traps, mainly in the disused area. Emergence traps were more effective than Malaise traps for some species. Seventy individuals of *Pherbellia ventralis* were collected in emergence traps but only two were caught in Malaise traps. No *Pteromicra angustipennis* were taken in Malaise traps, but 28 were collected in emergence traps on the farm in Co. Cork. P. Withers, (personal communication, 2003) found the uncommon *Colobaea bifasciella* and *P. angustipennis* in emergence traps in southeastern Ireland. Also see Section 10.3.

Larvae and puparia of many aquatic species, for example, *Dictya*, *Sepedon*, and *Tetanocera*, can be collected with various kinds of nets dragged over the water surface where vegetation is emerging (Frontispiece) or by treading the vegetation below the surface and watching for them to float to the surface. Floating snail shells collected in these ways or among flotsam or shoreline debris can be dissected or "candled," that is, held in front of a strong light to find puparia of some *Colobaea*, *Pherbellia*, *Pteromicra*, and *Sciomyza*. Flotsam and shoreline debris is examined effectively for floating larvae and puparia in white pans of water. Some of the first indications of associations between sciomyzids and snails resulted from finding puparia in floating shells (Lundbeck 1923).

Apparently, few attempts have been made to collect substratum material such as shoreline surface soil or leaf litter and to rear adults from larvae and puparia that may be hidden within. However, in this way Przhiboro (2001) obtained adults of 11 species in three genera, including the sole specimen of a new species of *Sciomyza*. The adults emerged from puparia in substrata samples taken in the "zone of water line", i.e., "from 10 cm above the water level to 5 cm below this level" and then held in the laboratory. His samples were made at five sites, ranging from 30–40 cm wide up to 50 m wide, on the shores of two small lakes near St. Petersburg, Russia.

In cold regions, overwintering larvae and puparia can be found floating around thick stems of plants projecting from melting ice. Larvae of semi-terrestrial and terrestrial species can be found in living or dead snails collected in nature and held in individual containers in the laboratory. The first discovery of a sciomyzid (*Salticella fasciata*) associated with snails (Perris 1850) was made in this way. Because aquatic, predaceous larvae stay with their prey for only a few hours and leave their prey when disturbed they are very rarely found in collections of living aquatic snails; the known records, including collecting methods, were summarized by Mc Donnell *et al.* (2005a). Further information on the natural prey of aquatic and semi-aquatic predators probably could be obtained by using the technique of Hope Cawdery & Lindsay (1977). They marked *Stagnicola palustris* by gluing small glass beads to the shells, released the snails, and recovered them by searching with a flashlight. In this way they found 28 first-instar and one third-instar larvae of *Hydromya dorsalis* feeding in the snails. Refuge traps, white plastic cups placed in pools so that the mouth of the trap was flush with the bottom of pool, captured 204 larvae, primarily third instars, of the aquatic predator, *Iliona albiseta*, between October and March in various temporary and permanent aquatic habitats, primarily on reclaimed blanket bog, in Ireland (Lindsay 1982). Eighteen of 21 larvae of *I. albiseta* collected in various aquatic habitats in Ireland during April were found in folds of discarded rubbish such as rags, milk cartons, etc. (Gormally 1987b).

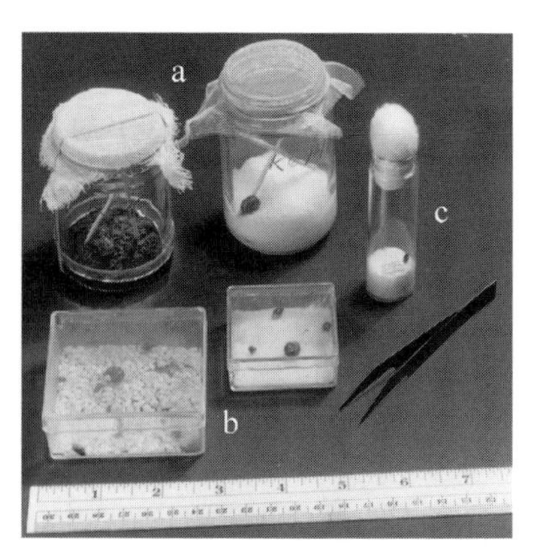

Fig. 20.2 Containers used in rearing Sciomyzidae. (a) adult breeding jars; (b) larval rearing containers; (c) pupal emergence vials. Photo by L. Knutson.

20.2 REARING AND RECORD KEEPING

What is not given is lost.

Indian Proverb.

Working out the life cycles of Sciomyzidae is best accomplished by a type of feed-back process, starting with observations of either field-collected larvae or with laboratory rearings initiated with adults collected in the field or emerged from puparia. Then, information gained from the laboratory is used to focus specific field observations, or vice versa. In other words, a life history is woven from the warp of nature and the woof of the laboratory. Field observations and experiments throughout the year, over a period of years, and at different latitudes and altitudes are important for studying some aspects (e.g., voltinism, overwintering). Many species have broad macrohabitat and microhabitat ranges, and some aspects of behavior might be linked with these differences and might be somewhat different in different situations. Rearing techniques have been described, generally very briefly, in many papers on sciomyzid life cycles. Depending upon the genus or suspected life style of the species under study, reference to the methods used in these publications should be useful. Most rearings have been conducted in the laboratory at room temperatures and humidities under natural light and room lighting but out of direct sunlight; see Section 9.3.2 for development under controlled conditions. Some usual types of rearing containers are shown in Fig. 20.2.

The methods and materials described here are primarily those used in conducting studies of the basic life cycles and when only a few dozen preserved specimens of each stage are required. Mass rearing of thousands of individuals is possible, however. Chock *et al.* (1961) described an outdoor, mass-rearing technique for the aquatic *Sepedomerus macropus*. Davis *et al.* (1961) described techniques for the hygrophilous species *Pherbellia dorsata*. Barnes (1976) further advanced large-scale rearing methods under controlled conditions for the aquatic predator *Sepedon fuscipennis*. McLaughlin & Dame (1989) described a mass-rearing method in the laboratory for the aquatic predator *Dictya floridensis*. Mass-rearing techniques have not been described for any terrestrial species.

Knowledge of the biology and immature stages of reared Sciomyzidae will be helpful in elucidating new life cycles. About 200 of the 539 described species have been reared. All have developed from hatching to puparium formation on snails, snail eggs, slugs, or fingernail clams except the aquatic oligochaete-feeding *Sepedonella nana*, *Sepedon knutsoni*, and the facultative *S. ruficeps* in West Africa. The only other exceptions are *Salticella fasciata*, which can be a facultative saprophage on other dead invertebrates, and the facultative snail killer *Atrichomelina pubera* which apparently can develop on dead, large bivalves. The major categories of hosts/prey of genera of Sciomyzidae are shown in Fig. 5.2. The reared genera of Sciomyzidae and the families of molluscs that they are known to feed on in nature or presumed to feed on from laboratory rearings are shown in Fig. 5.3. It is likely that most of the biologically unknown species also have some sort of malacophagous habits, but one should look for new prey and host relationships when rearing unknown species, and, especially, genera. Expect the unexpected in behaviors and hosts/prey.

Laboratory rearings can be started with any stage of the life cycle. As adults are the stage most frequently encountered in the field, they are usually used to start a culture.

But in starting with adults one might have difficulties in getting them to mate, getting females to oviposit, eggs to hatch, and, often the biggest problem, in selecting the proper food and conditions for the larvae. The most direct way to start a life cycle is with larvae found feeding in nature. The best way to find these is to collect terrestrial, semi-terrestrial, and stranded or otherwise exposed aquatic and semi-aquatic gastropods where adult Sciomyzidae have been taken. Then keep the gastropods in the laboratory until larvae might be seen. Because aquatic predators are essentially free living, remain with the prey for only a few hours and generally leave the prey when disturbed, collections of aquatic snails moving about freely in the water usually is not productive. Study of the truly aquatic, submersed, young larvae of clam killers and some snail feeders is best accomplished by keeping individual larvae and fingernail clams or snails in water in small (e.g., $3 \times 3 \times 3$ mm) transparent, plastic boxes that can be viewed under a dissecting microscope. Some of the fingernail clams live in waters that deposit ferrous or other obscuring coats of matter on the shells. It might be necessary to hold such food molluscs in clear, running water for some days to dissipate the obscuring matter or to remove it with a small brush in order to see the larvae within. If larvae are suspected of being aquatic oligochaete feeders, small pieces of floating vegetation bearing the tubes of the worms need to be isolated with them.

Adults collected with a sweep net are kept temporarily in small homeopathic vials with a piece of fresh vegetation secured between the cork and rim to maintain humidity. They may be left in such collecting vials for several hours before being placed in breeding jars in the laboratory. Some sort of transfer cage is useful when transferring adults. Their positive phototaxis should be taken advantage of by having a light behind the jar into which they are being placed. Also, adults can be anesthetized with ether or CO_2 for transfer. Usually only a female or a pair of adults is kept in each breeding jar. Wide-mouthed jars of various sizes may be used; ours range from 5×8 cm to 8×10 cm. Plastic jars can be used but glass jars do not become scratched and thus more easily permit one to observe mating, oviposition, and hatching. For aquatic Tetanocerini the breeding jars should have a substrate of about 2 cm of damp peat moss or cotton that has been packed down so that the surface is relatively flat and without projecting fibers that might entangle the flies or conceal newly hatched larvae. For terrestrial Tetanocerini and Sciomyzini which usually put their eggs directly on the substrate, slightly damp peat moss that has been packed so as to produce a flat surface permits one to see the eggs and the newly hatched larvae more easily and to remove them with a small wet brush. The water used to moisten the substrate should contain a mold inhibitor, e.g., 0.1% Tegosept M. In order to find and count eggs of species that oviposit on and in the substrate a bottomless jar or wide tube resting securely in a Petri dish of slightly damp peat moss is very useful. Breeding jars are provisioned with a small amount of a putty-like mixture of honey, dried milk, and brewer's yeast appressed to the inner rim; a snail crushed against the inner wall of the jar; a few small, living molluscs of various species to serve as possible oviposition sites; and a couple of small sticks or pieces of vegetation placed at an angle to serve as resting places and possible oviposition sites. The jar is covered with a piece of fine-mesh cloth secured with a rubber band. After putting the flies into the jars they are relatively easy to care for. A few drops or more of water should be added to the cotton or moss as it dries and fly food should be replaced as it becomes dry. The adults should be transferred to fresh breeding jars about once a week or more frequently when many eggs are being laid or hatching.

If eggs are laid on the gauze covers or on the included snails, sticks, or vegetation, they can be removed to small rearing containers for hatching. Do not pry the eggs loose if they are laid on the jar itself because the eggs of most species are affixed with a strong cement and the rather brittle chorion may be damaged if one tries to remove the eggs. If the eggs are laid on the bottom of the jar, they may be left to hatch in place, after which the first-instar larvae may be removed to rearing containers or allowed to start feeding on the living snails present. Infested snails subsequently may be removed to rearing containers. Newly hatched larvae of aquatic species may be recovered by flooding the substrate with water. Stereoscan studies of the eggs of Sciomyzidae (Gasc *et al.* 1984b) have indicated that those of most terrestrial species do not have aeropyles at the posterior pole, but in aquatic species both poles bear aeropyles; this feature should be useful in narrowing the search for the hosts/prey.

The following suggestions for rearing larvae apply to those collected in nature as well as to those obtained from eggs laid in the laboratory. The methods used for rearing larvae depend to a great extent on the type of prey or host involved; thus, an understanding of the microhabitat requirements of the food animals might be important in obtaining successful rearings. The following remarks pertain mainly to larvae that feed on hygrophilous, terrestrial, or exposed, aquatic snails.

Larvae of Sciomyzini (hygrophilous as well as terrestrial; e.g., *Pherbellia*, *Colobaea*, and *Atrichomelina*) are best kept in small plastic boxes (2 × 5 × 5 cm to 3 × 7 × 7 cm) or other shallow containers having a substrate of slightly damp cotton. The cotton should be packed down so that the surface is relatively smooth and there are no fibers sticking up in which the larvae might become entangled. For highly parasitoid species it is particularly important to follow closely the development of individually infested snails. This can be accomplished by isolating infested snails in narrow-diameter glass tubes, plugged with moist cotton, which are observed under the microscope. Also, wetting the surface of the shell with a small camel's hair brush while under observation might increase the translucency of the shell, enabling the larva within to be seen more easily. Larvae of aquatic Tetanocerini, e.g., most species of *Tetanocera*, *Sepedon*, *Elgiva*, etc., are placed in small plastic boxes which have a thin layer of fine, uniform, light-colored gravel covered with just enough water to fill the interstices. It is not necessary to leave or make an opening for air exchange in the rearing containers. For both types of rearing containers a few snails of various species are included so the larvae can select their preferred food. For newly hatched and other small larvae, very small snails, 1–3 mm in greatest dimension, should be used; larger larvae kill snails up to about 13.0 mm in greatest dimension. One dozen or more sciomyzid eggs or tiny larvae may be placed together initially in a small rearing container, but the culture should be subdivided as the larvae develop. Larvae of most aquatic predators will kill and eat 10–20 snails measuring from 1.0 to about 13.0 mm in greatest dimension during the 1 or 2 weeks required to complete development. If the larvae consume all the food in the container and search for food they might be able to crawl under the lid of the box or dish. For this reason it is important to keep the rearing containers stocked with a slight over-abundance of food. None of the reared sciomyzid larvae is cannibalistic. Rearing containers should be examined at least once daily to remove dead, partially eaten snails and fresh food should be added. Importantly, the fouled water in rearing containers with aquatic species should be removed with a bulb pipette; several pipettes of water should be squirted over the gravel and then removed and fresh water should be added. Exuviae, for microscopic study of cephalopharyngeal skeletons and integumental features, can be removed and saved at this time. Because terrestrial and hygrophilous larvae remain in dead snails for a long time and molt while immersed in the liquefied tissues, it is difficult to find the exuviae of these species.

Slug-feeding larvae are handled in nearly the same way as species that feed on terrestrial snails. Infested slugs should be kept in somewhat larger containers and the containers should be held in a dark, relatively cool place. A piece of carrot should be included for living slugs. The strong heat of a microscope lamp is very harmful to slugs, so they should not be held routinely under such a light for more than a few seconds at a time. Detailed biological features of slug-feeding species were discussed by Knutson *et al.* (1965) and Trelka & Foote (1970). A technique for rearing slugs individually was presented by Stephenson (1962).

Snail eggs are the only food source utilized during the first part of larval life by several species of *Anticheta* (Fisher & Orth 1964, Knutson 1966, Knutson & Abercrombie 1977, Robinson & Foote 1978). Females of these species routinely oviposit on the soft egg masses and egg capsules of Succineidae, Lymnaeidae, and Physidae, but none is known to oviposit routinely on or feed on the tougher egg capsules of any Planorbidae. A rather high degree of host specificity is displayed during the earlier stages of development; more mature larvae have a broader host range and also kill and feed on juvenile to mature snails.

Puparia of aquatic species are readily seen in the containers because these larvae pupariate on the surface of the wet gravel or on the sides or lids of containers. Hygrophilous and terrestrial species often form their puparia in snail shells or under or in the cotton or moss in the rearing containers. Their puparia should be looked for a little more carefully, and it might be necessary to flood containers having a moss substrate in order to recover the buried puparia. Puparia first should be gently washed in tap water containing a mold inhibitor and then placed singly in 2 × 8 cm homeopathic vials having a substrate of about 2 cm of slightly damp cotton and securely plugged with a wad of dry cotton. If puparia are kept too moist they will mold but if they are kept too dry they will desiccate. For best emergence results, vials of puparia should be kept in a temperature-controlled cabinet or in a humidity chamber maintained at 76% RH by a super-saturated solution of table salt. Laboratory-reared eggs and puparia in small containers covered with mesh cloth can be placed in the microhabitat to obtain parasitization by hymenopterous wasps; this is a relatively unexplored technique. The pupal period of most species lasts from several days to about 2 weeks. Some species have a pupal diapause, and these often require various exposures to low temperatures, e.g., 5 °C, and subsequent exposure to higher temperatures before emergence occurs.

Most newly emerged adults mate readily and several consecutive generations may be obtained in the laboratory. Adults to be pinned and kept for reference should be left alive and fed for about 48 hours before they are killed. This permits hardening of the exoskeleton and retention of diagnostic features in a more or less natural condition. For later identification of some species it is useful to spread the postabdomen of freshly killed males, exposing the surstyli with pins. Some special methods used in experimental studies of larval feeding behavior are mentioned in Section 7.3.

Pairs of adults and their progeny must be followed individually throughout their development to obtain information on fecundity, duration of stages, numbers of hosts/prey fed upon, etc.; thus, a precise record-keeping system is essential. Also, deposition of voucher specimens is essential (Knutson 1984). Our procedure and that used by many sciomyzid researchers at Cornell University has been to assign a unique Biological Note Number to each rearing of each species, chronologically, throughout a year. For example, K9903 was assigned to the third rearing initiated by Knutson in 1999 with a pair or an individual of field-collected adults, an individual egg, larva, or pupa, or a living or dead host/prey containing a larva or bearing an egg. The first group of eggs laid by the female was, for example, labeled K9903-A when the pair was transferred to a fresh breeding chamber. As the eggs hatch and larvae subsequently are transferred, individually or in small groups to rearing containers, these are labelled K9903-A-1, etc., in sequence. Puparia formed are placed in individual emergence tubes labeled K9903-A-1–1, etc. Individual field-collected larvae are also given unique codes. Individual field-collected puparia are usually labeled only with a field trip number; the adults that emerge are given a Biological Note Number if rearings subsequently are initiated with them. We keep laboratory rearing records on separate sheets of paper so that the sheets on a species can be put with others on that species at the end of the rearing. Observations are made at least daily, with the time noted. A Biological Note Number label is placed in each vial of specimens preserved in alcohol, on each slide-mounted specimen, and with each pinned adult. Shells of snails eaten by larvae are cleaned, dried, and glued to small cards that are pinned below reared adults, or kept in small vials associated with the adults. Soft parts of gastropods, preserved in formalin, have usually not been kept but it would be important to do so if the sciomyzid is suspected of being highly host specific or if studies are to be made of the specific tissues fed upon.

20.3 PRESERVING AND STUDYING ADULTS AND IMMATURE STAGES

Since many techniques are well described, e.g., by Ferrar (1987) and Smith (1989), only those especially suited or adapted for study of sciomyzids or not well documented are noted here. Eye-color patterns of freshly killed adults should be recorded. Some adults should be preserved in 70% ethanol and Bouin's fluid, as well as pinned, for study of non-sclerotized structures and internal anatomy. Especially, some should be preserved for molecular studies. The following techniques for collecting and preparing Sciomyzidae for DNA studies have been provided by E. G. Chapman (personal correspondence, 2007). Specimens can be collected a variety of ways for DNA studies. The highest quality extractions are taken from freshly collected specimens. However, that is often impractical. So, fresh tissue can be frozen at −20 °C (ideally at −70 °C) until DNA extraction. Preserving in near 100% ethanol or isopropanol is also useful for DNA work, and keeping such specimens at −20 °C or lower will preserve the DNA for years. Carvalho & Vieira (2000) compared a variety of preservation techniques (fresh specimens at −70 °C, 95% ethanol at −20 °C, −4 °C, and room temperature; silica gel at room temperature and in a buffer solution). They found that after 1 year, only specimens stored at room temperature in 95% ethanol showed any signs of DNA degradation.

E. G. Chapman had varying success with pinned specimens, the best being those that he collected, stored in >95% ethanol for a short time, and then transferred into hexamethyldisilazane (HMDS) under a fume hood. Specimens were soaked in HMDS for at least 24 hours, after which the excess liquid was poured off and the specimens were allowed to dry for 24 hours (under the hood). This method, which is excellent for allowing alcohol-collected material to dry without the wings becoming mangled, also allows for high-quality DNA extraction. DNA was extracted from specimens preserved this way after more than a year of being on a point at room temperature (specimens soaked in HMDS have to be pointed because the process completely dries them, making them too brittle to pin directly). Similarly, odonate workers have been preserving their specimens for years using acetone to dry them, which works to preserve their color. The acetone probably dissolves the fats that, if left in the insect body, cause the color to fade as they break down. Acetone also leaves the specimens very brittle, so they are stored in envelopes, where, if the specimen breaks apart, all of the parts remain together. Apparently this process also preserves the DNA and likely would work well for sciomyzids.

To study such features as head sutures, hind coxal bridge, mouth-parts, spermathecae, etc., pinned adults can be relaxed, removed from pins, processed entire in a weak NaOH (sodium hydroxide) solution, then in a water, acidic ethanol, ethanol series to 95% ethanol, and preserved in plastic microvials of glycerine. Adults preserved in ethanol can be transferred through a series of ethanol to 100%, then to xylol or Cellosolve ($C_4H_{10}O_2$), and then dried to produce specimens that can be glued to a pin or paper triangle, preferably with Gelva resin solution or some other transparent adhesive that requires a micro-amount for fixation. Dry specimens (pinned or not) can be relaxed by painting them with Barber's liquid (Matile 1993). This enables manipulation and exposure of the male postabdomen. Heraty & Hawks (1998) compared two methods of chemically drying soft-bodied insects preserved in ethanol: critical-point drying and use of HMDS.

Exceptionally fine slide preparations of wings can be made by placing a wing in a small drop of highly dilute Euparol or Canada balsam on a slide, carefully lowering a cover slip wet with xylol, Cellosolve, or LMRSOL (Labo-Moderne, this solution requires some weeks to dry) over the drop at an angle to expel air bubbles. Then during the next few days a few small drops of increasingly less dilute mountant are placed at the edge of the slide to fill air spaces as the first drop of more dilute mountant dries and shrinks. Entry into trapped air bubbles can be effected by breaking an opening with a minuten pin. Wing slides also can be prepared using clear fingernail polish (Matile 1993) and Histolemon Erba ($C_{10}H_{16}$), a clearing agent for histology and less toxic than xylol or LMRSOL.

Examination of minute details of the male genitalia by a compound microscope often is necessary, but standard slide preparations on flat slides usually cause distortions of these structures, as seen in some published drawings and photos (e.g., Verbeke 1967a, Rivosecchi 1992). Also, such slide mounts obviously do not permit manipulating the specimens in order to view them from all aspects. These deficiencies can be obviated by placing the structure in a small drop of glycerine in a very shallow depression slide and covering it with a glass coverslip; the specimen can be re-oriented as needed and eventually stored in glycerine in a plastic microvial.

Genitalia and other exoskeleton preparations usually are kept in glycerine in plastic or glass microvials with plastic stoppers, *never* with cork stoppers as over time the glycerine will creep upward to the cork, which then becomes hygroscopic, leading to erosion of the pin, discoloration of the genitalia, and deterioration of the labels. After placing the specimen in a microvial, a pin must be placed in the vial, the stopper twisted in, and then the pin removed allowing air to escape, thus ensuring an air-tight seal so that the stopper will not ease out due to changes in air pressure. Similarly, the stopper for an alcoholic vial must be "burped" with a thin plastic cord. Type specimens have been destroyed due to neglect of these simple procedures. Genitalia and other exoskeleton structures generally are prepared by soaking in NaOH solution at room temperature, washing in water, then soaking in slightly acidic 70% ethanol to stop the action of the caustic solution. Lactic acid solution (more than 5%) also can be used to clear/soften genitalia. An advantage of this procedure is that specimens can be left in the solution for several days without damage to the structures, and there is no need to wash the specimens in acidic alcohol or water. Genitalia and microscopic structures such as the cephalopharyngeal skeleton (cps) also can be placed in a drop of the water-soluble, synthetic resin DMHF (dimethylhydantoin formaldehyde) on a piece of plastic sheet pinned below the fly (Bameul 1990). Such short-term or moderately long-term preparations, washed in water and examined in glycerine facilitate the ease of re-examination. DMHF also can be used for permanent storage in microvials.

The cephalopharyngeal skeleton of the larva can be studied *in situ* under a dissecting microscope by making a few holes in the anterior end of an ethanol-preserved larva with a minuten pin, then soaking the larva in creosote to clear the integument and muscles. The cps of the molted third-instar larva becomes attached to the ventral cephalic cap of the puparium and can be removed for study after treating the cap with NaOH. Excellent slide mounts of the cps can be prepared from those removed from the cephalic cap and mounted as described above for preparation of slides of wings. Entire exuviae also can be mounted in this way. Also, an NaOH-treated cps can be mounted temporarily in glycerine as described above for genitalia for study under a compound microscope and drawing. Microchips of cover glass can be used to stabilize the position of the structure.

Proper preservation of larvae is one of the most neglected and seemingly difficult steps in the study of immature stages, but it is in fact relatively easy to accomplish. *Never* place living larvae in alcohol or a fixative. Boil water, remove container from heat, immerse larvae, and, most importantly, allow the water to cool. Then immediately put the larvae in Bouin's solution or a solution of kerosene–acetic acid–80% ethanol (in proportions of 1–2–10 parts) for 1–2 hours. The latter solution will cause the larvae to expand very slightly and straighten out. Then store in 70% ethanol. Bratt *et al.* (1969) noted,

> Numerous folds, wrinkles, pads, and welts of the transparent, white or nearly white integument become obvious when the larvae are dipped in a water-soluble stain (e.g., Delafield's hematoxylin) and placed in 70% alcohol. The stain coagulates and becomes trapped around the surface irregularities. After the excess is brushed off, the ventral spinule patches, encircling spinule bands, rows of spinules on the posterior spiracular disc, and other surface features appear accentuated.

A seldom-used technique for studying sensilla of larvae and their nerve supply was mentioned by Bhatia & Keilin (1937); a living larva is placed in a drop of water on a slide, covered with a cover slip, and examined under a compound microscope. Larvae that were preserved in alcohol and that have dried out sometimes can be reconstituted by placing them in a weak solution of NaOH, then passing them through a water, acidic ethanol, ethanol series back to 70% ethanol.

For study with a stereoscan electron microscope, eggs and larvae are prepared according to the usual technique. After cleaning in a water bath containing a few drops of household bleach (5% sodium hypochlorite), the specimens are dehydrated through a graded series of ethanol solutions, dried with CO_2 by the critical point technique, and coated with palladium gold. In a second technique, larvae are placed in a vial of water, then cryo-prepared in a Polaron LT 7400 system using a cryo-FESEM Hitachi 4200. The larvae are successively cryofixed by plunge-freezing at −210 °C into nitrogen, sublimated of superficial ice by progressively raising the temperature to −90 °C in a vacuum, and observed at −150 °C. The observations are made at 1 kV to reduce deep penetration of the electron beam (providing a very good sample surface), to minimize charging effects, and to permit direct scanning electron microscope (SEM) examination. Selected images are directly transferred onto a computer.

Puparia are best preserved dry in a gelatin capsule pinned beneath the adult fly; be sure to include the two puparial caps.

20.4 POPULATION STUDIES

Standard sweep net, pan trap, and emergence trap methods for making quantitative collections of adults to measure seasonal abundance are described in several publications, e.g., O. Beaver (1970), Eckblad (1973a), Vala & Brunel (1987), Vala (1984b), Vala & Manguin (1987), Appleton *et al.* (1993), and Speight (2001, 2004). Capture–mark–release–recapture methods and sweep-net methods were used to study adult populations of the aquatic predator *Sepedon fuscipennis* in open populations by Peacock (1973) and Eckblad & Berg (1972) and in relatively isolated populations by Arnold (1978b). Lincoln Indexes were used to estimate populations of adults by Eckblad & Berg (1972); Arnold (1978b) used the Jolly-Seber statistical model. Lynch (1965) estimated the relative densities of the aquatic predator *Dichetophora biroi* by dipping with a 10-cm sieve. McPhail traps baited with crushed *Lymnaea* snails were used successfully in Hawaii to capture adults of the introduced aquatic predator *Sepedomerus macropus* (Chock *et al.* 1961). Traps baited with live snails were used to sample larvae of the shoreline predator *Hydromya dorsalis* (Hope Cawdery & Lindsay 1977) and the aquatic predator *Ilione albiseta* (Lindsay 1982) in Ireland. Eckblad (1973a) used 1256 cm^2 chambers and 0.6 × 4.6 m quadrats enclosed by nylon cloth (110 μm openings) to study predation by three experimental population levels of the larvae of *S. fuscipennis* introduced into a typical habitat in nature. Arnold (1978a) developed sampling techniques and evaluated estimation formulae for determining the total numbers and absolute densities of eggs, larvae, and puparia of *S. fuscipennis* and several aquatic *Dictya* species in four 20 × 20 m experimental marshes in Ithaca, New York. He studied densities of immatures in 0.25 m^2 floating frame quadrats stratified by presence/absence of vegetation and suggested many improvements of the techniques he used. These studies were conducted jointly with estimation of adult populations by sweep-net collections and mark–release trials. For sampling methods used in biocontrol studies see Chapter 18.

20.5 NOTES ON MAJOR COLLECTIONS

In the World Checklist (Chapter 21), the depositories for type material of all currently valid species are given. In a world catalog being prepared the depositories for and nature of type material for all proposed species names, including synonyms, will be given. The type material of the 539 currently valid species in 61 genera is held in 64 collections in 27 countries, on all continents except Antarctica. The type material of at least 27 currently valid species is known or presumed to be lost, or the location is unknown or questionable. The USNM has the largest amount of type material, 110 species in 20 genera. In Europe, the BM(NH) has the largest amount of type material, 39 species in 17 genera. Other major type depositories are: MNHNP, 30 species; IRSNB, 29; NMW, 27; MCZ, 27; CU, 17. See Table 21.3 for the collection abbreviations.

Extensive information on type examinations over the past half century are documented in the publications of Fisher & Orth (North American collections), Steyskal and Knutson (North American and European collections, see especially Knutson *et al.* [1986] for MCZ and five other collections), Rozkošný (especially 1984b, 1987a) (European collections), and Verbeke (European collections).

Some problems exist in regard to many of the type specimens of the six Palearctic and 30 Afrotropical species in the IRSNB and MRACT described by J. Verbeke who routinely mounted the genitalia and other parts of types and paratypes on glass slides. He died suddenly in 1973, and thus the collection was left somewhat in disarray. Knutson curated the collection June 19–July 14, 1978, aided by preliminary lists of the species in the collections and of specimens bearing slide number labels, which were prepared by Verbeke's technician, W. Timmermans. The lists of slides were inscribed "Preparations microscopiques (dipteren in verzameling – micros. Prep. NIET)" [microscopic preparations (flies in the collection – no microscopic preparations)]. Some slides were found scattered through the collection, but many are presumed lost. A 15-page report including a revised list of the IRSNB and MRACT collections (box-by-box), a list of specimens (including type material) set aside to be returned to other museums, and Timmermans' lists was left with IRSNB. A copy is in Knutson's files and will be included in the Sciomyzidae web page.

A few collections include extensive material from diverse locations and long series of specimens. By far, the collection with the best representation of the world species and number of specimens is the USNM, holding 511 of the 539 currently valid species in 80 drawers. Other major collections are CU (Neartic, Palearctic, Neotropical, and Australian; 63 drawers), University of California, Riverside (California and nearby areas, about 25 000 specimens), CNC (Canada), and IRSNB and MRACT (Afrotropical species). It should be noted that the series of many of the species that were field-collected in many regions and reared at CU, are partly at CU and partly at the USNM as a result of the extensive collaboration between workers at these institutions. There are fairly large collections at other US institutions, and the North American material in many of these institutions has been studied and recorded over the past decade in preparation for a manual of North American Sciomyzidae. One of the most interesting "historical" specimens of Sciomyzidae we have seen is a female of *Protodictya chilensis*, collected in Valparaiso, Chile by Charles Darwin and in the BM(NH) (Plate 1b).

The only collections containing extensive amounts of immature stages are those of CU, USNM, and J-C. Vala, the latter of which will be deposited in MNHNP. The huge alcoholic collection at CU (mostly immature stages) consists of 883 vials stored in 68 jars. A representative collection of immature stages of 74 species in 24 genera reared at CU was prepared by Knutson and donated to USNM in 1968. Unfortunately, the extensive alcoholic collection, including, especially, species of North American *Tetanocera*, developed by B. A. Foote, was mistakenly put in the trash by a workman when Dr. Foote retired from Kent State University. Foote's 1961–1996 collection of pinned adults, many with puparia, was transferred to the Carnegie Museum, Pittsburgh, Pennsylvania; his 1954–1960 material is at Cornell University.

There is a large amount of reared adults from the Nearctic, Palearctic, Neotropical, and Australian regions with puparia in gelatine capsules pinned beneath them in CU, USNM, and the Carnegie Museum. This material is particularly important, not only because descriptions of the puparia of many species have not been published but also because males and females thus can be correctly associated and can be used for the neglected study of the females, especially the abdominal segments and spermathecae.

A very large collection of parts of immature stages mounted on about 2000 glass slides, from the rearing of worldwide species and used in the publications of C. O. Berg and his students, is maintained at CU. About

350 slides from that series, concerning mainly species of *Tetanocera*, *Coremacera*, *Salticella*, *Limnia*, *Colobaea*, and *Tetanura*, along with slides of wings of 46 species, are in the USNM.

It was required practice during rearings of Sciomyzidae at Cornell that the parent flies, and each reared adult, vial, or slide of immature stages bear a label: "[name or initial of researcher], Biological Note Number." This ensures the identity of the series, documents the rearing, and will be important in developing publications on undescribed immature stages. Field trip number labels also were added to some specimens by some Cornell sciomyzid researchers. In most cases, the biological notes were retained by the researchers, but some apparently were disposed of when Dr. Berg retired in 1978. Efforts are being made to prepare a directory of Biological Note and Field Trip numbers, to deposit the directory at several institutions and on the Sciomyzidae web page, and to preserve rearing notes in appropriate institutions.

Designation of voucher specimens has been grossly neglected in biological studies, in general (Knutson 1984). The specimens labeled with "Biological Note Number" and "Field Trip Number" mentioned above serve as voucher specimens for life-cycle studies. The need for voucher specimens persists. For example, J-C. Vala recently was able to determine, from examination of a few of their specimens, that the published report by Rondeleau *et al.* (2002) on "*Tetanocera arrogans*" actually pertains to a melange of *Tetanocera ferruginea* (one specimen) but mostly of species of Sarcophagidae and Muscidae.

21 • World checklist of Sciomyzidae and Phaeomyiidae

Au cours du temps, les espèces se sont différenciées en s'adaptant à la multiplicité des milieux offerts. La plupart de ces milieux ont été peu explorés par les naturalistes et des espèces restent encore à d'écouvrir et à d'écrire. [During the course of time, the species have become differentiated by their adaptation to a multiplicity of habitats. But, most of these habitats have been poorly explored by naturalists and there are many species to discover and describe].

Gruner & Riom (1977).

In the checklist (Table 21.1) we include subgenera where these have been proposed in seven genera. The subgeneric placement is shown by the abbreviation before the species name, corresponding to the abbreviation after the subgenus name. In some genera where subgenera have been named, they have not been designated for all species in the genus. Subgenera are included in the checklist because from a nomenclatorial point of view, the names are valid and available and from a biological point of view, they indicate relationships. Such indications are of potential aid in working out unknown biologies and in further cladistic analysis. "Groups" also have been proposed for *Dictya* (Steyskal 1954b), *Limnia* (Steyskal *et al.* 1978), *Pherbellia* (see Section 15.4.1), *Sepedon* (Knutson & Orth 2001), and *Tetanocera* (Cresson 1920, Boyes *et al.* 1972).

In Table 21.1 adventive species are in parentheses, e.g., N (NT), and broadly shared species are indicated by a dash, e.g., N-NT. Species included under H are not also included under P and N. After species entries, one asterisk indicates the type of the genus, two asterisks indicate the type of the subgenus.

Table 21.1 World checklist of Sciomyzidae and Phaeomyiidae, institutions where type-specimens are deposited, and zoogeographical distribution

Taxa	Type collections	Zoogeographical distribution
HUTTONININAE		
HUTTONININI		
Huttonina TONNOIR & MALLOCH 1928		
subg. *Huttonina* TONNOIR & MALLOCH (1928) (BARNES & KNUTSON 1989) (*Hina.*)		
subg. *Huttoninella* BARNES & KNUTSON 1989 (*Hella.*)		
Hina. abrupta TONNOIR & MALLOCH 1928*	CIN	SA
Hella. angustipennis TONNOIR & MALLOCH 1928**	CIN	SA
Hina. brevis MALLOCH 1930	USNM	SA
Hella. claripennis HARRISON 1959	CIN	SA
Hina. elegans TONNOIR & MALLOCH 1928	CIN	SA
Hina. furcata TONNOIR & MALLOCH 1928	CIN	SA
Hina. glabra TONNOIR & MALLOCH 1928	CIN	SA
Hina. scutellaris TONNOIR & MALLOCH 1928	CIN	SA
PROSOCHAETINI		
Prosochaeta MALLOCH 1935	DEIB	SA
prima MALLOCH 1935*		
SALTICELLINAE		
Salticella ROBINEAU-DESVOIDY 1830		
fasciata MEIGEN 1830*	MNHNP	P
stuckenbergi VERBEKE 1962	NMP	AF
SCIOMYZINAE		
SCIOMYZINI		
Apteromicra PAPP 2004		
parva PAPP 2004	MHNG	O
Atrichomelina CRESSON 1920		
pubera (LOEW 1862)*	MCZ	N (NT)
Calliscia STEYSKAL, in STEYSKAL & KNUTSON 1975		
callisceles (STEYSKAL 1963)*	USNM	NT
Colobaea ZETTERSTEDT 1837		
acuticerca CARLES-TOLRÁ 2008	MC-T	P
americana STEYSKAL 1954	CNC	N
beckeri (HENDEL 1902)	NMW	P
bifasciella (FALLÉN 1820)*	NRS	P
canadensis KNUTSON & ORTH, in Knutson *et al.* 1990	CNC	N
distincta (MEIGEN 1830)	MNHNP	P
eos ROZKOŠNÝ & ELBERG 1991	ZMMSU	P
flavipleura ROZKOŠNÝ & ELBERG 1991	ZMMSU	P
limbata (HENDEL 1933)	ZISP	P
montana KNUTSON & ORTH, in KNUTSON *et al.* 1990	USNM	N
nigroaristata ROZKOŠNÝ 1984	UZIL	P

Table 21.1 (*cont.*)

Taxa	Type collections	Zoogeographical distribution
pectoralis (Zetterstedt 1847)	UZIL	P (O)
punctata (Lundbeck 1923)	UZMC	P (O)
n. sp.	USNM	AF
Ditaeniella Sack 1939		
grisescens (Meigen 1830)*	MNHNP ? NMW ?	P-O
parallela (Walker 1853)	BM(NH)	N (NT)
patagonensis (Macquart 1851)	MNHNP	NT
trivittata (Cresson 1920)	CU	N
Neuzina Marinoni & Knutson, in Marinoni *et al.* 2004		
diminuta Marinoni & Zumbado, in Marinoni *et al.* 2004	INBio	NT
Oidematops Cresson 1920		
ferrugineus Cresson 1920*	MCZ	N
Parectinocera Becker 1919		
dissimilis (Malloch 1933)	BM(NH)	NT
inaequalis (Malloch 1933)	BM(NH)	NT
neotropica Becker 1919*	MNHNP	NT
Pherbellia Robineau-Desvoidy 1830		
subg. *Chetocera* Robineau-Desvoidy 1830 (Rozkošný 1964) (*C.*)		
subg. *Dictyomyza* Enderlein 1939 (*Dc.*) (Rozkošný 1964) (*Dc.*)		
subg. *Ditaenia* Hendel 1902 (Hendel 1910) (*Dt.*)		
subg. *Graphomyzina* Macquart 1835 (Cresson 1920) (*G.*)		
subg. *Oxytaenia* Sack 1939 (*O.*)		
subg. *Pherbellia* Robineau-Desvoidy 1830 (Steyskal 1949) (*P.*)		
C. albicarpa (Rondani 1868)	MLSF	P
C. albocostata (Fallén 1820)**	NRS	H
C. albovaria (Coquillett 1901)	USNM	N
aloea Orth 1983	USNM	N
C. alpina (Frey 1930)	ZMH	P
C. annulipes (Zetterstedt 1846)	UZIL	P
C. anubis Knutson, in Bratt *et al.* 1969	CU	N
argyra Verbeke 1967	IRSNB	H
C. argyrotarsis (Becker 1908)	DEIB	P
C. austera (Meigen 1830)	MNHNP	P
O. beatricis Steyskal 1949	UM	N
O. borea Orth 1982	USNM	N
brevistriata Li, Yang & Gu 2001	CAUB	P
O. brunnipes (Meigen 1838)**	MNHNP	P
O. bryanti Steyskal 1967	USNM	N
O. californica Orth 1982	USNM	N
causta (Hendel 1913)	DEIB	O
chiloensis (Malloch 1933)	BM(NH)	NT
Dt. cinerella (Fallén 1820)**	?	P (O)
G. cingulata (Verbeke 1950)	IRSNB	AF

Table 21.1 (*cont.*)

Taxa	Type collections	Zoogeographical distribution
Dc. clathrata (Loew 1874)**	DEIB	P
G. costata (Verbeke 1950)	IRSNB	AF
C. czernyi (Hendel 1902)	NMW	P
dentata Merz & Rozkošný 1995	NML	P
ditoma Steyskal 1956	UCLA	P
G. dives (Bezzi 1928)	BM(NH)	OC
C. dorsata (Zetterstedt 1846)	UZIL	P
C. dubia (Fallén 1820)	NRS	P
evittata (Malloch 1933)	BM(NH)	NT
C. fisheri Orth 1987	USNM	N
C. footei Steyskal 1961	USNM	N
frohnei Steyskal 1963	WSU	N
garganica Rivosecchi 1989	MCSNM	P
goberti Pandellé 1902	MNHNP	P
C. griseicollis (Becker 1900)	ZMH & MNB	H
C. griseola (Fallén 1820)	NRS	H
G. guttata (Coquillett 1901)	USNM	NT-N
guttipennis (Hendel 1932)	SMNS	NT
C. hackmani Rozkošný 1982	ZMH	H
C. hermonensis Knutson & Freidberg 1983	TAU	P
C. idahoensis Steyskal 1961	USNM	N
C. inclusa (Wollaston 1858)	BM(NH)	P
inflexa Orth 1983	USNM	N
G. javana (Meijere 1919)	ZMA	A-O
G. juxtajavana Knutson, Manguin & Orth 1990	ANIC	A
G. kivuana (Verbeke 1950)	IRSNB	AF
O. knutsoni Verbeke 1967	IRSNB	P
koreana Rozkošný & Kozánek 1989	SNMB	P
krivosheinae Rozkošný 1991	ZMMSU	P
kugleri Knutson 1985	TAU	P
G. limbata (Meigen 1830)**	MNHNP	P
luctifera (Loew 1861)	MCZ	N
lutheri Rozkošný 1982	ZMH	P
majuscula (Rondani 1868)	MLSF	P
O. marthae Orth 1982	USNM	N
C. melanderi Steyskal 1963	USNM	N
O. mikiana (Hendel 1900)	NMW	P
C. nana (Fallén 1820)		H (O)
nana s. str.	NRS	H
nana reticulata (Thomson 1869)	NRS	P (O)
C. obscura (Ringdahl 1948)	ZISP	H
C. obtusa (Fallén 1820)	NRS	P
C. oregona Steyskal 1961	UI	N

Table 21.1 (*cont.*)

Taxa	Type collections	Zoogeographical distribution
orientalis Rozkošný & Knutson 1991	AKMB	P
ozerovi Rozkošný 1991	ZMMSU	P
O. pallidicarpa (Rondani 1868)	MLSF	P
C. pallidiventris (Fallén 1820)	NRS	P
O. paludum Orth 1982	USNM	N
phela Steyskal 1963	USNM	N
philippii (Malloch 1933)	USNM	NT
C. pilosa (Hendel 1902)	NMW	P
O. prefixa Steyskal 1967	USNM	N
C. priscillae Knutson & Freidberg 1983	TAU	P
O. propages Steyskal 1967	USNM	N
C. quadrata Steyskal 1961	USNM	N
C. rozkosnyi Verbeke 1967	IRSNB	P
P. schoenherri (Fallén 1826)*		H
schoenherri s. str.	NRS?	P
schoenherri maculata (Cresson 1920)	ANSP	N
C. scutellaris (Roser 1840)	SMNS	P
C. seticoxa Steyskal 1961	USNM	N
shatalkini Rozkošný 1991	ZMMSU	P
silana Rivosecchi 1989	ISSR	P
C. similis (Cresson 1920)	USNM	N
C. sordida (Hendel 1902)	NMW	P
spectabilis Orth 1984	USNM	N
O. stackelbergi Elberg 1965	ZISP	P
C. steyskali Rozkošný & Zuska 1965	JZP	P
C. subtilis Orth & Steyskal, in Orth *et al.* 1980	USNM	N
C. suspecta Orth & Steyskal 1981	CNC	N
tenuipes (Loew 1872)	MCZ	N
terminalis (Walker 1858)	BM(NH)	O (P)
G. trabeculata (Loew 1872)	MCZ	N (NT)
tricolor Sueyoshi 2001	BLKUF	P
O. ursilacus Orth 1982	USNM	N
C. ventralis (Fallén 1820)	NRS	P
C. vitalis (Cresson 1920)	ANSP	N
vittigera (Malloch 1933)	BM(NH)	NT
Pseudomelina Malloch 1933		
apicalis Malloch 1933*	BM(NH)	NT
Pteromicra Lioy 1864		
albicalceata (Cresson 1920)	MCZ	N
angustipennis (Staeger 1845)	NMW	H
anopla Steyskal 1954	USNM	N
apicata (Loew 1876)	MCZ	N

Table 21.1 (*cont.*)

Taxa	Type collections	Zoogeographical distribution
glabricula (Fallén 1820)*	NRS	P
leucodactyla (Hendel 1913)	DEIB	O
leucopeza (Meigen 1838)	MNHNP	H
leucothrix Melander 1920	USNM	N
nigripalpis Rozkošný 1979	MNHB	P
oldenbergi (Hendel 1902)	NMW	P
pectorosa (Hendel 1902)	NMW	H
perissa Steyskal 1958	UCB	N
pleuralis (Cresson 1920)	ANSP	N
rudis Knutson & Zuska 1968	CU	N
similis Steyskal 1954	USNM	N
siskiyouensis Fisher & Orth 1966	CAS	N
sphenura Steyskal 1954	USNM	N
steyskali Foote 1959	CU	N
n. sp.	USNM	AF
Sciomyza Fallén 1820		
aristalis (Coquillett 1901)	USNM	N
dryomyzina Zetterstedt 1846	UZIL	H
pulchra Roller 1996	SNMB	P
sebezhica Przhiboro 2001	ZISP	P
simplex Fallén 1820*	NRS	H
testacea Macquart 1835	MNHNP	P
varia (Coquillett 1904)	MCZ	N
Tetanura Fallén 1820		
fallenii Hendel 1924	NMW, ? lost	P
pallidiventris Fallén 1820*	NRS	P
Tetanocerini		
Anticheta Haliday 1838		
subg. *Anticheta* Haliday 1839 (Steyskal 1960) (*A.*)		
subg. *Paranticheta* Enderlein 1936 (Steyskal 1960) (*P.*)		
A. analis (Meigen 1830)*	MNHNP	P
A. atriseta (Loew 1849)	MNB	P
P. bisetosa Hendel 1902**	NMW	P
A. borealis Foote 1961	CU	N
A. brevipennis (Zetterstedt 1846)	UZIL	P
A. canadensis (Curran 1923)	CNC	N
A. fulva Steyskal 1960	USNM	N
A. johnsoni (Cresson 1920)	USNM	N
A. melanosoma Melander 1920	USNM	N
A. nigra Karl 1921	ZMH	P
A. obliviosa Enderlein 1939	MNB	P
A. robiginosa Melander 1920	USNM	N
A. shatalkini Vikhrev 2008	ZMMU	P

Table 21.1 (*cont.*)

Taxa	Type collections	Zoogeographical distribution
A. testacea Melander 1920	USNM	N
A. vernalis Fisher & Orth 1971	CAS	N
Chasmacryptum Becker 1907		
seriatimpunctatum Becker 1907*	ZISP	P
Coremacera Rondani 1856		
amoena (Loew 1853)	MNB, ? lost	P
catenata (Loew 1847)	MNB	P
confluens Rondani 1877	MLSF	P
fabricii Rozkošný 1981	UZIL	P
halensis (Loew 1864)	MNB	P
marginata (Fabricius 1775)		P
*marginata s. str.**	UZMC	P
marginata pontica Elberg 1968	ZISP	P
obscuripennis (Loew 1845)	MNB	P
scutellata (Matsumura 1916)	HUS	P
turkestanica (Elberg 1968)	ZISP	P
ussuriensis (Elberg 1968)	ZISP	P
Dichetophora Rondani 1868		
subg. *Dichetophora* Rondani 1868 (Malloch 1928) (*D.*)		
subg. *Neosepedon* Malloch 1928 (*N.*)		
D. australis (Walker 1853)	BM(NH)	A
N. biroi (Kertész 1901)	MNHB	A
N. boyesi Steyskal, in Boyes *et al.* 1972	CNC	A
N. conjuncta Malloch 1928	USNM	A
finlandica Verbeke 1964	IRSNB	P
N. hendeli (Kertész 1901)	MNHB	A
intermedia Hendel 1912	DEIB	O
japonica Sueyoshi 2001	BLKUF	P
kumadori Sueyoshi 2001	BLKUF	P
meleagris (Hendel 1933)	NRS	P
D. obliterata (Fabricius 1805)*	MNHNP	P
N. punctipennis Malloch 1928**	USNM	A
Dictya Meigen 1803		
abnormis Steyskal 1954	USNM	N
adjuncta Valley, in Valley & Berg 1977	CU	N
atlantica Steyskal 1954	USNM	N
bergi Valley, in Valley & Berg 1977	CU	NT
borealis Curran 1932	AMNH	N
brimleyi Steyskal 1954	USNM	N
caliente Orth 1991	USNM	N
chihuahua Orth 1991	USNM	N
disjuncta Orth 1991	USNM	N
expansa Steyskal 1938	UM	N

Table 21.1 (*cont.*)

Taxa	Type collections	Zoogeographical distribution
fisheri Orth 1991	USNM	N
floridensis Steyskal 1954	USNM	N
fontinalis Fisher & Orth 1969	CAS	N
gaigei Steyskal 1938	UM	N
guatemalana Steyskal 1954	USNM	NT
hudsonica Steyskal 1954	CNC	N
incisa Curran 1932	AMNH	N (NT)
insularis Steyskal 1954	USNM	NT
jamaica Orth 1991	USNM	NT
knutsoni Orth 1991	USNM, lost?	N
laurentiana Steyskal 1954	CNC	N
lobifera Curran 1932	AMNH	N
matthewsi Steyskal 1960	CU	NT (N)
mexicana Steyskal 1954	CAS	NT (N)
montana Steyskal 1954	CAS	N
neffi Steyskal 1960	CU	NT
orion Orth 1991	SIVC	N
orthi Mathis, Knutson & Murphy 2009	USNM	N
oxybeles Steyskal 1960	USNM	N
pechumani Valley, in Valley & Berg 1977	CU	N (NT)
pictipes (Loew 1859)	MCZ	N
praecipua Orth 1991	USNM	N
ptyarion Steyskal 1954	USNM	N
sabroskyi Steyskal 1938	USNM	N (NT)
sinaloae Orth 1984	USNM	NT
steyskali Valley, in Valley & Berg 1977	CU	N
stricta Steyskal 1938	UM	N
texensis Curran 1932	AMNH	N (NT)
umbrarum (Linnaeus 1758)*	?	P
umbroides Curran 1932	CNC	N
valleyi Orth 1991	USNM	N
veracruz Orth 1991	USNM	N
zacki Orth & Fisher 1983	USNM	N
Dictyacium Steyskal 1956		
ambiguum (Loew 1864)*	MCZ	N
firmum Steyskal 1956	CNC	N
Dictyodes Malloch 1933		
dictyodes (Wiedemann 1830)*	?ZMHUB	NT
platensis Steyskal 1974	USNM	NT
Ectinocera Zetterstedt 1838		
borealis Zetterstedt 1838*	UZIL	P
Elgiva Meigen 1838		
connexa (Steyskal 1954)	USNM	N

Table 21.1 (*cont.*)

Taxa	Type collections	Zoogeographical distribution
cucularia (Linnaeus 1767)*	?	P
divisa (Loew 1845)	ZMHUB	H
elegans Orth & Knutson 1987	USNM	N
manchurica Rozkošný & Knutson 1991	AKMB	P
pacnowesa Orth & Knutson 1987	USNM,	N
solicita (Harris 1780)	? lost	H
Ethiolimnia Verbeke 1950		
brincki Verbeke 1961	UZIL	AF
geniculata (Loew 1862)	NRS	AF
lindneri Verbeke 1962	SMNS	AF
platalea Verbeke 1950*	IRSNB	AF
vanrosi Verbeke 1962	NRS	AF
vittipennis (Thomson 1869)	NRS	AF
zumpti Verbeke 1956	SAIMR	AF
Eulimnia Tonnoir & Malloch 1928		
milleri Tonnoir & Malloch 1928*	NZACA	SA
philpotti Tonnoir & Malloch 1928	NZACA	SA
Euthycera Latreille 1829		
alaris Vala 1983		
alaris s. str.	MNHNP	P
alaris sardoa Contini & Rivosecchi 1984	? UT	P
algira (Macquart 1849)	MNHNP	P
alpina (Mayer 1953)	NMW	P
arcuata (Loew 1859)	MCZ	N
atomaria (Linnaeus 1761)	LCL	P
chaerophylli (Fabricius 1798)*	? UZMC	P
cribrata (Rondani 1868)	? MLSF	P
flavostriata (Villeneuve 1911)	MNHNP	P
formosa (Loew 1862)	? MNB	P
fumigata (Scopoli 1763)	?	P
guanchica Frey 1936	ZMH	P
hrabei Rozkošný 1969	ZSBSM	P
korneyevi Rozkošný & Knutson 2006	MUB	P
merzi Rozkošný & Knutson 2006	MHNG	P
mira Knutson & Zuska 1968	CNC	N
seguyi Vala 1990	MNHNP	P
stichospila (Czerny 1909)	NMW	P
stictica (Fabricius 1805)	UZMC, lost?	P
sticticaria (Mayer 1953)	NMW	P
vockerothi Rozkošný 1988	CNC	P
zelleri (Loew 1847)	MNB	P
Euthycerina Malloch 1933		
pilosa Malloch 1933	BM(NH)	NT
vittithorax Malloch 1933*	BM(NH)	NT

Table 21.1 (*cont.*)

Taxa	Type collections	Zoogeographical distribution
Eutrichomelina Steyskal, in Steyskal & Knutson 1975		
albibasis (Malloch 1933)	BM(NH)	NT
fulvipennis (Walker 1837)*	BM(NH)	NT
Guatemalia Steyskal 1960		
nigritarsis Marinoni 1992	USNM	NT
straminata (Wulp 1897)*	BM(NH)	NT
Hedria Steyskal 1954		
mixta Steyskal 1954*	USNM	N
Hoplodictya Cresson 1920		
acuticornis (Wulp 1897)	BM(NH)	N
australis Fisher & Orth 1972	USNM	N (NT)
kincaidi (Johnson 1913)	MCZ	N
setosa (Coquillett 1901)*	USNM	N
spinicornis (Loew 1866)	MCZ	N (NT)
Hydromya Robineau-Desvoidy 1830		
dorsalis (Fabricius 1775)*	UZMC, ? lost	P (O, AF)
Ilione Haliday, in Curtis 1837		
subg. *Ilione* Haliday, in Curtis 1837 (*I.*)		
subg. *Knutsonia* Verbeke 1964 (*K.*)		
I. albiseta (Scopoli 1763)*	destroyed	P
I. corcyrensis (Verbeke 1964)	IRSNB	P
K. lineata (Fallén 1820)**	NRS	P
K. rossica (Mayer 1953)	NMW	P
I. trifaria (Loew 1847)	ZMHUB	P
I. truqui (Rondani 1863)	MLSF	P
I. turcestanica (Hendel 1903)	MNHB	P (O)
I. unipunctata (Macquart 1849)	MNHNP	P
Limnia Robineau-Desvoidy 1830		
boscii (Robineau-Desvoidy 1830)	? MNHNP	N
conica Steyskal, in Steyskal *et al.* 1978	USNM	N
fitchi Steyskal, in Steyskal *et al.* 1978	USNM	N
georgiae Melander 1920	USNM	N
inopa (Adams 1904)	UK	N
japonica Yano 1978	KUJ	P
lemmoni Fisher & Orth 1971	CAS	N
lindbergi Steyskal, in Steyskal *et al.* 1978	USNM	N
loewi Steyskal 1965	MCZ	N
lousianae Melander 1920	USNM	N
nambai Steyskal, in Steyskal *et al.* 1978	USNM	N
ottawensis Melander 1920	USNM	N
pacifica Elberg 1965	ZISP	P
paludicola Elberg 1965	ZISP	P
pubescens (Day 1881)	USNM, ? lost	N

Table 21.1 (*cont.*)

Taxa	Type collections	Zoogeographical distribution
sandovalensis Fisher & Orth, in Steyskal *et al.* 1978	USNM	N
septentrionalis Melander 1920	USNM	N
setosa Yano 1978	KUJ	P
severa Cresson 1920	CAS	N
shannoni Cresson 1920	USNM	N
sparsa (Loew 1862)	MCZ	N
unguicornis (Scopoli 1763)*	destroyed	P
Neodictya Elberg 1965		
jakovlevi Elberg 1965*	ZISP	P
Neolimnia Tonnoir & Malloch 1928		
subg. *Neolimnia* Tonnoir & Malloch 1928 (*N.*)		
subg. *Pseudolimnia* Tonnoir & Malloch 1928 (*P.*)		
subg. *Sublimnia* Harrison 1959 (*S.*)		
N. castanea (Hutton 1904)	CMC	SA
N. diversa Tonnoir & Malloch 1928	NZACA	SA
N. irrorata Tonnoir & Malloch 1928	NZACA	SA
N. minuta Tonnoir Malloch 1928	NZACA	SA
S. nitidiventris Tonnoir & Malloch 1928**	NZACA	SA
N. obscura (Hutton 1901)*	CMC	SA
N. pepekeiti Barnes 1979	NZACA	SA
N. raiti Barnes 1979	NZACA	SA
P. repo Barnes 1979	NZACA	SA
P. sigma (Walker 1849)**	BM(NH)	SA
N. striata (Hutton 1904)		SA
striata s. str.	CMC	SA
s. brunneifrons Tonnoir & Malloch 1928	NZACA	SA
P. tranquilla (Hutton 1901)	CMC	SA
P. ura Barnes 1979	NZACA	SA
S. vittata Harrison 1959	NZACA	SA
Oligolimnia Mayer 1953		
zernyi Mayer 1953*	NMW	P
Perilimnia Becker 1919		
albifacies Becker 1919*	MNHNP	NT
cineritia Zuska, in Kaczynski *et al.* 1969	CNC	NT
Pherbecta Steyskal 1956		
limenitis Steyskal 1956*	CU	N
Pherbina Robineau-Desvoidy 1830		
coryleti (Scopoli 1763)*	destroyed	P
intermedia Verbeke 1948	IRSNB	P
mediterranea Mayer 1953	NMW	P
testacea (Sack 1939)	DEIB	P
Poecilographa Melander 1913		
decora (Loew 1864)*	MCZ	N

Table 21.1 (*cont.*)

Taxa	Type collections	Zoogeographical distribution
Protodictya Malloch 1933		
apicalis Steyskal 1950	IOC	NT
bidentata Marinoni & Knutson 1992	USNM, ? lost	NT
brasiliensis (Schiner 1868)	NMW, ? lost	NT
chilensis Malloch 1933*	BM(NH)	NT
guttularis (Wiedemann 1830)	ZMHUB	NT
iguassu Steyskal 1950	IOC	NT
lilloana Steyskal 1953	IMLT	NT
nubilipennis (Wulp 1897)	BM(NH)	NT
Psacadina Enderlein 1939		
disjecta Enderlein 1939*	ZMHUB	P
kaszabi Elberg 1978	MNHB	P
verbekei Rozkošný, in Knutson *et al.* 1975	MUB	P
vittigera (Schiner 1862)	NMW	P
zernyi (Mayer 1953)	NMW	P
Renocera Hendel 1900		
amanda Cresson 1920	MCZ	N
johnsoni Cresson 1920	MCZ	N
longipes (Loew 1876)	MCZ	N
pallida (Fallén 1820)	NRS	P
striata (Meigen 1830)	NMW	H
stroblii Hendel 1900*	NMW	P
Sepedomerus Steyskal 1973		
bipuncticeps (Malloch 1933)		NT
bipuncticeps s. str.	BM(NH)	NT
bipuncticeps trinidadensis (Steyskal 1951)	USNM	NT
caeruleus (Melander 1920)	USNM	NT
macropus (Walker 1849)*	BM(NH)	NT(N)
Sepedon Latreille 1804		
subg. *Mesosepedon* Verbeke 1950 (*M.*)		
subg. *Parasepedon* Verbeke 1950 (*P.*)		
subg. *Sepedomyia* Verbeke 1950 (Steyskal 1973) (*Sm.*)		
subg. *Sepedon* Latreille 1804 (Verbeke 1950) (*S.*)		
P. acrosticta Verbeke, in Steyskal & Verbeke 1956	IRSNB	AF
S. aenescens Wiedemann 1830	UZMC	O (P)
Sm. alaotra Verbeke 1962	MNHNP	AF
P. albocostata Verbeke 1950	IRSNB	AF
americana Steyskal 1951	USNM	N
anchista Steyskal 1956	CU	N
armipes Loew 1859	MCZ	N (NT)
batjanensis Kertész 1899	MNHB	O-OC
bifida Steyskal 1951	CAS	N
borealis Steyskal 1951	USNM	N

Table 21.1 (*cont.*)

Taxa	Type collections	Zoogeographical distribution
capellei Fisher & Orth 1969	CAS	N
cascadensis Fisher & Orth 1974	CAS	N
chalybeifrons Meijere 1908	ZMA	O
M. convergens Loew 1862	?	AF
costalis Walker 1858	BM(NH)	A
crishna Walker 1859	BM(NH)	O (A)
M. dispersa Verbeke 1950	IRSNB	AF
P. edwardsi Steyskal, in Steyskal & Verbeke 1956	BM(NH)	AF
M. ethiopica Steyskal, in Steyskal & Verbeke 1956	BM(NH)	AF
femorata Knutson & Orth 1984	USNM	P
ferruginosa Wiedemann 1824	UZMC	O
floridensis Steyskal 1951	USNM	N
fuscipennis Loew 1859	MCZ	N
gracilicornis Orth 1986	USNM	N
hecate Elberg, Rozkošný & Knutson 2009	USNM	P
hispanica Loew 1862		AF (P)
hispanica s. str.	ZMHUB	P
hispanica ruhengeriensis Verbeke 1950	IRSNB	AF
P. iris Verbeke 1961	IRSNB	AF
P. ituriensis Verbeke 1950	MRACT	AF
P. katangensis Verbeke 1950	MRACT	AF
P. knutsoni Vala, Gbedjissi & Dossou 1994	MNHNP	AF
lata Bezzi 1928	BM(NH)	A-OC
lignator Steyskal 1951	UM	N
P. lippensi Verbeke 1950	IRSNB	AF
lobifera Hendel 1911	MNHB	O
P. maculifemur Verbeke 1950	IRSNB	AF
P. madecassa Verbeke 1961	MNHNP	AF
P. magerae Verbeke 1950	IRSNB	AF
mcphersoni Knutson & Orth 2001	INHS	N
melanderi Steyskal 1951	USNM	N
P. monacha Verbeke 1961	IRSNB	AF
P. nanoides Verbeke 1950	MRACT	AF
Sm. nasuta Verbeke 1950**	IRSNB	AF
neanias Hendel 1913	[DEIB? NMW]	O (P)
P. neavei Steyskal, in Steyskal & Verbeke 1956	BM(NH)	AF
neili Steyskal 1951	UM	N
P. notambe Speiser 1910**	? NRS	AF
noteoi Steyskal 1980	USNM	O-P
P. ochripes Verbeke 1950	IRSNB	AF
P. ophiolimnes Steyskal, in Steyskal & Verbeke 1956	BM(NH)	AF

Table 21.1 (*cont.*)

Taxa	Type collections	Zoogeographical distribution
P. ornatifrons ADAMS 1905	UK	AF
pacifica CRESSON 1914	ANSP	N
P. paranana VERBEKE 1950	MRACT	AF
P. pelex STEYSKAL, in STEYSKAL & Verbeke 1956	BM(NH)	AF
M. pleuritica LOEW 1862	?	AF
plumbella WIEDEMANN 1830	UZMC	O-OC-A
praemiosa GIGLIO-TOS 1893	UT	N (NT)
pseudarmipes FISHER & ORTH 1969	CAS	N
pusilla LOEW 1859	MCZ	N
relicta WULP 1897	BM(NH)	N
P. ruficeps BECKER 1922	NMW	AF (P)
P. saegeri VERBEKE 1950	IRSNB	AF
P. scapularis ADAMS 1903	UK	AF
M. schoutedeni VERBEKE 1950**	MRACT	AF
P. selenopa VERBEKE 1961	MRACT	AF
P. senegalensis MACQUART 1844	MNHNP	AF
senex WIEDEMANN 1830	NMW	O
P. simulans VERBEKE 1950	MRACT	AF
spangleri BEAVER 1974	USNM	O
S. sphegea (FABRICIUS 1775)*	BM(NH)	P (O)
S. spinipes (SCOPOLI 1763)	destroyed MLSF	P
P. straeleni VERBEKE 1963	IRSNB	AF
P. stuckenbergi VERBEKE 1961[1]	MNHNP	AF
tenuicornis CRESSON 1920	USNM, ? lost	N
P. testacea LOEW 1862	? NRS	AF
P. trichrooscelis SPEISER 1910	? NRS	AF
P. trochanterina VERBEKE 1950	IRSNB	AF
M. tuckeri BARRACLOUGH 1985	NMP	AF
P. uelensis VERBEKE 1950	IRSNB	AF
P. umbrosa VERBEKE 1950	MRACT	AF
Sepedonea STEYSKAL 1973		
barbosai KNUTSON & BREDT 1976	MZUSP	NT
canabravana KNUTSON & BREDT 1976	MZUSP	NT
giovana MARINONI & MATHIS 2006	USNM	NT
guatemalana (STEYSKAL 1951)	OSU	NT
guianica (STEYSKAL 1951)	BM(NH)	NT
incipiens FREIDBERG, KNUTSON & ABERCROMBIE 1991	USNM	NT
isthmi (STEYSKAL 1951)	USNM	NT
lagoa (STEYSKAL 1951)	IOC	NT
lindneri (HENDEL 1932)*	SMNS	NT
neffi FREIDBERG, KNUTSON & ABERCROMBIE 1991	USNM	NT
telson (STEYSKAL 1951)	IOC	NT
trichotypa FREIDBERG, KNUTSON & ABERCROMBIE 1991	USNM	NT

Table 21.1 (*cont.*)

Taxa	Type collections	Zoogeographical distribution
veredae Freidberg, Knutson & Abercrombie 1991	MZUSP	NT
Sepedonella Verbeke 1950		
bredoi Verbeke 1950	IRSNB	AF
nana Verbeke 1950*	IRSNB	AF
straeleni Verbeke, in Steyskal & Verbeke 1956	IRSNB	AF
wittei Verbeke 1950	IRSNB	AF
Sepedoninus Verbeke 1950		
curvisetis Verbeke 1950	IRSNB	AF
planifrons Verbeke 1950*	IRSNB	AF
n. sp.	BM(NH)	O
Shannonia Malloch 1933		
costalis (Walker 1837)*	BM(NH)	NT
meridionalis Zuska, in Kaczynski *et al.* 1969	CU	NT
Steyskalina Knutson, in Ghorpadé *et al.* 1999		
picta Ghorpadé & Marinoni, in Ghorpadé *et al.* 1999	NRS	O
Tetanocera Duméril 1800		
subg. *Chaetotelanocera* Mayer 1953 (*C.*)		
subg. *Tetanocera* Duméril 1800 (Mayer 1953) (*T.*)		
amurensis Hendel 1909	NMW	P
andromastos Steyskal 1963	CNC	N
annae Steyskal 1938	UM	N
arnaudi Orth & Fisher 1982	CAS	N
T. arrogans Meigen 1830	MNHNP	P
bergi Steyskal 1954	USNM	N
T. brevisetosa Frey 1924	ZMH	H
chosenica Steyskal 1951	CAS	O-P
clara Loew 1862	MCZ	N
T. elata (Fabricius 1781)*	MNHNP	P
T. ferruginea Fallén 1820	? NRS	H
T. freyi Stackelberg 1963	ZISP	H
fuscinervis (Zetterstedt 1838)	UZIL	H
T. hyalipennis Roser 1840	SMNS	P
ignota Becker 1907	ZISP	P
iowensis Steyskal 1938	USNM	N
T. kerteszi Hendel 1901	MNHNP	H
T. lapponica Frey 1924	ZMH	P
T. latifibula Frey 1924	ZMH	H
loewi Steyskal 1959	USNM	N
melanostigma Steyskal 1959	USNM	N
mesopora Steyskal 1959	USNM	N
T. montana Day 1881	? UCS	H
T. nigrostriata Li, Yang & Gu 2001	CAUB	O

Table 21.1 (*cont.*)

Taxa	Type collections	Zoogeographical distribution
obtusifibula Melander 1920	USNM	N
oxia Steyskal 1959	USNM	N
T. phyllophora Melander 1920	USNM	H
plebeja Loew 1862	MCZ	H
plumosa Loew 1847	MCZ	N (NT)
T. punctifrons Rondani 1868	MLSF	P
C. robusta Loew 1847**	ZMH	H
rotundicornis Loew 1861	MCZ	N
silvatica Meigen 1830	MNHNP	H
soror Melander 1920	USNM	N
spirifera Melander 1920	USNM	H
spreta Wulp 1897	BM(NH)	N
stricklandi Steyskal 1959	USNM	N
valida Loew 1862	MCZ	N
vicina Macquart 1844	MNHNP	N
Tetanoceroides Malloch 1933		
bisetosus (Thomson 1869)	NRS	NT
dentifer Zuska, in Zuska & Berg 1974	CU	NT
fulvithorax Malloch 1933	BM(NH)	NT
mendicus Zuska, in Zuska & Berg 1974	CU	NT
mesopleuralis Malloch 1933*	BM(NH)	NT
patagonicus (Thomson 1869)	NRS	NT
simplex Zuska, in Zuska & Berg 1974	CU	NT
Tetanoptera Verbeke 1950		
leucodactyla Verbeke 1950*	IRSNB	AF
Teutoniomyia Hennig 1952		
costaricensis Steyskal 1960	USNM	NT(N)
plaumanni Hennig 1952*	DEIB	NT
Thecomyia Perty 1833		
abercrombiei Marinoni & Steyskal, in Marinoni *et al.* 2003	BM(NH)	NT
autazensis Marinoni & Steyskal, in Marinoni *et al.* 2003	USNM	NT
bonattoi Marinoni & Steyskal, in Marinoni *et al.* 2003	AMNH	NT
chrysacra Marinoni & Steyskal, in Marinoni *et al.* 2003	USNM	NT
lateralis (Walker 1858)	BM(NH)	NT
limbata (Wiedemann 1819)	ZMHUB	NT
longicornis Perty 1833*	ZSBSM	NT
mathisi Marinoni & Steyskal, in Marinoni *et al.* 2003	USNM	NT
naponica Marinoni & Steyskal, in Marinoni *et al.* 2003	CNC	NT

Table 21.1 (*cont.*)

Taxa	Type collections	Zoogeographical distribution
papaveroi Marinoni & Steyskal, in Marinoni *et al.* 2003	USNM	NT
signorelli Marinoni & Steyskal, in Marinoni *et al.* 2003	MZUSP	NT
tricuneata Marinoni & Steyskal, in Marinoni *et al.* 2003	CNC	NT
Trypetolimnia Mayer 1953		
rossica Mayer 1953*	NMW	P
Trypetoptera Hendel 1900		
canadensis (Macquart 1844)	lost MNHNP	N
punctulata (Scopoli 1763)*	? destroyed MLSF	P
Verbekaria Knutson 1968		
punctipennis Knutson 1968*	CNC	AF
PHAEOMYIIDAE		
Akebono Sueyoshi, in Sueyoshi *et al.* 2009		
vernalis Sueyoshi, in Sueyoshi *et al.* 2009	NIAS	P
Pelidnoptera Rondani 1856		
fuscipennis (Meigen 1830)	NMW	P
leptiformis (Schiner 1864)	NMW	P
nigripennis (Fabricius 1794)*	? UZMC	P
triangularis Knutson & Ghorpadé, in Sueyoshi *et al.* 2009	NIAS	P(O)

Note: [1] Not in NMP as stated by Verbeke (1961).

Table 21.2. Fossil species of Sciomyzidae and Phaeomyiidae

Taxa	Collection	Geographical distribution (Age)
PHAEOMYIIDAE		
Prophaeomyia Hennig 1965		
loewi Hennig 1965*	BM(NH)	Baltic Region (Eocene/Oligocene)
SCIOMYZIDAE		
SALTICELLINAE		
Prosalticella Hennig 1965		
succini Hennig 1965*	UG	Baltic Region (Eocene/Oligocene)
SCIOMYZINAE		
Sciomyzini		
Palaeoheteromyza Meunier 1904		
crassicornis Meunier 1904*	UG	Baltic Region (Eocene/Oligocene)
curticornis Hennig 1965[1]	UG	Baltic Region (Eocene/Oligocene)
investiganda Hennig 1969	MNHNP	Baltic Region (Eocene/Oligocene)

Table 21.2. (*cont.*)

Taxa	Collection	Geographical distribution (Age)
Sciomyza Fallén 1820		
disjecta Scudder 1878[2]	?	Wyoming, USA (Eocene)
florissantensis Cockerell 1909	AMNH	Colorado, USA (Miocene) (= Ortalidae?)
manca Scudder 1878[2]	?	Wyoming, USA (Eocene)
revelata Scudder 1877[2]	?	British Columbia, Canada (Oligocene)
Tetanocerini		
Sepedonites Hennig 1965		
baltica Hennig 1965*	UG	Baltic Region (Eocene/Oligocene)
Tetanocera Duméril 1800		
alireticulata Théobald 1937[2]	?	France (Oligocene)
contenta Förster 1891[2]	?	Germany (Miocene)
preciosa Förster 1891[2]	?	Germany (Miocene)
variciliata Théobald 1937[2]	?	France (Oligocene)

Notes: [1] Questionably referred to Phaeomyiidae by Hennig (1969); [2] Not recognizable as Sciomyzidae by Hennig (1965).

Table 21.3. Abbreviations of collections in which type specimens of valid species are held

AKMB: Alexander Koenig Museum, Bonn, Germany
AMNH: American Museum of Natural History, New York, New York, USA
ANIC: Australian National Insect Collection, Canberra, Australia
ANSP: Academy of Natural Sciences, Philadelphia, Pennsylvania, USA
BLKUF: Biosystematics Laboratory, Kyushu University, Fukuoka, Japan
BM(NH): British Museum (Natural History), London, England
BMH: Bishop Museum, Honolulu, Hawaii, USA
CAS: California Academy of Sciences, San Francisco, California, USA
CAUB: China Agricultural University, Beijing, China
CIN: Cawthron Institute, Nelson, New Zealand
CMC: Canterbury Museum, Christchurch, New Zealand
CNC: Canadian National Collection, Ottawa, Ontario, Canada
CU: Cornell University, Ithaca, New York, USA
DEIB: Deutsches Entomologisches Institut, Berlin, Germany
HUS: Hokkaido University, Sapporo, Japan
IMLT: Instituto Miguel Lillo, Tucuman, Argentina
INB: Instituto Nacional de Biodiversidad de Costa Rica, Santo Domingo, Heredia, Costa Rica
INHS: Illinois Natural History Survey, Champaign-Urbana, Illinois, USA
IOC: Instituto Oswaldo Cruz, Rio de Janeiro, Brazil
IRSN: Institut royal des Sciences naturelles de Belgique, Brussels, Belgium
ISSR: Istituto Superiore di Sanità, Rome, Italy
JZP: Jan Zuska collection, Prague, Czech Republic
KUJ: Kyushu University, Kyushu, Japan

Table 21.3. (*cont.*)

LCL: Linnean Collection, London, England
MCSNM: Museo civico di Storia naturale di Milano, Milan, Italy
MC-T: Miguel Carles-Tolrá collection, Barcelona, Spain
MCZ: Museum of Comparative Zoology, Harvard College, Cambridge, Massachusetts, USA
MHNG: Museum d'Histoire naturelle, Geneva, Switzerland
MLSF: Museo Zoologica "La Specola," Florence, Italy
MNB: Museum für Naturkunde, Berlin, Germany
MNHB: Museum of Natural History, Budapest, Hungary
MNHNP: Museum national d'Histoire naturelle, Paris, France
MRACT: Musée royal Afrique central, Tervuren, Belgium
MUB: Masaryk University, Brno, Czech Republic
MZUSP: Museu de Zoologia da Universidade de São Paulo, São Paulo, Brazil
NIAS: National Institute of Agro-environmental Science, Tsukuba, Japan
NML: Natur-Museum, Luzern, Switzerland
NMP: Natal Museum, Pietermaritzberg, South Africa
NMW: Naturhistorisches Museum, Vienna, Austria
NRS: Naturhistoriska Riksmuseet, Stockholm, Sweden
NZACA: New Zealand Arthropod Collection, Auckland, New Zealand
OSU: Ohio State University, Columbus, Ohio, USA
SAIMR: South African Institute for Medical Research, Johannesburg, South Africa
SIUC: Southern Illinois University, Carbondale, Illinois, USA
SMNS: Staatliches Museum für Naturkunde, Stuttgart, Germany
SNMB: Slovak National Museum, Bratislava, Slovakia
TAU: Tel Aviv University, Israel
UCB: University of Colorado, Boulder, Colorado, USA
UCLA: University of California, Los Angeles, California, USA
UCS: University of Connecticut, Storrs, Connecticut, USA
UG: University of Göttingen, Göttingen, Germany
UI: University of Idaho, Moscow, Idaho, USA
UK: University of Kansas, Lawrence, Kansas, USA
UM: University of Michigan, Ann Arbor, Michigan, USA
USNM: United States National Museum of Natural History, Washington, DC, USA
UT: University of Turin, Italy
UZIL: Universitetets Zoologiske Institution, Lund, Sweden
UZMC: Universitetets Zoologiske Museum, Copenhagen, Denmark
WSU: Washington State University, Pullman, Washington, USA
ZISP: Zoological Institute, St. Petersburg, Russia
ZMA: Zoolögisch Museum, Amsterdam, the Netherlands
ZMH: Zoological Museum, Helsinki, Finland
ZMHUB: Zoologisches Museum, Humboldt University, Berlin, Germany
ZMMSU: Zoological Museum, Moscow State University, Russia
ZSBSM: Zoologische Sammlung des Bayerischen Staates, Munich, Germany

References

Longtemps, longtemps, longtemps, après que les poètes ont disparus,
leurs chansons courent encore dans les rues.
[Long, long, long after the poets have disappeared,
their songs still echo through the streets.]

Words and music by Charles Trenet, 1951.

PART A[1]

SCIOMYZIDAE AND PHAEOMYIIDAE

Abercrombie, J. (1970). *Natural History of Snail-killing Flies of South America (Diptera: Sciomyzidae: Tetanocerini).* Ph.D. thesis, Cornell Univ., Ithaca, NY, USA. 344 pp. Order No. 70–23, 095, Univ. Microfilms, Ann Arbor, MI. (*Diss. Abstr. Int., Ser.* **B. 31**, 3456–3457).

Abercrombie, J. (2000). Biology of *Sepedonea*, a Neotropical genus of snail-killing flies (Diptera: Sciomyzidae). In *XXI Int. Congr. Ent.*, Aug. 20–26, 2000, Brazil, ed. D. L. Gazzoni. *Abstracts Book I*, p. 925.

Abercrombie, J. & Berg, C. O. (1975). Natural history of *Thecomyia limbata* (Diptera: Sciomyzidae) from Brazil. *Proc. Ent. Soc. Wash.*, **77** (3), 355–368.

Abercrombie, J. & Berg, C. O. (1978). Malacophagous Diptera of South America: Biology and immature stages of *Dictyodes dictyodes*. *Rev. Bras. Ent.*, **22** (1), 23–32.

Acloque, A. (1897). *Faune de France, Contenant la Description des Espèces Indigènes Disposées en Tableaux Analytiques et*

[1] When two years are given for a publication, the actual year of publication, i.e., the year of distribution, is given first, followed in brackets by the year shown on the publication.

Illustrée de Figures Représentant les Types Caractéristiques des Genres. Vol. **2**. Orthoptères, etc. Paris: J-B. Baillière et Fils.

Adams, C. F. (1903). Dipterological contributions. *Kans. Univ. Sci. Bull.*, **2** (2) [=whole series **12** (2)], 21–47. (= *Kans. Univ. Bull.*, **4** [6]).

Adams, C. F. (1904). Notes on and descriptions of North American Diptera. *Kans. Univ. Sci. Bull.*, **2** [=whole series, **12** (14)], 433–455. (= *Kans. Univ. Bull.*, **4** [6]).

Adams, C. F. (1905). Diptera Africana, I. *Kans. Univ. Sci. Bull.*, **3** (6) [= whole series **13** (6)], 149–208. (= *Kans. Univ. Bull.*, **6** [2]).

Appleton, C. C., Miller, R. M. & Maharaj, R. (1993). Control of schistosomiasis host snails in South Africa – the case for biocontrol by predator augmentation using sciomyzid flies. *J. Med. Appl. Malacol.*, **5**, 107–116.

Arnold, S. L. (1978a). *Field and Simulation Studies of the Population Dynamics of* Sepedon fuscipennis *(Diptera: Sciomyzidae)*. Ph.D. thesis, Cornell Univ., Ithaca, NY, USA. 213 pp. Order No. 7817741, Univ. Microfilms, Ann Arbor, MI. (*Diss. Abstr. Int., Ser.* **B. 39**, 1614–1615).

Arnold, S. L. (1978b). Sciomyzidae (Diptera) population parameters estimated by the capture-recapture method. *Proc. Ent. Soc. Ont.*, **107** (1976), 3–9.

Ayatollahi, M. (1971). Importance of the study of Diptera and their role in biological control. *Ent. Phytopathol. Appl.*, **31**, 20–28.

Bailey, P. T. (1989). The millipede parasitoid *Pelidnoptera nigripennis* (F.) (Diptera: Sciomyzidae) for the biological control of the millipede *Ommatoiulus moreleti* (Diplopoda: Julida: Julidae) in Australia. *Bull. Ent. Res.*, **79**, 381–391.

Baker, G. H. (1985). Parasites of the millipede *Ommatoiulus moreleti* (Lucas) (Diplopoda: Iulidae) in Portugal, and their potential as biological control agents in Australia. *Aust. J. Zool.*, **33**, 23–32.

Ball, S. G. & McLean, I. F. G. (1986). Preliminary Atlas: Sciomyzidae. *Recording Scheme Newsletter*, No. 2.

Barker, G. M. (ed.) (2004). *Natural Enemies of Terrestrial Molluscs*. Wallingford: CABI Publishing.

Barker, G. M., Knutson, L., Vala, J-C., Coupland, J. B. & Barnes, J. K. (2004). Overview of the biology of marsh flies (Diptera: Sciomyzidae), with special reference to predators and parasitoids of terrestrial gastropods. In *Natural Enemies of Terrestrial Molluscs*, ed. G. M. Barker. Wallingford: CABI Publishing, pp. 159–225.

Barnes, J. K. (1976). Effect of temperature on development, survival, oviposition, and diapause in laboratory populations of *Sepedon fuscipennis* (Diptera: Sciomyzidae). *Environ. Ent.*, **5** (6), 1089–1098.

Barnes, J. K. (1979a). Revision of the New Zealand genus *Neolimnia* (Diptera: Sciomyzidae). *N. Z. J. Zool.*, **6**, 241–265.

Barnes, J. K. (1979b). Biology of the New Zealand genus *Neolimnia* (Diptera: Sciomyzidae). *N. Z. J. Zool.*, **6**, 561–576.

Barnes, J. K. (1979c). Bionomics of the New Zealand genera *Neolimnia* and *Eulimnia* (Diptera: Sciomyzidae). *J. N. Y. Ent. Soc.*, **86** (4), 277–278.

Barnes, J. K. (1979d). The taxonomic position of the New Zealand genus *Prosochaeta* Malloch (Diptera: Sciomyzidae). *Proc. Ent. Soc. Wash.*, **81** (2), 285–297.

Barnes, J. K. (1980a). Taxonomy of the New Zealand genus *Eulimnia*, and biology and immature stages of *E. philpotti* (Diptera: Sciomyzidae). *N. Z. J. Zool.*, **7**, 91–103.

Barnes, J. K. (1980b). Biology and immature stages of *Helosciomyza subalpina* (Diptera: Helosciomyzidae), an ant-killing fly from New Zealand. *N. Z. J. Zool.*, **7**, 221–229.

Barnes, J. K. (1980c). Immature stages of *Polytocus costatus* (Diptera: Helosciomyzidae) from the Snares Islands, New Zealand. *N. Z. J. Zool.*, **7**, 231–233.

Barnes, J. K. (1981). Revision of the Helosciomyzidae (Diptera). *J. R. Soc. N. Z.*, **11** (1), 45–72.

Barnes, J. K. (1984). Biology and immature stages of *Dryomya anilis* Fallén (Diptera: Dryomyzidae). *Proc. Ent. Soc. Wash.*, **86** (1), 43–52.

Barnes, J. K. (1988). Notes on the biology and immature stages of *Poecilographa decora* (Loew) (Diptera: Sciomyzidae). *Proc. Ent. Soc. Wash.*, **90** (4), 474–479.

Barnes, J. K. (1990). Biology and immature stages of *Sciomyza varia* (Diptera: Sciomyzidae), a specialized parasitoid of snails. *Ann. Ent. Soc. Am.*, **83** (5), 925–938.

Barnes, J. K. & Knutson, L. (1989). Family Sciomyzidae. In *Catalog of the Diptera of the Australasian and Oceanian Regions*, ed. N. L. Evenhuis. *Bishop Mus. Spec. Publ. 86*. Bishop Mus. Press and E. J. Brill, pp. 566–569.

Baronio, P. (1974). Gli insetti nemici dei molluschi gasteropodi. *Boll. Ist. Ent. Univ. Studi Bologna*, **32**, 169–187.

Barraclough, D. A. (1983). The biology and immature stages of some *Sepedon* snail-killing flies in Natal (Diptera: Sciomyzidae). *Ann. Natal Mus.*, **25** (2), 293–317.

Barraclough, D. A. (1985). Two new species of *Sepedon* (*Mesosepedon*) from southern Africa (Diptera: Sciomyzidae). *Ann. Natal Mus.*, **26** (2), 489–495.

Barraclough, D. A. (2007). Proboscis labellar hooks in southern African species of *Sepedon* (Diptera: Schizophora: Sciomyzidae). *Afr. Invert.*, **48** (2), 55–64.

Barraclough, D. A. & McAlpine, D. K. (2006). Natalimyzidae, a new African family of acalyptrate flies (Diptera: Schizophora: Sciomyzoidea). *Afr. Invert.*, **47**, 117–134.

Bay, E. C., Berg, C. O., Chapman, H. C. & Legner, E. F. (1976). Biological control of medical and veterinary pests. In *Theory and Practice of Biological Control*, ed. C. B. Huffaker & P. S. Messenger. London: Academic Press, pp. 457–479.

Beaver, O. (1970). *Studies on the Ecological Relationships of Certain Sciomyzid Flies.* Ph.D. thesis, Univ. Coll. North Wales, Bangor. 240 pp.

Beaver, O. (1972a). Notes on the biology of some British sciomyzid flies (Diptera: Sciomyzidae). I. Tribe Sciomyzini. *Entomologist* (*Lond.*), **105**, 139–143.

Beaver, O. (1972b). Notes on the biology of some British sciomyzid flies (Diptera: Sciomyzidae). II. Tribe Tetanocerini. *Entomologist* (*Lond.*), **105**, 284–299.

Beaver, O. (1973). Egg laying studies on some British sciomyzid flies (Diptera: Sciomyzidae). *Hydrobiologia*, **43** (1–2), 1–12.

Beaver, O. (1974a). A new species of *Sepedon* from Thailand (Diptera: Sciomyzidae). *Proc. Ent. Soc. Wash.*, **76** (1), 86–88.

Beaver, O. (1974b). Laboratory studies on competition for food of the larvae of some British sciomyzid flies (Diptera: Sciomyzidae). I. Intra-specific competition. *Hydrobiologia*, **44** (4), 443–462.

Beaver, O. (1974c). Laboratory studies on competition for food of the larvae of some British sciomyzid flies (Diptera: Sciomyzidae). II. Interspecific competition. *Hydrobiologia*, **45** (1), 135–153.

Beaver, O. (1989). Study of effect of *Sepedon senex* W. (Sciomyzidae) larvae on snail vectors of medically important trematodes. *J. Sci. Soc. Thailand*, **15**, 171–189.

Beaver, O., Knutson, L. & Berg, C. O. (1977). Biology of snail-killing flies (*Sepedon*) from southeast Asia (Diptera: Sciomyzidae). *Proc. Ent. Soc. Wash.*, **79** (3), 326–337.

Becher, E. (1882). Zur Kenntniss der Mundtheile der Dipteren. *Denkschr. Akad. Wiss. Wien*, **45**, 123–162, pls. 1–4.

Becker, T. (1900). Beiträge zur Dipteren-Fauna Sibiriens. Nordwest-Sibirische Dipteren. *Acta Soc. Sci. Fenn. Ser. B*, **26** (9), 1–66.

Becker, T. (1903). Aegyptische Dipteren. *Mitt. Zool. Mus. Berlin*, **2** (3), 67–195.

Becker, T. (1905). Sciomyzidae. In *Katalog der Paläarktischen Dipteren.* Vol. **IV**. *Cyclorrhapha Schizophora: Holometopa*, ed. T. Becker, M. Bezzi, K. Kertész & P. Stein. Budapest: Wesselényi, pp. 53–63.

Becker, T. (1907a). Die Ergebnisse meiner dipterologischen Frühjahrsreise nach Algier und Tunis. 1906. *Zschr. Hymenopterol. Dipterol.*, **7** (5), 369–407.

Becker, T. (1907b). Zur Kenntnis der Dipteren von Central-Asien. I. Cyclorrhapha Schizophora Holometopa und Orthorrhapha Brachycera. *Ann. Mus. Zool. Acad. Imp. Sci. St. Petersburg*, **12**, 253–317, 2 pls.

Becker, T. (1908). Dipteren der Kanarischen Inseln. *Mitt. Zool. Mus. Berlin*, **4** (28), 1–180.

Becker, T. (1919). Diptères. Brachycères. In *Ministère de l'Instruction publique, Mission du Service géographique de l'Armée pour la mesure d'un Arc de Méridien équatorial en Amérique du Sud, sous le Contrôle scientifique de l'Académie des Sciences, 1899–1906.* Vol. **10**, fasc. 2. Paris: Gauthier-Villard, pp. 163–215, pls. 14–17.

Becker, T. (1922). Wissenschaftliche Ergebnisse der mit Unterstützung der Akademie der Wissenschaften in Wien aus der Erbschaft Treitl von F. Werner unternommenen zoologischen Expedition nach dem Anglo-Ägyptischen Sudan (Kordofan) 1914. VI. Diptera. *Denkschr. K. Akad. Wiss. Wien. Math-Nat. K1.*, **98**, 57–82.

Bequaert, J. (1925). The arthropod enemies of mollusks, with description of a new dipterous parasite from Brazil. *J. Parasitol.*, **11**, 201–212.

Berg, C. O. (1953). Sciomyzid larvae that feed in snails. *J. Parasitol.*, **39** (6), 630–636.

Berg, C. O. (1961). Biology of snail-killing Sciomyzidae (Diptera) of North America and Europe. *Verhandl. XI. Int. Kongr. Ent. Wien, 1960*, **1** (1), 197–202.

Berg, C. O. (1964). Snail control in trematode diseases: the possible value of sciomyzid larvae, snail-killing Diptera. *Adv. Parasitol.*, **2**, 259–309.

Berg, C. O. (1966). The evolution of parasitoid relationships in malacophagous Diptera, especially the Sciomyzidae. In *Proc. 1st Int. Cong. Parasitol.*, 21–26 Sept., 1964, Rome, pp. 606–607.

Berg, C. O. (1971a). The fly that eats the snail that spreads disease. *Smithsonian*, **2** (6), 8–17.

Berg, C. O. (1971b). To kill a disease. *Intellectual Dig.*, **Dec.**, 38–41.

Berg, C. O. (1973a). Biological control of snail-borne diseases: a review. *Exp. Parasitol.*, **33**, 318–330.

Berg, C. O. (1973b). Two partially duplicated antennae in Sciomyzidae (Diptera) of Western Europe. *Norsk Ent. Tidsskr.*, **20** (2), 231–235.

Berg, C. O., Foote, B. A., Knutson, L., Barnes, J. K. Arnold, S. L. & Valley, K. (1982). Adaptive differences in phenology in sciomyzid flies. In *Recent Advances in Dipteran Systematics: Commemorative Volume in Honor of Curtis W. Sabrosky*, ed. W. N. Mathis & F. C. Thompson. *Mem. Ent. Soc. Wash.*, **10**, pp. 15–36.

Berg, C. O., Foote, B. A. & Neff, S. E. (1959). Evolution of predator-prey relationships in snail-killing sciomyzid larvae (Diptera). *Bull. Am. Malacol. Union*, **25**, 10–13.

Berg, C. O. & Knutson, L. (1978). Biology and systematics of the Sciomyzidae. *Ann. Rev. Ent.*, **23**, 239–258.

Berg, C. O. & Neff, S. E. (1959). Preliminary tests of the ability of sciomyzid larvae (Diptera) to destroy snails of medical importance. *Bull. Am. Malacol. Union*, **25**, 11–13.

Berg, C. O. & Valley, K. (1985a). Nuptial feeding in *Sepedon* spp. (Diptera: Sciomyzidae). *Proc. Ent. Soc. Wash.*, **87** (3), 622–633.

Berg, C. O. & Valley, K. (1985b). Further evidence of nuptial feeding in *Sepedon* (Diptera: Sciomyzidae). *Proc. Ent. Soc. Wash.*, **87** (4), 769.

Bertrand, H. I. (1954). Tetanoceridae. In *Les Insectes Aquatiques d'Europe*. Vol. **2**. In *Encycl. Ent., Ser. A.*, 31, 477–482. Paris: Lechevalier.

Bezzi, M. (1907). Ulteriori notizie sulla ditterofauna delle caverne. *Atti Soc. Ital. Sci. Nat., Milano*, **46**, 177–187.

Bezzi, M. (1926). Le "Stupide Mosche." *Rev. Sci. Nat.*, **27**, 1–19.

Bezzi, M. (1928). Sciomyzidae. In *Diptera, Brachycera and Athericera of the Fiji Islands Based on Material in the British Museum (Natural History)*. London: British Museum, pp. 84–86.

Bhatia, M. L. & Keilin, D. (1937). On a new case of parasitism of snail (*Vertigo genesii* Gredl.) by a dipterous larva. *Parasitology*, **29**, 399–408.

Bhuangprakone, S. & Areekul, S. (1973). Biology and food habits of the snail-killing fly, *Sepedon plumbella* Wiedemann (Sciomyzidae: Diptera). *Southeast Asian J. Trop. Med. Pub. Hlth.*, **4** (3), 387–394.

Blades, D. C. A. & Marshall, S. A. (1994). Terrestrial arthropods of Canadian peatlands: Synopsis of pan trap collections of four southern Ontario peatlands. In *Terrestrial Arthropods of Peatlands, with Particular Reference to Canada*, ed. A. T. Finnamore & S. A. Marshall. Symposium Volume on Peatland Arthropods. *Mem. Ent. Soc. Canada*, **169**, 221–284.

Blair, K. G. (1933). *Lucina fasciata* Mg. bred from *Helix pisana*. *Ent. Mon. Mag.*, **69**, 102.

Boray, J. C. (1964). Studies on the ecology of *Lymnaea tomentosa*. I. History, geographical distribution, and environment. *Austr. J. Zool.*, **12**, 217–230.

Boray, J. C. (1969). Experimental fascioliasis in Australia. *Adv. Parasitol.*, **7**, 96–210.

Boyes, J. W., Knutson, L. V., Jan, K. Y. & Berg, C. O. (1969). Cytotaxonomic studies of Sciomyzidae (Diptera: Acalyptratae). *Trans. Am. Microscop. Soc.*, **88** (3), 331–356.

Boyes, J. W., Knutson, L. V. & van Brink, J. M. (1972). Further cytotaxonomic studies of Sciomyzidae, with description of a new species, *Dichetophora boyesi* Steyskal (Diptera: Acalyptratae). *Genetica (The Hague)*, **43**, 334–365.

Bratt, A. D., Knutson, L. V., Foote, B. A. & Berg, C. O. (1969). Biology of *Pherbellia* (Diptera: Sciomyzidae). *N. Y. Agric. Exp. Stn. Ithaca. Mem.*, **404**, 1–247.

Brauer, F. (1880). Die Zweiflügler des Kaiserlichen Museums zu Wien. I. Bemerkungen zur Systematik der Dipteren. *Denkschr. Akad. Wiss., Wien*, **42**, 105–216, pls. 1–6.

Brauer, F. (1883). Die Zweiflügler des Kaiserlichen Museums zu Wien. III. Systematische Studien auf Grundlage der Dipteren-larven nebst einer Zusammenstellung von Beispielen aus der Literatur über dieselben und Beschreibung neuer Formen. *Denkschr. Akad. Wiss., Wien*, **47**, 1–100, 5 pls.

Bredt, A. & Mello, D. A. (1978). Nota sobre o ciclo biológico de duas espécies de dípteros da família Sciomyzidae. *Rev. Bras. Biol*, **38** (4), 767–770.

Brocher, F. (1913). *L'Aquarium de Chambre. – Introduction à l'Etude de l'Histoire Naturelle*. Paris: Librairie Payot.

Brunel, C. (1986). Etude éco-entomologique des zones humides de la Chaussée-Tirancourt (Vallée de la Somme). Similarités interstationnelles. *Acta Oecol., Oecol. Appl.*, **7** (4), 367–388.

Carles-Tolrá, M. (2001). Datos taxónomicós y ecológicós de 304 especies de dípteros acalípteros (Diptera, Acalyptrata). *Bull. Soc. Ent., Arag.*, **28**, 89–103.

Carles-Tolrá, M. (2003). Phaeomyiidae, p. 181 (by M. Carles-Tolrá); Sciomyzidae, pp. 185–186 (by M. Carles-Tolrá & M. Báez). In *Catálogo de los Diptera des España, Portugal y Andorra (Insecta)*, ed. M. Carles-Tolrá & H. Andersen. Zaragoza: Soc. Ent. Arag.

Carles-Tolrá, M. (2008). Four new acalyptrate dipterous species from Spain (Diptera: Camillidae, Carnidae and Sciomyzidae). *Heteropterus Rev. Ent.*, **8** (2), 125–130.

Carles-Tolrá, M., Garanto, O. & Checa, J. I. (1993). Nuevas citas y datos de dípteros acalípteros para España (Diptera: Acalyptrata). *Ses. Ent*, ICHN-SCL, 7 (1991), 83–89.

Čepelák, J. & Rozkošný, R. (1968). Zur Bionomie der Art *Angioneura cyrtoneurina* Zetterstedt, 1859 (Rhinophorinae, Diptera). *Acta. Zootech. Univ. Agric., Nitra*, **17**, 189–191.

Chandavimol, Y., Sucharit, S., Viraboonchai, S. & Tumarasvin, W. (1975). Biology of the snail-killing fly, *Sepedon spangleri* Beaver (Diptera: Sciomyzidae). **I**. Life history. *Southeast Asian J. Trop. Med. Pub. Hlth.*, **6** (3), 395–399.

Chandler, P. J. (1972). The distribution of snail-killing flies in Ireland. *Proc. Trans. Br. Ent. Soc.*, **5**, 1–21.

Chandler, P., Cranston, P., Disney, H. & Stubbs, A. E. (1978). Association with other animals and micro-organisms. Slugs, snails and bivalves (Mollusca). In *A Dipterists' Handbook*. ed. A. Stubbs & P. Chandler. *Amat. Ent.*, **15**, 190–192.

ChannaBasavanna, G. P. & Yano, K. (1969). Some observations on *Sepedon sauteri* Hendel (Diptera: Sciomyzidae) during the winter months in Fukuoka, Japan. *Mushi*, **42**, 181–187.

Chapman, E. G., Foote, B. A., Malukiewicz, J. & Hoeh, W. R. (2006). Parallel evolution of larval morphology and habitat in the snail-killing fly genus *Tetanocera*. *J. Evol. Biol.* **19** (5), 1459–1474.

Chock, Q. C., Davis, C. J. & Chong, M. (1961). *Sepedon macropus* (Diptera: Sciomyzidae) introduced into Hawaii as a control for the liver fluke snail, *Lymnaea ollula*. *J. Econ. Ent.*, **54** (1), 1–4.

Chong, M. (1968). Notes and exhibitions – September. *Proc. Hawaii. Ent. Soc.*, **20** (1), 15.

Ciprandi Pires, A., Marinoni, L. & Carvalho, C. J. Barros de (2008). Track analysis of the Neotropical genus *Sepedonea* Steyskal (Diptera: Sciomyzidae): a proposal based on the phylogenetic analysis of its species. *Zootaxa*, **1716**, 21–34.

Clarke, A. H., Jr. (1987). Clifford Osburn Berg 1912–1987. *Malacol. Data Net (Ecosearch Ser.)*, **1** (6), 137–140.

Cockerell, T. D. A. (1909). Fossil Diptera from Florissant, Colorado. *Bull. Am. Mus. Nat. Hist.*, **26**, 9–12.

Colless, D. H. & McAlpine, D. K. (1991). Diptera (Flies). In *The Insects of Australia: a Textbook for Students and Research Workers*. Vol. **2**, 2nd edn. Melbourne: CSIRO & Melbourne Univ. Press, pp. 717–786.

Contini, C. & Rivosecchi, L. (1984). Nuovi dati sugli Sciomyzidae della Sardegna. *Boll. Assoc. Rom. Ent.*, **38** (1983), 21–29.

Coquillett, D. W. (1901). New Diptera in the U. S. National Museum. *Proc. U.S. Natn. Mus.*, **23**, 593–618.

Coquillett, D. W. (1904). Several new Diptera from North America. *Can. Ent.*, **36**, 10–12.

Coupland, J. (1996a). The biological control of helicid snail pests in Australia: Surveys, screening and potential agents. In *Slug and Snail Pests in Agriculture*, ed. I. F. Henderson. *Br. Crop. Prot. Counc. Symp. Proc.*, **66**, pp. 255–261.

Coupland, J. B. (1996b). Influence of snail faeces and mucus on oviposition and larval behavior of *Pherbellia cinerella* (Diptera: Sciomyzidae). *J. Chem. Ecol.*, **22** (2), 183–189.

Coupland, J. & Baker, G. (1995). The potential of several species of terrestrial Sciomyzidae as biological control agents of pest helicid snails in Australia. *Crop Prot.*, **14** (7), 573–576.

Coupland, J. B., Espiau, A. & Baker, G. (1994). Seasonality, longevity, host choice, and infection efficiency of *Salticella fasciata* (Diptera: Sciomyzidae), a candidate for the biological control of pest helicid snails. *Biol. Contr.*, **4**, 32–37.

Courtney, G. W., Sinclair, B. J. & Meier, R. (2000). 1.4. Morphology and terminology of Diptera larvae. In *Contributions to a Manual of Palaearctic Diptera (with special reference to flies of economic importance)*. Vol. **3**. *Higher Brachycera*, ed. L. Papp & B. Darvas. Budapest: Science Herald, pp. 85–161.

Crampton, G. C. (1944). Suggestions for grouping the families of acalyptrate cyclorrhaphous Diptera on the basis of the male terminalia. *Proc. Ent. Soc. Wash.*, **46** (6), 152–154.

Cresson, E. T., Jr. (1914). Descriptions of new North American acalyptrate Diptera, I. *Ent. News*, **25** (10), 457–460.

Cresson, E. T., Jr. (1920). A revision of the Nearctic Sciomyzidae (Diptera, Acalyptratae). *Trans. Am. Ent. Soc.*, **46**, 27–89 + 3 pls.

Curran, C. H. (1923). New cyclorrhaphous Diptera from Canada. *Can. Ent.*, **55** (12), 271–279.

Curran, C. H. (1932). The genus *Dictya* Meigen (Tetanoceridae, Diptera). *Am. Mus. Novit.*, **517**, 1–7.

Curtis, J. (1837). *A Guide to an Arrangement of British Insects; Being a Catalogue of All the Named Species Hitherto Discovered in Great Britain and Ireland*. 2nd Edn, Greatly enlarged. London: J. Pigot & Co., vi + 294 cols.

Czerny, L. (1909). Spanische Dipteren. III. Cyclorrhapha Schizophora. Holometopa. *Verh. Zool.-Bot. Ges. Wien*, **59**, 247–290.

Darvas, B. & Fónagy, A. (2000). Postembryonic and imaginal development of Diptera. In *Contributions to a Manual of Palaearctic Diptera (with special reference to flies of economic importance)*, Vol. **1**. *General and Applied Dipterology*. ed. L. Papp & B. Darvas. Budapest: Science Herald, pp. 283–363.

Davis, C. J. (1959). Recent introductions for biological control in Hawaii. – IV. *Proc. Hawaii. Ent. Soc.*, **17** (1), 62–66.

Davis, C. J. (1960). Recent introductions for biological control in Hawaii. – V. *Proc. Hawaii. Ent. Soc.*, **17** (2), 244–248.

Davis, C. J. (1961a). Recent introductions for biological control in Hawaii. – VI. *Proc. Hawaii. Ent. Soc.*, **17** (3), 389–393.

Davis, C. J. (1961b). *Sciomyza dorsata Z. Proc. Hawaii. Ent. Soc.*, **17** (3), 329.

Davis, C. J. (1971). Recent introductions for biological control in Hawaii. – XVI. *Proc. Hawaii. Ent. Soc.*, **21** (1), 59–62.

Davis, C. J. (1972). Recent introductions for biological control in Hawaii. – XVII. *Proc. Hawaii. Ent. Soc.*, **21** (2), 187–190.

Davis, C. J. (1974). Recent introductions for biological control in Hawaii. – XVIII. *Proc. Hawaii. Ent. Soc.*, **21** (3), 355.

Davis, C. J., Chock, Q. C. & Chong, M. (1961). Introduction of the liver fluke snail predator *Sciomyza dorsata* (Sciomyzidae, Diptera) in Hawaii. *Proc. Hawaii. Ent. Soc.*, **17** (3), 395–397.

Davis, C. J. & Chong, M. (1968). Recent introductions for biological control in Hawaii – XIII. *Proc. Hawaii. Ent. Soc.*, **20** (1), 25–34.

Davis, C. J. & Chong, M. (1969). Recent introductions for biological control in Hawaii. XIV. *Proc. Hawaii. Ent. Soc.*, **20** (2), 317–322.

Davis, C. J. & Krauss, N. L. H. (1962). Recent introductions for biological control in Hawaii – VII. *Proc. Hawaii. Ent. Soc.*, **18** (1), 125–129.

Davis, C. J. & Krauss, N. L. H. (1967). Recent introductions for biological control in Hawaii – XII. *Proc. Hawaii. Ent. Soc.*, **19** (3), 375–380.

Day, L. T. (1881). Notes on Sciomyzidae with descriptions of new species. *Can. Ent.*, **13** (4), 85–89.

Denis, P. (1983). Diptères Sciomyzidae du nord de la France. Nouvel inventaire des espèces du Pas-de-Calais. *Bull. Soc. Ent. Mulhouse*, Jan.-March **1983**, 5–11.

Denis, P. & Leclercq, M. (1985). Cartes 1964–2030: Diptera Sciomyzidae. In *Atlas Provisoire des Insectes de Belgique (et des Régions Limitrophes)*, ed. J. Leclercq, C. Gaspar & C. Verstraeten. Cartographie des Invertébrés Européens, Ministère de la Région Wallonne pour l'Eau, l'Environnement et la Vie Rurale.

Desmarest, E. (1860). Table alphabétique des noms vulgaires et scientifiques de tous les sujets décrits et figurés dans cette Encyclopédie. Annelés. In *Encyclopédie d'Histoire Naturelle*, ed. J. Chenu. Paris: Maresq, 68 pp.

Didham, R. K. (1997). Diptera tree-crown assemblages in a diverse southern temperature rainforest. In *Canopy Arthropods*, ed. N. E. Store, J. Addis & R. K. Didham. London: Chapman & Hall, pp. 330–343.

Disney, R. H. L. (1964). A note on diet and habits of the larva and an ichneumonid parasitoid of the pupa of *Tetanocera ferruginea* Fall. (Diptera: Sciomyzidae). *Ent. Mon. Mag.*, **25**, 88–90.

Dowding, V. (1971). The feeding mechanism and its ecological importance in larvae of cyclorrhaphous Diptera. *Proc. XIII Int. Congr. Ent. Moscow, 1968*, **1**, 372.

Drake, C. M. (1988). Diptera from the Gwent Levels, South Wales. *Ent. Mon. Mag.*, **124**, 37–44.

Dufour, L. (1847a). Histoire des métamorphoses du *Tetanocera ferruginea*. *Compt. Rend. Acad. Sci. Paris*, **24**, 1030–1034.

Dufour, L. (1847b). Histoire des métamorphoses du *Tetanocera ferruginea*. *Ann. Soc. Ent. Fr.*, **7** (2), 67–79, pl. 3.

Dufour, L. (1851). Recherches anatomiques et physiologiques sur les Diptères, accompagnées de considérations relatives à l'histoire naturelle de ces insectes. *Mem. Math. Sav. Etrangers*, **11**, 171–360, 11 pls.

Duméril, A. M. C. (1800). Exposition d'une méthode naturelle pour la classification et l'étude des insectes, présentée à la société philomatique le 3 brumaire an 9. *J. Physiol. Chim. His. Nat. Arts*, **51**, 427–439, 5 pls.

Dyar, H. G. (1902). Illustrations of the early stages of some Diptera. *Proc. Ent. Soc. Wash.*, **5**, 56–59.

Eckblad, J. W. (1971). Weight-length regression models for three aquatic gastropod populations. *Am. Midl. Natural.*, **85**, 271–274.

Eckblad, J. W. (1973a). Experimental predation studies of malacophagous larvae of *Sepedon fuscipennis* (Diptera: Sciomyzidae) and aquatic snails. *Exp. Parasitol.*, **33** (2), 331–342.

Eckblad, J. W. (1973b). Population studies of three aquatic gastropods in an intermittent backwater. *Hydrobiologia*, **41**, 199–219.

Eckblad, J. W. (1976). Biomass and energy transfer by a specialized predator of aquatic snails. *Freshwater Biol.*, **6** (1), 19–21.

Eckblad, J. W. & Berg, C. O. (1972). Population dynamics of *Sepedon fuscipennis* (Diptera: Sciomyzidae). *Can. Ent.*, **104** (11), 1735–1742.

Eggleton, P. & Belshaw, R. (1992). Insect parasitoids: an evolutionary overview. *Phil. Trans. R. Soc. London B*, **337**, 1–20.

Eisenberg, R. M. (1966). The regulation of density in a natural population of the pond snail, *Lymnaea elodes*. *Ecology*, **47**, 889–906.

Elberg, K. J. (1965). New palaearctic genera and species of flies of the family Sciomyzidae (Diptera, Acalyptrata). *Ent. Rev.*, **44** (1), 104–109.

Elberg, K. J. (1968). Zur Fauna der Sciomyziden (Diptera) der USSR. *Akad. Nauk Estonskoi SSR Izv. Ser. Biol.*, **17** (2), 217–222.

Elberg, K. J. (1978). Sciomyzidae aus der Mongolei (Diptera). *Ann. Hist. Nat. Mus. Natn. Hung. Budapest*, **70**, 207–211.

Elberg, K. J. (1988). The extension of the distribution area of *Tetanocera amurensis* Hendel, 1909 (Diptera: Sciomyzidae) as far as the Baltic Sea. *Year-Book Estonian Natr. Soc.*, **72**, 116–121.

Elberg, K. J. & Rozkošný, R. (1978). Taxonomic and distributional notes on some Palaearctic Sciomyzidae (Diptera). *Scripta. Fac. Sci. Nat. Univ. Brunensis Biologia*, **2** (8), 47–53.

Elberg, K., Rozkošný, R. & Knutson, L. (2009). A review of the Holarctic *Sepedon fuscipennis* and *S. spinipes* groups with description of a new species (Diptera: Sciomyzidae). *Zootaxa*, **2288**, 51–60.

Elzinga, R. J. & Broce, A. B. (1986). Labellar modifications of Muscomorpha flies (Diptera). *Ann. Ent. Soc. Am.*, **79** (1), 150–209.

Enderlein, G. (1936). Ordnung: Zweiflügler, Diptera. In *Die Tierwelt Mitteleuropas. Bd. 6: Insekten 3 Teil, Lief. 2. 16*, ed. P. Brohmer, P. Ehrmann & G. Ulmer. Leipzig: Quelle & Meyer.

Enderlein, G. (1939). Zur Kenntnis der Klassifikation der Tetanoceriden (Diptera). *Veröff. Deutsch. Kolon. Mus. Bremen*, **2** (3), 201–210.

Englund, R. A., Arakaki, K., Preston, D. J., Evenhuis, N. L. & McShane, K. K. (2003). *Systematic Inventory of Rare and Alien Aquatic Species in Selected O'ahu, Maui, and Hawai'i Island Streams.* Contr. 2003–17, Hawaii Biol. Surv.

Evans, H. E. (1985). Marsh flies. In *The Pleasures of Entomology: Portraits of Insects and the People Who Study Them*. Washington, DC: Smithsonian Institution Press, pp. 125–132.

Evenhuis, N. L. (1994). Family Sciomyzidae, In *Catalogue of the Fossil Flies of the World (Insecta: Diptera)*, ed. N. L. Evenhuis. Leiden: Backhuys, pp. 430–431.

Fabricius, J. C. (1775). *Systema Entomologiae, Sistens Insectorum Classes, Ordines, Genera, Species Adiectis Synonymis, Locis, Descriptionibis, Observationibus.* Flensburgi et Lipsiae: Kortii.

Fabricius, J. C. (1781). *Species Insectorum Exhibentes Eorum Differential Specificas, Synonyma, Auctorum Loca Natalia, Metamorphosium.* Vol. **2**. Hamburgi et Kilonii: C. E. Bohnii.

Fabricius, J. C. (1794). *Entomologia Systematica Emendata et Aucta. Secundum. Classes, Ordines, Genera, Species Adjectis Synonimis, Locis, Observationibus, Kortii, descriptionibus.* Vol. **4**. Hafniae: Proft et Torch.

Fabricius, J. C. (1798). *Supplementum Entomologiae Systematicae.* Hafniae: Proft et Torch.

Fabricius, J. C. (1805). *Systema Antliatorum Secundum Ordines, Genera, Species.* Brunsvige: Carolum Reichard.

Faes, H. (1902). Myriapodes du Valais. *Rev. Suisse Zool.*, **10**, 138–140.

Falk, S. J. (1991). *A Review of the Scarce and Threatened Flies of Great Britain (Part 1). Research and Service in Nature Conservation 39.* Peterborough: Nature Conservancy Council.

Fallén, C. F. (1820a). *Opomyzides Sveciae.* Lundae: Litteris Berlingianis.

Fallén, C. F. (1820b). *Sciomyzides Sveciae.* Lundae: Litteris Berlingianis.

Fallén, C. F. (1826). *Supplementum Dipterorum Sveciae.* Lundae: Litteris Berlingianis.

Ferguson, F. F. [1977? undated]. *The Role of Biological Agents in the Control of Schistosome-bearing Snails.* U.S. Dept. Hlth. Educ. Welfare, Publ. Hlth. Serv., Cent. Dis. Contr., Bur. Labs., Atlanta, GA. N° 4H2129577.

Ferrar, P. (1987). *A Guide to the Breeding Habits and Immature Stages of Diptera. Cyclorrhapha.* (Part 1: text): Sciomyzidae, pp. 329–340; (Part 2: figures): Sciomyzidae, pp. 815–827. In *Entomonograph*, Vol. **8**, ed. L. Lyneborg. Leiden: E. J. Brill/Scandinavian Science Press.

Fisher, T. W. & Orth, R. E. (1964). Biology and immature stages of *Antichaeta testacea* Melander (Diptera: Sciomyzidae). *Hilgardia*, **36** (1), 1–29.

Fisher, T. W. & Orth, R. E. (1966). A new species of *Pteromicra* from western North America and resurrection of *Pteromicra pleuralis* (Cresson) (Diptera: Sciomyzidae). *Pan-Pac. Ent.*, **42** (4), 307–318.

Fisher, T. W. & Orth, R. E. (1969a). Two new species of *Sepedon* separated from *S. armipes* Loew in western North America (Diptera: Sciomyzidae). *Pan-Pac. Ent.*, **45** (2), 152–164.

Fisher, T. W. & Orth, R. E. (1969b). A new *Dictya* in California, with biological notes (Diptera: Sciomyzidae). *Pan-Pac. Ent.*, **45** (3), 222–228.

Fisher, T. W. & Orth, R. E. (1971a). A synopsis of the Nearctic species of *Antichaeta* Haliday with one new species (Diptera: Sciomyzidae). *Pan-Pac. Ent.*, **47** (1), 32–43.

Fisher, T. W. & Orth, R. E. (1971b). A new species of *Limnia* Robineau-Desvoidy from western North America (Diptera: Sciomyzidae). *Ent. News*, **82** (7), 169–175.

Fisher, T. W. & Orth, R. E. (1972a). Resurrection of *Sepedon pacifica* Cresson and redescription of *Sepedon praemiosa* Giglio-Tos with biological notes (Diptera: Sciomyzidae). *Pan-Pac. Ent.*, **48** (1), 8–20.

Fisher, T. W. & Orth, R. E. (1972b). Synopsis of *Hoplodictya* Cresson with one new species (Diptera: Sciomyzidae). *Ent. News*, **83** (7), 173–190.

Fisher, T. W. & Orth, R. E. (1974). A new species of *Sepedon* Latreille from Oregon (Diptera: Sciomyzidae). *Pan-Pac. Ent.*, **50** (3), 291–297.

Fisher, T. W. & Orth, R. E. (1975). Sciomyzidae of Oregon (Diptera). *Pan-Pac. Ent.*, **51** (3), 217–235.

Fisher, T. W. & Orth, R. E. (1983). The marsh flies of California (Diptera: Sciomyzidae). *Univ. Calif. Press. Bull. Calif. Insect Surv.*, **24**.

Fontana, P. G. (1972). Larvae of *Dichetophora biroi* (Kertész, 1901) (Diptera: Sciomyzidae) feeding on *Lymnaea tomentosa* (Pfeiffer, 1855) snails infected with *Fasciola hepatica* L. *Parasitology*, **64** (1), 89–93.

Foote, B. A. (1959a). Biology and life history of the snail-killing flies belonging to the genus *Sciomyza* Fallén (Diptera: Sciomyzidae). *Ann. Ent. Soc. Am.*, **52** (1), 31–43.

Foote, B. A. (1959b). A new species of *Pteromicra* reared from land snails, with a key to the Nearctic species of the genus. *Proc. Ent. Soc. Wash.*, **61** (1), 14–16.

Foote, B. A. (1961a). *Biology and Immature Stages of the Snail-killing Flies Belonging to the Genus Tetanocera (Diptera: Sciomyzidae)*. Ph.D. thesis, Cornell Univ., Ithaca, NY, USA. 190 pp. Order No. 62–105. Univ. Microfilms, Ann Arbor, MI. (*Diss. Abstr.* **22**, 3302–3303).

Foote, B. A. (1961b). The marsh flies of Idaho and adjoining areas (Diptera: Sciomyzidae). *Am. Midl. Natural.*, **65** (1), 144–167.

Foote, B. A. (1961c). A new species of *Antichaeta* Haliday, with notes on other species of the genus (Diptera: Sciomyzidae). *Proc. Ent. Soc. Wash.*, **63** (3), 161–164.

Foote, B. A. (1971). Biology of *Hedria mixta* (Diptera: Sciomyzidae). *Ann. Ent. Soc. Am.*, **64** (4), 931–941.

Foote, B. A. (1973). Biology of *Pherbellia prefixa* (Diptera: Sciomyzidae), a parasitoid-predator of the operculate snail *Valvata sincera*. *Proc. Ent. Soc. Wash.*,**75** (2), 141–149.

Foote, B. A. (1976). Biology and larval feeding habits of three species of *Renocera* (Diptera: Sciomyzidae) that prey on fingernail clams (Mollusca: Sphaeriidae). *Ann. Ent. Soc. Am.*, **69** (1), 121–133.

Foote, B. A. (1977). Biology of *Oidematops ferrugineus* (Diptera: Sciomyzidae), a parasitoid enemy of the land snail *Stenotrema hirsutum* (Mollusca: Polygyridae). *Proc. Ent. Soc. Wash.*, **79** (4), 609–619.

Foote, B. A. (1991). Sciomyzidae (Sciomyzoidea). In *Immature Insects*, ed. F. W. Stehr. Vol. **2**. Dubuque, IO: Kendall/Hunt, pp. 828–832.

Foote, B. A. (1996a). Biology and immature stages of snail-killing flies belonging to the genus *Tetanocera* (Insecta: Diptera: Sciomyzidae). I. Introduction and life histories of predators of shoreline snails. *Ann. Carneg. Mus.*, **65** (1), 1–12.

Foote, B. A. (1996b). Biology and immature stages of snail-killing flies belonging to the genus *Tetanocera* (Insecta: Diptera: Sciomyzidae). II. Life histories of predators of snails of the family Succineidae. *Ann. Carneg. Mus.*, **65** (2), 153–166.

Foote, B. A. (1999). Biology and immature stages of snail-killing flies belonging to the genus *Tetanocera* (Insecta: Diptera: Sciomyzidae). III. Life histories of predators of aquatic snails. *Ann. Carneg. Mus.*, **68** (3), 151–174.

Foote, B. A. (2004). Acalyptrate Diptera associated with stands of *Carex lacustris* and *C. stricta* (Cyperaceae) in northeastern Ohio. *Proc. Ent. Soc. Wash.*, **106** (1), 166–175.

Foote, B. A. (2007). Biology of *Pherbellia inflexa* (Diptera: Sciomyzidae), a predator of land snails belonging to the genus *Zonitoides* (Gastropoda: Zonitidae). *Ent. News*, **118** (2), 193–198.

Foote, B. A. (2008). Biology and immature stages of snail-killing flies belonging to the genus *Tetanocera* (Diptera: Sciomyzidae). IV. Life histories of predators of land snails and slugs. *Ann. Carneg. Mus.*, **77** (2), 301–312.

Foote, B. A. & Keiper, J. B. (2004). The snail-killing flies of Ohio (Insecta: Diptera: Sciomyzidae). *Kirtlandia*, **54**, 43–90.

Foote, B. A. & Knutson, L. V. (1970). Clam-killing fly larvae. *Nature (London)*, **226** (5244), 466.

Foote, B. A., Knutson, L. V. & Keiper, J. B. (1999). The snail-killing flies of Alaska (Diptera: Sciomyzidae). *Insecta Mundi*, **13** (1–2), 45–72.

Foote, B. A., Neff, S. E. & Berg, C. O. (1960). Biology and immature stages of *Atrichomelina pubera* (Diptera: Sciomyzidae). *Ann. Ent. Soc. Am.*, **53** (2), 192–199.

Förster, B. (1891). *Die Insekten der "Plattigen Steinmergels" von Brunstatt.* Abh. Geol. Specialk. Els.-Lothr. Vol. **3**. Strassburg: Strassburger Druckerei und Verlagsanstalt, pp. 333–594, pls. 11–16.

Freidberg, A., Knutson, L. & Abercrombie, J. (1991). A revision of *Sepedonea*, a neotropical genus of snail-killing flies (Diptera: Sciomyzidae). *Smithson. Cont. Zool.*, **506**.

Frey, R. (1921). Studien über den Bau des Mundes der niederen Diptera Schizophora. nebst Bemerkungen über die Systematik dieser Dipterengruppe. *Acta Soc. Fauna Flora Fenn.*, **48** (3), 1–246, 10 pls.

Frey, R. (1924). Die nordpälaarktischen *Tetanocera*-Arten (Diptera: Sciomyzidae). *Not. Ent.*, **4**, 47–53.

Frey, R. (1930). Neue Diptera brachycera aus Finnland und angrenzenden Ländern. *Not. Ent.*, **10**, 82–94.

Frey, R. (1936). Die Dipterenfauna der Kanarischen Inseln und ihre Probleme. *Comm. Biol. Soc. Sci. Fenn.*, **6** (1), 1–237.

Frömming, E. (1956). *Biologie der mitteleuropäischen Süsswasserschnecken*. Berlin: Duncker & Humblot.

Gaponov, S. P., Khitzovan L. N. & Sergeev A. S. (2006). Fine morphology of Sciomyzidae (Diptera) female head sensory system. In *6th Int. Congr. Dipterol., 23–28 Sept., 2006*, Fukuoka, Japan, ed. M. Suwa. Abstracts Vol., pp. 84–85.

Gasc, C., Vala, J-C. & Reidenbach, J. M. (1984a). Microstructures cuticulaires et récepteurs sensoriels des larves de *Sepedon sphegea* (F.) (Diptera: Sciomyzidae). *Int. J. Insect Morphol. Embryol.*, **13** (4), 275–281.

Gasc, C., Vala, J-C. & Reidenbach, J. M. (1984b). Etude comparative au microscope électronique a balayage des structures chorioniques d'oeufs de cinq espèces de Sciomyzidae à larves terrestres et aquatiques (Diptera). *Ann. Soc. Ent. Fr. (N.S.)*, **20** (2), 163–170.

Gbedjissi, L. G. (2003). *Relations mollusques – Diptères Sciomyzidae. Implications dans la Lutte Contre les Distomatoses au Bénin*. Ph.D. thesis, Université d'Avignon et des Pays du Vaucluse, France. 153 pp.

Gbedjissi, L. G., Vala, J-C., Knutson, L. & Dossou, C. (2003). Predation by larvae of *Sepedon ruficeps* (Diptera: Sciomyzidae) and population dynamics of the adult flies and their freshwater prey. *Rev. Suisse Zool.*, **110** (4), 817–832.

Geckler, R. P. (1971). Laboratory studies of predation of snails by larvae of the marsh fly *Sepedon tenuicornis* (Diptera: Sciomyzidae). *Can. Ent.*, **103** (5), 638–649.

Gercke, G. (1876). Ueber die metamorphose von *Sepedon sphegeus* und *spinipes*. *Verh. Ver. Nat. Unterhalt Hamburg*, **2**, 145–149.

Ghamizi, M. (1985). *Predation des Mollusques par les Larves de* Sepedon sphegea *Fab. (Diptera: Sciomyzidae) Aspects de la Dynamique Proie-prédateur*. Ph.D. thesis, Acad. Montpellier, Univ. Sci. Tech. Languedoc, Montpellier, France.

Ghorpadé, K., Marinoni, L. & Knutson, L. (1999). *Steyskalina picta*, new genus and species of Tetanocerini (Diptera, Sciomyzidae) from the Oriental Region. *Revta. Bras. Zool.*, **16** (3), 835–839.

Giglio-Tos, E. (1893). Diagnosi di nuovi generi e di nuove specie di Ditteri, IX. *Boll. Mus. Zool. Anat. Com. Reale Univ. Torino*, **8** (158), 1–14.

Glick, P. A. (1939). *The Distribution of Insects, Spiders, and Mites in the Air*. U.S. Dept. Agric. Tech. Bull. 673.

Glover, T. (1874). *Manuscript Notes from My Journal, or Illustrations of Insects, Native and Foreign. Diptera or Two-winged Flies*. Washington, DC: Gedney.

Godan, D. (1979). *Arthropoden, Schadschnecken und ihre Bekämpfung*. Stuttgart: Eugen Ulmer.

Gomez Pallerola, J. E. (1986). Nuevos insectos fósiles de las calizas litográficas del Cretácio inferior del Montsec (Lérida). *Bio. Geol. Min.*, **97**, 27–46.

Gormally, M. J. (1985a). *Ilione albiseta* – potential biological control agent of liver fluke? *J. Irish Grassl. Anim. Prod. Assoc.*, **19**, 82.

Gormally, M. J. (1985b). The effect of temperature on the duration of the egg stage of certain sciomyzid flies which predate *Lymnaea truncatula*. *J. Therm. Biol.*, **10** (4), 199–203.

Gormally, M. J. (1987a). Effect of temperature on the duration of larval and pupal stages of two species of sciomyzid flies, predators of the snail *Lymnaea truncatula*. *Ent. Exp. Appl.*, **43**, 95–100.

Gormally, M. J. (1987b). Notes on the larval habitat of *Ilione albiseta* with records of other Sciomyzidae (Diptera) collected at the same localities in Co. Sligo. *Ir. Nat. J.*, **22** (6), 217–264.

Gormally, M. J. (1988a). Studies on the oviposition and longevity of *Ilione albiseta* (Diptera: Sciomyzidae) – potential biological control agent of liver fluke. *Entomophaga*, **33** (4), 387–395.

Gormally, M. J. (1988b). Temperature and the biology and predation of *Ilione albiseta* (Diptera: Sciomyzidae) – potential biological control agent of liver fluke. *Hydrobiologia*, **166**, 239–246.

Gorokhov, V. V. (1971). On the possibility of using insects in the control of molluscs. *Tr. Vses. Inst. Gelmintol.*, **18**, 79–87.

Gouin, F. (1949). Recherches sur la morphologie de l'appareil buccal des Diptères. *Mem. Mus. nat. Hist. natl. Ser. A. Zool.*, **28** (4), 167–269.

Greathead, D. J. (1981). Arthropod natural enemies of bilharzia snails and the possibilities for biological control. *Biocontr. News Info., CIBC*, **2** (3), 197–202.

Green, T. (1977). A man's obsession reveals the riches of a hidden world. *Smithsonian*, **8** (8), 81–87.

Griffiths, G. C. D. (1972). *The Phylogenetic Classification of Diptera Cyclorrhapha with Special Reference to the Structure of the Male Postabdomen*. Ser. Ent. **8**. The Hague: Dr. Junk.

Griffiths, G. C. D. (1990). Book Review. Manual of Nearctic Diptera. Vol. **3**. *Quaest. Ent.*, **26**, 117–130.

Grist, D. H. & Lever, R. J. A. W. (1969). *Pests of Rice*. London: Longmans, Green & Co.

Haab, C. (1984). *Etude Expérimentale de la Biologie de* Sepedon sphegea *(Fabricius, 1775) et Aspects de sa Prédation Larvaire (Diptera: Sciomyzidae)*. Ph.D. thesis, Acad. Montpellier, Univ. Sci. Tech. Languedoc, Montpellier, France.

Hackman, W. & Väisänen, R. (1985). The evolution and phylogenetic significance of the costal chaetotaxy in the Diptera. *Ann. Zool. Fenn.*, **22**, 169–203.

Hågvar, S. & Greve, L. (2003). Winter active flies (Diptera, Brachycera) recorded on snow – a long-term study in south Norway. *Stud. dipterol.*, **10** (2), 401–421.

Hairston, N. G., Wurzinger, K. H. & Burch, J. B. (1975). Non-chemical methods of snail control. *WHO-VBC/* 75.573. *WHO-SCHISTO*-75.40.

Haliday, A. H. (1839) [1838]. XXII. New British insects indicated in Mr. Curtis's guide. *Ann. Nat. Hist.*, **2** (9), 183–190.

Harris, M. (1780) [1776]. *An Exposition of English Insects with Curious Observations and Remarks, Wherein Each Insect is Particularly Described; its Parts and Properties Considered; the Different Sexes Distinguished, and the Natural History Faithfully Related*. Decads III–V. London: Robson.

Harrison, R. A. (1959). Acalypterate Diptera of New Zealand. *N. Z. Dept. Sci. Indust. Res. Bull.*, **128**.

Haslett, J. R. (2001). Biodiversity and conservation of Diptera in heterogeneous land mosaics: A fly's eye view. *J. Insect Cons.*, **5**, 71–75.

Hendel, F. (1900a). Über eine neue *Sciomyza* (Diptera) aus dem österreichischen Litorale. *Wein. Ent. Ztg.*, **19** (4–5), 89–91.

Hendel, F. (1900b). Untersuchungen über die europäischen Arten der Gattung *Tetanocera* im Sinne Schiner's. Eine dipterologische Studie. *Verh. Zool.-Bot. Ges. Wein*, **50**, 319–358.

Hendel, F. (1901). Zur Kenntnis der Tetanocerinen (Dipt.). *Természt. Füzetek.*, **24**, 138–142.

Hendel, F. (1902). Revision der paläarktischen Sciomyziden (Dipteren-Subfamilie). *Abh. Zool.-Bot. Ges. Wien*, **2** (1), 1–94.

Hendel, F. (1903). Synopsis der paläarktischen *Elgiva*-Arten. *Ztg. System. Hym. Dipt.*, **3**, 213–215.

Hendel, F. (1909). Drei neue holometope Musciden aus Asien. *Wien. Ent. Ztg.*, **28** (3), 85–86.

Hendel, F. (1910). Über die Nomenklatur der Acalyptratengattungen nach Th. Beckers Katalog der paläarktischen Dipteren, Bd. 4. *Wien. Ent. Ztg.*, **29**, 307–313.

Hendel, F. (1911). Über die *Sepedon*-Arten der aethiopischen und indomalayischen Region. *Ann. Mus. Natn. Hung.*, **9**, 266–277.

Hendel, F. (1912). Neue Muscidae acalypteratae. *Wien. Ent. Ztg.*, **31** (1), 1–20.

Hendel, F. (1913). [H. Sauter's Formosa-Ausbeute.] Acalyptrate Musciden (Dipt.). II. *Suppl. Ent.*, **2**, 77–112.

Hendel, F. (1916). Beiträge zur systematik der Acalyptraten Musciden (Dipt.). *Ent. Mitt.*, **5**, 294–299.

Hendel, F. (1922). Die paläarktischen Muscidae acalyptratae Girsch. = Haplostomata Frey nach ihren Familien und Gattungen. – I. Die Familien. *Konowia*, **1** (4/5 & 6), 145–160.

Hendel, F. (1924) [1923]. Die paläarktischen Muscidae acalyptratae Girschn. = Haplostomata Frey nach ihren Familien und Gattungen. II. Die Gattungen. *Konowia*, **2** (5/6), 203–215.

Hendel, F. (1932). Die Ausbeute der deutschen Chaco-Expedition 1925/26. – Diptera. (Fortsetzung). XXX–XXXVI. Sciomyzidae, Lauxaniidae, Tanypezidae, Lonchaeidae, Tylidae, Drosophilidae, Milichiidae. *Konowia*, **11** (1), 98–114, (2), 115–145.

Hendel, F. (1933). Neue acalyptrate Musciden aus der paläarktischen Region (Diptera). *Deutsch. Ent. Ztsch.*, **1**, 39–40.

Hendel, F. (1934). Schwedisch-chinesische wissenschaftliche Expedition nach den nordwestlichen Provinzen Chinas, unter Leitung von Dr. Sven Hedin und Prof. Sü Ping-Chang. Insekten gesammelt vom schwedischen Arzt der Expedition Dr. David Hummel. 1927–1930. 13. Diptera. 5. Muscaria holometopa. *Ark. Zool.*, **25A** (21), 1–18.

Hendel, F. (1936–1937). Ordnung der Pterygogena (Dreissigste Ordnung der Insecta): Diptera = Fliegen. In *Handbuch der Zoologie. Band 4, 2 Hälfte, 2 Teil, Insecta 3, Lieferung 11*, ed. W. Kükenthal & T. Krumbach. Berlin: De Gruyter & Co., pp. 1729–1998.

Hennig, W. (1952a). Bemerkenswerte neue Acalpytraten in der Sammlung des Deutschen Entomologischen Institutes (Diptera: Acalyptrata). *Beitr. Ent.*, **2** (6), 604–618.

Hennig, W. (1952b). Family Sciomyzidae. In *Die Larvenformen der Dipteren.Teil 3*. Berlin: Akad. Verlag, pp. 241–243.

Hennig, W. (1958). Die Familien der Diptera Schizophora und ihre phylogenetischen Verwandtschaftsbeziehungen. *Beitr. Ent.*, **8**, 505–688.

Hennig, W. (1965). Die Acalyptratae des baltischen Bernsteins und ihre Bedeutung für die Erforschung der phylogenetischen Entwicklung dieser Dipteren-Gruppe. *Stuttg. Beitr. Naturkd.*, **145**, 1–212.

Hennig, W. (1969). Neue Übersicht über die aus dem Baltischen Bernstein bekannten Acalyptratae. (Diptera: Cyclorrhapha). *Stuttg. Beitr. Naturkd.*, **209**, 1–42.

Hennig, W. (1971). Neue Untersuchungen über die Familien der Diptera Schizophora (Diptera: Cyclorrhapha). *Stuttg. Beitr. Naturkd.*, **226**, 1–76.

Hennig, W. (1973). 31. Diptera (Zweiflügler). In *Handbuch der Zoologie. Bd. IV: Arthropoda – 2. Halfte: Insecta. Zweite Auflage.* M. Beier (Ed.-in-Chief). Berlin: De Gruyter, 254 pp.

Hinton, H. E. (1981). *Biology of Insect Eggs.* Vol. **2**. Oxford: Pergamon Press, pp. 475–778.

Hippa, H. (1986). Morphology and taxonomic value of the female external genitalia of Syrphidae and some other Diptera by new methodology. *Acta Zool. Fenn.*, **23** (3), 307–320.

Hope Cawdery, M. J. (1981). Changing temperatures and prediction models for the liver fluke (*Fasciola hepatica*). *J. Therm. Biol.*, **6**, 403–408.

Hope Cawdery, M. J. & Lindsay, W. (1977). Observations on the decline of the snail (*Lymnaea truncatula*, Linn.) and the liver fluke (*Fasciola hepatica*, Linn.) on reclaimed western blanket peat and its possible relationship to predation by *Hydromya dorsalis* (Fab). *Proc. R. Ir. Acad. Sem. Biol. Cont.*, 161–169.

Horsáková, J. (2003). Biology and immature stages of the clam-killing fly, *Renocera pallida* (Diptera: Sciomyzidae). *Eur. J. Ent.*, **100**, 143–151.

Hutton, F. W. (1901). Synopsis of the Diptera brachycera of New Zealand. *Trans. Proc. N. Z. Inst.*, **33**, 1–95.

Hutton, F. W. (1904). Two new flies. *Trans. Proc. N. Z. Inst.*, **36**, 153–154.

Johannsen, O. A. (1935). Aquatic Diptera. Part II. Orthorrhapha-Brachycera and Cyclorrhapha. *Mem. Cornell Univ. Agr. Exp. Sta.*, **177**, 1–62.

Johnson, C. W. (1913). The dipteran fauna of Bermuda. *Ann. Ent. Soc. Am.*, **6**, 443–452.

Johnson, A. W. & Hays, K. L. (1973). Some predators of immature Tabanidae (Diptera) in Alabama. *Environ. Ent.*, **2** (6), 1116–1117.

Jordan, P., Christie, J. D. & Unrah, G. O. (1980). Schistosomiasis transmission with particuliar reference to possible ecological and biological methods of control; a review. *Acta Trop.*, **37**, 95–135.

Juliano, S. A. (1981). *Trichogramma* spp. (Hymenoptera: Trichogrammatidae) as egg parasitoids of *Sepedon fuscipennis* (Diptera: Sciomyzidae) and other aquatic Diptera. *Can. Ent.*, **113**, 271–279.

Juliano, S. A. (1982). Influence of host age on host acceptability and suitability for a species of *Trichogramma* (Hymenoptera: Trichogrammatidae) attacking aquatic Diptera. *Can. Ent.*, **114**, 713–720.

Kaczynski, V. W., Zuska, J. & Berg, C. O. (1969). Taxonomy, immature stages, and bionomics of the South American genera *Perilimnia* and *Shannonia* (Diptera: Sciomyzidae). *Ann. Ent. Soc. Am.*, **62** (3), 572–592.

Kaltenbach, J. H. (1873). *Die Pflanzenfeinde aus der Klasse der Insekten.* Stuttgart: J. Hoffman.

Karl, O. (1921). Zwei neue Dipteren aus dem Kösliner Bezirk. *Stett. Ent. Ztg.*, **82**, 125–126.

Kassebeer, C. F. (1999a). Neue Nachweise von Netzfliegen (Diptera, Sciomyzidae) aus Marokko, mit einer Übersicht der Fauna des Maghreb. *Dipteron*, **2** (1), 1–10.

Kassebeer, C. F. (1999b). Die Netzfliegen (Diptera, Sciomyzidae) des Trentmoores bei Plön. *Dipteron*, **2** (8), 163–172.

Kassebeer, C. F. (2000). Die Hornfliegen (Diptera, Sciomyzidae & Phaeomyiidae) von Schleswig-Holstein und Hamburg. *Dipteron*, **3** (2), 179–216.

Kassebeer, C. F. (2001a). Die Hornfliegen (Diptera, Sciomyzidae & Phaeomyiidae) im Grossraum Berlin. *Dipteron*, **4** (1), 65–108.

Kassebeer, C. F. (2001b). Neue Nachweise von Hornfliegen (Diptera, Sciomyzidae) aus Island. *Dipteron*, **4** (2), 209–212.

Keilin, D. (1919). On the life history and larval anatomy of *Melinda cognata* Meigen parasitic in the snail *Helicella* (*Heliomanes*) *virgata* Da Costa, with an account of the other Diptera living upon mollusks. *Parasitology*, **11**, 430–455.

Keilin, D. (1921). Supplementary account of the dipterous larvae feeding upon mollusks. *Parasitology*, **13**, 180–183.

Keiper, J. B., Walton, W. E. & Foote, B. A. (2002). Biology and ecology of higher Diptera from freshwater wetlands. *Ann. Rev. Ent.*, **47**, 207–232.

Kertész, K. (1899). Verzeichniss einiger, von L. Biró in Neu-Guinea und am malayischen Archipel gesammelten Dipteren. *Termész. Füzet.*, **22**, 173–195.

Kertész, K. (1901). Neue und bekannte Dipteren in der Sammlung des ungarischen National-Museums. *Termész. Füzet.*, **24**, 403–432.

King, R. S. & Brazner, J. C. (1999). Coastal wetland insect communities along a trophic gradient in Green Bay, Lake Michigan. *Wetlands*, **19**, 426–437.

Knutson, L. V. (1965). Ecological notes on Tabanidae, Rhagionidae, and Xylophagidae in Europe (Diptera). *Proc. Ent. Soc. Wash.*, **67** (1), 59–60.

Knutson, L. V. (1966). Biology and immature stages of malacophagous flies: *Antichaeta analis, A. atriseta,*

A. brevipennis, and *A. obliviosa* (Diptera: Sciomyzidae). *Trans. Am. Ent. Soc.*, **92** (1), 67–101.

Knutson, L. V. (1968). A new genus and species of Sciomyzidae from Tanzania with a key to the genera of the Ethiopian region and distributional notes (Diptera: Acalyptratae). *J. Ent. Soc. S. Africa*, **31** (1), 175–180.

Knutson, L. V. (1970a). Biology and immature stages of *Tetanura pallidiventris*, a parasitoid of terrestrial snails (Diptera: Sciomyzidae). *Ent. Scand.*, **1** (2), 81–89.

Knutson, L. V. (1970b). Biology of snail-killing flies in Sweden (Diptera: Sciomyzidae). *Ent. Scand.*, **1** (4), 307–314.

Knutson, L. V. (1972). Description of the female of *Pherbecta limenitis* Steyskal (Diptera: Sciomyzidae), with notes on biology, immature stages, and distribution. *Ent. News*, **83** (1), 15–21.

Knutson, L. V. (1973). Biology and immature stages of *Coremacera marginata* F., a predator of terrestrial snails (Diptera: Sciomyzidae). *Ent. Scand.*, **4** (2), 123–133.

Knutson, L. (1976). Sciomyzidae. In *A Checklist of British Insects*. 2nd edn. *Handbooks for the Identification of British Insects*. Vol. **II**, Part 5, ed. G. S. Kloet & W. D. Hincks. London: R. Ent. Soc., pp. 76–78.

Knutson, L. V. (1977). Superfamily Sciomyzoidea, Family Sciomyzidae. In *A Catalog of the Diptera of the Oriental Region*. Vol. 3, *Suborder Cyclorrhapha*, ed. M. D. Delfinado & D. E. Hardy. Honolulu: Hawaii Univ. Press, pp. 168–172.

Knutson L. V. (1978). Sciomyzidae. In *Limnofauna Europaea*, 2nd edn, ed. J. Illies. Stuttgart: Gustav Fischer Verlag, pp. 485–488.

Knutson, L. V. (1980). 51. Family Sciomyzidae. In *Catalogue of the Diptera of the Afrotropical Region*, ed. R. W. Crosskey. London: British Museum (Nat. Hist.), pp. 597–600.

Knutson, L. (1985). *Pherbellia kugleri*, a remarkable new species from Mt. Hermon, with other new records of Sciomyzidae from Israel (Diptera: Acalyptratae). *Israel J. Ent.*, **19**, 111–117.

Knutson, L. (1987). Family Sciomyzidae. In *Manual of Nearctic Diptera*. Vol. **2**, ed. J. F. McAlpine. *Res. Br., Agric. Canada. Monogr.*, **28**, pp. 927–940.

Knutson, L. (1988). Life cycles of snail-killing flies: *Pherbellia griseicollis, Sciomyza dryomyzina, S. simplex*, and *S. testacea* (Diptera: Sciomyzidae). *Ent. Scand.*, **18**, 383–391.

Knutson, L. (2008 [2000]). Biology of two parasitoid snail-killing flies (*Sepedon trichrooscelis* and *S. hispanica ruhengeriensis*) attacking succineid snails in Africa, and proposal of a new category of larval food selection/behavior in the Sciomyzidae (Diptera). *Colemania*, **7**, 1–11.

Knutson, L. & Abercrombie, J. (1977). Biology of *Antichaeta melanosoma* (Diptera: Sciomyzidae), with notes on parasitoid Ichneumonidae (Hymenoptera). *Proc. Ent. Soc. Wash.*, **79** (1), 111–125.

Knutson, L. V. & Berg, C. O. (1963a). Biology and immature stages of a snail-killing fly, *Hydromya dorsalis* (Fabricius) (Diptera: Sciomyzidae). *Proc. R. Ent. Soc. London. Ser. A: Gen. Ent.*, **38** (4–6), 45–58.

Knutson, L. V. & Berg, C. O. (1963b). *Phaenopria popei* (Hymenoptera: Diapriidae) reared from puparia of sciomyzid flies. *Can. Ent.*, **95**, 724–726.

Knutson, L. V. & Berg, C. O. (1964). Biology and immature stages of snail-killing flies: the genus *Elgiva* (Diptera: Sciomyzidae). *Ann. Ent. Soc. Am.*, **57** (2), 173–192.

Knutson, L. V. & Berg, C. O. (1966). Parasitoid development in snail-killing sciomyzid flies. *Trans. Am. Microscop. Soc.*, **85** (1), 164–165.

Knutson, L. V. & Berg, C. O. (1967). Biology and immature stages of malacophagous Diptera of the genus *Knutsonia* Verbeke (Sciomyzidae). *Bull. Inst. r. Sci. nat. Belg.*, **43** (7), 1–60.

Knutson, L. V., Berg, C. O., Edwards, L. J., Bratt, A. D. & Foote, B. A. (1967a). Calcareous septa formed in snail shells by larvae of snail-killing flies. *Science*, **156** (3774), 522–523.

Knutson, L. & Bredt, A. (1976). Two new species of snail-killing flies from west-central Brazil (Diptera: Sciomyzidae). *Pap. Avul. Zool.*, **30** (7), 113–118.

Knutson, L. & Carvalho, C. J. Barros de. (1989). Seasonal distribution of a relatively rare and a relatively common species of *Thecomyia* at Belém, Pará, Brazil (Diptera: Sciomyzidae). *Mem. Inst. Oswaldo Cruz, Rio de Janeiro*, **89** (Suppl. IV), 287–289.

Knutson, L. & Freidberg, A. (1983). The Sciomyzidae (Diptera: Acalyptratae) of Israel, with notes on the distribution of the Near East species. *Ent. Scand.*, **14**, 371–386.

Knutson, L. & Ghorpadé, K. (2004). Family Sciomyzidae. In *Freshwater Invertebrates of the Malaysian Region*, ed. C. M. Yule & H. S. Yong. Kuala Lumpur, Malaysia: Acad. Sci. Malaysia, pp. 831–844.

Knutson, L. V. & Lyneborg, L. (1965). Danish acalypterate flies. 3. Sciomyzidae (Diptera). *Ent. Medd.*, **34** (1), 61–101.

Knutson, L., Manguin, S. & Orth, R. E. (1990). A second Australian species of *Pherbellia* Robineau-Desvoidy (Diptera: Sciomyzidae). *J. Aust. Ent. Soc.*, **29**, 281–286.

Knutson, L. V., Neff, S. E. & Berg, C. O. (1967b). Biology and immature stages of snail-killing flies from Africa and southern Spain (Sciomyzidae: *Sepedon*). *Parasitology*, **57** (3), 487–505.

Knutson, L. & Orth, R. E. (1984). The *Sepedon sphegea* complex in the Palearctic and Oriental regions: Identity, variation, and distribution (Diptera: Sciomyzidae). *Ann. Ent. Soc. Am.*, **77** (6), 687–701.

Knutson, L. & Orth, R. E. (2001). *Sepedon mcphersoni* n. sp., key to North American *Sepedon*, groups in *Sepedon s.s.*, and intra- and intergeneric comparison. *Proc. Ent. Soc. Wash.*, **103** (3), 620–635.

Knutson, L., Orth, R. E., Fisher, T. W. & Murphy, W. L. (1986). Catalog of Sciomyzidae (Diptera) of America North of Mexico. *Entomography*, **4**, 1–53.

Knutson, L., Orth, R. E. & Rozkošný, R. (1990). New North American *Colobaea*, with a preliminary analysis of related genera (Diptera: Sciomyzidae). *Proc. Ent. Soc. Wash.*, **92** (3), 483–492.

Knutson, L. V., Rozkošný, R. & Berg, C. O. (1975). Biology and immature stages of *Pherbina* and *Psacadina* (Diptera: Sciomyzidae). *Acta Sci. Nat. Acad. Sci. Bohem.-Brno*, **9** (1), 1–38.

Knutson, L. V., Shagudian, E. R. & Sahba, G. H. (1973). Notes on the biology of certain snail-killing flies (Sciomyzidae) from Khuzestan (Iran). *Iranian J. Pub. Hlth.*, **2** (3), 145–155.

Knutson, L. V. & Stephenson, J. W. (1970). The distribution of snail-killing flies (Diptera: Sciomyzidae) in the British Isles. *Ent. Mon. Mag.*, **106** (1268–1270), 16–21.

Knutson, L. V., Stephenson, J. W. & Berg, C. O. (1965). Biology of a slug-killing fly, *Tetanocera elata* (Diptera: Sciomyzidae). *Proc. Malacol. Soc. Lond.*, **36**, 213–220.

Knutson, L. V., Stephenson, J. W. & Berg, C. O. (1970). Biosystematic studies of *Salticella fasciata* (Meigen), a snail-killing fly (Diptera: Sciomyzidae). *Trans. R. Ent. Soc. Lond.*, **122** (3), 81–100.

Knutson, L., Steyskal, G. C., Zuska, J. & Abercrombie, J. (1976). Family Sciomyzidae. In *A Catalogue of the Diptera of the Americas South of the United States*, ed. N. Papavero. *Mus. Zool., Univ. São Paulo*, **64**, 1–24.

Knutson, L. & Vala, J-C. (1999). The male of *Tetanoptera leucodactyla* Verbeke (Diptera: Sciomyzidae). *Eur. J. Ent.*, **96** (4), 451–457.

Knutson, L. & Vala, J-C. (2002). An evolutionary scenario of Sciomyzidae and Phaeomyiidae (Diptera). *Ann. Soc. Ent. Fr., (n. s.)*, **38** (1–2), 145–162.

Knutson, L. & Valley, K. (1978). Biology of a neotropical snail-killing fly, *Sepedonea isthmi* (Diptera: Sciomyzidae). *Proc. Ent. Soc. Wash.*, **80** (2), 197–209.

Knutson, L. V. & Zuska, J. (1968). A new species of *Pteromicra* and of *Euthycera* from western North America (Diptera: Sciomyzidae). *Proc. Ent. Soc. Wash.*, **70** (1), 78–84.

Lammertsma, D. R. (1996). *Pherbellia stylifera* nieuw voor de Nederlandse fauna (Diptera: Sciomyzidae). *Ent. Ber.*, **56** (1), 12–13.

Latreille, P. A. (1804). Tableau Méthodique des Insectes. In *Société de Naturalistes et d'Agriculteurs, Nouveau Dictionnaire d'Histoire Naturelle, Appliqué aux Arts, Principalement à l'Agriculture et à l'Economie Rurale et Domestique*. Vol. **24** [sect. 3]: *Tableaux méthodiques d'histoire naturelle*. Paris: Déterville, pp. 129–200.

Latreille, P. A. (1825). *Familles Naturelles du Règne Animal, Exposées Succinctement et dans un Ordre Analytique, avec l'Indication de Leurs Genres*. Paris: J-B. Baillière.

Latreille, P. A. (1829). Les Crustacés, les Arachnides et les Insectes, Tome second [Vol. **2**]. In *Le Règne Animal . . .*, Vol. **5**. Edn. 2, ed. G. C. L. D. Cuvier. Paris: Déterville & Crochart.

Leclercq, M. & Schacht, W. (1986). The Sciomyzidae of Turkey (Diptera, Sciomyzidae). *Entomofauna*, **7** (4), 57–61.

Leclercq, M. & Schacht, W. (1987a). Additions to the Sciomyzidae of Turkey (Diptera, Sciomyzidae). *Entomofauna*, **8** (17), 269–270.

Leclercq, M. & Schacht, W. (1987b). The Sciomyzidae of Morocco (Diptera, Sciomyzidae). *Entomofauna*, **8** (30), 449–451.

Li, Z., Yang, D. & Gu, J. (2001). The new species of Sciomyzidae (Diptera: Sciomyzidae) from China. *Entomotaxonomia*, **23** (2), 137–140.

Lindroth, C. H., Andersson, H., Bödvarsson, H. & Richter, S. H. (1973). Surtsey, Iceland. The development of a new fauna, 1963–1970. Terrestrial invertebrates. *Ent. Scand.*, Suppl. **5**, 1–280.

Lindsay, W. (1982). *Unpublished Report on the Biology of* Knutsonia albiseta. 77 pp. + 8 figures. (Available at Library, AFT, Dunsinea Res. Ctr., Castleneck, Dublin 15, Ireland.)

Lindsay, W., Mc Donnell, R. J., Williams, C. D., Knutson, L. & Gormally, M. J. (2011). Biology of the snail-killing fly *Ilione albiseta* (Scopoli) (Diptera: Sciomyzidae). *Studia dipterol.*, **16**, 245–307.

Linnaeus, C. (1758). *Systema Naturae per Regna Tria Naturae*. Edn. 10, Vol. **1**. Holmiae: Salvii.

Linnaeus, C. (1761). *Fauna Svecica Sistens Animalis Sveciae Regni*. Edn. 2. Stockholmiae: Salvii.

Linnaeus, C. (1767). *Systema Naturae per Regna Tria Naturae*. Edn. 12 (rev.), Vol. **1**, Pt. 2, pp. 533–1327. Holmiae: Salvii.

Lioy, P. (1864). I ditteri distribuiti secondo un nuovo metodo di classificazione naturale. *Atti. Ist. Veneto Sci., Let., Arti, Ser. 3.*, **9**, 989–1027.

Loew, H. (1845). *Dipterologischer Beitrag.* [Erster Theil]. [Zu der] öffentlichen Prüfung der Schüler der Königlichen Friedrich-Wilhelms-Gymnasium zu Posen. Posen: W. Decker.

Loew, H. (1847a). Über *Tetanocera stictica* und die ihre nächsten Verwandten, nebst der Beschreibung zweier anderen neuen *Tetanocera*-Arten. *Stettin. Ent. Ztg.*, **8**, 114–124.

Loew, H. (1847b). Über *Tetanocera ferruginea* und ihre verwandten Arten. *Stettin. Ent. Ztg.*, **8**, 194–202.

Loew, H. (1847c). Dipterologisches. *Tetanocera trifaria* und Schlussbemerkungen über die Gattung *Tetanocera*. *Stettin. Ent. Ztg.*, **8**, 246–250.

Loew, H. (1849). Über *Sciomyza glabricula* Fall. und ihre nächsten Verwandten. *Stettin. Ent. Ztg.*, **10**, 337–341.

Loew, H. (1853). *Neue Beiträge zur Kenntnis der Dipteren.* Erster Beiträge. Berlin: Mittler & Sohn.

Loew, H. (1856). *Neue Beiträge zur Kenntniss der Dipteren.* Vierter Beiträge. Berlin: Mittler & Sohn.

Loew, H. (1859). Die nordamerikanische Arten der Gattungen *Tetanocera* und *Sepedon*. *Wien. Ent. Monat.*, **3** (10), 289–300.

Loew, H. (1861). Diptera Americae septentrionalis indigene. Centuria prima. *Berl. Ent. Ztsch.*, **5**, 307–359.

Loew, H. (1862a) [1859]. Über die europäischen Helomyzidae und die in Schlesien vorkommenden Arten derselben. *Ztsch. Ent. (Breslau)*, **13**, 3–80.

Loew, H. (1862b). Über griechischen Dipteren. *Berl. Ent. Ztsch.*, **6**, 69–89.

Loew, H. (1862c). *Monographs of the Diptera of North America.* Part I. Smithson. Inst. Misc. Coll., **6** (1 = [Pub. 141]), 1–221.

Loew, H. (1862d). Bidrag till kännedomen om Afrikas Diptera. *Öfvers. K. Vetensk. Akad. Förh.*, **19** (1), 3–14.

Loew, H. (1862e). Sechs neue europäischen Dipteren. *Wien. Ent. Monat.*, **6**, 294–300.

Loew, H. (1864a). Diptera Americae septentrionalis indigene. Centuria quinta. *Berl. Ent. Ztsch.*, **8**, 49–104.

Loew, H. (1864b). Ueber die in der zweiten Hälfte des Juli 1864 auf der Ziegelwiese bei Halle beobachteten Dipteren. *Ztsch. Ges. Naturwiss.*, **24** (11), 377–396.

Loew, H. (1866 [1865]). Diptera Americae septentrionalis indigene. Centuria sexta. *Berl. Ent. Ztsch.*, **9**, 127–186.

Loew, H. (1872). Diptera Americae septentrionalis indigene. Centuria decima. *Berl. Ent. Ztsch.*, **16**, 49–124.

Loew, H. (1874). Diptera nova a Hug. Theod. Christopho collecta. *Ztsch. Ges. Naturwiss.*, **43**, 413–420.

Loew, H. (1876). Beschreibungen neuer amerikanischer Dipteren. *Ztsch. Ges. Naturwiss.*, **48**, 317–340.

Lopes, H. d. S. (1940). Contribuição ao conhecimento do genero *Udamopyga* Hall e de outros Sarcophagideos que vivem em molluscos no Brasil (Diptera). *Rev. Ent. (Rio J.)*, **11**, 924–954.

Lundbeck, W. (1923). Some remarks on the biology of the Sciomyzidae, together with the description of a new species of *Ctenulus* from Denmark. *Vidensk. Medd. Dan. Naturhist. Foren. Kφbenhavn*, **76**, 101–109.

Lynch, J. J. (1965). The ecology of *Lymnaea tomentosa* (Pfeiffer 1855) in South Australia. *Aust. J. Zool.*, **13**, 461–473.

Macquart, J. (1834). *Histoire Naturelle des Insectes. Diptères, Tome première.* In *Collection des Suites à Buffon*, ed. N. E. Roret. Paris.

Macquart, J. (1835). *Histoire Naturelle des Insectes. Diptères, Tome deuxième.* In *Collection des Suites à Buffon*, ed. N. E. Roret. Paris.

Macquart, J. (1844) [1843]. Diptères Exotiques Nouveaux ou Peu Connus. Tome deuxième – 3ème partie. *Mém. Soc. R. Sci., Agr. Arts, Lille*, **1842**, 162–460.

Macquart, J. (1846). Diptères exotiques nouveaux ou peu connus. [Ier] Supplément. *Mém. Soc. R. Sci., Agr. Arts, Lille*, **1844**, 133–364.

Macquart, J. (1849). Diptères. In *Histoire Naturelle des Animaux Articulés. Pt. 3 – Insectes. Exploration Scientifique de l'Algérie Pendant les Années 1840, 1841, 1842 Publié par Ordre du Gouvernement et avec le Concours d'une Commission Académique. Sciences Physiques -Zoologie III*, ed. P. H. Lucas. Paris.

Macquart, J. (1851). Diptères exotiques nouveaux ou peu connus. Suite du 4e supplément. *Mém. Soc. Natl. Sci., Agr. Arts, Lille*, **1850**, 134–294. pls.15–28.

Madsen, H. (1990). Biological methods for the control of freshwater snails. *Parasitol. Today*, **6** (7), 237–240.

Maharaj, R. (1991). *Predator-Prey Interactions Between Sciomyzid Fly Larvae* (Sepedon *spp.) and Aquatic Snails.* M.S. thesis, University of Natal, Pietermaritzburg, South Africa. 150 pp.

Maharaj, R., Appleton, C. C. & Miller, R. M. (1992). Snail predation by larvae of *Sepedon scapularis* Adams (Diptera: Sciomyzidae), a potential biocontrol agent of snail intermediate hosts of schistosomiasis in South Africa. *Med. Vet. Ent.*, **6**, 183–187.

Maharaj, R., Appleton, C. C. & Miller, R. M. (2002). Bilharzia control by *Sepedon scapularis* larvae (Diptera: Sciomyzidae): a viable option? In *5th Int. Congr. Dipterol., 29 Sept. – 4 Oct., 2002, Brisbane, Australia.* Conf. chair D. Yeates. *Abstracts*, p. 141.

Malloch, J. R. (1928). Notes on Australian Diptera. No. xv. *Proc. Linn. Soc. New South Wales*, **53**, 319–335.

Malloch, J. R. (1930). New Zealand Muscidae Acalyptratae. Family Sciomyzidae (Suppl.). *Rec. Canterbury Mus.*, **3** (5), 343–344.

Malloch, J. R. (1933). Sciomyzidae. In *Diptera of Patagonia and South Chile, Part VI – Fasc. 4. Acalyptrata*. London: British Museum (Natural History), pp. 296–323.

Malloch, J. R. (1935). New Zealand Muscidae Acalyptratae. Sciomyzidae, Addendum. *Rec. Canterbury Mus.*, **4**, 95–96.

Manguin, S. (1989). Sexual dimorphism in size of adults and puparia of *Tetanocera ferruginea* Fallén (Diptera: Sciomyzidae). *Proc. Ent. Soc. Wash.*, **91** (4), 523–528.

Manguin, S. (1990). Population genetics and biochemical systematics of marsh flies in the *Sepedon fuscipennis* group (Diptera: Sciomyzidae). *Biochem. Syst. Ecol.*, **18** (6), 447–452.

Manguin, S. & Hung, A. C. F. (1991). Developmental genetics in larvae, pupae and adults of *Sepedon fuscipennis fuscipennis* (Dipt. Sciomyzidae). *Entomophaga*, **36** (2), 183–192.

Manguin, S. & Vala, J-C. (1989). Prey consumption by larvae of *Tetanocera ferruginea* (Diptera: Sciomyzidae) in relation to number of snail prey species available. *Ann. Ent. Soc. Am.*, **82** (5), 588–592.

Manguin, S., Vala, J-C. & Reidenbach, J. M. (1986). Prédation de mollusques dulçaquicoles par les larves malacophages de *Tetanocera ferruginea* Fallén, 1820 (Diptera, Sciomyzidae). *Can. J. Zool.*, **64**, 2832–2836.

Manguin, S., Vala, J-C. & Reidenbach, J. M. (1988a). Action prédatrice des larves de *Tetanocera ferruginea* (Diptera: Sciomyzidae) dans des systèmes à plusieurs espèces de mollusques-proies. *Acta Ecol./Ecol. Applic.*, **9** (3), 249–259.

Manguin, S., Vala, J-C. & Reidenbach, J. M. (1988b). Détermination des préférences alimentaires des larves de *Tetanocera ferruginea* (Diptera: Sciomyzidae), prédateur de mollusques dulçaquicoles. *Acta Ecol./Ecol. Applic.*, **9** (4), 353–370.

Marinoni, L. (1992). A new species of *Guatemalia* Steyskal (Diptera, Sciomyzidae). *Revt. Bras. Zool.*, **9** (3/4), 247–249.

Marinoni, L. & Carvalho, C. J. Barros de (1993). A cladistic analysis of *Protodictya* Malloch (Diptera, Sciomyzidae). *Proc. Ent. Soc. Wash.*, **95** (3), 412–417.

Marinoni, L. & Knutson, L. (1992). Revisão do gênero Neotropical *Protodictya* Malloch, 1933 (Diptera: Sciomyzidae). *Revt. Bras. Ent.*, **36** (1), 25–45.

Marinoni, L. & Knutson, L. (2010). Sciomyzidae. In *Manual of Central American Diptera*, vol. 2. ed. B. V. Brown, A. Borkent, D. M. Wood & M. A. Zumbado. National Research Council of Canada. Ottawa: Research Press.

Marinoni, L. & Mathis, W. N. (2000). A cladistic analysis of Sciomyzidae Fallén (Diptera). *Proc. Biol. Soc. Wash.*, **113** (1), 162–209.

Marinoni, L. & Mathis, W. N. (2006). A cladistic analysis of the Neotropical genus *Sepedonea* Steyskal (Diptera: Sciomyzidae). *Zootaxa*, **1236**, 37–52.

Marinoni, L., Steyskal, G. C. & Knutson, L. (2003). Revision and cladistic analysis of the Neotropical genus *Thecomyia* (Diptera: Sciomyzidae). *Zootaxa*, **191**, 1–36.

Marinoni, L., Zumbado, M. A. & Knutson, L. (2004). A new genus and species of Sciomyzidae (Diptera) from the Neotropical region. *Zootaxa*, **540**, 1–7.

Marshall, S. A. (2006). *Insects – Their Natural History and Diversity: with a Photographic Guide to Insects of Eastern North America*. Buffalo, NY: Firefly Books.

Mathis, W. N., Knutson, L. & Murphy, W. L. (2009). A new species of the snail-killing genus *Dictya* Meigen from the Delmarva States (Diptera: Sciomyzidae). *Proc. Ent. Soc. Wash.*, **111** (4), 785–794.

Matsumura, S. (1916). *Thousand Insects of Japan. Additamenta*. Vol. **2**, *Diptera*. Tokyo: Keisei-sha.

Mayer, H. (1953a). Beiträge zur Kenntnis der Sciomyzidae (Diptera Musc. acalyptr.). *Ann. Natur. Mus. Wien*, **59**, 202–219.

Mayer, H. (1953b). Bericht über das vorwiegend 1951 an den Ufern des Mauerbaches, Wien, NÖ., gesammelte Insektenmaterial, unter besonderer Berücksichtigung der Dipteren. *Wetter und Leben, Sonderheft*, **2**, 156–162.

McAlpine, D. K. (1991). Relationships of the genus *Heterocheila* (Diptera: Sciomyzoidea) with description of a new family. *Tijdsch. Ent.*, **134**, 193–199.

McAlpine, D. K. (1992). The earliest described species of Helosciomyzidae (Diptera: Schizophora). *Austr. Ent. Mag.*, **19** (3), 89–92.

McAlpine, J. F. (1963). Relationships of *Cremifania* Czerny (Diptera: Chamaemyiidae) and description of a new species. *Can. Ent.*, **95**, 239–253.

McAlpine, J. F. (1981). Morphology and terminology – adults. In *Manual of Nearctic Diptera*, Vol. **1**., ed. J. F. McAlpine, B. V. Peterson, G. E. Shewell, H. J. Teskey, J. R. Vockeroth & D. M. Wood. *Res. Br., Agric. Canada. Monogr.*, **27**, 9–63.

McAlpine, J. F. (1989). Phylogeny and classification of the Muscomorpha. In *Manual of Nearctic Diptera*, Vol. **3**, ed. J. F. McAlpine. *Res. Br., Agric. Canada. Monogr.*, **32**, 1397–1505.

McCoy, L. E. & Joy, J. E. (1977). Tolerance of *Sepedon fuscipennis* and *Dictya* sp. larvae (Diptera: Sciomyzidae) to the molluscicides Bayer 73 and sodium pentachlorophenate. *Environ. Ent.*, **6** (2), 198–202.

McCullough, F. S. (1981). Biological control of the snail intermediate hosts of human *Schistosoma* spp.: a review of its present status and future prospects. *Acta Trop.*, **38**, 5–13.

Mc Donnell, R. J. (2004). *The Biology and Behaviour of Selected Marsh Fly (Diptera: Sciomyzidae) Species, Potential Biological Control Agents of Liver Fluke Disease in Ireland*. Ph.D. thesis, National University of Ireland, Galway, Ireland.

Mc Donnell, R. & Gormally, M. J. (2000). *An Investigation of Marsh Flies (Diptera: Sciomyzidae) as Bioindicators of Wetland Habitats on Farms in the West of Ireland*. Report to the Heritage Council, Environmental Science Unit, National University of Ireland, Galway, Ireland.

Mc Donnell, R., Gormally, M. J. & Knutson, L. (2004). *Lymnaea truncatula* (Müller) the liver fluke snail: confirmed as alternate host of *Pherbellia s. schoenherri* in the wild (Diptera: Sciomyzidae). *Dipt. Dig.*, **10**, 69–71.

Mc Donnell, R., Knutson, L., Vala, J-C., Abercrombie, J., Henry, P-Y. & Gormally, M. J. (2005a). Direct evidence of predation by aquatic, predatory Sciomyzidae (Diptera, Acalyptrata) on freshwater snails from natural populations. *Ent. Mon. Mag.*, **141**, 49–56.

Mc Donnell, R. J., Mulkeen, C. J. & Gormally, M. J. (2005b). Sexual dimorphism and the impact of temperature on the pupal and adult stages *Sepedon spinipes spinipes*, a potential biological control agent of fascioliasis. *Ent. Exp. Appl.*, **115**, 291–301.

Mc Donnell, R. J., Paine, T. D. & Gormally, M. J. (2007a). Trail-following behaviour in the malacophagous larvae of the aquatic sciomyzid flies *Sepedon spinipes spinipes* and *Dictya montana*. *J. Ins. Behav.*, **20** (3), 367–376.

Mc Donnell, R. J., Paine, T. D., Orth, R. E. & Gormally, M. J. (2007b). Life history and biocontrol potential of *Dictya montana* Steyskal, 1954 (Diptera: Sciomyzidae), a snail-killing fly. *Pan-Pac. Ent.*, **83** (2), 101–109.

McLaughlin, H. E. & Dame, D. A. (1989). Rearing *Dictya floridensis* (Diptera: Sciomyzidae) in a continuously producing colony and evaluation of larval food sources. *J. Med. Ent.*, **26** (6), 522–527.

McLean, I. (1998). Sciomyzidae. In *Checklist of Insects of the British Isles (New Series). Part 1: Diptera (incorporating a list of Irish Diptera). Handbooks for the Identification of British Insects*, **12**, ed. P. J. Chandler. London: R. Ent. Soc., pp. 132–134.

Mead, A. R. (1961). *The Giant African Snail. A Problem in Economic Malacophagy*. Chicago: Univ. Chicago Press.

Meier, R. (1996). Adult characters versus larval characters in phylogenetic studies of the Sciomyzoidea (Cyclorrhapha). *Proc. XX Int. Congr. Ent.*, August 25–31, 1996, Firenze, Italy. *Abstracts*, p. 26.

Meigen, J. W. (1800). *Nouvelle Classification des Mouches à Deux Ailes (Diptera L.) d'Après un Plan Tout Nouveau*. Paris: Perronneau.

Meigen, J. W. (1803). Versuch einer neuen Gattungs-Eintheilung der europäischen zweiflügligen Insekten. *Mag. Insektenkd.*, **2**, 259–281.

Meigen, J. W. (1830). *Systematische Beschreibung der bekannten europäischen zweiflügeligen Insekten*. Vol. **6**. Hamm: H. W. Schultz.

Meigen, J. W. (1838). *Systematische Beschreibung der bekannten europäischen zweiflügeligen Insekten*. Vol. **7** *oder Supplementband*. Hamm: H. W. Schultz.

Meijere, J. C. H. de. (1902). Ueber die Prothorakalstigmen der Dipteren-puppen. *Zool. Jahrb. Abt. Anat. Ontog. Tiere*, **15**, 623–692.

Meijere, J. C. H. de. (1908). Studien über südostasiatische Dipteren. II. *Tijdschr. Ent.*, **51**, 105–180.

Meijere, J. C. H. de. (1919). Derde supplement op de Nieuwe Naamlijst van Nederlandsche Diptera. *Tijdschr. Ent.*, **62**, 161–195.

Melander, A. L. (1913). Note on two preoccupied muscid names. *Psyche*, **20**, 205.

Melander, A. L. (1920). Review of the Nearctic Tetanoceridae. *Ann. Ent. Soc. Am.*, **13** (3), 305–332.

Mello, D. A. & Bredt, A. (1978a). Distribuição geográfica de Sciomyzidae (Diptera: Insecta) no distrito federal e em algumas regiões de outros estados do Brasil. *Ciên. e Cult.*, **30** (2), 212–215.

Mello, D. A. & Bredt, A. (1978b). Estudos populacionais de cinco espécies de Sciomyzidae (Diptera-Insecta) no norte de Formosa, Goiās. *Ciên. e Cult.*, **30** (12), 1459–1464.

Mercier, L. (1921). Diptères de la côte du Calvados. IIe liste. *Ann. Soc. Ent. Belg.*, **61**, 162–164.

Merz, B. (1998). 65. Sciomyzidae. In *Fauna Helvetica. 1. Diptera-checklist*, ed. B. Merz, G. Bächli, J-P. Haenni & Y. Gonseth. Neuchâtel: Cent. Suisse Cart. Faun. und Schweiz. Ent. Gesell., pp. 251–253.

Merz, B. & Rozkošný, R. (1995a). A new *Pherbellia* (Diptera, Sciomyzidae) from Central Europe. *Bull. Soc. Ent. Suisse*, **68**, 435–440.

Merz, B. & Rozkošný, R. (1995b). New records of Sciomyzidae (Diptera) from Central Asia. *Stud. dipterol.*, **2** (2), 279–282.

Meunier, F. (1904). Contribution à la faune des Helomyzinae de l'ambre de la Baltique. *Feuil. jeun. Natural.*, **35**, 21–27.

Michelson, E. H. (1957). Studies on the biological control of schistosome-bearing snails. Predators and parasites of fresh-water Mollusca: a review of the literature. *Parasitology*, **47**, 413–426.

Miller, R. M. (1995). Key to genera of Afrotropical Sciomyzidae (Diptera: Acalyptratae) with new records, synonymies and biological notes. *Ann. Natal Mus.*, **36**, 189–201.

Miller, R. & Appleton, C. C. (2002). Biological and natural control of African schistosomiasis host snails by predatory sciomyzid fly larvae. In *5th Int. Congr. Dipterol.*, 29 Sept.–4 Oct., 2002, Brisbane, Australia. Conf. chair D. Yeates. *Abstracts*, p. 156.

Moor, B. (1980). Zur Biologie der Beziehung zwischen *Pherbellia punctata* (Diptera: Sciomyzidae) und ihrem Wirt *Succinea putris* (Pulmonata, Stylommatophora). *Rev. Suisse Zool.*, **87** (4), 941–953.

Munari, L. (1983). Lo studio degli Sciomyzidae (Diptera, Cyclorrhapha) per la lotta biologica ai molluschi vettori degli elminti parassiti, agenti eziologici delle bilharziosi e distomatosi umane e del bestiame. *Lav. Soc. Venezia Sci. nat.*, **8**, 9–30.

Munari, L. (1986). Nuovi dati sulle *Pherbellia* italiane (Diptera, Sciomyzidae). *Lav. Soc. Venezia Sci. nat.*, **11**, 35–40.

Munari, L. (1988). I ditteri Sciomyzidae, un esempio di radiazione adattativa. *Nat. e Mont.*, **35** (3), 35–40.

Munari, L. (1996). *Trixoscelis puncticornis* (Becker, 1907) (Trixoscelidae) e *Pherbellia cinerella* (Fallén, 1820) (Sciomyzidae): un esempio di mimetismo batesiano ? (Diptera: Acalyptrata). *Boll. Mus. civ. Stor. nat. Venezia*, **46**, 147–151.

Munari, L. & Cerretti, P. (2006). Some reflexions on the distributional patterns of two species of Euro-mediterranean Sciomyzidae (Diptera, Acalyptrata). *Boll. Mus. civ. Stor. nat. Venezia*, **57**, 117–122.

Nagatomi, A. & Kushigemachi, K. (1965). Life history of *Sepedon sauteri* Hendel (Diptera: Sciomyzidae). *Kontyû*, **33** (1), 35–38.

Nagatomi, A. & Tanaka, A. (1967). Egg of *Sepedon sauteri* Hendel (Diptera: Sciomyzidae). *Kontyû*, **35** (1), 31–33.

Needham, J. G. & Betten, C. (1901). Aquatic insects in the Adirondacks. *Bull. N. Y. State Mus.*, **47**.

Neff, S. E. (1964). Snail-killing sciomyzid flies: application in biological control. *Verhandl. Int. Ver. Limnol.*, **15** (2), 933–939.

Neff, S. E. & Berg, C. O. (1961). Observations on the immature stages of *Protodictya hondurana* (Diptera: Sciomyzidae). *Bull. Brooklyn Ent. Soc.*, **56** (2), 46–61.

Neff, S. E. & Berg, C. O. (1962). Biology and immature stages of *Hoplodictya spinicornis* and *H. setosa* (Diptera: Sciomyzidae). *Trans. Am. Ent. Soc.*, **88** (2), 77–93.

Neff, S. E. & Berg, C. O. (1966). Biology and immature stages of malacophagous Diptera of the genus *Sepedon* (Sciomyzidae). *Va. Agric. Exp. Stn. Bull.*, **566**, 1–113.

Newman, E. (1834). Entomological notes. *Ent. Mag.*, **2**, 379, 395.

Nielsen, P., Ringdahl, O. & Tuxen, S. L. (1954). Diptera 1 (exclusive of Ceratopogonidae and Chironomidae). In *The Zoology of Iceland*. Vol. **III**, Part 48 a, ed. Á. Friðriksson & S. L. Tuxen. Copenhagen and Reykjavik: Munksgaard.

Oldham, C. (1912). Report on land and freshwater Mollusca observed in Hertfordshire in 1910. *Trans. Hertford. Nat. Hist. Soc.*, **14**, 287–290.

Oldroyd, H. (1964). *The Natural History of Flies.* London: Weidenfeld & Nicholson.

O'Neill, W. L. (1973). Biology of *Trichopria popei* and *T. atrichomelinae* (Hymenoptera: Diapriidae), parasitoids of the Sciomyzidae (Diptera). *Ann. Ent. Soc. Am.*, **66** (5), 1043–1050.

O'Neill, W. L. & Berg, C. O. (1975). Reproduction by a grossly malformed fly (Diptera: Sciomyzidae). *Proc. Ent. Soc. Wash.*,**77** (1), 156–162.

Orth, R. E. (1982). Five new species of *Pherbellia* Robineau-Desvoidy, subgenus *Oxytaenia* Sack, from North America (Diptera: Sciomyzidae). *Proc. Ent. Soc. Wash.*, **84** (1), 23–37.

Orth, R. E. (1983). Two new species of *Pherbellia* from North America (Diptera: Sciomyzidae). *Proc. Ent. Soc. Wash.*, **85** (3), 537–542.

Orth, R. E. (1984a). A new species of *Pherbellia* from Montana (Diptera: Sciomyzidae). *Proc. Ent. Soc. Wash.*, **86** (3), 599–601.

Orth, R. E. (1984b). A new species of *Dictya* from Mexico (Diptera: Sciomyzidae). *Proc. Ent. Soc. Wash.*, **86** (4), 893–897.

Orth, R. E. (1986). Taxonomy of the *Sepedon fuscipennis* group (Diptera: Sciomyzidae). *Proc. Ent. Soc. Wash.*, **88** (1), 63–76.

Orth, R. E. (1987). A new species of *Pherbellia* from North America with range extensions for *P. hackmani* and *P. griseicollis* (Diptera: Sciomyzidae). *Proc. Ent. Soc. Wash.*, **89** (2), 344–350.

Orth, R. E. (1991). A synopsis of the genus *Dictya* Meigen with ten new species (Diptera: Sciomyzidae). *Proc. Ent. Soc. Wash.*, **93** (3), 660–689.

Orth, R. E. & Fisher, T. W. (1982). A new species of *Tetanocera* Duméril from Colorado (Diptera: Sciomyzidae). *Proc. Ent. Soc. Wash.*, **84** (4), 685–689.

Orth, R. E. & Fisher, T. W. (1983). A new species of *Dictya* from Idaho (Diptera: Sciomyzidae). *Proc. Ent. Soc. Wash.*, **85** (2), 217–221.

Orth, R. E. & Knutson, L. (1987). Systematics of snail-killing flies of the genus *Elgiva* in North America and biology of *E. divisa* (Diptera: Sciomyzidae). *Ann. Ent. Soc. Am.*, **80** (6), 829–840.

Orth, R. E. & Steyskal, G. C. (1981). A new species of *Pherbellia* Robineau-Desvoidy separated from a previously described North American species (Diptera: Sciomyzidae). *Proc. Ent. Soc. Wash.*, **83** (1), 99–104.

Orth, R. E., Steyskal, G. C. & Fisher, T. W. (1980). A new species of *Pherbellia* Robineau-Desvoidy with notes on the *P. ventralis* group (Diptera: Sciomyzidae). *Proc. Ent. Soc. Wash.*, **82** (2), 284–292.

Pandellé, L. (1902). Etudes sur les Muscides de France. 3e partie, Suite. *Rev. Ent. (Caen)*, **21**, 373–492.

Panov, A. A. (1994). Unusual structure of the retrocerebral complex in a representative of brachycerous cyclorrhaphous flies, *Sepedon sphegea* (Diptera, Sciomyzidae). *Zool. Zhur.*, **73** (9), 165–167.

Panzer, G. W. F. (1798). *Favne Insectorum Germanicae Initia Oder Devtschlands Insecten. H. 54.* Nürnberg: Felsecker Buchandlung.

Papp, L. (2004). Description of the first apterous genus of Sciomyzidae (Diptera), from Nepal. *Rev. Suisse Zool.*, **111** (1), 57–62.

Peacock, D. B. (1973). *Ecology of the Snail-Killing Fly,* Sepedon fuscipennis *Loew (Diptera: Sciomyzidae) in Lentic and Lotic Habitats.* M.S. thesis, Univ. Connecticut, Storrs, CT, USA. 61 pp.

Perris, E. (1850). Histoire des métamorphoses de quelques diptères. *Mem. Soc. Sci. Lille*, **1850**, 122–124.

Perty, M. (1833). *Insecta brasiliensia*. In *Delectus animalium articulatorum quae in itinere per Brasiliam annis MDCCCXVII-MDCCXX jussu et auspiciis Maximiliani Josephi I. Bavariae regis augustissimi peracto collegerunt Dr. J. B. de Sphix et Dr. C.F. Bh. de Martius.* Munich: F. S. Hübschmann & Leipzig: Fleischer, pp. 125–224, pls. 25–40.

Peterson, A. (1916). The head-capsule and mouth-parts of Diptera. *Ill. Biol. Monogr.*, **3** (2), 177–282.

Peterson, A. H. (1953). *Larvae of Insects. Part II*, 2nd edn. Ann Arbor: Edwards Bros.

Petitjean, M. (1966). Le contrôle biologique des mollusques nuisibles; revue des résultats essentiels, d'après les principaux travaux récents. *Ann. Biol.*, **5**, 271–295.

Platanova, T. A. (1985). Systematics of some closely related species of *Pseudocella* Filipov, 1927 (Nematoda, Enoplida). *Zool. Zhur.*, **64** (12), 1794–1801.

Platanova, T. A. (1988). New species of free living nematodes of the genus *Pseudocella* from coastal waters of Kamchatka. *Biol. Moriya*, **6**, 17–23.

Poinar Jr., G. & Poinar, R. (1999). *The Amber Forest.* Princeton: Princeton Univ. Press.

Poole, R. W. (1996). Sciomyzidae. In *Nomina Insecta Nearctica.* Vol. **3**. *Diptera, Lepidoptera, Siphonaptera*, ed. R. W. Poole & P. Gentili. Rockville: Ent. Inf. Serv., pp. 240–242.

Povolný, D. & Groschaft, J. (1959). Drei bedeutende Fliegenarten-Schmarotzer der Helicidae aus dem Gebiet der Tschechoslowakei. *Zool. Listy*, **8** (22), 131–136.

Pritchard, G., Harder, L. D. & Mutch, R. A. (1996). Development of aquatic insect eggs in relation to temperature and strategies for dealing with different thermal environments. *Biol. J. Linn. Soc.*, **58**, 221–244.

Przhiboro, A. A. (2001). A new species of the snail-killing fly genus *Sciomyza* (Diptera: Sciomyzidae), and a list of Sciomyzidae collected at Anninskoe and Anisimovo Lakes, Pskov Province. *Zoosyst. Rossica*, **10**, 183–188.

Reidenbach, J. M., Vala, J-C. & Ghamizi, M. (1989). The slug-killing Sciomyzidae (Diptera): potential agents in the biological control of crop pest molluscs, In *BCPC Symposium Proc. No. 41. Slugs and Snails in World Agriculture*, ed. I. Henderson. Thornton Heath: British Crop Protection Council, pp. 273–280.

Ringdahl, O. (1948). Dipterologische Notisen 3. Bemerkungen zu schwedischen Sciomyziden. *Opusc. Ent.*, **13** (2), 52–54.

Rivosecchi, L. (1984). Famiglia Sciomyzidae. In *Ditteri (Diptera)*. No. **28**. *Guide per il Riconoscimento delle Specie Animali delle Acque Interne Italiane*, ed. L. Rivosecchi. Consiglio Nazionale delle Richerche AQ/I/206. Verona: Valdonega, pp. 142–152.

Rivosecchi, L. (1989). Note sugli Sciomyzidae IX. Due nuove specie del genere *Pherbellia* provenienti dall'Italia meridionale (Diptera, Cyclorrhapha, Acalyptera). *Fragm. Ent.*, **21** (2), 153–161.

Rivosecchi, L. (1992). *Diptera Sciomyzidae.* Vol. **30**. In *Fauna D'Italia*. Bologna: Edizioni Calderini.

Rivosecchi, L. & Santagata, V. (1978). Introduzione allo studio degli Sciomyzidae (Diptera Acalyptera) d'Italia. *Parassitologia*, **20** (1–3), 113–130.

Rivosecchi, L. & Santagata, V. (1979). Note e osservazioni su qualche Sciomyzidae (Diptera Acalyptera) proveniente dall'Italia centrale. *Boll. Mus. Civ. Stor. Nat. Verona*, **6**, 469–489.

Robineau-Desvoidy, J. B. (1830). Essai sur les Myodaires. [Cl. des] Sci. Math. et Phys., Mém. présentés par divers Savants. *Acad. R. Sci., Inst. Fr.* [ser. 2] **2**, 1–813.

Robinson, W. H & Foote, B. A. (1978). Biology and immature stages of *Antichaeta borealis* (Diptera: Sciomyzidae), a predator of snail eggs. *Proc. Ent. Soc. Wash.*, **80** (3), 388–396.

Robinson, W. H & Turner, E. C. (1975). Insect fauna of some Virginia thermal streams. *Proc. Ent. Soc. Wash.*, **77** (3), 391–398.

Roller, L. (1995). Seasonal dynamics of Sciomyzids (Sciomyzidae, Diptera). *Biologia (Bratislava)*, **50**,171–176.

Roller, L. (1996). A new species of *Sciomyza* Fallén from Central Europe (Diptera: Sciomyzidae). *Dtsch. Ent. Zeit.*, **2**, 245–250.

Rombach, M. C. & Roberts, D. W. (1989). *Hirsutella* species (Deuteromycotina: Hyphomycetes) on Philippine insects. *Philippine Ent.*, **7**(5), 491–518.

Rondani, C. (1856). *Dipterologiae Italicae Prodromus.* Vol. **1**. *Genera Italica Ordinis Dipterorum Ordinatim Disposita et Distincta et in Familias et Stirpes Aggregata.* Parma: Stocchi.

Rondani, C. (1863). *Diptera Exotica Revisa et Annotata. Novis Nonnullis Descriptis.* Modena: Soliani.

Rondani, C. (1868). Sciomizinae [sic] Italicae collectae, distinctae et in ordinem dispositae. In *Dipterologiae Italicae. Prodromus. Pars VII, Fasc. II. 60 pp.* Vol. **7**. *Species italicae ordinis Dipterorum a Prof. Camillo Rondani collectae, distinctae, et in ordinem dispositae novis, vel minus cognitis descriptis, Pars sexta: Scatophaginae Sciomyzinae Ortalidinae. Atti Soc. Ital. Sci. Nat. Milano*, **11**, 199–256.

Rondani, C. (1877). Species Italicae ordinis dipterorum (Muscaria Rnd). Stirpis XIX Sciomyzinarum revisio. *Ann. Soc. Nat. Modena. Ann XI, Fasc. Primo, Ser. 2*, **11**, 7–79.

Rondelaud D., Vignoles, P. & Dreyfuss, G. (2002). The presence of predators modifies the larval developmenct of *Fasciola hepatica* in surviving *Lymnaea truncatula*. *J. Helminthol.*, **76**, 175–178.

Roser, C. L. F. von. (1840). Erster Nachtrag zu dem im Jahre 1834 bekannt gemachten Verzeichniss in Würtemberg vorkommender zweiflügliger Insekten. 1. Nachtrag. *Correspondenzbl. K. Württemberg. Landwirtsch. Ver. Stuttgart (n.s.)*, **17**, 49–64.

Rozkošný, R. (1962). Zur Verbreitung und Flugzeit der Hornfliegen (Sciomyzidae, Diptera) Südmahrens. *Publ. Fac. Sci. Univ. Purk. Brno*, **433**, 211–232.

Rozkošný, R. (1964). Zur Taxonomie der Gattung *Pherbellia* Robineau-Desvoidy (Diptera, Sciomyzidae). *Acta. Soc. Ent. Čechoslov.*, **61** (4), 384–390.

Rozkošný, R. (1965). Neue Metamorphosestadien mancher *Tetanocera*-Arten (Diptera: Sciomyzidae). *Zool. Listy*, **14**, 367–371.

Rozkošný, R. (1966). O životě a praktickém významw našich vláhomilek. *Živa*, **14** (5), 182–184.

Rozkošný, R. (1967). Zur Morphologie und Biologie der Metamorphosestadien mitteleuropäischer Sciomyziden (Diptera). *Acta Acad. Sci. Brno*, **1** (4), 117–160.

Rozkošný, R. (1968). Malacophagie als Lebensweise von Dipterenlarven in Mitteleuropa. *Abh. Ber. Naturkundemus. Görlitz*, **44** (2), 165–170.

Rozkošný, R. (1969). Zur Taxonomie und Verbreitung seltenerer palaärktischer Sciomyziden nebst Beschreibung der *Euthycera hrabei* sp. n. (Diptera). *Fol. Fac. Sci. Nat. Univ. Purk. Brun.*, **10** [*Biol.* **25** (*8*)], 107–114.

Rozkošný, R. (1970) [1969]. Beiträge zur Kenntnis der Fauna Afghanistans. (Sammelergebnisse von 0. Jakeš 1963–1964, D. Povolný 1965, D. Povolný und Fr. Tenora 1966, J. Šimek 1965–1966, D. Povolný, J. Gaisler, Z. Šebek und Fr. Tenora 1967). Sciomyzidae, Diptera. *Čas. Mor. Mus.*, **54**, Suppl., 287–292.

Rozkošný, R. (1979a). *Pteromicra nigripalpis* sp. n. from Mongolia and a world catalogue of the genus (Diptera, Sciomyzidae). *Acta Ent. Bohemoslov.*, **76** (3), 181–187.

Rozkošný, R. (1979b). Malacophagous larvae of Diptera as means of biological control. *Práce z oboru botan. zool.*, (1978–1979), 47–56.

Rozkošný, R. (1981). A new name and some new synonyms of Palearctic Sciomyzidae (Diptera). *Ent. Scand.*, **12** (2), 177–180.

Rozkošný, R. (1982). Three new species of *Pherbellia* and new synonyms of Holarctic and Palaearctic Sciomyzidae (Diptera). *Ann. Ent. Soc. Fenn.*, **48** (2), 51–56.

Rozkošný, R. (1984a). Review of *Colobaea* Zetterstedt (Diptera: Sciomyzidae), with a new species from northern Fennoscandia. *Ent. Scand.*, **15**, 85–88.

Rozkošný, R. (1984b). The Sciomyzidae (Diptera) of Fennoscandia and Denmark. *Fauna Ent. Scand.*, **14**. Leiden-Copenhagen: Scan. Sci. Press.

Rozkošný, R. (1985). New records of Sciomyzidae (Diptera) from Cyprus. In *Zbor. Organizmy a prostredie. Pedagogická Fakulta v Nitre*, pp. 153–157.

Rozkošný, R. (1987a). A review of the Palaearctic Sciomyzidae (Diptera). *Fol. Facul. Sci. Nat. Univ. Purkynanae Brun., Biol., Brno*, **86**. 100 pp. + 56 pls.

Rozkošný, R. (1987b). Sciomyzidae. In *Enumeration Insectorum Bohemoslovakiae. [Checklist of Czechoslovak Insects] II (Diptera)*, ed. J. Ježek. *Acta Faun. Ent. Mus. Nat. Prag*, **18**, 179–181.

Rozkošný, R. (1988). New records of Sciomyzidae (Diptera) from Spain, including the description of a new species of *Euthycera* Latreille. *Acta Ent. Bohemoslov.*, **85** (6), 457–463.

Rozkošný, R. (1991). A key to the Palaearctic species of *Pherbellia* Robineau-Desvoidy, with descriptions of three new species (Diptera, Sciomyzidae). *Acta Ent. Bohemoslov.*, **88**, 391–406.

Rozkošný, R. (1995). World distribution of Sciomyzidae based on the list of species (Diptera). *Stud. dipterol.*, **2**, 221–238.

Rozkošný, R. (1997a). Diptera Sciomyzidae, snail-killing flies. In *Aquatic Insects of North Europe – A Taxonomic Handbook*. Vol. **2**, ed. A. N. Nilsson. Stenstrup: Apollo, pp. 363–381.

Rozkošný, R. (1997b). Sciomyzidae. In *Check List of Diptera (Insecta) of the Czech and Slovak Republics*, ed. M. Chvála. Prague: Karolinum, Charles Univ. Press, pp. 73–74.

Rozkošný, R. (1998). Family Sciomyzidae. In *Contributions to a Manual of Palaearctic Diptera (with Special Reference to Flies of Economic Importance)*. Vol. **3**. *Higher Brachycera*, ed. L. Papp & B. Darvas. Budapest: Science Herald, pp. 357–376.

Rozkošný, R. (1999a). Sciomyzidae. In *Entomofauna Germanica 2. Checkliste der Dipteren Deutschlands*, ed. H. Schumann, R. Bährmann & A. Stark. *Stud. dipterol.*, Suppl. **2**, 188–189.

Rozkošný, R. (1999b). Phaeomyiidae, Sciomyzidae. In *Diptera of the Pálava Biosphere Reserve of UNESCO*, **II**, ed. R. Rozkošný & J. Vaňhara. *Folia Fac. Sci. Nat. Univ. Masaryk., Brun., Biol.*, **100**, 281–286.

Rozkošný, R. (2002). Insecta: Diptera: Sciomyzidae. **21** (23). In *Süsswasserfauna von Mitteleuropa*, ed. J. Schwoerbel & P. Zwick. Spektrum, Heidelberg: Akad. Verl., pp. 15–122.

Rozkošný, R. & Elberg, K. (1984). Family Sciomyzidae (Tetanoceridae). In *Catalogue of Palaearctic Diptera*. Vol. **9**. *Micropezidae–Agromyzidae*, ed. A. Soós & L. Papp. Budapest: Akad. Kiadó, pp. 167–193.

Rozkošný, R. & Elberg, K. (1991). Two new species of *Colobaea* Zetterstedt (Diptera, Sciomyzidae) from Palaearctic Asia. *Aquatic Ins.*, **13** (1), 55–63.

Rozkošný, R. & Knutson, L. V. (1970). Taxonomy, biology, and immature stages of Palearctic *Pteromicra*, snail-killing Diptera (Sciomyzidae). *Ann. Ent. Soc. Am.*, **63** (5), 1434–1459.

Rozkošný, R. & Knutson, L. (1972). The male of *Pherbellia terminalis* (Walker), comb. n. and distributional notes on Sciomyzidae of Afghanistan (Diptera). *Acta Ent. Bohemoslov.*, **69** (6), 417–422.

Rozkošný, R. & Knutson, L. (1991). Two new species of Sciomyzidae (Diptera) from Palaearctic China. *Entomotaxonomia*, **13** (1), 65–69.

Rozkošný, R. & Knutson, L. (2005a). Sciomyzidae. In *Fauna Europaea: Diptera, Brachycera*, version 1.2, ed. T. Pape. http//www.faunaeur.org.

Rozkošný, R. & Knutson, L. (2005b). Phaeomyiidae. In *Fauna Europaea: Diptera, Brachycera*, version 1.2, ed. T. Pape. http//www.faunaeur.org.

Rozkošný, R. & Knutson, L. (2006). Two new species of *Euthycera* Latreille with records of Central Asian Sciomyzidae (Diptera). *Fol. Heyrovskyana, Ser. A.*, **13** (4), 155–169.

Rozkošný, R. & Kozánek, M. (1989). *Pherbellia koreana* sp. n. and a list of the Korean Sciomyzidae (Diptera). *Biológia (Bratislava)*, **44** (10), 1020–1026.

Rozkošný, R. & Vaňhara, J. (1992). Diptera (Brachycera) of the agricultural landscape in southern Moravia. *Acta. Sci. Nat. Brno*, **26** (4), 1–64.

Rozkošný, R. & Vaňhara, J. (1993). Diptera Brachycera of a forest steppe near Brno (Hády Hill). *Acta Sci. Nat. Brno*, **27** (2–3), 1–76.

Rozkošný, R. & Vaňhara, J. (1995). Monitoring Diptera in southern Moravia by pitfall traps. *Dipterol. Bohemoslov.*, **7**, 143–157.

Rozkošný, R. & Zuska, J. (1965). Species of the family Sciomyzidae (Diptera) new to Central Europe and description of a new *Pherbellia* from Czechoslovakia. *Entomologist*, **98** (1228), 197–206.

Ryder, C., Moran, J., Mc Donnell, R. J. & Gormally, M. J. (2005). Conservation implications of grazing practices on the plant and dipteran communities of a turlough in Co. Mayo, Ireland. *Biodivers. Cons.*, **14**, 187–204.

Sabrosky, C. W. (1997). Obituary. George C. Steyskal 1909–1996. *Proc. Ent. Soc. Wash.*, **99** (2), 379–398.

Sabrosky, C. W. (1999). *Family-Group Names in Diptera. An Annotated Catalog*, ed. F. C. Thompson. *Myia*, **10**, 3–360.

Sack, P. (1939). Sciomyzidae. In *Die Fliegen der palaearktischen Region, Lief.* **125**, Parts 1, 2, 3, ed. E. Lindner. Stuttgart: Schweitzerbart, pp. 1–87.

Schalie, H. van der. (1969). Man meddles with nature – Hawaiian style. *The Biologist*, **51** (3), 136–14.

Schiner, I. R. (1862). Vorläufiger Commentar zum dipterologischen Theile der "Fauna Austriaca," mit einer näheren Begründung der in derselben aufgenommen neuen Dipteren-Gattungen. IV. *Wien. Ent. Monatsch.*, **6**, 143–152.

Schiner, I. R. (1864) [1862]. Die Fliegen (Diptera). Vol. **2**, Part 8. In *Fauna Austriaca*, ed. L. Redtenbacher & I. R. Schiner. Wien: Karl Gerold's Sohn, pp. 1–80.

Schiner, I. R. (1868). Diptera. [Art. 1]. In *Reise der österreichischen Fregatte Novara um die Erde in den Jahren*

1857, 1858, 1859. Zoologischer Theil, Vol. **2**, Abt. 1, [Sect.] (in charge) B. von Wullerstorf-Urbair. Wien: B. K. Gerold's Sohn.

Schmitz, H. (1917). Biologische Beziehungen zwischen Dipteren und Schnecken. *Biol. Zentralbl.*, **37**, 24–43.

Schneider, M. R. (1988). Die Sciomyzidae-Fauna auf den Nordseeinseln Mellum und Memmert (Diptera). *Drosera*, **88**, 293–301.

Schumann, H., Bâhrmann, R. & Stark, A, ed. (1999). Entomofauna Germanica 2. Checkliste der Dipteren Deutschlands. *Stud. Ent.*, Suppl. **2**, 1–354.

Scopoli, J. A. (1763). *Entomologia Carniolica Exhibens Insecta Carnioliae Indigene et Distributa in Ordines, Genera, Species, Varietates, Methodo Linnaeana*. Vindobonae: Otto Guglia.

Scudder, S. H. (1877). The first discovered traces of fossil insects in the American Tertiaries. *Bull. U. S. Geol. Geogr. Surv. Terr.*, **3**, 741–762.

Scudder, S. H. (1878a). Additions to the insect-fauna of the Tertiary beds at Quesnel (British Columbia). *Rep. Prog. Geol. Surv. Can.*, 1875–1876, 266–280.

Scudder, S. H. (1878b). An account of some insects of unusual interest from the Tertiary rocks of Colorado and Wyoming. *Bull. U. S. Geol. Geogr. Surv. Terr.*, **4** (2), 519–543.

Scudder, S. H. (1878c). The fossil insects of the Green River shales. *Bull. U. S. Geol. Geogr. Surv. Terr.*, **4**, 747–776.

Séguy, E. (1934). Diptères. (Brachycères) (Muscidae Acalypterae et Scatophagidae). *Faune Fr.*, **28**, 1–832.

Séguy, E. (1937). Fasc. 8. Diptères. In *La Faune de la France en Tableaux Synoptiques Illustrés*, ed. R. Perrier. Paris: Delagrave, pp. 1–16.

Sivinski, J., Aluja, M., Dodson, G., Freidberg, A., Headrick, D., Kaneshiro, K. & Landolt, P. (1999). Topics in the evolution of sexual behavior in Tephritidae. In *Fruit Flies (Tephritidae): Phylogeny and Evolution of Behavior*, ed. M. Aluja & A. L. Norrbom. Boca Raton: CRC Press, pp. 751–792.

Soós, A. (1958). Ist das Insektenmaterial der Museen für ethologische und ökologische Untersuchungen verwendbar? Angaben über die Flugzeit und Generationszahl der Sciomyziden (Diptera). *Acta Ent. Mus. Natl. Pragae*, **32** (493), 101–150.

Speight, M. C. D. (1969). The prothoracic morphology of Acalypterates (Diptera) and its uses in systematics. *Trans. R. Ent. Soc. Lond.*, **121** (9), 325–421.

Speight, M. C. D. (1994). Some Liechtenstein records of snail-killing flies and their allies (Diptera: Phaeomyiidae & Sciomyzidae). *Ber. Bot.-Zool. Ges. Liechtenstein – Sargans – Werdenberg*, **21**, 185–190.

Speight, M. C. D. (2001). Farms as biogeographical units: 2. The potential role of different parts of the case-study farm in maintaining its present fauna of Sciomyzidae and Syrphidae (Diptera). *Bull. Ir. Biogeogr. Soc.*, **25**, 248–278.

Speight, M. C. D. (2004). Predicting impacts of changes in farm management on Sciomyzids (Diptera, Sciomyzidae): a biodiversity case study from southern Ireland. *Dipt. Dig.*, **11**, 147–166.

Speight, M. C. D. & Chandler, P. J. (1995). *Paragus constrictus, Pteromicra pectorosa* and *Stegana similis*: insects new to Ireland and *Stegana coleoptrata*, presence in Ireland confirmed (Diptera). *Ir. Nat. J.*, **25**, 28–32.

Speiser, P. (1910). Sciomyzidae. In *Wissenschaftliche Ergebnisse der schwedischen zoologischen Expedition nach dem Kilimandjaro, dem Meru und den Umgebenden Massaisteppen Deutsch-Ostafrikas 1905–1906*. Vol. **2**, Abt. 8–14, Pt. 10. *Diptera*. 5. *Cyclorrhapha, Aschiza*, ed. Y. Sjöstedt. Stockholm: Palmquist, pp. 168–171.

Stackelberg, A. A. (1963). Species of the genus *Tetanocera* Dum. (Diptera: Sciomyzidae) in the European part of the USSR. *Ent. Rev. USSR*, **42** (4), 492–497.

Staeger, C. (1845). Bemaerkninger til synonymien af *Sciomyza glabricula* Fall. *Naturhist. Tidsskr.*, (1844–1845). Ser. 2. **1**, 38–40.

Stephenson J. W. & Knutson, L. V. (1966). A resumé of recent studies of invertebrates associated with slugs. *J. Econ. Ent.*, **59** (2), 356–360.

Stevens, N. M. (1908). A study of the germ cells of certain Diptera with reference to the heterochromosomes and the phenomena of synapsis. *J. Exp. Zool.*, **5**, 359–374.

Steyskal, G. C. (1938). New Stratiomyidae and Tetanoceridae (Diptera) from North America. *Occas. Pap. Mus. Zool. Univ. Mich.*, **386**, 1–10.

Steyskal, G. C. (1949). New Diptera from Michigan (Stratiomyidae, Sarcophagidae, Sciomyzidae). *Pap. Mich. Acad. Sci., Arts, Lett.*, **33** (2), 173–180.

Steyskal, G. C. (1950). The genus *Protodictya* Malloch (Diptera, Sciomyzidae). *Proc. Ent. Soc. Wash.*, **52** (1), 53–59.

Steyskal, G. C. (1951a). [1950]. The genus *Sepedon* Latreille in the Americas (Diptera: Sciomyzidae). *Wasmann J. Biol.*, **8** (3), 271–297.

Steyskal, G. C. (1951b). A new species of *Tetanocera* from Korea (Diptera: Sciomyzidae). *Wasmann J. Biol.*, **9** (1), 79–80.

Steyskal, G. C. (1953). Notes on a collection of Argentine Sciomyzidae. *Acta Zool. Lilloana*, **13**, 73–75.

Steyskal, G. C. (1954a). *Colobaea* and *Hedria*, two genera of Sciomyzidae new to America (Diptera: Acalyptrate). *Can. Ent.*, **86** (2), 59–65.

Steyskal, G. C. (1954b). The American species of the genus *Dictya* Meigen (Diptera, Sciomyzidae). *Ann. Ent. Soc. Am.*, **47** (3), 511–539.

Steyskal, G. C. (1954c). The genus *Pteromicra* Lioy (Diptera: Sciomyzidae) with especial reference to the North American species. *Pap. Mich. Acad. Sci., Arts, Lett.*, **39**, 257–269.

Steyskal, G. C. (1954d). The Sciomyzidae of Alaska (Diptera). *Proc. Ent. Soc. Wash.*, **56** (2), 54–71.

Steyskal, G. C. (1956). New species and taxonomic notes in the family Sciomyzidae (Diptera: Acalyptratae). *Pap. Mich. Acad. Sci., Arts, Lett.*, **41**, 73–87.

Steyskal, G. C. (1958). A new species of the genus *Pteromicra* associated with snails (Diptera: Sciomyzidae). *Proc. Ent. Soc. Wash.*, **59** (6), 271–272.

Steyskal, G. C. (1959). The American species of the genus *Tetanocera* Duméril (Diptera). *Pap. Mich. Acad. Sci., Arts, Lett.*, **44**, 55–91.

Steyskal, G. C. (1960a). New North and Central American species of Sciomyzidae (Diptera: Acalyptratae). *Proc. Ent. Soc. Wash.*, **62** (1), 33–43.

Steyskal, G. C. (1960b). The genus *Antichaeta* Haliday, with special reference to the American species (Diptera: Sciomyzidae). *Pap. Mich. Acad. Sci., Arts, Lett.*, **45**, 17–26.

Steyskal, G. C. (1961). The North American Sciomyzidae related to *Pherbellia fuscipes* (Macquart) (Diptera: Acalyptratae). *Pap. Mich. Acad. Sci., Arts, Lett.*, **46**, 405–415.

Steyskal, G. C. (1963). Taxonomic notes on Sciomyzidae (Diptera, Acalyptratae). *Pap. Mich. Acad. Sci., Arts, Lett.*, **48**, 113–125.

Steyskal, G. C. (1965a). The subfamilies of Sciomyzidae of the world (Diptera: Acalyptratae). *Ann. Ent. Soc. Am.*, **58** (4), 593–594.

Steyskal, G. C. (1965b). Family Sciomyzidae (Tetanoceridae). In *A Catalog of the Diptera of America North of Mexico*, ed. A. Stone *et al.* U.S. Dept. Agric. Handbook No. 276, pp. 685–695.

Steyskal, G. C. (1967) [1966]. The Nearctic species of *Pherbellia* Robineau-Desvoidy, subgenus *Oxytaenia* Sack (Diptera: Sciomyzidae). *Pap. Mich. Acad. Sci., Arts, Lett.*, **51**, 31–36.

Steyskal, G. C. (1973). A new classification of the *Sepedon* group of the family Sciomyzidae (Diptera) with two new genera. *Ent. News*, **84** (5), 143–146.

Steyskal, G. C. (1974a). A gynandromorphic specimen of the genus *Limnia* (Diptera: Sciomyzidae). *J. Wash. Acad. Sci.*, **64** (1), 11.

Steyskal, G. C. (1974b) [1973]. The genus *Dictyodes* Malloch (Diptera: Sciomyzidae). *Proc. Ent. Soc. Wash.*, **75** (4), 427–430.

Steyskal, G. C. (1980). Family Sciomyzidae. In *Insects of Hawaii*. Vol. **13**, ed. D. E. Hardy & M. D. Delfinado. Honolulu: Univ. Press of Hawaii, pp. 108–125.

Steyskal, G. C. & El-Bialy, S. (1968) [1967]. A list of Egyptian Diptera with a bibliography and key to families. *United Arab Republic. Min. Agric. Tech. Bull.*, **3**. 87 pp.

Steyskal, G. C., Fisher, T. W., Knutson, L. & Orth, R. E. (1978). Taxonomy of North American flies of the genus *Limnia* (Diptera: Sciomyzidae). *Univ. Calif. Publs. Ent.*, **83**, 1–48.

Steyskal, G. C. & Knutson, L. V. (1975a). The cochleate vesicle, a highly specialized device for sperm transfer in male sciomyzid flies (Diptera). *Ann. Ent. Soc. Am.*, **68** (2), 367–370.

Steyskal, G. C. & Knutson, L. (1975b). Key to the genera of Sciomyzidae (Diptera) from the Americas south of the United States, with descriptions of two new genera. *Proc. Ent. Soc. Wash.*, **77** (3), 274–277.

Steyskal, G. C. & Knutson, L. (1978). Helosciomyzinae in Australia (Diptera: Sciomyzidae). *Aust. J. Zool.*, **26** (4), 727–743.

Steyskal, G. C. & Verbeke, J. (1956). Sepedoninae (Sciomyzidae, Diptera) from Africa and southern Arabia. *Bull. Inst. r. Sci. natr. Belg.*, **32** (7), 1–14.

Stratton, L. W. (1963). An ecological study. *J. Conchol.*, **25** (5), 174–179.

Stratton, L. W. (1970). Numbers of shells. *J. Conchol.*, **27**, 171–176.

Strickland, E. H. (1953). The ptilinal armature of flies (Diptera, Schizophora). *Can. J. Zool.*, **31**, 263–299.

Stuke, J.-H. (2005). Die Sciomyzoidea (Diptera: Acalyptratae) Niedersachsens und Bremens. *Drosera*, **2005** (2), 135–166.

Sucharit, S., Chandavimol, Y. & Sornmani, S. (1976). Biology of the snail-killing fly, *Sepedon spangleri* Beaver (Diptera: Sciomyzidae). II. Ability of the larvae to kill snails of medical importance in Thailand. *Southeast Asian J. Trop. Med. Publ. Hlth.*, **7** (4), 581–585.

Sueyoshi, M. (2001). A revision of Japanese Sciomyzidae (Diptera), with description of three new species. *Ent. Sci.*, **4** (4), 485–506.

Sueyoshi, M. (2002). Phylogeny of Dryomyzidae (Diptera: Acalyptrata). In *5th Int. Congr. Dipterol.* 29 Sept.–4 Oct., 2002, Brisbane, Australia. Conf. chair D. Yeates. *Abstracts*, p. 231.

Sueyoshi, M. (2005). Sciomyzidae. In *Aquatic Insects of Japan: Manual with Keys and Illustrations*, ed. T. Kawai & K. Tanida. Kanagawa: Tokai Univ. Press, pp. 1229–1256.

Sueyoshi, M., Knutson, L. & Ghorpadé, K. (2006). Discovery of new species of *Pelidnoptera* Rondani and related new genus from Asia and their implications to the basal lineage of Sciomyzidae. In *6th Int. Congr. Dipterol.*, 23–28 Sept., 2006, Fukuoka, Japan, ed. M. Suwa. *Abstracts*, pp. 242–243.

Sueyoshi, M., Knutson, L., & Ghorpadé, K. (2009). A taxonomic review of *Pelidnoptera* Rondani (Diptera: Sciomyzoidea), with discovery of a related new genus and species from Asia. *Ins. Syst. Evol.*, **40** (4), 389–409.

Szadziewski, R. (1983). Flies (Diptera) of the saline habitats of Poland. *Pol. Pismo Ent.*, **53**, 31–76.

Teskey, H. (1981). Morphology and terminology – larvae. In *Manual of Nearctic Diptera*. Vol. **1**, ed. J. F. McAlpine *et al. Res. Br., Agric. Canada. Monogr.*, **27**, 65–88.

Theobald, N. (1937). *Les insectes fossiles des terrains oligocènes de France*. Nancy: O. Thomas.

Thomas, J. D. (1973). Schistosomiasis and the control of molluscan hosts of human schistosomes with particular reference to possible self-regulatory mechanisms. *Adv. Parasitol.*, **11**, 307–395.

Thomson, C. G. (1869) [1868]. 6. Diptera. Species Nova Descripsit. (= h. 12, no. 2). In *K. Svenska Vetenskaps-Akademien. Kongliga Svenska Fregatten Eugenies Resa Omkring Jorden under Befäl af C. A. Virgin, åren 1851–1853*. Pt. 2: *Zoologi*, [sec.] 1, *Insekter*. Stockholm: P. A. Norstedt & Söner, pp. 443–614, pl. 9.

Tirgari, S. (1977). Theoretical and experimental evidences of biological control of aquatic snails by snail-killing flies in relation to schistosomiasis (Diptera: Sciomyzidae) (*Sepedon sphegea*). *First Medit. Conf. Parasitol.*, 5–10 Oct., 1977, Izmir, Turkey. *Summaries*, pp. 103–04.

Tirgari, S. (1986). On the biology and mass production procedure of snail-killing flies (*Sepedon sphegea* F.) (Diptera, Sciomyzidae). *3rd European Congr. Ent.*, 24–29 Aug., 1986, Amsterdam, the Netherlands. *Abstracts*, p. 51.

Tirgari, S. & Massoud, J. (1981). Study on the biology of snail-killing flies and prospect of biological control of aquatic snails *Sepedon sphegea* (Fabricius) (Insecta, Diptera, Sciomyzidae). *Sci. Publ. 2051*. Tehran: Sch. Publ. Hlth., Inst. Publ. Hlth. Res., Tehran Univ.

Tonnoir, A. L. & Malloch, J. R. (1928). New Zealand Muscidae Acalyptratae. Part IV. Sciomyzidae. *Rec. Canterbury Mus.*, **3** (3), 151–179.

Trelka, D. G. (1973). *The Behavior of Predatory Larvae of* Tetanocera plebeia *(Diptera: Sciomyzidae) and Toxicological and Neurological Aspects of a Toxic Salivary Secretion Used to Immobilize Slugs*. Ph.D. thesis, Cornell Univ., Ithaca, NY, USA. 125 pp. Order No. 73–6673. Univ. Microfilms, Ann Arbor, MI. (*Diss. Abstr.*, **33** (9), 4322-B).

Trelka, D. G. & Berg, C. O. (1977). Behavioral studies of the slug-killing larvae of two species of *Tetanocera* (Diptera: Sciomyzidae). *Proc. Ent. Soc. Wash.*, **79** (3), 475–486.

Trelka, D. G. & Foote, B. A. (1970). Biology of slug-killing *Tetanocera* (Diptera: Sciomyzidae). *Ann. Ent. Soc. Am.*, **63** (3), 877–895.

Tscheuschner, M. H., Appleton, C. C. & Miller, R. M. (1993). Prospects for bilharzia snail control by sciomyzid flies. In *Proc. Ninth Ent. Soc. So. Afr. Congr.*, 28 June–1 July, 1993, Johannesburg, South Africa. *Abstracts*, p. 163.

Tscheuschner, M. H., Appleton, C. C. & Miller, R. M. (1994). Prospects for bilharzia snail control by sciomyzid flies. In *3rd Int. Congr. Dipterol.*, 15–19 August, 1994, Univ. Guelph, Canada, ed. J. E. O'Hara. *Abstracts*, pp. 228–229.

Tscheuschner, M. H., Appleton, C. C. & Miller, R. M. (2002). Prospects for bilharzia control by sciomyzid flies. In *5th Int. Congr. Dipterol.*, 29 Sept.–4 Oct., 2002, Brisbane, Australia. Conf. chair D. Yeates. *Abstracts*, p. 243.

Vala, J-C. (1983). Description de *Euthycera alaris*, n. sp. et désignation du lectotype et du paralectotype de *Euthycera flavostriata* (Villeneuve, 1911) (Diptera: Sciomyzidae). *Rev. Fr. Ent. (N.S.)*, **5** (4), 166–170.

Vala, J-C. (1984a). Diptères Sciomyzidae: classification supragénérique et détermination pratique des genres pour l'Europe. *Bull. Soc. Linn. Lyon*, **5**, 167–175.

Vala, J-C. (1984b). Phenology of Diptera Sciomyzidae in a Mediterranean forestry biotop. *Ent. Basil.*, **9**, 432–440.

Vala, J-C. (1986). Description des stades larvaires et données sur la biologie et la phénologie de *Trypetoptera punctulata* (Diptera, Sciomyzidae). *Ann. Soc. Ent. Fr. (N.S.)*, **22** (1), 67–77.

Vala, J-C. (1989a). *Diptères Sciomyzidae Euro-méditerranéens. Faune de France. France et Régions limitrophes. N° 72.* Paris: Féd. Franc. Soc. Sci. Nat.

Vala, J-C. (1989b). Les mouches Sciomyzidae: des insectes utiles. *O.P.I.E. Insectes*, **74** (3), 16–20.

Vala, J-C. (1990). Une nouvelle espèce d'*Euthycera* Latreille récoltée en France (Diptera: Sciomyzidae). *Ann. Soc. Ent. Fr. (N.S.)*, **26** (3), 451–455.

Vala, J-C., Bailey, P. T. & Gasc, C. (1990). Immature stages of the fly *Pelidnoptera nigripennis* (Fabricius) (Diptera: Phaeomyiidae), a parasitoid of millipedes. *Syst. Ent.*, **15**, 391–399.

Vala, J-C. & Brunel, C. (1987). Diptères Sciomyzides capturés dans le Département de la Somme. *Bull. mem. Soc. linn. Lyon*, **56** (6), 187–191.

Vala, J-C. & Caillet, C. (1985). Description des stades immatures et biologie de *Euthycera leclercqi* (Diptera, Sciomyzidae). *Rev. fr. Ent.*, **7** (1), 19–26.

Vala, J-C., Caillet, C. & Gasc, G. (1987). Biology and immature stages of *Dichetophora obliterata*, a snail killing fly (Diptera–Sciomyzidae). *Can. J. Zool.*, **65**, 1675–1680.

Vala, J-C. & Gasc, C. (1990a). *Pherbina mediterranea*: immature stages, biology, phenology and distribution (Diptera: Sciomyzidae). *J. Nat. Hist.*, **24**, 441–451.

Vala, J-C. & Gasc, C. (1990b). Ecological adaptations and morphological variation in the posterior disc of larvae of Sciomyzidae (Diptera). *Can. J. Zool.*, **68** (3), 517–521.

Vala, J-C., Gasc, C., Gbedjissi G. & Dossou, C. (1995). Life history, immature stages and sensory receptors of *Sepedon* (*Parasepedon*) *trichrooscelis* an Afrotropical snail-killing fly (Diptera: Sciomyzidae). *J. Nat. Hist.*, **29**, 1005–1014.

Vala, J-C., Gbedjissi, G. & Dossou, C. (1994). Les Sciomyzidae du Bénin, description de *Sepedon* (*Mesosepedon*) *knutsoni* n. sp. (Diptera, Sciomyzidae). *Bull. Soc. Ent. Fr.*, **99** (5), 497–504.

Vala, J-C., Gbedjissi, G., Knutson, L. & Dossou, C. (2000a). Not all Sciomyzidae kill molluscs: *Sepedonella nana* eats worms. In *XXI Int. Congr. Ent.*, 20–26 Aug., 2000, Brazil, ed. D. L. Gazzoni. *Abstract Book I*, p. 973.

Vala, J-C., Gbedjissi, G., Knutson, L. & Dossou, C. (2000b). Extraordinary feeding behavior in Diptera Sciomyzidae, snail-killing flies. *C. R. Acad. Sci. Paris, Sci. Vie, Sér. III*, **323** (3), 245–340.

Vala, J-C., Gbedjissi, G., Knutson, L. & Dossou, C. (2002). *Sepedon knutsoni*, a second oligochaete feeding sciomyzid from Africa. In *5th Int. Congr. Dipterol.*, 29 Sept.–4 Oct., 2002, Brisbane, Australia. Conf. chair D. Yeates. *Abstracts*, p. 250.

Vala, J-C. & Ghamizi, M. (1991). Sciomyzidae du Maroc. *L'Entomologiste*, **47** (4), 205–208.

Vala, J-C. & Ghamizi, M. (1992). Aspects de la biologie de *Pherbellia schoenherri* parasitoïde de *Succinea elegans* (Mollusca) (Diptera, Sciomyzidae). *Bull. Soc. Ent. Fr.*, **97** (2), 145–154.

Vala, J-C., Greve, L. & Knutson, L. (2000c). Description of the male of *Verbekaria punctipennis* (Diptera: Sciomyzidae). *Stud. dipterol.*, **7** (1), 247–255.

Vala, J-C. & Haab, C. (1984). Etude expérimentale du développement larvaire de *Tetanocera ferruginea* Fallén 1820. Influences de la température et de la photopériode, diapause pupale, biomasse alimentaire. *Bull. Ann. Soc. r. Belg. Ent.*, **120**, 165–178.

Vala, J-C. & Knutson, L. (1990). Stades immatures et biologie de *Limnia unguicornis* (Scopoli), Diptère Sciomyzidae prédateur de mollusques. *Ann. Soc. Ent. Fr. (N. S.)*, **26** (3), 443–450.

Vala, J-C., Knutson, L. V. & Gasc, C. (1999). Stereoscan studies with descriptions of new characters of the egg and larval instars of *Salticella fasciata* (Meigen) (Diptera: Sciomyzidae). *J. Zool.*, **247**, 531–536.

Vala, J-C. & Manguin, S. (1987). Dynamique et relations Sciomyzidae-Mollusques d'un biotope aquatique asséchable dans le sud de la France (Diptera). *Bull. Ann. Soc. r. Belg. Ent.*, **123** (2), 153–164.

Vala, J-C., Pineau, X. & Bro, E. (2006). Discrimination entomologique des types de jachères. *Symbioses, Bull. Mus. Hist. natr. Reg. Centre (N. S.)*, **17**, 1–10.

Vala, J-C., Reidenbach, J. M. & Gasc, C. (1983). Biologie des stades larvaires de *Euthycera cribrata* (Rondani 1868), parasitoide de gastéropodes terrestres. Premier cycle d'une espèce du genre *Euthycera* Latreille 1829 (Dipt., Sciomyzidae). *Bull. Soc. Ent. Fr.*, **88**, 250–258.

Valley, K. R. (1974). *Biology and Immature Stages of Snail-killing Diptera of the Genus* Dictya *(Sciomyzidae)*. Ph.D. thesis, Cornell Univ., Ithaca, NY, USA. 174 pp. Order No. 75–1630, Univ. Microfilms, Ann Arbor, MI. (*Diss. Abstr.*, **35** (7), 3372-B).

Valley, K. R. & Berg, C. O. (1977). Biology and immature stages of snail-killing Diptera of the genus *Dictya* (Sciomyzidae). *Search. Agric. Ent. (Ithaca)*, **18**, 7 (2), 1–44.

Vaňhara, J. (1981). Lowland forest Diptera (Brachycera, Cyclorrhapha). *Acta Sci. Nat. Acad. Sci. Bohemoslov., Brno*, **15** (1), 1–32.

Vaňhara, J. (1986). Impact of man-made moisture changes on floodplain forest Diptera. *Acta Sci. Nat. Acad. Sci. Bohemoslov., Brno*, **20** (7), 1–36.

Vaňhara, J. & Rozkošný, R. (1997). Long-term development of floodplain forest Diptera (Brachycera) in the lower part of the Dyje River. In *Dipterol. Bohemoslov.* Vol. **8**, ed. J. Vaňhara & R. Rozkošný. *Fol. Fac. Sci. Nat. Univ. Masaryk. Brun., Biol.*, 95, 193–199.

Verbeke, J. (1948). Contribution à l'étude des Sciomyzidae de Belgique (Diptera). *Bull. Mus. r. Hist. natr. Belg.*, **24** (3), 1–31.

Verbeke, J. (1950). Sciomyzidae (Diptera Cyclorrhapha). Fasc. 66. In *Exploration Parc natl. Albert, Mission G. F. de Witte, (1933–35)*. Brussels: Inst. Parcs natn. Congo Belge.

Verbeke, J. (1956). Contributions à l'étude de la faune entomologique du Ruanda-Urundi. CV. Diptera

Micropezidae, Sciomyzidae et Psilidae. *Ann. Mus. r. Congo Belg. (Ser. 8, Sci. Geol.)*, **51**, 475–488.

Verbeke, J. (1960). Révision du genre *Pherbina* Robineau-Desvoidy (Diptera: Sciomyzidae). *Bull. Inst. r. Sci. natr. Belg.*, **36** (34), 1–15.

Verbeke, J. (1961). Sciomyzidae (Diptera Brachycera Malacophaga). Fasc. 61. In *Explor. Parc natl. Upemba. Mission G. F. de Witte*. Brussels: Inst. Parcs natn. Congo, Rwanda-Burundi.

Verbeke, J. (1962a). Contribution à l'étude des diptères malacophages. I. Sciomyzidae nouveaux ou peu connus d'Afrique du Sud et de Madagascar. *Bull. Inst. r. Sci. natr. Belg.*, **38** (54), 1–16.

Verbeke, J. (1962b). Sciomyzidae africains (Diptera) (Ergebnisse der Forschungsreise Lindner 1958–59 – Nr. 14). *Stuttg. Beitr. Naturk.*, **93**, 1–4.

Verbeke, J. (1963). Sciomyzidae Sepedoninae (Diptera: Brachycera). *Explor. Parc natn. Garamba, Mission H. de Saeger*, **39** (3), 51–86.

Verbeke, J. (1964a). Contribution à l'étude des diptères malacophages. II. Données nouvelles sur la taxonomie et la répartition géographique des Sciomyzidae paléarctiques. *Bull. Inst. r. Sci. natr. Belg.*, **40** (8), 1–27.

Verbeke, J. (1964b). Contribution à l'étude des diptères malacophages. III. Révision du genre *Knutsonia* nom. nov. (= *Elgiva* Auct.). *Bull. Inst. r. Sci. natr. Belg.*, **40** (9), 1–44.

Verbeke, J. (1967a). Contribution à l'étude des diptères malacophages. IV. L'identité de *Pherbellia obtusa* (Fallén) (1820) et la description d'une espèce paléarctique nouvelle *Pherbellia argyra* sp. n. (Diptera: Sciomyzidae). *Bull. Inst. r. Sci. natr. Belg.*, **43** (6), 1–9.

Verbeke, J. (1967b). Contribution à l'étude des diptères malacophages. V. Trois espèces paléarctiques nouvelles du genre *Pherbellia* Robineau-Desvoidy et quelques données sur l'identité de *P. scutellaris* (von Roser) (Diptera: Sciomyzidae). *Bull. Inst. r. Sci. natr. Belg.*, **43** (18), 1–12.

Verbeke, J. & Knutson, L. V. (1967). Sciomyzidae. In *Limnofauna Europaea*, ed. J. Illies. Stuttgart: Gustav Fischer Verlag, pp. 417–421.

Vikhrev, N. E. (2008). New species of *Anticheta* Haliday, 1838 (Diptera: Sciomyzidae) from Russian Far East. *Russian Ent. J.*, **17** (2), 241–242.

Villeneuve, J. (1911). Diptères nouveaux recueillis en Syrie par M.H. Gadeau de Kerville et décrits. *Bull. Soc. Sci. natr. Rouen*, **47**, 40–55.

Vimmer, A. 1925. *[Larvae and pupae of Central-European Diptera with special reference to pests of cultivated plants]*. Prague: Nakledem Česke.

Wagner, R., Barták, M., Borkent, A., Courtney, G., Goddeeris, B., Haenni, J.-P., Knutson, L., Pont, A., Rotheray, G. E., Rozkošný, R., Sinclair, B., Woodley, N., Zatwarnicki, T. & Zwick, P. (2008). Global diversity of dipteran families (Insecta Diptera) in freshwater (excluding Simulidae, Culicidae, Chironomidae, Tipulidae and Tabanidae). *Hydrobiologia*, **595**, 489–519.

Walker, F. (1837). Notes on Diptera. *Ent. Mag. (Lond.)*, **4**, 226–230.

Walker, F. (1849). *List of the Specimens of Dipterous Insects in the Collection of the British Museum*. Part 4, pp. 689–1172. London: British Museum.

Walker, F. (1853). Diptera. Part IV. cont. In *Insecta Saundersiana: or Characters of Undescribed Insects in the Collection of W. W. Saunders, Esq.* Vol. **1**, ed. W. W Saunders. London: Van Voorst, pp. 253–414, pls. 7–8.

Walker, F. (1858). Catalogue of the dipterous insects collected in the Aru Islands by Mr. A. R. Wallace, with descriptions of new species. [concl]. *J. Proc. Linn. Soc. Lond. (Zool.)*, **3**, 111–131.

Walker, F. (1859). Catalogue of the dipterous insects collected at Makessar in Celebes, by Mr. A. R. Wallace, with descriptions of new species. [part]. *J. Proc. Linn. Soc. Lond. (Zool.)*, **4**, 90–144.

Webb, J. R. & Lott, D. A. (2006). The development of ISIS: a habitat-based invertebrate assemblage classification system for assessing conservation interest in England. *J. Insect Conserv.*, **10**, 179–188.

Wesché, W. (1904). The mouth parts of the Nemocera and their relations to the other families in Diptera. *J. R. Microscop. Soc.*, **1904**, 28–47.

Wesenberg-Lund, C. (1943). *Biologie der Süsswasserinsekten*. Copenhagen: Gyldendalske Boghandel, Nordisk Forlag.

Whalley, P. E. S. & Jarzembowski, E. A. (1985). Fossil insects from the lithographic limestone of Montsech (Late Jurassic–Early Cretaceous) Lérida Province, Spain. *Bull. Br. Mus. Nat. Hist. (Geol.)*, **38**, 381–412.

Whiles, M. R. & Goldowitz, B. S. (2001). Hydrologic influences on insect emergence production from Central Platte River wetlands. *Ecol. Applic.*, **11** (6), 1829–1842.

Wiedemann, C. R. W. (1819). Brasilianische Zweiflügler. *Zool. Mag. (Wiedemann's)*, **1** (3), 40–56.

Wiedemann, C. R. W. (1824). *Munus Rectoris in Academia Christiana Albertina Aditurus Analecta Entomologica ex Museo*. Kiliae: Regio Typographeo scholarum.

Wiedemann, C. R. W. (1830). *Aussereuropäische zweiflügelige Insekten* [concl.] Vol. **2**. Hamm: Schulz.

Wild, S. V. & Lawson, A. K. (1937). Enemies of the land and freshwater Mollusca of the British Isles. *J. Conchol.*, **20** (12), 351–361.

Williams, C. D., Gormally, M. J., & Knutson, L. V. (2010). Very high population estimates and limited movement of snail-killing flies (Diptera: Sciomyzidae) on an Irish turlough (temporary lake). *Proc. R. Irish Acad. Biol. Env.*, **1108** (2), 81–94.

Williams, C. D., Moran, J., Doherty, O., Mc Donnell, R. J., Gormally, M. J., Knutson, L. & Vala, J-C. (2009). Factors affecting Sciomyzidae (Diptera) across a transect at Skealoghan Turlough (Co. Mayo, Ireland). *Aquatic Ecol.*, **43** (1), 117–133.

Willomitzer, J. (1970). Some observations on experimental breeding of sciomyzid flies (Diptera) under laboratory conditions. *Acta Vet. Brno*, **39**, 307–313.

Willomitzer, J. & Rozkošný, R. (1977). Further observation on the rearing of sciomyzid larvae (Diptera) for the control of intermediate host snails. *Acta Vet. Brno*, **46**, 315–322.

Wollaston, T. V. (1858). Brief diagnostic characters of undescribed Madieran insects [concl.]. *Ann. Mag. Nat. Hist., Ser.* **3** (1), 18–28, 113–125.

Wulp, F. M. van der. (1897). Family Muscidae, Sciomyzinae. In *Biologia Centrali-Americana. Insecta Diptera*. Vol. **2**, ed. F. D. Godman & O. Salvin. London: Taylor & Francis, pp. 344–360, pl. IX.

Yano, K. (1968). Notes on Sciomyzidae collected in paddy field (Diptera). I. *Mushi*, **41** (15), 189–200.

Yano, K. (1975). Bionomics of *Sepedon sphegeus* (Fabricius) (Diptera: Sciomyzidae). In *Approaches to Biological Control, JIBP Synthesis*. Vol. **7**, ed. K. Yasumatsu & H. Mori. Tokyo: Univ. Tokyo Press, p. 85.

Yano, K. (1978). Faunal and biological studies on the insects of paddy fields in Asia. Part I. Introduction and Sciomyzidae from Asia (Diptera). *Esakia*, **11**, 1–27.

Yano, K. (1984). Biology of marsh flies (Diptera: Sciomyzidae). *Insectarium*, **21** (2), 4–7.

Yoneda, Y. (1981). The effect of temperature on development and predation of marsh fly, *Sepedon aenescens* Wiedemann (Diptera: Sciomyzidae). *Jpn. J. Sanit. Zool.*, **32** (2), 117–123.

Yoneda, Y. (1984). Malformation of *Sepedon aenescens* Wiedemann (Diptera: Sciomyzidae). *Makunagi*, **12**, 11–19.

Zaitzev, V. F. (1992). *Evolutionary Pathways of Morphofunctional Adaptations.* St. Petersburg: Nauka.

Zetterstedt, J. W. (1837). *Conspectus Familiarum, Generum et Specierum Dipterorum*. In *Fauna Insectorum Lapponica Descriptorum. Isis.(Oken's)*, **21**, 28–67.

Zetterstedt, J. W. (1838). *Dipterologis Scandinaviae. Sectio Tertia. Diptera*. In *Insecta Lapponica*. Lipsiae, pp. 477–868.

Zetterstedt, J. W. (1846). *Diptera Scandinaviae. Disposita et Descripta*. Vol. **5**. Lund: Officina Lundbergiana, pp. 1739–2162.

Zetterstedt, J. W. (1847). *Diptera Scandinaviae. Disposita et Descripta*. Vol. **6**. Lund: Officina Lundbergiana, pp. 2163–2580.

Zherichin, V. V. (1985). Insects of the Cretaceous Period in Europe. *Priroda*, **1985** (10), 119–120. [in Russian].

Zuska, J. & Berg, C. O. (1974). A revision of the South American genus *Tetanoceroides* (Diptera: Sciomyzidae), with notes on colour variations correlated with mean temperatures. *Trans. R. Ent. Soc. Lond.*, **125** (3), 329–362.

PART B[2]

REFERENCES CITED OTHER THAN SCIOMYZIDAE AND PHAEOMYIIDAE

Addison, J. (1770). Cited by Stearn, W. T. (1981). *The Natural History Museum at South Kensington: a History of the British Museum (Natural History) 1953–1980.* London: Heinemann.

Aditya, G. & Raut, S. K. (2001). Predation of water bug *Sphaerodema rusticum* Fabricius on the snail *Pomacea bridgesi* (Reeve), introduced in India. *Curr. Sci.*, **81**, 1413–1414.

Aditya, G. & Raut, S. K. (2002). Predation of water bug *Sphaerodema rusticum* on the freshwater snails *Lymnaea (Radix) luteola* and *Physa acuta*. *The Veliger*, **45**, 267–269.

Aldrich, J. M. (1916). *Sarcophaga and Allies in North America*. Thomas Say Publ. No. I. Lafayette, IN: Ent. Soc. Am.

Andrewartha, H. G. & Birch, L. C. (1954). *The Distribution and Abundance of Animals.* Chicago: Univ. Chicago Press.

Arditi, R. (1983). A unified model of the functional response of predators and parasitoids. *J. Anim. Ecol.*, **52**, 293–303.

Aristotle, S. (384–322 BC). *De Historia Animalium. Book 5.* Oxford: Clarendon. (translated by D. W. Thompson).

Aristotle, S. (384–322 BC). *Traité de l'âme.* **I**, I, 2.

Aristotle, S. (384–322 BC). *Politique.* **I**, 1, 2.

Armitage, P. D. (1968). Some notes on the food of the chironomid larvae of a shallow woodland lake in South Finland. *Ann. Zool. Fenn.*, **5**, 6–13.

[2] Quotations cited at the beginning of Chapters or Sections are attributed below, except for some taken from published collections of quotations where no original date of publication was given.

Armitage, P. D., Cranston, P. S. & Pinder, L. C. V. (1995). *The Chironomidae. The Biology and Ecology of Non-biting Midges*. London: Chapman & Hall.

Askew, R. R. (1971). *Parasitic Insects*. London: Heinemann.

Ayal, Y., Broza, M. & Pener, M. P. (1999). Geographical distribution and habitat segregation of bush crickets (Orthoptera: Tettigoniidae) in Israel. *Isr. J. Zool.*, **45**, 49–64.

Bacon, F. (1605). *Advancement of Learning*. Book 1.

Baer, J. G. (1953). Notes de faunistiques éburnéenne. III. Contribution à l'étude morphologique et biologique de *Wandolleckia achatinae* Cook, Phoridae (Diptera) commensal d'Achatines de la forêt tropicale. *Acta Tropica*, **10**, 73–79.

Bailey, N. T. J. (1951). On estimating the size of mobile populations from recapture data. *Biometrika*, **38**, 293–306.

Bailey, N. T. J. (1952). Improvements in the interpretation of recapture data. *J. Anim. Ecol.*, **21**, 120–127.

Balzac, H. de. (1832). Ch. **16**. *Louis Lambert*. In *La Comédie humaine*. Œuvres complètes de H. de Balzac. Paris: Edt. Furne, 1846.

Bameul, F. (1990). Le DMHF: un excellent milieu de montage en entomologie. *L'Entomologiste*, **46** (5), 233–239.

Bardach, J. E. (1972). Some ecological implications of Mekong River development plans. In *The Careless Technology*, ed. M. T. Farven & J. P. Milton. Garden City, NY: The Natural History Press.

Barker, J. F. (1969). Notes on the life cycle and behaviour of the drilid beetle *Selasia unicolor* (Guerin). *Proc. R. Ent. Soc. Lond. A*, **44**, 169–172.

Barnard, K. H. (1911). Chironomid larvae and watersnails. *Ent. Mon. Mag.*, **47**, 76–78.

Bates, M. (1949). *The Natural History of Mosquitoes*. New York: Harper & Row.

Beaver, R. A. (1972). Ecological studies on Diptera breeding in dead snails. I. Species found in *Cepaea nemoralis* (L.). *Entomologist (Lond.)*, **105** (1305), 41–52.

Beaver, R. A. (1977). Non-equilibrium "island" communities: Diptera breeding in dead snails. *J. Anim. Ecol.*, **46**, 783–798.

Beddington, J. R., Hassell, M. P. & Lawton, J. H. (1976). The components of arthropod predation. II. Predator rate of increase. *J. Anim. Ecol.*, **45**, 165–185.

Beedham, G. E. (1966). A chironomid (Dipt.) larva associated with the lamellibranchiate mollusk, *Anodonta cygnea* L. *Ent. Mon. Mag.*, **101**, 142–143.

Beedham, G. E. (1970). A further example of an association between a chironomid (Diptera) larva and a bivalve mollusc. *Ent. Mon. Mag.*, **105**, 3–5.

Benthem-Jutting, T. van. (1938). A freshwater pulmonate (*Physa fontinalis* (L.)) inhabited by the larva of a non-biting midge (*Tendipes* (*Parachironomus*) *varus* Goetg.). *Arch. Hydrobiol.*, **32** S, 693–699.

Bergson, H. (1907). *L'Evolution créatrice*, ed. PUF. ed 2007. Paris: Collection Quadrige textes.

Bernard, C. (1865). *Introduction à l'étude de la médecine expérimentale*. edn. 1966. Paris: Eds. Garnier-Flammarion.

Berthold, A. A. (1827). *Latreille's natürliche Familien des Thierrichs. Aus dem Französichen. mit Anmerkungen und Zusätzen*. Weimar: Gr. H. S. priv. Landes = Industrie.

Bigot, L. & Bodot, P. (1972–1973). Contribution à l'étude biocénotique de la garrigue à *Quercus coccifer*. II. Composition biotique du peuplement des invertébrés. *Vie Mil.*, **23** (2), 229–249.

Bliss, C. I. & Owen, A. R. G. (1958). Negative binomial distributions with a common *k*. *Biometrika*, **45**, 37–58.

Borgmeier, T. (1963). Revision of the North American phorid flies. Part I. The Phorinae, Aenigmatiinae and Metopininae, except *Megaselia* (Diptera, Phoridae). *Stud. ent. (N.S.)*, **6**, 1–256.

Börner, C. (1909). Die Verwandlungen der Insekten (Vorl. Mitt.). *Sitzungsber. Ges. Naturforsch. Freunde Berlin*, **7** (1903), 290–311.

Borradaile, L. A., Potts, F. A., Eastham, L. E. S. & Saunders, J. T. (1951). *The Invertebrata*, 2nd edn. London: Cambridge Univ. Press.

Bourguignat, J. R. (1864). *Malacologie de l'Algérie. Histoire Naturelle des Animaux. Mollusques Terrestres et Fluviatiles*. Vol. **2**. Paris: Challamel Bastid.

Brauer, F. (1863). *Monographie der Oestriden*. Wien: C. Ueberreuter.

Bray, J. R. & Curtis, J. T. (1957). An ordination of upland forest communities of southern Wisconsin. *Ecol. Monog.*, **27**, 325–349.

Brown, B. V. (1992). Generic revision of Phoridae of the Nearctic Region and phylogenetic classification of Phoridae, Sciadoceridae, and Ironomyiidae (Diptera: Phoroidea). *Mem. Ent. Soc. Can.*, **164**, 1–144.

Brown, B. V. (1993). Taxonomy and preliminary phylogeny of the parasitic genus *Apocephalus*, subgenus *Mesophora* (Diptera: Phoridae). *Syst. Ent.*, **18**, 191–230.

Brown, D. S. (1980). *Freshwater Snails of Africa and their Medical Importance*. London: Taylor & Francis.

Brown, D. S. & Kristensen, T. K. (1993). *A Field Guide to African Freshwater Snails. 1. West African Species*. Charlottenlund: Danish Bilharziasis Lab.

Bryant, E. H. & Hall, A. E. (1975). The role of medium conditioning in the population dynamics of the housefly. *Res. Pop. Ecol.*, **16**, 188–197.

Buck, J. B. & Keister, M. L. (1953). Cutaneous and larval respiration in the *Phormia* larva. *Biol. Bull.*, **105** (3), 402–411.

Buffon, L. L. de. (1749). *Histoire Naturelle Générale et Particulière*. **13** vols. Paris: Plonteaux.

Burch, J. B. (1982). *Freshwater Snails (Mollusca: Gastropoda) of North America*. Cincinnati, OH: U. S. Env. Prot. Ag., Env. Monit. & Sup. Lab.

Burger, J. F., Anderson, J. A. & Knudson, M. F. (1980). The habitats and life history of *Oedoparena glauca* (Diptera: Dryomyzidae), a predator of barnacles. *Proc. Ent. Soc. Wash.*, **82**, 360–377.

Buschman, L. L. (1984). Biology of the firefly *Pyractomera luctifera* (Coleoptera: Lampyridae). *Florida Ent.*, **67**, 529–542.

Butler, S. (1912). Notebooks.

Campos, R. E. & Lounibos, L. P. (2000). Natural prey and digestion times of *Toxorhynchites rutilis* (Diptera: Culicidae) in southern Florida. *Ann. Ent. Soc. Am.*, **93** (6), 1280–1287.

Carvalho, A. O. R. & Vieira, L. G. E. (2000). Comparison of preservation methods of *Atta* spp. (Hymenoptera: Formicidae) for RAPD analysis. *Ann. Soc. Ent. Brasil*, **29** (3), 489–496.

Cheatum, E. P. (1934). Limnological investigations on respiration, annual migratory cycle, and other related phenomena in freshwater pulmonate snails. *Trans. Am. Microscop. Soc.*, **53**, 348–407.

Coates, J. (2003). The last word. *Anim. Health Centre Diagn. Diary*, **13** (1), 15.

Cock, M. J. W. (1978). The assessment of preference. *J. Anim. Ecol.*, **47**, 805–816.

Combes, C. (1982). Trematodes: antagonism between species and sterilizing effects on snails in biological control. *Parasitology*, **84**, 151–175.

Combes, C. (1995). Interactions durables. *Ecologie et Evolution du parasitisme*. Collection écologie n°26. Paris: Masson.

Cooke, A. H. (1895). Molluscs. *Cambridge Nat. Hist.*, **3**, 1–459.

Copeland, J. (1981). Effects of larval firefly extracts on molluscan cardiac activity. *Experimentia*, **37**, 1271–1272.

Coupland, J. B. & Baker, G. (1994). Host distribution, larviposition behaviour and generation time of *Sarcophaga penicillata* (Diptera: Sarcophagidae), a parasitoid of conical snails. *Bull. Ent. Res.*, **84**, 185–189.

Coupland, J. B. & Barker, G. (2004). Diptera as predators and parasitoids of terrestrial gastropods, with emphasis on Phoridae, Calliphoridae, Sarcophagidae, Muscidae and Fanniidae. In *Natural Enemies of Terrestrial Molluscs*, ed. G. M. Barker. Wallingford: CABI Publishing, pp. 85–158.

Crampton, G. C. (1944). A comparative morphological study of the terminalia of male calyptrate cyclorrhaphous Diptera and their acalypterate relatives. *Bull. Brook. Ent. Soc.*, **39**, 1–31.

Cranston, P. S. (1990). Biomonitoring and taxonomy. *Environ. Monit. Assess.*, **14**, 265–273.

Cros, A. (1926). Moeurs et èvolution de *Drilus mauritanicus* Lucas. *Bull. Soc. Hist. Nat. Afr. Nord*, **17** (7), 181–206.

Daget, J. (1979). *Les modèles mathèmatiques en écologie*, 2nd edn. Paris: Masson.

Danks, H. V. (1991). Life cycle pathways and the analysis of complex life cycles in insects. *Can. Ent.*, **123**, 23–40.

Danks, H. V. (1993). Seasonal adaptations in insects from the high arctic. In *Seasonal Adaptation and Diapause in Insects*, ed. M. Takeda & S. Tanaka. Tokyo: Bun-ichi-Sogo Publ., pp. 54–66.

Danks, H. V. (1999a). Life cycles in polar arthropods – flexible or programmed? *Eur. J. Ent.*, **96**, 83–102.

Danks, H. V. (1999b). The diversity and evolution of insect life cycles. *Ent. Sci.*, **2** (4), 651–660.

Danks, H. V. (2000). Measuring and reporting life-cycle duration in insects and arachnids. *Eur. J. Ent.*, **97**, 285–303.

Danks, H. V. (2001). The nature of dormancy responses in insects. *Acta Soc. Zool. Bohemoslov.*, **65**, 169–179.

Danks, H. V. (2002). The range of insect dormancy responses. *Eur. J. Ent.*, **99**, 127–142.

Danks, H. V., Kukal, O. & Ring, R. A. (1994). Insect cold-hardiness: insights from the Arctic. *Arctic*, **47** (4), 391–404.

Darwin, C. (1859). *On the Origin of Species by Means of Natural Selection*. Facsimile of 1st edn. with introduction by E. Mayer. Cambridge, MA: Harvard Univ. Press, 1964.

Daudet, A. (1866). *Lettres de mon Moulin, Le rouge et le blanc*. Paris: Ed. Librio, (no. 12, 2006).

Deeming, J. C. (1996). The Calliphoridae (Diptera: Cyclorrhapha) of Oman. *Fauna Saudi Arabia*, **15**, 264–279.

DeLong, D. M. (1932). Some problems encountered in the estimation of insect populations by the sweeping method. *Ann. Ent. Soc. Am.*, **25**, 13–17.

Delpino, F. (1869). Ulteriori osservazioni sulla dicogamia nel regno vegetale. Parte 2. *Atti Soc. Ital. Sci. Nat. Milano*, **12**, 314.

Dendy, A. (1914). *Outlines of Evolutionary Biology*, 2nd edn. London: Constable.

Denlinger, D. L. & Ždárek, J. (1994). Metamorphosis behavior of flies. *Ann. Rev. Ent.*, **39**, 243–266.

DeQuieroz, A. & Wimberger, P. H. (1993). The usefulness of behavior for phylogeny estimation: levels of homoplasy in behavioral and morphological characters. *Evolution*, **47**, 46–60.

DeWitt, R. M. (1954). The intrinsic rate of natural increase in a pond snail (*Physa gyrina* Say). *Am. Nat.*, **88**, 353–359.

DeWitt, T. J., Robinson, B. W. & Wilson, D. S. (2000). Functional diversity among predators of a freshwater snail imposes an adaptative trade-off for shell morphology. *Evol. Ecol. Res.*, **2**, 129–148.

DeWitt, T. J., Sih, A. & Hucko, J. A. (1999). Trait compensation and cospecialization: size, shape, and antipredator behavior. *Anim. Behav.*, **58**, 397–407.

Disney, R. H. L. (1977). A further case of a scuttlefly (Dipt. Phoridae) whose larvae attack slug eggs. *Ent. Mon. Mag.*, **112**, 174 (1976).

Disney, R. H. L. (1979). Natural history notes on some British Phoridae (Diptera) with comments on a changing picture. *Ent. Gaz.*, **30**, 141–150.

Disney, R. H. L. (1982). A scuttle fly (Diptera: Phoridae) that appears to be a parasitoid of a snail (Stylommatophora: Zonitidae) and is itself parasitised by a Braconid (Hymenoptera). *Ent. Rec. J. Var.*, **94**, 151–154.

Disney, R. H. L. (1994). *Scuttle Flies: The Phoridae.* London: Chapman & Hall.

Dowding, V. M. (1967). The function and ecological significance of the pharyngeal ridges occuring in the larvae of some cyclorrhaphous Diptera. *Parasitology*, **57**, 371–388.

Dowding, V. M. (1968). The formation of the cuticular ridges in the larval pharynx of the blow fly (*Calliphora vicina* R-D.). *Parasitology*, **58**, 683–690.

Drake, C. M. (2001). The British species of *Notophila* Fallén (Diptera, Ephydridae), with the description of a new species. *Dipt. Dig.*, **8**, 91–126.

Dunn, D. W., Crean, C. S. & Gilburn, A. S. (2001). Male mating preference for female survivorship in the seaweed fly *Gluma musgravei* (Diptera: Coelopidae). *Proc. R. Soc. Lond. B.*, **268**, 1255–1258.

Eggleton, P. & Gaston, K. (1990). Parasitoid species and assemblages: convenient definitions or misleading compromises? *Oikos*, **59**, 417–421.

Egglishaw, H. J. (1960a). The life-history of *Helcomyza ustulata* Curt. (Dipt., Dryomyzidae). *Ent. Mon. Mag.*, **96**, 39–42.

Egglishaw, H. J. (1960b). Studies on the family Coelopidae (Diptera). *Trans. R. Ent. Soc. Lond.*, **112**, 109–140.

Eiseley, L. (1958). *Darwin's Century.* Garden City, NY: Doubleday.

Elton, C. S. (1966). *The Pattern of Animal Communities.* London: Methuen.

Emden, F. I. van. (1957). The taxonomic significance of the characters of immature insects. *Ann. Rev. Ent.*, **2**, 91–106.

Emlen, J. M. (1966). The role of time and energy in food preference. *Am. Nat.*, **100**, 611–617.

Fabre, J.-H. (1921). *La vie des Insectes. Extraits des Souvenirs Entomologiques.* Paris: Delagrave.

Fage, L. (1933). A propos du parasitisme des phorides. *Bull. Soc. Zool. Fr.*, **58**, 90.

Fairweather, I. & Boray, J. C. (1999). Mechanisms of fasciolicide action and drug resistance in *Fasciola hepatica*. In *Fasciolosis*, ed. J. P. Dalton. Wallingford: CABI Publishing.

Falkner, G., Obrdlík, P., Castella, E. & Speight, M. C. D. (2001). *Shelled Gastropoda of Western Europe.* Munich: Friedrich-Held-Gesellschaft.

Farris, J. S. (1988). *Hennig86 Reference Manual, Version 1.5.* Port Jefferson Station, published by the author.

Feener, D. H. & Brown, B. V. (1997). Diptera as parasitoids. *Ann. Rev. Ent.*, **42**, 73–97.

Felsenstein, J. (2004). *Inferring Phylogenies.* Sunderland, MA: Sinauer Associates.

Ferrar, P. (1975). Life-history and larviparous reproduction of *Musca fergusoni* J. & B. (Diptera, Muscidae). *Bull. Ent. Res.*, **65**, 187–189.

Ferris, T. (1997). Human dignity, racism and political correctness. *Zadok Perspectives*, **57**, 18.

Finkelstein, J. L, Schleinitz, M. D., Carabin, H. & McGarvey, S. T. (2008). Decision-model estimation of the age-specific disability weight for Schistosomiasis Japonica: A systematic review of the literature. *PloS Neglect. Trop. Dis.*, **2** (3), e158.

Fischer, P. H. (1951). Causes de destruction des mollusques; maladies et mort. *Jour. Conchyliol., Paris*, **91**, 29–59.

Fisher, T. W., Moore, I., Legner, E. F. & Orth, R. E. (1976). *Ocypus olens*, a predator of Brown Garden Snail. *Calif. Agric.*, **March**, 20–21.

Force, D. C. (1975). Succession of *r* and *K* strategists in parasitoids. In *Evolutionary Strategies of Parasitic Insects and Mites*, ed. P. W. Price. New York: Plenum, pp. 112–129.

Fougeret de Monbron, L-C. (1750). *Le Cosmopolite ou le Citoyen du Monde.*

Fountain, H. (2005). Despite worries, push to build big dams is strong. *New York Times*, June 11, 2005 (*Le Monde* edn., p. 5).

Futuyma, D. J. & Moreno, G. (1988). The evolution of ecological specialization. *Ann. Rev. Ent.*, **19**, 207–233.

Gain, W. A. (1896). Enemies of our land and fresh-water molluscs. *Naturalists' J.*, **1896**, 78–79.

Gardner, J. W. (1961). *Excellence – Can We Be Equal and Excellent – Too?* New York: Harper Row.

Germain, L. (1908). Etude sur les mollusques, etc. In *Voyage Zoologique en Khroumérie (Tunisie) par Henri Gadeau de Kerville*. Paris: Baillière, pp. 129–297.

Ghamizi, M. (1998). *Les Mollusques des Eaux Continentales du Maroc: Systématique, Biologie et Malacologie Appliquée*. Doctoral thesis. Univ. Cadi Ayyad, Marrakech, Morocco.

Ghysen, A. (1980). The projection of sensory neurons in the central nervous system of *Drosophila*: choice of the appropriate pathway. *Dev. Biol.*, **78**, 521–541.

Gilbert, F., Rotheray, G., Emerson, P. & Zafar, R. (1994). The evolution of feeding strategies. In *Phylogenetics and Ecology. Linn. Soc. Symp. Ser. No. 17*, ed. P. Eggleton & R. Vane-Wright. London: Academic Press, pp. 323–343.

Gilby, A. R. & McKeller, J. W. (1976). The calcified puparium of a fly. *J. Insect. Physiol.*, **22**, 1465–1468.

Godfray, H. C. J. (1994). *Parasitoids. Behavioral and Evolutionary Ecology*. Princeton, NJ: Princeton Univ. Press.

Grandcolas, P. (1997). Preface. In *The Origin of Biodiversity in Insects: Phylogenic Tests of Evolutionary Scenarios*, ed. P. Grandcolas. *Mem. Mus. natn. Hist. natr.*, **173**, 9–10.

Griffiths, G. C. D. (1980). Preface. In *Flies of the Nearctic Region*, Vol. **1**, Part 1. ed. G. C. D. Griffiths. Stuttgart: E. Schweizerbart'sche, pp. v–vii.

Griffiths, G. C. D. (1994). Relationships among the major subgroups of Brachycera (Diptera): a critical review. *Can. Ent.*, **126**, 861–886.

Gruner, L. & Riom, J. (1977). *Insectes et papillons des Antilles*. Papeete-Tahiti: Les Editions du Pacifique.

Guibé, J. (1942). Chironomes parasites de mollusques gastéropodes. *Chironomus varus limnaei* Guibé espèce jointive de *Chironomus varus* Goetgh. *Bull. Biol. Fr. Belg.*, **76**, 283–297.

Haase, E. (1885). Ein neuer Schmarotzer von *Julius*. *Zool. Beitr.*, **1**, 252–256.

Harman, W. N. & Berg, C. O. (1971). The freshwater snails of central New York with illustrated keys to the genera and species. *Search Agric. (Geneva, N. Y.)*, **1** (4), 1–68.

Hartley, J. C. (1963). The cephalopharyngeal apparatus of syrphid larvae and its relationship to other Diptera. *Proc. Zool. Soc. Lond.*, **141**, 261–280.

Hassell, M. P. & Southwood, T. R. E. (1978). Foraging strategies of insects. *Ann. Rev. Ecol. Syst.*, **9**, 75–98.

Hawkins, C. P. & MacMahon, J. A. (1989). Guilds: the multiple meanings of a concept. *Ann. Rev. Ent.*, **34**, 423–451.

Hengeveld, R. & Walter, G. H. (1999). The two coexisting ecological paradigms. *Acta Biotheo.*, **47**, 141–170.

Hennig, W. (1948–1952). *Die Larvenformen der Dipteren*. **3** vols. Berlin: Akademie Verlag.

Hennig, W. (1950). *Gründzüge einer Theorie der phylogenetischen Systematik*. Berlin: Deutsches Entomologisches Institut.

Hennig, W. (1972). Eine neue Art der Rhagionidengattung *Litoleptis* aus Chile, mit Bemerkungen über Fühlerbildung und Verwandtschaftbezeihungen einiger Brachycerenfamilien (Diptera: Brachycera). *Stuttg. Beitr. Naturk.*, **242**, 1–18.

Heraty, J. & Hawks, D. (1998). Hexamethyldisilazane – a chemical alternative for drying insects. *Ent. News*, **109**, 369–379.

Hess, A. D. & Hall, T. F. (1943). The interaction line as a factor in anopheline ecology. *J. Nat. Malaria Soc.*, **2** (2), 93–98.

Holland, W. J. (1903). *The Moth Book*. New York: Doubleday, Page & Company.

Holling, C. S. (1959). Some characteristics of simple types of predation and parasitism. *Can. Ent.*, **91**, 385–398.

Holling, C. S. (1961). Principles of insect predation. *Ann. Rev. Ent.*, **6**, 163–182.

Hopper, K. R. (1995). Potential impacts on threatened and endangered insect species in the continental United States from introductions of parasitic Hymenoptera for the control of insect pests. In *Biological Control: Benefits and Risks*, ed. H. Hokkanen & J. M. Lynch. Cambridge: Cambridge Univ. Press, pp. 64–74.

Horvitz, L. A. (2000). *The Quotable Scientist*. New York: McGraw-Hill.

Hüber, F. (1985). Approaches to insect behaviour of interest to both neurobiologists and behavioural ecologists. *Fla. Ent.*, **68**, 52–78.

Huston, M. A. (1994). *Biological Diversity. The Coexistence of Species on Changing Landscapes*. Cambridge: Cambridge Univ. Press.

Hutton, J. (1795). *Theory of the Earth with Proofs and Illustrations*. **2** vols. Edinburgh & New York: Reprint Haffner Pub. Co. (1959).

Huxley, T. H. (1868). On the classification and distribution of the *Alecotromorphae* and *Heteromorphae*. *Proc. Zool. Soc. Lond.*, 294–319.

Isidoro, N., Romani, R., Vinson, S. B. & Bin, F. (2000). Spiracular glands of *Drosophila melanogaster* puparia elicit recognition response by *Trichopria drosophilae*. In *XXI Int. Congr. Ent.*, 20–26 Aug., 2000, Brazil. ed. D. L. Gazzoni. *Abstracts*, p. 367.

Jackson, R. R. & Barrion, A. (2004). Heteropteran predation on terrestrial gastropods. In *Natural Enemies of Terrestrial Molluscs*, ed. G. M. Barker. Wallingford: CABI Publishing, pp. 483–496.

Jacobs, J. (1974). Quantitative measurement of food selection. A modification of the forage ratio and Ivlec's electivity index. *Oecologia* (Berlin), **14**, 413–417.

Jammes, L. (1904). *Zoologie Pratique. Basée sur la Dissection des Animaux les Plus Répandus*. Paris: Masson et Cie.

Jobin, W. R. & Michelson, E. H. (1967). Mathematical simulation of an aquatic snail population. *Bull. Wld Hlth Org.*, **37**, 657–664.

Johnson, C. G. (1969). *Migration and Dispersal of Insects by Flight*. London: Methuen.

Jokinen, E. H. (1983). *The Freshwater Snails of Connecticut*. State Geol. Nat. Hist. Surv. Conn., Dept. Env. Prot. Bull.

Jolly, G. M. (1965). Explicit estimates from capture-recapture data with both death and immigration-stochastic model. *Biometrika*, **52**, 225–247.

Jordan, P., Christie, J. D. & Unrao, G. O. (1980). Schistosomiasis transmission with particular reference to possible ecological and biological methods of control. A review. *Acta Tropica*, **37**, 95–135.

Keilin, D. (1912). Structure du pharynx en fonction du régime chez les larves des Diptères cyclorrhaphes. *Comp. Rend. Hebd. Séanc. Acad. Sci., Paris*, **155**, 1546–1550.

Keilin, D. (1915). Recherches sur les larves de Diptères Cyclorrhaphes. *Bull. Scient. Fr. Belg.*, **49**, 15–198.

Keilin, D. (1917). Recherches sur les Anthomyides à larves carnivores. *Parasitology*, **9**, 325–430.

Kesler, D. H. & Munns, Jr., W. R. (1989). Predation by *Belostoma flumineum* (Hemiptera): an important cause of mortality in freshwater snails. *J. N. Am. Benthol. Soc.*, **8**, 342–350.

Kirby, W. & Spence, W. (1846). *An Introduction to Entomology; or, Elements of the Natural History of Insects*, 6th edn. Philadelphia: Lea & Blanchard.

Knab, F. (1913). Some earlier observations on the habits of *Aphiochaeta juli* Brues. *Insec. Inscit. Menstr.*, **1**, 24.

Kneidel, K. A. (1984). Competition and disturbance in communities of carrion breeding Diptera. *J. Anim. Ecol.*, **53**, 849–865.

Knutson, L. V. (1973). Taxonomic revision of the aphid-killing flies of the genus *Sphaerophoria* in the Western Hemisphere (Syrphidae). *Misc. Publ. Ent. Soc. Am.*, **9** (1), 1–50.

Knutson, L. (1984). Voucher material in entomology: A status report. *Bull. Ent. Soc. Am.*, **30** (11), 8–11.

Knutson, L. & Coulson, J. R. (1997). Procedures and policies in the USA regarding precautions in the introduction of classical biological control agents. *EPPO Bull.*, **27**, 133–142.

Krivosheina, N. P. (1991). Larval morphology and the classification of the Diptera. *Proc. Soc. Int. Congr. Dipterol.*, Bratislava, Czechoslovakia, 27 Aug.–1 Sept., 1990, ed. L. Weisman, I. Országh & A. C. Pont. The Hague: SPB Academic Publishing, pp. 161–181.

Kubinyi, H. (1999). Chance favors the prepared mind – from serendipity to rational drug design. *J. Recept. Signal Transduct. Res.*, **19** (1–4), 15–39.

La Rochefoucauld, F. (1665). *Réflexions ou Sentences et Maximes Morales et Réflexions Diverses*. Edn. 2002, ed. L. Plazenet. Paris : Champion.

Leonardo da Vinci. In Sertilanges, A. D. (undated). *The Thoughts of Leonardo Da Vinci* (translated by R. S. Walker). Amboise, France: Le Clos-Lucé.

Lim, S. G. & Meier, R. (2006). Conflict between larval and adult data: Which has more homoplasy and which provides more information for phylogenetics ? In *6th Int. Congr. Dipterol.*, 23–28 Sept., 2006, Fukuoka, Japan, ed. M. Suwa. *Abstracts*, p. 154.

Loden, M. S. (1974). Predation by chironomid (Diptera) larvae on oligochaetes. *Limnol. Ocean*, **19**, 156–159.

Lopes, H. de S. (1979a). Redescription of the holotypes of some Neotropical Sarcophagidae (Diptera) described by C. H. T. Townsend. *Can. Ent.*, **111**, 149–160.

Lopes, H. de S. (1979b). Contribution to the knowledge of the tribe Johnsoniini (Diptera, Sarcophagidae). *Revt. Bras. Biol.*, **39**, 919–942.

Luck, R. F. (1984). Principles of arthropod predation. *Ecol. Ent.*, **16**, 497–529.

Lunau, K. & Knüttel, H. (1998). Fly's vision through coloured eyes. In *Fourth Int. Congr. Dipterol.*, 6–13 Sept., 1998, Oxford, ed. J. W. Ismay. *Abstracts*, p. 126.

Lydekker, R. (1896). *A Geographical History of Mammals*. Cambridge: Cambridge Univ. Press.

Magnan, A. (1934). *La locomotion chez les animaux*. Vol. **I**. *Le Vol des Insectes*. Paris: Hermann et Cie.

Mathias, P. & Boulle, L. (1933). Sur une larve de chironomide (Diptère) parasite d'un mollusque. *Comp. Rend. Acad. Sci. Paris*, **196**, 1744–1746.

Matile, L. (1990). Recherches sur la systématique et l'évolution des Keroplatidae (Diptera, Mycetophiloidea). *Mem. Mus. natn. Hist. natr. (A)*, **148**, 1–682.

Matile, L. (1993). *Les Diptères d'Europe Occidentale*. Vol. **I**. Paris: Soc. Nouv. Eds. Boubée.

Matile, L. (1997). Phylogeny and evolution of larval diet in the Sciaroidea (Diptera, Bibionomorpha) since the Mesozoic. In *The Origin of Biodiversity in Insects: Phylogenetic Tests of Evolutionary Scenarios*, ed. P. Grandcolas. *Mem. Mus. natn. Hist. natr.*, **173**, 273–303.

Mayr, E. (1944). Wallace's line in the light of recent zoogeographic studies. *Q. Rev. Biol.*, **19**, 1–14.

McKillup, S. C. & McKillup, R. V. (2000). The effects of two parasitoids on the life history and metapopulation structure of the intertidal snail *Littoraria filosa* in different-sized patches of mangrove forest. *Oecologica*, **123**, 525–534.

McKillup, S. C., McKillup, V. & Pape, T. (2000). Flies that are parasitoids of a marine snail: the larviposition behaviour and life cycles of *Sarcophaga megafilosa* and *Sarcophaga meiofilosa*. *Hydrobiologica*, **439**, 141–149.

Meier, M. (1987). Lebenszyklus und Parasit-Wirt-Beziehungen von *Parachironomus varus* (Diptera: Chironomidae) und *Radix ovata* (Pulmonata: Lymnaeidae) in einem Weiher in Süddeutschland. *Arch. Hydrobiol.*, **109**, 367–376.

Meier, R. (1996). Larval morphology of the Sepsidae (Diptera: Sciomyzoidea), with a cladistic analysis using adult and larval characters. *Bull. Am. Mus. Nat. Hist.*, **228**, 1–147.

Mellini, E. & Baronio, P. (1971). Superparassitismo sperimentale e competizioni larvali del parassitoide solitario *Macquartia chalconota* Meig. *Boll. Ist. Ent. Univ. Bologna*, **30**, 133–152.

Mendis, V. W. (1997). *A Study of Slug Egg Predation using Immunological Techniques.* Ph.D. thesis, Cardiff University, Wales.

Miller, R. M. (1977). Ecology of Lauxaniidae (Diptera: Acalyptratae) I. Old and new rearing records with biological notes and discussion. *Ann. Natal Mus.*, **23** (1), 215–238.

Miller, R. M. & Foote, B. A. (1975). Biology and immature stages of eight species of Lauxaniidae (Diptera). I. Biological observations. *Proc. Ent. Soc. Wash.*, **77** (3), 308–328.

Miller, R. M. & Foote, B. A. (1976). Biology and immature stages of eight species of Lauxaniidae (Diptera). II. Description of immature stages and discussion of larval feeding habits and morphology. *Proc. Ent. Soc. Wash.*, **78** (1), 16–37.

Mitter, R. M., Farrell, B. & Wiegmann, B. (1988). The phylogenetic study of adaptative zones: has phytophagy promoted insect diversification? *Am. Nat.*, **132**, 107–128.

Miyan, J. A. (1989). The thoracic mechanism for eclosion and digging during the extrication behaviour of Diptera. *Physiol. Ent.*, **14**, 309–317.

Monchadsky, A. S. (1937). The larvae evolution and its connection with the adult mosquitoes evolution in family Culicidae. *Izv. Akad. Nauk. SSSR, Ser. Biol.*, **4**, 1130–1351.

Morelet, A. (1880). La faune malacologique du Maroc en 1880. *J. Conchyliol. Paris*, **20** (28), 1–79.

Morgan, T. H., Bridges, C. B. & Sturtevant, A. H. (1925). The genetics of *Drosophila*. *Biblio. Gen.*, **2**, 1–262.

Müller, P. (1974). *Aspects of Zoogeography.* The Hague: Dr. W. Junk.

Müller, S. (1846). Über den character der Thierwelt auf Inseln des indischen Archipels. *Arch. Naturg.*, **12**, 109–128.

Munroe, E. (1977). Summary and appraisal of the symposium. In *Beltsville Symposia in Agricultural Research. [2] Biosystematics in Agriculture*, ed. J. A. Romberger. Montclair, NJ: Allanhold, Osmun, pp. 313–332.

Murdoch, W. W. (1969). Switching in general predators. Experiments on predator specificity and stability of prey populations. *Ecol. Monogr.*, **39**, 335–354.

Nabokov, V. (1943). On discovering a butterfly. *New Yorker*, **19** (13), 26.

Nei, M. (1978). Estimation of average heterozygosity and genetic distance from a small number of individuals. *Genetics*, **89**, 583–590.

Nijhout, H. F. (2000). The evolutionary and ecological significance of metamorphosis and its hormonal regulation. *XXI Int. Congr. Ent., 20–26 Aug., 2000, Brazil. Abstract Book 1.*

Noble, E. R. & Noble, G. A. (1961). *Parasitology. The Biology of Animal Parasites.* Philadelphia: Lea & Febiger.

Norrbom, A. L., Carroll, L. E. & Freidberg, A. (1998). Status of knowledge. In *Fruit Fly Expert Identification System and Systematic Information Database*, ed. F. C. Thompson. *Myia*, **9**, 9–47.

Ochiai, A. (1957). Zoogeographic studies on the soleid fishes found in Japan and its neighbouring regions. *Bull. Jap. Soc. Sci. Fish.*, **22**, 526–530.

Oldroyd, H. (1970). *Diptera. I. Introduction and Key to Families. Vol. IX, Part I. Handbooks for the Identification of British Insects.* London: R. Ent. Soc. Lond.

Orth, R. [E.], Moore, I., Fisher, T. W. & Legner, E. F. (1975a). Biological notes on *Ocypus olens* a predator of brown garden snail, with descriptions of the larva and pupa (Coleoptera: Staphylinidae). *Psyche*, **82** (3–4), 292–298.

Orth, R. [E.], Moore, I., Fisher, T. W. & Legner, E. F. (1975b). A rove beetle, *Ocypus olens*, with potential for biological control of the Brown Garden Snail, *Helix aspersa*, in California, including a key to the Nearctic species of *Ocypus*. *Can. Ent.*, **107**, 1111–1116.

Pape, T. (1990a). Two new species of *Sarcophaga* Meigen from Madeira and mainland Portugal (Diptera: Sarcophagidae). *Tijdschr. Ent.*, **133**, 39–42.

Pape, T. (1990b). Revisionary notes on American Sarcophaginae (Diptera: Sarcophagidae). *Tijdschr. Ent.*, **133**, 43–74.

Papp, L. (2002). Dipterous guilds of small-sized feeding sources in forests of Hungary. *Acta Zool. Acad. Sci. Hung.*, **48** (Suppl.), 197–213.

Pearl, R. (1922). Experimental studies on the duration of life. VI. A comparison of the laws of mortality in *Drosophila* and man. *Am. Nat.*, **56**, 398–405.

Pelseneer, P. (1928). Les parasites des mollusques et les mollusques parasites. *Bull. Soc. Zool. Fr.*, **53** (3), 158–189.

Pesigan, T. P., Hairston, N. G., Jauregui, J. J., Garcia, E. G., Santos, B. C. & Besa, A. A. (1958). Studies on *Schistosoma japonicum* infection in the Philippines. *WHO Bull.*, **18**, 481–578.

Picard, F. (1930). Sur le parasitisme d'un phoride (*Megaselia cuspidata* Schmitz) aux dépens d'un myriapode. *Bull. Soc. Zool. Fr.*, **55**, 180–183.

Pimentel, D. & White, Jr., P. C. (1959). Biological environment and habits of *Australorbis glabratus*. *Ecology*, **40**, 541–550.

Plate, H-P. (1951). Die ökologischen Beziehungen zwischen Arthropoden und Mollusken. *Ztschr. Ang. Ent.*, **32**, 406–432.

Poincaré, H. (1902). Chapitre IX. Les hypothèses en physique. In *La Science et l'Hypothèse*. Paris: Flammarion.

Poincaré, H. (1908). Chapitre III. L'Invention mathématique. In *Science et Méthode*. Paris: Flammarion.

Prado, A. P. do. (1965). Segunda contribuição ao conhecimento da familía Rhopalomeridae (Diptera, Acalyptratae). *Stud. Ent. (N.S.)*, 8, 209–268.

Price, P. W. (1975). *Insect Ecology*. New York: J. Wiley & Sons.

Price, P. W. (1980). *Evolutionary Biology of Parasites*. New Haven CT: Princeton Univ. Press.

Price, P. W. (1997). *Insect Ecology*, 3rd edn. New York: J. Wiley & Sons.

Pringle, G. (1960). Invasion of the egg masses of the mollusc *Bulinus* (*Physopsis*) *globosus* by larval chironomids. *Parasitology*, **50**, 497–499.

Raut, S. K. & Saha, T. C. (1989). The role of the water bug *Sphaerodema annulatum* in the control of disease transmitting snails. *J. Med. Appl. Malacol.*, **1**, 97–106.

Renn, C. E. (1941). The food economy of *Anopheles quadrimaculatus* and *A. crucians* larvae. Relationships of the air–water interface and the surface feeding mechanism. In *Symposium on Hydrobiology*. Univ. Wisconsin Press, pp. 329–342.

Renn, C. E. (1943). Emergent vegetation, mechanical properties of the water surface, and distribution of *Anopheles* larvae. *J. Nat. Malaria Soc.*, **2** (1), 47–52.

Reuter, O. M. (1913). *Lebensgewohnheiten und Instinkte der Insekten*. Berlin: R. Friedländer & Sohn.

Ribera, I. & Vogler, A. P. (2000). Habitat type as a determinant of species range sizes: the example of lotic-lentic differences in aquatic Coleoptera. *Biol. J. Linn. Soc.*, **71**, 33–52.

Richard, D. S., Saunders, D. S., Egan, V. M. & Thomson, R. C. K. (1986). The timing of larval wandering and puparium formation in the flesh fly *Sarcophaga argyrostoma*. *Physiol. Ent.*, **11**, 53–60.

Riley, C. V. (1882). Darwin's work in entomology. *Proc. Biol. Soc. Wash.*, **1**, 71–72.

Roback, S. S. & Moss, W. W. (1978). Numerical taxonomic studies on the congruence of classifications for the genera and subgenera of *Macropelopiini* and *Anatopyniini* (Diptera: Chironomidae: Tanypodinae). *Proc. Acad. Nat. Sci. Phil.*, **129**, 125–150.

Roberts, M. J. (1971). The structure of the mouthparts of some calypterate dipteran larvae in relation to their feeding habitats. *Acta Zool.*, **52**, 171–188.

Robinson, W. H & Foote, B. A. (1968). Biology and immature stages of *Megaselia aequalis*, a phorid predator of slug eggs. *Ann. Ent. Soc. Am.*, **61** (6), 1587–1594.

Roethke, T. (1964). *The Far Field*. Garden City, NY: Doubleday.

Rogers, D. (1972). Random search and insect population models. *J. Anim. Ecol.*, **41**, 369–383.

Rohdendorf, B. B. (1960). Special characters of ontogenesis and their significance in the insect evolution. In *The Ontogeny of Insects. Acta Symposii de Evolutione Insectorum*, Prague 1959. Prague: Publ. House Czechoslovak Acad. Sci., pp. 56–60.

Rohdendorf, B. B. (1964). The historical development of Diptera. *Trudy Paleontol. Inst.*, **100**. Moscow: Nauka.

Rondelaud, D. & Mage, C. (1988). Limnée tronquée et molluscicides. *Bull. Group. Tech. Vét.*, **335**, 69–76.

Root, R. B. (1967). The niche exploitation pattern of the blue-gray gnatcatcher. *Ecol. Monogr.*, **37**, 317–350.

Roubaud, E. (1916). Nouvelles observations de phorésie chez les Diptères du groupe des Borboridae. *Bull. Soc. Zool. Fr.*, **41**, 43.

Rozeboom, L. E. & Hess, H. D. (1944). The relation of the intersection line to the production of *Anopheles quadrimaculatus*. *J. Nat. Malaria Soc.*, **3**, 169–179.

Russell, P. F. & Rao, T. R. (1941). On surface tension of water in relation to behavior of *Anopheles* larvae. *Am. J. Trop. Med.*, **21**, 767–777.

Russell-Hunter, W. (1961). Annual variations in growth and density in natural populations of freshwater snails in the west of Scotland. *Proc. Zool. Soc. Lond.*, **136**, 219–253.

Russell-Hunter W. D. (1975). Variations in populations of *Lymnaea palustris* in upstate New York. *Am. Midl. Nat.*, **94**, 401–420.

Samways, M. J. (1981). *Biological Control of Pests and Weeds.* London: Edward Arnold.

Saunders, D. S. (2000). Phenology and diapause in Diptera. In *Contributions to a Manual of Palaearctic Diptera (with Special Reference to Flies of Economic Importance).* Vol. **1**. *General and Applied Dipterology*, ed. L. Papp & B. Darvas. Budapest: Science Herald, pp. 365–404.

Schelle, P., Collier, U., & Hoock, P. (2004). Rivers at risk – dams and the future of freshwater ecosystems. *7th Int. River Symp.*, Brisbane, Australia.

Schickel, R. (2005). *Time Magazine.* 13 June.

Schilthuizen, M. T., Kemperman, C. M. & Gittenberger, E. (1994). Parasites and predators of *Albinaria* (Gastropoda, Pulmonata: Claussiliidae). *Bios*, **2**, 177–186.

Schindewolf, O. H. (1950). *Der Zeitfaktor in Geologie und Palaeöntologie.* Stuttgart: Schweizerbart.

Schmitz, F. H. (1939). A new species of Phoridae (Diptera) associated with millipedes, from the Yemen. *Proc. R. Ent. Soc. Lond. Ser. B*, **8** (3), 43–45.

Schwalb, H. H. (1961). Beiträge zur Biologie der einheimschen Lampyriden, *Lampyris noctiluca* (Geoffr.) und *Phausis splendidula* (LeC.) und experimentelle Analyses ihres Beutefang und Sexual-verhaltens. *Zool. Jahrb., Abt. Allg. Zool. Physiol. Tiere*, **88**, 399–550.

Scudder, S. H. (1877). The first discovered traces of fossil insects in the American Tertiaries. *Bull. U.S. Geol. Geogr. Surv. Terr.*, **3**, 741–762.

Scudder, S. H. (1878). An account of some insects of unusual interest from the Tertiary rocks of Colorado and Wyoming. *Bull. U.S. Geol. Geogr. Surv. Terr.*, **4** (2), 519–543.

Seber, G. A. F. (1965). A note on the multiple-recapture census. *Biometrika*, **52**, 249–259.

Sen, S. K. (1921). Notes on the life-history of two species of Celyphidae. *Report and Proc. 4th Ent. Mtng, Pusa*, 1921, pp. 293–297.

Shakespeare, W. (1564). *The Most Lamentable Romaine Tragedy of Titus Andronicus.*

Shannon, C. E. & Weaver, W. (1948). *The Mathematical Theory of Communication.* Urbana, IL: Univ. of Illinois Press.

Simpson, P. (1990). Lateral inhibition and the development of the sensory bristles of the adult peripheral nervous system of *Drosophila. Development*, **109**, 509–519.

Simpson, P., Woehl, R. & Usui, K. (1999). The development and evolution of bristle patterns in Diptera. *Development*, **126**, 1349–1364.

Sinclair, B. J. (1992). A phylogenetic interpretation of the Brachycera (Diptera) based on the larval mandible and associated mouthpart structures. *Syst. Ent.*, **17**, 233–252.

Singh, K. R. P. & Micks, D. W. (1957). The effects of surface tension on mosquito development. *Mosq. News*, **17** (2), 70–73.

Skidmore, P. (1985). *The Biology of the Muscidae of the World. Ser. Ent.*, **29**. Dordrecht: Dr. W. Junk.

Skingsley, D. R., White, A. T. & Weston, A. (2000). Analysis of pulmonate mucus by infrared spectroscopy. *J. Moll. Stud.*, **66**, 363–371.

Sluss, T. P. & Foote, B. A. (1971). Biology and immature stages of *Leucopis verticalis* (Diptera: Chamaemyiidae). *Can. Ent.*, **103**, 1427–1434.

Sluss, T. P. & Foote, B. A. (1973). Biology and immature stages of *Leucopis pinicola* and *Chamaemyia polystigma* (Diptera: Chamaemyiidae). *Can. Ent.*, **105**, 1443–1452.

Smith, K. G. V. (1989). *An Introduction to the Immature Stages of British Flies: Diptera Larvae, with Notes on Eggs, Puparia and Pupae. Handbooks for the Identification of British Insects.* Vol. **10**, Part 14. London: R. Ent. Soc. Lond.

Smith, W. (1817). *Strata Identified by Organised Fossils. Part 3.* London: Arding & Merrett.

Snodgrass, R. E. (1930). *Insects: Their Ways and Means of Living.* Washington, DC: Smithsonian Institution Series.

Sokal, R. R. & Sneath, P. H. A. (1963). *Principles of Numerical Taxonomy.* San Francisco & London: W. H. Freeman & Co.

Solem, A. & Yochelson, E. L. (1979). North American Paleozoic Land Snails with a Summary of other Paleozoic Nonmarine Snails. *U.S. Geol. Surv. Prof. Paper*, **1072**, 1–42.

Speight, M. C. D. (2000). Syrph the Net: A database of biological information about European Syrphidae (Diptera) and its use in relation to the conservation of biodiversity. In *Biodiversity: The Irish Dimension. Proceedings of Seminar 4–5 March, 1998*, ed. B. S. Rushton. Dublin: R. Ir. Acad., pp. 156–171.

Speight, M. C. D., Castella, E. & Obrlik, P. (2000). Use of the Syrph the Net database 2000. In *Syrph the Net, the Database of European Syrphidae.* Vol. **25**, ed. M. C. D. Speight, E. Castella, P. Orbrlik & S. Ball. Dublin: Syrph the Net Publications.

Stearn, W. T. (1981). *The Natural History Museum at South Kensington: A History of the British Museum (Natural History) 1753–1980.* London: Heinemann.

Stehr, F. W. (ed.) (1991). *Immature Insects*. Vols 1 & 2. Dubuque, IO: Kendall/Hunt.

Stephenson, J. W. (1962). A culture method for slugs. *Proc. Malacol. Soc. Lond.*, **35**, 43–45.

Steyskal, G. C. (1952). Review of "Grundzüge einer Theorie der phylogenetischen Systematik" by Dr. Willi Hennig. *Ann. Ent. Soc. Am.*, **45**, 502–503.

Steyskal, G. C. (1953). On the nature of systematics. *Syst. Zool.*, **2**, 41.

Steyskal, G. C. (1964). Principles of numerical taxonomy. Sokal, R. R. & P. H. A. Sneath (book review). *Ann. Ent. Soc. Am.*, **57**, 390.

Steyskal, G. C. (1965). Trend curves of the rate of species description in zoology. *Science*, **149**, 880–882.

Sunderland, K. D. (1996). Progress in quantifying predation using antibody techniques. In *The Ecology of Agricultural Pests, Biochemical Approaches*, ed. W. O. C. Symondson & J. E. Liddell. *Syst. Assoc. Spec.* Vol. **53**. London: Chapman & Hall, pp. 419–455.

Swofford, D. L. (1998). *PAUP**, *Phylogenetic Analysis Using Parsimony (* and other methods). Version 4*. Sunderland, MA: Sinauer Associates.

Symondson, W. O. C. (2002a). Diagnostic techniques for determining carabid diets. In *The Agro-ecology of Carabid Beetles*, ed. J. Holland. Andover: Intercept, pp. 137–164.

Symondson, W. O. C. (2002b). Molecular identification of prey in predator diets. *Mol. Evol.*, **11**, 627–641.

Symondson, W. O. C. (2004). Coleoptera (Carabidae, Staphylinidae, Lampyridae, Drilidae and Silphidae) as predators of terrestrial gastropods. In *Natural Enemies of Terrestrial Molluscs*, ed. G. M. Barker. Wallingford: CABI Publishing, pp. 37–84.

Tauber, M. J., Tauber, C. A. & Masaki, S. (1984). Adaptations to hazardous seasonal conditions: dormancy, migration, and polyphenism. In *Ecological Entomology*, ed. C. B. Huffaker & R. L. Rabb. New York: J. Wiley & Sons, pp. 149–183.

Tauber, M. J., Tauber, C. A., Nyrop, J. P. & Villani, M. G. (1998). Moisture, a vital but neglected factor in the seasonal ecology of insects: hypotheses and tests of mechanisms. *Environ. Ent.*, **27** (3), 523–530.

Thompson, D. W. (1910). *Historia Animalium. Vol. IV of the Works of Aristotle*, ed. J. A. Smith & W. D. Ross. Oxford: Oxford Univ. Press.

Thompson, D. W. (1917). *On Growth and Form*. Cambridge: Cambridge Univ. Press. Original edition 1917; abridged and edited edition 1961.

Thompson, F. C. (1999). Data dictionary and standards [for Fruit Fly Systematic Information Database]. *Myia*, **9**, 49–63.

Thorpe, W. H. & Crisp, D. J. (1947). Studies on plastron respiration. I. The biology of *Aphelocheirus*, Hemiptera, Aphelocheidae (Naucoridae) and the mechanism of plastron retention. *J. Exp. Biol.*, **24**, 227–269.

Tinbergen, L. (1960). The natural control of insects in pine woods. I: Factors influencing the intensity of predation by songbirds. *Arch. Neerl. Zool.*, **13**, 265–344.

Tod, M. E. (1973). Notes on beetle predators of molluscs. *The Entomologist*, **106**, 196–201.

Tokeshi, M. (1991). On the feeding habits of *Thienemannimyia festiva* (Diptera, Chironomidae). *Aquat. Ins.*, **13** (1), 9–16.

Truman, J. W. & Riddiford, J. M. (2002). Endocrine insights into the evolution of metamorphosis in insects. *Ann. Rev. Ent.*, **47**, 467–500.

Ulrich, H. (1991). The present state of comparative morphology of thoracic musculature in Diptera. In *Proc. Second Int. Congr. Dipterol.*, 27 Aug–1 Sept., 1990, Bratislava, ed. I. Országh & A. Pont. The Hague: SPB Academic Publishing, pp. 315–326.

Ulrich, H. & Schmelz, R. M. (2001). Enchytraeidae as prey of Dolichopodidae, recent and in Baltic amber (Oligochaeta; Diptera). *Bonn. Zool. Beitr.*, **50** (1–2), 89–101.

Voelker, J. (1966). Wasserwanzen als obligatorische Schneckenfresser im Nildelta (*Limnogeton fieberi* Mayr, Belostomatidae, Hemiptera). *Ztschr. Tropenmed. Parasitol.*, **17**, 155–165.

Voelker, J. (1968). Unterschungen zu Ernährung, Fortpflansungsbiologie und Entwicklung von *Limnogeton fieberi* Mayr (Belostomatidae, Hemiptera) als Beiträg zur Kenntnis von natürlichen Feinden tropischer Süsswasserschnecken. *Ent. Mitt. Zool. Mus. Hamburg*, **3** (60), 1–31.

Voltaire, F-M. (1759). *Candide ou l'Optimisme*. Paris: Classiques Larousse Collection (1990).

Wade, C. M., Mordan, P. B. & Clarke, B. (2000). A phylogeny of the land snails. *Proc. R. Soc. Lond. B*, **268**, 413–422.

Wallace, A. R. (1869). *The Malay Archipelago: The Land of the Orang-utan, and the Bird of Paradise*. New York: Dover. Republication of 10th edn., published (1890) by Macmillan & Company.

Walter, G. H. & Hengeveld, R. (2000). The structure of the two ecological paradigms. *Acta Biotheo.*, **48**, 15–46.

Walter, H. (1985). *Vegetation of the Earth and Ecological Systems of the Geo-biosphere*. 3rd edn., transl. from 5th German edn. by O. Muise. Berlin: Springer.

Walter, H. J. (1969). Illustrated biomorphology of the "*angulata*" lake form of the basommatrophoran snail *Lymnaea catascopium* Say. *Malacol. Rev.*, **2**, 1–102.

Wardaugh, K. G. (1986). Diapause strategies in the Australian plague locust (*Chortoicetes terminifera* Walker). *Aust. J. Ecol.*, **5**, 187–191.

Waters, W. E. (1959). A quantitative measure of aggregation in insects. *J. Econ. Ent.*, **52**, 1180–1184.

Watt, K. E. F. (1959). A mathematical model of the effect of densities of attacked and attacking species on the number of attacked. *Can. Ent.*, **91**, 129–144.

Weber, M. (1902). *Der Indo-australische Archipel und die Geschichte seiner Tierwelt*. Jena: G. Fischer.

Wheater, C. P. (1987). Observations on the food of *Staphylinus olens* (Col., Staphylinidae). *Ent. Mon. Mag.*, **123**, 116.

Wheater, C. P. (1989). Prey detection by some predatory Coleoptera (Carabidae and Staphylinidae). *J. Zool.*, **218**, 171–185.

Wheeler, W. M. (1923). *Social Life Among the Insects*. New York: Harcourt, Brace.

Whitfield, J. B. (1998). Phylogeny and evolution of host–parasitoid interactions in Hymenoptera. *Ann. Rev. Ent.*, **43**, 129–151.

WHO, World Health Organization (1961). Molluscicides: second report of the expert committee on bilharziasis. *Tech. Rept. Ser.*, **214**, 50 pp.

WHO, World Health Organization (1984). Report of an informal consultation on research on the biological control of snail intermediate hosts. Unpublished Document. *TDR/BCV-SCH/SIH/84.3*.

Wiegmann, B. M., Mitter, C. & Farrell, B. (1993). Diversification of carnivorous parasitic insects: extraordinary radiation or specialised dead end? *Am. Nat.*, **142** (5), 737–754.

Wiegmann, B. M., Yeates, D. K., Thorne, J. L. & Kishino, H. A. (2003). Time flies, a new molecular time-scale for Brachyceran fly evolution without a clock. *Syst. Biol.*, **52** (6), 745–756.

Wiggins, G. B. & Mackay, R. J. (1978). Some relationships between systematics and trophic ecology in Nearctic aquatic insects, with special reference to Trichoptera. *Ecology*, **59** (6), 1211–1220.

Wiktor, A. (2000). Agrolimacidae (Gastropoda: Pulmonata) – A systematic monograph. *Ann. Zool. (Warszawa)*, **49** (3), 347–590.

Wilken, G. B. & Appleton, C. C. (1991). Avoidance responses of some indigenous and exotic freshwater pulmonate snails to leech predation. *S. Afr. J. Zool.*, **26**, 6–10.

Williams, F. X. (1951). Life-history studies of East African *Achatina* snails. *Bull. Mus. Comp. Zool., Harvard*, **105** (3), 295–317.

Williston, S. W. (1896). *Manual of the Families and Genera of North American Diptera*, 2nd edn. New Haven, CT: Hathaway.

Woodley, N. E. (1989). Phylogeny and classification of the "Orthorrhaphous" Brachycera. In *Manual of Nearctic Diptera*. Vol. 3, ed. J. F. McAlpine. *Res. Br., Agric.* Canada. Mono. No. 32, pp. 1371–1395.

Wright, C. A. (1968). Some views on biological control of trematode diseases. *Trans. R. Soc. Trop. Med. Hyg.*, **62**, 320–329.

Wright, C. A. (1971). *Flukes and Snails. Science of Biology Series No. 4*. London: G. Allen & Unwin.

Yeates, D. K. & Greathead, D. (1997). The evolutionary pattern of host use in the Bombyliidae (Diptera): a diverse family of parasitoid flies. *Biol. J. Linn. Soc.*, **60**, 149–185.

Yeates, D. K. & Wiegmann, B. M. (1999). Congruence and controversy: toward a higher-level phylogeny of Diptera. *Ann. Rev. Ent.*, **44**, 397–428.

Zetterstedt, J. W. (1842). Diptera Scandinaviae. *Disposita et Descripta*. Vol. **1**. Lund: Officina Lundbergiana.

Equations

Equation 5.1 (Section 5.2.1, Table 5.8, Figs 5.5–5.8)

(a) $c = (P1/P2)/(N1/N2)$ (Murdoch's 1969 index)
(b) $D = (P1/P2 - N1/N2)/(P1/P2 + N1/N2)$ (Jacobs' 1974 index)

In the case of strong preference only one species is consumed, thus $P2 = 0$, and use of Murdoch's index is impossible. So, a modification was proposed by Manguin *et al.* (1988b):

(c) $M = (P1/N1 - P2/N2)/(P1/N1 + P2/N2)$

To determine the value of c, Murdoch (1969) proposed the following equation:

(d) $Y = (100cX)/(100 - X + cX)$

As the slope is different in comparison with Murdoch's c, Manguin *et al.* (1988b) used c' and transformed Y to the form $y = ax + b$:

(e) $1/Y = (100 - X + c'X)/100c'X$, which can also be written $1/Y = 1/c'X + 1/100(1-1/c')$

so, $1/Y = y$, $1/X = x$ and $1/c' = a$
c = Murdoch's (1969) index; c' = new index of Manguin *et al.* (1988b); $N1$, $N2$ = number of prey present; $P1$, $P2$ = number of prey consumed; X = percentage for one of the prey species; Y = percentage consumed of the same species. From Manguin *et al.* (1988b).

Equation 5.2 (Fig. 5.5, Table 5.8)

Eq. 5.2a. $1/Y = 0.30, 1/X + 0.008$, $(r = 0.73)$***, $(c' = 3.30)$
Eq. 5.2b. $1/Y = 0.13, 1/X + 0.010$, $(r = 0.63)$***, $(c' = 7.96)$
Eq. 5.2c. $1/Y = 0.60, 1/X + 0.004$, $(r = 0.69)$***, $(c' = 1.69)$
Eq. 5.2d. $1/Y = 0.98, 1/X + 0.001$, $(r = 0.61)$***, $(c' = 1.02)$
Eq. 5.2e. $1/Y = 0.95, 1/X + 0.001$, $(r = 0.51)$**, $(c' = 1.06)$

Significant differences at * 0.5% ($P < 0.05$); **1 % ($P < 0.01$); ***, 1 ‰ ($P < 0.001$). From Manguin *et al.* (1988b).

Equation 7.1 (Section 7.1)

x, in 7.1a,b, snail size; in 7.1c,d, larval weight; y, in 7.1a,c, total feeding time; in 7.1b,d, post-establishment time.
Eq. 7.1a. $y = -78.87 + 47.86\,x$ with $r = 0.91$, d.f. = 48, $P < 0.01$
Eq. 7.1b. $y = -85.83\,x$ with $r = 0.91$, d.f. = 48, $P < 0.01$
Eq. 7.1c. $y = 112.8 - 0.08\,x$ with $r = -0.16$, d.f. = 48, $P > 0.1$
Eq. 7.1d. $y = 109.7 - 0.08\,x$ with $r = -0.19$, d.f. = 48, $P > 0.1$
From Lindsay (1982).

Equation 7.2 (Fig. 7.6, Table 7.4 Section 7.2.4)

$R = (n/N)100$
R = Rate of successful infestation, n = number of infesting larvae, N = number of larvae used in experiment. From Reidenbach *et al.* (1989).

Equation 10.1 (Fig. 10.6, Table 10.3)

$Fi = (ni/N)100$
$Pi = (qi/Q)100$
Fi = frequency of species i; N = total number of collections; ni = number of sweeps with species i; Pi = relative abundance of species i; qi = number of captures with species i; Q = number of individuals captured of all of the species. From Vala & Manguin (1987).

Equation 13.1 (Fig. 13.1, Section 13.1)

$N = m(n + 1)/(r + 1)$
m = number of marked flies; N = calculated population size; n = number of individuals in the second sample; r = total number of recaptured marked flies. From Eckblad & Berg (1972).

Equation 19.1 (Fig. 19.1)

Eq. 19.1a. $y = 0.7416x^2 + 0.9597x - 5.4377$ ($R2 = 0.98$)
Eq. 19.1b. $y = 0.0076x^2 + 1.6431x$ ($R2 = 0.99$)
x, decades for species (19.1a) and specific years for genera (19.1b); y, tendency curve equation (solid line); $R2$, determination coefficient.

Index 1: Subject

Index 2: Genera and species cited other than Sciomyzidae, Phaeomyiidae, and Mollusca

Index 3: Group names cited other than Sciomyzidae, Phaeomyiidae, and Mollusca

Index 4: Suprageneric names in Sciomyzidae and Phaeomyiidae (see especially Table 15.14)

Information about the DVD

The enclosed CD was prepared from a 14.5-minute, 16-mm color film produced by Professor C. O. Berg in 1973 at Cornell University, Ithaca, New York, with the assistance of his then-standing crop of graduate students in the Department of Entomology, J. W. Eckblad, W. L. O'Neill, D. J. Trelka, and K. R. Valley. The filming was the exquisite work of Kjell Sandved (creator of the "Butterfly Alphabet"), who was the lead natural history research cinemaphotographer at the U.S. National Museum of Natural History at the Smithsonian Institution, 1960-1989. The commentary was written and narrated by Professor Berg; he appears in opening and closing footage.

The film centers around the possibilities for biological control of freshwater snails that are obligate intermediate hosts of trematode flatworms that cause schistosomiasis of man and fascioliasis of man and domestic animals (see Chapter 18). Main aspects of the lifecycle of a sciomyzid, *Sepedon* sp., a freshwater predator, are shown. Remarkable footage is featured of *Sepedon* sp. larvae swimming at the water surface and attacking and killing *Biomphalaria* sp. and *Lymnaea* sp. snails (see Chapter 7), as well as various aquatic insect predators killing *Sepedon* larvae and parasitoid Ichneumonidae wasps ovipositing into and emerging from *Sepedon* puparia (see Chapter 11). Dr. Eckblad is shown examining the floating cylinders used in his experimental studies of natural populations of *Sepedon fuscipennis* at Bool's Backwater, near Ithaca (see Chapter 13). An overview is presented by Professor Berg of the basic principles of biological control by natural enemies and possibilities for use of Sciomyzidae.

L. V. Knutson